# HANDBOOK OF APPLIED MULTIVARIATE STATISTICS AND MATHEMATICAL MODELING

# HANDBOOK OF APPLIED MULTIVARIATE STATISTICS AND MATHEMATICAL MODELING

Edited by

**Howard E. A. Tinsley**

*Department of Psychology*
*University of Florida*
*Gainesville, Florida*

**Steven D. Brown**

*Department of Leadership,*
*Foundations, and Counseling Psychology*
*Loyola University of Chicago*
*Wilmette, Illinois*

**ACADEMIC PRESS**
A Harcourt Science and Technology Company

San Diego   San Francisco   New York   Boston   London   Sydney   Tokyo

Academic Press
*A Harcourt Science and Technology Company*
525 B Street, Suite 1900, San Diego, California 92101-4495, USA
http://www.academicpress.com

Academic Press
Harcourt Place, 32 Jamestown Road, London NW1 7BY, UK
http://www.hbuk.co.uk/ap/

Library of Congress Catalog Card Number: 99-68685

International Standard Book Number: 0-12-691360-9

PRINTED IN THE UNITED STATES OF AMERICA
00  01  02  03  04  05  SB  9  8  7  6  5  4  3  2  1

# CONTENTS

## I INTRODUCTION

## 1 Multivariate Statistics and Mathematical Modeling

HOWARD E. A. TINSLEY AND STEVEN D. BROWN

## 2 Role of Theory and Experimental Design in Multivariate Analysis and Mathematical Modeling

JOHN HETHERINGTON

# 3 Scale Construction and Psychometric Considerations

RENE V. DAWIS

# 4 Interrater Reliability and Agreement

HOWARD E. A. TINSLEY AND DAVID J. WEISS

# 5 Interpreting and Reporting Results

MARK HALLAHAN AND ROBERT ROSENTHAL

# II MULTIVARIATE ANALYSIS

## 6 Issues in the Use and Application of Multiple Regression Analysis

ANRE VENTER AND SCOTT E. MAXWELL

## 7 Multivariate Analysis of Variance and Covariance

CARL J HUBERTY AND MARTHA D. PETOSKEY

## 8 Discriminant Analysis

MICHAEL T. BROWN AND LORI R. WICKER

## 9 Canonical Correlation Analysis

ROBERT M. THORNDIKE

# 10 Exploratory Factor Analysis

ROBERT CUDECK

## 13 Time-Series Designs and Analyses

MELVIN M. MARK, CHARLES S. REICHARDT, AND LAWRENCE J. SANNA

## 14 Poisson Regression, Logistic Regression, and Loglinear Models for Random Counts

PETER B. IMREY

# III EVALUATION OF MATHEMATICAL MODELS

## 15 Structural Equation Modeling: Uses and Issues
LISABETH F. DILALLA

## 16 Confirmatory Factor Analysis
RICK H. HOYLE

## 17 Multivariate Meta-analysis

BETSY J. BECKER

## 18 Generalizability Theory

GEORGE A. MARCOULIDES

## 19 Item Response Models for the Analysis of Educational and Psychological Test Data

RONALD K. HAMBLETON, FREDERIC ROBIN, AND DEHUI XING

## 20 Multitrait–Multimethod Analysis

LEVENT DUMENCI

## 21 Using Random Coefficient Linear Models for the Analysis of Hierarchically Nested Data

ITA G. G. KREFT

## 22 Analysis of Circumplex Models

TERENCE J. G. TRACEY

## 23 Using Covariance Structure Analysis to Model Change over Time

JOHN B. WILLETT AND MARGARET K. KEILEY

# CONTRIBUTORS

*Numbers in parentheses indicate the pages on which the author's contribution begins.*

**Betsy J. Becker** (499), College of Education, Michigan State University, East Lansing, Michigan 48824

**Michael T. Brown** (209), Graduate School of Education, University of California, Santa Barbara, Santa Barbara, California 93016

**Steven D. Brown** (3), Department of Leadership, Foundations, and Counseling Psychology, Loyola University of Chicago, Wilmette, Illinois 60091

**Robert Cudeck** (265), Department of Psychology, University of Minnesota, Minneapolis, Minnesota 55455

**Mark L. Davison** (323), Department of Educational Psychology, University of Minnesota, Minneapolis, Minnesota 55455

**Rene V. Dawis** (65), University of Minnesota, Minneapolis, Minnesota 55414; and The Ball Foundation, Glen Ellyn, Illinois 60137

**Lisabeth F. DiLalla** (439), School of Medicine, Southern Illinois University, Carbondale, Illinois 62901

**Levent Dumenci**[1] (583), Department of Psychiatry, University of Arkansas, Little Rock, Arkansas 72204

**Paul A. Gore, Jr.** (297), Department of Psychology, Southern Illinois University, Carbondale, Illinois 62901

**Mark Hallahan**[2] (125), Department of Psychology, Clemson University, Clemson, South Carolina 29634

**Ronald K. Hambleton** (553), College of Education, University of Massachusetts, Amherst, Massachusetts 01003

**John D. Hetherington** (37), Department of Psychology, Southern Illinois University, Carbondale, Illinois 62901

**Rick H. Hoyle** (465), Department of Psychology, University of Kentucky, Lexington, Kentucky 40506

**Carl J Huberty** (183), Department of Educational Psychology, University of Georgia, Athens, Georgia 30602

**Peter B. Imrey** (391), Departments of Statistics and Medical Information Science, University of Illinois at Urbana-Champaign, Champaign, Illinois 61820

**Margaret K. Keiley** (665), Department of Child Development and Family Studies, Purdue University, West Lafayette, Indiana 47907

**Ita G. G. Kreft** (613), School of Education, California State University at Los Angeles, Los Angeles, California 90032

**George A. Marcoulides** (527), Department of Management Science, California State University at Fullerton, Fullerton, California 92834

**Melvin M. Mark** (353), Department of Psychology, Pennsylvania State University, University Park, Pennsylvania 16802

**Scott E. Maxwell** (151), Department of Psychology, University of Notre Dame, Notre Dame, Indiana 46556

**Martha D. Petoskey** (183), Department of Educational Psychology, University of Georgia, Athens, Georgia 30602

**Charles S. Reichardt** (353), Department of Psychology, University of Denver, Denver, Colorado 80208

**Frederic Robin** (553), College of Education, University of Massachusetts, Amherst, Massachusetts 01003

---

[1] Current address: Department of Psychiatry, University of Vermont, Burlington, Vermont 05401.

[2] Current address: Psychology Department, College of the Holy Cross, Worcester, Massachusetts 01610.

**Robert Rosenthal**[3] (125), Department of Psychology, Harvard University, Cambridge, Massachusetts 02138

**Lawrence J. Sanna** (353), Department of Psychology, Washington State University, Pullman, Washington 99164

**Stephen G. Sireci** (323), Department of Educational Policy, Research, and Administration, University of Massachusetts, Amherst, Massachusetts 01003

**Robert M. Thorndike** (237), Department of Psychology, Western Washington University, Bellingham, Washington 98225

**Howard E. A. Tinsley** (3, 95), Department of Psychology, University of Florida, Gainesville, Florida 32611

**Terence J. G. Tracey**[4] (641), School of Education, University of Illinois at Urbana-Champaign, Champaign, Illinois 61820

**Anre Venter** (151), Department of Psychology, University of Notre Dame, Notre Dame, Indiana 46556

**David J. Weiss** (95), Department of Psychology, University of Minnesota, Minneapolis, Minnesota 55455

**Lori R. Wicker** (209), Graduate School of Education, University of California, Santa Barbara, Santa Barbara, California 93016

**John B. Willett** (665), Graduate School of Education, Harvard University, Cambridge, Massachusetts 02138

**Dehui Xing** (553), College of Education, University of Massachusetts, Amherst, Massachusetts 01003

[3] Current address: Department of Psychology, University of California, Riverside, Riverside, California 92502.

[4] Current address: Division of Psychology in Education, Arizona State University, Tempe, Arizona 85287.

# PREFACE

Our objective in the *Handbook of Applied Multivariate Statistics and Mathematical Modeling* is to provide a comprehensive, readable overview of the multivariate statistical and mathematical modeling procedures that are used by research scientists in the physical, biological, and social sciences and in the humanities. The *Handbook* differs from other textbooks on multivariate statistics and mathematical modeling in three respects. First, the *Handbook's* coverage of both multivariate statistical procedures and mathematical modeling procedures in a single volume is unique. Although no single volume could provide a completely exhaustive coverage of these procedures, the approaches covered by the *Handbook* account for over 95% of the multivariate statistics and mathematical modeling procedures reported in the research literature.

Second, the *Handbook* emphasizes the appropriate uses of multivariate statistics and mathematical modeling techniques; it prescribes practices that will enable applied researchers and scholars to use these procedures effectively. Each chapter provides an overview of (a) the types of questions that can best be answered using the technique, (b) the data requirements of the technique (e.g., assumptions and robustness, properties of data, measurement requirements, and requirements for subject selection and sample size), (c) the decisions necessary to perform the analysis, and (d) considerations to keep in mind in interpreting the results. Common mistakes and pitfalls are reviewed, and each chapter provides an in-depth example that illustrates the application of the procedure in addressing issues of concern in the biological and social sciences.

Third, the *Handbook* is intended for applied researchers who have no need to know about the mathematical bases of these procedures nor about the mathematical manipulations involved in the analysis. Throughout, we have assumed that the intended reader will use a computer and readily available software to perform the necessary statistical computations. For the most part, therefore, the presentations are conceptual in nature. Emphasis is given to the rationales, applications, and interpretations of the techniques, rather than to their mathematical, computational, and theoretical aspects. Formulas are used only when they are necessary to achieve conceptual clarity or to describe essential computations that are not performed by the readily available computer programs.

Nevertheless, the *Handbook* is not completely devoid of formulas. The coverage of both multivariate statistics and the analysis of mathematical modeling in a single volume required that we tolerate some variability in the level of mathematical treatment. Some of the procedures described in the *Handbook* are more complex than others, and user-friendly computer software is not readily available for some of the newer procedures. In those instances it was necessary to rely more heavily on statistical formulas to explain the computations involved in performing the analysis. These chapters will require greater persistence on the part of the reader. Nevertheless, we are confident that careful study of these chapters will provide applied researchers with a conceptual understanding of the purposes and limitations of the procedure and allow them to understand the statistical computations necessary to perform the analysis.

Each chapter is written by a knowledgeable expert in language that can be understood by the average bright applied researcher. The authors are prolific contributors to the professional literature who use multivariate statistics and mathematical modeling procedures in their own work and possess expert knowledge of their technique. Each has great experience in teaching these procedures to their students and colleagues. The authors conducted exhaustive reviews and critical analyses of the theoretical and empirical literature on the informed use of their techniques and they provide authoritative recommendations, taking care to distinguish clearly among those that are firmly grounded in empirical research, those for which some empirical information is available, and those "rules of thumb" that are made in the absence of any empirical evidence. These distinctions also provide a valuable research agenda for further study of these multivariate techniques and mathematical modeling procedures. The authors emphasize the practical application of their techniques and provide an objective, critical evaluation of the technique, taking care to avoid the temptation to serve as an enthusiastic booster. In short, the *Handbook* provides simple, direct, practical advice, based on the theoretical and empirical literature, that will enable readers to make informed use of each technique.

All readers will benefit from a close reading of the first section of the *Handbook,* which provides the general background necessary to understand and effectively use multivariate statistics and mathematical modeling procedures. We recommend that those who are not familiar with multiple regression also read Chapter 6 before reading any of the subsequent chapters. Many of the procedures covered in subsequent chapters build on basic multiple regression concepts and procedures. Beyond that, it is not necessary to read the *Handbook* chapters in any particular order. The authors have written their chapters as self-contained treatments of their topics, taking care to cross-reference related chapters to help readers gain a more complete understanding of the technique. Indeed, we anticipate that many readers will use the *Handbook* as a reference work and will read selected portions only as needed.

The *Handbook* is intended for a wide audience of students and applied researchers who make use of these procedures in their professional work and are seeking expert advice on how to use these techniques to best advantage.

- The *Handbook* can be used as a required text in introductory multivariate statistics classes in which the emphasis is on the practical application of multivariate statistics and mathematical modeling procedures.
- Graduate students in all academic disciplines will find the *Handbook* useful as a supplement to required multivariate texts, which often are too technical for many students enrolled in such courses.
- Applied researchers and scholars who deal in their professional lives with the design, analysis, and interpretation of research will find the *Handbook* useful in helping them (a) understand what multivariate statistics and mathematical modeling procedures can do and how they might be applied in their own research, (b) become more insightful collaborators with multivariate analysts and statistical consultants, and (c) supervise employees or students who are using these procedures.
- Instructors, counseling and clinical psychology practitioners, psychiatrists, physicians, program consultants, and other practitioners who have not had the time or the need to learn to perform these procedures in their own professional work will find the *Handbook* useful in helping them to become more insightful consumers of the

ever-increasing proportion of research in which these procedures are used.

Although both of us have reviewed and edited all the chapters in this book, Brown took primary responsibility for the chapters on interrater reliability and agreement, multiple regression, exploratory and confirmatory components and factor analysis, multidimensional scaling, time-series analysis, and loglinear analysis (i.e., Chapters 4, 6, 10, 12–14, and 16). Tinsley assumed primary responsibility for the chapters on theory and experimental design, scale construction and psychometric properties, interpreting results, multivariate analysis of variance and covariance, discriminant analysis, canonical analysis, cluster analysis, structural equation modeling, multivariate meta-analysis, generalizability theory, item response theory, multitrait–multimethod analysis, hierarchical models, circumplex models, and growth modeling (i.e., Chapters 2, 3, 5, 7–9, 11, 15, and 17–23).

We are indebted to the contributing authors, who gave so generously of their time and expertise to the *Handbook*. Their chapters were a genuine pleasure to read and they greatly enhanced our knowledge of the design issues, multivariate statistics, and mathematical modeling techniques covered in this volume. Their responsiveness to our editorial suggestions greatly eased our work as editors.

We are indebted to Nikki Levy, Executive Editor at Academic Press, who encouraged to pursue this project and offered valuable advice as we progressed through the process. We also are indebted to Eileen Favorite, Production Editor at Academic Press, for her skillful editing, which improved the clarity and readability of this volume, and for her patience, persistence, and skill in shepherding the *Handbook* through the production process.

# PART I

# INTRODUCTION

# I

# MULTIVARIATE STATISTICS AND MATHEMATICAL MODELING

## HOWARD E. A. TINSLEY

*Department of Psychology, University of Florida, Gainesville, Florida*

## STEVEN D. BROWN

*Department of Leadership, Foundations, and Counseling Psychology, Loyola University of Chicago, Wilmette, Illinois*

The use of multivariate statistics and mathematical models has increased dramatically during the last 30 years. In one analysis of research published in the *Journal of Consulting and Clinical Psychology* (JCCP), the number of articles that used at least one multivariate statistical procedure increased 744% during the 16-year period from 1976 to 1992 (Grimm & Yarnold, 1995). An increase of 356% occurred in the *Journal of Personality and Social Psychology* (JPSP) during the same period (see Figure 1.1). The use of multivariate statistics is now standard practice in contemporary social science research.

Several factors have contributed to this dramatic increase, and to the need for the *Handbook of Multivariate Statistics and Mathematical Modeling*. First, scholars have come to realize that most important phenomena in the physical, biological, and social sciences and the humanities have causes and effects that are inherently multivariate in nature. For example, most human behaviors are influenced by the interactions of individual differences in attributes such as gender, age, goals, and expectations, and with differences in environmental factors associated with diurnal, seasonal, and geographic variations. Gaining an understanding of these phenomena requires the simultaneous consideration of the interactions of a great many

*Handbook of Applied Multivariate Statistics and Mathematical Modeling*

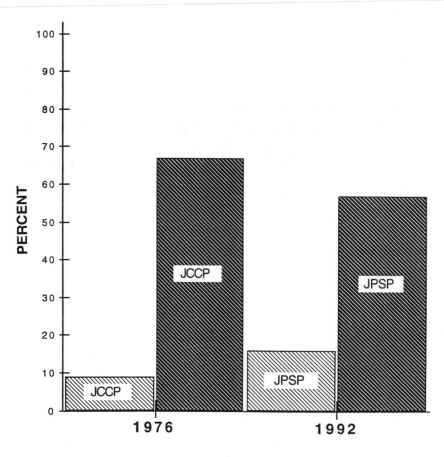

**YEAR**

**FIGURE 1.1**    Percent of articles published in two American Psychological Association research journals in 1976 and 1992 that used multivariate statistics. JCCP, *Journal of Consulting Psychology;* JPSP, *Journal of Personality and Social Psychology.*

variables; the sheer complexity of the task is overwhelming without the assistance of some form of statistical analysis. Univariate statistics that examine the effects of one variable at a time are of little help in this endeavor because they provide such a limited glimpse of the dynamic interactions that are occurring. For this reason, the ability to understand multivariate statistics and mathematical modeling procedures and to use them effectively is now an essential skill that scholars and practitioners in the sciences and humanities must master.

Second, computers have become far more powerful and widely available than most could have imagined in the 1960s. The mathematical bases of modern multivariate statistical techniques were developed in the early decades of the twentieth century, but these techniques were seldom used because the necessary calculations are so complex. The development of computers in the years following World War II enabled a select group of scholars to begin experimenting with the use of multivariate statistics, and by the early-1950s statistically erudite professors at major universities were able to make use of selected multivariate procedures. Completion of such an analysis virtually guaranteed publication in a reputable journal, but their use was rare because a procedure such as factor analysis took months to complete (see Cattell, 1988). Today even high school students have access to a computer capable of performing the necessary computations in seconds.

A third factor that has contributed to the increased use of multivariate statistics and mathematical modeling procedures is the increased availability of menu-driven software. Comprehensive packages of computer programs have been developed to the point that today's users need virtually no knowledge of computer programming or the computations involved in the statistical procedures. The combination of more powerful and widely available computers and menu-driven software packages now makes it possible for scholars in the sciences and humanities to move beyond the limitations of univariate analysis and seek a deeper understanding of the complex relations among multiple sets of variables. Today multivariate statistics and mathematical modeling procedures are applied regularly to problems arising in the physical sciences (e.g., environmental studies, meteorology, and geology), biological sciences (e.g., medicine, biology, and health psychology), social sciences (e.g., sociology, economics, education, psychology, political science, and sports), and humanities (e.g., art history, literary analysis, and criticism).

Given the present trends, it is now essential for scholars, practitioners, and students in the sciences and humanities to develop a basic understanding of the use and interpretation of these procedures. Consumers of research articles who fail to attain at least a basic knowledge of multivariate statistics and mathematical modeling run the risk of becoming scientifically illiterate.

Unfortunately, our impressions after years of reviewing manuscripts and editing journals is that many practicing researchers lack an adequate knowledge of these statistical techniques, and many of those who obtained a solid grounding in these procedures during graduate school have found it difficult to keep abreast of developments. Several recent studies support our impressions. For example, a survey of the corresponding authors of articles published in American Psychological Association (APA) research journals found that over 40% of the answers to a series of five "factual" questions about the use of statistics were incorrect (Zuckerman, Hodgins, Zuckerman, & Rosenthal, 1993). There are signs that this generally low level of competence will persist into the next generation of scholars. Aiken, West, Sechrest, and Reno (1990) surveyed 186 psychology departments and concluded that few advances had occurred in the statistical and methodological curriculum in the last 20 years, and that the respondents were

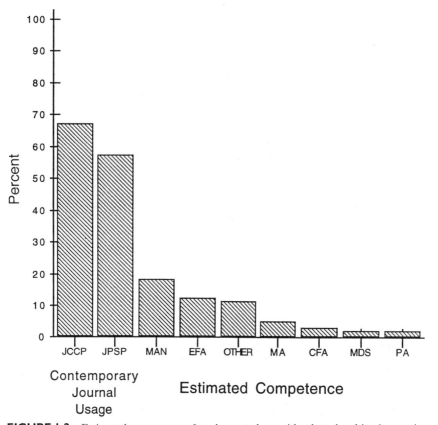

**FIGURE 1.2**   Estimated competence of graduate students with selected multivariate statistics compared to contemporary usage of multivariate statistics. JCCP, *Journal of Consulting and Clinical Psychology;* JPSP, *Journal of Personality and Social Psychology;* MAN, multivariate analysis of variance; EFA, exploratory factor analysis; OTHER, other multivariate analysis; MA, meta-analysis; CFA, confirmatory factor analysis; MDS, multidimensional scaling analysis; PA, path analysis.

pessimistic about their graduate students' ability to apply multivariate statistics in their own research. A comparison of the usage of multivariate statistics in articles published in contemporary psychology journals with the percent of graduate students judged competent to use various procedures reveals a grim picture (see Figure 1.2).

The situation may be no better with practicing researchers. The ever expanding array of multivariate statistics and mathematical modeling techniques increases the amount of time needed to develop technical proficiency with the available options, and most researchers cannot afford the time. In response to this situation, many active researchers tend to use a limited range of techniques that are commonly used in their research area and familiar to the editors of the journals in which they publish. The result is that less than optimal procedures are used, and enthusiasm for specific multivariate procedures sometimes obscures the substantive issue under investigation (see Tinsley, 1995).

This situation need not persist. Early treatments of multivariate techniques were understandable only by learned mathematicians, but over the years scholars have developed simple explanations of the concepts underlying these procedures. Gaining a basic understanding of these techniques is now a reasonable objective for the average graduate student. Moreover, a systematic body of research is now available to guide the practical application of multivariate statistics. For example, empirical research has revealed that it sometimes does no practical harm to violate some of the restrictive theoretical assumptions of these statistical procedures. There has never been a better time for scholars in the sciences and humanities to achieve an informed use of these statistical techniques.

The objective of the *Handbook* is to explain the appropriate uses of those multivariate statistics and mathematical modeling techniques that applied researchers and research consumers are most likely to encounter in the research literature and to use in their own research. The *Handbook* prescribes practices that will enable applied researchers in the sciences and humanities to use these procedures effectively without needing to concern themselves with their mathematical basis or with arcane statistical issues surrounding their use. The *Handbook* emphasizes multivariate statistics and mathematical models as tools; the objective is to inform readers about which tools to use to accomplish the various tasks that confront them. Researchers generally turn their attention to the selection of a statistical procedure when they have a question they need to answer. Therefore, each chapter dealing with a statistical technique includes a discussion of the kinds of questions the technique can (and cannot) be used to answer. These chapters focus on the decisions necessary to use the technique effectively, and they provide examples that illustrate its use. The emphasis is on providing simple, direct, practical advice based on the theoretical and empirical literature to enable readers to make informed use of each technique.

In the remainder of this chapter we briefly introduce five topics in roughly the order users encounter them in the data analysis process. This general overview sets the stage for the authors' more detailed treatments of these issues as they relate to their particular technique. We begin by discussing aspects of the critical but often overlooked process of preparing your data for analysis. Then we consider some preliminary steps you should take to begin the interpretation of your data prior to starting your formal statistical analysis. Next we discuss some of the factors to consider in selecting the best statistical technique to accomplish your research objective; each author provides a more detailed discussion of the uses and limitations of their technique. In selecting a statistical technique it is essential that users take into account any properties of the data that may limit the applicability of the alternative statistical procedures. We conclude with a discussion of critical issues to consider in interpreting the results of your statistical analyses.

Often throughout the remainder of this chapter we will refer in a generally inclusive sense to all of the multivariate statistical techniques and mathematical modeling testing procedures covered in the *Handbook*. We will use terms such as "statistical techniques" or "multivariate procedures" for ease of communication when making general references to the procedures covered in the *Handbook*.

## I. DATA PREPARATION

"Garbage in, garbage out," has been used as a cautionary metaphor for so long that it is now cliché. Nevertheless, the underlying message that it conveys is even more important when relying on computers to perform multivariate procedures than it was at midcentury when researchers performed univariate analyses by hand using mechanical desk calculators. The answers that will emerge from a statistical analysis are dependent on the quality of the data submitted to analysis, so it is critical that the data be as accurate as possible. For this reason, the first step in the analysis process is a close inspection of the data to correct any errors and to deal with missing data and outliers. It is ironic that the painstaking preparation of your data for analysis may take days or even weeks, while the primary analysis often takes only a few hours.

### A. Accuracy

It is likely that all data files of the size submitted to multivariate statistical analysis or used in testing mathematical models contains some errors. These can occur as a result of respondent error (e.g., responding in the wrong place on a scan sheet or using the wrong set of response alternatives), an error by data-entry personnel, or even an error in scanning the answer sheet. Although it is not widely known, scanning errors occur as a consequence of stray marks on the form, dust on the glass, and other idiosyncrasies of the scanning process.

With small data files it is prudent to proofread the data by comparing the computerized data file against the original answers. However, that will seldom be possible with the large data files used in multivariate procedures. Instead, begin by preparing a frequency distribution for each item in the data file and inspect these meticulously for out-of-range, unusual, and missing responses. Check possible errors by comparing original answers given by the respondent against the number appearing in the data file. For example, a missing response may have been recorded on the answer sheet too lightly for the scanner to record it. An out-of-range response may be the result of a stray mark on the answer sheet, despite the presence of a valid answer for that question. The necessary remedial action is clear when these types of errors are identified, and making the necessary corrections is unlikely to introduce significant error into the data. Make all corrections in the data file and on the original (answer sheet or questionnaire) simultaneously to avoid the possibility of future confusion about which is the "correct" response.

Other errors that are due to more subtle causes may still be correctable. For example, the respondent might have answered each question faithfully but skipped one circle on a scan sheet. That would be apparent if the response to one item appears to be missing and the highest item number answered is one greater than the number of questions. Another possibility is the presence of out-of-range responses (e.g., an answer of "4" when only three response alternatives were available). Depending on the pattern of responses, you might decide to change such a response to "3" under the assumption that the respondent intended to use the highest response alternative and made a clerical error.

Also examine combinations of responses to determine whether the pattern of responses is logically consistent. For example, respondents might describe themselves as an only child with two older siblings, as a 15-year-old having a doctorate, as a school teacher having only a high school education, or as having been in a relationship longer than they have been alive (see Behrens, 1997). In instances such as these the likelihood of an error is high, but the way to correct the error is not obvious. You may decide to leave the data unchanged, delete the response that seems most likely to be wrong, or even to delete the respondent from the data file.

Other errors involving combinations of items can occur when item responses are summed to create a composite score. Dawis (chapter 3, this volume) provides a thorough review of scaling methods and the potential problems that can occur. When all of the responses that are aggregated to produce the composite are within the acceptable range, then the summary score of the individual must be within the acceptable range. Nevertheless, the composite score may be quite atypical for the sample of respondents. (We will return to this issue in the section on outliers.) Also examine the distribution of composite scores for the entire sample and the mean and standard deviation to determine whether they are plausible. Both the shape of the distribution and the mean and standard deviation of the sample are important. The proper use of multivariate procedures requires more than a knowledge of the techniques themselves; it also requires a thorough knowledge of the theory and research pertinent to the substantive issues under investigation (Comrey & Lee, 1992). Hetherington (chapter 2) discusses the critical role of theory as a guide to the informed use of multivariate statistics and model-testing procedures.

Given the likelihood that errors will occur in most large data files, it is best to think carefully about these possibilities before beginning the data preparation and to develop an a priori strategy for taking remedial action when such errors are detected. Develop a consistent strategy for dealing with such errors and record all editing decisions for future reference. It is important to avoid the introduction of alternative errors into the data file under the guise of correcting errors.

Two particularly troublesome problems when performing multivariate statistics and testing mathematical models are missing data and outliers.

## B. Missing Data

Venter and Maxwell (chapter 6, this volume), Huberty and Petoskey (chapter 7, this volume), and DiLalla (chapter 15, this volume) discuss alternative procedures for dealing with the problem of missing data. They explain the advantages and disadvantages of such procedures as pairwise and listwise deletion of missing data, briefly describe and evaluate alternative procedures for substituting estimated values in the place of the missing data, and provide references to more extensive treatments of the missing data options. Although we touch upon some of the procedures they review, we have structured this discussion to minimize redundancy. We recommend you turn to the sections of those chapters dealing with missing data after reading this chapter to gain a more complete understanding of the advantages and disadvantages of the alternative strategies for dealing with missing data.

The presence of a small amount of randomly distributed missing data in a large data file is not likely to cause serious problems, and most of the standard procedures for dealing with missing data will be adequate. On the other hand, the presence of systematic missing data can be serious. Systematic missing data occurs when the variable for which information is missing is related to other variables included in the data file (e.g., information about political affiliation is missing when you believe that political affiliation is related to other important variables, such as income level and attitudes towards social issues).

Although it is tempting to assume that any missing data is randomly distributed, we recommend that you test to determine whether the missing data is randomly distributed vis-à-vis the other variables of interest. Create a dummy variable to divide the sample into those who provided the requested information (coded 0) and those who did not (coded 1) and compare the two subgroups statistically on those variables you believe are related to the missing data. You can conclude that the missing information is not systematically related to the other variables if the means of the two groups do not differ significantly. Dropping the cases having missing data will most likely address the problem adequately when the missing data is random and only a few respondents have missing data. However, the missing data is not random if there is a significant mean difference, and the problem is especially troublesome when the significant mean difference accounts for a substantial amount of variance. Under those circumstances dropping the cases having missing data would introduce a distortion into the data.

Another possibility is to drop variables from the data file. This option is most practical when the missing values occur primarily for one or a relatively few variables and those variables are not critical to the research, or the data file contains another variable for which complete data is available that correlates highly with the variable having missing data.

Numerous strategies have been used for estimating (imputing) missing values. Tabachnick and Fidell (1996) argue that it is reasonable to make a well-educated guess if the sample is large, the number of missing values is small, and the researcher has extensive knowledge of the research area. This recommendation again underscores the importance of possessing a thorough knowledge of the substantive area under investigation. The effective use of multivariate procedures requires both a familiarity with the techniques themselves and with the substantive area under investigation.

More typically, missing values are replaced with a mean value calculated on the basis of the entire sample or the subgroup to which the person belongs. Using the mean for the entire sample is a conservative procedure that does not change the overall mean, but it does reduce the variance in scores on that variable, and therefore the correlation of that variable with other variables. Using the mean for the subgroup to which the respondent belongs does not change the mean of the group, but it does reduce the within-group heterogeneity on the variable. Because some tests of significance examine the relation of the variance between groups to the variance within groups, this procedure introduces a slight bias in favor of finding a statistically significant difference between groups and increases the likelihood of a Type I error.

We recommend a variation of that strategy when one of the values in a composite is missing. For example, suppose a respondent failed to answer one of the items on the 10-item Self-Esteem Scale (SES; Rosenberg, 1965). Calculate that respondent's mean score on the nine SES items that were answered and substitute that mean in place of the missing value. This strategy is conservative in that it does not change the respondent's overall mean on the scale, but it inflates the internal consistency of the scale slightly.

Another procedure is to use regression to calculate predicted values for the missing data. Use the other variables in the data file as the independent variables and calculate a regression equation based on the data of all those respondents who do not have missing data for the (dependent) variable. Impute the resulting predicted values in place of the missing values. This process can be iterated; the predicted value can be inserted in place of the missing value and the data of all the respondents used to generate a second predicted missing data value for each respondent having missing data. Then compare the second missing data value to the first missing data value and repeat this process until the predicted values stabilize.

Despite it's seeming sophistication, this procedure is feasible only when the data file contains at least one variable having no missing data that is a reasonable predictor of the variable with missing data. Furthermore, this procedure reduces the variance in the variable having missing data and introduces greater consistency among the variables. Moreover, the regression equation will sometimes yield out-of-range estimated scores that must be discarded or adjusted to fall within the acceptable range.

A point often overlooked by researchers is that the presence of missing data may itself be useful as a predictor in your research. The respondent's decision not to answer certain items represents a sample of behavior that may have a meaningful relation to the issues under investigation. You can test this possibility by creating a dummy variable that distinguishes cases having complete data from cases having missing data. Including this information about the respondent in your analysis might convert the liability of missing data to an asset.

Finally, whenever missing data are imputed, it is good practice to rerun your analyses using only those cases having complete data. Compare the results obtained using those cases having complete data results to those based on the entire sample when missing data is imputed. You can have confidence in the results based on the entire sample if they are essentially the same as those based on the subgroup having complete data. If the results of the two analyses are different, however, you need to investigate the reasons for the difference.

## C. Outliers

Outliers are values on one or a combination of variables that are so extreme they exert a disproportionate influence on the analysis. As such, they distort the statistics produced by the analysis and lead to results that do not generalize. Outliers can be univariate (i.e., cases with an extreme value on a single variable) or multivariate (i.e., cases with an unusual combination of scores on two or more variables). The proper handling of outliers involves a three-step process of detecting the outliers, determining the cause of the outliers, and reducing the influence of the outliers. Venter and Maxwell (chapter 6, this volume) and Huberty and Petoskey (chapter 7, this volume) describe diagnostic procedures for detecting the presence of outliers and remedial procedures for nullifying their negative effects. As with our treatment of missing data, we have structured our discussion of outliers to minimize redundancy. For example, Venter and Maxwell distinguishes among outliers, leverage, and influence, a distinction we do not draw in our discussion. After reading this chapter we recommend you read the sections of chapters 6 and 7 dealing with outliers to gain a more complete overview of procedures for detecting outliers and nullifying their influence.

Outliers can occur as a result of an error in the data file (e.g., entry of an incorrect value), a programming error (e.g., an error in recoding or transforming variables or a failure to identify missing data values correctly), or the presence of a valid but exceptional data point. Outliers that result from an error in the data file should be identified and corrected while editing the data. Outliers that occur as a result of programming errors should be detected and remedied during the process of debugging the analysis. The presence of valid but extreme data points is the most troublesome to remedy.

Univariate outliers for dichotomous variables have very uneven distributions between the two categories (i.e., 90% of the cases in one category). In this instance, scores in the category with 10% of the cases have more influence on the analysis than scores in the category with 90% of the cases. Univariate outliers for continuous variables are those cases that have large standardized scores (i.e., $z$-scores $> 3.29$; $p < .001$ two-tailed). Multivariate outliers are cases that have an unusual pattern of scores on two or more variables. These cannot be detected as easily as univariate outliers because each of the individual scores may be within the typical range. For example, it would not be unusual to find a respondent who was 14 years old in a data file nor to find a respondent who had a college education, but finding a 14-year-old respondent who had completed a college education would be quite unusual.

One statistical procedure for screening for multivariate outliers is to compute the distance of each case from the centroid (i.e., mean of all the variables) of the remaining cases. This value, referred to as the Mahalanobis distance, is available from most of the leading computer statistics packages. We recommend you screen for both univariate outliers and multivariate outliers so that you will have an understanding of the full range of the problem before deciding what corrective action to take. Also, it is best to screen for outliers again after taking corrective action because the extreme nature of some cases might have been camouflaged by the presence of even more extreme cases. Once the most extreme outliers have been nullified, the extreme nature of some of the remaining cases may become more apparent.

The nature of univariate outliers is clear; the case has an extreme score on one or more variables. A large Mahalanobis distance signifies a multivariate outlier, but it does not indicate the variables on which the case is deviant. You can identify these variables by creating a dummy variable to separate the rest of the cases (coded 0) from the case having the outlier (coded 1) and then using the dummy variable as the dependent (grouping) variable in a discriminant analysis (see M. Brown & Wicker, chapter 8, this volume).

Several procedures are available for reducing the influence of outliers. One obvious strategy is to make sure the data values were entered into the data file accurately for the case in question. If several outliers have been identified, check to determine whether one variable is responsible for most or all of them. If so, consider whether that variable could be eliminated from the analysis. As in the case of missing data, that option will be most feasible if the variable is not critical to the substantive issue under investigation or the data file contains another variable that correlates highly with it. In the case of univariate outliers, it may be possible to transform the relevant variables to lessen their effect (see Hallahan & Rosenthal, chapter 5, and Venter & Maxwell, chapter 6, this volume). However, this strategy may not have the desired effect on multivariate outliers because it is the combination of scores on two or more variables that causes the problem.

If it is feasible to conclude that the cases that are outliers are not part of the population from which you intended to sample, another possibility is to drop those cases from the analysis. Do not make this decision lightly; the cases dropped represent the type of cases to which your results do not generalize. If you decide to drop cases from the analysis, we recommend that you run your analyses with and without the excluded cases and compare the results. You can have more confidence in the generalizability of the results based on the trimmed sample if they are essentially the same as those based on the entire sample. If the results of the two analyses are different, however, the excluded outliers represent the kinds of cases to which your results do not apply.

## II. STUDY YOUR DATA

The first—irreplaceable—steps in interpreting data are to calculate basic descriptive statistics for the primary variables under investigation (e.g., means, standard deviations, and correlation coefficients) and to develop a visual depiction of the relations among the variables (cf. Anscombe, 1973; Behrens, 1997; Tufte, 1997; Tukey, 1969). It is tempting to skip this step, for these procedures can be time-consuming, and the quick and easy answers provided by the computer create a strong enticement to pick a statistical technique and apply it to your data forthwith. Data analysis is always an exciting part of the research process because it marks the point at which the answers to the questions that drove the research begin to emerge. Nevertheless, we urge you to remember that your objective is not to "analyze data," but to obtain meaningful answers to your research questions.

The perusal of descriptive information is important in helping you gain an understanding of the richness and complexity of your data, and in alerting you to any aberrations that do not conform to the general rules emerging from the analysis. These basic descriptive procedures provide an early glimmering of your findings, and they are useful in directing your selection of inferential statistics. In practice, data analysis often involves an iterative procedure in which analysts try a variety of strategies and options for extracting meaning from their data. Hallahan and Rosenthal (see chapter 5, this volume) suggest a number of specific procedures that you can use in beginning the analysis process. The statistical techniques described in the *Handbook* will bring greater objectivity, precision, and explanatory power to the analysis, but your prior study of the data may provide critical information that will allow you to make decisions about the analysis in an insightful manner.

It is important to avoid the prevailing practice of testing the null hypothesis and making dichotomous pronouncements of statistical significance that ignore fundamental aspects of the data (Cohen, 1994; Estes, 1997; Gonzalez, 1994; Loftus, 1996; Tinsley, 2000). Gonzalez has criticized significance testing as a "ritual" of the social sciences, done more as a matter of routine than as a way to understand the meaning of the data. The unthinking use of multivariate procedures can be criticized on this basis. Testing the null hypothesis that the patterns of association underlying your data are not "completely random" will not contribute significantly to your understanding of the meaning of the data. The best solution is to base conclusions on the skillful integration of descriptive and inferential information. This can happen only if you take the time to extract the descriptive information.

Once your analyses are completed, we recommend you revisit the descriptive statistics and graphical representations of the data. Many of the statistical techniques described in the *Handbook* provide an option for a graphical display of the relations among the variables included in the analysis. Careful inspection of these displays may reveal important information not apparent in the *p*-values, fit indices, or other statistics that summarize the results. In some instances the information gained by reexamining the descriptive statistics or graphical displays may lead you to repeat the analyses after making important modifications in the options. They could lead you to use another method of analysis to avoid critical artifacts of the procedure initially selected, or to transform your data to lessen the influence of artifacts on the analysis. In still other instances it could lead you to modify or qualify the decisions suggested by a preliminary interpretation of the analysis. For example, a common oversight in the literature is to fail to consider whether "statistically significant" results are of practical importance.

## III. SELECTING A STATISTICAL TECHNIQUE

Conceptually, the task of selecting the best technique for analyzing your data is simple, requiring only a consideration of the question(s) you want to answer and the data requirements of each of the applicable techniques. Putting this strategy into operation can be tricky. The statistical techniques described in the *Handbook* overlap in their potential applications; often there is more than one technique that can be applied to the solution of a problem. Furthermore, many of these statistical techniques are flexible procedures that can be used to address more than one type of question, so often more than one analysis strategy is available. In the face of this complexity it is comforting to know that the set of statistical procedures covered in the *Handbook* is both finite and reasonably comprehensive.

It would be nice to have a concise table or decision tree that would direct you to the proper procedure after answering a few simple questions, and Hetherington has undertaken just such an effort (see chapter 2, this volume). His decision tree takes into account (a) whether multiple predictors (independent variables), multiple outcomes (criteria or dependent variables) or both multiple predictors and multiple outcomes are to be analyzed; (b) whether these data were measured at a nominal level or at a higher level of measurement; (c) whether the purpose of the analysis is to determine degree of association, group differences, or group membership; and (d) whether the analysis is exploratory (i.e., guided by general hypotheses and focused on interpreting the emergent characteristics of the data) or confirmatory (i.e., tests of empirical support for explicit hypotheses).

Hetherington's table will be helpful in identifying technique(s) that are likely to satisfy your analysis needs, but the statistical procedures covered in the *Handbook* are so flexible and interrelated that no brief tabular presentation could do justice to the full range of their potential applications. For example, Hetherington does not list any techniques for testing hypotheses when multiple predictors and multiple outcomes are involved, but specifically stated hypotheses could be tested using canonical analysis (listed under exploratory procedures). No confirmatory procedures are listed for testing group membership on multiple outcomes using a single predictor, but creative applications of multidimensional scaling could be used to address this issue. Furthermore, Hetherington cautions that the distinction between exploratory and confirmatory analyses depicted in his table is a convenient heuristic that actually depicts the end points of a continuum: a single study could test both exploratory and confirmatory components.

Additional information about the relations among the statistical techniques is provided throughout the *Handbook*. Table 1.1 provides a brief summary, but repeated references to other procedures for addressing questions appear in many of the chapters. Furthermore, some procedures such as regression, exploratory factor analysis, confirmatory factor analysis and structural equation modeling are so basic and flexible that most chapters contain one or more references to them. Furthermore, many of the procedures used to test mathematical models actually make use of basic multivariate procedures such as multiple regression, so repeated references and comparisons appear throughout the chapter.

## IV. DATA REQUIREMENTS

### A. Level of Measurement

Tinsley and Weiss (chapter 4, this volume) note that the level of measurement used in obtaining ratings is a critical determinant of the proper approach to the estimation of interrater reliability and agreement; that is no less true for many of the other statistical procedures described in the *Handbook*. Some of the data submitted to multivariate analysis possess the properties of nominal-level measurement. Nominal scales involve simple classification without order; the numbers function as shorthand names for qualitatively different categories. Many of the descriptive characteristics of respondents are measured at the nominal level (e.g., gender, race, religious affiliation, and political affiliation), as are attributes such as clinical diagnosis (e.g., borderline and bipolar disorder) and occupation (e.g., plumber and school teacher).

Measurements in the social sciences and humanities more frequently possess the properties of ordinal-level measurement. Ordinal measurement implies the ability to rank objects on a continuum, but without implying equal intervals between the ranks. Ordinal scales, in the form of the simple graphic rating scale, are among the most frequently used scales in counseling research (see Dawis, chapter 3, this volume). The two dimensions along which these scales most commonly vary are the number of scale levels (e.g., 3-, 6-, or 11-point scales) and the manner in which the scale levels are defined (e.g., numerical values, adjective descriptions, or graphic intervals).

Data from the social sciences and humanities seldom achieve the interval level of measurement, but data from the biological sciences may do so. True measures of the distance between objects on a continuum are possible at the interval level of measurement. The application of statistical procedures that assume an interval level of measurement to data that do not possess those properties may distort the true relations among the cases and variables if the assumption of equal intervals is grossly inappropriate.

### B. Assumptions

The multivariate techniques described in the *Handbook* require a number of assumptions, some of which may not be apparent to users. Each author has identified those that are most relevant to the techniques they are

**TABLE 1.1    Relations among Multivariate Statistical and Mathematical Modeling Techniques**

| Techniques | Author | Chapter | Pages |
|---|---|---|---|
| Analysis of covariance | | | |
|   Canonical analysis | Thorndike | 9 | 240–241 |
|   Hierarchically nested models | Kreft | 21 | 613, 618–619 |
|   Logistic regression | Imrey | 14 | 392 |
|   Poisson regression | Imrey | 14 | 392 |
| Analysis of variance | | | |
|   Hierarchically nested models | Kreft | 21 | 613 |
|   Logistic regression | Imrey | 14 | 392 |
|   Loglinear models | Imrey | 14 | 392 |
|   Multitrait–multimethod models | Dumenci | 20 | 586–593 |
|   Poisson regression | Imrey | 14 | 392 |
|   Regression | Thorndike | 9 | 240–241 |
| Canonical analysis | | | |
|   Analysis of covariance | Thorndike | 9 | 240–241 |
|   Discriminant function analysis | Thorndike | 9 | 240–241 |
|   Factor analysis (exploratory) | Thorndike | 9 | 241–242 |
|   Multivariate analysis of covariance | Thorndike | 9 | 240–241 |
|   Multivariate analysis of variance | Thorndike | 9 | 240–241 |
|   Principal components analysis | Thorndike | 9 | 241–242 |
| Circumplex models | | | |
|   Factor analysis (exploratory) | Tracey | 22 | 644–647 |
|   Loglinear modeling | Tracey | 22 | 659 |
|   Multidimensional scaling | Davison[a] | 12 | 347 |
| | Tracey | 22 | 644–647, 649–650 |
|   Principal components analysis | Tracey | 22 | 644–647 |
|   Structural equation modeling | Tracey | 22 | 650–655 |
| Cluster analysis | | | |
|   Multidimensional scaling | Davison[a] | 12 | 344 |
| Discriminant function analysis | | | |
|   Canonical analysis | Thorndike | 9 | 240–241 |
| Factor analysis (confirmatory) | | | |
|   Factor analysis (exploratory) | Hoyle | 16 | 469 |
|   Multitrait–multimethod models | Dumenci | 20 | 586, 593–597 |
|   Structural equation modeling | DiLalla | 15 | 442–443 |
| | Hoyle | 16 | 465–466 |
| Factor analysis (exploratory) | | | |
|   Canonical analysis | Thorndike | 9 | 241–242 |
|   Circumplex models | Tracey | 22 | 644–647 |
|   Factor analysis (confirmatory) | Hoyle | 16 | 469 |
|   Multidimensional scaling | Davison[a] | 12 | 345–346 |
|   Partial correlation | Cudeck | 10 | 267 |
|   Structural equation modeling | DiLalla | 15 | 442 |
|   Regression | Cudeck | 10 | 269 |
| Generalizability theory | | | |
|   Item response theory | Marcoulides | 18 | 542–545 |
|   Multitrait–multimethod models | Dumenci | 20 | 597 |
| Hierarchically nested models | | | |
|   Analysis of covariance | Kreft | 21 | 618–619 |
|   Analysis of variance | Kreft | 21 | 613 |
|   Structural equation modeling | Kreft | 21 | 620 |
|   Regression | Kreft | 21 | 613, 618–619, 635–637 |

*(continued)*

**TABLE 1.1** (*continued*)

| Techniques | Author | Chapter | Pages |
|---|---|---|---|
| Item response theory | | | |
|   Generalizability theory | Marcoulides | 18 | 542–545 |
|   Multidimensional scaling | Davison[a] | 12 | 346 |
| Logistic regression | | | |
|   Analysis of covariance | Imrey | 14 | 391–392 |
|   Analysis of variance | Imrey | 14 | 391–392 |
| Loglinear modeling | | | |
|   Analysis of variance | Imrey | 14 | 392 |
|   Circumplex models | Tracey | 22 | 659 |
| Moderator and mediator models | | | |
|   Structural equation modeling | DiLalla | 15 | 444 |
| Multidimensional scaling | | | |
|   Circumplex models | Davison[a] | 12 | 347 |
| | Tracey | 22 | 644–647, 649–650 |
|   Cluster analysis | Davison[a] | 12 | 344 |
|   Factor analysis (exploratory) | Davison[a] | 12 | 345–346 |
|   Item response theory | Davison[a] | 12 | 346 |
|   Moderator and mediator models | DiLalla | 15 | 444 |
| Multitrait-multimethod models | | | |
|   Analysis of variance | Dumenci | 20 | 586–593 |
|   Covariance component analysis | Dumenci | 20 | 586, 597–601 |
|   Factor analysis (confirmatory) | Dumenci | 20 | 586, 593–597 |
|   Generalizability theory | Dumenci | 20 | 597 |
|   Structural equation modeling | Dumenci | 20 | 586, 587–591 |
| Multivariate analysis of covariance | | | |
|   Canonical analysis | Thorndike | 9 | 240–241 |
| Multivariate analysis of variance | | | |
|   Canonical analysis | Thorndike | 9 | 240–241 |
| Partial correlation | | | |
|   Factor analysis (exploratory) | Cudeck | 10 | 267 |
| Poisson regression | | | |
|   Analysis of covariance | Imrey | 14 | 391–392 |
|   Analysis of variance | Imrey | 14 | 391–392 |
| Principal components analysis | | | |
|   Canonical analysis | Thorndike | 9 | 241–242 |
|   Circumplex models | Tracey | 22 | 644–647 |
|   Multitrait-multimethod models | Dumenci | 20 | 586, 597–601 |
| Regression | | | |
|   Analysis of variance | Thorndike | 9 | 240–241 |
|   Exploratory factor analysis | Cudeck | 10 | 269 |
|   Hierarchically nested models | Kreft | 21 | 613, 618–619, 635–637 |
|   Logistic regression | Imrey | 14 | 392 |
|   Poisson regression | Imrey | 14 | 392 |
|   Structural equation modeling | DiLalla | 15 | 443–444 |
| Structural equation modeling | | | |
|   Circumplex models | Tracey | 22 | 650–655 |
|   Factor analysis (confirmatory) | DiLalla | 15 | 442–443 |
| | Hoyle | 16 | 465–466 |
|   Factor analysis (exploratory) | DiLalla | 15 | 442 |
|   Hierarchically nested models | Kreft | 21 | 620 |
|   Multitrait–multimethod models | Dumenci | 20 | 586, 587–591 |
|   Regression | DiLalla | 15 | 443–444 |

[a] Davison & Siceri.

describing. Most common are the assumptions of independence, multivariate normality, homoscedasticity, homogeneity of variance, homogeneity of the variance–covariance matrices, and linearity. Venter and Maxwell (chapter 6, this volume) provide the most extended discussion in the *Handbook* of the assumptions underlying multivariate procedures, diagnostic procedures for detecting violations of the assumptions, and procedures for correcting violations of the assumptions.

These assumptions are in place because they were necessary to allow psychometricians to derive the formulas used to perform the multivariate computations. Research has shown that violations of some of these assumptions has only a small effect on the power of the analysis and the Type I error rate (e.g., Baker, Hardyck, & Petrinovich, 1966; Everitt, 1979). The violation of other assumptions can greatly reduce the power of the analysis or inflate the probability of a Type I error. Therefore, it is critical when planning a multivariate study to consider the assumptions required by the statistical procedures you plan to use, the robustness of these procedures for violations of their assumptions, and the extent to which your data satisfy the assumptions.

The assumption of independence, by far the most important assumption, is that every observation is statistically independent of every other observation. Even small violations of this assumption produce a substantial effect on the power and Type I error rate of the analysis (see Hallahan and Rosenthal, chapter 5, this volume, for an example.) You might think that independence is assured if the data are collected from each case without overt cooperation among the respondents, but that is not necessarily the case. Observations are most likely independent when the treatment is administered individually (Glass & Hopkins, 1984). However, the observations may violate the assumption of independence whenever interactions occur among people (e.g., members of the same class or discussion group, marital or dating partners, or members of the same team or club). M. Brown and Wicker (chapter 8, this volume) underscore the importance of the assumption of independence and discuss one heuristic proposed for testing violations of the assumption of independence.

The assumption of multivariate normality is that each variable and all possible linear combinations of the variables are distributed normally. This assumption is difficult to test because of the practical difficulty of testing all possible linear combinations of the variables. Although some multivariate procedures appear to be robust for violations of this assumption, a blanket assurance of robustness in not possible. Robustness of the multivariate procedure for violations of the assumption of normality depend on both the specific technique and features of the design (e.g., whether the data are grouped or ungrouped and whether the purpose of the analysis is descriptive or inferential). Hallahan and Rosenthal (chapter 5), Venter and

Maxwell (chapter 6), Huberty and Petoskey (chapter 7), M. Brown and Wicker (chapter 8), DiLalla (chapter 15), and Hoyle (chapter 16) discuss tests of multivariate normality, the robustness of their statistical procedures for violations of this, and procedures that can be used when violations of this assumption are detected.

The assumption of homoscedasticity is that for any given pair of continuous variables the variability in scores for one of the variables is roughly the same for all values of the other continuous variable. Violations of the assumption of homoscedasticity are likely to result in weaker but still valid results. Correcting the heteroscedasticity by transformation of the variables is desirable, if possible, because the strength of the linear relations among the variables will be increased.

The more widely known assumption of homogeneity of variance is that the variability in the dependent (continuous) variable is roughly the same for all values of the independent (discrete) variable. This is simply the assumption of homoscedasticity applied to the special case when one of the two variables, usually the independent (or grouping) variable is measured at the nominal level. The assumption of homogeneity of the variance–covariance matrices is the multivariate extension of the homogeneity of variance assumption. Violations of the assumptions of homogeneity of variance and homogeneity of variance–covariance matrices inflate Type I error rates when the sample sizes are unequal. This is most likely to occur when the larger variance is associated with the smaller cell size. Violations of these assumptions are controlled by variable transformations or the use of a more stringent criterion for statistical significance. Hallahan and Rosenthal (chapter 5), Venter and Maxwell (chapter 6), Huberty and Petoskey (chapter 7), and Brown and Wicker (chapter 8) describe diagnostic procedures for testing these assumptions and suggest strategies for dealing with violations when they are detected.

The assumption of linearity is that a linear (i.e., monotonic or straight line) relation exists between pairs of variables, where one or both of the variables can be composites of several of the variables under investigation. This assumption is of practical importance because it has implications for the measure used to determine the strength of association between variables. For example, Pearson's $r$ only accounts for the linear relations among the variables. This assumption also has theoretical importance in that most models in the social sciences and humanities assume a linear relation among population variables. Therefore, the presence of nonlinearity in the data file may signify that the sample does not provide a valid representation of the population. Venter and Maxwell (chapter 6, this volume) describe diagnostic procedures for evaluating the linearity assumption and transformations that can be used when violations of this assumption are detected.

## V. INTERPRETING RESULTS

Many of the statistical techniques described in the *Handbook* are maximization procedures that "peek" at the data file and identify an atheoretical solution that maximizes the amount of variance explained. This feature is desirable when you are interested only in the sample on which the analysis is performed, for it provides the fullest possible accounting of the variance in that sample, but this property has drawbacks when you intend to generalize the results of your analysis. The accuracy of these generalizations is contingent on the extent to which the sample is representative of the population. Unfortunately, most samples have idiosyncrasies that are incorporated into the solution by the statistical procedure. The tendency of the analysis to capitalize on sample-specific variations introduces a positive bias into the results. Therefore, it is essential to evaluate the generalizability of the results to an independent sample.

Four procedures have been suggested for evaluating the generalizability of multivariate results, the use of shrinkage formulas (used with multiple regression), cross-validation, and the use of bootstrapping and jackknife procedures. We will describe each of these procedures briefly, then conclude this section with a brief consideration of the distinction between practically significant and statistically significant results.

### A. Shrinkage Estimates

Shrinkage formulas (e.g., Cohen & Cohen, 1983) are used with multiple regression to estimate the extent to which the multiple correlation (multiple-**R**) would shrink if the regression weights were applied to a new sample of respondents. These formulas are applied to the sample used to calculate multiple-**R**, so access to a second sample is not required. In general, these formulas consider the relation between the number of predictors and the number of cases, and return large shrinkage estimates when the number of predictors is relatively large. Venter and Maxwell (chapter 6, this volume) provide a more detailed discussion of the use of the shrunken multiple-$R^2$ in prediction and its relation to multiple-$R^2$ and the adjusted multiple-$R^2$.

Wherry (1951) criticized shrinkage estimates because they require a number of statistical assumptions that are not likely to be satisfied in practice, and Mosier (1951) and others advocated cross-validation as superior to shrinkage estimates. However, subsequent research has shown that estimates of multiple-**R** derived from shrinkage formulas are more accurate than estimates derived from cross-validation studies when predicting from a fixed set of predictors (Herzberg, 1969).

Often the purpose of the analysis is to select a reduced set to predictor variables from the larger set of variables under investigation. For this purpose, it is essential to use some form of cross-validation to estimate the generalizability of the results to independent samples. In addition to the positive bias in the estimate of the multiple-**R**, it is also possible that the predictor variables that are selected for retention in the reduced set may not be the best set of predictor variables in other samples. Shrinkage formulas do not provide an estimate of the shrinkage that will occur as a consequence of selecting a subset of the predictors.

## B. Cross-Validation

Numerous strategies exist for cross-validation (see Mosier, 1951; Norman, 1965). We will briefly comment on two such strategies; Dawis (chapter 3, this volume), M. Brown and Wicker (chapter 8, this volume), and Thorndike (chapter 9, this volume) provide additional information about cross-validation. In both procedures, you begin by randomly dividing your sample into two parts. Perform your analysis on one of the two subsamples; this subsample is called the developmental, derivation, training, or screening sample because the initial solution (or equation) is developed (or derived) from an analysis of this group. Then apply this solution to the second subsample, called the cross-validation, calibration, or hold-out sample. The results provide an indication of how well the solution actually works when applied to an independent sample from the same population.

The hold-out group procedure is one common cross-validation method. In this procedure you randomly split your sample into a larger group (typically containing 2/3 to 3/4 of the cases) that is used as the developmental sample and a smaller group that serves as the cross-validation group. The likelihood that an analysis will take advantage of sample-specific idiosyncrasies is positively related to the number of variables and negatively related to the number of cases, so assigning most of the cases to the developmental sample maximizes the reliability (i.e., stability) of the results and the likelihood of obtaining generalizable results. An alternative cross-validation strategy is the double-split cross-validation procedure, in which you divide the total sample into half and use each half as a derivation (i.e., developmental) sample. Then apply the results from subsample A to subsample B and the results from subsample B to subsample A. In essence, you are performing two cross-validation studies simultaneously. If the results of the two analyses are similar and both sets of results cross-validate, it is then permissible to perform the analysis on the entire sample, thereby taking advantage of all of the information at your disposal and assuring yourself of the greatest possible stability for your results.

## C. Bootstrapping and Jackknifing

Two other methods for estimating the generalizability of multivariate results are bootstrapping and jackknifing. Like the other procedures, bootstrapping and jackknifing are concerned primarily with the reliability of results across samples drawn from the same population. Unlike cross-validation and shrinkage analyses, bootstrapping and jackknifing are resampling procedures in which samples of various sizes are taken from the same large data set and relevant statistics (e.g., structural coefficients or regression coefficients) are calculated on the reduced data set and then summarized over the entire larger data set.

The difference between the two procedures revolves around how samples are drawn from the large data set, how estimates are calculated, and how overall summary estimates are obtained. In bootstrapping, the researcher randomly draws a large number (e.g., 1000) of subsamples from the full data set, and with replacement, calculates the relevant statistic (i.e., bootstrap values) in each sample and builds a distribution of bootstrap values. The mean of this sampling distribution is the estimate of the population statistic (called a bootstrap estimate), and its standard deviation is an estimate of the population standard error. The bootstrap distribution can be used to estimate a variety of confidence intervals and to test null hypotheses about the value of the test statistic in the population.

In using the jackknife procedure, researchers would first calculate the relevant statistic (e.g., structure coefficient, regression coefficient, multiple or bivariate correlation) using the full sample. They would then randomly divide the data set into small (e.g., $n = 1$ or 2) subsamples and repeatedly recalculate the relevant statistic on the reduced full sample data. Pseudo-values of the statistic are then obtained by subtracting weighted values of the sample reduced statistic from a weighted value of the statistic obtained from the full sample, and a distribution of pseudo-random values is obtained. The mean of the distribution provides an estimate of the population parameter (called a jackknife estimate), and the standard deviation of the distribution provides an estimate of the standard error. As in the bootstrap procedure, the jackknife distribution can be used to develop various confidence intervals and to test null hypotheses about the value of the test statistic in the population.

A growing body of research has focused on further developing each method (e.g., comparing different methods for sampling and creating confidence intervals in the bootstrap method) and on comparing methods, but these studies have not always yielded consistent results. For example, Lederer (1996) found that the bootstrap method yielded positively biased estimates of the multiple correlation in regression analyses, especially in small samples, when compared to conventional cross-validation procedures. Thompson (1995) found that bootstrapping yielded positively biased canonical correlation coefficients. He argued that a positive bias is to be expected because resampling is carried out on the sample in hand and will be influenced by commonalities among cases in the sample. Thompson (1995) concluded that "inflated empirical evaluations of reliability are often superior to a mere presumption of reliability [we would assume under conditions where cross-validation is impossible], especially when the researcher can take this capitalization into account during interpretation." (p. 92). However, Thorndike (chapter 9, this volume) concludes that standard cross-validation procedures are still the method of choice for testing cross-sample reliability and estimating population multiple correlations until we learn more about the properties of the bootstrapping procedure.

Direct comparisons of jackknifing and bootstrapping procedures suggest that bootstrapping procedures yield positively biased estimates of relevant population values, but that the jacknifing procedure may provide even larger underestimates and may yield inflated type I error rates, especially for canonical correlation estimates and hypothesis tests (e.g., Dalgleish, 1994). Dalgleish (1994) concluded that bootstrap methods are to be preferred over jackknife tests in discriminant analyses, but recent efforts to correct for the positive bias associated with bootstrap methods yielded inflated type I errors in simulation tests of the canonical correlation (Efron, 1987).

There are obviously differing opinions in the literature about the value of bootstrapping, jackknifing, and other procedures for estimating the cross-sample reliability of multivariate results, and research on these methods continues to be a vigorous area of inquiry. All of the authors in this volume provide some coverage of these issues as they relate to the specific analytic techniques covered in the chapter. The authors and we agree with Thompson (1995) that "the business of science is formulating *generalizable* insight" (p. 92) and that direct tests of reliability, even though somewhat biased, are superior to nonempirical assumptions of reliability.

Meta-analysis also provides a method of assessing cross-sample replication by allowing investigators to estimate the population effect size and to test the degree to which sampling error and other artifacts account for effect size variability in a body of literature. Meta-analytic methods also can be used to investigate the function of both substantive and methodological variables as moderators of effect size variability that limit the generalizability of findings. Meta-analysis has the advantage of using an entire literature to estimate cross-sample replication rather than having to estimate generalizability from a single sample drawn for the purpose of testing hypotheses in a single study (see Becker, chapter 17, this volume; Schmidt, 1996).

## VI. STATISTICAL VERSUS PRACTICAL SIGNIFICANCE

Much has been written in the social sciences in recent years (e.g., Cohen, 1990, 1994; Harris, 1997; Schmidt, 1996; Wilkinson and the Task Force on Statistical Inference, 1999) on the meaning of statistical significance and what it does and does not say about the practical meaning of the obtained results. The conventional alpha levels used in most social science research are estimates of the probability of making a type I error (i.e., rejecting the null hypothesis when it is, in fact, true). Alpha levels are not (as is often implied) estimates of the reliability, replicability, or practical importance of the effect. Instead, estimates of the practical importance of an effect must be based on direct estimates of effect size (i.e., all else being equal, larger effects are more important than smaller effects) and effect sizes and associated confidence intervals must be presented in written reports of the research (see Wilkinson et al., 1999) in addition to alpha levels. (Hallahan & Rosenthal, chapter 5, this volume, provide a more complete discussion of this issue.)

However, researchers must guard against dismissing seemingly small effects as having no practical significance for several reasons. First, human behavior, thought, and emotion are complexly, multiply, and interactively determined. Second, single-study effect size estimates are sample-based estimates that were obtained using less than perfectly reliable measures. Therefore, they tend to underestimate population effect sizes. Large sample studies produce less biased estimates of population effects than small sample studies, but they also yield negatively biased estimates of population effects because measurement error is always present. Thus, accounting for a small amount of variance in a multiply determined, fallibly measured dependent variable using a single (or small set) of independent variables that also have been fallibly measured may be both theoretically and practically important.

Conversely, obtaining seemingly large effect sizes should alert investigators to the possibility that something may be amiss in their data file (coding, transcription, or recording errors), that common method variance may be producing inflated effect size estimates, or that measures bearing different names are actually measuring the same constructs. In our view, Cohen's (1988) guidelines for determining effects obtained in the social sciences as small (e.g., $r = .10$, $d = .20$), medium (e.g., $r = .30$, $d = .50$), or large (e.g., $r = .50$, $d = .80$) need another category reflecting that the obtained effect size estimate may be too large ($r > .50$, $d > .80$).

Third, the context must be considered in interpreting the practical importance of the obtained effect size, a point discussed further by Hallahan and Rosenthal (chapter 5, this volume). For example, biomedical researchers routinely terminate studies when effect sizes reach what social science guidelines would consider to be small (i.e., $r = .10$ to .20) on the grounds that it would be unethical to withhold such a demonstrably effective treatment (Rosenthal, 1990). The same might hold for some findings from other areas of applied science if standard effect sizes were recast into a Binomial Effect Size Display (BESD: Rosenthal & Rubin, 1982) that expresses correlation coefficients as indices of increases in (for example) survival, success, and improvement rates. Thus, for example, a correlation of .23 between AZT treatment and AIDS survival rates suggests that AZT treatment accounts for only 5% of the variance in survival rates, but the BESD reveals that 23% more people will survive if given AZT treatment than if the treatment was not available to them. The chapters by Hetherington (chapter 2) and Hallahan and Rosenthal (chapter 5) provide more detailed discussions of the issues to be considered in interpreting the practical importance of the results obtained from multivariate statistics and mathematical modeling procedures. Hetherington explains the influence of theory on the determination of the research question, sampling of variables, selection of statistical technique (all of which affect the generalizability of the results), and interpretation of results. Hallahan and Rosenthal also emphasize the importance of assessing the extent to which the data correspond to specific theoretically meaningful predictions, and they caution against assuming that statistically significant results are important. They also provide additional examples of how seemingly trivial effects can have practical importance.

## VII. OVERVIEW

The *Handbook* is organized into three parts for ease of use by practitioners and researchers. Although the *Handbook* would make an excellent required text in multivariate statistics classes, we anticipate that many applied researchers will use it as a reference work to be read as needed. For that reason, we have separated the more familiar multivariate statistics (Part 2) and the newer applications of these procedures in model testing (Part 3) for ease of use. Each chapter in Part 2 and Part 3 is written as a relatively free-standing treatment that does not rely on a prior knowledge of information in the other chapters. In contrast, part 1 (Introduction) provides the general background necessary to understand and effectively use multivariate statistics and mathematical modeling procedures. We anticipate that all users of multivariate statistics will benefit from a close reading of these chapters.

### A. Part I: Introduction

In chapter 2 Hetherington reviews the relations among multivariate statistical techniques and emphasizes the importance of theory in the effective use of multivariate statistics and the testing of mathematical models. Multivariate techniques are tools; just because the computer makes a tool available does not mean it should be used. A careful consideration of theory in the selection of variables and the formulation of specific hypotheses and research questions should inform the use of these techniques. Hetherington reviews the major theoretical perspectives that have guided research in the twentieth century and advocates the use of strong theory and critical multiplism to guide scientific inquiry.

In chapter 3 Dawis provides a comprehensive overview of measurement issues and procedures that are likely to have an important influence on the outcomes of multivariate research. Most researchers fail to grasp this essential point: fully 64% of the experienced researchers who participated in one survey incorrectly thought that measurement error could cause spuriously significant results (Zuckerman et al., 1993). Dawis identifies important differences in scales constructed for theoretical as opposed to practical purposes, and evaluates the Thurstone, Likert, Guttman, paired comparisons, multiple rank orders and ratings methods of measuring constructs. His treatments of reliability, validity, and the attenuation paradox highlight critical features of the data that can influence the outcomes obtained from multivariate statistics and the evaluation of mathematical models.

In chapter 4 Tinsley and Weiss discuss the use of ratings data in multivariate analysis and the analysis of mathematical models. The rating scale is one of the most widely used measuring instruments in the biological sciences, social sciences, and humanities, but few applied researchers seem to understand the design issues that are unique to the collection of ratings data nor the proper procedures for evaluating ratings data. Tinsley and Weiss explain the classic distinction between interrater reliability and agreement they introduced (Tinsley & Weiss, 1975) and review the appropriate statistical procedures for evaluating each.

In chapter 5 Hallahan and Rosenthal provide guidelines to help researchers avoid common sources of confusion about statistical procedures and obtain a more complete understanding of their data using procedures such as graphical displays, exploratory data analysis, and contrast analysis. There is a growing sentiment among quantitative psychologists that the current emphasis on null hypothesis testing should be replaced with an informed consideration of effect sizes, point estimates, and confidence intervals (e.g., Cohen, 1994; Meehl, 1978; Rosenthal & Rubin, 1994; Schmidt, 1992). Hallahan and Rosenthal review the distinction between practical and statistical importance, and discuss procedures for interpreting statistical significance levels and interpreting effect sizes. They emphasize that the data analysis plan should be developed prior to data collection and advocate replacing simple dichotomous judgments about statistical significance with a thoughtful consideration of the meaningfulness or practical importance of statistical results.

### B. Part 2: Multivariate Statistics

In this section each author provides cautious, qualified, informed recommendations about the use of their technique based on their expertise as a leading authority on the technique and their exhaustive review and critical analysis of the theoretical and empirical literature that bears on the informed use of their technique. They provide recommendations for the best way to use the technique and distinguish among those recommendations that are firmly grounded in empirical research, those for which some empirical information is available, and those "rules of thumb" that are made in the absence of any empirical evidence. These distinctions provide a research agenda for the development of further empirical evidence to guide the use of multivariate techniques and mathematical modeling procedures.

In chapter 6 Venter and Maxwell review the use of multiple regression for prediction and for explanation and explain the implications of this distinction for issues in multiple regression. They explain diagnostic procedures for evaluating the validity of the assumptions underlying multiple regression and corrective steps to take when the assumptions are violated. They discuss critical issues such as the use of data transformations, interaction and moderator effects, the calculation of sample size requirements, and procedures for dealing with missing data.

In chapter 7 Huberty and Petoskey review multivariate analysis of variance (MANOVA) and multivariate analysis of covariance (MANCOVA). They cover design aspects beginning with the purposes of these techniques, the selection of grouping and response variables, the sampling of respondents, and sample size requirements. Preparation of data for analysis is covered in detail, including the use of transformations, treatment of outliers, and dealing with problems of missing data. Huberty and Petoskey discuss at length the analysis sequence, suggest guidelines for data analysis strategies, and offer tips on interpreting and reporting MANOVA results.

In chapter 8, M. Brown and Wicker explain the use of discriminant analysis to identify latent dimensions that distinguish among groups of interest to the applied researcher and to classify new cases into preexisting groups. They describe in detail the data requirements for descriptive discriminant analysis and the steps necessary to conduct an analysis and interpret the results. They explain the importance of evaluating classification accuracy and the critical role of cross-validation in that process. M. Brown and Wicker briefly compare predictive discriminant analysis with descriptive discriminant analysis, recommend the use of discriminant analysis following MANOVA, and explain why stepwise discriminant analysis should not be used. They conclude with a discussion of reporting requirements for publishing discriminant analysis results.

In chapter 9 Thorndike begins with a discussion of the relations among multivariate statistical procedures and, in particular, the relation of canonical analysis to factor analysis and principle components analysis. He introduces the reader to the vocabulary of canonical analysis, reviews significance tests, redundancy analysis, and rotation of canonical components. Methodological issues are discussed, including stepwise procedures and estimating bias in canonical analysis results, and he emphasizes the importance of cross-validation.

In chapter 10 Cudeck describes the conceptual underpinnings of exploratory factor analysis and its relation to classical regression. He explains the factor analysis model, the conceptual rationale and problems underlying factor rotation, and the relation of factor and components analysis. Cudeck reviews the primary issues that applied researchers must resolve to perform an exploratory factor analysis, including the method of factor estimation, number of factors to extract, and the problems associated with drawing inferences about population loadings from a sample factor loading. He explains the theoretical rationale underlying factor rotation, reviews alternative factor rotation algorithms, and discusses the purpose and procedures for performing target rotations.

In chapter 11 Gore explains that cluster analysis refers to a large and diverse family of procedures for grouping objects according to latent structure. He explains the relation of cluster analysis to discriminant function and exploratory factor analysis, describes both exploratory and hypothesis testing uses of cluster analysis, and reviews the major approaches to cluster analysis. Gore explains the differences between similarity and distance matrices, and provides guidelines for selecting the type of matrix best suited to the objectives of the analysis. He reviews the agglomerative hierarchical methods that have been shown to be most effective in recovering a known cluster structure, and describes procedures for deciding on the number of clusters, establishing the external validity of the clusters, and reporting the cluster analysis results.

In chapter 12 Davison and Siceri discuss applications of exploratory multidimensional scalling (MDS) to proximity matrices containing distance measures. They describe the selection of stimuli, participants, and a proximity measure, and review issues to consider when gathering proximity data using paired comparisons and sorting approaches. The merits of weighted and unweighted scaling models are considered, and strategies for determining the dimensionality of the matrix and interpreting the MDS solution are explained. Davison and Siceri conclude with a consideration of the relation of MDS to other multivariate techniques, including cluster analysis and exploratory factor analysis, and its relation to mathematical modeling testing procedures, such as item response theory and the analysis of circumplex structures.

In chapter 13 Mark, Reichardt, and Sanna discuss time series designs for the analysis of patterns of change over time, forecasting (i.e., predicting future values of a variable that has been measured in a time series), assessing the effect of a treatment or intervention, and assessing the relation between two or more time-series variables. They explain the problem of auto-correlation (i.e., the likelihood that error scores are correlated because the same individuals are tested across time), introduce the family of ARIMA modeling approaches to control for auto-correlation, and describe the iterative ARIMA modeling process. Applications to multiple cases, threats to internal validity, and time-series studies of covariance are covered, and design issues, such as the number of data collection points and number of cases, are discussed.

In chapter 14 Imrey introduces readers to three procedures for analyzing categorical (i.e., nominal) dependent variables, Poisson regression (used most frequently to analyze rates), logistic regression (used most commonly to analyze proportions), and loglinear modeling (used to model dichotomous and polytomous dependent variables as a function of multiple independent categorical variables). He explains the differences between rates, probabilities, and odds ratios, and the measures of association used with each, then considers these techniques as special cases of the generalized linear model that evaluate associations among count data. Practical issues such as sample size and variable selection are discussed.

## C. Part 3: Mathematical Models

Part 3, a feature unique to the *Handbook* begins with a coverage of structural equation modeling, confirmatory components and factor analysis, meta-analysis, Generalizability theory, and item response theory. These flexible procedures can be used to evaluate a wide variety of models. Part 3 continues with a coverage of multitrait–multimethod (MTMH) models, hierarchical models, circumplex models, and growth models. Although these procedures are equally flexible, they are used for evaluating specific models that are widely used throughout the sciences and humanities

A challenge confronted by each *Handbook* author was to achieve an optimal balance between mathematical precision and conceptual simplicity in the treatment of their topic. The chapters in this section, of necessity, rely more heavily on mathematical/statistical explanations than the chapters in the first two sections because programs to perform the analyses are not widely available in the leading software packages (e.g., SPSS, SAS). In may instances those packages can be used to compute statistical components that will be used in the analysis, but programs that will do the entire analysis are not available. Nevertheless, the authors in this section made every effort to shift their treatment as far in the direction of a conceptual, step-by-step treatment as possible. Each author describes the types of questions that can be answered using the procedures they were describing, how to perform the analysis, and how to interpret the results.

In chapter 15 DiLalla describes the flexibility of structural equation modeling (SEM) and its potential for use in performing exploratory factor analyses, confirmatory factor analyses (CFA), and analyses of moderator and mediator hypotheses. She reviews the data requirements for SEM and explains procedures for evaluating the assumption of multivariate normality and for dealing with missing data. She explains the interpretation of fit indices and the procedures for assessing model fit. DiLalla highlights important checks users should perform on the output to insure that the analysis has been performed correctly and explains the procedure for interpreting parameter estimates, using modification indices, and comparing nested models.

In chapter 16 Hoyle introduces CFA as a special application of SEM that focuses on the structural model (i.e., the directional relations among constructs). The prototypical use of CFA is to evaluate the correspondence between hypothesized and observed patterns of association among the variables. Hoyle describes the use of CFA to address questions of dimensionality and evaluate higher order models, growth models, measurement invariance, and content validity. He reviews the data requirements for CFA, provides a detailed description of the elements of a CFA (i.e., model specification, parameter identification and estimation, and respecification), reviews absolute and comparative fit indices, and details the use of the latter in model comparisons.

In chapter 17 Becker describes flexible procedures for the synthesis of multivariate data that can be applied regardless of the form of the data (i.e., effect size, correlation, proportion, or other index) to estimate the magnitude of effects across studies and to evaluate the variations in outcome patterns. In contrast to the more familiar univariate meta-analysis, multivariate meta-analysis takes into account the magnitude of within-study differences in effects, and the differential relations of alternative predictors to the outcomes under investigation. She describes procedures for determining effect sizes when data are available on several outcome variables, for the same variables across multiple times, and for multiple treatment and control groups. Becker explains the analysis of fixed effect models, random effects models, and mixed models.

In chapter 18 Marcoulides provides an overview of generalizability theory and illustrates its use as a comprehensive method for designing, assessing, and improving the dependability of measurement procedures. His objective is to enable readers to begin to use generalizability theory procedures in their own research. He describes the distinction between relative and absolute decisions, and the calculation of error variances and generalizability coefficients for each, and he clarifies the often misunderstood distinction between generalizability and decision studies. Marcoulides explains the application of generalizability theory procedures to one-facet crossed designs, one-facet nested designs, multifaceted designs, and multivariate designs and concludes with a summary of recent work relating generalizability theory and item response theory.

In chapter 19 Hambleton, Robin, and Xing introduce item response theory (IRT) and the process of estimating person-ability and item-difficulty parameters using IRT models. They describe models for use with discrete or continuous responses formats that are dichotomously or polytomously scored, models for use with ordered or unordered item score categories, and models that can be used with homogeneous or heterogeneous latent abilities. The concepts of item and test characteristic functions are introduced and the applications of those in test development and computer-adaptive testing described. The authors provide a brief overview of software for IRT analysis and conclude with a summary of new areas in which applications of IRT models are receiving increased attention.

In chapter 20 Dumenci explains that Campbell and Fiske's (1959) original procedure for analyzing multitrait–multimethod matrices lacks an explicit statistical model, requires the restrictive (and generally violated) assumption that all the correlations in the matrix were generated by only four correlations, and allows for great subjectivity in interpretation. These shortcomings led to the development of a second generation of procedures (e.g., exploratory factor analysis, nonparametric ANOVA, partial correlation methods, and smallest space analysis) that assume the validity of the measurement model without providing any means of testing that assumption. Recently, psychometricians have advocated the use of the three-way random ANOVA model, CFA model, covariance components analysis model, and composite direct product model for modeling the trait-method interactions underlying MTMM matrices. Dumenci evaluates the adequacy of these procedures for testing the convergent and discriminant validity of measurement procedures and determining the amount of variance attributable to method effects and provides guidelines for their use.

In chapter 21 Kreft describes models that can be used in the analysis of hierarchically nested data. Multilevel data files consist of data in which measurements have been made at two or more levels of a hierarchical system (e.g., measurements of employees nested within firms or patients nested within hospitals). Multilevel data also can involve repeated measurements, in which case the levels involve measurements at two or more times nested within students. Multilevel data can have as few as two levels, but many more levels are possible, such as prisoners nested within prisons nested within level of security nested within geographic region nested within country. Kreft explains the use of the random coefficient model for modeling multilevel data and describes the advantages of the multilevel model over the fixed linear and traditional regression models. She provides guidelines for the use of the random coefficient model in modeling multilevel data.

In chapter 22 Tracey describes the analysis of circumplex models. The correlations between variables can be represented in two-dimensional space by plotting their location on the two primary dimensions underlying the correlation matrix. A circumplex, the resulting circular pattern formed by the correlations (Guttman, 1954), provides a parsimonious description of the relations among the variables in which the distance of the variables from each other on the circumplex is inversely proportional to the relations among the variables. Tracey distinguishes among three major variations of the circumplex model, the circulant (also referred to as the circumplex), the geometric circulant (a more restrictive version of the circulant model), and the quasi-circumplex (also referred to as the circular order) models. He describes exploratory approaches to the analysis of circumplex models based on visual inspection (e.g., visual inspection of correlation matrices and plotted factor scores) and statistical tests (e.g., the Selkirk-Neave Gap test and Kupier-Stephens K test) of the circular distribution, and confirmatory approaches that make use of constrained MDS, SEM, and the randomization test.

In chapter 23 Willett and Keiley introduce the concept of individual growth modeling, a flexible tool for investigating individual change over time in which repeated measures are modeled as functions of time in a statistical model. Individual growth modeling permits the investigation of interindividual differences in change by allowing the parameters of the individual growth model to reflect individual differences on the predictors (e.g., gender, environment, and experiences) under investigation. They explain the process of mapping multilevel models of change onto the general covariance structure model and illustrate its application when introducing time-invariant (e.g., gender or race) and time-varying (e.g., peer pressure) predictors of change.

## REFERENCES

Aiken, L. S., West, S. G., Sechrest, L., & Reno, R. R. (1990). Graduate training in statistics, methodology, and measurement in psychology: A survey of Ph.D. programs in North America. *American Psychologist, 45,* 721–734.

Anscombe, F. J. (1973). Graphs in statistical analysis. *The American Statistician, 17,* 17–21.

Baker, B. O., Hardyck, C. D., & Petrinovich, L. F. (1996). Weak measurement vs. strong statistics: An empirical critique of S. S. Stevens' proscriptions on statistics. *Educational and Psychological Measurement, 26,* 291–309.

Behrens, J. T. (1997). Principles and procedures of exploratory data analysis. *Psychological Methods, 2,* 131–160.

Campbell, D., & Fiske, D. (1959). Convergent and discriminant validation by the multitrait-multimethod matrix. *Psychological Bulletin, 56,* 81–105.

Cattell, R. B., (1988). The meaning and strategic use of factor analysis. In J. R. Nesselroade & R. B. Cattell (Eds.). *Handbook of multivariate experimental psychology* (2nd ed.) (pp. 131–203). New York: Plenum Press.

Cliff, N. (1987). *Analyzing multivariate data.* San Diego, CA: Harcourt-Brace Jovanovich.

Cohen, J. (1988). *Statistical power analysis for the behavioral sciences* (2nd ed.). Hillsdale, NJ: Erlbaum.

Cohen, J. (1990). Things I have learned (so far). *American Psychologist, 45,* 1303–1312.

Cohen, J. (1994). The earth is round ($p < .05$). *American Psychologist, 49,* 997–1003.

Cohen, J., & Cohen, P. (1983). *Applied multiple regression/correlation analysis for the behavioral sciences.* (2nd ed.). Hillsdale, NJ: Erlbaum.

Comrey, A. L., & Lee, H. B. (1992). *A first course in factor analysis (2nd ed.)*. Hillsdale, NJ: Lawrence Erlbaum.

Dalgleish, L. I. (1994). Discriminant analysis: Statistical inference using jackknife and bootstrap procedures. *Psychological Bulletin, 116*, 498–508.

Efron, B. (1987). Better bootstrap confidence intervals. *Journal of the American Statistical Association, 82*, 171–185.

Estes, W. K. (1997). Significance testing in psychological research: Some persisting issues. *Psychological Sciences, 8*, 18–20.

Everitt, B. S. (1979). A Monte Carlo investigation of the robustness of Hotelling's one and two sample $T^2$ tests. *Journal of the American Statistical Association, 74*, 48–51.

Glass, G. V., & Hopkins, K. (1984). *Statistical methods in education and psychology*. Englewood Cliff, NJ: Prentice-Hall.

Gonzalez, R. (1994). The statistics ritual in psychological research. *Psychological Science, 6*, 325–328.

Grimm, L. G., & Yamold, P. R. (1995). *Reading and understanding multivariate statistics*. Washington, DC: American Psychological Association.

Guttman, L. R. (1954). A new approach to factor analysis: The radix. In P. F. Lazarsfeld (Ed.), *Mathematical thinking in the social sciences* (pp. 258–348). New York: Columbia University Press.

Harris, R. J. (1985). *A primer of multivariate statistics (2nd ed.)* San Diego, CA: Academic Press.

Harris, R. J. (1997). (Guest Editor). Special section on significance testing. *Psychological Science, 8*, 1–20.

Herzberg, P. A. (1969). The parameters of cross-validation. *Psychometrika Monograph Supplement, 34*, (Whole No. 16).

Lederer, M. (1996). *Cross-validation, formula estimation, and a bootstrap approach to estimating the population cross-validity of multiple regression equations*. Unpublished M.S. Thesis, Western Washington University.

Loftus, G. R. (1996). Psychology will be a much better science when we change the way we analyze data. *Current Directions in Psychological Science, 5*, 161–171.

Meehl, P. E. (1978). Theoretical risks and tabular asterisks: Sir Karl, Sir Ronald, and the slow progress of soft psychology. *Journal of Consulting and Clinical Psychology, 46*, 806–834.

Mosier, C. I. (1951). Problems and designs of cross-validation. *Educational and Psychological Measurement, 11*, 5–11.

Norman, W. T. (1965). Double-split cross-validation: an extension of Mosier's design, two undesirable alternatives, and some enigmatic results. *Journal of Applied Psychology, 49*, 348–357.

Rosenberg, M. (1965). *Society and the adolescent self-image*. Princeton, NJ: Princeton University Press.

Rosenthal, R. (1990). How are we doing in soft psychology? *American Psychologist, 45*, 775–777.

Rosenthal, R., & Rubin, D. B. (1982). A simple general purpose display of magnitude of experimental effect. *Journal of Educational Psychology, 74*, 166–169.

Rosenthal, R., & Rubin, D. B. (1994). The counternull value of an effect size. A new statistic. *Psychological Science, 5*, 329–334.

Schmidt, F. L. (1992). What do data really mean? Research findings, meta-analysis, and cumulative knowledge in psychology. *American Psychologist, 47*, 1173–1181.

Schmidt, F. L. (1996). Statistical significance testing and cumulative knowledge in psychology: Implications for training of researchers. *Psychological Methods, 1*, 115–129.

Tabachnick, B. G., & Fidell, L. S. (1996). *Using multivariate statistics (3rd ed.)*. New York: Harper Collins.

Thompson, B. (1995). Exploring the reliability of a study's results: Bootstrap statistics for the multivariate case. *Educational and Psychological Measurement, 55*, 84–94.

Tinsley, H. E. A. (1995). The Minnesota counseling psychologist as a broadly trained applied psychologist. In. D. Lubinski & R. V. Dawis, *Assessing individual differences in human behavior: New concepts, methods and findings.* Palo Alto, CA: Davies-Black Publishing.

Tinsley, H. E. A. (2000). The congruence myth: An analysis of the efficacy of the person–environment fit model. *Journal of Vocational Behavior, 56.*

Tinsley, H. E. A., & Weiss, D. J. (1975). Interrater reliability and agreement of subjective judgments. *Journal of Counseling Psychology, 22,* 358–376.

Tufte, E. R. (1997). *Visual explanations: Images and quantities, evidence and narrative.* Cheshire, CT: Graphics Press.

Tukey, J. W. (1969). Analyzing data: Sanctification or detective work? *American Psychologist, 24,* 83–91.

Wherry, R. J. (1951). Comparison of cross-validation with statistical inference of betas and multiple R from a single sample. *Educational and Psychological Measurement, 11,* 23–28.

Wilkinson, L., & the Task Force of Statistical Inference (1999). Statistical methods in psychology journals: Guidelines and explanations. *American Psychologist, 54,* 594–604.

Zuckerman, M., Hodgins, H. S., Zuckerman, A., & Rosenthal, R. (1993). Contemporary issues in the analysis of data: A survey of 551 psychologists. *Psychological Science, 4,* 49–53.

# 2

# ROLE OF THEORY AND EXPERIMENTAL DESIGN IN MULTIVARIATE ANALYSIS AND MATHEMATICAL MODELING

**JOHN HETHERINGTON**

*Department of Psychology, Southern Illinois University, Carbondale, Illinois*

## I. THE IMPORTANCE OF THEORY IN SCIENTIFIC METHODOLOGY

Imagine the task before you is to help solve a jigsaw puzzle. Most of the pieces are laid out face-up on a large table, but some have been previously joined to form small patches of the design. You glance around for a picture of the completed puzzle from which to work, but apparently none is available. You examine the already combined pieces in an attempt to discern the picture, but upon closer inspection you discover that the pieces do not all appear to be from the same puzzle. In fact, some of the already solved patches include several pieces that have been forced to fit into their current positions. Furthermore, it appears to you that all of the puzzle pieces may not be present. With no guiding picture, possible missing pieces, and ostensibly extraneous pieces, you might think that solving this particular jigsaw puzzle would be an impossible task worthy of Sisyphus. As social scientists striving to explain the complexity and nuance of human behavior, however, we continually engage in this very activity.

The scientific enterprise portrayed as a game of solving jigsaw puzzles is admittedly a simple analogy, but the task of constructing representations or *theories* about the world is one—if not arguably the only—objective of the game of science. The premise that generating an explanatory theory is the primary goal of scientific inquiry is central to many practitioners of the scientific method, as evidenced by Kerlinger's (1986) statement: "The basic aim of science is theory" (p. 8). From this perspective, a substantive theory that explains the phenomenon of interest by identifying critical variables and specifying their interrelations is considered the culmination of scientific activities (i.e., a teleological end). In the jigsaw puzzle analogy, for example, the completion of the entire puzzle is reason to begin the project in the first place.

Other social scientists have emphasized a more instrumental, pragmatic view of the role of theory in scientific inquiry. Lipsey (1990) asserts that, "Theory orientation may have its greatest advantage in the domain of practical research design" (p. 35). In the jigsaw puzzle analogy, the picture (or representation) of the completed pattern both guides the placement of pieces, and suggests what additional pieces are needed. Similarly, theory serves a fundamental role in formulating research design, outlining strategy for data analysis, interpreting the meaningfulness of research findings, and generating additional hypotheses to spur subsequent research. Both the explanatory and instrumental roles of theory are important, and even though much discussion has been generated emphasizing one over the other, they are not inherently incongruous (Cattell, 1988a). As scientists we should strive to construct well-defined theories, and conscientiously use them to direct our research. Kurt Lewin's (1951) oft-cited sentiment regarding the pragmatic utility of theory clearly illustrates the complementary nature of these positions:

> The greatest handicap of applied psychology has been the fact that, without proper theoretical help, it had to follow the costly, inefficient, and limited method of trial and error. Many psychologists working today in an applied field are keenly aware of the need for close cooperation between theoretical and applied psychology. This can be accomplished in psychology, as it has been accomplished in physics, if the theorist does not look toward applied problems with highbrow aversion or with a fear of social problems, and if the applied psychologist realizes that there is nothing so practical as a good theory. (p. 169)

Plainly stated, theory is both a useful component and a desired goal state of modern scientific inquiry, and these two roles are complementary rather than mutually exclusive.

The proliferation of multivariate statistics in the sociobehavioral sciences over the last half of this century is a direct result of more widespread access to increasingly powerful computers and statistical software packages (see Tinsley and Brown, chapter 1, this volume). Modern multivariate analytic techniques in the span of 50 years have progressed from laborious, time-consuming computations by hand to readily obtainable results from a personal microcomputer (Judd, McClelland, & Culhane, 1995). Cattell

(1988b) opened a window to this not-so-distant past by reminding us that in the early 1950s it took approximately 6 months to calculate an exploratory factor analysis and obtain simple structure! In fact, many current multivariate techniques (e.g., structural equation modeling) and many parametric estimation algorithms (e.g., maximum likelihood estimation) could not be realized without the widespread availability of sophisticated computer technology (Nunnally & Bernstein, 1994).

Data analysis via statistical software has proliferated in tandem with the rapid progress of computer hardware. Today a veritable cornucopia of software programs is available to the social scientist, which offers an impressive array of univariate and multivariate analyses in convenient, unified statistical packages (Stevens, 1996; Tabachnick & Fidell, 1996). Some of these programs now no longer require the user to know computer programming syntax to request particular statistical operations, as many increasingly support the familiar Windows point-and-click interface. Never before have so many statistical procedures been available to so many researchers, regardless of the user's level of knowledge—or lack thereof—about statistical assumptions or operations.

Suffice to say, the possibilities have never been greater for the social scientist to analyze complex, multivariate data. Rather than slavishly relying on experimental manipulations and the controls of classical bivariate designs, researchers can—and increasingly choose to—statistically partition variance without the onus of creating artificial laboratory situations (Aiken, West, Sechrest, & Reno, 1990; Cattell, 1988a; Pedhazur & Schmelkin, 1991). Researchers can concurrently examine common and specific variance components of large, multiple variable data sets, which better affords the opportunity to model complex causality, measurement error, and the effects of mediating variables (Cattell, 1988b; Nunnally & Bernstein, 1994). Furthermore, the partitioning of hierarchically nested variance components allows the researcher to better examine correlated and uncorrelated multilevel data (Bryk & Raudenbush, 1992). In the face of this impressive array of quantitative data analysis options, however, it is essential for sociobehavioral scientists to think carefully about the role of theory in conducting multivariate statistics and mathematical modeling. Are these seemingly omnipotent multivariate techniques primary or secondary to theory in scientific inquiry?

The purpose of this chapter is to discuss the role of theory and experimental design in the analysis of multivariate data. A variety of topics is explored, from the importance of theory for the scientific method in general to multivariate data analysis in particular. Throughout this discourse, the practical, instrumental value of theory is overtly emphasized, while implicit recognition of the importance of the teleological value of theory development to sociobehavioral science research is assumed. The evolution of modern, postpositivistic scientific methodology is briefly examined to identify the more prominent, modern criticisms of scientific inquiry, and the research strategy of critical multiplism (Cook, 1985; Shadish, 1989) is presented as a comprehensive approach that specifically addresses such concerns. Simply put, the intent of this chapter is to convince the reader of the undeniable necessity of explicit, substantive theory throughout the entire research enterprise and especially for multivariate data analysis.

## A. Strong Theory versus Law of the Instrument

This chapter is based on the proposition that the role of theory in the scientific enterprise must be periodically entertained, thoroughly examined, and explicitly reaffirmed by scientific practitioners. As a case in point, Kaplan (1964) criticized the predisposition of individual scientists to exclusively conceptualize research only in relation to those techniques in which he or she is specifically trained. Kaplan referred to this tendency as the Law of the Instrument, where a particular statistical or methodological technique directs scientific practice—not theory. The anecdote of "give a kid a hammer and everything looks like a nail" aptly illustrates this tendency. The effort and time spent learning research skills constrain the researcher to a limited range of tactics; Kaplan argued, "The price of training is always a certain 'trained incapacity': the more we know how to do something, the harder it is to learn to do it differently" (p. 29). Consequently, the Law of the Instrument undermines the essence of scientific inquiry, because methods and techniques are mechanically applied without deliberate, critical appraisal.

In this chapter, Kaplan's idea of the Law of the Instrument is expanded to refer to the uncritical application of any data analytic technique and to the practice of allowing the technique to determine the particular elements of the research design. Blind obedience to the Law of the Instrument is neither good practice nor good science, because as Kaplan (1964) comments, "Statistics is [sic] never in itself [sic] a source of knowledge" (p. 220). Theory should be used in the research process to establish guidelines for data analysis and to provide an essential frame of reference for understanding the merits of the results. Slavish adherence to the Law of the Instrument leaves the researcher amid an ocean of unappraisable facts, or as Bordens and Abbott (1991) state: "Experimentation conducted without the guidance of theory produces a significant amount of irrelevant information that is likely to obscure important observations" (p. 499). The familiar adage of "Garbage In-Garbage Out" (GIGO) is a similar principle, which warns that statistics do not by themselves impart meaning to vacuous data. Although multivariate statistics can generate an impressive array of information, they may nevertheless produce nothing more than "well-dressed" GIGO without the guidance of substantive theory (Aiken, 1994; Bernstein, 1988; Gorsuch, 1983).

Adherence to the Law of the Instrument also makes researchers more susceptible to the nominalistic fallacy (i.e., considering illusory, empirically generated artifacts as factual), because they are less likely to effectively discriminate between target and nontarget results in the absence of a theory (Cliff, 1983). For example, the advent of common variance statistical models—in particular exploratory factor analysis in the 1950s—has allowed researchers to better examine latent, theoretical constructs and quantify measurement error, but it has also enabled researchers to engage in indiscriminate, atheoretical research (McDonald, 1986; Muliak, 1972). These "blind" or "shotgun" exploratory factor analyses belie the original intent underlying the development of factor analytic techniques: hypothesis testing guided by explicit theory (Thurstone, 1948). According to Gorsuch (1983), the numerous criticisms of exploratory factor analysis as practiced in the social and behavioral sciences indict not the procedure *per se,* but the failure to use "a theoretical approach that integrates the data collection, factor analysis, and interpretation, and which leads to the future use of the results" (p. 372).

Unwitting adherence to the Law of the Instrument can result in numerous, undesirable consequences. Yet, fifteen of the leading multivariate statistics textbooks, written during the last 25 years for use in graduate training in the sociobehavioral sciences, provide no more than a cursory review of the role of theory in the scientific enterprise (Cliff, 1987; Dillon & Goldstein, 1984; Grimm & Yarnold, 1995; Hair, Anderson, Tatham, & Black, 1992; Hirschberg & Humphreys, 1982; Johnson & Wichern, 1988; Kachigan, 1982; Marcoulides & Hershberger, 1997; Morrison, 1990; Overall & Klett, 1972; Srivastava & Carter, 1983; Stevens, 1996; Tabachnick & Fidell, 1996; Tacq, 1997; Tatsuoka, 1988). The authors of these volumes generally emphasize the statistical derivation of multivariate techniques and the procedural aspects of their computation. Three notable exceptions to this disheartening state of affairs are those written by Bernstein (1988), Nesselroade and Cattell (1988), and Pedhazur and Schmelkin (1991). The emergence of greater computational power and more sophisticated multivariate statistical techniques over the past several decades seems only to have exacerbated these problems, as Cattell (1988a) observes: "Alas, one must also recognize that the computer has brought dangers of magnifying design mistakes, of spreading vulgar abuses, and favoring uninspired collection of data" (p. 9). Technology and statistics *per se* are tools of thought, but they do not—and cannot—substitute for its faculty.

Conversely, the opposite of the Law of the Instrument is what I refer to as Strong Theory, or the explicit, intentional application of theory to guide observation, design, and analysis by constructing precise, testable and meaningful hypotheses about measurement or causal relationships. Similarly, Meehl (1978) suggested the concept of substantive theory as a "theory about the causal structure of the world, the entities and processes underlying the phenomena" (p. 824). Moreover, Platt (1964) offered "strong inference" as a means of ruling out alternative explanations by designing crucial experiments that place multiple hypotheses in juxtaposition so that one or the other can be eliminated from further consideration. Strong Theory does not require the pre-existence of a single, generally accepted theoretical framework, but that the relevant theoretical landscape be thoroughly explored and explicitly incorporated in designing the research, and collecting and analyzing the data. In contrast, Weak Theory is the identification of theoretically relevant variables without further specifying the nature, direction, or strength of their interrelations. Hence, Weak Theory is strikingly similar to the Law of the Instrument, because the statistical algorithm or model determines the meaningfulness of the results, and therefore, Weak Theory can offer only weak corroboration or rejection of the research question at hand (Meehl, 1986, 1990).

In sum, the tension between Strong Theory, Weak Theory, and the Law of the Instrument is ubiquitous to scientific practitioners: Should an explicitly stated theory or the statistical procedure drive the research process? As if in answer, Poincaré (1913/1946) skillfully employs a metaphor, one which has been revitalized several times in the social and behavioral sciences (e.g., Forscher, 1963; Lipsey, 1997), regarding the necessity of theory: "Science is built up of facts, as a house is with stones. But a collection of facts is no more a science than a heap of stones is a house" (p. 127). This sentiment is echoed by Thurstone (1948) for psychologists when he writes: "(Statistics) should be our servants in the investigation of psychological ideas. If we have no psychological ideas, we are not likely to discover anything interesting" (p. 277). Theory is a necessary cornerstone of the scientific method, without which investigators are unlikely to accomplish their research objectives. There are no statistical adjustments to correct for having asked the wrong research question research.

### B. Need for Theory in Multivariate Research

Although it is easy to wax poetic and list the salubrious effects of theory for scientific inquiry in general, a more difficult task is to draw out the specific implications for multivariate analysis in practice. This is in part because the scientific method is readily adaptable to the possibly infinite combinations of possible variables across different research designs, but also in part because of the impressive array of multivariate statistics currently available. Three concrete applications of theory to the research process are discussed: the role of theory in (a) the selection of variables, (b) the specification of relations among variables, and (c) the interpretation of results.

## 1. Variable Selection

Theory plays an essential part in the selection of variables for research, because it specifies not only *what* should be observed but also *how* it should be measured. Popper (1959) makes a strong case that observations are necessarily guided by theory:

> Experiment is planned action in which every step is guided by theory. We do not stumble upon our experiences, nor do we let them flow over us like a stream. Rather, we have to be active: We have to make our experiences. (p. 280)

Because we cannot observe everything, some theory—whether explicitly stated or implicitly implied—delimits the boundaries of the investigation. This fact has led some philosophers of science to denounce the objectivity of the scientific method (Feyerabend, 1975/1988; Kuhn, 1962/1970). Other philosophers of science (e.g., Popper, 1959, 1963) have used the ubiquity of theory-laden observations to argue for making theory explicit before data collection as a means of improving scientific practice. In either case, most postpositivist approaches to the scientific method accept that theory plays an essential role in determining the selection of variables and constructs for any given inquiry. Hence, at least two types of errors can occur when modeling complex, multivariate phenomena: exclusion of relevant variables and inclusion of irrelevant variables (Pedhazur, 1982; Pedhazur & Schmelkin, 1991).

Simon and Newell (1963) refer to the exclusion of relevant variables as errors of omission—or Type I errors—in theory development and corroboration. These errors are a problem for both common variance models (e.g., exploratory and confirmatory factor analyses) and predictive models (e.g., path analysis and multiple regression correlation), because estimates of the variance attributable to those constructs are not included in the model (Cohen & Cohen, 1983; Gorsuch, 1988; Pedhazur, 1982). Furthermore, the parameters in predictive models may be either suppressed or augmented because of a relation with the excluded variable, or what is referred to as the third variable problem (Bernstein, 1988; Tabachnick & Fidell, 1996). Although not every variable can be realistically included in any one statistical model, it is necessary to specify a set of variables that exhaust a substantial segment of the variance lest one misrepresent the phenomena of interest (Huberty, 1994). Theory can identify a parsimonious set of salient variables in contrast to the inefficient trial-and-error approach to modeling.

Simon and Newell (1963) refer to the inclusion of irrelevant variables as errors of commission—or Type II errors—in theory development and corroboration. Inclusion of irrelevant variables is a problem when using specific variance component models (e.g., analysis of variance and multiple regression correlation), which do not effectively manage high intercorrelations among supposedly independent variables (viz., multicollinearity). For example, Figueredo, Hetherington, and Sechrest (1992) have criticized the practice of some investigators of entering large numbers of independent variables (e.g., 100–500) into a single linear equation. Strategies to avoid such situations—like prespecifying common variables or hierarchically entering variables sequentially into the model—should be based on guidance from theory (Cohen & Cohen, 1983; Pedhazur, 1982).

## 2. Explication of Relations among Variables

Theory also is essential in specifying exactly what relations (viz., associative or causal) exist and do not exist among variables and their underlying constructs. As a result, theory has definite implications for the formulation of the research design, the operational definition of the constructs, the measurement of the manifest variables, and the analysis of the data. Three general dimensions are relevant to the specification of a theoretical model and selection of an analytic procedure: (a) the number of predictor and outcome variables, the type of relation postulated, and the theoretical emphasis (see Table 2.1). First, multivariate statistics can be categorized by whether they include multiple predictors (or independent variables), multiple criteria (or outcome or dependent variables), or both (see Tinsley & Brown, chapter 1, this volume). Multivariate statistical techniques also can be distinguished by the statistical relations hypothesized. Table 2.1 illustrates that different multivariate procedures are used to test hypotheses about the degree of association among variables, differences among groups, and the prediction of group differences. Examples of the multivariate procedures used to test each of these three statistical relations, fully crossed by the number-of-variables dimension, are also provided. Additional information about each technique is discussed in greater detail throughout subsequent chapters in this volume.

The third dimension represented in Table 2.1 is the distinction between exploratory and confirmatory research objectives (Cattell, 1988a; Tukey, 1977, 1980). Exploratory analyses typically are guided by vague, general hypotheses, and they focus on interpreting the emergent characteristics of the data after analysis (Gorsuch, 1988; Hair et al., 1992). Confirmatory analyses typically test the empirical support for a number of explicit hypotheses about the expected properties of the data (Muliak, 1988). However, a single multivariate study could involve both exploratory and confirmatory components. Therefore, exploratory and confirmatory multivariate statistical techniques should not be considered dichotomous, because they represent the polemic endpoints of a continuum reflecting how and when theory is emphasized in the research process (Bernstein, 1988).

## 3. Interpreting Results

Theory also plays an essential role in the interpretation of results from multivariate analyses. Theory is the primary schema, or frame of reference,

**TABLE 2.1  Taxonomy of Common Multivariate Statistical Techniques**

| | Multiple predictors | | Multiple outcomes | | Multiple predictors and outcomes | |
|---|---|---|---|---|---|---|
| | Exploratory | Confirmatory[a] | Exploratory | Confirmatory[a] | Exploratory | Confirmatory[a] |
| **Degree of association** | | | | | | |
| Multiple regression | Hierarchical multiple regression | | Factor analysis (unconstrained factor extraction) | Factor analysis (specific factor extraction) | Canonical correlation | None[b] |
| Logistic regression | Hierarchical logistic regression | | Multidimensional scaling (unspecified dimensionality) | Multidimensional scaling (specified dimensionality) | Confirmatory factor analysis (maximization of fit indices) | Confirmatory factor analysis (nested models) |
| **Group differences[c]** | | | | | | |
| ANOVA (post hoc comparisons) | ANOVA (planned comparisons) | | One-way MANOVA (post hoc comparisons) | One-way MANOVA (stepdown) | Factorial MANOVA (post hoc comparisons) | Factorial MANOVA (stepdown and/or planned comparisons) |
| ANCOVA (post hoc comparisons) | ANCOVA (planned comparisons) | | One-way ANCOVA (post hoc comparisons) | One-way MANCOVA (stepdown) | Factorial MANCOVA (post hoc comparisons) | Factorial MANCOVA (stepdown and/or planned comparisons) |
| **Group membership** | | | | | | |
| One-way discriminant analysis | Hierarchical one-way discriminant analysis | | Cluster analysis | None[b] | Factorial discriminant analysis | Hierarchical factorial discriminant analysis |

[a] All listed techniques could conceivably be "confirmed" through cross-validation analysis using multiple samples. The confirmatory procedures listed assume one-sample designs.

[b] There are no unique confirmatory applications for these inherently exploratory statistical techniques.

[c] ANOVA, analysis of variance; MANOVA, multiple analysis of variance; ANCOVA, analysis of covariance; MANCOVA, multiple analysis of covariance.

by which the researcher understands the content and implications of the research findings. This is true regardless of whether a theory is explicitly recognized, wholly comprehensive, or well understood (Chen & Rossi, 1983; Phillips, 1992). The emphasis and role that theory plays is different between confirmatory and exploratory approaches to the research process. For confirmatory procedures, specific hypotheses based on theory are generated before the collection of data (i.e., a priori), and for exploratory procedures, theory plays greater role in the interpretation of the results after data analysis (i.e., post hoc). Regardless of the adopted approach, the comparison of results to expected hypotheses and possible subsequent modification of theory is necessary to complete one iteration of the inductive-hypothetico-deductive cycle (Cattell, 1988a; Kerlinger, 1986). Popper employed a quote from the Romantic poet Novalis to illustrate the explanatory role of scientific theory: "Theories are nets: Only he who casts will catch" (1959, p. 11). When interpreting results, Popper argued, the object is to make the mesh of the net increasingly finer in order to "catch" (i.e., explain or account for) as many phenomena as possible. This continual refinement of the theoretical net is accomplished through the iterative, recursive relation between theory, method, and analysis of the inductive-hypothetico-deductive scientific method: "Even the careful and sober testing of our ideas by experience is in turn inspired by ideas" (p. 280).

## C. Design Implications for Multivariate Analysis

Over the last few decades, Cattell (1988a), Kerlinger (1986), and others noted that applied social scientists have become increasingly aware of the multivariate nature of social phenomena. Whether this is attributable primarily to a theoretical interest in modeling more complex phenomena (i.e., Strong Theory) or to technological advances that afford the opportunity (i.e., Law of the Instrument), most modern scholars agree that the use of multivariate models in the sociobehavioral sciences is increasing, and that this trend will continue in the future (Grimm & Yarnold, 1995; Tabachnick & Fidell, 1996). It is essential, therefore, to consider the implications of design issues such as the type of experimental control, the collection of data from multiple sources, and the determination of sampling frame for the analysis of multivariate data.

A common distinction made in sociobehavioral research is between experimental, quasi-experimental, and nonexperimental designs (Cook & Campbell, 1979; Kerlinger, 1986; Pedhazur & Schmelkin, 1991). Although the primary intent underlying both experimental and quasi-experimental designs is to identify causal relations between independent and dependent variables, experimental designs use random assignment to equate treatment conditions and use direct manipulations to control the levels of one or more of the independent variables (Cattell, 1988a). Quasi-experimental designs generally use multiple observations over time or the matching of groups on identified characteristics in place of random assignment of subjects, and such designs attempt to capitalize on naturally occurring variation in (or the indirect manipulation of) the independent variables (Cook & Campbell, 1979). Nonexperimental or correlational designs do not involve random assignment nor the direct or indirect manipulation of the independent variables. Instead, they rely exclusively on statistical control to partition measured variance (Tabachnick & Fidell, 1996).

The type of control employed in a particular design has a number of implications for data analysis. First, the veracity of the causal conclusions drawn from the research findings is dependent upon the ability of the researcher to rule out alternative hypotheses through either experimental or statistical control. However, the researcher must know or suspect *a priori* the existence and influence of these variables to be able to use either of these types of control (Kerlinger, 1986; Pedhazur & Schmelkin, 1991). Furthermore, the addition of variables to traditional bivariate experimental research designs exponentially increases the number of possible relations among the variables that must be either experimentally incorporated into the research design or statistically accounted for in the data analysis. Second, the choice of a particular methodological design limits to some extent the statistical procedures that are appropriate to address the research question (Cattell, 1988b). For example, ANOVA (and MANOVA) require that independent variables be truly orthogonal for accurate between-group testing (i.e., to avoid serious inflation of alpha level), but this assumption is very difficult to maintain in most sociobehavioral research (Stevens, 1996). Finally, other factors, such as level of measurement (continuous vs. categorical) and theoretical emphasis (exploratory vs. confirmatory) also suggest appropriate multivariate statistical techniques.

For these reasons, decisions about multivariate analysis ultimately involve questions of sampling. The researcher must decide how to sample from a population of predictors, outcomes, occasions, and persons (Keppel, 1982; Pedhazur & Schmelkin, 1991). Typically, researchers systematically sample to increase the probability that subsequent inferences are correct (i.e., increase generalizability, representativeness, or applicability of particular results). Although a variety of different sampling techniques may be used, those that establish selection criteria a priori are preferred because of their relative freedom from systematic bias (Keppel, 1982). Cattell's (1988c) Data Box and Rummel's (1970) Data Cube are models used to identify all of the possible individual elements of a particular multivariate research design and conceptually organize the possible relations among them (see Figure 2.1). These heuristic models depict three dimensions representing the most commonly varied factors in research designs: persons, characteristics, and occasions. No one study can include all of these possible components (i.e., every person across every characteristic across every

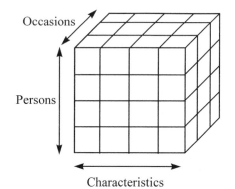

Characteristics

**FIGURE 2.1**    The Data Box (adapted from Cattell, 1988c; Rummel, 1970).

occasion), so researchers must determine what cells of the cube are meaningful and feasible within the pragmatic limitations of the study (Keppel, 1982; Kerlinger, 1986). In either case, the Data Box illustrates the relation between the sampled characteristics of the design to the universe of generalization (Petrinovich, 1979).

In sum, multivariate designs involve the collection of a considerable amount of information that is challenging to interpret; therefore, it is vital to plan multivariate analyses strategically before data collection (Pedhazur & Schmelkin, 1991; Tacq, 1997). The sheer volume of data collected in multivariate designs creates more opportunities for things to go wrong. Multivariate designs and analyses have the potential to discover wondrous amounts of information, but that will occur only when the results can be meaningfully understood and effectively communicated to an audience (Bernstein, 1988).

## II. THE EVOLUTION OF POSTPOSITIVIST SCIENTIFIC METHOD

Most researchers are introduced to the rudimentary inductive-hypothetico-deductive cycle of the scientific method, but the continuously iterative nature of this model is more difficult to grasp. Science is a dynamic enterprise in which both knowledge and the methods used to obtain it change. The role of theory in the scientific method also has evolved throughout the development of modern scientific inquiry (Phillips, 1992). Applied researchers can learn much by studying the prominent criticisms of previous incarnations of the scientific method. By necessity, this brief portrayal of the main ideologies of the scientific method will omit a considerable amount of information, so the reader is directed to Bechtel (1988), Giere (1988), Phillips (1992), and Slife and Williams (1995) for more in-depth treatment.

In the following section, I distill the defining characteristics of historically influential approaches to the scientific method, identify the most telling criticisms, and explain how response to those criticisms have changed current approaches to the scientific method.

## A. Naive Inductivism

The origin of modern scientific inquiry began with naive inductivism, which provided the epistemological foundation of empiricism (Bechtel, 1988; Giere, 1988). The maturation of naive inductivism over a period of several thousand years claims a rich intellectual heritage, including Aristotle, Francis Bacon, Sir Isaac Newton, and John Stuart Mill. According to naive inductivism, the process of obtaining knowledge begins with an observation based on experience. The inductive method involves a simple progression from singular statements about particular occurrences at a particular place and time to universal statements or generalizations about all events of a particular kind at all places and times. A scientific theory derived using the naive inductive model is simply a statement of the common elements distilled from a compendium of observations.

Bacon (1620/1956) contended that simple observation under a variety of different conditions allows one to rule out alternative explanations or unknown confounds. However, naive inductivism does not eschew deduction, because the derivation of a universal law or theory and the establishment of given certain initial conditions enables one to deduce consequences that could serve as explanations and predictions of phenomena: "For our road does not lie on a level, but ascends and descends; first ascending to axioms, then descending to works" (Bacon, 1620/1956, p. 97). The ability to predict future conditions was considered the real power of the scientific method for naive inductivism.

Various problems with the method of naive inductivism have been noted, but the limitation of induction—originally identified by David Hume (1748/1963) and subsequently elaborated by Popper (1959)—is the most serious. Hume questioned whether experience alone can establish the truth or falsity of universal statements. Specifically, there is no assurance that one can predict future outcomes based solely on inductive reasoning, because such natural laws or generalizations are based exclusively on repeated observations and there is no guarantee that a contradictory fact will not be discovered in the future. Hence, inductive laws or theories are tentative and cannot be empirically verified by relying exclusively on induction. Similarly, Sahlin (1991) argued that induction does not provide any knowledge of the causal mechanisms underlying phenomena, because it only provides models of prediction, not models of understanding or explanation.

In spite of these criticisms, naive inductivism has contributed much to the development of scientific methodology. For example, naive inductivism stressed the importance of empirical observation in the development of knowledge about the world. In addition, Bacon (1620/1956) recognized the fallibility of single observations, and attempted to overcome this limitation through repeated observation across a number of circumstances. These insights today enjoy the status of basic tenets upon which current postpositivism scientific inquiry is built (Chalmers, 1982).

### B. Logical Positivism (Sophisticated Inductivism)

Naive inductivism is "naive" because of an excessive reliance on the inductive method. Philosophical attempts to address the problem of induction, as identified by Hume and the others, led to the development of sophisticated inductivism and logical positivism during the latter part of the 19th century and the early half of the 20th century. Influential in its development were scholars such as Auguste Comte, Carl Hempel, Hans Reichenbach, Herbert Feigl, and Rudolf Carnap. Sociobehavioral researchers are generally familiar with the most common extension of positivistic principles (e.g., logical positivism or logical empiricism), but logical positivism and classical positivism are not identical philosophical positions (Bechtel, 1988; Phillips, 1992). Logical positivism was derived from the principles of classical positivism, and they both share many similarities (e.g., the inductive-hypothetico-deductive approach to the scientific method). However, several of the additional assumptions underlying logical positivism (e.g., verifiability principle) resulted in untenable ontological positions (e.g., antirealism), whereas the legacy of classical positivism still survives today in postpositivistic science. Nevertheless, I review briefly the tenets of logical positivism rather than those of classical positivism, because of the importance of logical positivism in the development of modern scientific practice.

Logical positivism affirms that the origin of knowledge is grounded on experience or direct observation, but scientific theories and laws are not the simple summation of individual propositions. Rather, theories are an axiomatic network of statements from which singular propositions can be derived and subsequently verified. This modification in the structure of knowledge is a direct response to the problem of induction. According to logical positivism, a theory is a tool or instrument for prediction, and explanatory theories are nothing more than premises for deducing accurate and specific predictions. To this end, logical positivism sought to elevate scientific discourse by using a rigorous procedural language (e.g., symbolic logic) to establish a hierarchical unification of all scientific disciplines under the most fundamental science: physics.

Logical positivism used a procedure known as the hypothetico-deductive method to acquire knowledge (Bechtel, 1988). Cattell (1988a) more precisely refers to it as the inductive-hypothetico-deductive method, because it still relies on inductive inference to modify the original theory.

Logical positivism asserted that induction is used to formulate general laws and theories. Those then provide a basis for formulating scientific hypotheses and establishing the initial conditions of the research design. If the predictions derived from the theory are corroborated by direct experience, then the hypothesis is confirmed. The positivistic inductive-hypothetico-deductive method, therefore, attempted to avoid the problem of sole reliance on induction by explicitly stating theoretically based hypotheses prior to observation.

For logical positivism, the validity of knowledge is evaluated through a strict process of verification. All hypotheses must be phrased in such a way that the phenomena under investigation can be observed empirically, and the verification of statements can be corroborated by experience. Hence, theoretical statements about facts must be stated in ways that are directly and completely testable, and the truth or falsity of the statements is revealed to the observer by empirical observation (Gjertsen, 1989, Weinberg, 1960). Phillips (1992) noted that a direct result of this manifesto was the use of operational definitions in scientific inquiries. However, this strict criterion of verification caused the eventual downfall of logical positivism. Verification requires that statements be made such that they are able to be confirmed or disconfirmed by experience, but abstract constructs and ideas cannot be so stated. Logical positivism was designed to establish scientifically based universal laws and theories, so its inability to deal with abstract phenomena was devastating.

The inductive-hypothetico-deductive method of logical positivism provided the methodological template for modern scientific inquiry. Instead of a simple compendium of facts from which one could make predictions, as in naive inductivism, the outcome of predictions or hypotheses in the inductive-hypothetico-deductive method is used to modify the initial theory. The inductive-hypothetico-deductive method is a successful defense against the problem of induction, and provides scientists the opportunity to originate and modify general theories or universal statements about empirical phenomena.

### C. Popper's Falsificationism

In response to logical positivism, Sir Karl Popper (1959, 1963) proposed a different account of the inductive-hypothetico-deductive scientific method: falsificationism. Rather than constantly seeking confirmation, Popper argues that scientists should expose theories as clearly as possible to refutation. Falsificationism disputes the inductivist assumption that science is a cumulative process and replaces it with an account of science based on a Darwinian metaphor of natural selection: survival of the fittest (Mayr, 1989).

Falsificationism is empirical, but unlike naive inductivism, it asserts that observations are necessarily guided by theory, and it denies the inductivist claim that theories can be established as true. Falsificationism recognizes the limitations of induction and considers theories as speculations that are accepted only on a tentative basis. Hypotheses must be tested rigorously and ruthlessly by observation and experience, and scientists should focus on demonstrating that some hypotheses are false, rather than attempting to prove their truth. Those hypotheses that fail are eliminated and replaced, but those that survive are not confirmed or true, as naive inductivism would claim. Rather, falsificationism demands that hypotheses and theories must be tested repetitively under different conditions, and a hypothesis or theory is never accepted as a universal truth even if it survives a large number of tests.

Falsificationism implies that empirical investigation must proceed by disproving hypotheses, and therefore, any hypothesis or theory proposed must be able to be falsified through empirical observation. Good scientific theories are ones that can be put to critical tests where one can specify in advance what would count against the theory. The more possibilities a theory rules out, the more powerful that theory is because it tells the scientist more about the world. Furthermore, Popper (1959) argued that falsification can be used as a criterion for separating scientific theories that run the risk of being falsified from nonscientific theories that do not.

Falsificationism stipulates that theories and hypotheses must be stated clearly and precisely, and that science must rule out ambiguity in the interpretation of experimental results. Although the objective of logical positivism was to reduce ambiguity in scientific language, the objective of falsificationism is to reduce ambiguity in producing scientific knowledge. Therefore, Popper (1959) rules out ad hoc modifications to protect theories or hypotheses from falsification unless those modifications are testable independently from the theory or hypothesis. To Popper (1959), scientific progress is a competitive and evolutionary process that produces theories of a higher level of universality (viz., they are more testable and have absorbed older, well-corroborated theories).

However, falsificationism is not without a number of criticisms that potentially undermine the approach. Popper (1959) argued scientists should not use observations to confirm theories. Observations are fallible, and scientists using fallible observations can make mistakes in falsifying theories just as they make mistakes in confirming theories (Chalmers, 1982). Hence, theories cannot be either conclusively confirmed or falsified.

In addition, philosophers of science have disputed the implication of falsificationism and inductivism that testing theory merely involves testing a single statement or hypothesis that directly reflects the theory from which it is derived. These philosophers argue that testing a statement or hypothesis really involves testing a complex number of statements, which include auxiliary assumptions regarding the method used and the initial conditions of the testing situation (Duhem, 1954/1982). Therefore, theories cannot be conclusively falsified because of the attending auxiliary statements and initial conditions that surround all theories and all observations. The tenets of falsificationism advanced by Karl Popper have provided current scientific practice with a number of unique benefits. Popper's recognition that empirical inquiry is based on and guided by theory is an important element in the development of current scientific methodologies. Furthermore, the stipulations that theories and hypotheses—as well as any ad hoc modifications—should be clearly stated and falsifiable have proven invaluable in testing and comparing scientific theories. Finally, the recognition that disconfirmatory evidence is as important as confirmatory evidence—even though both are fallible—has had a profound effect on the interpretation of research findings in the sociobehavioral sciences.

### D. Postpositivism (Modern Inductive–Hypothetico-deductive Method)

Unlike the previous scientific movements, postpositivism is not a unified, well-organized approach to scientific practice championed by an individual or group, but rather a rough, patchwork quilt incorporating many of the effective strategies of inductivism, positivism, and falsificationism. Popper's (1959) idea of theoretical universality provides an apt characterization of postpositivism, because well-corroborated and highly effective scientific tactics from the other three philosophies of science have been absorbed, recycled, and incorporated in postpositivism (Kerlinger, 1986; Pedhazur & Schmelkin, 1991). The development of postpositivism as a loose conglomeration of tenets and principles of scientific inquiry is a direct response to decades of criticism from philosophers of science. Nevertheless, many modern critics of the scientific method erroneously continue to believe that the one received view of scientific inquiry is logical positivism (Shadish, 1995).

### III. CRITICISMS OF THE MODERN INDUCTIVE–HYPOTHETICO-DEDUCTIVE METHOD

The scientific method is inherently a critical enterprise; theories, methods, and facts are all subject to active scrutiny, critique, and challenge by the scientific community. In this section I review three contemporary criticisms of postpositivism—Historicism, Anarchism, and Constructivism—to characterize the emergent nature of postpositivistic scientific inquiry.

### A. Historicism (Kuhn's Paradigms)

Historicism refers to philosophies of science that are grounded in the history of scientific investigation (Chalmers, 1982). These philosophical accounts are concerned with the actual character of scientific inquiry, and they generally attempt to provide a descriptive rather than a normative account of scientific inquiry. Historicism generally focuses on the larger framework in which science operates rather than on individual theories. For example, Kuhn (1962/1970) challenged the position of inductivism, positivism, and falsificationism that science is a steadily accumulating body of knowledge that is based on purely rational methods: "There is no logic of justification. There is only the historical process of competition among segments of a scientific community" (Giere, 1988, p. 33).

In place of the realist account of knowledge of naive inductivism and Popper's falsificationism, Kuhn (1962/1970) posited an instrumentalist epistemological view that the goal of science is neither to confirm nor falsify theories, but to concentrate on fitting theory to nature by solving problems. Kuhn (1962/1970) presented an original epistemological structure distinct from inductivism and falsificationism by introducing the concept of a *paradigm*. A paradigm is a framework that a particular discipline uses for characterizing the phenomena it takes as its subject matter. It includes the general theoretical assumptions, laws, and methodological techniques that members of a particular scientific community adopt. In other words, paradigms can be viewed as a basic model or general (meta) theory that instructs how scientific theories or models are to be developed and applied in further research.

Kuhn argued that scientists are unable to compare and evaluate differences between competing paradigms on any rational grounds because each paradigm's empirical knowledge is based on the theoretical assumptions and methodological techniques of that paradigm. No neutral observational language exists between two rival paradigms. Furthermore, each paradigm establishes empirical and theoretical standards for observations and their interpretation, so researchers cannot compare two different paradigms because the standards by which to evaluate them are different. Kuhn (1962/1970, 1977) referred to this inability to compare paradigms on either theoretical or empirical grounds as the incommensurability of paradigms. Therefore, Kuhn argued, scientists must resort to extrarational means to choose between paradigms.

One concern with Kuhn's earlier (1962/1970) paradigmatic account of the scientific method is that it offers nothing more than relativistic knowledge. Relativism, the view that knowledge is relative and may vary from individual to individual, group to group, or time to time (Runes, 1984), is potentially fatal to any theory of science because scientific knowledge is relegated to the position of opinion or belief. However, Phillips (1992) argued that a lens is a better metaphor for understanding the form of relativism Kuhn (ca. 1977) advocated in his later works. A paradigm is a lens through which scientists look to gain knowledge and an understanding of the world. Disagreement in perception occurs because different lenses (paradigms) provide different levels of magnification, and not altogether different realities. For this reason, knowledge is understandable primarily within the originating paradigm, and incommensurability between paradigms is neither absolute or certain.

Kuhn's (1957, 1962/1970, 1977) work challenged the notion that scientists operate in a purely rational manner and that science is a cumulative process. Kuhn repeatedly observed that science is undeniably a social process that is influenced by a myriad of factors, and that scientific development does not necessarily proceed under strict rational guidelines. Kuhn's conceptualization of scientific paradigms further underscored the irrefutable interdependency of theory, method, and fact.

### B. Anarchism (Feyerabend's Relativism)

Feyerabend's (1975/1988, 1981) examination of the methods by which knowledge is obtained and tested for validity provides another perspective on scientific inquiry. Drawing from Kuhn's and Popper's arguments about the theory-ladenness of observations, Feyerabend (1975/1988) concluded that scientists should consider a number of different theories and methodologies:

> A scientist who wishes to maximize the empirical content of the views he holds and who wants to understand them as clearly as he possibly can, must therefore introduce other views; that is he must adopt a pluralist methodology. He must compare ideas with other ideas rather than "experience" and he must improve rather than discard the views that have failed in competition. (p. 21)

Using this counter-inductive, pluralist method provides for a variety of opinion, which consequently leads to progressively better objective knowledge. Feyerabend strongly urged consideration of a theory even if it has no supporting empirical evidence, because accurate scientific knowledge depends on maintaining a theoretical and methodological plurality.

Feyerabend emphasized Kuhn's idea of theoretical and empirical incommensurability at the theoretical—rather than paradigmatic—level. Because the fundamental principles of two rival theories may be radically different, they may not agree on what is considered to be an empirical observation. Choice between theoretical explanations is necessarily subjective when there are no mutually shared rational elements on which to compare them. Furthermore, "hardly any theory is *consistent with* [italics in original] the facts. The demand to admit only those theories which are consistent with the available and accepted facts leaves us without any theory" (Feyerabend, 1975/1988, p. 51). In other words, an infinite number of possible theories could explain the same set of empirical facts, thereby implying that no rational basis exists for choosing between theories.

Because no scientific theory can completely explain all of the known facts in a specific domain or discipline, Feyerabend (1975/1988) denied the existence of any sound, rational basis for imposing methodological principles on scientific inquiry. In support of this argument for methodological anarchism, Feyerabend described a number of famous scientific programs (e.g., Galileo) that succeeded only because they violated rational rules or principles about how science should be conducted. The only principle of scientific methodology acceptable, according to Feyerabend, is "anything goes." In other words, any theory or methodology, whether defunct or not, offers some contribution to scientific knowledge regarding worldly phenomena.

Critics have countered that Feyerabend's maxim of "anything goes" in regard to theory and methodology implies that "everything stays"(Krige, 1980), because scientific or objective knowledge cannot improve if every theory or method is legitimate in one sense or another. In addition, Laudan (1990) argued that the relativist position regarding the underdetermination of theory fails to distinguish between facts that post hoc "fit" the theory in question and facts that were predicted *a priori* to support the theory in question. The problem of underdetermination of theory implicates rampant exploratory inductivism but not research guided by Strong Theory. Although Feyerabend's idea of methodological pluralism is too arbitrary for realistic scientific practice, modern postpositivistic sociobehavioral research has adopted his emphasis on the fallibility of observation and measurement (Cook; 1985; Cronbach & Meehl, 1955; Webb, Campbell, Schwartz, & Sechrest, 1966).

## C. Constructivism

A number of alternative transactional methods of inquiry have developed in response to positivistic scientific methodology, including neo-Marxism, feminism, and materialism (Guba & Lincoln, 1989). Transactional methods posit that scientific data are literally created from the interactive interchange between an investigator and participant. Constructivism, the predominant contemporary transactional approach, shares many epistemological and methodological similarities with other transactional views of the scientific enterprise, but it is unique in its denial of the existence of a single reality. Constructivism argues that multiple realities are socially constructed through intra- and interpersonal dialectic conflict and consensus (Greene, 1990; Guba & Lincoln, 1997). None of these constructed realities is more or less a true reflection of reality, but some may be more or less informed or sophisticated.

Knowledge is built through interactive subjectivity, and not the clinically distant objectivity claimed by positivism. Information is collected using conventional hermeneutical (explanatory) techniques, and interpreted by means of a dialectical interchange (i.e., a systematic examination—and debate about—the meaning of facts or ideas to resolve any real or apparent contradictions). Data collection and interpretation typically occurs in natural contexts in order to capture holistically the implicit meaning of the discourse. Lincoln (1990) notes:

> The aim of constructivist science is to create idiographic knowledge, usually expressed in the form of pattern theories, or webs of mutual and plausible influence expressed as working hypotheses, or temporary, time- and place-bounded knowledge. (p. 77)

In other words, constructivists seek to elicit individually constructed realities, and to compare and contrast these multiple realities dialectically to identify a smaller number of constructions having substantial consensus (Guba & Lincoln, 1989; Schwandt, 1990).

Although constructivism subscribes to an individually based, relativistic view of knowledge, the quality of the information collected during an inquiry still must be evaluated. For example, constructivism replaces validity with participants' assessment of the believability of the results after data collection and analysis. Transferability (i.e., generalizability) of the results implies that the description of the particular context is sufficient to allow other investigators to judge the applicability of the information collected. Likewise, dependability replaces reliability, and confirmability is substituted for objectivity (Guba & Lincoln, 1997; Schwandt, 1990).

Transactional approaches like constructivism have been criticized on two grounds. First, the epistemological stance of relativism necessarily implies that all knowledge is incommensurable (Laudan, 1990). As with the incommensurability of Kuhn's paradigms and Feyerabend's theories, constructivism embraces incommensurability by positing that multiple and fundamentally different realities coexist at the same time. This casts serious doubt on the uniqueness of scientific knowledge, leaving no basis on which to separate knowledge from uninformed opinion. Furthermore, critics have rejected several of the criteria used by constructivists to warrant knowledge claims. For example, Phillips (1992) argued: "Credibility is a scandalously weak and inappropriate surrogate for truth or veracity—under appropriate circumstances any nonsense at all can be judged 'credible'" (p. 117). Nevertheless, constructivism's stress (along with historicism's and anarchism's) on the importance of recognizing socially constructed phenomena, identifying the potential biases of the investigator and acknowledging the influence of social institutions have influenced contemporary thinking about scientific practice.

## IV. CRITICAL MULTIPLISM (POSTPOSITIVIST INDUCTIVE–HYPOTHETICO-DEDUCTIVE METHOD)

The sociobehavioral sciences have been comparatively sensitive to the limitations that result from an overreliance on single methods of measurement and unitary theoretical perspectives (Pedhazur & Schmelkin, 1991). For example, Campbell and Fiske (1959) advocated the use of multiple methods to distinguish between trait (construct) and method variance. Webb et al. (1966) advanced operational definitions beyond the strict limitations of positivism by using multiple operationalism in measurement. Both of these advancements reflect the progressive spirit of postpositivism in the sociobehavioral sciences. More recently, a number of researchers (e.g., Cook, 1985; Houts, Cook, & Shadish, 1986; Shadish, 1989) have expanded upon this previous work and developed the strategy of critical multiplism to address modern criticisms and concerns regarding the scientific enterprise.

Critical multiplism encourages scientists to use an array of (unavoidably imperfect) methods and theories in a strategic manner that minimizes systematic and constant biases (Shadish, 1989). This research strategy requires a "critical" examination of the assumptions and biases inherent in each method or approach to measurement, as opposed to the "uncritical" or relativistic reliance on convenient and available methods (i.e., Law of the Instrument or "Anything goes"). Critical multiplism advocated a deliberately planned approach in which the principles of multiple operationalism are systematically extended to all elements of the scientific enterprise, including the selection of theoretical models, formulation of research questions, operationalization of constructs, application of research methodologies, analyses of data, interpretation of data and replication (Sechrest & Figueredo, 1992).

Rather than selecting multiple indicators across multiple disciplines with no particular purpose (i.e., methodological pluralism), critical multiplism seeks to identify the strengths, biases, and assumptions associated with each measuring instrument, theory, or approach. By selecting complementary operations, Cook (1985) and Shadish (1989) argue, one is better able to ensure that constant biases are avoided and no plausible bias is overlooked. Critical multiplism, however, does not embrace the maxim "anything goes": "Multiplism by itself is like blind variations, generating many novel options with little sense of which option is worth retaining" (Shadish, 1989, p. 14). Instead, emphasis is placed on the judicious use of methods to ensure a heterogeneity of biases of both omission and commission in all facets of a research program. Under these circumstances, confidence in the resulting empirical knowledge is increased when multiple operations produce similar results across research designs having different biases.

Critical multiplism advocated searching at each stage of a scientific inquiry for constant, systematic biases in the operations that *have* and *have not* been employed previously. Question formation should involve a consideration of the types of questions that have and have not been asked within the content domain, and of the ways in which the questions have been operationalized. Theories and models should be selected to consider a heterogeneity of plausible, alternative perspectives to ensure that inadvertent systematic biases are not introduced, a procedure sometimes referred to as the method of multiple working hypotheses (Chamberlain, 1890) or strong inference (Platt, 1964). Research design should incorporate multiple outcomes, multiple indicators, multiple situations, and multiple occasions in order to (a) address issues of construct and external validity, (b) model complex causality, and (c) assess measurement error and bias. Data analysis should involve multiple analytic techniques that discriminate between exploratory and confirmatory approaches and between linear and nonlinear trends. Interpretation of research results should attempt to narrow down the number of working hypotheses to those that seem more credible on the basis of available evidence (Sechrest & Figueredo, 1992). Finally, replication is essential to continually increase the heterogeneity of different sampling dimensions (e.g., subjects, instruments or contexts) and to map out domains of generalizability (Houts et al., 1986).

Planned critical multiplism has enjoyed considerable success across a variety of disciplines in the sociobehavioral sciences. Houts et al. (1986) investigated the person–situation debate in personality by reanalyzing data using a variety of working hypotheses. Donaldson (1995) used planned critical multiplism to summarize the current state of research on worksite health-promotion programs. Lösel (1995) used critical multiplism as a general rubric to review and synthesize the numerous meta-evaluations of the efficacy of correctional treatments.

Critical multiplism is a constructive response to modern criticisms of the scientific method. Rather than resigning to the fact that no one theory or method can explain all of the known facts of a particular domain and adopting a position of epistemological relativism, proponents of critical multiplism have used these limitations to develop a systematic strategy for reducing uncertainty about scientific knowledge. The key element of critical multiplism is the recognized need to identify the assumptions, capitalize on the strengths, and compensate for the biases inherent in all plausible methods and theories.

## V. CONCLUSION

The role of theory and experimental design in the analysis of multivariate data is as complex and multifaceted as it is foundational. Whatever the substantive question or methodological approach, theory provides an essential guide for research design, measurement operations, and data analysis. The use of Strong Theory to guide an inquiry improves the quality of the research and alleviates many of the philosophical criticisms of postpositivistic approaches reviewed herein. Conducting research in the absence of an explicit Strong Theory condemns researchers to the use of inefficient methods and forces them to accept indeterminate conclusions, especially when multivariate analyses are used. To return to the jigsaw puzzle analogy with which I began this chapter, two-, four- or eight-piece jigsaw puzzles do not require a detailed representation to solve. However, the task of selecting the correct twenty-four, forty-eight, or ninety-two puzzle pieces from a confusing agglomeration of thousands of puzzle pieces necessitates the use of some strategy for determining which pieces are relevant. We call these strategies *theories*.

## REFERENCES

Aiken, L. R. (1994). Some observations and recommendations concerning research methodology in the behavioral sciences. *Educational and Psychological Measurement, 54,* 848–860.

Aiken, L. S., West, S. G., Sechrest, L., & Reno, R. R. (1990). Graduate training in statistics, methodology, and measurement in psychology. *American Psychologist, 45,* 721–734.

Bacon, F. (1956). *Novum organum.* New York: Anchor. (Original work published 1620)

Bechtel, W. (1988). *Philosophy of science: An overview for cognitive science.* Hillsdale, NJ: Lawrence Erlbaum Associates.

Bernstein, I. H. (1988). *Applied multivariate analysis.* New York: Springer-Verlag.

Bordens, K. S., & Abbott, B. B. (1991). *Research design and methods: A process approach.* Mountain View, CA: Mayfield.

Bryk, A. S., & Raudenbush, S. W. (1992). *Hierarchical linear models: Applications and data analysis methods.* Newbury Park, CA:Sage.

Campbell, D. T., & Fiske, D. W. (1959). Convergent and discriminant validity by the multitrait-multimethod matrix. *Psychological Bulletin, 56,* 81–105.

Cattell, R. B. (1988a). Psychological theory and scientific method. In J. R. Nesselroade & R. B. Cattell (Eds.), *Handbook of multivariate experimental psychology* (2nd ed., pp. 3–20). New York: Plenum Press.

Cattell, R. B. (1988b). The meaning and strategic use of factor analysis. In J. R. Nesselroade & R. B. Cattell (Eds.), *Handbook of multivariate experimental psychology* (2nd ed., pp. 131–203). New York: Plenum Press.

Cattell, R. B. (1988c). The data box: Its ordering of total resources in terms of possible relational systems. In J. R. Nesselroade & R. B. Cattell (Eds.), *Handbook of multivariate experimental psychology* (2nd ed., pp. 69–130). New York: Plenum Press.

Chalmers, A. F. (1982). *What is this thing called science?* St. Lucia, Qld: University of Queensland Press.

Chamberlain, T. C. (1890). The method of multiple working hypotheses. *Science, 15,* 92.

Chen, H. T., & Rossi, P. H. (1983). Evaluating with sense: The theory-driven approach. *Evaluation Review, 7,* 283–302.

Cliff, N. (1983). Some cautions concerning the application of causal modeling methods. *Multivariate Behavioral Research, 18,* 115–126.

Cliff, N. (1987). *Analyzing multivariate data.* New York: Harcourt Brace Jovanovich.

Cohen, J., & Cohen, P. (1983). *Applied multiple regression/correlation analysis for the behavioral sciences* (2nd ed.). Hillsdale, NJ: Lawrence Erlbaum Associates.

Cook, T. C., & Campbell, D. T. (1979). *Quasi-experimentation: Design and analysis issues for field settings.* Dallas, TX: Houghton Mifflin Co.

Cook, T. C. (1985). Postpositivist critical multiplism. In L. Shotland & M. Mark (Eds.), *Social science and social policy* (pp. 21–62). Beverley Hills, CA: Sage Publications.

Cronbach, L. J., & Meehl, P. E. (1955). Construct validity in psychological tests. *Psychological Bulletin, 52,* 281–302.

Dillon, W. R., & Goldstein, M. (1984). *Multivariate analysis: Methods and applications.* New York: Wiley.

Donaldson, S. I. (1995). Worksite health promotion: A theory-driven, empirically based perspective. In L. R. Murphy, J. J. Hurrell, S. L. Sauter, & C. P. Keita (Eds.), *Job stress interventions* (pp. 73–90). Washington, DC: American Psychological Association.

Duhem, P. (1982). *The aim and structure of physical theory.* Princeton: Princeton University Press. (Original work published 1954)

Feyerabend, P. (1988). *Against method.* New York: Verso. (Original work published in 1975)

Feyerabend, P. (1981). *Philosophical papers* (Vols. 1–2). Cambridge: Cambridge University Press.

Figueredo, A. J. , Hetherington, J., & Sechrest, L. (1992). Water under the bridge: A response to Bingham, R. D., Heywood, J. S., & White, S. B. Evaluating schools and teachers based on student performance: Testing an alternative methodology. *Evaluation Review, 16,* 40–62.

Forscher, B. K. (1963). Chaos in the brickyard. *Science, 142,* 399.

Giere, R. N. (1988). *Explaining science: A cognitive approach.* Chicago: University of Chicago Press.

Gjertsen, D. (1989). *Science and philosophy: Past and present.* London: Penguin Books.

Gorsuch, R. L. (1983). *Factor analysis* (2nd ed.). Hillsdale, NJ: Lawrence Erlbaum Associates.

Gorsuch, R. L. (1988). Exploratory factor analysis. In J. R. Nesselroade & R. B. Cattell (Eds.), *Handbook of multivariate experimental psychology* (2nd ed., pp. 231–258). New York: Plenum Press.

Greene, J. (1990). Three views on the nature and role of knowledge in social science. In E. Guba (Ed.), *The paradigm dialogue* (pp. 227–245). Newbury Park, CA: Sage.

Grimm, L. G., & Yarnold, P. R. (1995). *Reading and understanding multivariate statistics.* Washington, DC: American Psychological Association.

Guba, E. G., & Lincoln, Y. S. (1989). *Fourth generation evaluation.* Newbury Park, CA: Sage.

Guba, E. G., & Lincoln, Y. S. (1997). Competing paradigms in qualitative research. In N. K. Denzin & Y. S. Lincoln (Eds.), *The landscape of qualitative research* (pp. 195–220). Thousand Oaks, CA: Sage.

Hair, J. F., Anderson, R. E., Tatham, R. L., & Black, W. C. (1992). Multivariate data analysis with readings. New York: Macmillan Publishing Company.

Hirschberg, N., & Humphreys, L. G. (1982). *Multivariate applications in the social sciences.* Hillsdale, NJ: Lawrence Erlbaum.

Houts, A. C., Cook, T. D., & Shadish, W. R. (1986). The person-situation debate: A critical multiplist perspective. *Journal of Personality, 54,* 52–105.

Huberty, C. J. (1994). Why multivariate analyses? *Educational and Psychological Measurement, 54,* 620–627.

Hume, D. (1963). *An enquiry concerning human understanding* (E. Mossner, Ed.). New York: Washington Square Press. (Original work published in 1748)

Johnson, R. A., & Wichern, D. W. (1988). *Applied multivariate statistical analysis* (2nd ed.). Englewood Cliffs, NJ: Prentice Hall.

Judd, C. M., McClelland, G. H., & Culhane, S. E. (1995). Data analysis: Continuing issues in the everyday analysis of psychological data. *Annual review of psychology* (Vol. 46, pp. 433–465). New York: Annual Reviews, Inc.

Kachigan, S. K. (1982). *Multivariate statistical analysis.* New York: Radius Press.

Kaplan, A. (1964). *The conduct of inquiry.* New York: Chandler Publishing.

Keppel, G. (1982). *Design and analysis: A researcher's handbook* (2nd ed.). Englewood Cliffs, NJ: Prentice-Hall, Inc.

Kerlinger, F. N. (1986). *Foundations of behavioral research.* (3rd ed.). New York: Harcourt Brace College Publishers.

Krige, J. (1980). *Science, revolution, and discontinuity.* Brighton, Sussex: Harvester.

Kuhn, T. S. (1957). *The copernican revolution.* Chicago: University of Chicago Press.

Kuhn, T. S. (1970). *The structure of scientific revolutions.* Chicago: University of Chicago Press. (Original work published 1962)

Kuhn, T. S. (1977). *The essential tension.* Chicago: University of Chicago Press.

Laudan, L. (1990). *Science and relativism.* Chicago: University of Chicago Press.

Lewin, K. (1951). *Field theory in social science* (D. Cartwright, Ed.). New York: Harper & Brothers Publishers.

Lincoln, Y. S. (1990). The making of a constructivist: A remembrance of transformations past. In E. Guba (Ed.), *The paradigm dialogue* (pp. 46–87). Newbury Park, CA: Sage.

Lipsey, M. W. (1990). Theory as method: Small theories of treatments. In L. Sechrest, E. Perrin, & J. Bunker (Eds.), *Research methodology: Strengthening causal interpretations of non-experimental data* (pp. 33–51). Washington, DC: U.S. Department of Health and Human Services.

Lipsey, M. W. (1997). What can you build with thousands of bricks? Musings on the culmination of knowledge in program evaluation. In D. J. Rog & D. Fournier (Eds.), *Progress and future directions in evaluation: Perspectives on theory, practice and methods. New Directions for Program Evaluation, 76* (pp. 7–23). San Francisco: Jossey-Bass.

Lösel, F. (1995). The efficacy of correctional treatment: A review and synthesis of meta-evaluations. In J. McGuire (Ed.), *What works: Reducing reoffending-guidelines from research and practice* (pp. 79–111). New York: John Wiley and Sons.

Marcoulides, G. A., & Hershberger, S. L. (1997). *Multivariate statistical methods: A first course.* Mahwah, NJ: Lawrence Earlbaum Associates.

Mayr, E. (1989). *Toward a new philosophy of biology.* Cambridge, MA: Harvard University Press.

McDonald, R. P. (1986). Describing the elephant: Structure and function in multivariate data. *Psychometrika, 51,* 513–534.

Meehl, P. E. (1978). Theoretical risks and tabular asterisks: Sir Karl, Sir Ronald, and the slow progress of soft psychology. *Journal of Consulting and Clinical Psychology, 46,* 806–834.

Meehl, P. E. (1986). What social scientists don't understand. In D. W. Fiske & R. A. Shweder (Eds.), *Metatheory on social science: Pluralisms and subjectivities* (pp. 315–338). Chicago: University of Chicago Press.

Meehl, P. E. (1990). Why summaries of research on psychological theories are often uninterpretable. *Psychological Reports, 66,* 195–244.

Morrison, D. F. (1990). *Multivariate statistical methods* (3rd ed.). New York: McGraw-Hill.

Muliak, S. A. (1972). *The foundations of factor analysis.* New York: McGraw-Hill.

Muliak, S. A. (1988). Confirmatory factor analysis. In J. R. Nesselroade & R. B. Cattell (Eds.), *Handbook of multivariate experimental psychology* (2nd ed., pp. 259–288). New York: Plenum Press.

Nesselroade, J. R., & Cattell, R. B. (1988). *Handbook of multivariate experimental psychology.* New York: Plenum Press.

Nunnally, J. C., & Bernstein, I. H. (1994). *Psychometric theory* (3rd ed.). New York: McGraw-Hill.

Overall, J. E., & Klett, C. J. (1972). *Applied multivariate analysis.* New York: McGraw-Hill.

Pedhazur, E. J. (1982). *Multiple regression in behavioral research* (2nd ed.). New York: Holt, Rinehart and Winston.

Pedhazur, E. J., & Schmelkin, L. P. (1991). *Measurement, design, and analysis: An integrated approach.* Hillsdale, NJ: Lawrence Erlbaum Associates.

Petrinovich, L. (1979). Probabilistic functionalism: A conception of research method. *American Psychologist, 34,* 373–390.

Phillips, D. C. (1992). *The social scientist's bestiary.* New York: Pergamon Press.

Platt, J. R. (1964). Strong inference. *Science, 146,* 347–353.

2. ROLE OF THEORY

Poincaré H. (1946). *The foundations of science* (G. B. Halsted, Trans.). Lancaster, PA: The Science Press. (Original work published 1913)

Popper, K. (1959). *Logic of scientific discovery.* New York: Basic Books.

Popper, K. (1963). *Conjectures and refutations.* London: Routledge & Kegan Paul.

Rummel, R. J. (1970). *Applied factor analysis.* Evanston, IL: Northwestern University Press.

Runes, D. D. (Ed.). (1984) *Dictionary of philosophy.* Totowa, NJ: Littlefield, Adams & Co.

Sahlin, N. (1991). Baconian inductivism in research on human decision-making. *Theory and Psychology, 1,* 431–450.

Schwandt, T. R. (1990). Paths to inquiry in social disciplines: Scientific, constructivist and critical theory methodologies. In E. Guba (Ed.), *The paradigm dialogue* (pp. 258–276). Newbury Park, CA: Sage.

Sechrest, L., & Figueredo, A. J. (1992). Approaches used in conducting outcomes and effectiveness research. In P. Budetti (Ed.), *A research agenda for outcomes and effectiveness research.* Alexandria, VA: Health Administration Press.

Shadish, W. R. (1989). Critical multiplism: A research strategy and its attendant tactics. In L. Sechrest, H. Freeman, & A. Mulley (Eds.), *Health services research methodology: A focus on AIDS* (pp. 5–28). Washington, DC: U. S. Department of Health and Human Services.

Shadish, W. R. (1995). Philosophy of science and the quantitative-qualitative debates: Thirteen common errors. *Evaluation and Program Planning, 18,* 63–75.

Simon, H. A., & Newell, A. (1963). The uses and limitations of models. In M. Marx (Ed.), *Theories in contemporary psychology* (pp. 89–104). New York: MacMillan.

Slife, B. D., & Williams, R. N. (1995). *What's behind the research? Discovering hidden assumptions in the behavioral sciences.* Newbury Park, CA: Sage.

Srivastava, M. S., & Carter, E. M. (1983). *An introduction to applied multivariate statistics.* New York: North-Holland.

Stevens, J. (1996). *Applied multivariate statistics for the social sciences* (3rd ed.). Mahwah, NJ: Lawrence Erlbaum Associates, Inc.

Tabachnick, B. G., & Fidell, L. S. (1996). *Using multivariate statistics* (3rd ed.). New York: Harper Collins Publishers.

Tacq, J. (1997). *Multivariate analysis techniques in social science research.* Thousand Oaks, CA: Sage.

Tatsuoka, M. M. (1988). *Multivariate analysis* (2nd ed.). New York: Macmillan.

Thurstone, L. L. (1948). Psychological implications of factor analysis. *American Psychologist, 3,* 402–408.

Tukey, J. W. (1977). *Exploratory data analysis.* Reading, MA: Addison-Wesley.

Tukey, J. W. (1980). We need both exploratory and confirmatory. *The American Statistician, 34,* 23–25.

Webb, E. J., Campbell, D. T., Schwartz, R. D., & Sechrest, L. (1966). *Unobtrusive measures: Nonreactive research in the social sciences.* Chicago: Rand McNally.

Weinberg, J. R. (1960). *An examination of logical positivism.* Paterson, NJ: Littlefield, Adams & Company.

# 3

# SCALE CONSTRUCTION AND PSYCHOMETRIC CONSIDERATIONS

## RENE V. DAWIS

*University of Minnesota, Minneapolis, Minnesota; and
The Ball Foundation, Glen Ellyn, Illinois*

## I. INTRODUCTION

This chapter is about the construction of scales—the instruments social scientists use to collect data. Scales are common features of social science research, reflecting the belief that data in the social sciences (as, indeed, in all science) should be quantitative in form. This belief is a fundamental presupposition for this chapter, but no rationale for it will be presented here (however, see Meehl, 1998).

This chapter is not about scaling theory or scaling models. Granted, any discussion of scale construction cannot avoid mention of scaling theory or scaling models. Nevertheless, the main concern here is with the scale itself—the actual measuring instrument—and how it is designed, constructed, and evaluated as a measuring instrument. Regardless of the subsequent use to which a scale may be put, its first use is measurement.

Measurement can be defined, following Stevens (1946), as the assignment of numbers ("numerals") to objects or events according to rules. The numbers so assigned are often called a "scale" (e.g., as in the Richter scale of earthquake magnitude). However, in this chapter, the term *scale* will be

*Handbook of Applied Multivariate Statistics and Mathematical Modeling*
Copyright © 2000 by Academic Press. All rights of reproduction in any form reserved.

used to refer to the instrument that produces the numbers (e.g., as in "weighing scale"), and *scale scores* will be used to refer to the numbers. Numbers have several properties that reflect different levels of measurement, and different scale scores may incorporate different properties of numbers. At the lowest level of measurement, the *nominal* level, it is simply the naming or labeling property of numbers that is used. At the next level, the *ordinal* level, we use the property that numbers are sequenced in an invariant order, allowing us to rank order the objects being measured. At the next higher level, the *interval* level, we use the property that equal intervals theoretically separate the numbers, allowing us to use the basic arithmetic operations of addition and subtraction with the scale scores. At the highest level of measurement, the *ratio* level, the introduction of a zero point in addition to equal intervals allows us to calculate ratios and use the arithmetic operations of multiplication and division. Each level of measurement includes the properties of the levels below it.

Numbers are assigned to "objects" that typically are tangible entities in the physical and biological sciences, but in the social sciences, numbers are assigned to entities that are intangible for the most part. Variables such as ability, motive, need, attitude, preference, value, and personality trait, among others, cannot be observed directly but rather must be inferred from behavior. These variables are theoretical constructions ("constructs") that have to be defined operationally by the methods used to measure them—which is why it is important to understand how scales are constructed.

Scale construction methodology originated primarily in the discipline of psychology. In the beginning, the first generation of psychological scientists used physical instruments to obtain their measurements (and many psychologists today still do). For example, the early psychophysicists discovered that a scale of sensation (e.g., of visual discrimination) could be derived as a mathematical function of the intensity of the stimulus (e.g., light intensity) that produces the sensation—Weber's Law. However, measurement with physical instruments required (among other things) the operation of "concatenation" (linking or combining empirically, e.g., putting two weights together to produce a heavier weight), whereas most of the interesting social and psychological phenomena could not be concatenated. Measurement of such phenomena required a different approach.

The new approach was found in the use of statistical models, whose introduction into psychology can be credited to Francis Galton. Galton (1869/1961) originated the method of scaling individual differences based on the normal curve, with the standard deviation as the unit. The best known of the early individual-differences scales, the Binet scale of intelligence (Binet & Simon, 1905–1908/1961), was based on differences between age groups. Alfred Binet introduced several innovations that have become characteristic of scale construction methodology ever since: (a) use of multiple indicators (items), (b) item selection (based on item analysis), (c) aggregation of item scores to create a total score, (d) standardized administration, and (e) validation (the innovation with the most far-reaching consequences not just for scale construction methodology but for psychological theorizing as well).

A mathematical foundation for scale construction methodology was made possible by another of Galton's monumental contributions: the invention of correlation (Galton, 1888/1961). Karl Pearson, Galton's disciple, developed the correlation to such a degree that it has become the basic analytic tool of the social sciences. It is no coincidence, therefore, that almost all of the multivariate analyses and mathematical modeling procedures discussed in this volume are based on the correlation.

Charles Spearman laid the groundwork for modern psychometrics with two landmark papers that discussed the concept of "reliability" and the measurement of an unobserved (inferred) variable. Spearman's (1904a) work on the correction of a correlation for attenuation led to the development of reliability theory, which in turn became the basis for classical test theory. His demonstration that correlational analysis could be used to infer and measure a postulated latent variable (Spearman, 1904b) set the stage for factor theory and eventually latent trait theory.

These (and other) developments led to the increasing use of scales in the applied social sciences. In education, for example, Thorndike (1910) introduced a handwriting scale consisting of graded handwriting specimens ordered according to the average "scale value" assigned by expert judges. A teacher could then grade a student's handwriting by using as a score the scale value of the most similar scale specimen. In industry, rating scales were used to evaluate work performance (Viteles, 1932; Landy & Farr, 1980). The construction and publication of ability tests, interest inventories, personality inventories, and other kinds of scales for use in the school, the clinic, and the workplace, became a new and booming business that continues to this day (DuBois, 1970).

The last fifty years have witnessed the rapid development of scaling theory and scaling models. Topics such as axiomatic measurement theory, conjoint measurement, order theory, multidimensional scaling, latent trait theory, and item response theory, among others, have progressed to unanticipated levels of sophistication. Yet disappointingly, these developments have had little influence on scale construction. As Ozer and Reise (1994) have noted, scale construction technology has remained "fundamentally unchanged for decades" (p. 368). More than two decades ago, Cliff (1973) observed the same disjunction between scaling theory and scaling practice, and explained it with observations that seem still to apply: If scaling theory is to be useful in scale construction, "empirical realizability is as important as mathematical elegance" (p. 475); that is, to be useful, scaling theory should make it possible to construct scales that actually work in practice.

However, this has not happened because the new approaches appear unable to "handle the difficulties posed by the inconsistencies inherent in fallible data" (p. 478). Human inconsistency (i.e., variability) seems effectively to have set an upper limit on scale construction technology, at least for the time being.

## II. SCALE DEFINITION

### A. A Psychological View

From a psychological standpoint, the typical scale may be described as consisting of a set of stimuli (items) designed to elicit particular kinds of responses, a set of response alternatives, and a set of directions on how to respond to the items. The responses that the scale is designed to elicit are taken either as being the construct of interest (e.g., a preference) or, more frequently, as indicative of the construct of interest (e.g., a preference as indicative of an attitude or a value). To effect measurement, numbers are assigned to the response alternatives according to rules dictated by some scaling method. These numbers are used to "score" responses. "Item scores" are typically aggregated into a "total score" (or "scale score" or "total scale score").

Each response to a scale item is the result of an input-mediation-output behavior sequence. The input (directions and the item) elicits and initiates the behavior sequence. Mediation—the cognitive-affective behavior that occurs in the respondent—results in the choice of a response alternative, which in turn triggers the output, the response behavior itself. Unfortunately, the mediation that actually occurs is not always the mediation for which the scale was designed. Extraneous mediation can occur if the scale stimulates affect that is not part of the construct (e.g., suspicion of the assessment process), if the respondent engages in unthinking compliant behavior, or if random behavior occurs as a result of boredom or lack of motivation. Furthermore, the response that occurs is not always the response that was intended by the respondent. Unintended response can occur if the respondent is careless or distracted, or if some feature of the scale instrument itself promotes unintended response (e.g., a confusing format or the use of too fine a print).

The mediation segment of the behavior sequence is ordinarily what is of most interest to the scale constructor and scale user. Scales are typically named after the psychological process(es) presumed to occur in the mediation segment. For example, interest inventories are intended to stimulate cognitions about interests, and ability scales are intended to call into play the targetted abilities. The scale (items, directions, and response alternatives) constitutes the core of an operational definition of the construct at issue.

## B. Categories of Scales

Social scientists construct a wide variety of scales, as reflected in the following categorizations:

### 1. Dimensionality

Scales can be classified into those consisting of only one item and those consisting of multiple items. A one-item scale must measure only one dimension to be interpretable. Multiple-item scales are sometimes described as "homogeneous" or "heterogeneous," depending on whether they include one kind of content (e.g., a homogeneous scale of arithmetic items) or multiple contents (e.g., a heterogeneous scale of arithmetic, vocabulary, and reading items). Factor theory uses these terms to refer to the number of dimensions underlying the scale. By definition, homogeneous scales are unidimensional (or unifactorial) and heterogeneous scales are multidimensional (multifactorial). The more the correlations among items tend to approach 1.00, the more "homogeneous" the scale is likely to be; the more the correlations tend to approach 0.00, the more "heterogeneous."

A multiple-item scale can be partitioned into "subscales," each consisting of a subset of items that are homogeneous. Subscale scores should be unidimensional to be interpretable.

### 2. Inferential Basis

Cook and Selltiz (1964) distinguished among the following five types of scales according to the basis on which inferences are drawn: (a) self-report, (b) observation of overt behavior, (c) reaction to, or interpretation of, partially structured stimuli, (d) performance of tasks, and (e) physiological reactions. These five scale types can be subsumed under two basic categories: scales involving the direct observation of behavior without verbal mediation (i.e., Cook and Selltiz's [b] and [e] type scales), and scales that require verbal mediation (i.e., Cook and Selltiz's types [a], [c], and [d]). Because these latter scales require verbal directions, a certain level of proficiency in the language is presumed for the use of such scales.

### 3. Test versus Self-Report

Verbally mediated scales, in turn, can be classified into two groups according to the directions used. In one group, respondents are directed to perform as well as they possibly can (Cook and Selltiz's performance of tasks); hence these scales are appropriately called "tests." In the other group, respondents are directed to report on their mediational processes—

their feelings, attitude, preferences, opinions, judgments, self-conceptions, or perceptions of their own or others' attributes or behavior. This group combines Cook and Selltiz's (a) and (c) scale types, and may collectively be called "self-report" scales. The two groups correspond to Cronbach's (1970) dichotomy of "maximum performance" versus "typical performance" measures. Tests (maximum performance scales) are intended to probe the limits of human performance, whereas self-reports (typical performance scales) attempt to portray the usual behavior of people.

Tests are premised on maximal respondent motivation and optimal performance conditions; maximum performance is not possible in the absence of one of these conditions. Responses to test items consist of perceptual, cognitive, and motor components in that they all start with the respondent's perception of the task stimuli (items), which stimulates cognitive processing that leads to some motor response. Tests can therefore be classified according to which component(s) contributes most to scale score variance. For example, perceptual tests are so called because they are designed so that most of the test score variance is due to individual differences in the perceptual component of response to the items, with the contributions of the cognitive and motor components of response being kept to a minimum.

Self-report scales—usually in questionnaire or rating format (including checklist)—presume both the ability and the willingness of respondents to report the "truth" about their typical behavior. Both factors, ability and motivation, have to be considered when evaluating the veridicality of self-report scale scores. Self-report scales also involve perceptual, cognitive, and motor components, but in contrast to tests, only the cognitive component is of interest. The nature of the cognitive component is what differentiates the different types of constructs and variables measured by self-report scales.

Tests and self-report scales differ in other ways. Test items are typically scored "right" or "wrong," and their basic parameter is level of difficulty (proportion of respondents failing the task). There are no "right" or "wrong" response alternatives for self-report scales. Rather, they are scored for "direction" (e.g., attitude scales are scored on an accept–reject direction, interest scales on a like–dislike direction, and value scales on an important–unimportant direction) and for "intensity" (zero to high). Typically, tests are timed, whereas self-report scales are not. On tests, respondents can only "fake" in one direction—downwards (deliberately choosing to fail), whereas self-report scales can be faked in either direction (reporting "more" or "less" than the "truth").

In this chapter, I will focus on the construction of self-report scales, which include personality scales, interest inventories, measures of attitudes, opinions, needs and values, and many of the ad hoc measures constructed or used by scholars in the social sciences for various purposes other than the testing of ability and achievement. (For another, though not altogether different, view on scale construction, see Hough and Paullin, 1994).

## III. SCALE CONSTRUCTION

Scale construction proceeds in five phases: scale design, item construction, item selection, scale validation, and scale evaluation.

### A. Scale Design

Scales are constructed for either a theoretical purpose (where the focus is on the construct being measured) or a practical purpose (the focus is on predicting a criterion). Either purpose requires that the scale measure a construct or a variable, so the first step in scale design is to develop a detailed description of the construct or variable to be measured. This description must have enough detail to guide the selection of items that are appropriate for the scale ("construct representation," Embretson, 1983) and to indicate the kinds of evidences that are required to demonstrate construct validity (Cronbach & Meehl, 1955). In effect, scale construction must begin with a "theory of the scale" that shows how the scale is related to the construct it measures.

The design of a scale should specify expectations about the internal structure of the scale (bearing on reliability) and the external relations that the scale should have with other measures (bearing on validity). For example, you should specify: (a) whether the internal structure of the scale should stress internal consistency (hence, item homogeneity) or stability (which allows for item heterogeneity); (b) whether the scale is to be used in studies that are exploratory (where internal structure is not constrained) or confirmatory (where internal structure is constrained); (c) the relations to specified external variables that are needed to demonstrate construct validity and/or social utility; and (d) the external variables to which the scale should *not* be related.

Finally, the scale design should address pragmatic concerns that will affect the scale construction effort. These include such matters as the projected size of the scale (target number of items), any limitations imposed by the population of interest (e.g., reading level, format), the conditions under which the scale will be administered (which of these are under the scale administrator's control), the sources of samples for scale development (and what is required to gain access to them), the time-line and time-frame for scale development, and of course, the cost. One helpful practice is to use a small lay advisory committee to serve as a sounding board throughout the scale construction process. This committee can be especially helpful in advising you about matters that involve sociocultural and ethical questions (e.g., about items that are offensive to a demographic group or invasion-of-privacy questions).

## B. Item Construction

All scales—including nonverbally mediated scales—require communication with the respondents, so the language proficiency of the scale's target population is an important initial consideration. The reading ability level of the target population should be ascertained, particularly when developing verbally mediated scales. Writing items for a heterogeneous target population requires a careful balance between being understood by the lower levels and avoiding "talking down" to the upper levels of the population.

I recommend conducting exploratory interviews on the subject matter with representative persons from the target population before writing items (Dawis & Weitzel, 1979). These interviews should provide information about how the construct of interest is "operationalized" in the verbal expression of this population. Writing items using the idiom of the target population can promote communication, but you also run the risk that the items can become obsolete with time.

A useful aid to writing items is Guttman's *facet analysis* (Guttman, 1959; Tziner, 1987). The facet approach requires the "mapping" (i.e., parsing) of items according to facets identified by theory or practice. For example, Elizur and Guttman (1976) "mapped" the domain of attitudes toward work and technological change according to three facets: behavior modality, referent group, and organizational component. Three behavior modalities (affect, cognition, action), five referent groups (computer staff, management, supervisor, colleagues, self), and two organizational components (computer, work) were of interest in the study. Respondents' attitudes were hypothesized to depend on their perception of the {behavior modality} of the {referent group} in regard to {organizational component}. In combination, these facets yielded $3 \times 5 \times 2 = 30$ unique item "types." For example, the item "How satisfied were you with the introduction of computers?" comprised the combination of {behavior modality: affect (satisfaction)}, {referent group: self}, and {organizational component: computer}. You can use the facet approach beforehand to ensure that item writing is systematic, and afterwards, to check on the item pools' coverage of the content domain. After the data are in, facet analysis can identify the facet(s) that contribute the most to item characteristics (item parameters) and scale score variance.

Item writing also requires you to decide on the number of response alternatives to use. The options range from two alternatives at one extreme to an unstructured response at the other extreme. The unstructured response alternative can be scored according to some scoring system (as, e.g., in scoring the Thematic Apperception Test [TAT]), but it is probably best used in exploratory studies to ascertain what structured response alternatives might be best suited for use in subsequent scales. As it is, many scale constructors unthinkingly opt for the "Likert 5-point rating" as the response alternative set, relying on its wide usage as justification for the choice. I recommend you think carefully about this issue instead of blindly adopting a conveniently available and popular option. For instance, there is reason to believe that the optimal number of response alternatives is related linearly to the cognitive level of the respondents.

Using an odd number of response alternatives typically results in the inclusion of a "middle ground" response alternative (such as "?" or "Undecided") that tends to attract a significant number of responses (in part, reflecting a "central tendency" response bias). You can eliminate this phenomenon by using an even number of alternatives. This approach confers the added advantage of allowing you to "collapse" categories into a defensible dichotomy, thereby fulfilling the commonly required statistical assumption of linearity (a 2-point scale is always linear). For this reason, it is psychometrically preferable to use an even number of alternatives. On the other hand, an even number of alternatives may not mirror reality faithfully, inasmuch as ordinarily there may be many truly "Undecided" or "middle ground" respondents. I recommend you include the "middle ground" alternative only if you are interested in this respondent category (e.g., as in the use of "Undecided" when polling voters). Otherwise an even number of alternatives is more advantageous.

You may sometimes need to include alternatives such as "No Opinion" or "Not Applicable" to account for other logical possibilities. Significant numbers of respondents may choose such out-of-scale alternatives, so I recommend you pretest the scale format to evaluate this possibility. If the proportions of "No Opinion" or "Not Applicable" respondents are small (say, 1 or 2%), you may drop the out-of-scale alternatives. However, if you include them, you will have to decide how they are to be scored in light of your theory of the construct. For example, you might conclude that the "No Opinion" and "Not Applicable" responses should be treated differently. Your alternatives are not scoring them; giving them either a zero score or the lowest nonzero score on the scale; and scoring them with the "?" or "Undecided" middleground category scoring weight.

Do not forget to check both the items and the directions for grammar and language usage. The presence of grammatical errors may cause the respondents to take the scale less seriously or reduce their motivation to participate in the study. Furthermore, the discovery of grammatical errors after completing the scale construction process, and especially after collecting normative data at considerable expense, will present you with the Hobson's choice of continuing with a flawed scale or starting all over again.

Items can be written to produce a homogeneous or a heterogeneous scale. Homogeneous scales are better for studying respondent (individual) differences, whereas heterogeneous scales are more useful when studying a stimulus (content) domain. To elaborate: Suppose you wanted to differentiate among individuals (say, in personnel selection). A heterogeneous scale would allow different individuals with different response patterns to obtain the same or similar scores, thereby making differentiation difficult. In contrast, a homogeneous scale would rank order individuals more reliably, thus making for better differentiation. On the other hand, you would be better served by a heterogeneous scale if you were in the early stages of articulating a complex construct. In this case, a homogeneous scale would focus your attention prematurely on one dimension, whereas a heterogeneous scale would reveal other dimensions for you to consider.

The homogeneity–heterogeneity question may be conceptualized as a two-way analysis of variance (ANOVA) design problem, with items and respondents as the two factors. The best course of action is to maximize the variance for the factor you wish to study and minimize it for the other factor, unless your purpose is to study the interaction.

Your initial guide in developing homogeneous or heterogeneous scales is item content: writing items of similar content yields more homogeneous items, whereas items dissimilar in content form heterogeneous items. After the initial item writing stage, however, do not rely on the appearance of similarity or dissimilarity. You can ascertain the actual homogeneity or heterogeneity of the scale only by analyzing actual respondent data. Never forget that the decision to write homogeneous or heterogeneous items should be dictated by your theory of the construct and your scale design.

I recommend you pretest the item pool on a small sample, with follow-up interviews, to uncover any remaining problems with the wording of the items or the directions. Do this pretest using the same kind of groups you use to prepare for item writing.

Following all precautions and exercising all due care will not guarantee that all the items you write will meet the criteria for item selection. Therefore, you must write more items than you intend to have in the final, developed scale. The number of items to write depends on the nature of the construct, the nature of the respondents, and your skill as an item writer. In ability testing, writing four or five items is often required to obtain one item that meets the selection criteria. For self-report scales, however, writing twice as many items as desired for the final scale probably might suffice.

## C. Item Selection

You can select items on the basis of internal criteria, external criteria, or both, depending on the scaling method you choose. I will discuss specific scaling methods later in its own section. In this section, I discuss the general methods that have come into widespread use because of rapid advances in personal computer technology and in statistical analysis programs.

## 1. Internal Criterion Methods

The original internal criterion method, introduced by Rensis Likert (1932), was to select items that correlate the most with total score. A variant of this method is to select items that contribute the most to reliability (information routinely provided by computer statistical analysis programs). These two methods provide practically identical results.

Factor analysis (see Cudeck, chapter 10, this volume) is often used to select items. Selecting items by factor analysis (setting aside for the moment issues about the choice of factor solution) requires a large sample ($N = 100$ is minimal, $N = 400$ to 500 is optimal) and the adoption of some criterion for item selection. The simplest method you could use for selecting items is to rank order the items according to their factor loadings and decide the cutoff on other (e.g., pragmatic) grounds. For example, you might decide to select the 10 best items from the pool of items submitted to analysis. Another approach is to specify a minimum factor loading to use as a cutoff for item selection. One option is to use an absolute cutoff criterion, on the premise that a factor loading is the correlation between the item and the factor. For example, setting a factor loading of .50 as the cutoff criterion for item selection implies that you accept a minimum of 25% (i.e., $.50^2$) shared variance between item and factor as sufficient evidence that the item measures the factor. Alternatively, you could use a relative cutoff criterion. For example, the factor loading squared divided by the communality of the item indicates the proportion of the item's communality (i.e., the common variance of the item) accounted for by the item's factor loading. Use of that criterion is consistent with the premise that an item should go to a scale (or subscale) with which it shares the largest proportion of its common variance. For example, an item with a factor loading of .50 ($r^2 = .25$) shares 31% of its common variance with the factor if the item's communality is .80, and 83% if its communality is .30. Hence, if you require a shared common variance of at least 50% as the criterion for item selection, an item with factor loading of .50 would not be selected if its communality were .80 but would be selected if its communality were .30.

How many items should constitute a scale? Factor theory requires three items to identify a factor, so in theory this number should be the minimum size of a scale. However, you usually need at least four or five items for scales to achieve an internal consistency reliability of .70. One factor to consider is the well-known relation between the number of items and degree of reliability, given in the Spearman-Brown prophecy formula, as follows:

$$r_X = \frac{Nr_Y}{1 + (N - 1)r_Y}, \tag{1}$$

where $X$ and $Y$ are two scales of different lengths, $r$ is the reliability coefficient, and N is the ratio of scale X's number of items to scale Y's number of items. Which is better: a 6-item scale with reliability of .77, an 8-item scale with reliability of .82, or a 10-item scale with reliability of .85? According to the Spearman-Brown formula, all three scales would have the same reliabilities if they had the same number of items (i.e., the 6-item scale with reliability of .77 would have a reliability of .85 if boosted to 10 items, using items of comparable quality that measured the same construct).

Cluster analysis (see Gore, chapter 11, this volume) is also used to select items and is most useful when your item pool is heterogeneous and you want to construct subscales. Typically, hierarchical clustering solutions are used, and the problem is deciding at what grouping level to stop the analysis (i.e., how many item groups should be formed). You can use two criteria. The average intercorrelation within the item group can be compared with the average correlation of the items in the group with nonmember items. A ratio of 2.0 is typically used as a minimum criterion; the higher the ratio, the more stringent the criterion. Another possible criterion can be based on the number of items in the groups. For example, you might decide on subscales with a minimum of 5 items, or that the longest subscale should not exceed 10 items. Or, you could just calculate the internal consistency reliability of each potential subscale and stop the analysis when the least reliable subscale (usually the one with the fewest items) is at the threshold you set as minimally acceptable.

Regardless of the internal criterion method used, the set of items selected for a scale (or a subscale) should be subjected to the basic test of the method: internal consistency reliability. Also, you might need to assess test–retest stability depending on how you conceptualize your construct. The particulars for stability assessment (e.g., length of intertest interval, characteristics of sample) should flow from your theory of the construct and should be incorporated in your scale design. It is important when assessing stability to distinguish between test–retest score differences and test–retest correlation. You can obtain a high test–retest correlation and still have large differences (positive or negative) between test scores and retest scores. This difference will be discernable in the mean difference between test and retest scores. When such a difference occurs, you will have to account for this change in scale scores—unless such change is expected from your theory of the construct. (See Tinsley & Weiss, chapter 4, this volume, for more on the distinction between reliability and agreement.)

A word of warning: Internal consistency reliability may be the *sine qua non* for item selection by internal criterion methods, but you should not use it in an automatic fashion, without examining the items you do select. Your scale may turn out to consist of items that are merely paraphrases of each other. Such a scale will yield very little information because the items essentially ask the same question and get the same response over and over again. You will be better served by a scale of lower but acceptable reliability having items of more varied content.

## 2. External Criterion Methods

Item selection by the external criterion method requires you to specify one or preferably more external variables that theory or practice says should be substantively related to the scale's construct. You then select items to the degree that they correlate with the external variables. This procedure (used in the empirical scaling method, described later) has the added advantage of simultaneously validating the scale as scale construction proceeds.

You may select items singly, as in Strong's (1943) empirical method, or in combination, using multiple regression or other multivariate techniques. Using Strong's method, you select items one at a time, based on the usefulness of each in differentiating between an external criterion group and a base-line or reference group (this is equivalent to the item correlating with the external variable). Strong recommended the use of large samples ($N \geq 400$) to ensure the validity of the items. The single-item approach does not take account of item intercorrelation, so you run the risk of selecting redundant (i.e., highly correlated) items. The multiple-items-in-combination approach has the virtue of using more information than the single-item method, specifically, information about item covariation. Redundancy is not necessarily a disadvantage, however, as is shown by the robustness of Strong's Occupational Scales (Hansen & Campbell, 1985).

Regardless of which external criterion method you use to select items, you should cross-validate, because all such methods capitalize on chance factors (see Tinsley & Brown, chapter 1, this volume). The results you obtain will be uniquely tailored to the sample used and will almost always differ, sometimes drastically, for another (ostensibly similar) sample. Cross-validation provides information about the extent to which the results are generalizable to different samples.

External criterion methods select items that tend to be uncorrelated or lowly correlated, so scales consisting of such items will tend to have lower internal consistency reliabilities. That is not troublesome because internal consistency is not the goal for such scales. It is more appropriate to assess the stability (test–retest reliability) of these scales rather than the internal consistency reliability. Relatively longer scales of 30 items or more typically are necessary to obtain acceptable levels of stability reliability.

You can use both internal criterion and external criterion methods by combining the results of the two analyses and selecting the items that meet both criteria. For this approach to be practicable, you may have to relax the decision rules used for each method. Relaxing decision rules may allow marginal items to be selected. I recommend you specify an item-selection algorithm beforehand (algorithms make replication easier). The procedure can also be reversed: you could look at the data analysis results, select the final set of items on the basis of your experience, intuition, and theory, then devise an algorithm that would "select" the selected items—and report this as the algorithm to use. Such a post hoc algorithm undoubtedly capitalizes on chance factors, so you should cross-validate it.

Some final comments on item selection: You should obtain item-selection data on samples representative of the target population, using administration procedures that approximate the conditions you anticipate for regular scale use. Obtain data on administrative matters at the same time, such as how long it took to complete the scale, how comfortable the respondents were with the task, how clear the directions were, what questions were asked, what other comments were made, and how many respondents one administrator could handle comfortably.

### D. Scale Validation

Some scale constructors validate their scales with the same data set that they use to select their items. When they do this, they are assuming that the responses to the selected items in the context of the reduced set of items (i.e., the final scale) will be the same as responses to those items in the context of the larger item pool. This assumption is not always justifiable because item context can have an influence on response (e.g., as in the "halo" effect, wherein preceding high ratings tend to "inflate" ratings on subsequent items). When resources are limited, you may have no other choice, but it is always preferable to collect new data for the purpose of validating a scale, separate from the data used to select items.

Data obtained for scale validation will permit you to assess scale reliability as well. Reliability is a precondition for validity: a scale cannot be valid unless it is reliable. Put another way, reliability is one type of evidence used to evaluate a scale's construct validity. (Recall also that different constructs may require different kinds of reliability.)

When you assess scale validity, you essentially test hypotheses about the scale and the scale's construct (Landy, 1986; Loevinger, 1957). The hypotheses are all of the form, "If this scale is a measure of the construct, then. . ." The more consequents you can generate and test, the more potential evidence you can have for the scale's validity. These consequents can include the practical use of the scale as well as the demonstration of theoretical expectations. (Specific validity issues are discussed in a later section.)

### E. Scale Evaluation

The final step in scale construction is to evaluate the scale. You can do this from a number of points of view, including a theoretical, psychometric, administrative, cultural, and ethical perspective. Theoretical evaluation addresses the question, "Is this scale a measure of the construct, and under what conditions?" Psychometric evaluation focuses on measurement issues, principally of reliability and validity (and it overlaps with theoretical evaluation). Administrative evaluation takes the form of a cost–benefit analysis of the scale's production and use. Among the aspects considered are the costs of production, ease of administering and scoring, and feasibility of use in mail surveys and other data collection venues. Cultural (or sociopolitical) evaluation examines the place of the scale and its construct in the larger cultural context: will it be socially acceptable and beneficial, or controversial and divisive? Finally, ethical evaluation covers a range of issues from the competence of scale construction to the scale's potential to do harm to people. As I recommended earlier, it is helpful to assemble a small advisory committee of people representing the constituencies that would have interest in or dealings with the scale. This committee should be involved from scale design through all construction stages, but especially in scale evaluation.

Fortunately, there are aids to help you evaluate a scale. Prominent among these is the *Standards for Educational and Psychological Testing* (American Educational Research Association, American Psychological Association & National Council on Measurement in Education, 1985). Although the *Standards* refer to "psychological testing," they are applicable to all social science scales.

## IV. SCALING METHODS

Following are brief descriptions of some common scaling methods. The major concerns here are how items are selected and how scale scores are generated from these items.

### A. Thurstone Scaling

Leon Thurstone brought rigorous quantitative methods to bear on the measurement of previously "unmeasurable" constructs, such as attitudes and values (Thurstone, 1928, 1959). In scale construction, he is best known for the Thurstone scaling method. This method is used to develop a scale designed to differentiate among stimulus objects (as opposed to differentiating among respondents). You can use it, for example, to develop a scale of occupations according to social status, which you can then use to "score" other occupations in terms of their social status.

The object in Thurstone scaling is to select a set of benchmark stimuli that span equal intervals on a scale dimension. You accomplish this by asking a group of judges to assign a scale value to each stimulus (item) in a large item pool, using scale values from 1 to 11. You can improve within-judge reliability by doing the assignment of scale values in a three-stage sorting process. First, assign each item a preliminary scale value of 1, 2, or 3 (i.e., sort the items into three groups in rank order of relative scale value). Then, take each group of items in turn and sort the items into three rank-ordered subgroups. This results in nine subgroups of items in rank order from lowest to highest scale value. Finally, take the lowest and highest item subgroups and sort each into two: the "extreme" items and the "less extreme." This sorting procedure yields 11 subgroups of items, which can then be assigned scale values of 1 to 11, respectively. Judges are not required to assign equal numbers of items to the subgroups; the determining factor should be a judgment of whether the items are similar or different in scale value.

Typically, Thurstone scaling requires a large number of items (perhaps 200 to 300) and a small group of judges (20 to 30). For each item then, you would be able to calculate a mean scale value and a standard deviation (which indicates judge consensus). This allows you to select benchmark items to represent each integer on the 11-point scale dimension by picking the item with the nearest mean scale value and smallest standard deviation. Because two items are customarily picked for each scale point, a 22-item scale is often a tip-off that the Thurstone scaling method was used.

To use a Thurstone scale as a self-report measure, you ask the respondents to pick the item or items that best represent their position on the construct. A respondent's scale score is the scale value of the item picked or the average scale value if more than one item is picked. If more than one item is picked, they should be close to each other in scale value, otherwise the respondent's response is suspect.

The Thurstone scale is easy to construct and has the important measurement advantage of being an interval scale (or more correctly, an "equal-appearing interval scale"). From the point of view of item selection, it has the advantage that judges tend to give the same scale value to an item regardless of their own personal opinions about the item's subject matter, so you do not need to represent the whole range of opinions among your judges (although it is always better if you do). One difficulty with the Thurstone scale is that you have to ascertain the scale's reliability by the test–retest method because internal consistency methods cannot apply.

### B. Likert Scaling

Rensis Likert (1932) pioneered the use of aggregation ("summated ratings") in the measurement of attitudes. Where Thurstone used psychophysical methods to scale stimuli, Likert used psychometric methods to scale responses. Where Strong used an external criterion to select items, Likert used an internal criterion. The Likert scale is arguably the most frequently encountered type of scale in the contemporary social sciences.

The method discussed in Section III. Scale Construction is essentially the Likert method, so description of the method's specific steps will not be repeated here. The object of Likert scaling is to select a set of items that constitutes an internally consistent scale. Originally, Likert (1932) used two approaches to accomplish this objective: correlation between item and total score, and mean score difference between upper and lower scoring groups (both approaches reflecting the use of an internal criterion). Results were practically the same. Currently, as discussed in the Scale Construction section, we use item-total score correlation or contribution to reliability as the criterion for item selection. Likert (1932) also compared different weighting schemes for the response alternatives, and he found that results were just as good when the five response alternatives were scored simply 1 through 5 as when a more sophisticated scoring scheme was used. The 5-point scoring scheme has become so popular that any scale using it is often called a Likert scale, even if the Likert scaling method was not used in its construction.

## C. Guttman Scaling

Guttman scaling (Guttman, 1944) begins with the idea of a perfect unidimensional scale, which can be depicted in a "scalogram." A scalogram is a matrix in which respondents are represented as rows and items as columns (items are dichotomously scored: endorsed vs. not endorsed). In a perfectly unidimensional scale, the pattern of responses is completely predictable from a respondent's total score and an item's number of endorsements. In such a scale, if respondents were ordered according to total score and items according to number of endorsements, an orderly stepwise progression of endorsement for both respondents and items would appear, with the region of endorsement clearly separated from the region of nonendorsement by the diagonal. Given two respondents who differ in total score by one, the higher scorer would have endorsed all the items endorsed by the lower scorer, plus one other item. Given two items that differ by one endorsement, the higher item would have been endorsed by all respondents endorsing the lower item, plus one other respondent.

The Guttman scale can be illustrated in the classic Bogardus (1933) scale of social distance. In this scale (abbreviated for illustration purposes), you are asked whether you would be willing to admit members of a group (say, a nationality group) into each of the following social relationships:

1. Entry into your country as aliens
2. Citizenship in your country
3. Neighbors on your street
4. Close kinship in marriage

If you answer "yes" to item 3, you will most likely answer "yes" to items 2 and 1. If you have a score of "2," you most probably would have answered "yes" to items 1 and 2, and "no" to items 3 and 4. Results like these allow you to infer that the scale is unidimensional.

Use the scalogram to ascertain if your items constitute a unidimensional scale. From the proportion of deviations from perfect unidimensionality, you can calculate an index called the "coefficient of reproducibility," which should be .85 or better for a scale to be deemed functionally unidimensional. You can improve the coefficient of reproducibility by throwing out nonconforming items and respondents (items and respondents with significant deviations from perfect unidimensionality). You can examine the characteristics of the retained items and respondents to learn more about the nature of your construct.

A Guttman scale is often difficult to realize in practice, especially when the construct at issue is complex and thus probably not unidimensional. Furthermore, respondents frequently do not conform to the unidimensional model required by the Guttman scale. (Respondents give "inconsistent" answers, for example, on the Bogardus scale, answering "yes" to item 3 but "no" to item 2). For these reasons, Guttman scaling is not frequently used. Nevertheless, Guttman's ideas about scaling, especially about unidimensionality, have had great impact on subsequent developments in scaling theory and scaling models.

### D. Paired Comparisons

This classic scaling method originated in the psychophysical study of judgment and choice, and was given mathematical form by Thurstone (1927) in his famous Law of Comparative Judgment. The paired comparisons method has become a favorite method in the study of any kind of human preference. In this method, you pair the objects to be scaled, each object with every other object, and present the pairs to the respondents whose task is to choose one in each pair according to the scale's construct. You then analyze the judgment data to yield "scale values" for each object by using one of several methods of analysis that are available (see, e.g., Guilford, 1954). Because the number of pairings increases rapidly with each additional object, this method becomes impractical when the number of objects to be paired is greater than 20 to 25.

A special problem for the paired comparisons scaling method is the phenomenon of "intransitivity." This occurs when you choose A over B, B over C, but C over A—when logic expects you to choose A over C (termed "transitivity"). Intransitivity can result from an error in responding or because it is difficult for you to discriminate among the three objects. The first kind of intransitivity speaks to the reliability of the scale (the scale may need revision to minimize responding errors); the second kind of intransitivity speaks to the nature of the construct underlying the scale (the construct may be multidimensional). The study of intransitivity can be a fruitful research direction.

## E. Other Ranking Methods

### 1. Multiple Rank Orders

When the number of objects to pair is too large, you can use the method of multiple rank orders (Gulliksen & Tucker, 1961) to accomplish the same objective. In this method, you present the objects to be judged in groups of 3, 4, 5, or more, and the respondents' task is to rank order the objects within each group according to the scale's construct. You can use grouping schemes known as "balanced incomplete block" designs (see Gulliksen & Tucker, 1961) to create groupings such that each object is in a group just once with every other object. This is equivalent to each object being paired once with every other object, and therefore you can derive a paired comparisons matrix from multiple rank orders data and then analyze the matrix to produce scale values.

### 2. Simple Ranking

Assigning ranks in simple rank order is a straight-forward scaling method that assumes a single underlying dimension. You can assign ranks to either objects (items) or persons and use the ranks (or better, their complements) as scores. To improve the reliability of the ranking, you can start at the ends (i.e., rank the highest, then the lowest, then the next highest, followed by the next lowest, etc.) on the premise that discrimination is better at the ends than in the middle of the distribution. You can also convert ranks to z-scores (if you can assume a normal distribution), or to some other suitable transformation (see Guilford, 1954).

## F. Rating Methods

Rating scales have been used in the social sciences since at least the beginning of the 20th century, and there is an extensive but scattered literature on it (for an introduction, see Guilford, 1954, pp. 263–301; or Landy & Farr, 1980, pp. 82–94). There are a number of issues you have to deal with when constructing rating scales, such as, whether to use discrete points or a continuum as the rating scale, how many rating points to use, how to protect against rating bias (e.g., leniency, strictness, and halo), how to deal with rating errors (e.g., errors of central tendency and dispersion), whether to use verbal descriptions to "anchor" the rating points or use unlabeled numbers alone, whether two or more different rating scales can be used for each item, and whether the rating scales should all point in the same direction for all items or should alternate in direction for successive items.

Underlying many of these issues is the question of psychometric versus practical considerations (e.g., the optimal number of rating points is 5 to 9 [Cox, 1980], but 3 points may be more practical in some situations).

### 1. Simple Rating

A simple rating scale—in effect, a one-item rating scale—is the easiest way to construct a scale. Over the years, however, research has shown several problems with simple rating scales, and their use has been supplanted by other methods, notably by the multi-item Likert scale. For one thing, the reliability of one-item rating scales is usually low—and very frequently it is not ascertained at all by its users. You can only ascertain the reliability of one-item rating scales by the test–retest method, although in theory you can estimate it from a larger group of items by using the Spearman-Brown prophecy formula.

### 2. Semantic Differential

This is a special way of using rating scales, originated by Osgood, Suci, and Tannenbaum (1957) in their well-known study of meaning. In this method, you use several 7-point simple rating scales to measure the respondent's perception or judgment (e.g., the meaning of an object to the respondent). Each rating scale is anchored at each end by antonyms (e.g., "good" vs. "bad," "strong" vs. "weak"). Rating profiles can be compared for respondents rating the same object, and for objects rated by the same respondent or group of respondents. Three dimensions consistently have been found to underlie the simple rating scales used in semantic differential studies: (a) evaluation, (b) potency, and (c) activity.

### G. Empirical Scaling

In all of the scaling methods discussed above, you assume that responses to the items will reflect the construct that the scale is designed to measure. Therefore, you also can assume that an aggregated score is a "best estimate" of the underlying construct, and thus you can use total score as the (internal) criterion for selecting items for the scale. However, such use of an internal criterion can be criticized as being based on circular logic. That is, you select items as measuring a construct because they correlate with the criterion, but the criterion you use is the total score derived from the items.

You can avoid this problem of logical circularity by resorting to an external criterion when selecting items. Edward Strong, Jr. (1927) reintroduced the idea of using an external criterion in scale construction (first introduced by Binet) when he selected items on which he found sizable differences in response between an occupational group and a reference or baseline group to construct an occupational interest scale. Strong's "empirical approach" yielded reliable (stable) and valid scales that performed better than contemporaneous scales produced by the "rational approach," whose scoring was based entirely on psychological theory. Empirical scaling was subsequently adopted in personality scale construction (notably in the construction of the Minnesota Multiphasic Personality Inventory [MMPI]) with similarly striking results. The case for the "rational approach," however, is not without support (Ashton & Goldberg, 1973; Jackson, 1975), and in recent times, has gathered strong support (Hough & Paullin, 1994).

In empirical scaling, the external criterion often takes the form of groups of respondents who are known or assumed with good reason to differ (if possible, radically) on the construct of interest. The scale construction process involves selecting items that discriminate substantially between these contrasting groups. (This is mathematically equivalent to selecting items that correlate with a dichotomous external criterion.) Obviously, the quality of the scale you construct by this method depends, in part, on the quality of the criterion. Furthermore, constructing scales by this method can capitalize on chance or irrelevant factors, so you should cross-validate your scale on new samples.

A major difficulty with empirical scaling is that it is not always easy to identify suitable contrasting groups. Many constructs are complex in composition, often being correlated with other constructs, so that sometimes you will find it impracticable to identify groups that differ only on the construct at issue and no other. Furthermore, the items you select can be heterogeneous in nature, making interpretation of the scale score problematical. You can overcome this difficulty by improving the internal consistency of your scale through the method of reciprocal averages (Horst, 1935), which revises the weights assigned to the item-response alternatives to maximize internal consistency reliability.

If your purpose is to illuminate the nature of a construct, empirical scaling most likely will not be a fruitful approach to take. However, if you want to predict an external practical criterion (such as success in training or employment), it will be difficult to improve on a cross-validated empirical scale.

## H. Q-Sort

Stephenson (1953) proposed to identify suitable contrasting groups via an internal criterion by the use of Q-methodology. In this procedure, respondents rather than items are correlated (Q-correlation), and the resulting respondent correlation matrix is factor analyzed. The patterns of factor loadings can then be used to identify suitably contrasting groups of respondents.

Stephenson adopted a sorting procedure similar to the Thurstone procedure described above to facilitate the process, and this procedure came to be called the "Q-sort" because it was part of the Q-methodology. In time, the Q-sort became a favored technique in the study of intraindividual differences.

## V. PSYCHOMETRIC CONSIDERATIONS

### A. The Attenuation Paradox

The mathematical relation between item scores and total score is described by the *variance theorem,* which states that the variance of a sum (total score) is equal to the sum of the item variances plus twice the sum of the item covariances. The variance theorem has two important consequences for scale construction:

1. When item covariance is at a maximum (i.e., item intercorrelations approach 1.00), total score variance is also at a maximum. Because reliability depends on variance, the higher the intercorrelation of the scale's items, the greater the reliability of the scale scores is likely to be.

2. When item covariance is at a minimum (i.e., item intercorrelations tend toward 0.00), common variance among items is at a minimum. No two items measure the same thing, thus the content domain coverage of the items is at a maximum.

This second condition gives the scale the best chance of correlating with an external variable. In a scale with high item intercorrelations, the items all measure the same thing, which in turn minimizes the content domain coverage of the scale and diminishes the scale's chance of correlating with an external variable. There is a parallel here with multiple regression in that $R^2$ is maximized to the degree that intercorrelation among the independent variables tends toward zero, and is minimized by high intercorrelation among the independent variables, the condition called collinearity. (See Maxwell & Venter, chapter 7, this volume.)

Thus, there appears to be a built-in trade-off between reliability and validity, a trade-off known as the "attenuation paradox" (Loevinger, 1954). One condition, high item intercorrelation, favors reliability, whereas the opposite condition, low item intercorrelation, favors validity. Optimizing both reliability and validity requires sacrificing the maximization of each. Understanding this paradox is a key to understanding the lower reliabilities of social science scales when compared with physical science scales. Constructs in the social sciences (e.g., anxiety, social class) are very broad (i.e., heterogeneous) compared with those in the physical sciences (e.g., temperature) and they are used as explanatory constructs in a much broader array of situations.

### B. Reliability

"Reliability" is interpreted as "precision" in the physical science sense when the reliability coefficient is converted into a standard error of measurement. However, the concept of precision requires repeated measurement, but repeated measurement in the social sciences can change the respondent ("learning" takes place). For this reason, social scientists interpret reliability differently depending on the operations used to obtain the reliability estimate.

"Internal consistency," reflected in the high correlation among items or subsets of items, signifies that the items on the scale behave equivalently as though they were a single measure. Obviously, the internal consistency idea of reliability cannot apply to heterogeneous scales. You have to use the idea of stability.

The test–retest correlation, used to indicate temporal stability, depends on the interval between measurements, among other factors. Immediate retest would be the condition most like the concept of precision (repeated measurement). However, immediate retest reliability coefficients reflect not only the stability of the scale scores but also the ability of the respondents to remember previous answers, changes in the respondents' physical situation (e.g., fatigue), and changes in the respondents' motivation. Retest intervals of two weeks or more are often used to minimize these effects. With such intervals, and even more so with longer periods (months, years), the reliability coefficient becomes crucial evidence about the "trait or state" nature of the construct being measured. Trait variables require high stability over time, whereas state variables require evidence of susceptibility to changing conditions or to intervention.

Reliability should not be thought of only as a scale characteristic; it is also affected by the respondent sample. That is, reliability is correctly seen as an attribute—not of the scale—but of the scale scores, and scale scores are the product of scale-by-respondent interaction. It therefore behooves scale users to estimate reliability for *each use* of the scale. Citing reliability coefficients reported in the manual does not suffice; these are not estimates of reliability for the scale as used in the particular instance at issue.

What level of reliability is acceptable? Reliabilities reported for most self-report scales rarely rise above .90. Reported internal consistency reliabilities are typically in the .80s; stability coefficients are usually lower. Reliabilities have not improved much since World War II, improvements in scale construction theory and technology notwithstanding. This asymptote probably reflects a realistic ceiling imposed on scale reliability by human inconsistency (Cliff, 1973).

## C. Validity

Reliability and validity are the scale construction counterparts of precision and accuracy in physical measurement. As with reliability, the parallel for validity is true only up to a point. Validity is similar to accuracy when we think of validity as the degree to which scale scores, as sampling estimates, approximate the (usually unknown) population parameter. The accuracy of such estimation is described via the standard error of estimate, the formula for which incorporates the validity coefficient. But the concept of validity—from the Latin word meaning strong, effective, hence supportable—goes beyond just hitting the target.

Scales have two major kinds of validity: validity for theory and validity for practice. Validity for theory, commonly called construct validity, is the degree to which a scale is supportable as a measure of a theoretical construct. Validity for practice, called criterion-related validity, is the degree to which a scale is useful in predicting a practical (i.e., socially significant) criterion variable.

## 1. Validity for Theory

Construct validity includes any and all evidences that the scale measures the theory-defined construct (Messick, 1989). Evidence may be positive or negative (i.e., supporting what the construct is, as well as what it is not). The statistical evidence of validity may take the form of mean differences, variance differences, or correlations.

Embretson's (1983) distinction between construct representation and nomothetic span is useful (cf. Campbell & Stanley's, 1966, "internal" and "external" validity). Construct representation is a showing that the items (content and form) derive from the construct's theoretical definition (they represent the construct). Nomothetic span (cf. Cronbach and Meehl's, 1955, "nomological net") is the degree to which the construct's relations with other variables is articulated and supported. The more relations with other variables you can state about a construct, the more evidences you can gather about a scale as a measure of the construct. Thus, you can even include evidence of reliability and of the scale's usefulness in professional practice to demonstrate construct validity.

Construct validity often has been demonstrated through the Campbell and Fiske (1959) multitrait-multimethod matrix (MTMM) approach (see Dumenci, chapter 20, this volume). The MTMM format has formalized two lines of evidence in construct validation: a showing that different scales measure the same thing, and a showing that a scale does not measure the same constructs that other scales measure. The "multimethod" portion of MTMM is often very difficult to implement, in part, because data collection methods other than the self-report method are often impracticable. More and more, the last "M" in "MTMM" has stood for "Measure"—in lieu of different methods, researchers have been using different self-report measures. Use of an MTMM format as a conceptual framework for systematizing your evaluation of construct validity is commendable so long as you do not allow it to constrain your plans for construct validation.

Another favored approach to construct validation has been the use of factor analysis, either exploratory or confirmatory (see Cudeck, chapter 10, and Hoyle, chapter 16, this volume). Exploratory factor analysis has the appearance of objectivity in that no preconceptions are imposed on the data—the data are allowed to "speak for themselves." Confirmatory factor analysis is championed for precisely the opposite reason: that the role of data is to test explicitly stated hypotheses, not to provide them after the fact. There is no reason you cannot apply both approaches to the same data.

You can conduct exploratory factor analysis at the item level (with scale items as the variables) to explore the dimensionality of the scale, or at the scale level (perhaps including demographic variables and scales from other instruments) to provide evidences of convergent and discriminant validity. In the latter instance, the more rigorous way would be to conduct an MTMM analysis using confirmatory factor analysis to "confirm" the postulated "traits" (constructs) and to ascertain how much scale score variance is due to "method factors."

### 2. Validity for Practice

The "criterion problem" is a well-known problem of long standing in the applied social sciences (see Austin & Villanova, 1992, for a summary). Initially, "criterion" was used (correctly) to mean a "standard" or model scale by which to evaluate a new scale (did it measure up to the "standard"?). However, "standards" were hard to come by for many kinds of scales, so suitable substitutes were sought. At the same time, social scientists became more interested in predicting socially significant variables (e.g., school performance, work adjustment, crime and delinquency), and thus in scales that could predict these variables. In time, "criterion" came to mean a socially significant variable that was predicted by a scale, and conversely, the meaning or interpretation of a scale score came to be couched in terms of the criterion variable(s) it could predict. For practitioners, "validity" had become criterion-related validity.

Two other aspects of the "criterion problem" are worth noting: (a) some criteria may not be appropriate for a given scale. You cannot choose a criterion just on the basis of convenience; you must show its logical link to the scale's construct; and (b) a criterion must be measured as rigorously as any scale, and that means a showing of reliability and validity. Do not measure criteria using one-item rating scales that can conveniently be inserted in a multipurpose questionnaire. You must allot sufficient effort to criterion measurement if practical validity is to be of any consequence. A scale can be found to be "invalid" if the criterion measure is itself invalid or unreliable.

### D. Cross-Validation

Science deals with "facts," which comes from the same root word as does *manufacture* and *factory,* that is, from the Latin word meaning *to make.*

Researchers construct scales to produce or educe (i.e., to make) facts. To achieve the status of "fact," however, your finding (outcome, result) has to be "proved," that is, put to the test. In law, proving a fact requires the corroboration of witnesses; in journalism, two independent sources are required. In scale construction, you require cross-validation.

Cross-validation is a form of replication (Lykken, 1968), that is, seeing if you can get the same results on repetition of the experiment. In item selection, cross-validation demonstrates whether you would select the same items when you try the same procedures on a second sample. After you have formed the scale, cross-validation will show if the scale performs in the same manner (i.e., if you obtain the same levels of reliability and validity) when you use the scale with another sample. You would do even better with a double cross-validation, in which the items you select and the scale you develop in each of two samples are cross-validated on the other sample.

You can extend the procedure of cross-validation to doing other kinds of replications to learn more about your scale. For example, you could divide your total sample into two groups according to gender and see if you get the same results for either group—which brings up the question of bias.

### E. Bias and Equivalence between Groups

A common user question concerns whether a scale constructed for one group is appropriate for use with another group that differs from the original group on some important characteristic, such as age, gender, ethnicity, or geographic location. Essentially, the question is whether the scale is valid or biased for the new group. The short answer is that the appropriateness of using a scale with a group that is different from the scale construction group should always be evaluated empirically. To be appropriate, the scale must function equivalently for the new group, as shown by similarity in means, variances, item covariances (including factor structures), reliabilities, and validities.

"Bias" can be taken as the opposite of "equivalence." Predictive bias, the bias of most interest to social scientists, is systematic error in prediction for a group that is different from the base or norm group. Predictive bias is indicated by nonequivalence (i.e., a statistically significant difference obtained when based on the groups' estimated "true" scores) in any of three regression parameters: intercept, slope, or standard error of estimate (Jensen, 1980). This definition is tricky because it is based on true scores, whose estimation (by formula) depends on the scale's reliability. Thus, it is mathematically possible to find no evidence of predictive bias for a scale with low reliability, yet paradoxically to find bias when the scale's reliability is improved. Also, it is not mathematically necessary to have equivalent validity coefficients and standard deviations for the groups to show that there is no predictive bias. The regression parameters (i.e., intercept, slope, and standard error of estimate) for these groups may be equivalent even when validity coefficients or standard deviations are not equivalent. The moral is: you must do the calculations before you reach any conclusions. Furthermore, because "nonequivalence" is defined as a "statistically significant" difference, a conclusion of predictive bias can depend on sample size. Thus, a trivial difference that does not make any difference in practice can still lead to a conclusion of predictive bias if your samples are large enough (say, in the thousands).

There are two basic sources of bias: the scale and the respondent. The scale is a source of bias when it does not function as a measure of the construct equivalently for the two groups. This may come about, for example, when the scale content is more familiar to one group or when the scale format favors one group. The respondent is a source of bias when predispositions in responding introduce systematic error into the scale scores. Among these are tendencies (a) to respond leniently or strictly (i.e., to use only one side of the response alternative range); (b) to use the middle alternative(s) and avoid the extremes ("central tendency bias"), or vice versa; and (c) to respond similarly across different rating dimensions, better known as the "halo effect" when the responses are on the positive or high end, and the "horn effect" when they are on the negative or low end. These tendencies correspond respectively to response bias in the means, variances, and covariances (correlations) of scale scores.

Bias is to be distinguished from fairness. Bias is an attribute of the scale scores and is the concern of the scale constructor. Fairness is the concern of the scale user and refers to decisions based on scale scores (in whole or in part) that affect members of two (or more) groups unequally or disproportionately. At bottom, fairness has to do with equity (equivalence of the outcome-to-input ratios). An unbiased scale can be used in an unfair manner, but you are likely to be unfair without intending to if you use a biased scale.

Your first responsibility as a scale constructor is to minimize bias. This means, at a minimum, that you should examine the influence of "unchangeable" demographic variables (such as gender, age, and ethnicity) on the item and scale scores. Put in another way, you should examine differences between gender groups, age groups, ethnic groups, and the like, in the means, standard deviations, correlations, factor structures, and criterion-related validities of both item and scale scores.

Your second responsibility as a scale constructor is to promote the fair use of the scale. This includes alerting the user, especially in the manual, to known and potential sources of bias in the scale scores, and being vigilant about reports of unfair use of the scale. As a scale constructor, you have a social responsibility to educate the public about the proper use of the scale you have constructed.

## VI. A FINAL WORD

I conclude this chapter with two observations and a prescription. First, the list of published self-report scales is much too long, and growing longer at an alarming rate. Many reasons can explain the surfeit of scales. Scale construction is an appealing way to meet the doctoral dissertation and "publish or perish" requirements. Using published scales in research is sometimes too expensive; constructing your own can be more economical. Published scales often fall short of what is required in specific studies. Many scales appear only in the "fugitive literature," and tracking them down is too much work. Finally, there is also the "positive" reason that social scientists are creative and energetic.

Second, the .50 barrier on validity coefficients, first noted by Hull (1928), appears to be alive and well. Loevinger (1991) explains this phenomenon in terms of Brunswik's "lens model." There are many pathways from a focal light source (i.e., a construct) through a lens (i.e., the scale) to a focal point behind the lens (i.e., the criterion). Thus, each separately constructed scale can capture only a fraction of the construct–criterion relation. Viewed in this context, explaining 25% of the criterion variance is an accomplishment rather than a frustration.

Finally, to understand better the correlation between construct scale and criterion scale, I recommend that validity be examined at the item level. That is, the correlation at the scale level should be explicated in terms of the correlations at the item level. Besides reporting a validity coefficient, you would do well to report the cross-correlations between construct scale items and criterion scale items, even if only the significant correlations. When constructing a new scale, you should search out "convergent" scales that purportedly measure the same or similar constructs and examine their items' cross-correlations with yours. Item cross-correlation analysis might cut down on unneeded duplication in scale construction, but even more important, it might help you understand better the nature of your scale's validity.

In scale construction—to paraphrase a saying from another field of endeavor—validity is not the only thing, it is everything.

## REFERENCES

AERA, APA, & NCME (1985). *Standards for educational and psychological testing.* Washington, DC: Author.

Ashton, S. G., & Goldberg, L. R. (1973). In response to Jackson's challenge: The comparative validity of personality scales constructed by the external (empirical) strategy and scales developed intuitively by experts, novices, and laymen. *Journal of Research in Personality, 7,* 1–20.

Austin, J. T., & Villanova, P. (1992). The criterion problem: 1917–1992. *Journal of Applied Psychology, 77,* 836–874.

Binet, A., & Simon, T. (1961). The development of intelligence in children. In J. J. Jenkins & D. G. Paterson (Eds.), *Studies in individual differences* (pp. 81–111). New York: Appleton-Century-Crofts. (Original work published 1905–1908).

Bogardus, E. S. (1933). A social distance scale. *Sociology and Social Research, 17,* 265–271.

Campbell, D. T., & Fiske, D. W. (1959). Convergent and discriminant validation by the multitrait-multimethod matrix. *Psychological Bulletin, 56,* 81–105.

Campbell, D. T., & Stanley, J. C. (1966). *Experimental and quasi-experimental designs for research.* Chicago: Rand McNally.

Cliff, N. (1973). Scaling. *Annual Review of Psychology, 24,* 473–506.

Cook, S. W., & Selltiz, C. (1964). A multiple-indicator approach to attitude measurement. *Psychological Bulletin, 62,* 36–55.

Cox, E. P. (1980). The optimal number of response alternatives for a scale: A review. *Journal of Marketing Research, 17,* 407–422.

Cronbach, L. J. (1970). *Essentials of psychological testing* (3rd ed.) New York: Harper & Row.

Cronbach, L. J., & Meehl, P. E. (1955). Construct validity in psychological tests. *Psychological Bulletin, 52,* 281–302.

Dawis, R. V., & Weitzel, W. (1979). Worker attitudes and expectations. In D. Yoder & H. G. Heneman, Jr. (Eds.), *ASPA handbook of personnel and industrial relations* (Vol. 6, pp. 23–49). Washington, DC: Bureau of National Affairs.

DuBois, P. H. (1970). *A history of psychological testing.* Boston: Allyn & Bacon.

Elizur, D., & Guttman, L. (1976). The structure of attitudes toward work and technological change within an organization. *Administrative Science Quarterly, 21,* 611–622.

Embretson (Whitely), S. (1983). Construct validity: construct representation versus nomothetic span. *Psychological Bulletin, 93,* 179–197.

Galton, F. (1961). Classification of men according to their natural gifts. In J. J. Jenkins & D. G. Paterson (Eds.), *Studies in individual differences* (pp. 1–16). New York: Appleton-Century-Crofts. (Original work published 1869).

Galton, F. (1961). Co-relations and their measurement, chiefly from anthropometric data. In J. J. Jenkins & D. G. Paterson (Eds.), *Studies in individual differences* (pp. 17–26). New York: Appleton-Century-Crofts. (Original work published 1868).

Guilford, J. P. (1954). *Psychometric methods* (2nd ed.). New York: McGraw-Hill.

Gulliksen, H., & Tucker, L. R. (1961). A general procedure for obtaining paired comparisons from multiple rank orders. *Psychometrika, 26,* 173–183.

Guttman, L. (1944). A basis for scaling qualitative data. *American Sociological Review, 9,* 139–150.

Guttman, L. (1959). A structural theory of intergroup beliefs and action. *American Sociological Review, 24,* 318–328.

Hansen, J. C., & Campbell, D. P. (1985). *Manual for the SVIB-SCII* (4th ed.). Stanford, CA: Stanford University Press.

Horst, P. (1935). Measuring complex attitudes. *Journal of Social Psychology, 6,* 369–374.

Hough, L. M., & Paullin, C. (1994). Construct-oriented scale construction: The rational approach. In G. S. Stokes, M. D. Mumford, & W. A. Owens (Eds.), *Biodata handbook: Theory, research and use of biographical information in selection and performance prediction* (pp. 109–145). Palo Alto: Consulting Psychologists Press.

Hull, C. L. (1928). *Aptitude testing.* Yonkers, NY: World Book.

Jackson, D. N. (1975). The relative validity of scales prepared by naive item writers and those based on empirical methods of personality scale construction. *Educational and Psychological Measurement, 35,* 361–370.

Jensen, A. R. (1980). *Bias in mental testing.* New York: Free Press.

Landy, F. J. (1986). Stamp collecting versus science: Validation as hypothesis testing. *American Psychologist, 41,* 1183–1192.

Landy, F. J., & Farr, J. L. (1980). Performance rating. *Psychological Bulletin, 87,* 72–107.

Likert, R. (1932). A technique for the measurement of attitudes. *Archives of Psychology,* No. 140.

Loevinger, J. (1954). The attenuation paradox in test theory. *Psychological Bulletin, 51,* 493–504.

Loevinger, J. (1957). Objective tests as instruments of psychological theory. *Psychological Reports, 3,* 635–694.

Loevinger, J. (1991). Personality structure and the trait-situation controversy: On the uses of low correlations. In W. M. Grove & D. Cicchetti (Eds.), *Thinking clearly about psychology* (Vol. 2, pp. 36–53). Minneapolis: University of Minnesota Press.

Lykken, D. T. (1968). Statistical significance in psychological research. *Psychological Bulletin, 70,* 151–159.

Meehl, P. E. (1998). *The power of quantitative thinking.* (Cattell Award Address). Washington, DC: American Psychological Association.

Messick, S. (1989). Validity. In R. L. Linn (Ed.), *Educational measurement* (3rd ed., pp. 13–103). New York: American Council on Education/Macmillan.

Osgood, C. E., Suci, G. J., & Tannenbaum, P. H. (1957). *The measurement of meaning.* Urbana, IL: University of Illinois Press.

Ozer, D. J., & Reise, S. P. (1994). Personality assessment. *Annual Review of Psychology, 45,* 357–388.

Spearman, C. (1904a). The proof and measurement of association between two things. *American Journal of Psychology, 15,* 72–101.

Spearman, C. (1904b). "General intelligence," objectively determined and measured. *American Journal of Psychology, 15,* 201–292.

Stephenson, W. (1953). *The study of behavior.* Chicago: University of Chicago Press.

Stevens, S. S. (1946). On the theory of scales of measurement. *Science, 103,* 677–680.

Strong, Jr., E. K. (1927). A vocational interest test. *Educational Record, 8,* 107–121.

Strong, Jr. E. K. (1943). *Vocational interests of men and women.* Stanford, CA: Stanford University Press.

Thorndike, E. L. (1910). Handwriting. *Teachers College Record, 11,* No. 2.

Thurstone, L. L. (1927). A law of comparative judgment. *Psychological Review, 34,* 273–286.

Thurstone, L. L. (1928). Attitudes can be measured. *American Journal of Sociology, 33,* 529–554.

Thurstone,L. L. (1959). *The measurement of values.* Chicago: University of Chicago Press.

Tziner, A. E. (1987). *The facet analytic approach to research and data processing.* New York: Peter Lang.

Viteles, M. S. (1932). *Industrial psychology.* New York: Norton.

# INTERRATER RELIABILITY AND AGREEMENT

**HOWARD E. A. TINSLEY**

*Department of Psychology, University of Florida, Gainesville, Florida*

**DAVID J. WEISS**

*Department of Psychology, University of Minnesota, Minneapolis, Minnesota*

The rating scale is a ubiquitous measuring instrument, enjoying widespread usage in popular culture, the physical, biological and social sciences, and the humanities. Rating scales occur in a great variety of forms; most use numbers to reflect gradations in the attribute being judged, but some use pictures or other symbols (i.e., thumbs up vs. thumbs down, one to five stars, one to five mice, or one to five cows). Regardless of the cosmetic differences in appearance, rating scales typically require a rater, judge, or observer (we use the terms interchangeably in this chapter) to evaluate some characteristic of an object and assign it to some point on the rating scale. The object to be rated can be a person (e.g., an employee or counselor), a thing (a book, music CD, or movie), a process (e.g., types of counseling or recovery from anesthesia), an outcome (e.g., recovery of function or assessment of apraxia), an idea (e.g., quality of life or contribution to society) or a datum (e.g., time since last update or occurrence of target behavior).

Despite the widespread popularity of rating scales, few applied researchers seem to understand the design issues unique to the collection of ratings nor the proper procedures for evaluating ratings. The quality of ratings can be critical to the success of your research, however, because

*Handbook of Applied Multivariate Statistics and Mathematical Modeling*
Copyright © 2000 by Academic Press. All rights of reproduction in any form reserved.

the effect size and power of your analyses are contingent on maintaining measurement error within acceptable limits. Failure to maintain the reliability of your ratings data (and, possibly, the level of agreement among your raters) within an acceptable level may make it impossible for your research to yield noteworthy effect sizes. Yet Thompson (1994) and Fabrigar, Wegenar, MacCallum, and Strahan (1999) estimate that reliability is given insufficient attention in 40 to 50% of the published literature, and Pedhazer and Schmelkin (1991) refer to the lack of attention given to measurement as the Achilles heel of research in the social and biological science.

This problem is even more serious when working with rating scales because the datum recorded on a rating scale is the subjective judgment of an individual (i.e., the rater). It is always necessary to establish the reliability of a set of ratings to demonstrate that the variance in the ratings is due to a systematic rather than a random ordering of the objects. Knowledge of the interrater reliability of a set of ratings is crucial in evaluating the validity and generalizability of the results, yet the failure to report the interrater reliability of ratings is not uncommon. Another common practice is to report the interrater reliability as having been determined in a previous investigation. Those who do so apparently do not grasp the critical point that reliability is not a property of the rating scale, but a property of a specific set of data that result from a specific use of the scale. Knowledge of the interrater agreement of a set of ratings also may be important, yet the interrater agreement of ratings seldom is reported. These practices are unacceptable.

In this chapter we explain the critical distinction between interrater reliability and agreement, a distinction that is given little attention in other areas of measurement (see Dawis, chapter 3, this volume). We review design issues such as the type of replication and the level of measurement (see Tinsley & Brown, chapter 1, this volume and Dawis, chapter 3, this volume) that researchers must consider in determining how to collect and analyze ratings data. However, the primary purpose of this chapter is to provide guidelines for selecting the appropriate statistical procedure for analyzing ratings data. Determining the proper procedure for calculating interrater reliability and interrater agreement requires a consideration of the level of measurement achieved by the rating scale and the intended use of the ratings. We also consider briefly the desirability of using weighting schemes for transforming nominal scale ratings to ordinal scale ratings.

## I. AGREEMENT VERSUS RELIABILITY

One critical distinction that applied researchers must understand to properly evaluate the psychometric quality of ratings data is that between interrater reliability and interrater agreement. This distinction is analogous to coming to a T-shaped intersection while on a Sunday drive; it controls the direction your analysis will take. Making the wrong decision most likely means that your data will not be analyzed properly no matter what decisions you make later.

Procedures for analyzing interrater reliability were suggested as early as the 1930s (e.g., Burt, 1936; Hoyt, 1941), but formal introduction of the intraclass correlation $(R)$ did not occur for another decade (e.g., Bartke, 1966; Ebel, 1951; Guilford, 1954; Lindquist, 1953). Widespread confusion regarding the difference between interrater reliability and agreement continued until the mid-1970s, however, with most researchers using those terms incorrectly as synonyms. For example, K. Lu (1971) proposed a measure of interrater agreement that is conceptually identical to a measure of interrater reliability proposed earlier by Finn (1970). Recognition that interrater reliability indices and interrater agreement indices provide fundamentally different information about a set of ratings has greatly improved since we clarified their quintessential difference (Tinsley & Weiss, 1975). Sadly, however, failures to understand this distinction and to use interrater reliability and interrater agreement indices properly still are not uncommon. A quick scan of the literature published in medicine and psychology since 1997 using the Medline database revealed over 30 instances in which interrater agreement indices were represented as measuring interrater reliability. The consequence of these errors is wastefulness.

One noteworthy illustration of the problems that can occur when researchers confuse interrater reliability and interrater agreement is James, Demaree, and Wolf's (1984) proposal of an index of interrater reliability $(r_{wg})$ for use when a single stimulus has been rated. Schmidt and Hunter (1989) criticized Kozlowski and Hults's (1987) use of this index on the grounds that, "it is not conceptually possible within standard measurement theory to compute a legitimate estimate of the interrater reliability coefficient when raters rate only one stimulus" (p. 370). Schmidt and Hunter (1989) expressed concern that $(r_{wg})$, "may have no meaningful interpretation" (p. 369). The issue was resolved only after Kozlowski and Hattrup (1992) clarified that $r_{wg}$ is actually a measure of interrater agreement, a point that James, Demaree, and Wolf (1993) subsequently acknowledged.

In classical psychometric theory, reliability is defined as the ratio of true score variance to total variance (Crocker & Algina, 1986). In the typical measurement situation in which internal consistency reliability is assessed, a sample of persons complete an instrument consisting of multiple items at a single point in time. Because the status of each person is measured only once, and it is assumed that all of the items are measuring the same construct, any variance in the scores on the instrument should be the result of differences among the persons being tested. Therefore, the variance among persons is interpreted as true score variance and any nonsystematic variance is interpreted as error variance.

Essentially the same situation exists when a rating scale is used, except that the multiple raters take the place of the multiple items. The attribute on which the objects are rated is assumed to be constant at the time of measurement, all of the raters are assumed to be rating the same object, and the variability in the ratings should be the result of differences among the rated objects. Therefore, the variance attributable to differences among the rated objects is interpreted as true score variance, and any additional variance is interpreted as error variance.

Interrater reliability, therefore, provides an indication of the extent to which the variance in the ratings is attributable to differences among the objects rated. As with other forms of reliability, it is not necessary that the rated objects be assigned exactly the same rating for interrater reliability to be high. This is because, just like internal consistency reliability, interrater reliability is calculated using correlational (or analysis of variance [ANOVA]) indices. To understand the significance of that, consider that the pattern of ratings a judge assigns to a set of objects can be thought of as a profile of scores, just as an individual's scores on the 10 clinical scales of the Minnesota Multiphasic Personality Inventory (MMPI) form a profile of scores. Two profiles can differ in level (i.e., the overall mean score of one person is higher than the overall mean score of the second person), scatter (i.e., the two persons' scores differ in the extent to which they vary around the overall mean), and shape (i.e., the two persons' scores have the same overall mean and variability, but their high and low scores occur on different scales). Correlational indices of the degree of association between two profiles are insensitive to differences in level and scatter because they standardize the scores by setting the mean of each set to zero and the standard deviation of each set to 1.0. Therefore, Interrater reliability indicates the degree to which the ratings of different judges are the same when expressed as deviations from their means. In practice, a high interrater reliability means that the relation of one rated object to other rated objects is the same across judges, even though the absolute numbers used to express this relation may differ from judge to judge. Interrater reliability is sensitive only to the relative ordering of the rated objects.

In contrast, interrater agreement represents the extent to which the different judges tend to assign exactly the same rating to each object. As such, interrater agreement indices are sensitive to differences in the level, scatter, or shape of the profile of ratings assigned by the judges. When judgments are made using a numerical scale, a high interrater agreement signifies that the judges assigned precisely the same numerical values when rating the same person.

Table 4.1 shows hypothetical data (assuming interval-level measurement) in which three judges have rated 10 job applicants on a 10-point applicant rating scale. Case 1 illustrates a situation in which all three judges assign exactly the same ratings to each of the 10 applicants. These ratings

**TABLE 4.1   Hypothetical Ratings of Job Applicants Illustrating Different Levels of Interrater Reliability and Interrater Agreement for Interval-Scaled Data**

|  | Case 1: High interrater reliability and high interrater agreement | | | Case 2: High interrater reliability and low interrater agreement | | | Case 3: Low interrater reliability and high interrater agreement | | |
|---|---|---|---|---|---|---|---|---|---|
|  | Rater | | | Rater | | | Rater | | |
| Candidate | 1 | 2 | 3 | 1 | 2 | 3 | 1 | 2 | 3 |
| A | 1 | 1 | 1 | 1 | 3 | 6 | 5 | 6 | 5 |
| B | 2 | 2 | 2 | 1 | 3 | 6 | 5 | 4 | 4 |
| C | 3 | 3 | 3 | 2 | 4 | 7 | 6 | 4 | 6 |
| D | 4 | 4 | 4 | 2 | 4 | 7 | 4 | 5 | 6 |
| E | 5 | 5 | 5 | 3 | 5 | 8 | 5 | 4 | 4 |
| F | 6 | 6 | 6 | 3 | 5 | 8 | 6 | 6 | 5 |
| G | 7 | 7 | 7 | 4 | 6 | 9 | 4 | 4 | 5 |
| H | 8 | 8 | 8 | 4 | 6 | 9 | 5 | 5 | 4 |
| I | 9 | 9 | 9 | 5 | 7 | 10 | 4 | 5 | 3 |
| J | 10 | 10 | 10 | 5 | 7 | 10 | 6 | 6 | 6 |
| | | | | Descriptive statistics | | | | | |
| Mean | 5.5 | 5.5 | 5.5 | 3.0 | 5.0 | 8.0 | 5.0 | 5.1 | 4.7 |
| SD | 3.0 | 3.0 | 3.0 | 1.5 | 1.5 | 1.5 | .8 | .9 | .8 |
| | | | | Interrater reliability | | | | | |
| $R_{di}$ | | 1.00 | | | 1.00 | | | .38 | |
| $R_{gi}$ | | 1.00 | | | .02 | | | .40 | |
| $R_{dc}$ | | 1.00 | | | 1.00 | | | .65 | |
| $R_{gc}$ | | 1.00 | | | .06 | | | .67 | |
| $r_i$ | | 1.00 | | | .23 | | | .94 | |
| $r_c$ | | 1.00 | | | .47 | | | .98 | |
| | | | | Interrater agreement | | | | | |
| $T_0$ | | 1.00 | | | .00 | | | .00 | |
| $T_1$ | | 1.00 | | | .00 | | | .67 | |
| $T_2$ | | 1.00 | | | .00 | | | 1.00 | |

have both high interrater agreement and high interrater reliability, and the distinction between these indices is blurred.

Case 2 shows ratings having high interrater reliability but low interrater agreement. These data have high reliability because the ratings assigned to the job applicants by the three judges are the same when expressed as deviations from their overall mean. You can verify this for yourself by subtracting the judge's mean from each rating. For example, the ratings given the first candidate by judge #1 (i.e., $1 - 3 = -2$), Judge #2 (i.e., $3 - 5 = -2$), and Judge #3 (i.e., $6 - 8 = -2$) are identical when expressed as deviations from each judge's overall mean rating.

Despite the perfect interrater reliability of the ratings in Case 2, these ratings have low interrater agreement because no two raters gave the same rating to any job applicant. We explained that agreement indices are sensitive to differences in the level (i.e., mean), scatter (i.e., variability), and shape of the ratings. In Case 2 the pattern of scores and the variability of the ratings are equal across judges, but you can see the cause of the low agreement among the raters reflected in their mean ratings of 3.0, 5.0, and 8.0, respectively.

Given the similarity of the pattern and variability of the judges' ratings, you might be tempted to dismiss the differences in the mean level of the ratings. To understand the importance of this difference, pretend for a moment that you are applying for a job and could pick one of these judges to evaluate your suitability for employment. Would you care which judge rated you? Of course you would; we all would prefer to be evaluated by rater #3 because it seems likely that the ratings assigned by the three judges will translate into meaningfully different statements about the acceptability of these job applicants. In other situations (e.g., when rating degree of disability or carotid artery plaque morphology) the specific interpretations attached to the different levels of the rating scale may be quite important.

A critical distinction to understand, therefore, is that interrater reliability is a satisfactory index of the quality of the ratings when you are interested only in the relative ordering of the rated objects (i.e., a norm-referenced interpretation). However, interrater reliability indices are insensitive to differences in the mean and variability of the ratings given the rated objects. Whenever the absolute value of the ratings or the meaning of the ratings as defined by the points on the scale is of concern (i.e., criterion referenced interpretation), the interrater agreement index provides critical information about the ratings.

The interrater agreement of the ratings is always critical when judgments about the rated objects are based on information from different sets of judges. Suppose, for example, that you actually have 99 applicants for 10 positions. Two scenarios typically occur in situations such as this. In Scenario One each candidate is rated by only one judge. If each judge rated 33 of the applicants, none of the candidates interviewed by judge 1 or judge 2 would be offered a position. In Scenario Two each candidate is rated by two judges in an effort to provide a broader perspective on each candidate. In this scenario the candidates rated by Judges #1 and #2 and the candidates rated by Judges #1 and #3 would not be offered positions. Only those candidates interviewed by Judges #2 and #3 would have a chance at employment. Although the fallacy of these scenarios may seem obvious, they describe the situation often faced in applied research. Despite best intentions, the rating task is found to exceed the available resources and Scenario One or Two is adopted as an expedient. In situations such as this it is mandatory that interrater agreement be assessed and reported in addition to interrater reliability.

Finally, Case 3 depicts ratings that have high interrater agreement but low interrater reliability. The high interrater agreement results from the fact that the ratings assigned to each job applicant by the three judges are quite similar. The low interrater reliability is the result of the restricted range of ratings given by the judges. This occurs because reliability is defined as the proportion to true score variance attributable to total variance. Therefore, as total variance (the denominator in the equation) approaches zero, reliability approaches zero. Ratings such as these may signify that the job applicants are homogeneous on the trait of interest, or they could indicate that the judges did not use the rating scale properly. This issue can be investigated by further study of the same subjects or by having the judges rate a sample of subjects known to be heterogeneous on the trait of interest. Regardless of the cause, however, interrater reliability will be low when the variability of the ratings is low, as often occurs in applied research.

In summary, interrater reliability and interrater agreement refer to two important and meaningfully different attributes of a set of ratings. High reliability is no indication that the raters agree in an absolute sense on the degree to which the objects rated possess the characteristic being judged (Case 2). On the other hand, low reliability does not necessarily indicate that the raters are in disagreement (Case 3). Both types of information are necessary to evaluate ratings. When both interrater reliability and agreement are low, a case not illustrated in Table 4.1, the ratings have no validity and should not be used for research or applied purposes.

## II. LEVEL OF MEASUREMENT

Tinsley and S. Brown discuss the differences in nominal, ordinal, and interval level measurements in chapter 1 (this volume). That distinction is as important in assessing interrater reliability and interrater agreement as it is in using the other multivariate statistics and model testing procedures covered in the *Handbook*. Interval-level statistics often can be used appropriately on ratings that result from ordinal level measurement, but keep in mind that such applications can result in interrater reliability and interrater agreement statistics that distort the true relations among raters when the assumption of equal intervals is grossly inappropriate.

Nevertheless, the critical distinction for interrater reliability and interrater agreement is that between nominal level measurement and higher levels of measurement. The distinction between interrater reliability and interrater agreement ceases to exist when ratings are made at the nominal level of measurement. With nominal ratings scales the rating categories do not differ quantitatively, so the disagreements in categorization do not differ in their severity. Therefore, the concept of "proportionality of ratings," which is central to interrater reliability, ceases to make sense at the nominal-scale level. Agreement becomes an absolute; ratings are either in agreement or disagreement. As a result, the concept of interrater reliability ceases to make sense for nominal scales. The term interrater agreement is more appropriate, because the attribute being measured is the extent to which the judges assigned the same rating to each rated object.

## III. TYPE OF REPLICATION

The concept of reliability involves some type of replication or repetition of the measuring process. In classical psychometric theory the measurements are replicated by having the same persons complete two different measures of the same construct (i.e., parallel-forms reliability), by having the same persons complete the instrument at two points in time (i.e., test–retest reliability), or by having the same persons complete multiple items that are presumed to measure the same construct (i.e., internal consistency reliability). In much the same manner, determination of interrater reliability also requires some form of replication. We are not aware of a parallel-forms analog in assessing interrater reliability, but analogs to the test–retest and internal consistency data collection procedures exist.

One method of replicating a set of ratings is the rate–rerate method, the analog of the test–retest procedure for determining test reliability. In this method the judges rate a group of subjects on some characteristic, then rerate the same subjects on the same characteristic at a later date. This replication has been used recently by Dreessen and Arntz (1998) and Chorpita, Brown, and Barlow (1998).

The rate–rerate method is an inappropriate procedure for demonstrating interrater agreement or interrater reliability for several reasons. First, this measure of reliability provides no information about interrater agreement because it treats the raters as a constant and ignores between-judges variability in the ratings. The rate–rerate method could yield a high rate–rerate reliability for three raters who were in complete disagreement, so long as each rater was consistent across time. Second, the rated objects may change over time on the characteristics being rated. The more accurately the judges amend their previous ratings to reflect this change, the lower the reliability will appear. Third, rate–rerate methods may give spuriously high reliabilities when the between-ratings interval is short and a nonchanging database is used (e.g., behavioral ratings of audio- or videotape excerpts). In this situation the second set of ratings will likely be contaminated by the raters' recall of their previous judgments. Finally, the rate–rerate method requires that the judges rate each subject on two different occasions. This is seldom feasible in the field situations in which much applied research is conducted.

For these reasons we advocate replication of ratings between judges (i.e., have several judges rate the same objects on the same attribute), as illustrated in Table 4.1. This procedure is the analogue of internal consistency reliability with the multiple judges taking the place of the multiple items.

## IV. INTERRATER RELIABILITY

Students of classical psychometric theory know that approaches, to the calculation of reliability based on the correlation coefficient (Kuder & Richardson, 1937) and ANOVA (Hoyt, 1941) offer psychometrically different starting points for arriving at the same destination. Hoyt's (1941) reliability and Cronbach's (1951) coefficient alpha reliability are algebraically equivalent, and both produce the same result as the Kuder-Richardson-20 (Kuder & Richardson, 1937) formula when applied to dichotomous data. A similar phenomenon occurred in the development of measures of interrater reliability. Numerous procedures were proposed for calculating interrater reliability prior to Ebel's (1951) now classic work on the intraclass correlation ($R$; e.g., Gulliksen, 1950; Horst, 1949; Peters & Van Voorhis, 1940), and additional procedures continue to be recommended (e.g., Finn, 1970), some of which are represented as measures of interrater agreement (e.g., K. Lu, 1971). These procedures yield values that are identical to or approximate a value obtained from one of the forms of $R$. Although we will describe one situation in which Finn's $r$ is useful, we advocate the use of $R$ in most instances.

Generalizability theory (Cronbach, Gleser, Nanda, & Rajaratnam, 1972; see Marcoulides, chapter 18, this volume) provides a conceptual framework within which to understand the intraclass correlation. Generalizability theory introduced the notion that test scores (and ratings) are influenced by multiple forms of error, and that different measures of reliability can be calculated to assess the influence of these sources of error. Multiple formulations of $R$ are possible for precisely this reason (see Ebel, 1951; Hoyt, in press; Shrout & Fleiss, 1979; Tinsley & Weiss, 1975); in this chapter we discuss four variations that have the greatest applicability in applied research. Generalizability theory provides a structured framework within which investigators can compare the alternative forms of the intraclass correlation in terms of the (a) type of error assessed and conceptual meaning of the reliability index, (b) assumptions underlying the index, (c) relevance of the index to their research objectives, and (d) costs of the index.

Although technically, $R$ requires the assumption of interval-level ratings, we recommend that $R$ be used with ordinal scales that assume interval properties unless the variance in the ratings is severely restricted (as in Case 3). We will discuss the use of Finn's (1970) $r$ in that situation later in this chapter. $R$ can be interpreted as the proportion of the total variance in the ratings due to variance in the persons being rated (i.e., as the proportion of variance attributable to true score variance). As such, $R$ is consistent with the classical definition of reliability. In actual practice, $R$ can be expected to range from 1.00 to zero, with high values indicating high interrater reliability (i.e., most or all of the variance in the ratings is attributable to differences among the objects rated). Values close to zero indicate a complete lack of reliability (viz. almost all of the variability in the ratings is due to error). Although negative values of $R$ are mathematically possible, they are rarely observed in actual practice. Negative values are most likely to occur when the variance in the ratings is severely restricted, or when a Rater × Object interaction is present (e.g., Judge 1 rates men high and women low, whereas Judge 2 does the reverse).

Two decisions are necessary to decide on the proper approach to calculating $R$. Ebel (1951) pointed out that $R$ permits investigators to treat the between-rater variance as error variance or to exclude it from the error term. In order to make this decision, researchers must decide whether differences in the level (i.e., mean) or scatter (i.e., variance) in the ratings of the judges represent error or inconsequential differences. In addition, investigators must decide whether they want to know the average reliability of the individual judge or the reliability of the composite rating of the panel of judges. The product of these two dichotomous decisions yields four versions of $R$. We will discuss each of these decisions in turn.

## A. Between-Raters Variance

### 1. Interpretation of Between-Raters Variance

Case 2 in Table 4.1 illustrates a situation in which the judges assigned identical rankings to job candidates, but they differed in their average ratings (i.e., the mean ratings were 3.0, 5.0, and 8.0 for judges 1, 2, and 3, respectively). These "level" differences account for all of the difference in the judges ratings. It stands to reason, therefore, that elimination of these differences in the average rating given by the judges (i.e., the between-judges level differences) would yield data having perfect reliability (i.e., $R = 1.00$). Some writers have suggested that this formulation of interrater reliability be interpreted as a measure of interrater agreement (e.g., Suen & Ary, 1989), a point to which we will return shortly. On the other hand, the decision to treat the between-judges variance as error results in an intraclass correlation of .017. Whenever differences are apparent in the mean or variance of the judges' ratings, the decision regarding the interpretation of between-raters variance will influence the interrater reliability of the data. If sizable differences in mean or variance are present, the exclusion of the between-raters variance from the error term will cause the reliability coefficient to be substantially higher than if it were included.

The desirability of removing the interjudge variance from the error term when estimating the interrater reliability depends on the way in which the ratings are to be used (Ebel, 1951). Do not include the between-raters variance in the error term if these differences do not influence the interpretation of the ratings. That will be the case when decisions are based on the mean or sum of the ratings assigned by the set of observers, or on ratings that have been adjusted (e.g., by conversion to ranks or $Z$ scores) to eliminate the between-rater differences. Include the between-raters variance in the error term if (a) decisions are made by comparing objects rated by different judges or different sets of judges, or (b) the investigator wants to generalize the results to other samples of judges using the same scale with a similar sample of subjects.

## 2. Generalizability of $R$

Several writers suggest or imply that the ability to generalize the interrater reliability is desirable (Shrout & Fleiss, 1979; Suen & Ary, 1989). We maintain that applied researchers should make every effort to design their research so that between-judges variation can be removed from the error term; as can be seen in Case 2, the cost of treating between-judges variance as error can be great. Excluding between-judges variation from the error term means that the sample of raters used in the computation of $R$ must be regarded as "fixed" (i.e., they represent the population of judges rather than a sample of judges from some population; Burdock, Fleiss, & Hardesty, 1963; Tinsley & Weiss, 1975). The cost of this decision is that the calculated $R$ is a descriptive statistic, applicable to only the data on which it was calculated, and therefore it cannot be generalized to any other set of judges using the same rating scale on a similar sample of persons.

This cost is not as great as it may appear, however, for it is seldom feasible to generalize an interrater reliability coefficient for both conceptual and practical reasons. Conceptually, as we noted earlier, it is now widely agreed that reliability is not a generalizable property of a rating scale, but a descriptive property of a specific set of data that result from a specific use of the scale. Beyond that, researchers would need to satisfy several rather difficult requirements to generalize the calculated $R$ to another sample of judges. First, a random sample of judges is necessary so that the judges can be considered representative of the population of judges to which you wish to generalize. Second, the new judges must use the same rating scale. Third, the judges must be trained to the same level of proficiency as the original judges. Fourth, the objects to be rated must be the original objects or a random sample from the same population as the original objects. Finally, the equation for $R$ must incorporate an estimation of the between-judges variance so that the degree to which $R$ will vary across samples of judges can be estimated.

It seems unlikely that the requirement of a random sample of judges will be satisfied in applied research, and the original investigator has no control over the selection of the judges, training of the judges, or selection of objects to be rated in the subsequent research unless he or she is conducting that research. Therefore, despite your best intentions and your willingness to accept a lower reliability, it seems unlikely that the reliability obtained in your research will generalize to other studies. Ebel (1951) provided formulas for establishing confidence intervals around $R$, but in most applied research studies the lower limit of the confidence interval falls below zero because of the small number of raters used. This further emphasizes the folly of attempting to generalize interrater reliability coefficients. For all of these reasons, we contend that interrater reliability is best treated as a descriptive statistic that describes properties of a specific set of data that were obtained from a specific use of the scale.

### 3. Computation of $R$

Operationally, the decision about whether to include the between-judges variance in the error term involves the choice between computing a one-way analysis of variance (ANOVA) or a two-way ANOVA (Ebel, 1951; Guilford, 1954). Use the two-way procedure when you do not want to include the between-judges variance in the error term (i.e., when you plan to make a norm-referenced interpretation about the ordering of the rated objects relative to each other). Given the assumption of no interaction between persons and judges, the analysis involves the use of standard ANOVA procedures to compute the mean square for persons ($MS_p$), mean square for judges ($MS_j$), mean square for error ($MS_e$), and total mean square ($MS_t$). These components are then inserted into the standard equation for the intraclass correlation:

$$R_i = \frac{MS_p - MS_e}{MS_p + MS_e(K - 1)} \tag{1}$$

where $i$ indicates that the average reliability of a single (i.e., individual) rater is estimated, a point we will explain in the next section, and $K$ = the number of judges ratings each person. Use of this approach to calculate the interrater reliability for Cases 1, 2, and 3 in Table 4.1 yields $R_{di}$ = 1.00, 1.00, and 0.38, respectively (see Table 4.1), where $d$ signifies that the reliability is descriptive of the data on which it is calculated.

Suen and Ary (1989) suggested that $R_{di}$ be interpreted as a measure of interrater agreement, but Case 2 illustrates the problem with that approach (see Table 4.1). Interpreting $R_{di}$ as a measure of agreement leads to the conclusion that the Case 2 raters were in perfect agreement (i.e., $R_{di} = 1.00$) when in fact the raters did not agree on the rating to be assigned to a single job candidate. Instead, $R_{di}$ indicates the extent to which the judges ranked the objects in the same order.

You may wonder why the $MS_j$ was computed but then not used in Equation 1 to compute $R$. In ANOVA the error term is the residual that remains after all of the desired components have been subtracted from the $MS_t$. Computationally, therefore, calculation of the $MS_j$ component in the two-way ANOVA is tantamount to removing it from the error term.

The easiest way to compute the components for Equation 1 when the between-judges variance is to be included in the error term is to use the standard one-way ANOVA procedure, in which only mean square for persons $MS_p$ and mean square of error $MS_e$ are calculated (Ebel, 1951; Tinsley & Weiss, 1975). The one-way ANOVA procedure does not extract a variance component to estimate the between-judges variability, thereby leaving the mean square for judges in the residual that is used as an estimate of error. Following the computation of its components, the interrater reliability is calculated using Equation 1. When the between-judges variance is interpreted as error, the intraclass correlations for Cases 1, 2, and 3 in Table 4.1 are $R_{gi}$ = 1.00, .017, and .40, respectively (see Table 4.1), where $_g$ indicates that the intraclass correlation is generalizable to the population of judges from which the sample of judges was drawn, and $_i$ indicates that the average reliability of a single (i.e., individual) rater is estimated. The difference between the two estimates of reliability is greatest for Case 2 (see Table 4.1), where substantial differences were observed in the means of the raters.

Frequently, applied researchers are confronted with incomplete data. When the objects have been rated by varying numbers of raters and the investigator plans to use the one-way ANOVA procedure, the following equation form Snedecor (1946) yields an average value of $K$ that can be used in Equation 1:

$$\overline{K} = \frac{1}{N-1}\left(\sum K - \frac{\sum K^2}{\sum K}\right), \qquad (2)$$

where $\overline{K}$ = the average value of $K$ to be inserted in Equation 1 for $K$, $N$ = the number of subjects, and $K$ = the number of judges rating each subject (this value will vary from subject to subject).

## B. Reliability of Composite Ratings

When measuring a construct using a multiple-item scale, researchers typically obtain a composite score that reflects the respondents' scores on all of the items (i.e., a mean item score or sum of the item scores). Generally, they are concerned about the reliability of this composite because judgments about the relative ordering of the respondents are based on the composite. In an analogous manner, we have recommended that researchers use the mean rating or sum of the ratings of the panel of judges for making decisions.

Therefore, it is the reliability of the composite rating that is of interest, but the value of $R$ yielded by Equation 1 actually indicates the average reliability of the single judge. As such, Equation 1 underestimates the interrater reliability of the composite rating.

Researchers should use a measure of reliability that is consistent with the way in which the ratings data are interpreted. If conclusions typically are drawn from the ratings of a single judge, the average reliability of the individual judge (i.e., $R_{di}$ or $R_{gi}$) is appropriate even if the ratings of several judges are available because decisions are based on the ratings of a single judge. Conversely, when decisions are based on a composite formed by combining information from a group of observers, it is necessary to know the reliability of the composite. This reliability will be higher than the reliability yielded by Equation 1; it can be estimated by applying the Spearman-Brown Prophecy formula (see Crocker & Algina, 1986, p. 119) to the result obtained from Equation 1, but it is quicker to calculate it directly using the following equation:

$$R_c = \frac{MS_p - MS_e}{MS_p}, \tag{3}$$

where $_c$ signifies that the intraclass correlation estimates the reliability of the composite rating.

In a manner analogous to adding items to a scale to increase its reliability, increasing the number of raters yields gains in interrater reliability when the initial reliability is low (Guilford, 1954). Interrater reliability will increase rapidly as a result of adding the first several raters, but decreasing increments in reliability occur with the addition of each new rater. Use of the variance components derived from the two-way ANOVA in Equation 3 yields $R_{dc} = 1.00$, $1.00$, and $.65$ for Cases 1, 2, and 3 in Table 4.1, respectively, where $_d$ indicates the interrater reliability is descriptive of the sample on which it is calculated. Use of the variance components derived from the one-way ANOVA yields $R_{gc} = 1.00$, $.06$, and $.67$ for Cases 1, 2, and 3 in Table 4.1, respectively, where $_g$ signifies that the reliability is generalizable to other situations, given the restrictive conditions specified previously.

### 1. Differential Weighting

Equation 3 and the Spearman-Brown formula assign unit weights to the ratings of each judge (i.e., weights them equally) in estimating the reliability of the composite rating. This is the same as the procedure used in calculating a composite score from the responses to a multiple-item scale (i.e., you add the score for item one to the score for item two, and so on). In this instance, however, you might wonder whether differentially weighting the ratings based on some a priori information about the quality of the judges would yield a higher estimate of interrater reliability than

Equation 3. For example, Kelley (1947) suggested calculating the pairwise correlation among the judges and assigning weights to the ratings of the different observers based on their estimated reliability. When calculating a multiple-item composite score, the unit weighting scheme (i.e., simply adding the item scores) actually weights the items according to their item variances. That is because the variance of a composite is equal to the sum of the variances of the items forming the composite plus the sum of their interitem covariances. The same general principle holds when ratings are combined to produce a composite rating; the ratings are differentially weighted by the variances of the individual judges. In the extreme case, the ratings of a judge who rated every object the same would receive zero weight in a composite score.

We recommend the use of unit weights (i.e., Equation 3) in most situations. The use of differential weights can actually decrease the reliability of the composite rating under some circumstances (see Lawshe & Nagle, 1952; Overall, 1965). Furthermore, a set of optimal weights is applicable only to a specific set of observers. Any modification of the rating scale, change in the membership of the group of judges, or change in the reliability or variance of the judges' ratings will require the calculation of a new set of optimal weights. For example, in many rating situations judges receive consistent feedback about the reliability of their ratings. This information may well be used by the low-reliability raters to improve their reliability. For this reason, optimal weights cannot be assumed to be stable across rating occasions, even if the same group of judges uses the same rating scale in rating the same or a similar group of objects.

## C. Reliability When Variance Is Low

We previously noted that $R$ can be greatly reduced when the variance in the ratings is low (see Case 3). The question remains about what to do in such situations. Rather than immediately discard the intraclass correlation for another statistic, we recommend you first determine the cause of the low variance. It may be that the judges are using the rating scale improperly. When that is the case it is best to provide additional training until the judges' performance meets the standard you have established for an acceptable level of proficiency. Another possibility is that a homogenous sample of objects is being rated. When that occurs and the sample of objects is not representative of the population, the best strategy is to obtain a more representative sample for rating. However, it is possible in applied research that the judges are performing competently and the sample, although rather homogeneous, is representative of the population or is of interest in its own right. In those circumstances it may be necessary to use an alternative estimate of interrater reliability.

Finn (1970) developed the following index of interrater reliability for use with ordinal-scaled ratings when within-judge variance is low:

$$r_i = 1.0 - \frac{MS_e}{(K^2 - 1)/12},\qquad(4)$$

where $MS_e$ = mean square for error obtained from a one-way ANOVA (unless only one object has been rated, in which cases $MS_e$ = the variance of the assigned ratings), $K$ = the number of scale categories, and $_i$ indicates that the average reliability of a single (i.e., individual) rater is estimated. Computationally, the denominator is the variance expected if the ratings were assigned at random; conceptually, it provides an estimate of the total variance. Therefore, the ratio indicates the proportion of total variance that is attributable to error variance and $r_i$ indicates the proportion of total variance that is attributable to true score variance (i.e., to genuine differences among the rated objects on the attribute being rated). An $r_i$ of 1.00 indicates perfect reliability; an $r_i$ of 0 indicates that the observed ratings varied as much as chance ratings. For the data in Table 4.1, $r_i$ = 1.00, .23, and .94 for Cases 1, 2, and 3, respectively. Comparison of Finn's $r_i$ with the four methods of computing $R$ is informative. The results for Case 3 reveal that $r_i$ is, indeed, relatively independent of the variance in the ratings, whereas Case 2 illustrates that $r_i$ treats between-judges variance as error.

Finn's $r_i$, like $R_{gi}$, indicates the average reliability of a single rater. Application of the Spearman-Brown formula to the values obtained from Equation 4 reveals that the reliability of the composite rating (i.e., $r_c$) for Cases 1–3 in Table 4.1 is 1.00, .47, and .98, respectively.

Finn's $r_i$ is a descriptive statistic; in actual practice the random assignment of ratings by a set of observers could result in an observed variance that is smaller than the expected variance purely by chance. Researchers can test the hypothesis that the observed variance is equal to the chance variance using the following chi-square test:

$$\chi^2 = \frac{N(k - 1)MS_e}{(K^2 - 1)/12},\qquad(5)$$

where $N$ = number of subjects, $N(k - 1)$ = the degrees of freedom, and $MS_e$ and $K$ are defined in Equation 4. A chi-square value lower than the lower critical value for the one-tailed chi-square test (i.e., the .99 level for a test at $p \le .01$) would indicate that the observed variance is significantly less than the chance variance. The interrater reliability is significantly greater than zero only if the null hypothesis is rejected.

We recommend a stringent critical value (e.g., $p < .01$) because this use of chi-square assumes that the ratings will be normally distributed and that both the subjects and raters will be randomly selected. Typically, these assumptions will be violated in applied research. Violation of the normality assumption are quite serious when inferences are made about variances (Hays, 1994, p. 359). Also keep in mind that a significant chi-square indicates only that the pattern of ratings is not consistent with the hypothesis of completely random responding. Finally, we caution that computation of the chance variance requires the assumption that the ratings of the judges are purely random and that every rating has the same probability. This assumption is violated whenever the judges have a response set to avoid the extreme categories of the rating scale. Whenever that occurs, the estimated chance variance used in Equation 4 will exceed the "true" chance variance, thereby causing $r_i$ (and $r_c$) to be spuriously high.

## V. INTERRATER AGREEMENT

Berk (1979) identified 16 different interrater agreement indices and numerous indices have been proposed since that time (e.g., James et al., 1984; Lindell & Brandt, 1997; Szalai, 1998). These measures include simple approaches such as calculating the pairwise correlation between observers ratings, approaches based on ANOVA and information theory (K. Lu, 1971), a variety of chi-square indices, occurrence–nonoccurrence indices (e.g., Hawkins & Dotson, 1975), and the proportion or percentage of agreement ($p$).

Many of the errors in assessing interrater agreement are attributable to the use of statistical indexes designed for other purposes as measures of interrater agreement. For example, pairwise correlations would be better suited for use as a measure of interrater reliability than as a measure of interrater agreement. Numerous chi-square indices have been used as measures of interrater agreement, but it is not possible to draw any inference about the degree of agreement from a chi-square test alone. As we noted previously (Tinsley & Weiss, 1975), chi-square tests the hypothesis that the proportion of objects assigned to the various rating categories by the different judges do not differ significantly. A nonsignificant chi-square merely indicates that the observed disagreement is not greater than the disagreement that could be expected on the basis of chance, whereas a significant chi-square indicates that the amount of agreement is greater than or less than expected on the basis of chance.

One deceptively simple statistic is the percentage (or proportion) of agreements, defined as

$$\rho = \frac{N_a}{N_a + N_d}, \tag{6}$$

where $N_a$ = the number of agreements and $N_d$ = the number of disagreements. $\rho = 1.00$, .00 and .30, respectively, for the three cases in Table 4.1. Despite repeated criticisms (e.g., Cohen, 1960; Kelly, 1977; Mitchell, 1979;

Robinson, 1957; Tinsley & Weiss, 1975), variations of $\rho$ continue to be the most widely used measures of interrater agreement. This is most likely attributable to its ease of calculation and its misleadingly simple interpretation.

Several problems limit the usefulness of $\rho$ as an index of interrater agreement. First, $\rho$ treats interrater agreement in an absolute, all-or-none fashion. For example, $\rho = 0$ for the data in Case 2, despite the fact that the judges were in perfect agreement on the relative merits of the job applicants. Another problem is that some agreements among observers can be expected on the basis of chance alone. For example, if two judges made a series of dichotomous ratings at random, their ratings would be expected to agree 25% of the time purely on the basis of chance. $\rho$ ignores the likelihood of chance agreements and thereby overestimates the true agreement by an amount related to the number of raters and number of points on the scale. Suen and Lee (1985) reanalyzed data from a sample of published studies and concluded that correcting for chance agreements would have led to the conclusion that the quality of the ratings data was unacceptable in 25% to 75% of these studies. Nevertheless, these difficulties have been partially avoided in the better measures of interrater agreement, and $\rho$ is the foundation upon which these measures of agreement are constructed.

## A. Nominal Scales

Numerous writers have suggested measures of interrater agreement for nominal scales based on the percentage or proportion of agreements among judges ($p$) (Cohen, 1960, 1968; Everitt, 1968; Fleiss, 1971; Fleiss, Cohen, & Everitt, 1969; Goodman & Kruskal, 1954; Scott, 1955). Although $p$ treats agreement as an absolute, that is not a disadvantage if all of the disagreements are regarded as equally serious, as is typically the case when dealing with nominal ratings. Nevertheless, the problem of chance agreement remains. Guttman, Spector, Sigal, Rakoff, and Epstein (1971) concluded that there was "tacit" consensus that 65% agreement represented the minimum acceptable standard, but fortunately more discriminating measures are available that represent observed agreement as a function of chance agreement and provide statistical tests of the significance of $\rho$.

### 1. Two Judges

Cohen's (1960) coefficient kappa ($\kappa$) indicates the proportion of agreements between two raters after adjusting for chance agreements. $\kappa$ is calculated as follows:

$$\kappa = \frac{P_o - P_c}{1 - P_c}, \tag{7}$$

where $P_o$ = the proportion of ratings in which the two judges agree, and $P_c$ = the proportion of ratings for which agreement is expected by chance. The expected chance agreement can be obtained as follows:

$$P_c = \sum_{j=1}^{k} \hat{P}_{jj}, \tag{8}$$

where $\hat{P}_{jj}$ = the product of the marginal proportions. For the data in Table 4.2, which shows two judges' hypothetical classifications of 100 interview statements, $P_c = [(.20 \times .30) + (.30 \times .20) + (.30 \times .40) + (.20 \times .10)] = .26$, $P_o = [.18+.18+.24+.10] = .70$, and $\kappa$ is .59.

$\kappa$ can vary from 1.00 (which indicates perfect agreement among the judges) to $-1.00$ when the marginal frequencies are identical. A $\kappa$ value of 0 indicates that the observed agreement is equal to the agreement that would be expected by chance, whereas a negative value indicates that the observed agreement is less than the expected chance agreement.

Several measures of interrater agreement have been suggested that are identical in form to $\kappa$ but differ in their definition of chance agreement (e.g., Goodman & Kruskal, 1954; Scotts, 1955), but each makes more restrictive assumptions that $\kappa$ (e.g., the assumption of equal distribution of ratings). In contrast, $\kappa$ recognizes that judges distribute their ratings differently across the categories and the only assumptions required are that (a) the objects to be rated are independent, (b) the judges assign their ratings independently, and (c) the categories of the nominal scale are independent, mutually exclusive, and exhaustive. Formulas for the standard error of $\kappa$, for testing the significance of the difference between two kappa coefficients, and for testing the hypothesis that $\kappa = 0$ are provided by Cohen (1960) and Fleiss et al. (1969). Fleiss (1971) has formulated an extension of kappa ($\kappa_v$) for measuring interrater agreement when subjects are rated by different sets of judges, but the number of judges per subject is constant, and for evaluating the interrater agreement regarding assignments to a particular rating category.

**TABLE 4.2  Proportion of 100 Interview Statements Classified by Two Reviewers as Indicative of Four Psychological Motivations**

| | Reviewer A | | | | |
|---|---|---|---|---|---|
| **Reviewer B** | **Autonomy** | **Affiliation** | **Competence** | **Status** | **Total** |
| Autonomy | .18 | .00 | .02 | .00 | .20 |
| Affiliation | .00 | .18 | .12 | .00 | .30 |
| Competence | .06 | .00 | .24 | .00 | .30 |
| Status | .06 | .02 | .02 | .10 | .20 |
| Total | .30 | .20 | .40 | .10 | — |

## 2. Weighted Kappa

$\kappa$ is based on the assumption that all disagreements in classification are equally serious, but in some instances researchers may consider some disagreements among judges to be more serious than others. For example, if two therapists were diagnosing patients as normal, neurotic, psychotic, or suicidal, disagreement as to whether a patient was normal as opposed to psychotic or suicidal might be regarded as more serious than disagreement over whether the patient was normal or neurotic. Cohen (1968) developed weighted kappa ($\kappa_w$) for use as an index of interrater agreement when researchers want to differentially weight the various disagreements that occur in assigning objects to nominal scale categories. $\kappa$ is actually a special case of $\kappa_w$ in which all agreements (i.e., the diagonal entries in Table 4.2) are assigned weights of 1.0 and all disagreements (the off-diagonal values in Table 4.2) are assigned weights of 0. However, $\kappa_w$ permits the assignment of variable weights to the various types of disagreements (i.e., the off-diagonal cells). Cohen (1968), Everitt (1968), and Fleiss et al. (1969) have provided computational formulas and examples.

Although this option may sound attractive, our prediction that there would be few valid uses of $\kappa_w$ has proven to be accurate (Tinsley & Weiss, 1975). Assigning weights to order nominal rating categories along a unidimensional continuum transforms nominal ratings to ratings possessing the properties of ratio measurement. Under these circumstances, $\kappa_w$ is identical to the intraclass correlation (Fleiss & Cohen, 1973), and $\kappa_w$ actually measures interrater reliability, not interrater agreement, but it lacks the flexibility and ease of computation of the intraclass correlation. The other possibility is to weight the nominal rating categories using some nonlinear (i.e., multi-dimensional) scheme. That is most likely to occur when the researcher is operating on the basis of an explicit theory about the relations among the rating categories. When $\kappa_w$ is used the investigator must specify the weights used in the analysis and explain in detail the basis upon which the weights were assigned.

### B. Ordinal and Interval Scales

A central tenet of decision theory (Cronbach & Gleser, 1965) is that errors of measurement differ greatly in their importance. Some very small errors can be quite serious (e.g., a mistake of one gram in preparing a prescription or a mistake of one degree in computing the trajectory of the space shuttle), whereas in other instances quite large differences are of no appreciable consequences (e.g., a mistake of one hundred pounds when measuring 100 metric tons of grain or a mistake of one mile when estimating the distance from New York to Los Angeles). This essential truth is often ignored in the sciences and humanities, where a premium is placed on achieving the greatest precision possible. For example, measures of interrater reliability, by their nature, emphasize precision, and departures from the linear model underlying these measures are interpreted as serious deficiencies in the ratings. Nevertheless, failure to distinguish between important and inconsequential errors can be limiting and even self-defeating. The assessment of interrater agreement differs from the assessment of interrater reliability in this critical aspect. It is not possible to assess interrater agreement adequately without considering the seriousness or cost of the various types of disagreements that can occur.

Lawlis and Lu (1972) developed a measure of interrater agreement that allows the researcher the option of defining agreement as identical ratings, as ratings that differ by no more than 1 point, or as ratings that differ by no more than 2 points. Their flexible model allows researchers to judge the seriousness of disagreements among the raters and to distinguish between serious and unimportant disagreements. Agreement is tallied one object at a time, by determining whether the total set of ratings given that object satisfies the criterion. For example, three interviewers from a multinational corporation who gave a job applicant ratings of 1, 2, and 3 on the scale illustrated in Table 4.1 would have disagreed in an absolute sense, but essentially all would have rated the applicant as low in the qualities viewed as desirable for employment. If agreement were defined as ratings that differed by not more than two scale categories, the Lawlis and Lu index of agreement would score the set of ratings as an agreement.

Lawlis and Lu (1972) proposed a nonparametric chi-square test of the significance of interrater agreement, but neglected to provide a descriptive index of agreement. Tinsley and Weiss (1975) proposed the T-index, patterned after Cohen's (1960) $\kappa$, as a measure of agreement:

$$T = \frac{N_a - N\rho_c}{N - N\rho_c}, \tag{9}$$

where $N_a$ = the number of agreements, $N$ = the number of individuals rated, and $\rho_c$ = the probability of chance agreement on an individual. The Appendix indicates the probability of chance agreement when agreement is defined as 0, 1, and 2 points discrepancy.

Positive values of $T$ indicate that the observed agreement is greater than chance agreement, negative values indicate that the observed agreement is less than chance agreement, and $T$ is zero when the observed agreement is equal to the expected chance agreement. The ratings for Case 3 of Table 4.1 can be used to illustrate the flexibility of this approach and the critical importance of the operational definition of agreement adopted. When agreement is defined as 0, 1, and 2 points discrepancy, respectively, the number of agreements equals 1 (applicant J), 7 (all applicants except C, D, and I), and 10 (all of the applicants), and the corresponding $T$ values are $T_0 = .00$, $T_1 = .67$, and $T_2 = 1.00$.

The statistical significance of $T$ can be tested using Lawlis and Lu's (1972) nonparametric chi-square, with 1 degree of freedom:

$$\chi^2 = \frac{(N_a - N\rho_c - .5)^2}{N\rho_c} + \frac{(N_d - N(1 - \rho_c) - .5)^2}{N(1 - \rho_c)}, \tag{10}$$

where $N_a$, $N$, and $\rho_c$ are defined as in Equation 9, $.5$ = a correction of continuity, and $N_d$ = the number of disagreements. A significant chi-square indicates that the observed agreement is greater than the agreement that could be expected on the basis of chance, whereas a nonsignificant chi-square means that the hypothesis that the ratings were assigned at random cannot be rejected. Several factors could be responsible for the failure to find significant agreement, including characteristics of the scale (e.g., vaguely defined or overlapping categories), the judges (e.g., carelessness or lack of adequate training), or the objects (e.g., homogeneity of the sample or inconsistent expression of the attribute being rated). Continued use of the scale in the manner in which the data were obtained is ill advised when a nonsignificant chi-square is obtained.

Researchers must study their rating scales and think carefully about the implications of the alternative definitions of agreement for their research, for the $T$-index and the results of Lawlis and Lu's (1972) chi-square are specific to the definition of agreement. The advantages of being able to distinguish disagreements that have no serious consequences from those that are of practical significance carries with it the temptation to manipulate the definition of agreement so that the results will reach the desired level of agreement. Consequently, the definition of agreement should be determined on theoretical, empirical, or practical grounds before data analysis has begun, and preferably before data collection has begun. Furthermore, the definition of agreement and the rationale for its adoption should be stated when reporting the interrater agreement of the ratings.

One further consideration is that Lawlis and Lu's (1972) estimate of chance agreement is based on the assumption that every judgment has the same probability. We pointed out earlier that under some circumstances (e.g., when raters do not use the end categories of a rating scale), the true probability of chance agreement $(\rho_c)$ is greater than $N\rho_c$. In practice, therefore, $N\rho_c$ is the lower limit for the unknown probability of chance agreement, $(\rho_c)$ is often underestimated, and the significance of the observed agreement is overstated. Careful investigators can avoid underestimating $(\rho_c)$ by adjusting the required level of significance from the traditional $.05$ to a more stringent level, or by calculating the probability of chance agreement on fewer scale categories than are actually available (e.g., use the probability of chance agreement for a 5- or 6-point scale when a 7-point scale was used).

## VI. SUMMARY OF RECOMMENDATIONS

Evidence regarding both interrater reliability and interrater agreement should be reported whenever rating scales are used. Interrater reliability and agreement are functions of the objects rated, the scale used, and the observers making the ratings, so they are descriptive of a specific set of data. Attempts to generalize interrater reliability and agreement from other research or other groups of raters is misleading. All of the recommended measures of interrater reliability and agreement are one-trial estimates, so calculation of these indices should pose no logistic problems if the study has been designed properly.

The intraclass correlation (Equation 1) is the best measure of interrater reliability for ordinal and interval level measurement; nominally scaled data permit an analysis only of interrater agreement. Use Finn's $r_c$ (Equation 4) only when the within-raters variance is so severely restricted that the intraclass correlation is inappropriate (e.g., Case 3 in Table 4.1). Interpret Finn's $r_c$ cautiously because chance variance may be overestimated, thereby spuriously inflating $r_c$. Be explicit in explaining whether between-raters' variance was included or excluded from the error term, the implications of that decision for the possibility of generalizing your results, and whether the reported reliability represents the average reliability of a single rater or the reliability of the composite rating.

The reliability of composite scores based on unit weights will be satisfactory for most situations. Use optimal weights only when the decision to be made requires the greatest possible measurement precision and the ratings of the judges differ in variance or reliability. The procedure developed by Overall (1965) is appropriate in this situation, but take care to report the variance and estimated reliability of each judge, the weights assigned the ratings of each judge, and the estimated reliability of the composite based on unit weights and optimal weights. Also keep in mind that the optimal weights were tailored to your unique rating situation and cannot be generalized across rating occasions, even if the same judges used the same rating scale with the same or a similar sample of objects.

Use the $T$-Index (Equation 9) and Lawlis and Lu's chi-square (Equation 10) to estimate the interrater agreement of ordinal and interval scaled ratings. Use a stringent level of significance in evaluating agreement or a value of $\rho_c$ based on fewer scale categories than were actually available to correct for the possibility that the raters were predisposed to avoid the extreme categories of the scale. Report all decisions made in conducting the analysis, including the definition of agreement adopted and the rational for the definition.

With nominal scaled ratings use Cohen's $\kappa$ (Equation 7) when the same two judges rate each object and Fleiss's $\kappa_v$ when the objects are rated by different judges but the same number of judges rate each object. We do not recommend the use of weighted kappa ($\kappa_w$).

## REFERENCES

Bartke, J. J. (1966). The intraclass correlation coefficient as a measure of reliability. *Psychological Report, 19*, 3–11.

Berk, R. A. (1979). Generalizability of behavioral observations; A classification of interobserver agreement and interobserver reliability. *American Journal of Mental Deficiency, 83*, 460–472.

Burdock, E. I., Fleiss, J. L., & Hardesty, A. S. (1963). A new view of interobserver agreement. *Personnel Psychology, 16*, 373–384.

Burt, C. (1936). An analysis of examination marks. In P. Hartog & E. C. Rhodes (Eds.), *The marks of examiners* (pp. 245–314). London: Macmillan.

Chorpita, B. F., Brown, T. A., & Barlow, D. H. (1998). Diagnostic reliability of the DSM-III-R anxiety disorders: Mediating effects of patient and diagnostician characteristics. *Behavior Modification, 22*, 307–320.

Cohen, J. (1960). A coefficient of agreement for nominal scales. *Educational and Psychological Measurement, 20*, 37–46.

Cohen, J. (1968). Weighted kappa: Nominal scale agreement with provision for scaled disagreement or partial credit. *Psychological Bulletin, 70*, 213–220.

Crocker, L., & Algina, J. (1986). *Introduction to classical and modern test theory.* New York: Holt, Rinehard and Winston.

Cronbach, L. J. (1951). Coefficient alpha and the internal structure of tests. *Psychometrika, 16*, 297–334.

Cronbach, L. J., & Gleser, G. C. (1965). *Psychological tests and personnel decisions.* Urbana: University of Illinois Press.

Cronbach, L. J., Gleser, G. C., Nanda, H., & Rajaratnam, N. (1972). *The dependability of behavioral measurements: Theory of generalizability for scores and profiles.* New York: John Wiley.

Dreessen, L., & Arntz, A. (1998). Short-interval test-retest interrater reliability of the Structured Clinical Interview for DSM-III-R personality disorders (SCID-II) in outpatients. *Journal of Personality Disorders, 12*, 138–148.

Ebel, R. L. (1951). Estimation of the reliability of ratings. *Psychometrika, 16*, 407–424.

Everitt, B. S. (1968). Moments of the statistics kappa and weighted kappa. *British Journal of Mathematical and Statistical Psychology, 21*, 97–103.

Fabrigar, L. R., Wegener, D. T., MacCallum, R. C., & Strahan, E. J. (1999). Evaluating the use of exploratory factor analysis in psychological research. *Psychological Methods, 4*, 272–299.

Finn, R. H. (1970). A note on estimating the reliability of categorical data. *Educational and Psychological Measurement, 30*, 71–76.

Fleiss, J. L. (1971). Measuring nominal scale agreement among many raters. *Psychological Bulletin, 76*, 378–382.

Fleiss, J. L., & Cohen, J. (1973). The equivalence of weighted kappa and the intraclass correlation coefficient as measures of reliability. *Educational and Psychological Measurement, 33*, 613–619.

Fleiss, J. L., Cohen, J., & Everitt, B. S. (1969). Large sample standard errors of kappa and weighted kappa. *Psychological Bulletin, 72*, 323–327.

Goodman, L. A., & Kruskal, W. H. (1954). Measures of association for cross classifications. *Journal of the American Statistical Association, 49*, 732–764.

Guilford, J. P. (1954). *Psychometric methods* (2nd ed.). New York: McGraw-Hill.

Gulliksen, H. (1950). *Theory of mental tests*. New York: Jn Wiley.

Guttman, H. A., Spector, R. M., Sigal, J. J., Rakoff, V., & Epstein, N. B. (1971). Reliability of coding affective communication in family therapy sessions: Problems of measurement and interpretation. *Journal of Consulting and Clinical Psychology, 37*, 397–402.

Hawkins, R. P., & Dotson, V. A. (1975). Reliability scores that delude: An Alice in Wonderland trip through the misleading characteristics of interobserver agreement scores in interval recording. In E. Ramp & G. Semb (Eds.), *Behavior analysis: Areas of research and application.* (pp. 359–376). Englewood Cliffs, NJ: Prentice-Hall.

Hays, W. L. (1994). *Statistics* (5th ed.). Orlando, FL: Harcourt Brace.

Horst, P. (1949). A generalized expression for the reliability of measures. *Psychometrika, 14*, 21–31.

Hoyt, C. J. (1941). Test reliability estimated by analysis of variance. *Psychometrika, 6*, 153–160.

Hoyt, W. T. (in press). Rater bias in psychological research: When is it a problem and what can we do about it? *Psychological Methods.*

James, L. R., Demaree, R. G., & Wolf, G. (1984). Estimating within-group interrater reliability with and without response bias. *Journal of Applied Psychology, 69*, 85–98.

James, L. R., Demaree, R. G., & Wolf, G. (1993). An assessment of within-group interrater agreement. *Journal of Applied Psychology, 78*, 306–309.

Kelley, T. L. (1947). *Fundamentals of statistics.* Cambridge, Mass.: Harvard University Press.

Kelly, M. B. (1977). A review of the observational data-collection and reliability procedures in *The Journal of Applied Behavior Analysis. Journal of Applied Behavior Analysis, 10*, 97–101.

Kozlowski, S. W. J., & Hattrup, K. (1992). A disagreement about within-group agreement: Disentangling issues of consistency versus consensus. *Journal of Applied Psychology, 77*, 161–167.

Kozlowski, S. W. J., & Hults, B. M. (1987). An exploration of climates for technical updating and performance. *Personnel Psychology, 40*, 539–563.

Kuder, G. F., & Richardson, M. W. (1937). The theory and estimation of test reliability. *Psychometrika, 2*, 151–160.

Lawlis, G. F., & Lu, E. (1972). Judgment of counseling process: Reliability, agreement, and error. *Psychological Bulletin, 78*, 17–20.

Lawshe, C. H., & Nagle, B. F. (1952). A note on the combination of ratings on the basis of reliability. *Psychological Bulletin, 49*, 270–273.

Lindell, M. K., & Brandt, C. J. (1997). Measuring interrater agreement for ratings of a single target. *Applied Psychological Measurement, 21*, 271–278.

Lindquist, E. F. (1953). *Design and analysis of experiments in psychology and education.* Boston: Houghton-Mifflin.

Lu, K. H. (1971). A measure of agreement among subjective judgments. *Educational and Psychological Measurement, 31*, 75–84.

Mitchell, S. K. (1979). Interobserver agreement, reliability, and generalizability of data collected in observational studies. *Psychological Bulletin, 86*, 376–390.

Overall, J. E. (1965). Reliability of composite ratings. *Educational and Psychological Measurement, 25*, 1011–1022.

Pedhazer, E. J., & Schmelkin, L. P. (1991). *Measurement, design and analysis: An integrated approach.* Hillsdale, NJ: Lawrence Erlbaum.

Peters, C. C., & Van Voorhis, W. R. (1940). *Statistical procedures and their mathematical bases.* New York: McGraw-Hill.

Robinson, W. S. (1957). The statistical measurement of agreement. *American Sociological Review, 22*, 17–25.

Schmidt, F. L., & Hunter, J. E. (1989). Interrater reliability coefficients cannot be computer when only one stimulus is rated. *Journal of Applied Psychology, 74*, 368–370.

Scott, W. A. (1955). Reliability of content analysis: The case of nominal scale coding. *Public Opinion Quarterly, 19,* 321–325.

Shrout, P. E., & Fleiss, J. L. (1979). Intraclass correlations: Uses in assessing rater reliability. *Psychological Bulletin, 86,* 420–428.

Snedecor, G. W. (1946). *Statistical methods* (4th ed.). Ames: Iowa State College Press.

Suen, H. K., & Ary, D. (1989). *Analyzing quantitative behavioral observation data.* Hillsdale, NJ: Lawrence Erlbaum.

Suen, H. K., & Lee, P. S. C. (1985). Effects of the use of percentage agreement on behavioral observation reliabilities: A reassessment. *Journal of Psychopathology and Behavioral Assessment, 7,* 221–234.

Szalai, J. P. (1998). Kappa-sub(sc): A measure of agreement on a single rating category for a single item or object rated by multiple raters. *Psychological Reports, 82,* 1321–1322.

Thompson, B. (1994). *Inappropriate statistical practices in counseling research: Three pointers for readers of research literature.* (Report No. EDO_CG_95–33). Greensboro, NC: ERIC Clearinghouse on Counseling and Student Services.

Tinsley, H. E. A., & Weiss, D. J. (1975). Interrater reliability and agreement of subjective judgements. *Journal of Counseling Psychology, 22,* 358–376.

## APPENDIX  Probability of Chance Agreement for Three Definitions of Agreement[a]

| Points on scale | Number of judges | | | | | | | | | | | | |
|---|---|---|---|---|---|---|---|---|---|---|---|---|---|
| | 2 | 3 | 4 | 5 | 6 | 7 | 8 | 9 | 10 | 11 | 12 | 13 | 14 |

Agreement defined as zero points discrepancy in ratings

| Points on scale | 2 | 3 | 4 | 5 | 6 | 7 | 8 | 9 | 10 | 11 | 12 | 13 | 14 |
|---|---|---|---|---|---|---|---|---|---|---|---|---|---|
| 2 | .500 | .250 | .125 | .063 | .031 | .106 | .008 | .004 | .002 | .001 | .001 | — | — |
| 3 | .333 | .111 | .037 | .012 | .004 | .001 | .001 | — | — | — | — | — | — |
| 4 | .250 | .063 | .016 | .004 | .001 | — | — | — | — | — | — | — | — |
| 5 | .200 | .040 | .000 | .002 | — | — | — | — | — | — | — | — | — |
| 6 | .167 | .028 | .005 | .001 | — | | | | | | | | |
| 7 | .143 | .020 | .003 | — | — | | | | | | | | |
| 8 | .125 | .016 | .002 | — | — | | | | | | | | |
| 9 | .111 | .012 | .001 | — | — | | | | | | | | |
| 10 | .100 | .010 | .001 | — | — | | | | | | | | |
| 11 | .091 | .003 | .001 | — | | | | | | | | | |
| 12 | .083 | .007 | .001 | — | | | | | | | | | |
| 13 | .077 | .006 | .001 | — | | | | | | | | | |
| 14 | .071 | .005 | — | — | | | | | | | | | |
| 15 | .067 | .004 | — | — | | | | | | | | | |
| 16 | .063 | .004 | — | | | | | | | | | | |
| 17 | .059 | .004 | — | | | | | | | | | | |
| 18 | .056 | .003 | — | | | | | | | | | | |
| 19 | .053 | .003 | — | | | | | | | | | | |
| 20 | .050 | .003 | — | | | | | | | | | | |
| 21 | .048 | .002 | — | | | | | | | | | | |
| 22 | .046 | .002 | — | | | | | | | | | | |
| 23 | .044 | .002 | — | | | | | | | | | | |
| 24 | .042 | .002 | — | | | | | | | | | | |
| 25 | .040 | .002 | — | | | | | | | | | | |
| 26 | .039 | .002 | — | | | | | | | | | | |
| 27 | .037 | .001 | — | | | | | | | | | | |
| 28 | .036 | .001 | — | | | | | | | | | | |
| 29 | .035 | .001 | — | | | | | | | | | | |
| 30 | .033 | .001 | — | | | | | | | | | | |
| 31 | .032 | .001 | — | | | | | | | | | | |
| 32 | .031 | .001 | — | | | | | | | | | | |
| 33 | .030 | .001 | — | | | | | | | | | | |
| 34 | .029 | .001 | — | | | | | | | | | | |
| 35 | .029 | .001 | — | | | | | | | | | | |

*(continued)*

**APPENDIX** *(continued)*

| Points on scale | Number of judges | | | | | | | | | | | | |
|---|---|---|---|---|---|---|---|---|---|---|---|---|---|
| | 2 | 3 | 4 | 5 | 6 | 7 | 8 | 9 | 10 | 11 | 12 | 13 | 14 |
| Agreement defined as $\leq$ one point discrepancy in ratings[b] | | | | | | | | | | | | | |
| 3 | .778 | .556 | .383 | .259 | .174 | .117 | .078 | .052 | .035 | .023 | .015 | .010 | .007 |
| 4 | .625 | .344 | .180 | .092 | .046 | .023 | .012 | .006 | .003 | .002 | .001 | — | — |
| 5 | .520 | .232 | .098 | .040 | .0162 | .007 | .003 | .001 | — | — | — | — | — |
| 6 | .444 | .167 | .059 | .020 | .007 | .002 | .001 | — | | | | | |
| 7 | .388 | .125 | .038 | .011 | .003 | .001 | — | — | | | | | |
| 8 | .343 | .098 | .026 | .007 | .002 | — | — | — | | | | | |
| 9 | .309 | .078 | .018 | .004 | .001 | — | — | — | | | | | |
| 10 | .280 | .064 | .014 | .003 | .001 | — | — | — | | | | | |
| 11 | .256 | .053 | .010 | .002 | — | | | | | | | | |
| 12 | .236 | .045 | .008 | .004 | — | | | | | | | | |
| 13 | .219 | .039 | .006 | .001 | — | | | | | | | | |
| 14 | .204 | .034 | .005 | .001 | — | | | | | | | | |
| 15 | .191 | .029 | .004 | .001 | — | | | | | | | | |
| 16 | .180 | .026 | .003 | — | | | | | | | | | |
| 17 | .170 | .023 | .003 | — | | | | | | | | | |
| 18 | .161 | .021 | .002 | — | | | | | | | | | |
| 19 | .152 | .019 | .002 | — | | | | | | | | | |
| 20 | .145 | .017 | .002 | — | | | | | | | | | |
| 21 | .138 | .015 | .002 | — | | | | | | | | | |
| 22 | .132 | .014 | .001 | — | | | | | | | | | |
| 23 | .127 | .013 | .001 | — | | | | | | | | | |
| 24 | .122 | .012 | .001 | — | | | | | | | | | |
| 25 | .117 | .011 | .001 | — | | | | | | | | | |
| 26 | .112 | .010 | .001 | — | | | | | | | | | |
| 27 | .108 | .009 | .001 | — | | | | | | | | | |
| 28 | .105 | .009 | .001 | — | | | | | | | | | |
| 29 | .101 | .008 | .001 | — | | | | | | | | | |
| 30 | .098 | .008 | .001 | — | | | | | | | | | |
| 31 | .095 | .007 | .001 | — | | | | | | | | | |
| 32 | .092 | .007 | — | — | | | | | | | | | |
| 33 | .089 | .006 | — | — | | | | | | | | | |
| 34 | .087 | .006 | — | — | | | | | | | | | |
| 35 | .084 | .006 | — | — | | | | | | | | | |

*(continues)*

## APPENDIX *(continued)*

| Points on scale | Number of judges | | | | | | | | | | | | |
|---|---|---|---|---|---|---|---|---|---|---|---|---|---|
| | **2** | **3** | **4** | **5** | **6** | **7** | **8** | **9** | **10** | **11** | **12** | **13** | **14** |
| Agreement defined as ≤ two points discrepancy in ratings[c] | | | | | | | | | | | | | |
| 4 | .875 | .719 | .570 | .443 | .340 | .259 | .196 | .148 | .117 | .084 | .063 | .047 | .036 |
| 5 | .760 | .520 | .338 | .213 | .132 | .081 | .049 | .030 | .02 | .11 | .007 | .004 | .002 |
| 6 | .667 | .389 | .213 | .113 | .059 | .030 | .015 | .008 | .004 | .002 | .001 | .001 | — |
| 7 | .592 | .300 | .142 | .065 | .029 | .013 | .006 | .002 | .001 | — | — | — | — |
| 8 | .531 | .238 | .099 | .040 | .016 | .006 | .002 | .001 | — | — | — | — | — |
| 9 | .482 | .193 | .072 | .026 | .009 | .003 | .001 | — | — | — | — | — | — |
| 10 | .440 | .160 | .054 | .017 | .005 | .002 | .001 | — | — | — | — | — | — |
| 11 | .405 | .135 | .041 | .012 | .003 | .001 | — | | | | | | |
| 12 | .375 | .115 | .032 | .009 | .002 | .001 | — | | | | | | |
| 13 | .349 | .099 | .026 | .006 | .002 | — | — | | | | | | |
| 14 | .327 | .086 | .021 | .005 | .001 | — | — | | | | | | |
| 15 | .307 | .076 | .017 | .004 | .001 | — | — | | | | | | |
| 16 | .289 | .067 | .014 | .003 | .001 | — | | | | | | | |
| 17 | .273 | .060 | .012 | .002 | — | — | | | | | | | |
| 18 | .259 | .054 | .010 | .002 | — | — | | | | | | | |
| 19 | .247 | .048 | .009 | .002 | — | — | | | | | | | |
| 20 | .235 | .044 | .008 | .001 | — | — | | | | | | | |
| 21 | .225 | .040 | .006 | .001 | — | | | | | | | | |
| 22 | .215 | .037 | .006 | .001 | — | | | | | | | | |
| 23 | .206 | .034 | .005 | .001 | — | | | | | | | | |
| 24 | .198 | .031 | .004 | .001 | — | | | | | | | | |
| 25 | .190 | .029 | .004 | .001 | — | | | | | | | | |
| 26 | .183 | .026 | .003 | — | | | | | | | | | |
| 27 | .177 | .025 | .003 | — | | | | | | | | | |
| 28 | .171 | .023 | .003 | — | | | | | | | | | |
| 29 | .165 | .021 | .003 | — | | | | | | | | | |
| 30 | .160 | .020 | .002 | — | | | | | | | | | |
| 31 | .155 | .019 | .002 | — | | | | | | | | | |
| 32 | .150 | .018 | .002 | — | | | | | | | | | |
| 33 | .146 | .017 | .002 | — | | | | | | | | | |
| 34 | .142 | .016 | .002 | — | | | | | | | | | |
| 35 | .138 | .015 | .001 | — | | | | | | | | | |

*(continues)*

**APPENDIX** *(continued)*

| Points on scale | Zero Points | | | ≤ One Point | | | ≤ Two Points | | | |
|---|---|---|---|---|---|---|---|---|---|---|
| | **2** | **3** | **4** | **2** | **3** | **4** | **2** | **3** | **4** | **5** |
| 36 | .028 | .001 | — | .082 | .005 | — | .134 | .014 | .001 | — |
| 37 | .027 | .001 | — | .080 | .005 | — | .131 | .013 | .001 | — |
| 38 | .026 | .001 | — | .078 | .005 | — | .127 | .013 | .001 | — |
| 39 | .026 | .001 | — | .076 | .005 | — | .124 | .012 | .001 | — |
| 40 | .025 | .001 | — | .074 | .004 | — | .121 | .011 | .001 | — |
| 41 | .024 | .001 | — | .072 | .004 | — | .118 | .011 | .001 | — |
| 42 | .024 | .001 | — | .070 | .004 | — | .006 | .010 | .001 | — |
| 43 | .023 | .001 | — | .069 | .004 | — | .113 | .010 | .001 | — |
| 44 | .023 | .001 | — | .067 | .004 | — | .111 | .010 | .001 | — |
| 45 | .022 | .001 | — | .066 | .003 | — | .108 | .010 | .001 | — |
| 46 | .022 | .001 | — | .064 | .003 | — | .106 | .009 | .001 | — |
| 47 | .021 | .001 | — | .063 | .003 | — | .104 | .008 | .001 | — |
| 48 | .021 | — | — | .062 | .003 | — | .102 | .008 | .001 | — |
| 49 | .020 | — | — | .060 | .003 | — | .010 | .008 | .001 | — |
| 50 | .020 | — | — | .059 | .003 | — | .010 | .007 | .001 | — |

[a] Tables computed from formulas appearing in Lawlis and Lu (1972).

[b] Probability of chance agreement for a 3-point scale when agreement defined as ≤1 point discrepancy equals .005 (15 judges), .003 (16 judges), .002 (17 judges), and .001 (≥18 judges).

[c] Probability of chance agreement for a 4-point scale when agreement defined as ≤2 points discrepancy equals .027 (15 judges), .020 (16 judges), .015 (17 judges), .011 (18 judges), .009 (19 judges), .006 (20 judges), .005 (21 judges), .004 (22 judges), .003 (23 judges), .002 (24 judges), .002 (25 judges), and .001 (≥26 judges). Probability of chance agreement for a 5-point scale when agreement defined as ≤2 points discrepancy equals .001 (≥15 judges).

# 5

# INTERPRETING AND REPORTING RESULTS

**MARK HALLAHAN**

*Department of Psychology, Clemson University, Clemson, South Carolina*

**ROBERT ROSENTHAL**

*Department of Psychology, Harvard University, Cambridge, Massachusetts*

## I. INTRODUCTION

This chapter presents guidelines to help researchers (a) avoid some common sources of confusion about statistical procedures, (b) achieve a more thorough and accurate understanding of their data, and (c) communicate research results more effectively. With their characteristic wisdom, John Tukey (1969) and Jacob Cohen (1990, 1994) emphasized that researchers should approach data with flexibility and exercise their own judgment rather than mechanically follow prescribed rules. In that spirit, we do not intend to present prescriptions for data analysis, but rather general advice that may be useful to consider when working with data. We believe that the best decisions about data are made by researchers who see data analysis as an inextricable part of the larger process of doing research and whose decisions are informed by an in-depth understanding of theory, measurement, and other scientific and practical concerns specific to their research. Data analysis should advance two primary scientific goals: identifying patterns of regularity in the social or natural world, and communicating about these patterns in a way that it is accessible to people who are interested. Rather than asking decontextualized questions such as "am I allowed to

*Handbook of Applied Multivariate Statistics and Mathematical Modeling*
**125**

analyze my data with this procedure?" (Abelson, 1995, p. xii), researchers should be concerned with these two goals.

This chapter discusses the general process of data analysis as well as specific procedures. Of course, researchers need to use and interpret statistical procedures appropriately, but their approach to the data analysis process is also crucial. Scientific understanding should be the basic organizing principle motivating the research process. Any practice that reduces researchers' direct contact with their data, fosters rote decision making, or obviates the need for judgment and interpretation is a potential barrier to scientific understanding. This chapter has six major sections. We discuss graphical displays and exploratory data analysis, contrast analysis, interpreting significance levels, interpreting effect sizes, the assumptions that underlie statistical procedures, and general advice about the process of working with data.

## II. GRAPHICAL DISPLAYS AND EXPLORATORY DATA ANALYSIS

### A. Graphical Displays

A good picture is often worth more than a thousand words. Graphical displays of data yield many benefits, such as identifying outlying cases, catching data entry errors, and checking that the assumptions of statistical tests are met. However, conceptual understanding is the most important benefit. Graphical displays reveal information that is difficult to discern from statistical summaries and clearly represent complex patterns in large data sets. As Tukey (1969) said, "'plot and eye' is the most diverse channel to the human mind. Not that it transmits more bits per second, but rather that it will transmit a greater variety of messages on unexpected topics easily and rapidly" (p. 83).

Anscombe (1973) provided an excellent illustration of how a simple graphical display can show trends that are lost in quantitative summaries of data. Table 5.1 has four different sets of 11 $x$, $y$ pairs. Based on typical statistical summaries, they appear to be identical. For each, the average $x = 9.0$, the average $y = 7.5$, the correlation between $x$ and $y$ is $r = .82$, and the regression equation is $y = 3 + .5x$. However, plots of $x$ and $y$ (see Figure 5.1) reveal obvious differences. There is a straightforward linear relation between variables $x$ and $y$ in data set $A$, but a curvilinear relation in data set $B$. Data sets $C$ and $D$ highlight the value of identifying outlying cases. If one outlier were removed from these datasets, the remaining scores in data set $C$ would perfectly fit a regression line with a less steep slope, whereas in data set $D$, there would be no variability.

Plotting the relation between two variables is good practice when computing a correlation. The value of $r$ is sensitive to even one erroneous or

**TABLE 5.1    Four Sets of 11 x, y Pairs**[a]

| Data set A | | Data set B | | Data set C | | Data set D | |
|------|------|------|------|------|------|------|------|
| *x* | *y* | *x* | *y* | *x* | *y* | *x* | *y* |
| 10.0 | 8.04 | 10.0 | 9.14 | 10.0 | 7.46 | 8.0 | 6.58 |
| 8.0 | 6.95 | 8.0 | 8.14 | 8.0 | 6.77 | 8.0 | 5.76 |
| 13.0 | 7.58 | 13.0 | 8.74 | 13.0 | 12.74 | 8.0 | 7.71 |
| 9.0 | 8.81 | 9.0 | 8.77 | 9.0 | 7.11 | 8.0 | 8.84 |
| 11.0 | 8.33 | 11.0 | 9.26 | 11.0 | 7.81 | 8.0 | 8.47 |
| 14.0 | 9.96 | 14.0 | 8.10 | 14.0 | 8.84 | 8.0 | 7.04 |
| 6.0 | 7.24 | 6.0 | 6.13 | 6.0 | 6.08 | 8.0 | 5.25 |
| 4.0 | 4.26 | 4.0 | 3.10 | 4.0 | 5.39 | 19.0 | 12.50 |
| 12.0 | 10.84 | 12.0 | 9.13 | 12.0 | 8.15 | 8.0 | 5.56 |
| 7.0 | 4.82 | 7.0 | 7.26 | 7.0 | 6.42 | 8.0 | 7.91 |
| 5.0 | 5.68 | 5.0 | 4.74 | 5.0 | 5.73 | 8.0 | 6.89 |

[a] From Anscombe (1977). Reprinted with permission of the American Statistical Association.

extreme datum. Researchers need to identify and understand genuinely unusual cases. However, outliers also may be errors that, if corrected, would produce more accurate results (and if not, may distort results seriously). For example, Wainer and Thissen (1976) illustrated how a correlation of height and weight for 25 people went from $r = .83$ to $r = -.26$ when a single person's height and weight values were reversed.

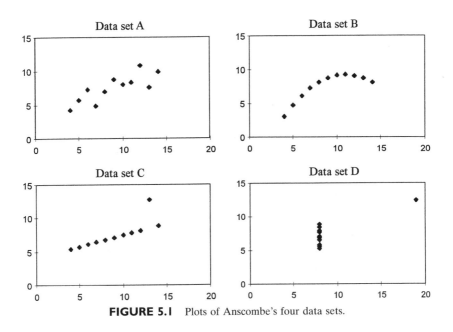

**FIGURE 5.1**    Plots of Anscombe's four data sets.

For complex research designs, graphical displays far surpass textual descriptions in their ability to express the full pattern of the data. For example, imagine an experiment examining the effectiveness of fear-arousing communications. Cigarette smokers were assigned to hear a message that would elicit low, moderate, or high levels of fear about the dangers of smoking, either by itself or accompanied by specific instructions about how to quit smoking. The dependent variable, cigarettes smoked per day, was measured four times, once before the message and then at 2, 4, and 6 weeks after the message. A typical journal article might report this experiment's results something like this:

> Results of a $2 \times 3 \times 4$ factorial ANOVA showed a significant main effect of instructions, $F_{(1, 66)} = 4.15$, $p = .046$, and a significant main effect of message type, $F_{(2, 66)} = 10.56$, $p = .0001$. The main effect of time, the repeated measures factor, also was significant, $F_{(3, 198)} = 7.13$, $p = .0001$. The interaction of time by instructions was significant, $F_{(3, 198)} = 4.80$, $p = .003$, as was the interaction of time by message type, $F_{(6, 198)} = 4.61$, $p = .0002$. Neither the interaction of instruction by message type, nor the interaction of time by instruction by message type were significant ($F$'s $< 1.0$).

This passage is virtually incomprehensible, but a graphical display (Figure 5.2) nicely depicts the patterns in these data. Figure 5.2 clearly shows moderately fear-arousing messages are most effective in reducing smoking, especially when given in conjunction with instructions, and high and low fear-arousing messages are effective when given with instructions, albeit to a lesser extent than moderate messages, and are ineffective when given without instructions. Figure 5.2 contains 95% confidence intervals around each condition mean, using Loftus and Masson's (1994) procedure for repeated measures data. Including confidence intervals or standard error bars around condition means provides a sense of how precisely the sample means estimate population values and greatly improves viewers' understanding of the data.

Journal editors may reasonably encourage tabular rather than graphical presentation of data. Tables of means are less costly to print and more precise than figures. However, readers may struggle to discern interactions and other complex trends from a table for a complex design with many condition means (e.g., our example has 24 condition means). No matter what presentation is employed for publication, researchers should use graphical displays to enhance their own understanding of their data.

Times have changed since Tukey (1977) wrote about "scratching down numbers" on graph paper. The ability to displays data visually is now a standard feature of statistical software. From basic plots and displays, which can be created with amazing ease and speed, to powerful interactive graphing software, this capacity makes it increasingly easy to understand and to communicate about data. Behrens (1997) discusses the availability of graphing and exploratory data analysis software, and many excellent books

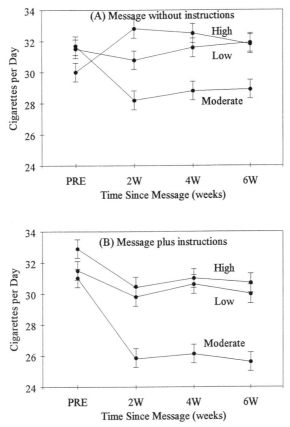

**FIGURE 5.2**    Smoking by instructions, message type, and time.

provide guidance about using graphical displays (e.g., Cleveland, 1993; Tufte, 1983, 1990, 1997).

## B. Exploratory Data Analysis

Exploratory data analysis (Hoaglin, Mosteller, & Tukey, 1983; Tukey, 1977), which Tukey (1977) characterized as "numerical detective work" (p. 1), emphasizes data analysis as a process of discovery and understanding. Tukey (1977) argued that exploratory data analysis, which provides insight into data and generates new hypotheses, should be used alongside confirmatory data analysis, which provides formal tests of a priori hypotheses. In general, exploratory techniques emphasize data description using statistics that are resistant to odd or extreme data, re-expression to place data on a more meaningful scale, and analysis of residuals in an iterative process of seeking the best model for the data. We cannot fully review exploratory data analysis here (see Behrens, 1997, for a comprehensive review), but we will briefly discuss a few of its classic techniques.

### 1. Data Description

The stem-and-leaf display summarizes data while preserving enough detail so that irregular and unexpected features are easily detected. Table 5.2 shows the following set of 23 scores presented in a stem-and-leaf display: 54, 36, 82, 51, 58, 41, 73, 97, 56, 19, 65, 67, 42, 76, 57, 23, 57, 52, 48, 34, 62, 56, 68. In this case, the "stems" represent the *10*'s digit and the "leaves" represent the *1*'s digit. The beauty of the stem-and-leaf display is that it organizes data in a way that provides information about the shape of the distribution, its central tendency and variability, extreme cases, and distributional anomalies like gaps or clusters of scores, without losing any of the original information.

The 5-number summary includes the median value (56 in these data); the upper and lower "hinges" or fourths, defined as the .25 $(n + 1)$th score from the top and bottom of a set of scores (42 and 67); and the extreme scores (19 and 97). The $H$-spread, defined as the difference between the upper and lower hinges, provides a measure of variability that is not influenced by extreme cases. The $H$-spread also can be used to identify outlier values. Tukey (1977) considered observations more than 1.5 times the $H$-spread beyond the hinges to be outliers. The box plot represents the five-number summary visually. A rectangular box extends between the hinges, with a horizontal line at the median. Lines extend from the box to the extreme values (or to the most extreme values within the outlier cutoff range, with outliers represented individually). Figure 5.3 shows five sets of scores represented with box plots. Note that this display conveys informa-

**TABLE 5.2 Stem-and-Leaf Display**

| Stems | Leaves (unit = 1) |
|-------|-------------------|
| 9 | 7 |
| 8 | 2 |
| 7 | 3 6 |
| 6 | 2 5 7 8 |
| 5 | 1 2 4 6 6 7 7 8 |
| 4 | 1 2 8 |
| 3 | 4 6 |
| 2 | 3 |
| 1 | 9 |

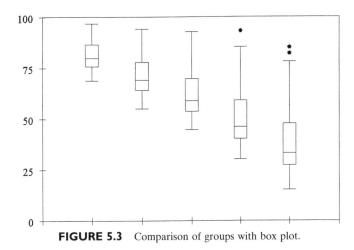

**FIGURE 5.3**    Comparison of groups with box plot.

tion about central tendency, variability and skew, as well as identifying out-liers.

### 2. Re-expression

Re-expression of raw scores on different scales (e.g., the square root or log of the raw scores) enhances conceptual understanding because data then often fit a more parsimonious or more interpretable model. Additionally, re-expression tends to lessen problems, such as asymmetrical distributions, unequal variance, and outlying scores. An appropriate re-expression of the data in Figure 5.3 would likely pull the extreme scores within outlier cutoffs, reduce the skew, and make the variability more similar across the five groups. Tukey (1977) discussed how to select an appropriate transformation.

### 3. Resistant Statistics

Exploratory data analysis also emphasizes resistant statistics, or statistics whose values are not affected by extreme or unusual data. Commonly used statistics, such as the mean, standard deviation, and correlation coefficient, can be highly sensitive to even a few extreme values. Examples of resistant statistics include measures of central tendency based on medians, measures of variability based on the *H*-spread, and measures of the relation between two variables that are not based on correlations or ordinary least squares regression. These statistics are useful for their ability to describe sets of data, regardless of how these data are distributed.

## III. CONTRASTS

Researchers should assess the extent to which their data correspond to specific, theoretically meaningful predictions. Although this can be accomplished easily by computing contrasts or planned comparisons (see e.g., Rosenthal, & Rosnow, 1985; Rosenthal, Rosnow, & Rubin, 1999), researchers too often use omnibus comparisons (e.g., chi-square with more than one degree of freedom or $F$ with more than one numerator degree of freedom), which address less specific and usually less interesting questions. Abelson (1995) described omnibus testing as "like playing the guitar with mittens on" (p. 105), an apt metaphor for these procedures' imprecision. A low $p$-value for $F$ from one-way analysis of variance (ANOVA) implies only that the observed sample means did not likely come from a population in which the null is true, that is, in which all condition means are identical ($\mu_1 = \mu_2 = \mu_3 \ldots = \mu_n$). This may be a basis to infer that the condition means differ from each other in some way, but it provides no insight into how they differ.

Researchers can almost always specify a more precise prediction for their data than "there will be some difference somewhere" or "not all conditions will be identical." Therefore, their data analysis should be similarly precise. Contrast analysis expresses a prediction as a set of numeric weights and then computes effect size estimates and significance levels that reflect the degree of correspondence between data and prediction. The computation of contrast effect sizes and significance levels is straightforward and is discussed in many other places (e.g., Hays, 1994; Rosenthal & Rosnow, 1985; Rosenthal et al., 1999; Snedecor & Cochran, 1989).

### A. Contrast Weights

Expressing a prediction as contrast weights is the most important step in contrast analysis. These numeric weights assigned to condition means must accurately reflect the intended prediction, and as a set sum to zero. Sources that discuss contrast weights might distinguish between orthogonal and nonorthogonal sets of weights (sets of weights that are independent of and related to each other, respectively), or mention well-known sets of orthogonal contrast weights (e.g., orthogonal polynomials or Helmert weights). However, when selecting contrast weights, the only important consideration should be that they reflect the intended prediction accurately and with appropriate precision, regardless of whether those predictions are independent of each other or whether they correspond to well-known sets of weights like the orthogonal polynomials. Contrast weights are flexible; they can express any prediction, no matter how complex. Hallahan and Rosenthal (1996a) present a simple algorithm for generating contrast weights.

## B. Benefits of Contrasts

Computing an omnibus comparison followed by some type of post hoc comparison is an alternative to pairwise comparisons of multiple group means (Miller, 1981; Shaffer, 1995). Although this is more informative than an omnibus test alone, contrasts have two advantages over this approach. First, many post hoc procedures rely on a series of pairwise comparisons, with the number of necessary comparisons quickly becoming quite large as the number of conditions in one's research design increases. However, when a prediction specifies a pattern in a set of means, it is easier and more informative to compute a single contrast that directly assesses how well the means correspond to the predicted pattern. Second, a planned contrast has more statistical power than a set of post hoc comparisons, which must employ a stricter criterion for obtaining $p$-values adjusted for post hoc analysis.

Contrasts are easy to use, they provide specific information about the extent to which data fit conceptually meaningful predictions, and they have greater statistical power than omnibus and post hoc comparisons. The process of generating contrast weights provides an additional benefit: it requires researchers to think about theory and to specify precisely the pattern in their data that theory predicts.

## IV. INTERPRETING SIGNIFICANCE LEVELS

The 1990s have witnessed vigorous debate about the role of significance tests in the analysis of empirical data. Some observers argue that significance tests should be abandoned because of serious problems associated with their use (and misuse) that impede researchers' ability to understand their data (e.g., Hunter, 1997; Loftus, 1996; Schmidt, 1996). Others, who do not advocate banning significance tests outright, voice concern that significance tests are overemphasized, misused, and misinterpreted (e.g., Abelson, 1997; Estes, 1997; Gonzalez, 1994; Scarr, 1997). Despite this controversy, significance tests are used widely and receive heavy emphasis in methodology courses. Significance tests to reject a null hypothesis receive the heaviest criticism, namely that (a) they are used mechanically without concern for their meaning or appropriateness, (b) they are misinterpreted due to widely held misconceptions about their meaning, and (c) they are overemphasized at the expense of more important information about effect sizes.

## A. Criticisms of Null Hypothesis Significance Testing

### 1. Applied Mechanically

Significance testing has been described as a "ritual" of social science research (Gonzalez, 1994). Many researchers seem to compute and report significance levels more as a matter of routine than as a way to understand their data. Significance levels are too often computed without careful consideration of their meaning and appropriateness for a given research question (Cleveland, 1993; Gigerenzer, 1993; Gonzalez, 1994).

Gonzalez (1994) and Loftus (1996) discuss the problem of "off-the-shelf" assumptions. Commonly used statistics assume that data have certain properties such as symmetric distributions, equal variance, and independent observations. Although these statistics often can be used reasonably when these assumptions are not met, in some cases, the deviations from this basic statistical model are critical to a researcher's theory. Rather than force data into analyses whose basic assumptions directly contradict their theory, thoughtful researchers use models that fit the underlying theory more closely. For example, Kenny (1994) sought to investigate the patterns of nonindependence in observations from within dyads. Instead of using standard analyses, which assume independence, he developed new procedures to incorporate the substantively interesting nonindependence into a model of the data (Kenny, 1994). However, when researchers mechanically use standard statistical procedures whose assumptions contradict their theory, they decrease the accuracy of those procedures and are likely to have difficulty addressing theoretically interesting questions.

The convention to use $p = .05$ as the criterion for rejecting the null, though arbitrary, has near universal acceptance and is often employed mechanically. As a result of this convention, some researchers seem to take a dichotomized view of their results, in which $p < .05$ is taken to mean there is an effect and $p > .05$ is taken to mean there is no effect (Rosenthal & Gaito, 1963). Results with $p = .06$ and $p = .05$ are extremely similar in terms of the likelihood of the data coming from a population in which the null were true, but often they are interpreted very differently by virtue of falling on different sides of the arbitrary .05 criterion.

## 2. Common Misconceptions

Many researchers believe the $p$-value associated with a significance test provides information about the probability that the null hypothesis is true, $P(H_0)$ (see Oakes, 1986). Other prevailing beliefs are that $1 - p$ provides information about the probability the alternative hypothesis is true, $P(H_1)$ and that $p$-values provide information about the probability of an observed result replicating in subsequent research. None of these interpretations are correct. The $p$-value associated with a significance test provides information about the probability of observing the data if the null hypothesis were true, $P(D|H_0)$.

Not understanding what information is (and is not) provided by $p$-values may lead researchers to misinterpret them. Inferring that failure to reject the null hypothesis implies its truth is one serious, common error. In other words, many researchers assume that if the observed $p$-value is not less than .05, then there is "no effect." This may be due to poor understanding of two related concepts, Type II error and statistical power (Cohen, 1988). Type II error is the failure to reject a null that is actually false. Statistical power is the probability of not making a Type II error, or the probability of obtaining a statistically significant result when a given size effect exists. A recent survey (Zuckerman, Hodgins, Zuckerman, & Rosenthal, 1993), in which a majority of psychologists answered questions about power and Type II error incorrectly, reflects the generally poor understanding of power, as does psychology research's generally poor level of statistical power. For example, in a reputable journal, the median level of power to detect a typical medium size effect (i.e., $d = .50$) was less than .50 (Cohen, 1962; Sedlmeier & Gigerenzer, 1989). This implies researchers will not obtain significance more than half of the time when a medium size effect truly exists. Therefore, assuming $p > .05$ implies the null's truth is a serious mistake, especially with low statistical power.

In contrast to the apparent neglect of power and Type II error, researchers take great pains to avoid making Type I errors, or rejecting a null hypothesis that is actually true, as evidenced by the thought and effort devoted to multiple comparisons (e.g., Miller, 1981; Shaffer, 1995). However, with $\alpha = .05$, Type I error occurs only 5% of the time when the null is actually true—not 5% of all significance tests, but 5% of the significance tests in which the null is actually true. It has been argued that the null hypothesis is almost never true (e.g., Cohen, 1990; Hunter, 1997). For example, in Hunter's (1997) review of 302 meta-analyses, less than 1% reported a meta-analytic effect size of zero. This should not be surprising; based on previous research, theory, or scientific intuition, researchers typically have good reason to expect there will be differences in their data. Thus, only a very small percentage of significance tests that are done are likely to result in Type I error.

Serious misinterpretations of significance levels often reflect multiple misconceptions occurring together. For example, consider a study comparing the means of 36 people who received a treatment and 36 people in a control group. The mean difference was in the predicted direction and significant, $t_{(70)} = 2.09$, $p = .04$. Suppose a second study was done as a replication of Study 1 with 28 people in each condition. The means were in the same direction as Study 1 with $t_{(54)} = 1.84, p = .07$. Some researchers would misinterpret these two results to mean that Study 2 "failed to replicate" Study 1. This misinterpretation reflects many of the problems we have discussed. The $p$-values for Study 1 and Study 2 are quite similar, both indicating that the observed difference between treatment and control would not have been very likely if the null were true. Interpreting these two studies as yielding conflicting results because their $p$-values fall on opposite sides of the .05 line makes the mistake of assuming that a failure to reject the null implies its truth, and it shows a lack of awareness of statistical power. Because Study 2 has fewer observations and, therefore, lower power, it has less chance of achieving significance than does Study 1. In fact, the size of the difference between treatment and control is exactly the same in Studies 1 and 2, (half a standard deviation). By ignoring effect size, overemphasizing significance levels, and not interpreting them correctly, these two studies may be seen to be different when their results are, in fact, remarkably similar.

### 3. *p*-Values Do Not Provide Useful Information

The most serious criticism of null hypothesis significance testing is that little useful information is gained by rejecting a null. The null is rarely a conceptually interesting hypothesis, and, many believe, it is rarely, if ever, true (e.g., Cohen, 1990; Hunter, 1997). If the null is not conceptually interesting, and there is little reason to believe it to be true, what value is there in rejecting it?

Focusing on estimating the size of effects and providing information about how precisely these effects are estimated is a better approach. Rather than a dichotomous rejection or failure to reject the null hypothesis, estimating effect sizes provides more useful information that researchers can use in judgments of an effect's practical and scientific importance. Tukey (1969) noted how social science would benefit from focusing more on the estimation of effect size: "The physical sciences have learned so much by storing up amounts, not just directions. If, for example, elasticity had been confined to 'when you pull on it, it gets longer!' Hooke's law, the elastic limit, plasticity, and many other important topics could not have appeared" (p. 86).

There are many ways to express an effect size (see Rosenthal, 1994). Effect size statistics that are independent of the scale in which raw scores were measured are useful when different measuring instruments have been used. Effect size is frequently represented by standardized differences between condition means or by correlations, and one effect size statistic can be translated to another quite easily. In some form or another, researchers should report the size of the effects obtained in their research.

### B. Alternatives to Null Hypothesis Significance Testing

### 1. Confidence Intervals

The practice of reporting effect sizes along with confidence intervals is widely suggested as a meaningful alternative to the way in which null hypothesis significance testing is currently used (e.g., Abelson, 1997; Cohen, 1990, 1994; Gonzalez, 1994; Loftus, 1996; Scarr, 1997; Tukey, 1991). Reporting confidence intervals instead of *p*-values places emphasis more appropriately on the size of the observed effect. The width of the confidence intervals provides information about how precisely the existing effect can be estimated from the sample data while still providing all of the information that is contained in a significance test of the null hypothesis (i.e., the $p$-value is less than the specified $\alpha$ if zero does not fall within the range of the confidence interval, and greater than the specified $\alpha$ if zero is in that range).

## 2. The Counternull

Reporting confidence intervals around measures of effect size still requires users to specify the $\alpha$ level to use in creating the confidence interval. The counternull statistic (Rosenthal & Rubin, 1994) is an alternative that avoids this requirement. The counternull value of an effect size is the non-null magnitude of the effect size that is supported by just the same amount of evidence as supports the null value of the effect size. For effect sizes that are distributed symmetrically, such as $d$, the counternull effect size can be obtained with the following equation:

$$ES_{(\text{counternull})} = 2ES_{(\text{obtained})} - ES_{(\text{null})}. \tag{1}$$

For example, imagine observing a sample effect size $d = .25$, with $p = .30$. This $p$-value means that for samples from a population in which the null were true (i.e., population value of $d = .00$), only 3 samples in 10 would be likely to produce a sample effect size as large as $d = .25$. The counternull is an alternative population effect size value that is equally likely as the null to produce the observed sample effect size. Here the value of the counternull is $d = .50$, computed by $(2 * .25 - .00)$. If $d = .50$ were the true value of the population effect size, only 3 samples in 10 would be likely to produce a sample effect size as small as $d = .25$. In other words the counternull illustrates that populations with $d = .00$ and $d = .50$ are equally likely to produce a sample effect size $d = .25$. Computing the counternull points out that samples are just as likely to underestimate the actual population value of an effect as overestimate it. For nonsymmetric effect sizes, like the Pearson $r$, the counternull formula can be applied after the effect size has been transformed to a symmetric scale, as Fisher's $z$ transformation does for the Pearson $r$.

## 3. Meta-analysis

A positive development is the increasing emphasis on meta-analytic procedures (e.g., Cooper & Hedges, 1994; Hunter & Schmidt, 1990; Rosenthal, 1991) to cumulate the results of multiple studies. Statistical significance quickly becomes a nonissue with meta-analysis because there is so much more data than in individual studies even the smallest effects will be statistically significant. Therefore, the estimation of effect sizes becomes a more central focus. Effect sizes are also estimated much more precisely because of the large amount of data.

## V. INTERPRETING THE SIZE OF EFFECTS

### A. Problems

It is a mistake to assume that any statistically significant result is also practically important. With large samples, even the smallest effects will be statistically significant, but they may not be large enough to be practically important. Therefore, it is important to interpret effect sizes in terms of their scientific importance.

There are three obstacles to understanding the practical importance of effect sizes. First, the variables used in social science research often are not scaled in a way that has intrinsic meaning. For example, how does one know whether a three-point difference on 50-point paper-and-pencil attitude or personality tests is a large enough effect to be practically important? Second, even experienced researchers do not seem to have a good intuitive sense for the degree of association that is represented by standard effect size measures such as correlations (Oakes, 1982). Third, an effect of a given size may be trivial in one context and quite important in another. The interpretation of effect size requires an exercise of judgment, informed by knowledge of the content area and the practical implications of the result.

### B. Solutions

We suggest three solutions to address these problems. First, use variables that are scaled in a way that has intrinsic meaning whenever possible. People will have a much easier time interpreting results that are presented in a metric they understand well, such as the increase in income associated with additional years of school, the increase in the percentage of children reading at grade level associated with a new teaching method, or the decrease in the probability of a heart attack associated with not smoking.

The binomial effect size display (BESD, Rosenthal & Rubin, 1982) expresses correlation coefficients in terms that are easily understood. For example, Altemeyer (1981) reported that right-wing-authoritarianism was correlated ($r = .43$) with the amount of shock people administered in an experimental learning task. The BESD clarifies the meaning of a correlation of this size by representing two dichotomous groups in the rows and columns of a $2 \times 2$ table. For example, the table's rows might represent high and low right-wing-authoritarianism and the columns might represent people who gave above and below average levels of shock. The BESD assumes each row and column sum to 100. Table 5.3 displays a correlation $r = .43$ with the BESD. The cells on one diagonal of the table contain the value $100(.5 + r/2)$ and the cells on the other diagonal contain the value $100(.5 - r/2)$. Thus, a correlation $r = .43$ would correspond with the difference between 71.5% of high vs. 28.5% of low right-wing-authoritarian people giving more than the average level of shock.

**TABLE 5.3**  Binomial Effect Size Display for Correlation between Right-Wing Authoritarianism and Level of Shock Administered

|  | Give less than average shock level | Give more than average shock level | Total |
|---|---|---|---|
| Low right-wing authoritarianism | 71.5 | 28.5 | 100 |
| High right-wing authoritarianism | 28.5 | 71.5 | 100 |
| Total | 100 | 100 | 200 |

It is also useful to interpret an effect in the context of other known predictors of a similar outcome. The correlation between right-wing-authoritarianism and level of shock could be compared to Milgram (1974), which reported for many different experimental conditions the levels of shock people administered in a similar learning task. From Milgram's data, we computed correlations $(r)$ that reflect the difference in the amount of shock administered between an experimental condition and Milgram's baseline condition (in the baseline condition, conducted at Yale University, a single participant administered shock at a level determined by an experimenter to a victim in another room whose voice could be heard). The relation between right-wing-authoritarianism and level of shock $(r = .43)$ is similar in size to the difference between the baseline and a condition that required participants physically to touch the victim's hand to a metal plate to deliver shock $(r = .37)$. Right-wing-authoritarianism predicted shock level substantially better than the difference between whether the research was conducted at Yale or in Bridgeport, Connecticut $(r = .20)$, but not as well as the difference between whether the participant or the experimenter chose the level of shock to administer $(r = .81)$.

### C. How Large an Effect Is Important?

Reporting effects in metrics that have intrinsic meaning, using the BESD to illustrate the size of effects, and reporting effects in the context of other known predictors are all useful for understanding an effect size. However, the question of how large an effect must be to be considered important needs to be addressed. There is no standard answer to this question. Effect sizes must be interpreted in the context of the specific research and the costs and benefits associated with the result. Some effects that generally would be considered "small" are practically important (Abelson, 1985; Prentice & Miller, 1992; Rosenthal, 1990). For example, the correlation between taking aspirin and having a heart attack $(r = .034$; Steering Committee of the Physicians' Health Study Research Group, 1988) is an important effect because the outcome is so consequential and the cost of implementing a daily aspirin treatment is trivial. In other cases, such as when assessing the reliability of a measuring instrument, a correlation 10 times larger (i.e., $r = .34$) may be considered too small to be of practical importance.

## VI. UNDERSTANDING ASSUMPTIONS

### A. Between-Subjects Comparisons

### 1. Normality and Homogeneity of Variance

Many statistical procedures are intended to be used when data have certain properties. For example, $F$ and $t$ statistics for between-subjects comparisons of means assume normally distributed data, homogenous variance across groups, and that observations are independent. Lacking these assumptions, obtained $p$-values may be inaccurate, sometimes drastically so. To use any statistic thoughtfully, researchers should be aware of the assumptions underlying that statistic, the extent to which their data violate those assumptions, and the statistic's robustness against violations of those assumptions.

The accuracy of $p$-values associated with $F$s and $t$s are not especially sensitive to deviations from normality, even fairly large deviations, unless sample sizes are small (i.e., $n < 30$). $F$s and $t$s are generally robust to violations of homogeneity of variance if the sample sizes are equal. Under many conditions, violations of these two assumptions are not problematic for $p$-value accuracy (Glass, Peckham, & Sanders, 1972). When there is reason to be concerned, data transformations effectively produce more homogenous variability and more symmetrical distributions. If transformation is not an effective remedy, nonparametric tests that have different underlying assumptions can be used. Judd, McClelland, and Culhane (1995) provides an in-depth review of these assumptions.

### 2. Independence

Violations of the independence assumption are more consequential (Kenny & Judd, 1986) and more often overlooked when analyzing data. When observations are collected in groups, the observations within a group may be more closely related to each other than they are to observations from other groups. Nonindependence may occur for many reasons. For example, sometimes treatments are administered to groups, such as classrooms, or data are collected from groups of people in different experimental sessions or by different experimenters. Observations from people within the same classroom, the same experimental session, or the same experimenter may not be independent of each other, a problem referred to as "hidden nesting."

Consider a study conducted in a school in which five classrooms of 10 students each were assigned to the treatment group and five classrooms of 10 students each were assigned to the control group. The left and right sides of Table 5.4 show, respectively, the analyses that would result if the fact that children came from different classrooms were ignored and considered. When classroom is ignored (i.e., the left side of Table 5.4), computing the $F$ for the main effect of treatment is straightforward, $MS_{\text{Treatment}}/MS_{\text{Error}} = F_{(1, 98)} = 25.67, p = .0000019, r = .456$. The obtained significance level assumes observations are independent and is not correct if that assumption is erroneous.

The ANOVA on the right side of Table 5.4 accounts for the fact that children come from different classrooms. With this information, the intraclass correlation ($r_{\text{ic}}$) can be computed, which assesses the degree of nonindependence within the classrooms:

$$r_{\text{ic}} = \frac{MS_{\text{Class(Treatment)}} - MS_{\text{Error}}}{MS_{\text{Class(Treatment)}} + (c - k)\,MS_{\text{Error}}} = \frac{30.73 - 4.29}{30.73 + (10 - 2)\,4.29} = .406, \quad (2)$$

with $c$ being the total number of classrooms distributed over $k$ levels of the treatment variable. Computing the $F$ for the effect of classroom ($MS_{\text{Class(Treatment)}}/MS_{\text{Error}} = F_{(8, 90)} = 7.16, p = .0000003$, also suggests that in this example there is a substantial classroom effect. Because of the lack of independence within classrooms, treating classroom as the unit of analysis will obtain a more accurate result. Thus, the classroom variance should be the error term when computing $F$ for the main effect of treatment, $MS_{\text{Treatment}}/MS_{\text{Class(Treatment)}} = F_{(1,8)} = 5.39, p = .049, r = .634$. Any factor that might create nonindependence should be measured and included in analyses. If examination of $r_{\text{ic}}$ suggests that observations within groups are independent, the individual and group sources of variance can be pooled (Green & Tukey, 1960).

Ignoring hidden structure in data can have a more insidious consequence, though it is not necessarily a problem of independence. Between-group and within-group effects may possibly "go in different directions."

**TABLE 5.4** Tables of Variance for Analyses That Ignore and Include the Effect of Classroom[a]

| I. Ignoring classroom | | | | II. Including classroom | | | |
|---|---|---|---|---|---|---|---|
| Source | SS | df | MS | Source | SS | df | MS |
| Treatment | 165.54 | 1 | 165.54 | Treatment | 165.54 | 1 | 165.54 |
| Error | 632.07 | 98 | 6.45 | Class (Treatment) | 245.82 | 8 | 30.73 |
| | | | | Error | 386.26 | 90 | 4.29 |

[a] SS, Sum of squares; MS, mean square.

How analyses account for the structure in the data may seriously affect results. Consider a study that found a weak negative relation between anxiety and alcohol consumption ($r = -.12$ with the 95% confidence interval ranging from $r = -.27$ to $r = .04$) in a group of 160 male college students. In these data, anxious people drank somewhat less alcohol. However, when it became known that the data consisted of 20 men each from 8 different fraternities, it made sense to look at the relation between anxiety and drinking separately at the between- and within-group levels.

Table 5.5 shows the average levels of drinking and anxiety for each fraternity and the correlation between these two variables for the men in each fraternity. Across groups there is a very strong negative relation between average level of anxiety and average drinking, $r = -.90$. Less anxious groups tend to drink more. However, within groups, there is a generally positive relation between anxiety and drinking. Within groups, the more anxious group members tend to drink more (median within group $r = .28$). In this case, failure to account for this structure in the data produces a substantively different answer. This analysis can be handled many ways if the unit of analysis of interest is the individual; including a regression analysis that codes for fraternity membership, subtracting the fraternity mean out of each anxiety and drinking score before correlating anxiety and drinking, or computing correlations separately within each fraternity and using meta-analytic procedures to combine the data. If the group is the unit of interest, of course, we simply correlate the means obtained for the set of groups.

## B. Within-Subjects Comparisons

Omnibus $F$-tests for repeated measures ANOVA assume sphericity (Huynh & Feldt, 1970), and are not robust to violations of this assumption. However,

**TABLE 5.5    Average Anxiety, Average Drinking, and the Anxiety–Drinking Correlation for Eight Fraternities[a]**

| Fraternity | Mean anxiety | Mean drinking | $r_{\text{anx-drink}}$ |
|---|---|---|---|
| A | 21.40 | 9.90 | .096 |
| B | 23.85 | 10.45 | .060 |
| C | 18.25 | 10.80 | .245 |
| D | 18.55 | 13.45 | .447 |
| E | 17.60 | 14.50 | −.054 |
| F | 15.75 | 15.85 | .310 |
| G | 13.90 | 18.15 | .548 |
| H | 14.75 | 18.40 | .378 |

[a] $n = 20$ per fraternity.

computing contrasts for the pattern in the repeated measures variable and using a contrast-specific error term avoids this assumption. This can be done quite easily by using individual contrast scores, sometimes called $L$-scores (Hallahan & Rosenthal, 1996a; Rosenthal & Rosnow, 1985). These scores, which are computed for each individual participant, assess the correspondence between an individual's responses across the levels of a repeated measures variable and a specified prediction.

For example, consider an experiment in which participants were measured three times and theory predicted that scores would increase linearly from *Time 1* to *Time 2* to *Time 3*. Contrast weights of $-1$, $0$, and $+1$, respectively, express this prediction. The following equation computes individual contrast scores ($L$):

$$L = \Sigma \, \lambda x, \qquad (3)$$

in which, $\lambda$ = the contrast weights expressing the prediction and $x$ = the person's scores at each level of the repeated measures variable. Thus, for a person with scores of *3, 5,* and *6* at *Times 1, 2,* and *3, L* = $(-1)(3)$ + $(0)(5)$ + $(+1)(6)$ = *3.*

If the null were true (i.e., if there were no relation between the predicted and actual patterns in the data) then the expected value of $L = 0$. Examining the extent to which the average $L$-score differs from zero assesses the strength of the relation between the data and the prediction (with positive and negative $L$-scores indicating, respectively, positive and negative relations between data and prediction). Comparing $L$-scores across different levels of between-subjects variables provides information about moderating variables.

Using $L$-scores to compute contrasts in repeated measures data addresses a more specific, conceptually meaningful question than does an omnibus comparison. This approach eliminates concern over the sphericity assumption as well. Also, recasting a repeated measures analysis as a comparison of $L$-scores requires substantially fewer computer resources, which may aid the analysis of complex designs.

## VII. PROCESS

Good data analysis requires not only a thorough understanding of statistical procedures, but also that researchers approach their data in a way that allows them to perceive the scientifically meaningful patterns in their data. Questions about data analysis are inextricably linked to theory, measurement, and design (see chapter 2 by Hetherington and chapter 3 by Dawis, this volume). Statistical procedures, no matter how sophisticated, cannot overcome fundamental problems in these areas. This section presents general advice about how to connect data analysis more closely to these basic scientific objectives.

## A. Formulate Data Analysis Plan Prior to Data Collection

It is useful to formulate a data analysis plan in the early stages of a research project. First, the plan should list the conceptual questions to be addressed, prioritized in order of theoretical and practical importance. Next, the plan should specify what analyses will address these hypotheses. Each analysis should be planned in detail, including precisely specifying contrast weights that will be used. Review the plan to ascertain that the proposed analyses actually address the intended hypotheses. This process may motivate important changes in the proposed research design. For example, thinking carefully about how to quantify a conceptual hypothesis might suggest conditions that need to be added to the design or might reveal conditions that could be dropped because they are not relevant to the specified hypotheses.

Researchers inevitably face trade-offs because of constraints on time, participant availability, and material resources. For example, a researcher might have the option of giving the short form of two different measures or the more reliable long form of one. A prioritized list of hypotheses is useful for resolving such trade-offs. By recognizing which of their hypotheses are most important, researchers can strategically concentrate their resources to obtain better information about their most important questions at the expense of peripheral ones.

In planning, researchers may confront the problem of deciding among possible ways to analyze their data (see Abelson, 1995). There may not be a principled reason to choose one procedure over another; sometimes there is not a single "best" option. Excessive deliberation, or "analysis paralysis," is unproductive. Instead, plan to compute each possible alternative. If the results converge to a similar conclusion, then the question of which procedure to use does not warrant further consideration. A reassuring footnote can report that different analyses obtained similar results. When different procedures yield divergent results, researchers can view the strongest and weakest results as tentative upper and lower bounds on the size of the true effect, but also need to devote more effort to understanding the cause of these differences, perhaps by consulting someone with more technical expertise.

If planning to compute significance tests, researchers should consult power tables (Cohen, 1988) to determine the sample size that will provide an acceptable level of power. If sample size is constrained because participants are costly or difficult to obtain, statistical power can be increased in many ways (Hallahan & Rosenthal, 1996b). For example, contrasts provide more power than omnibus comparisons. Design changes also can increase power, such as the use of more reliable measuring instruments, stronger treatments, blocking variables, or within-subjects designs. For any given number of research participants, the relative number allocated to each condition also can affect power; equal $n$ is not always the most efficient design (McClelland, 1997).

## B. Use Statistical Consultants Appropriately

The improper use of statistical consultants can be problematic. Most importantly, a researcher absolutely never should hand over data to a consultant with the idea that "the consultant will analyze the data for me." Such "outsourcing" removes the data analysis from the larger context of the research project. However knowledgeable about statistical procedures, statistical consultants who do not know the substantive content area are probably not qualified to make informed judgments about the data. Investigators who do not work closely with their data lose a precious opportunity to see unexpected trends that might signal unanticipated problems or important serendipitous findings. Researchers may fail to detect a consultant's errors and misinterpretations if too far removed from the data analysis. Researcher, not consultants, are ultimately responsible for the results that are reported, and therefore, need to understand their data thoroughly.

Statistical consultants should serve as a guide to an investigator's judgment rather than be a substitute for it. We recommend that researchers plan the analyses of their own data and then, if desirable, have a statistical consultant review their plan. Consultants can answer specific technical questions, such as the consequences of violating certain assumptions; help decide among alternative analyses; or suggest strategies for analyzing data that are not well suited to well-known statistical procedures. Involve consultants early in the research process, even before data have been collected, and allow time for consultants to learn about the content area. Be wary of consultants who do not ask questions to educate themselves about the content area.

## C. Use Statistical Software Appropriately

Although statistical software has brought great benefits in the speed, efficiency, and accuracy of data analysis, it also has brought its own set of problems. With a few mouse clicks through easy-to-use menus, users can generate output for extremely complex analyses without understanding their meaning. Without understanding what is being done and whether it is appropriate for one's research hypotheses, churning out complex analyses is useless; even worse, it is damaging to the scientific goals of the research.

Researchers need to understand how their software computes different statistics. For example, with unequal $n$ per condition, there are different ways to compute sums of squares that can produce substantially different results when there are large disparities in sample size. Some software, such as SAS, explicitly indicates which sums of squares are being reported; however, most software does not. If the manual does not provide information about the computation method, create a small sample dataset that can be analyzed easily by hand. Comparing hand-calculated results to computer output should reveal the software's computation method.

Be wary of procedures that "make decisions" about data analysis. Examples include procedures to determine how many predictors to include in a stepwise regression model, or how many factors represent the "best" solution in factor analysis. These arbitrary algorithms do not necessarily produce the best understanding of the patterns in a specific set of data. For example, Henderson and Vellerman (1981) showed that a stepwise regression procedure using a software's defaults produced a very different regression model than did an iterative process that employed exploration of the data and human judgment.

### D. Focus on Scientific Understanding

When working with empirical data, remain focused on the important scientific questions that inspired the collection of those data. View data analysis as a means to conceptual understanding, never as an end in itself. Roger Brown's (1989) self-described "laborious, low-tech, minimally mathematical research style" (p. 49) epitomizes this approach to data analysis:

> I have an almost Talmudic taste for poring over data, for hands-on solitary contact unmediated by machine, involving no human interaction, uncommitted to particular statistical analyses—involving nothing but the free exercise of the principles of induction.... The strength of my research method is that it aims at understanding, not at $p$ values, and understanding is the more demanding test. (pp. 49–50)

The product of Brown's method was a body of research remarkable for the depth and breadth of its scientific contribution and its good record of replicability.

### VIII. CONCLUSION

This chapter has presented many ideas that may be useful to researchers for understanding their data. Graphical displays and exploratory data analysis are useful for representing the full pattern in a dataset, for checking assumptions, and for identifying outliers and data entry errors. Contrasts examine specific, conceptually meaningful trends, unlike omnibus tests that address less specific and often less interesting questions. When using significance tests, strive to (a) understand what information is (and is not) provided by the $p$-value, (b) understand the implications of statistical power, and (c) avoid common misconceptions about the interpretation of significance levels. Researchers should emphasize the estimation of effect sizes and report them accompanied by confidence intervals or counternulls, and try to interpret effects in terms of their practical importance. Facilitate the conceptual interpretation of effect sizes by using variables that are intrinsically meaningful, displaying correlations by means of the binomial effect size display, and reporting effect sizes in the context of other known predictors. Check that the assumptions of statistical procedures are met and understand how robust a statistic is to violation of its underlying assumptions. More generally, approach data analysis in a way that is primarily focused on conceptual understanding and make decisions about data analysis that are informed by an understanding of the content area.

## REFERENCES

Abelson, R. P. (1985). A variance explanation paradox: When a little is a lot. *Psychological Bulletin, 97,* 129–133.

Abelson, R. P. (1995). *Statistics as principled argument.* Hillsdale, NJ: Erlbaum.

Abelson, R. P. (1997). On the surprising longevity of flogged horses: Why there is a case for the significance test. *Psychological Science, 8,* 12–15.

Altemeyer, R. A. (1981). *Right wing authoritarianism.* Winnipeg: University of Manitoba Press.

Anscombe, F. J. (1973). Graphs in statistical analysis. *The American Statistician, 27,* 17–21.

Behrens, J. T. (1997). Principles and procedures of exploratory data analysis. *Psychological Methods, 2,* 131–160.

Brown, R. W. (1989). Roger Brown. In G. Lindzey (Ed.), *A history of psychology in autobiography* (Vol. 8, pp. 36–60). Stanford, CA: Stanford University Press.

Cleveland, W. S. (1993). *Visualizing data.* Summit, NJ: Hobart Press.

Cohen, J. (1962). The statistical power of abnormal-social psychological research: A review. *Journal of Abnormal and Social Psychology, 65,* 145–153.

Cohen, J. (1988). *Statistical power analysis for the behavioral sciences* (2nd ed.). Hillsdale, NJ: Erlbaum.

Cohen, J. (1990). Things I have learned so far. *American Psychologist, 45,* 1304–1312.

Cohen, J. (1994). The earth is round ($p < .05$). *American Psychologist, 49,* 997–1003.

Cooper, H., & Hedges, L. V. (Eds.) (1994). *The handbook of research synthesis.* New York: Russell Sage Foundation.

Estes, W. K. (1997). Significance testing in psychological research: Some persisting issues. *Psychological Science, 8,* 18–20.

Gigerenzer, G. (1993). The superego, the ego, and the id in statistical reasoning. In Keren, G. & Lewis, C. (Eds.), *A handbook for data analysis in the behavioral sciences: Vol. 1, Methodological issues* (pp. 311–339). Hillsdale, NJ: Erlbaum.

Glass, G. V., Peckham, P. D., & Sanders, J. R. (1972). Consequences of failure to meet assumptions underlying analysis of variance and covariance. *Review of Educational Research, 42,* 237–288.

Gonzalez, R. (1994). The statistics ritual in psychological research. *Psychological Science, 6,* 321, 325–328.

Green, B. F., Jr., & Tukey, J. W. (1960). Complex analysis of variance: General problems. *Psychometrika, 25,* 127–152.

Hallahan, M., & Rosenthal, R. (1996a). Contrast analysis in education research. *Journal of Research in Education, 6,* 3–17.

Hallahan, M., & Rosenthal, R. (1996b). Statistical power: Concepts, procedures, and applications. *Behaviour Research and Therapy, 34*, 489–499.

Hays, W. L. (1994). *Statistics* (5th ed.). New York: Harcourt Brace.

Henderson, H. V., & Vellerman, P. F. (1981). Building multiple regression models interactively. *Biometrics, 37*, 391–411.

Hoaglin, D. C., Mosteller, F., & Tukey, J. W. (Eds.) (1983). *Understanding robust and exploratory data analysis.* New York: Wiley.

Hunter, J. E. (1997). Needed: A ban on the significance test. *Psychological Science, 8*, 3–7.

Hunter, J. E., & Schmidt, F. L. (1990). *Methods of meta-analysis: Correcting error and bias in research findings.* Newbury Park, CA: Sage.

Huynh, H., & Feldt, L. S. (1970). Conditions under which mean square ratios in repeated measurements designs have exact $F$-distributions. *Journal of the American Statistical Association, 65*, 1582–1589.

Judd, C. M., McClelland, G. H., & Culhane, S. E. (1995). Data analysis: Continuing issues in the everyday analysis of psychological data. *Annual Review of Psychology, 46*, 433–465.

Kenny, D. A. (1994). *Interpersonal perception.* New York: Guilford.

Kenny, D. A., & Judd, C. M. (1986). Consequences of violating the independence assumption in analysis of variance. *Psychological Bulletin, 99*, 422–431.

Loftus, G. R. (1996). Psychology will be a much better science when we change the way we analyze data. *Current Directions in Psychological Science, 5*, 161–171.

Loftus, G. R., & Masson, M. E. J. (1994). Using confidence intervals in within-subjects designs. *Psychonomic Bulletin and Review, 1*, 476–490.

McClelland, G. H. (1997). Optimal design in psychological research. *Psychological Methods, 2*, 3–19.

Milgram, S. (1974). *Obedience to authority: An experimental view.* New York: Harper and Row.

Miller, R. G., Jr. (1981). *Simultaneous statistical inference.* New York: Springer-Verlag.

Oakes, M. (1982). Intuiting the strength of association from a correlation coefficient. *British Journal of Psychology, 73*, 51–56.

Oakes, M. (1986). *Statistical inference: A commentary for the social and behavioral Sciences.* New York: Wiley.

Prentice, D. A., & Miller, D. T. (1992). When small effects are impressive. *Psychological Bulletin, 112*, 160–164.

Rosenthal, R. (1990). How are we doing in soft psychology? *American Psychologist, 45*, 775–777.

Rosenthal, R. (1991). *Meta-analytic procedures for social research* (rev. ed.). Newbury Park, CA: Sage.

Rosenthal, R. (1994). Parametric measures of effect size. In H. Cooper & L. V. Hedges (Eds.), *The handbook of research synthesis* (pp. 231–244). New York: Russell Sage.

Rosenthal, R., & Gaito, J. (1963). The interpretation of levels of significance by psychological researchers. *Journal of Psychology, 55*, 33–38.

Rosenthal, R., & Rosnow, R. L. (1985). *Contrast analysis: Focused comparisons in the analysis of variance.* New York: Cambridge University Press.

Rosenthal, R., Rosnow, R. L., & Rubin, D. B. (1999). *Contrasts and effect sizes in behavioral research: A correlational approach.* New York: Cambridge University Press.

Rosenthal, R., & Rubin, D. B. (1982). A simple general purpose display of magnitude of experimental effect. *Journal of Educational Psychology, 74*, 166–169.

Rosenthal, R., & Rubin, D. B. (1994). The counternull value of an effect size: A new statistic. *Psychological Science, 5*, 329–334.

Scarr, S. (1997). Rules of evidence: A large context for the statistical debate. *Psychological Science, 8*, 16–17.

Schmidt, F. L. (1996). Statistical significance testing and cumulative knowledge in psychology: Implications for training researchers. *Psychological Methods, 1*, 115–129.

Sedlmeier, P., & Gigerenzer, G. (1989). Do studies of statistical power have an effect on the power of studies? *Psychological Bulletin, 105,* 309–316.

Shaffer, J. P. (1995). Multiple hypothesis testing. *Annual Review of Psychology, 46,* 561–584.

Snedecor, G. W., & Cochran, W. G. (1989). *Statistical methods* (8th ed.). Ames, IA: Iowa State University Press.

Steering Committee of the Physicians' Health Study Research Group (1988). Preliminary report: Findings from the aspirin component of the ongoing physicians' health study. *New England Journal of Medicine, 318,* 262–264.

Tufte, E. R. (1983). *The visual display of quantitative information.* Cheshire, CT: Graphics Press.

Tufte, E. R. (1990). *Envisioning information.* Cheshire, CT: Graphics Press.

Tufte, E. R. (1997). *Visual explanations: Images and quantities, evidence and narrative.* Cheshire, CT: Graphics Press.

Tukey, J. W. (1969). Analyzing data: Sanctification or detective work? *American Psychologist, 24,* 83–91.

Tukey, J. W. (1977). *Exploratory data analysis.* Reading, MA: Addison-Wesley.

Tukey, J. W. (1991). The philosophy of multiple comparisons. *Statistical Science, 6,* 110–116.

Wainer, H., & Thissen, D. (1976). When jackknifing fails (or does it?). *Psychometrika, 41,* 9–34.

Zuckerman, M., Hodgins, H. S., Zuckerman, A., & Rosenthal, R. (1993). Contemporary issues in the analysis of data: A survey of 551 psychologists. *Psychological Science, 4,* 49–53.

# PART II

# MULTIVARIATE ANALYSES

# 6

# ISSUES IN THE USE AND APPLICATION OF MULTIPLE REGRESSION ANALYSIS

ANRE VENTER AND SCOTT E. MAXWELL

*Department of Psychology, University of Notre Dame, Notre Dame, Indiana*

It is widely acknowledged that statistical methods are best used within a broader theoretical context and appropriate research design. Statistical methods, in and of themselves, are not concerned with the validity of the underlying assumptions, how the data were gathered, or whether the results they provide answer the questions the researchers are asking. These issues should be addressed during the design phase of the research process. Although no one complains about the ease by which modern computer packages perform sophisticated analyses, it is important to note that one byproduct has been an increase in the thoughtless and incorrect application of statistical analyses to inappropriate data.

Given the importance of theory in the appropriate use of statistical techniques, this chapter attempts to blend the underlying theory (and assumptions) with the practical aspects of multiple regression. Throughout we assume among readers a basic working knowledge of multiple regression. Our intent is to highlight several critical but often underappreciated issues that arise in the use and interpretation of multiple regression analyses. Specifically, we review (a) the theoretical basis for the use of multiple regression, (b) the underlying assumptions and how their validity may be assessed, (c) transformations, (d) interaction and moderator effects, (e)

*Handbook of Applied Multivariate Statistics and Mathematical Modeling*

sample size requirements, and finally, (f) how missing data can be approached.

## I. OVERVIEW OF MULTIPLE REGRESSION

Generally speaking, multiple regression examines the relation between a single dependent (or criterion) variable and a set of independent (or predictor) variables to best represent the relation in the population. The technique is used for both predictive and explanatory purposes within experimental or nonexperimental designs. Setting aside for the moment differences in the design phase and the incompatibility between the stepwise method of entering variables and the purpose of explanation, the mechanics of the analyses are the same. Ultimately, the differences lie in the interpretation of the results.

### A. The Prediction–Explanation Dichotomy

Much ambiguity and controversy surrounds the debate on the explanation–prediction dichotomy. Some philosophers of science see these as separate and distinct concepts (Scriven, 1959), whereas others argue that they are structurally and logically identical (Hempel, 1965). Although this philosophical debate has been complex and controversial, it is much simpler to delineate research intended primarily to fulfill one or the other purpose. Predictive research emphasizes practical applications, whereas explanatory research focuses on achieving a theoretical understanding of the phenomenon of interest.

### 1. Multiple Regression for Prediction

Industrial and organizational psychologists attempt to predict job performance (Campbell, 1990), educational psychologists use Scholastic Aptitude Test (SAT) or Graduate Record Examination (GRE) scores to predict later academic performance (Sternberg & Williams, 1997), and counseling psychologists use communication factors to predict marital satisfaction (Smith, Vivian, & O'Leary, 1990). The goals here are primarily predictive with little to no emphasis given to understanding or explicating underlying relations between the variables. The critical issue is identifying the variables most useful in predicting future success; in other words, those variables that maximize the predictive power of the model. Typically this is done via the stepwise or all-subsets methods of entry. Consequently, all that can be said about the final regression model is simply that it contains the set of variables that best predicts the criterion given the specific design constraints. As a result the regression coefficients cannot be interpreted as indices of the effects of the predictors on the criterion variable (Pedhazur, 1982).

Prediction uses the regression weights obtained in a sample to predict the criterion in the population (or in other samples drawn from the population) in which the criterion scores are unknown (Darlington, 1990; Raju, Bilgic, Edwards, & Fleer, 1997). From this perspective it does not matter why the predictor works; whether it is a cause or symptom of the criterion, or is simply correlated with other causal factors is irrelevant. All prediction does is identify the most useful predictors. From a broader perspective, however, these issues may be relevant even for prediction, because they can have important implications for generalizability.

## 2. Multiple Regression for Explanation

Explanatory research tries to explain the relations between variables and emphasizes understanding the phenomena of interest. Thus, one goal might be to explain academic achievement in college as opposed to predicting future academic performance. Here the theory describing the relations between the variables of interest should determine the choice of variables, the analytic approach, and the interpretation of the results. Explanatory research is "inconceivable without theory" (Pedhazur, 1982, p. 174).

Typically this research measures the variability in the dependent variable accounted for by the independent variable (i.e., the multiple $R$-squared). Variance partitioning (or hierarchical regression analysis), partitions the multiple $R^2$ into portions attributable to the different independent variables separately. Although the order of entry of regressors into the equation and their intercorrelation can influence the results (Pedhazur, 1982), an independent variable's effect on the dependent variable controlling for the remaining independent variables can still be evaluated within the framework of a regression model.

Another multiple regression analytic approach involves the use of regression coefficients as indices of the effects of independent variables upon dependent variables. Here the regression coefficient obtained when a dependent variable is regressed onto a set of independent variables indicates the expected change in the dependent variable associated with a unit change in a given independent variable controlling for the other independent variables. Ideally, controlling for the effects of the other independent variables allows us to interpret the regression weights as partial slopes for the specific independent variables at a fixed value of the remaining independent variables. However, this interpretation holds only when a set of restrictive assumptions (to be presented shortly) is valid.

## 3. Experimental versus Nonexperimental Designs

It is important to realize that the research design within which the multiple regression is employed does not influence the actual mechanics of the analyses. Pedhazur (1982) notes that the major differences arise in the manner in which the regression equations obtained within each design are interpreted. Within the experimental paradigm, one can reasonably conclude that implementing certain changes in an independent variable would produce a desired effect in a dependent variable. However, the situation is much more ambiguous and complex within the nonexperimental paradigm. Here regression equations simply reflect the average relation between the dependent variable and the set of independent variables; inferences about the causal nature of the underlying relation are inherently more difficult than in experimental designs.

### 4. Multiple $R$-Squared, Adjusted $R$-Squared, and Shrunken $R$-Squared

In both predictive and explanatory research, multiple regression estimates how well a set of predictors correlates with the criterion in the population. There is, however, a subtle yet critical difference between the two purposes. Predictive research evaluates how well the sample-specific regression weights predict the criterion in the population as a whole or in other random samples drawn from the population. Explanatory research, on the other hand, focuses on the true predictive power of the variables in the population. In other words, how well do the actual population regression weights predict the criterion in the population?

Most researchers know that (a) the multiple $R^2$ calculated in a sample overestimates the true multiple $R^2$ in the population, and (b) adding predictor variables necessarily increases the sample multiple $R$. Given that the size of the correlation between a criterion and predictor variable is indicative of predictive validity (Nunnally & Bernstein, 1994), it is tempting to assume that more predictors are always "better." Many researchers might respond that the adjusted $R^2$ provided by most statistical packages avoids the problems of overestimation and excessive predictors as, after all, the adjusted $R^2$ squared is always lower than the sample multiple $R^2$.

It turns out that more predictors are not always better and that the purpose of the research determines which statistic should be evaluated. Recall that prediction evaluates how well the sample-specific regression weights predict the criterion in the population as a whole or in other random samples drawn from the population. In this case, the parameter of interest should be the population cross-validity parameter or shrunken $R$ (Darlington, 1990), and not the multiple $R^2$ or adjusted $R^2$. Conversely, explanatory research estimates the true predictive power of the variables in the population, which is measured by the adjusted $R^2$. This statistic estimates the multiple correlation based on the optimal population weights and thus evaluates the accuracy of the set of predictor variables by allowing the weights to be recalculated in the population. This provides a "more nearly unbiased" estimate of the population $R^2$ (Darlington, 1990, p. 121).

The distinction between the adjusted and shrunken $R$ is subtle yet important. Although both estimate how well a set of predictors correlate with the criterion in the population, they differ in their weighting of the predictors. The shrunken $R$ estimates the correlation based on the sample weights; the adjusted $R$ estimates the correlation based on the optimal population weights. Why is this distinction important? Suppose that in a sample with $n = 36$, a regression model with five predictors produces a multiple $R$ of .55 and thus an $R$-squared of .30. Darlington (1990) provides a series of equations that allow for the estimation of the shrunken $R$. The final equation in the series is

$$RS = \sqrt{\frac{(N - P - 3)RHO4 + ARS}{(N - 2P + 2)ARS + P}} \, , \qquad (1)$$

where $P$ is the number of predictors, and $RS$ estimates the shrunken $R$, $ARS$ the adjusted $R$-squared, and $RHO4$ the population $R^4$. Although space limitations prevent the presentation of the first two equations in the series (Darlington, 1990, p. 161), in this case we find that ARS = .18 and RHO4 = .02. Substituting these values into Equation 1 produces a shrunken $R$ of .26 and thus a shrunken $R$-squared of .07.

The difference between the shrunken and multiple $R$-squared is termed validity shrinkage (Darlington, 1990) which, in this example, reduces the space proportion of variance in the criterion variable accounted for by the predictor variables from 30% to 7%. Although the adjusted $R$-squared (.18) does adequately estimate the squared correlation based on the optimal population weights, it does not make enough of an adjustment for the purpose of prediction in another sample. Clearly our interpretation about the usefulness of the regression model changes depending on which estimate is examined. Thus it is critical that the correct statistic be examined, and for prediction this is the shrunken $R$-squared. This parameter can be assessed in a number of ways, ranging from the traditional empirical methods (cross-validation), formula-based methods, to the use of equal weights. The alternative methods become especially attractive when small sample sizes prevent utilization of the traditional cross-validation method. Readers are referred to Raju, Bilgic, Edwards, and Fleer (1997) for a detailed overview of these methods. Note that sample size is one factor that may influence the amount of shrinkage, an issue that is discussed later in the sample size section.

### 5. Restriction of Range

Another point of practical value concerns the issue of restriction of range in selected samples. For example, consider the prediction of success in graduate school (in terms of GPA) for those students already admitted. If, in this case, GPA is predicted from those variables upon which admission was based (e.g., undergraduate GPA and GRE scores), the lack of variability in the predictors may distort the true relationships between the variables. Prediction would be optimized if students were admitted randomly, their performance measured, and then predictors are identified and evaluated, but such a sequence is frequently impractical and possibly unethical.

## B. Conclusion

This section provided a brief overview of the two major purposes of multiple regression and illustrated that, ultimately, the major difference in answering the two types of questions lies in the design of the project and the interpretation of the results. As a result, unless explicitly stated otherwise, the technical discussions about the mechanics of multiple regression that follow do not distinguish between the two purposes.

## II. ASSUMPTIONS AND ROBUSTNESS

Darlington (1990) lists the regression assumptions as follows: (a) random sampling, (b) linearity, (c) homoscedasticity, and (d) normality. Although random sampling is generally assumed by any inferential statistical procedure, the other assumptions concern the nature of the distribution of criterion values at each vector of predictor values in the population (hereafter referred to as the conditional $Y$ distributions). These assumptions are presented here in terms of the way they influence the various uses of multiple regression. In other words, what assumptions are necessary when multiple regression is used as a descriptive technique, an inferential technique, or to enable causal inferences? Before describing these assumptions it is necessary to present a theoretical framework for the discussion.

Judd and McClelland (1989) show that data analysis techniques in general, and multiple regression in particular, can be conceptualized in the following manner

$$\text{DATA} = \text{MODEL} + \text{ERROR} \tag{2}$$

The MODEL component reflects how well the selected regression model mirrors the true functional relation between the variables in the population. The ERROR component measures the amount by which the MODEL fails to represent the data accurately. In other words, it contains any variance not explained by the MODEL term.

Multiple regression can be viewed as a methodology that establishes a functional relation between $Y$ and a vector of $X$'s. At each fixed value of this vector of $X$'s $Y$ has a certain probability distribution. Multiple regression models some parameter of this $Y$ distribution, most often the mean of the distribution $(\mu Y)$. Although $\mu Y$ is often the appropriate parameter, instances may arise when the conditional distributions of $Y$ (i.e., the distributions of $Y$ at each value of $X$) are sufficiently skewed that the mean is not suitable. This raises two issues. First, researchers should know their data well enough to be aware of violations of the normality assumption. Second, options are available should the conditional $Y$ distributions be sufficiently skewed to preclude the use of the mean as the parameter of interest. Possible strategies include transforming Y in an attempt to normalize the distribution or selecting alternative techniques that model other parameters such as, for example, the trimmed mean or median. Readers are referred to the section in this chapter dealing with transformations and re-expressions of the regression model. The remainder of this section assumes that $\mu Y$ has been chosen as the parameter of interest and discusses the assumptions and robustness of multiple regression for the MODEL and ERROR components separately, as well as those required for the drawing of causal inferences.

## A. Accuracy of Description

At issue here is whether the relation between the dependent and independent variables is being described accurately. Does the regression model match the nature of the $X–Y$ relation in the population? This question concerns the MODEL component (Judd & McClelland, 1989) and is relevant whether the goal is prediction or explanation.

The assumption of linearity becomes relevant in this realm. The relation between the $X$ variables and $Y$ is assumed to be linear in nature, such that if we were to plot the population mean values of $Y$ for each value of $X$, they would fall in a straight line or plane. The interpretation of the regression coefficients as partial slopes controlling for the remaining independent variables depends on this assumption. Note that this meaning of linearity does not preclude the use of multiple regression when the relation between the vector of $X$s and $Y$ is nonlinear. Nonlinearity can be addressed in multiple regression models through the use of transformations, which may produce an expression of either or both the $X$ variables and $Y$ such that a relation that is originally nonlinear is re-expressed in a linear fashion. Although polynomial terms may be applied a priori to deal with any curvilinearity indicated by theory, such terms are most often fit post hoc as a result of regression diagnostics.

Linearity can have an entirely different meaning in the context of multiple regression analysis. The fundamental issue here is whether the functional relation between the vector of $X$s and $Y$ in the population is "linear in the parameters." Although there are instances in psychology where specific theories exist suggesting the need for a particular mathematical nonlinear function (for example, the model describing the learning curve, Estes, 1950), this is quite rare in psychological research. However, it may be that the functional relation in the population is intrinsically nonlinear, meaning that no transformation of either or both X and Y is able to produce a model in which the variables relate in a linear fashion.

The problem that arises when describing relations is that if the model selected does not reflect the functional relation between $X$ and $Y$ in the population, the regression coefficients produced may be biased. For example, the effects of an omitted predictor that is correlated with those predictors in the MODEL are manifested in the ERROR term. As a result the ERROR term is no longer independent of those predictors in the model. This specification error has two consequences. First, power is reduced and the estimates are less precise. Second, even with the reduced power, Type I errors are more likely as the bias that is introduced means that, even if the regression coefficients are zero in the population, the confidence intervals around the coefficients obtained in the sample may be too unlikely to include zero. Although regression diagnostics may, in certain situations, indicate a specification error, they usually only indicate whether the addition of a different form of a predictor already in the equation is required to improve the fit of the model. For example, the diagnostics may suggest that the addition of a quadratic term would produce a model that better fits the data, but they usually do not provide any information about omitted variables. More detail is provided in the section on diagnostics.

It appears that the selection of the regression model should, as far as possible, be based upon a strong theoretical framework. Although theoretical support for intrinsically nonlinear models is seldom found in psychology, theory may well indicate the presence of some form of nonlinear relation in the population. Depending on the specific theory, options are to either transform some or all the variables or to include higher order polynomial terms in the model.

## B. Inferential Estimation and Testing

Statistical inference has to do with making statements (typically through hypothesis tests or confidence intervals) about a population based upon the collection of sample data. To enable this, a certain set of assumptions about the population data are required. Recall that the MODEL component of the multiple regression procedure attempts to explain the mean of the conditional distributions of $Y$ and involves the fit between the regression model and the population function. The ERROR component, on the other hand, allows for the fact that there is some variability in the distribution of $Y$ at each fixed vector of $X$ values and provides a measure of the variance not explained by the MODEL component. Although purely descriptive purposes require only that the MODEL be specified accurately, inferential purposes require that additional assumptions about the ERROR term be met for the statistical tests to be accurate.

The ERROR term is assumed to be identically and independently distributed. Although the independence portion of this assumption can usually be assured by the sampling design of the research, the identical aspect is based upon three assumptions that all concern the conditional distribution of $Y$ in the population. First, the conditional distributions of $Y$ are assumed to be normally distributed. Second, these distributions are assumed to have the same variances (the assumption of homoscedasticity). Third, the regression model is assumed to adequately reflect the population function between the dependent and independent variables. This is simply the assumption described earlier for accurate description illustrating that description is a prerequisite for inference.

This illustrates the area where the two components of multiple regression, the MODEL and ERROR components, are interdependent. Whether all the assumptions for the ERROR term have been met depends to some degree on how well the specified model fits the true population model. Thus if a linear model is fit to a relation that is in fact quadratic, one of the assumptions about the ERROR term would be violated and these would no longer be identically distributed, though they may still be independent. If, however, all three of the assumptions (i.e., normality, linearity, and homoscedasticity) are met, the error terms for each conditional $Y$ distribution will be identically distributed. Assuming then that the cases are independent of each other, the assumption of independent and identical error distributions is met and the use of hypothesis tests and confidence intervals for inferential purposes is valid.

## C. For Causal Inferences

If multiple regression is to be used as the basis for causal inferences, some additional assumptions that flow from those described above become necessary (Kenny, 1979). Recall that the point was made in the MODEL assumption section that, unless a predictor variable that is correlated with those in the regression equation is omitted from the analysis, the residuals are uncorrelated with the predictor variables. This assumption of uncorrelated errors has three implications (Kenny, 1979). First, that there is no reverse causation such that the dependent variable has any causal effect on the independent variables. Second, that the predictor variables are measured without error and with perfect validity. Third, that there are no common causes, otherwise known as the third variable or excluded variable problem. An additional assumption that comes into play when the variables are measured cross-sectionally is that of stationarity.

Kenny notes that these are stringent assumptions. Ruling out reverse causal effect is theoretically or logically based and often relates to the time precedence principle of causation. Thus if $X$ precedes $Y$ and is measured before $Y$, $Y$ cannot be said to cause $X$. The assumption of the perfect measurement of the $X$ variables is more problematic. Although some variables may be measured perfectly (such as race and sex), many of the variables of interest in psychology are clearly not measured perfectly (motivation, attitudinal, and behavioral variables). Latent variable modeling (DiLalla, chapter 15, this volume) often provides the most useful solution to this problem. The third variable problem can be addressed by attempting to identify and measure all the relevant variables. However, Kenny points out that whenever the multiple $R$ is $< 1$, it is logically impossible to demonstrate that all excluded variables have been included. Omitting relevant variables or measuring them imperfectly results in sample regression coefficients that are biased estimators of the population parameters. Because these problems are usually present in observational designs, causal inferences based on regression analyses tend to be a more tenuous venture. In contrast, true experimental designs may produce stronger inferences.

Predictive and explanatory research make the same assumptions, and their achievement is enhanced when the appropriate model is fit. However, recall that the coefficients they examine differ. Thus prediction attempts to maximize the multiple $R$-squared, whereas explanation focuses on the accurate estimation of the regression coefficients. Prediction is compromised most by the omission of an uncorrelated predictor, whereas explanation is compromised most when correlated independent variables (predictors) are omitted.

## III. REGRESSION DIAGNOSTICS AND TRANSFORMATIONS

### A. Introduction

The preceding section presents a theoretical overview of the assumptions that apply to the MODEL and ERROR components in multiple regression analyses. The next section deals with (a) the manner in which the validity of the assumptions of linearity, normality, and homoscedasticity can be assessed within a given data set, and (b) what action can be taken should the data indicate that any of these assumptions have been violated.

Regression diagnostics and transformations are related strategies. Diagnostics are generally used to address two basic issues: first, whether the assumptions underlying multiple regression have been met, and, second, whether any influential data points that unduly affect the regression results are present in the data. This is accomplished by examining plots, graphs, and diagnostic statistics. Transformations are typically performed for four purposes related to the underlying assumptions (Ryan, 1997). First, models that are "nonlinear in the parameters" (but not *intrinsically* nonlinear) can be transformed so that they become "linear in the parameters." Second, should the relation between $X$ and $Y$ be nonlinear, either $X$ and/or $Y$ may be transformed so that the relation between the new $X$ and $Y$ variables is linear. Third, transformations can produce relatively constant error variances (i.e., can address violations of the homoscedasticity assumption).

Finally, transformations can obtain approximate normality for the distribution of the error term.

There are a number of ways in which the question of diagnostics and the need for transformations can be approached. Note that while a variety of transformations exist, we limit our discussion here to two particular methods reviewed by Ryan (1997). These two methods are intended to transform the data so as to produce a model that fits the data well and remedies any nonlinearity, nonnormality, and/or heteroscedasticity in the data. First, the Box–Tidwell method provides a means of determining the most appropriate transformation of the $X$ variables. Second, the Box–Cox method provides a means of determining the most appropriate transformation of $Y$. In addition, Ryan presents a combined Box–Cox and Box–Tidwell approach, which can be used in a two–stage process. In contrast to the somewhat subjective visual nature of the diagnostic process, the approaches presented by Ryan can be thought of as objective methods. Here the need for any transformation is indicated by the particular procedure and is not based on the examination of any graphs. In this chapter, space limitations constrain us to describing the Box–Cox transformation only.

### B. Box–Cox Transformation

In situations where the dependent variable, $Y$, fails to meet necessary assumptions, it may be possible to transform $Y$ by raising it to some power, which we will denote $\lambda_j$. Ryan (1997) notes that the Box–Cox procedure attempts to estimate lambda ($\lambda$) such that the transformed $Y$ variable, $Y^\lambda$, is a linear function of the regressors, with approximate normality, and a constant error variance. Note that although a $\lambda$ value of 1.0 indicates that no transformation is necessary, a positively skewed Y distribution can often be remedied by choosing $\lambda$ to be some value less than 1.0, and a negatively skewed $Y$ distribution can be addressed by choosing $\lambda$ to be $> 1.0$. The family of Box-Cox transformations is as follows

$$Y^{(\lambda)} = \begin{cases} \dfrac{Y^\lambda - 1}{\lambda} & \lambda \neq 0 \\ \log(Y) & \lambda = 0 \end{cases} \qquad (3)$$

The regression equation would then model $Y^\lambda$ as a linear function of the regressors and is based upon the same assumptions as the original model for Y, namely linearity, normality, and homoscedasticity. This is important, as, even though the transformation is intended to produce a model that fits the data well, remedying any nonlinearity, nonnormality, and heteroscedasticity in the data, there are no guarantees that it will succeed in all these areas. Thus even if this type of procedure is used, it is still wise to plot and examine the relevant diagnostics to assess the validity of the assumptions in the transformed data as well.

A number of options exist to estimate $\lambda$ so as to identify the $\lambda$ value that best remedies skewness and heteroscedasticity. These range from those presented by Draper and Smith (1981), Ruppert and Aldershof (1989), Cook and Weisberg's inverse response plot (1994a,b), and the nonparametric alternative presented by Ryan (1997). As none of these approaches are uniformly best, readers are referred to Ryan, who presents a thorough review of these approaches.

It is possible to begin with the diagnostics on the original untransformed regression model or perform the suggested transformations first. For example, we could perform the combined Box-Cox and Box-Tidwell approach on the data, implement any indicated transformations, and subsequently run the diagnostics. Alternatively, we could fit the model we believe to be appropriate, run the diagnostics to assess whether any transformations are indicated, implement any indicated transformations, and rerun the diagnostics. Our approach in this chapter is a blend of both approaches. We fit the model we think appropriate and present the diagnostics while, at the same time, presenting the diagnostics for the transformation indicated by the data. This approach allows us to clearly demonstrate the effects of the transformations upon the data. Thus the figures and tables compare the model statistics and diagnostics of the model with the original variables to the model of the transformed variables. Much has been written on this topic (Darlington, 1990; Fox, 1997; Hair, Anderson, Tatham, & Black, 1995; Ryan, 1997), and the following discussion is not exhaustive; rather, we illustrate how the diagnostics and transformations can be used in assessing and addressing violations of the assumptions.

### C. Data Set and Regression Model Description

The data are drawn from an ongoing study of the correlates of depression in grade school children. The dependent variable in this example is the children's score on the Child Depression Inventory (CDI) (Kovacs, 1982, 1985) and the independent variables are measures of the children's perception of their own competence in five domains: academic $(X_1)$, appearance $(X_2)$, behavioral $(X_3)$, social $(X_4)$, and sports $(X_5)$ competence. Note that scores of 0 are possible on the CDI denoting the presumed absence of depression.

Because there is no theoretical reason to specify a nonlinear model, we have chosen to fit a model of the form

$$Y = \beta_0 + \beta_1 X_1 + \beta_2 X_2 + \beta_3 X_3 + \beta_4 X_4 + \beta_5 X_5 + \varepsilon \tag{4}$$

At this stage it is worth noting that we do have some a priori reason to expect that the scores on the dependent variable may be positively skewed. Typically a log or square root transformation is used to correct this skewness (Cohen & Cohen, 1983). However, rather than just implementing either of these transformations, we used the Box-Cox method to evaluate which transformation would be most suitable.

Because zero values in the data complicate the transformation (Maxwell & Delaney, 1990), one was added to each score on the dependent variable. The specific Box–Cox transformation was performed using STATA 5.0 (Intercooled Stata 5.0 for Windows 95). The point estimate of $\lambda$ for these data is 0.14, and the 95% confidence interval around $\lambda$ does not include 1.0 (ranging as it does from $-.03$ to $0.32$), suggesting that some transformation is necessary. However, the confidence interval does contain zero, which, if used as $\lambda$, produces a transformation equivalent to taking the log of the original dependent variable. As a result, we chose the log transformation and report those results in this chapter.

## D. Diagnostics and Transformations

The residual term, the difference between the actual dependent variable and its predicted value, comprises the principal means of evaluating the validity of the assumptions. We proceed by evaluating the assumptions of homoscedasticity, normality, and linearity. Finally, the issue of influential observations and the use of diagnostic plots and statistics in identifying such cases are presented.

### 1. Evaluating the Homoscedasticity Assumption

Because the residuals in the data (e) do not have a constant variance even when the population ERROR term ($\varepsilon$) does, this assumption is not evaluated by plotting the raw residuals against either the predicted $Y$ or $X$ (Ryan, 1997). We follow the example of Fox (1997) in plotting the studentized residuals against the values of the predicted dependent variable. Figure 6.1 presents these plots for the original dependent variable and its log transformation.

Examination of the plot of the original $Y$ variable clearly indicates that the homoscedasticity assumption has been violated. There is little doubt that the variability of the residuals increases as the predicted values of $Y$ increase. The plot for the log transformation suggests that this transformation succeeded in remedying the nonconstant error variances, producing error variances that appear approximately constant across the values of the predicted $Y$ variable.

### 2. Evaluating the Normality Assumption

Next we evaluated whether the residuals (or, alternatively, the conditional distributions of Y) are normally distributed. In terms of identifying any skewness present in the dependent variable, our initial reaction may be to examine the frequency distribution of the raw $Y$ scores. This may,

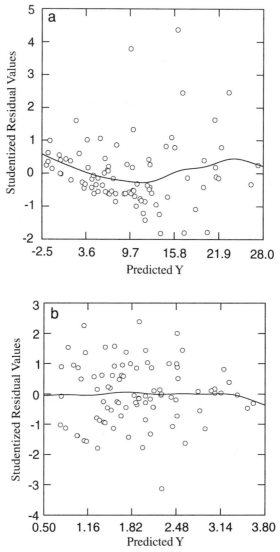

**FIGURE 6.1**    (a) Studentized residual plots for the original $Y$. (b) Studentized residual plots for the log of $Y$.

however, lead to the wrong conclusion, as the raw score distribution may be skewed, whereas the studentized residuals are normally distributed. This can occur if one or more of the $X$ variables are also positively skewed, so the marginal distribution of $Y$ is skewed even when all the conditional distributions are normal. Thus we examine the normal probability plot of the regression-studentized residuals.

Figure 6.2 presents this plot for the original dependent variable and the log transformed dependent variable. If the data were normally distributed, the residuals should fall along the diagonal with no substantial or systematic differences (Hair et al., 1995). It is apparent that the residuals for the original dependent variable deviate from the diagonal in a systematic manner, supporting our expectation that this distribution is skewed.

The plot of the log of $Y$ suggests that the transformation had the effect

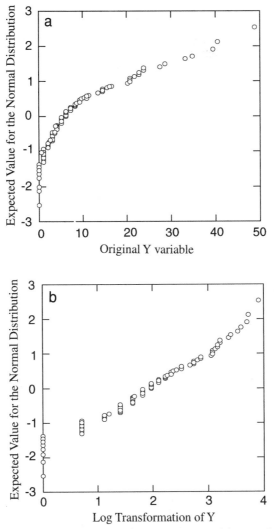

**FIGURE 6.2**   (a) Normal probability plot for the original $Y$. (b) Normal probability plot for the log of $Y$.

of drawing the residuals closer to the diagonal as compared to the original dependent variable. Although the positive skew has been reduced, this plot still indicates that a number of scores cluster around the left tail of the distribution as a result of a floor effect.

Given the results of the diagnostics presented thus far, it seems that the original dependent variable evidences both heteroscedasticity and non-normality, and the log transformation is effective in addressing both of these violations. However, it is important to acknowledge a subtle point concerning the use of transformations in correcting these violations. Both of these assumptions are subsumed by that part of the ERROR-term assumption requiring that the ERROR term be identically distributed. The assumptions are about the population values, not the sample data. If the distribution of $Y$ at each vector of the $X$s is normally distributed in the population, then the ERROR term in the model (i.e., $\varepsilon$, an unknown) is normally distributed. Similarly, if the homoscedasticity assumption holds then these conditional distributions of Y share equal variances, and the ERROR term also exhibits equal variances. In sample data we test both assumptions by examining the distribution of the sample regression residuals. Thus the difference between the observed and predicted $Y$ values in the data (i.e., $e$) is used to test whether the unknown $\varepsilon$ is normally distributed with equal variances. As the empirical distributions of $\varepsilon$ and e can differ considerably, testing the effectiveness of transforming $Y$ in the sample to induce normality in the population necessarily involves a certain degree of uncertainty (Ryan, 1997).

## 3. Evaluating the Linearity Assumption

At this point we turn to the assumption that a linear model can adequately describe the relation between the dependent and independent variables. The assumption of linearity is assessed at two levels. First, by examining the plots of the studentized residuals presented in Figure 6.1 we can establish whether there is evidence of a nonlinear pattern for the overall equation. Second, by examining the partial residual plots (Fox, 1997) we can assess the nature of the partial relation between each $X$ and $Y$ and can detect departures from linearity. Note that theoretically these individual relations will not be linear in both forms of the dependent variable (original and transformed). If these relations are linear in the original dependent variable, then the log transformation of $Y$ should have the effect of making them nonlinear. Due to space limitations, we only present the partial regression plots for one of the five independent variables.

The pattern of the residuals in the plot for the original $Y$ in Figure 6.1 suggest a nonlinear component to the overall relation between this original $Y$ and the independent variables. The plot for the log of Y suggests that the transformation successfully remedied the nonlinear pattern. Figure 6.3 presents the partial residual plots for the academic competence independent variable and the two forms of the $Y$ variable.

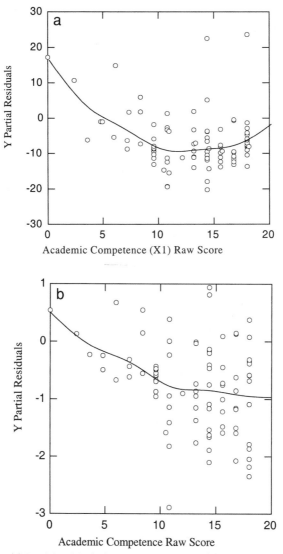

**FIGURE 6.3** (a) Partial residual plot for the original $Y$. (b) Partial residual plot for the log of $Y$.

The results are very similar to those illustrated in Figure 6.1. Again, nonlinearity is present in the plot for the original Y and appears to have been remedied by the log transformation. Very similar patterns were found for the partial residual plots for the remaining four independent variables. Thus it appears that the log transformation also successfully eliminated the nonlinearity present in the relation between the original dependent variable and the independent variables.

In conclusion, it is clear that the data in their original form contained unequal variances, skewness, and nonlinearity, which the log transformation appears to have successfully remedied. Table 6.1 presents the regression results for the original $Y$ and its log transformation.

The analysis of the original Y variable produces the following equation:

$$Y = 36 - .50X_1 - .45X_2 - .75X_3 - .59X_4 + .19X_5 + \varepsilon \qquad (5)$$

In order for our interpretation of, for example, the coefficient of $-.50$ as the partial slope for $X_1$ at a fixed value of the remaining regressors to hold, the assumption of linearity must be valid. Given the presence of curvilinearity evidenced by the diagnostics, the decrease in $Y$ as a result of any unit change in $X_1$ depends on the value of $X_1$. Furthermore, the value of $-.50$ may not be an unbiased estimate of the partial slope for $X_1$ because the other predictors may not relate linearly to Y. In addition, the presence of nonnormality and heteroscedasticity in the data challenges the validity of the inferential statistics produced. Thus, if the study is replicated 100 times and 100 confidence intervals are produced for the $X_1$ coefficient, we cannot be sure that, in the long run, 95 of these would contain the population parameter for $X_1$.

In addition, note the effect of the transformation on the results of the significance tests in Table 6.1. Both sets of results indicate that the partial slopes for the academic, appearance, and behavioral competence variables are significant, although that of the sports competence variable is nonsignificant. Notice that the transformed data produces larger $t$ values for the former three variables as well as a smaller $t$ value for the latter. Of particular note in this example, however, are the results for the variable of social competence. Although the untransformed data suggest that there is some relation between social competence and depression at fixed values of the other independent variables, $t(88) = -2.65, p = .01$, the transformed data suggest that this may not be the case, $t(88) = -1.73, p > .05$. Recall the point made earlier that to interpret the regression coefficients as partial slopes only, the assumption of linearity must be valid. The evidence of nonlinearity in the untransformed data raises concerns about the appropriateness of interpreting the regression coefficient for social competence as a partial slope. On the other hand, the transformation appears to have addressed the nonlinearity in the data, making this interpretation of the regression coefficient more plausible in these data. These data would then lead us to conclude that there is no evidence that social competence is related to depression at fixed values of the remaining independent variables.

### E. To Transform or Not to Transform?

Reviewing the diagnostic information presented earlier, it is clear that a transformation of the dependent variable was necessary to address viola-

**TABLE 6.1  Multiple Regression Results for the Different Regression Models**

**1. Original dependent variable**

Overall regression model results

| | |
|---|---|
| Multiple $R$ | 0.67 |
| Multiple $R^2$ | 0.45 |
| Standard error of estimate | 7.89 |

Coefficients

| Variable | Coefficient | Standard error of coefficient | Standardized coefficient | $t$-value | Significance |
|---|---|---|---|---|---|
| Intercept | 36.034 | 3.625 | | 9.941 | .000 |
| Academic competence | −.497 | .243 | −.200 | −2.042 | .044 |
| Appearance competence | −.451 | .209 | −.225 | −2.164 | .033 |
| Behavioral competence | −.746 | .217 | −.305 | −3.441 | .001 |
| Social competence | −.589 | .223 | −.276 | −2.645 | .010 |
| Sports competence | 0.190 | .217 | 0.092 | 0.873 | .385 |

**2. Log transformation**

Overall regression results

| | |
|---|---|
| Multiple $R$ | 0.69 |
| Multiple $R^2$ | 0.47 |
| Standard error of estimate | 0.75 |

Coefficients

| Variable | Coefficient | Standard error of coefficient | Standardized coefficient | $t$-value | Significance |
|---|---|---|---|---|---|
| Intercept | 4.623 | .344 | | 13.459 | .000 |
| Academic competence | −.058 | .023 | −.239 | −2.497 | .014 |
| Appearance competence | −.049 | .020 | −.252 | −2.476 | .015 |
| Behavioral competence | −.082 | .021 | −.345 | −3.985 | .000 |
| Social competence | −.037 | .021 | −.176 | −1.729 | .087 |
| Sports competence | 0.010 | .021 | 0.053 | 0.510 | .612 |

tions of the underlying assumptions. In addition, it appears that the transformation was successful in remedying the violations producing a model that fit the data better. There is, however, the fundamental issue of how the regression results for a model with a transformed dependent variable are interpreted. There is a trade-off between improving the regression model through transformations while at the same time making the results less interpretable. For example, interpreting the regression coefficients of the log transformation of the CDI may be less straightforward than interpreting the original CDI. However, not performing required transformations may result in failing to identify important predictors or incorrectly identifying predictors as important when they might not be.

### F. Identifying Influential Observations

Due to space constraints, we simply highlight statistics and plots that aid this process and refer readers to more detailed discussions of the procedure. Influential observations wield a disproportionate amount of influence on the regression results. Three types of influence are typically identified as (a) distance from the regression line (outliers), (b) distinctiveness in terms of the pattern of independent variable scores (leverage), or (c) influence, which is a combination of distance and leverage (Fox, 1997). Outliers and leverage points alone are not necessarily influential observations. Influential observations can be data errors, valid but exceptional data points due to extraordinary reasons, exceptional data points without such reason, and data points that are ordinary in their individual characteristics but exceptional based on their pattern of regressor scores (Hair et al., 1995). These observations can distort the sample results such that they fail to represent accurately the relations between the variables in a population.

The process of identifying influential observations is one of convergence rather than reliance on a single measure of an observation's influence, and, as with many of the statistical tools at our disposal, the interpretive aspect of the process is important. Fox (1997) and Hair et al. (1995) present detailed examples of the process and the kinds of options that exist in terms of statistics that evaluate influence. We present a brief overview of one method of identifying influential observations. First, the studentized residuals are examined to identify outliers. These residuals correspond to $t$ values (Hair et al.xs 1995). Therefore, the upper and lower thresholds against which the actual residuals are compared can be selected to correspond to specific significance values. For example, one might select limits of plus or minus 2, which corresponds roughly to a 95% confidence interval. Most computer packages provide a means of identifying specific cases that exceed a specified value (and allow you to set the specified value).

Second, the leverage points are examined to assess how the observations influence the predicted values of the dependent variable. Recall that the leverage points measure the atypicalness of each observation's pattern of regressor scores. The best-known measure of leverage is denoted $h_I$, and readers are referred to Darlington (1990) and Fox (1997) for a detailed discussion of this measure. The leverage points are compared to a threshold value that is obtained by the formula $2p/n$, where $p$ is the number of predictors and $n$ is the sample size (Hair et al., 1995; Fox, 1997). Third, the studentized deleted residuals, the residual from the regression equation estimated without that specific case's influence, are examined. Again we would select cutoffs at approximately 2 standard deviations above and below the mean to identify outlying cases.

Fourth, we examine the DFBETA values, which measure the relative effect of each single observation on each regression coefficient individually. This is calculated as the change in the regression coefficient when that specific observation is deleted. Hair et al. (1995) note that as exact thresholds are not available, the goal is to identify observations that change the coefficients in the regression equation in a substantial manner. Another option is the COVRATIO measure, which, though similar to DFBETA, estimates the effect of each observation on all the coefficients as a group. Finally, Cook's Distance is used to identify those observations that exhibit both large residuals and high leverage influence. Hair et al. (1995) note that in the case that no single observation exceeds the rule of thumb threshold of 1.0, the plot should still be evaluated to assess whether any small group of observations has values substantially higher than the rest.

Computer packages generally allow the option of saving these measures to the data file, which is useful in creating plots to examine the data. However, care is required in eliminating any cases that are identified as influential, as this process can be abused when data are trimmed solely for the purpose of obtaining good results. Our view is that the identification of influential observations should be pursued with the goal of understanding why they are so exceptional, not simply to delete them thoughtlessly. Unless scores are clearly erroneous, deletion may not be justified. Instead, we recommend either of two alternatives. First, transformations of relevant variables may lessen the impact of influential observations. Second, methods that are robust to the effects of influential observations are also available (see Wilcox, 1997).

## IV. INTERACTIONS AND MODERATOR EFFECTS

Many psychological hypotheses stipulate that two or more independent variables interact with one another in influencing a dependent variable. Although such interaction effects have long been a focal point of analysis of variance, they have historically received less attention in regression analysis.

We will begin our discussion of interactions by returning to our illustrative data set examining depression and competence. After briefly presenting the rationale and mechanics of performing an interaction test, we will describe how one might interpret the results of the test. Finally, we will highlight two further complications that may arise in testing interactions with multiple regression: deficient statistical power and spurious results.

## A. Rationale and Mechanics

Our analysis to this point of the relation between competencies and childhood depression has implicitly assumed an additive model. However, in reality, Cole's (1990) model of depression suggests that the influence of any single competence may depend on whether a child is high or low on the other competence domains. Examining whether such interaction effects truly exist requires a more complex model.

To illustrate how to assess the possible presence of interaction effects, we will consider only whether there is evidence that Academic Competence and Behavioral Competence interact for the log transformed measure of depression. With five predictors, one could examine 5 two-way interactions, 10 three-way interactions, 5 four-way interactions, and 1 five-way interaction, but in the interest of simplicity and brevity, we will consider only a possible two-way interaction of Academic and Behavioral Competences, ignoring any effects of the other three variables (see Aiken & West, 1991, for consideration of higher-order interactions).

To establish a context for interpreting the interaction model, it is useful to know the regression coefficients for the additive model that includes only Academic and Behavioral Competences. For the sake of illustration, we will focus our attention on Academic Competence, which has a coefficient of $-0.094$ in the additive model when only these two predictors are included. In the absence of an interaction, this value can be interpreted as the slope relating Academic Competence and the log of depression at every value of Behavioral Competence. However, if an interaction exists, it may be misleading to describe this relation with any single slope coefficient, because an interaction implies that the value of the slope differs at different values of Behavioral Competence.

The most common type of interaction examined in regression models is a "bilinear" interaction, which is the single degree of freedom portion of the interaction corresponding to the linear effects of the predictors. A bilinear interaction exists if the slope for Academic Competence is itself linearly related to Behavioral Competence, so that the slope consistently either increases or decreases at higher values of Behavioral Competence. It would be possible to examine other components of a possible interaction, such as an interaction between the linear effect of Academic Competence and the quadratic effect of Behavioral Competence, which would allow the slope for Academic Competence first to increase and then to decrease as Behavioral Competence increases, but we will follow common practice in considering only the bilinear interaction.

The bilinear interaction can be estimated and tested simply by adding an additional predictor variable to the model containing the additive predictors. In our case, the resultant model has three predictors: Academic Competence, Behavioral Competence, and a third variable that is literally the product of Academic Competence and Behavioral Competence. There is a statistically significant interaction in the depression data, as shown by a $t$-value of $-1.986$ ($p = .050$) for the product term. It is often useful to center the variables being multiplied before calcuting the product term. Such centering is not mathematically necessary, because the unstandardized regression coefficient for the interaction, its estimated standard error, and hence associated $p$-value will be the same regardless of whether centering is performed. Nevertheless, centering is often recommended because it can facilitate interpretation and reduce the likelihood of severe collinearity problems.

## B. Interpretation of the Interaction

The nature of the interaction can best be understood through the regression coefficients, which are 0.0174, 0.0166, and $-0.0085$ for Academic Competence, Behavioral Competence, and the product term, respectively. These coefficients imply that the slope of the log of depression on Academic Competence is given by

$$0.0174-0.0085 \text{ (Behavioral Competence)} \tag{6}$$

The fact that $-0.0085$ is significantly different from 0 implies that the slope for Academic Competence depends (linearly, for the bilinear model) on the value of Behavioral Competence. For example, the mean value of Behavioral Competence in these data is 13.20, so the slope of Academic Competence at the mean value of Behavioral Competence can be found simply by substituting 13.20 for Behavioral Competence in Equation 6, yielding a slope of $-0.095$. At higher levels of Behavioral Competence, Academic Competence is more strongly related to depression. For example, the slope of Academic Competence when Behavioral Competence is one standard deviation above the mean (in these data, a value of 17.42) can be found by substituting 17.42 for Behavioral Competence in Equation 6, yielding a slope of $-0.131$. On the other hand, when Behavioral Competence is one standard deviation below the mean (a score of 8.98 for these data), the corresponding slope is $-0.059$, obviously weaker than at higher values of Behavioral Competence. Further insight into such interactions can often be gained by graphing these regression slopes, much as is typically done through interaction plots in analysis of variance (ANOVA) (see Aiken & West, 1991, for details).

## C. Further Considerations

Interactions have historically been reported much less in regression analysis than in ANOVA. Although one reason for this difference may relate to the types of questions typically addressed by the two methods, another possible reason for this discrepancy is that interactions may be more difficult to detect in many regression applications. Two factors contribute to frequent inadequacies of statistical power for detecting interactions in the types of situations where regression analysis is commonly used. First, Busemeyer and Jones (1983) showed that the product terms used to test interaction effects are usually less reliable than the corresponding main effect variables. Thus, measurement error in the predictors will often reduce power more for testing interaction than for testing main effects. Second, McClelland and Judd (1993) showed that the distributional properties of typical variables in field studies lead to less power than would the distributions typically used in experimental studies. Specifically, extreme values of the predictors often do not co-occur in field studies, but such co-occurrences are crucial for detecting interactions. Thus, although there is nothing inherent in regression analysis that lowers the power to detect interactions, the use of variables measured with error to detect interactions in observational field settings may lack sufficient statistical power unless very large sample sizes are available.

Although statistical power may often be problematic, Lubinski and Humphreys (1990) ironically showed that the standard approach for testing interactions as described above can also lead to the detection of spurious interaction effects. For example, suppose two predictors, $X$ and $Z$, are highly correlated. Further suppose that the product term $XZ$ explains additional variance in $Y$ beyond that accounted for by $X$ and $Z$. Although this would seem to suggest an interaction, another possible explanation is that $X$ and/ or $Z$ have a quadratic nonlinear relation with $Y$. In this case, $XZ$ may appear important simply because it happens to correlate highly with $X^2$ or $Z^2$. A suggested solution for addressing this problem is to include quadratic terms of the predictors in the regression model along with the product term. For example, in our illustrative analysis, the square of Academic Competence and the square of Behavioral Competence would be included, along with the linear effects and the product term, yielding a total of five predictor variables. In this particular case, neither quadratic term is statistically significant, whereas the interaction term remains significant ($p = .020$), but in some data the interaction term may no longer be significant, suggesting that what originally appeared to be an interaction effect may in reality be a higher order effect but nevertheless an additive effect.

MacCallum and Mar (1995) discuss further issues in distinguishing interaction effects from quadratic effects.

## V. SAMPLE SIZE REQUIREMENTS FOR MULTIPLE REGRESSION ANALYSES

The determination of sample size requirements is related to statistical power and forms an integral part of the design phase of any research project. In fact, determining sample size depends on the purpose underlying the specific research project. Recall that multiple regression can be used for either prediction or explanation, and that different statistics are used to evaluate each of these goals. Predictive research tries to identify predictors that optimize the predictions, thus it focuses on the multiple $R^2$ as measured by the shrunken $R$. Explanatory research attempts to identify the underlying causes of phenomena and their relative degrees of importance. Here the statistical influence of the regression coefficient and the manner in which it influences the multiple $R^2$ is typically evaluated. This section illustrates how sample sizes can be determined for each of these two purposes separately.

### A. Sample Size Requirements for Prediction

Identifying the most effective set of predictors is primarily determined by evaluating the multiple $R^2$ coefficient. Sample size is most often obtained by determining the sample size required to achieve a desired degree of power in detecting a particular effect size. Although we briefly overview power analysis in terms of the multiple $R^2$, we suggest that its utility is limited for predictive research. Readers are encouraged to become more conversant with the details of power analysis provided by Cohen and Cohen (1983) and Cohen (1988).

Cohen (1988) proposes the following procedure. First, select the significance level. Cohen's tables provide significance levels of .05 and .01. For this example, we select an alpha of .05. Second, .80 is selected as the desired degree of power. Third, we need to determine the population effect size, or $f^2$, which Cohen defines as

$$f^2 = \frac{R^2}{1 - R^2}. \tag{7}$$

This $R^2$ may represent a probable population value indicated by previous research, a minimum value of theoretical or practical importance, or some conventional value (Cohen & Cohen, 1983). Let's suppose that we expect five predictors to account for 30% of the variance in our criterion variable. Substituting this value into Equation 7 produces an $f^2$ value of 0.43. Referring to the $L$ tables in Cohen and Cohen (1983, Tables E1 and E2, pp. 526–527), we find that with alpha of .05, $k_B$ of 5 (here $k_B$ is the number of independent variables), and power of .80, $L$ equals 12.83. This value is then substituted into Equation 8 (Cohen & Cohen, 1983)

$$n = \frac{L}{f^2} + k + 1 \qquad (8)$$

where $k$ is equal to the number of independent variables. Thus we find that we need 36 cases to detect a significant population $R^2$ of .30 with 80% probability using an alpha of .05.

Recall that when regression is used for prediction, predictors that maximize the multiple $R^2$ are chosen, and the parameter of interest should be the cross-validity parameter, or shrunken $R$ (Darlington, 1990). Although the power analysis suggests that $n = 36$ is sufficient, this sample size is inadequate if our goal is to develop a prediction equation based on the shrunken $R$. As described earlier, the multiple $R^2$ of .30 in this example produces a shrunken $R^2$ of .07, which would suggest that our prediction equation is not very effective. What effect would increasing the sample size have? Suppose that as before, we have a model with five predictors, a multiple $R^2$ of .30, and an $n$ of 200. Substituting these values in the relevant equations (Darlington, 1990, and Equation 1) produces a shrunken multiple $R^2$ of .26 which closely approximates the multiple $R^2$ of .30. Thus, for the same population, deriving the regression coefficients in a sample of 200 instead of 36 is expected to produce an equation that will explain 26% of the variance in $Y$, instead of only 7%.

### B. Sample Size Requirements for Explanation

With explanation as our goal, we are more likely to be interested in the amount of unique variance each independent variable (IV) accounts for in the dependent variable. Cohen notes that because the standardized and unstandardized regression coefficients, as well as the semipartial and partial correlation coefficients, must all have identical significance test results, we could use any of these in determining the $f^2$ value to obtain the required sample size. Thus the null hypothesis in this instance states that any partial correlation or regression coefficient for a given independent variable (in a set of $k$-independent variables) is zero. The sum of the squared semipartial coefficients for a set of independent variables will typically be lower than the multiple $R^2$ value due to the effects of collinearity. The more correlated the IVs are, the more pronounced this difference will tend to be.

Suppose that we expect the set of IVs to produce a multiple $R^2$ of .30 (as in the first example). Further, assume that we expect the specific IV of interest to make a unique contribution of .03. In other words, we expect the squared semipartial correlation coefficient $(sr_j^2)$ to equal .03. For $k_B =$

1 (as we are evaluating the effect of a single IV), alpha (.05), and power of .80, $L$ is equal to 7.85 (Table E2, Cohen & Cohen, 1983, p. 527). Combining the formula for $f^2$, which is now written as

$$f^2 = \frac{sr_i^2}{1 - R^2} \qquad (9)$$

with Equation 8, the required sample size whenever alpha is .05 and power is .80 is provided by

$$n = \frac{7.85(1 - R^2)}{sr_i^2} + k + 1 \qquad (10)$$

Substituting $R^2 = .30$, $sr_i^2 = .03$, and $k = 5$ into Equation 9 yields $n = 189$. Thus 189 subjects are necessary to test the null hypothesis that $sr_i^2 = 0$ with power of .80. Note that the value of $k$ in Equation 10 remains 5, as this represents the total number of independent variables in the regression equation.

The sample size required in testing the effects of the individual IVs will typically be substantially larger than that required to test the multiple $R^2$ when the regressors are correlated. In this example, for instance, we would need approximately 153 more subjects to test the effects of an individual IV. In performing power analyses as a means of obtaining the required sample size to attain a certain power level, it is important that the analysis is done with the correct coefficient in mind. Recall that, when obtaining the tabled $L$ value during the power analysis process, $k$ represents the degrees of freedom of the numerator in the $F$ test. Care needs to be taken that our choice of $k$ accurately reflects whether we are testing the effects of a single IV or a set of IVs.

## VI. HANDLING MISSING DATA

Complete data sets are rare in most types of psychological or social science research. Missing data can affect the estimation and inferential process adversely. Our primary concern when faced with missing data should be to examine the underlying reasons so that an appropriate course of action can be selected. Unfortunately, the typical reaction appears to be the selection of the listwise or pairwise deletion options offered by most statistical packages without thinking about the processes behind the missing data. However, as this section shows, neither of these options is efficient and both can be biased. Again readers are provided with specific resources that provide greater detail in this regard.

## A. Listwise Deletion

This option is viable only when a small proportion of the participants in the sample are missing on any variable to be used in the analysis. The omission of these participants from the analysis reduces the sample size and thus statistical power is also reduced. Although it would seem that this practice is fairly innocuous when the sample size is large and the proportion of individuals with any missing data is small, it is difficult to specify "large" and "small" in practical terms (Cohen & Cohen, 1983). In addition, this method requires the stringent assumption that the data are missing completely at random. Practically speaking then, the observed values must be a true random sample of all the possible values, both observed and missing.

If the data are missing completely at random, then parameter estimates are not biased. However, efficiency is lowered because non-missing data are excluded from the analysis. If the data are not missing completely at random (as is usually the case), the picture becomes even gloomier because estimates are biased (i.e., because the remaining observations may be a "nonrepresentative" subset of all the observations), so the results do not generalize to the population from which the sample was drawn.

## B. Pairwise Deletion

In this method the computed correlation matrix used in the multiple regression analysis is constructed based on all cases with data available for each pair of variables being correlated. As a result, each correlation coefficient in the matrix can be based on a different number and subsample of observations. Although this procedure may appear to be more efficient than the listwise deletion method, it too has flaws that threaten the validity of any inferences.

As with listwise deletion, pairwise deletion also assumes that the data are missing completely at random. Once again, if this assumption is met, estimates are unbiased. However, if this assumption is violated, the generalizability of the results is questionable, as the various computed correlation coefficients may not represent a single population. Cohen and Cohen (1983) point out that this correlation matrix may contain values that are mutually inconsistent with each other. In other words, this computed correlation matrix need not be positive definite and may produce values that are mathematically impossible were the data complete. For example, the multiple $R^2$ value may exceed 1.0. Even when the results are not explicitly false or impossible, their meaningfulness may be questionable.

Both methods described above depend on the assumption that the data are missing completely at random. Although both methods produce valid and generalizable results if this assumption holds, neither makes the most efficient use of the data. Two points are worth making in this regard. First, we do not have any method of testing the validity of this assumption. Second, to make matters worse, Cohen and Cohen (1983) argue that in most regression applications where data are missing, this assumption is "imprudent, if not patently false" (p. 279).

## C. Traditional Imputation Methods

A number of imputation alternatives are available. Again, space constraints prevent us from presenting these methods in any detail. Rather we provide a brief description and the relevant citations. One option simply replaces the missing data points with the mean for the specific variable based on those cases not missing this variable (Cohen & Cohen, 1983). However, this procedure biases the estimated variances and covariances toward zero, which influences the regression coefficients in an unpredictable manner. Another method substitutes the missing $X_i$ values with the predicted values obtained by regressing $X_i$ on the remaining $X$ variables. This method tends to bias the correlations away from zero, as any such substituted values will be perfectly correlated with the predicted value obtained in the regression on the new complete data. Adding some random error back once the imputation is complete is one way of addressing this bias.

Both methods have an additional flaw that needs to be considered. The final regression analysis performed once the particular method of imputation is performed fails to distinguish the actual data from the imputed data resulting in an overstatement of precision (Little & Rubin, 1990). In other words, the probability values and confidence intervals obtained from the analysis may be misleading, and we have no means of measuring the effects of the imputed data on the results.

## D. Possible Solutions

We present two alternatives to the traditional methods of imputation. The first option can be used when the data contain a few predominant patterns of missing data. This approach creates groups of subjects based upon their pattern of missing data, and models the means and covariances of the groups through multisample structural equation modeling. Allison (1987) provides a detailed description of the conceptual basis for this approach and notes that this method does not require the stricter assumption that the data be missing completely at random. The second alternative is the method of multiple imputation as proposed by Rubin (1987) and described in detail by Schafer (1997). This method retains the advantages of imputation while also allowing valid large-sample inferences about all parameters. Multiple imputed values for each missing value are created based upon the predictive distribution of the specific value, and are combined to create one inference. Ultimately, with multiple imputations and large samples, this method produces interval estimates with their nominal coverages and tests with their nominal significance levels (Little & Rubin, 1990).

In conclusion, both of the alternatives presented here require the less stringent assumption that the data be missing at random as opposed to missing completely at random. The missing-at-random assumption allows for missing data when such "missingness" is conditioned upon specific values of one or more $X$ variables. In addition, both of these methods are more efficient than the options listed earlier. We suggest that researchers faced with missing data at least consider the possible ramifications of the missing data, and explore the options available to them rather than instantly deciding to delete cases via the listwise or pairwise options.

## VII. CONCLUSION

We have attempted to address both the theoretical and practical aspects of multiple regression, highlighting a number of points that we believe should be considered when using this technique. Briefly these points were as follows. Researchers should be clear as to whether the goal of their research is predictive or explanatory in nature, as this determines which statistic is of interest. The assumptions required for multiple regression analyses were presented, together with an illustration of the effect of violations of these assumptions and how such violations may be identified through diagnostics and remedied through transformations. The use of multiple regression to detect interactions and moderator effects was discussed. Sample size requirements were presented. Within this topic, the importance of being clear about the purpose for which multiple regression is used was again highlighted, as sample size requirements can vary for prediction and explanation. Finally, the issue of missing data was described and discussed.

Space constraints have influenced the scope of the discussion in terms of both potential topics and the amount of detail presented. Finally, it should be acknowledged that multiple regression forms the foundation for a number of other techniques that are becoming more widely used in psychological research. For example, the techniques of structural equation modeling (DiLalla, chapter 15, this volume) and those used for modeling change over time (Willett & Keiley, chapter 23, this volume) come to mind.

## ACKNOWLEDGMENTS

Our appreciation to David Cole for allowing us access to the data. The 89 cases analyzed here are a small subset from a larger longitudinal dataset (Cole, 1990) and are used simply to illustrate technical points in terms of diagnostics and transformations. The analysis and results presented are not intended to represent the results of the actual research project.

# REFERENCES

Aiken, L. S., & West, S. G. (with Reno, R. R.). (1991). *Multiple regression: Testing and interpreting interactions.* Newbury Park, CA: Sage.

Allison, P. D. (1987). Estimation of linear models with incomplete data. In C. Clogg (Ed.) *Sociological methodology* (pp. 71–103). San Francisco: Jossey-Bass.

Busemeyer, J. R., & Jones L. E. (1983). Analysis of multiplicative combination rules when the causal variables are measured with error. *Psychological Bulletin, 93* (3), 549–562.

Campbell, J. P. (1990). Modeling the performance prediction in industrial and organizational psychology. In M. D. Dunnette & L. M. Hough (Eds.), *Handbook of industrial and organizational psychology,* Vol. 1 (pp. 687–732). Palo Alto, CA: Consulting Press, Inc.

Cohen, J. (1988). *Statistical power analysis for the behavioral sciences.* New York: Academic Press.

Cohen, J., & Cohen, P. (1983). *Applied multiple regression/correlation analysis for the behavioral sciences.* Hillsdale, NJ: Lawrence Erlbaum Associates.

Cole, D. A. (1990). Preliminary support for a competency-based model of depression in children. *Journal of Abnormal Psychology, 100,* (2) 181–190.

Cook, R. D., & Weisberg, S. (1994a). Transforming a response variable for linearity. *Biometrika, 81,* 731–737.

Cook, R. D., & Weisberg, S. (1994b). *An introduction to regression graphics.* New York: Wiley.

Darlington, R. B. (1990). *Regression and linear models.* New York: McGraw-Hill, Inc.

Draper, N. R., & Smith, H. (1981). Applied regression analysis (2nd ed.) New York: Wiley.

Estes, W. K. (1950). Toward a statistical theory of learning. *Psychological Review, 57,* 94–107.

Fox, J. (1997). Applied *regression analysis, linear models, and related methods.* Thousand Oaks, CA: Sage Publications.

Hair, J. F., Anderson, R. E., Tatham, R. L., & Black, W. C. (1995). *Multivariate data analysis.* Englewood Cliffs, NJ: Prentice Hall.

Hempel, C. G. (1965). *Aspects of scientific explanation.* New York: Free Press.

Judd, C. M., & McClelland, G. H. (1989). *Data analysis: A model comparison approach.* San Diego: Harcourt Brace Jovanovich.

Kenny, D. A. (1979). *Correlation and causality.* New York: John Wiley & Sons.

Kovacs, M. (1982). *The Children's Depression Inventory: A self-rating scale for school-aged youngsters.* Unpublished manuscript, University of Pittsburgh, Pittsburgh, PA.

Kovacs, M. (1985). The Children's Depression Inventory (CDI). *Psychopharmacology Bulletin, 21,* 995–998.

Little, R. J. A., & Rubin, D. B. (1990). The analysis of social science data with missing values. In J. Fox & J. S. Long (Eds.), *Modern methods of data analysis* (pp. 374–409). Newbury Park, CA: Sage Publications.

Lubinski, D., & Humphreys, L. G. (1990). Assessing spurious "moderator effects": Illustrated substantively with the hypothesized ("synergistic") relation between spatial and mathematical ability. *Psychological Bulletin, 107* (3), 385–393.

MacCallum, R. C., & Mar, C. M., (1995). Distinguishing between moderator and quadratic effects in multiple regression. *Psychological Bulletin, 118* (3), 405–421.

Maxwell, S. E., & Delaney, H. D. (1990). *Designing experiments and analyzing data: a model comparison perspective.* Belmont, CA: Wadsworth Publishing Company.

McClelland, G. H., & Judd, C. M. (1989). Statistical difficulties in detecting interactions and moderator effects. *Psychological Bulletin, 114,*(2), 376–390.

Nunnally, J. C., & Bernstein, I. H. (1994). *Psychometric theory.* New York: McGraw-Hill, Inc.

Pedhazur, E. J. (1982). *Multiple regression in behavioral research.* Fort Worth, TX: Holt, Rhinehart, and Winston, Inc.

Raju, N. S., Bilgic, R., Edwards, J. E., & Fleer, P. F. (1997). Methodology review: Estimation of population validity and cross-validity, and the use of equal weights in prediction. *Applied Psychological Measurement, 21* (4), 291–305.

Rubin, D. B. (1987). *Multiple imputation for nonresponse in surveys.* New York: John Wiley & Sons.

Ruppert, D., & Aldershof, B. (1989). Transformations to symmetry and homoscedasticity. *Journal of the American Statistical Association, 84,* 437–446.

Ryan, T. P. (1997). *Modern regression methods.* New York: John Wiley & Sons.

Schafer, J. L. (1997). *Analysis of incomplete multivariate data.* New York: Chapman and Hall.

Scriven, M. (1959). Explanation and Prediction in Evolutionary Theory. *Science, 130,* 477–482.

Smith, D. A., Vivian, D., & O'Leary, K. D. (1990). Longitudinal prediction of marital discord from premarital expressions of affect. *Journal of Consulting and Clinical Psychology, 58* (6), 790–798.

Sternberg, R. J., & Williams, W. M. (1997). Does the Graduate Record Examination predict meaningful success in the graduate training of psychologists? *American Psychologist, 52,* (6) 630–641.

Wilcox, R. R. (1997). *Introduction to robust estimation and hypothesis testing.* San Diego, CA: Academic Press.

# 7

# MULTIVARIATE ANALYSIS OF VARIANCE AND COVARIANCE

**CARL J HUBERTY AND MARTHA D. PETOSKEY**

*Department of Educational Psychology, University of Georgia, Athens, Georgia*

## I. OVERVIEW

This chapter provides a review of some of the conceptual details related to multivariate analysis of variance (MANOVA) and multivariate analysis of covariance (MANCOVA) using a design involving one or two grouping variables and a collection of response variables that may include some concomitant variables (i.e., covariates). More of the chapter is devoted to MANOVA than to MANCOVA, which may be viewed as a "special case" of the former. We begin with a brief discussion of the purposes of multivariate analyses and describe research situations found in the behavioral science literature that call for the use of a MANOVA or a MANCOVA. Next we cover some MANOVA design aspects with a focus on the initial choice of a response variable system and on sampling. We discuss at length a number of suggested guidelines for data analysis strategies and for reporting and interpreting MANOVA results and illustrate these guidelines using a research example. We conclude this chapter with some recommended practices regarding the typical use of MANOVA and MANCOVA by applied researchers.

*Handbook of Applied Multivariate Statistics and Mathematical Modeling*

Readers should note that we do not discuss in this chapter repeated measures designs, which involve multiple outcome variable measures and may call for the use of a MANOVA. Generally speaking, there are two types of repeated measures designs that may be considered multivariate in nature, those involving multiple measures (e.g., over time) on a single outcome variable, and those involving multiple measures on multiple outcome variables (sometimes called a *doubly multivariate* design). With either design, a MANOVA may be conducted to test for "factor" effects. Discussions of using repeated measures design MANOVAs are given in books (e.g., Crowder & Hand, 1990, pp. 60–70; Harris, 1985, pp. 190–195; Keselman & Algina, 1996, pp. 56–58; Keselman & Keselman, 1993, pp. 118–122; Rencher, 1995, pp. 226–243) and in journal articles (e.g., Algina & Keselman, 1997; Looney & Stanley, 1989; O'Brien & Kaiser, 1985). Readers will find discussions of topics related to repeated measures designs in this *Handbook* in Mark et al.'s coverage of time series analysis (chapter 13) and Willett and Keiley's discussion of modeling growth or change (chapter 23).

## A. Research Example

Throughout this chapter we illustrate the use of MANOVA and MAN-COVA using data described in Stevens (1990, 1996). The data were obtained from an evaluation by the Educational Testing Service of the *Sesame Street* television series, which aspired to teach preschool level cognitive skills (e.g., knowledge about body parts) to children in the 3–5-year age range. The sample of 240 children was designed to be representative of five populations of interest: (a) 3- to 5-year-old disadvantaged children from the inner city, (b) 4-year-old advantaged suburban children, (c) advantaged rural children, (d) disadvantaged rural children, and (e) disadvantaged Spanish-speaking children. We use two grouping variables: the frequency with which children viewed the show, assessed using a scale that varied from 1 (rarely watched) to 4 (watched more than five times per week), and the setting in which they watched it (1 = at home or 2 = at school).

The children completed the following six tests of cognitive skills both before and after viewing the series: (a) knowledge of the names and functions of body parts, (b) ability to recognize, name, and match letters in words, (c) understanding of amount, size, and position relationships, (d) classification by size, form, number, and function, (e) ability to recognize and name forms, and (f) recognition of numbers, counting, addition, and subtraction.

## B. Computer Programs

The two computer program packages we used for analyses in this chapter are SAS (1997, Version 6.9e), which offers the GLM procedure, and SPSS

(1996, Version 7.5.1 for Windows), which offers the MANOVA procedure and the GLM procedure. Information related to MANOVA/MANCOVA may also be obtained via SAS CANDISC, SAS DISCRIM, and SAS STEP-DISC, SPSS DISCRIMINANT, and BMDP 5M and 7M (see Huberty, 1994, pp. 20–21).

## II. PURPOSE OF MULTIVARIATE ANALYSIS OF VARIANCE

There are three views of the purpose for conducting a MANOVA. Suppose, using the research example above, as an applied researcher you wished to analyze data on the effect of viewing *Sesame Street* on children's knowledge. One view of the purpose is that of *comparing* viewing frequency in terms of the six outcome variables (i.e., cognitive skills). A second view is that of studying the *relation* between viewing frequency and one or more linear composites of the six outcome variables. A third view of the analysis purpose pertains to the *effect* of viewing frequency on one or more of the linear composites. We favor the third view because it is more generally applicable, as when discussing interaction effects in the context of a multiple factor MANOVA.

Now suppose you also are interested in including the consideration of a second grouping variable, setting (i.e., at home or at school). This suggests the use of a two-factor MANOVA to study the effects of the interaction of viewing frequency and setting on a number of linear composites of the six outcome variables. Furthermore, in either the one-factor situation or the two-factor situation, you might want to remove (or partial out) the influence of, for example, pretest knowledge from the six outcome variables when assessing the effects of the grouping variable(s). These "extra" (con-comitant) variables are considered *covariates,* and you would conduct a MANCOVA. Heuristically, *but not analytically,* a MANCOVA is com-pleted (under certain correlation restrictions) via a MANOVA on residuals obtained by regressing the outcome variable(s) on the covariate(s).

Applied researchers may conduct a MANOVA when they have a col-lection of outcome variables and one or more grouping variables. MAN-COVA requires the addition of one or more concomitant response vari-ables. Two examples of research situations are Harrist, Zaia, Bates, Dodge, and Pettit's (1997) two-factor MANOVA and Simon and Werner's (1996) MANCOVA.

## III. DESIGN

### A. Grouping Variable(s)

Grouping variables or "factors" in an ANOVA or a MANOVA context should be well defined in the sense that the membership of an analysis unit in any group or factor level should be determined unambiguously at the start of a study. [It should be noted that the expression "grouping variable" is *not* used herein to describe a categorical response variable.] Choice of the grouping variable(s) will depend upon the research question(s), and, unless obvious, group definitions should be made explicit by the researcher. Assessment (and generalization) of group differences is not possible without a sound group definition. Suppose as an applied researcher you want to compare college students who perform at "high," "middle," and "low" levels in mathematics. These group designations have no reliable, unambiguous definitions. If you simply impose two cutoffs, the presence of measurement error ensures that students near the high-middle and middle-low cutoffs will be classified differently if they were to go through the evaluation process again. Another problematic situation occurs when group definitions change over time. For example, suppose you plan to compare three types of universities, with university type being defined in the year 2000. Generalizations of the type differences you discover may not be applicable in 2010 because type definition in 2010 may be different from the definition in 2000.

Applications of MANOVA and MANCOVA are possible using a variety of factorial designs involving both single and multiple grouping variables. The latter designs may involve "blocking" factors or a combination of "between" and "within" factors, as with repeated measures studies. Huberty and Lowman (1996) found some 52 applications of MANOVA and MANCOVA in six behavioral science journals in 1994 and 1995. A scan of these same journals through 1996 and most of 1997 yielded 156 additional MANOVA and MANCOVA applications, resulting in a total of 208 applications during the review period (Keselman et al., 1998).[1] Many (134 or approximately 64%) of the 208 MANOVA and MANCOVA studies used designs with more than one grouping variable.

## B. Response Variables

We use "response variable" to refer to any variable that is *not* a grouping variable (e.g., a predictor variable, criterion variable, outcome variable, or covariate, to list a few). A response variable may be continuous (e.g., intelligence) or categorical (e.g., occupational aspiration). The multiple response variables in a MANOVA context are herein labeled "outcome variables." So, in a MANCOVA situation we have multiple outcome variables and one or more covariates.

The *initial* choice of the outcome variables—and covariate(s) if applicable—is very important. It is conceptually desirable that the collection of outcome variables constitutes a "system" (i.e., a pattern of relations should

---

[1] Henceforth in this chapter this survey of applications will be referred to as "The Review."

exist among the outcome variables that can be explained in terms of a *construct*). The expectation of arriving at meaningful constructs that underlie group differences is, therefore, dependent upon the initial choice of outcome variables. Applied researchers should consider conducting multiple ANOVAs when the collection of outcome variables does not constitute a "system" of conceptually related variables (relations that are based on well-developed theories and beliefs) or multiple MANOVAs when more than one outcome variable system is of theoretical interest.

The importance of the choice of initial variables also applies to the choice of covariate(s). Applied researchers should base the decision to include one or more covariates (theoretically or empirically supported) on substantive reasons. Do not use covariates in an attempt to equate what may be judged to be nonequivalent groups. Attempting to equate groups via a covariance analysis may yield misleading results because the adjustments that the covariates yield do not compensate for initial group inequality—see Porter and Raudenbush (1987) for a discussion of this issue in a single-outcome-variable context. Group equivalence is a judgment call to be made (and discussed) by the researcher.

The use of covariates in multivariate group comparison studies has been somewhat limited, at least in the behavioral sciences. The Review indicated that only 39 out of 208 studies (approximately 19%) used MANCOVA.

## C. Sampling

There are at least four ways to sample children for a study involving multiple groups. For simplicity, suppose you want to compare the means of three groups on several outcome variables. One approach is to select a group of analysis units from each of the three well-defined populations using proportional random sampling. A second approach is to randomly select one sample of units and then randomly assign units to the three groups. A third approach is to obtain information from an available set of analysis units—a convenience or grab sample—and then randomly assign those units to the three groups. The final approach is to obtain information from three convenience samples.

We will not elaborate upon the pros and cons of these four sampling designs, but we offer four brief comments. First, drawing a random sample is not possible unless the researcher has access to the population to be analyzed. Second, the sampling design depends on group definition. If the grouping variable is manipulable, then one can assign analysis units to the groups; otherwise, random assignment is not feasible. If the research situation involves naturally existing groups, random assignment is out of the question. Third, the random selection of units does *not* ensure representativeness. Finally, the random assignment of analysis units to groups does *not* ensure group equivalence with respect to all relevant characteristics. The best sampling design is that which yields the greatest likelihood that the analysis groups will be representative of their respective populations. Applied researchers may have greater confidence in the likelihood of representativeness and equivalence when the number of analysis units involved in the randomization procedures is "large." Always use the best sampling procedure possible, describe the sampling process used, and use your best professional judgment and common sense to ensure the highest degree of representativeness possible.

The Review revealed that investigators seldom use either form of randomization (i.e., sample selection or group assignment). In 18 (approximately 9%) of the 208 applications, the investigators reported that they used random selection (all from restricted "populations"), and 37 investigators (approximately 18%) reported use of random assignment. These findings do not describe condemnable or laudatory practices; they merely reflect what may be common practice in the behavioral sciences.

## D. Sample Size

There are a number of factors to consider when determining the desired sample size to use with MANOVA. Perhaps most importantly, applied researchers must consider what is necessary to ensure the *representativeness* of the sample. Five other considerations when determining necessary group size for the conduct of a MANOVA are (a) the probability of committing a Type I error in the statistical testing process; (b) the probability of committing a Type II error; (c) the number of design groups/cells; (d) the number of outcome variables ($p$); and (e) the size of the effect of interest. Although sample-size determination is based on a complex array of factors, some simple-minded recommendations are available. Kres (1983, pp. 432–451) provides sample-size tables that call for specifying some "parameters" (e.g., $\alpha$, $p$, effect size) and will yield desirable sample sizes. The Kres (1983) table information suggests that the size of the smallest group range from six to ten times the number of outcome variables (i.e., from $6p$ to $10p$). Whether one leans toward $6p$ or $10p$ would depend on the five considerations listed above. For example, if $p$ is "small" and a "large" group separation is expected, then one could "get by" by using $6p$ for the smallest group size ($n$). On the other hand, $10p$ is preferable when $p$ is "large" and little group separation exists.

A researcher's resources often are limited, and those limitations usually restrict the size of the total sample somewhat. Therefore, it behooves researchers to think carefully about their design(s) and particularly about the number of groups and the number of outcome variables initially to be chosen. Our research example includes four levels of viewing frequency, with all group sizes ($n_1 = 54, n_2 = 60, n_3 = 64, n_4 = 62$) of approximately $10p$.

## IV. ANALYSIS GUIDELINES

### A. Data Characteristics

The data set to be analyzed using MANOVA may be visualized as a (total number of analysis units, $N$)-by-(number of outcome variables, $p$) matrix. We discuss four aspects of entries in the $N \times p$ data matrix.

### 1. Independence

The $p$-vector of scores in one row (i.e., the scores for one analysis unit on all of the variables) is assumed to be independent of the score vectors in all other rows. This condition should not present a problem because it is under the control of the researcher.

### 2. Data Transformation

Scores for any outcome variable must appropriately represent the analysis unit. This typically is not a problem with a continuous outcome variable that is measured using an ordinal, interval, or ratio scale, but use of a nominal scale may present some problems. At least three uses of a nominal scale are possible. One situation involves a two-category or binomial variable (e.g., gender, drop-out versus stay-in); simple 0–1 scoring is acceptable in this instance. Integer scaling is appropriate for a variable having three or more *ordered* categories (e.g., "1" for assistant professor, "2" for associate professor, and "3" for full professor). Integer scaling also is appropriate for a continuous outcome variable upon which three or more categories have been imposed (e.g., trichotomizing mathematics achievement scores into "low," "middle," and "high" categories). With such imposed categories, researchers must consider carefully the measurement error around the two cutoffs that constitute the category definitions.

The most complicated variable scoring situation occurs when the researcher uses a categorical variable having three or more *unordered* categories. In this situation investigators typically use dummy or indicator variables to represent the categorical variable. We do not favor this approach. Use of dummy variables may increase the number of outcome "variables" used in the final analysis; thus inferences made may be questionable because of estimation problems (see Huberty, 1994, p. 152). Instead we recommend that researchers scale the unordered categorical variable using the Fisher-Lancaster (F-L) method (Huberty, 1994, pp. 153–154). The F-L method yields scale values that are relative to one another and have much more utility than dummy variable values (Huberty, Wisenbaker, Smith, & Smith, 1986).

Yet another data transformation is appropriate in some situations. Suppose a large number of scores were available for each analysis unit, as might result when a 40-item survey instrument is used to obtain outcome variable measures. Assume that the 40 items use the same measurement scale. In this situation, some *data reduction* may be appropriate. For example, it may be desirable to group the items into subsets, each of which measures a common construct, and sum these item scores to produce an outcome "variable" score. One way to group the items into subsets is on the basis of a subjective analysis of the item content. Another way is to conduct a principal component analysis (see Cudeck, chapter 10, this volume) to obtain a set of component scores that would be considered outcome variable scores.

### 3. Outliers

An initial step in any data analysis sequence is to examine the data matrix for aberrant measures (i.e., misrecordings, outliers) on the $p$ outcome variables. An outlier is an individual analysis unit whose vector of outcome variable scores is relatively distant from the centroid of the unit's group. [There is no formal, specific, general, agreed-upon empirical definition of an outlier for all data sets.] Identifying multivariate outliers may be somewhat involved because a univariate outlying measure may not necessarily indicate a multivariate outlier. One way to detect multivariate outliers is to run the SAS OUTLIER macro. Friendly (1991, pp. 570–572) illustrates the OUTLIER command statements.

Using our research example, we examined the 240 × 6 data matrix for missing and aberrant scores; none were apparent. Then we used the SAS OUTLIER macro and detected one outlier using a $P$ value of .01. We judged this child's scores on the six outcome variables to be a "large" distance from the centroid of her or his group, so we excluded this child's data from the analysis.

Another way to identify outliers is to calculate the squared (Mahalanobis) distance between each complete score vector and the respective group centroid. Assuming that it makes sense to use the $p \times p$ pooled (i.e., error) covariance matrix based on all complete score vectors from all $k$ groups, list the $N$-squared distances in rank order and look for a "gap" in the distribution. If the $p \times p$ covariance matrices for the $k$ groups are not "in the same ballpark," then use separate-group covariance matrices to calculate the Mahalanobis distances; technically, this may present a mixing of distance metrices across all $N$ analysis units, but "weird" unit vectors may still be identified in the form of a "gap" in the distribution of $N$-squared distances.

Every approach to identifying outliers requires some judgment calls. Applied researchers should examine each suspect score profile and make a common-sense judgment as to whether to retain it in the final analysis. In addition, we recommend that you run analyses with outliers included and with outliers excluded and compare the two sets of results, again making a common-sense decision. Finally, sometimes it may be reasonable to delete identified outliers one at a time; all steps would call for some judgments.

## 4. Missing Data

Roth (1994) provides a very readable discussion of the missing data problem. The more adept quantitative researcher may want to refer to Little and Schenker (1995), Rencher (1998, pp. 23–27), and Schafer (1997). We offer some guidelines here. First, examine the columns of the data matrix and delete any outcome variable if a "large" proportion of the entries in its column are empty. Make a similar examination of the data matrix rows, and delete any analysis unit if the proportion of empty cells is "large" for that unit. What constitutes "large" in either case is a judgment call.

The best strategy for replacing missing data is an issue among methodologists. We discuss two general strategies that have been proposed. Suppose analysis unit $u_i$ has a single missing score for outcome variable $Y_j$. The first strategy is to replace that missing entry with either the mean on $Y_j$ of the group of which $u_i$ is a member or the mean on $Y_j$ based on all $N$ units. There is disagreement as to which mean is more appropriate. Proponents of the first mean argue that it is more reasonable because we are dealing with a missing score for a unit in a particular group and that mean represents a better "fit" if a grouping variable effect is present in the data. Others argue that the second mean is preferable because it is more conservative in that it favors the MANOVA null hypothesis. We favor the first mean. Data imputation using means may be accomplished via the SAS STANDARD procedure.

A second data imputation method, termed "hot-deck imputation," is easy to apply, especially when the number of outcome variables ($p$) is reasonably small. Suppose $p = 4$ and a $Y_3$ score is missing for a particular analysis unit. Identify another unit (in the same group as the unit with the missing score) that has a complete score vector and has ($Y_1$, $Y_2$, $Y_4$) scores reasonably close to those of the unit for which the $Y_3$ score is missing, and use the $Y_3$ score in the complete vector to replace the missing $Y_3$ score.

## B. Analysis Sequence

### 1. Descriptives

The reporting of descriptive information is essential to inform the reader about the data set and to help the reader (and the researcher) judge the representativeness of the sample. Descriptive information may be reported in tabular or graphical form. The best format for a table is a group-by-group summary for each variable that provides the maximum score, the three quartiles, and the minimum score (see Table 7.1 for numerical descriptives for our data set). A set of box-and-whisker plots or boxplots may represent the same numerical information (see Figure 7.1 for graphical descriptives for our data set). Such plots provide less precise information

**TABLE 7.1** Five-Point Summaries for Four Viewer Frequency Groups on Four Outcome Variables

| Outcome variables | Minimum score | $Q_1$ | $Q_2$ | $Q_3$ | Maximum score |
|---|---|---|---|---|---|
| Rarely watch ($n = 53$) | | | | | |
| Body parts | 11 | 17.5 | 21.0 | 26.5 | 32 |
| Letters | 0 | 13.0 | 15.0 | 17.5 | 46 |
| Forms | 0 | 8.5 | 11.0 | 13.0 | 19 |
| Letters | 0 | 14.0 | 19.0 | 26.5 | 44 |
| Sometimes watch ($n = 60$) | | | | | |
| Body parts | 11 | 21.3 | 26.0 | 28.0 | 32 |
| Letters | 9 | 15.0 | 19.5 | 28.8 | 52 |
| Forms | 4 | 11.0 | 13.0 | 15.8 | 20 |
| Numbers | 7 | 19.0 | 26.0 | 36.0 | 52 |
| Often watch ($n = 64$) | | | | | |
| Body parts | 11 | 22.3 | 28.0 | 31.0 | 32 |
| Letters | 11 | 19.0 | 32.0 | 43.0 | 53 |
| Forms | 3 | 12.0 | 15.0 | 17.0 | 20 |
| Numbers | 8 | 22.0 | 35.5 | 43.8 | 53 |
| Watch >5 times per week ($n = 62$) | | | | | |
| Body parts | 17 | 27.0 | 29.0 | 31.0 | 32 |
| Letters | 13 | 22.8 | 35.0 | 46.0 | 54 |
| Forms | 11 | 15.0 | 17.0 | 19.0 | 20 |
| Numbers | 13 | 29.0 | 40.0 | 47.0 | 54 |

than a table, and they are more costly to print, but we recommend them for communicating with lay audiences.

Another type of descriptive information essential to report is the correlations among the outcome variables (see Table 7.2 for additional descriptives for our data set). If the covariance matrix homogeneity condition (discussed below) is reasonably satisfied, report the $p \times p$ error (i.e., within-groups) correlation matrix. When homogeneity cannot be concluded report the separate $p \times p$ correlation matrices for each group.

## 2. Covariance-Matrix Equality and $p$-Variate Normality

Two data requirements pertain to the $k$ covariance matrices and to outcome variable score distribution form. Conducting a MANOVA, technically, should be done only if the $k$ covariance matrices are approximately equal and the $p$ outcome variable scores approximate a $p$-variate normal probability distribution. Maxwell (1992) has reported some evidence that the MANOVA test is robust with respect to relatively minor distortions from $p$-variate normality, provided the group sizes are "respectable" (see the Sample Size subsection). Consideration of these issues may lead applied researchers to a paradoxical dilemma. Increasing sample size to improve the robustness of the MANOVA test for a lack of normality also increases the power of the test for covariance matrix equality, thereby increasing the likelihood that this test will spuriously suggest that significant differences exist among the covariance matrices.

Two tests of covariance matrix equality are available in the statistical computer packages. The SAS DISCRIM procedure yields a chi-squared statistic; the SAS GLM procedure does not yield homogeneity test information. The SPSS MANOVA procedure yields an approximate chi-squared (Bartlett) statistic and an approximate $F$ (Box) statistic. We prefer the information provided with the Box test.

Both the Bartlett test and the Box test are extremely powerful, so we recommend the following procedure. Proceed with the MANOVA test if the $P$ value for the Box $F$ test is higher than .005. If the $P$ value is less than .005, check the logarithms of the error covariance matrix and of the $k$ group covariance matrix. [SPSS MANOVA outputs with the Box test the values of the (natural) logarithms of all covariance matrices.] Proceed with the MANOVA test if the $k + 1$ logarithms are "in the same ballpark," (a judgment call made by the researcher). Another alternative is to compare the sum of the main diagonal elements for each of the $k + 1$ covariance

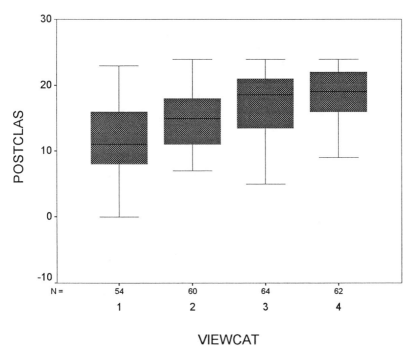

VIEWCAT

**FIGURE 7.1**   Box-and-whisker plot for postclas ($N = 239$).

**TABLE 7.2** Mean, Standard Deviation, and Error Correlations among Six Outcome Variables for Total Sample ($N = 239$)

| Outcome variable | VIEWCAT mean/(SD) | | | | Error correlations | | | | |
|---|---|---|---|---|---|---|---|---|---|
| | 1 | 2 | 3 | 4 | Letters | Forms | Numbers | Relational terms | Classification |
| Body parts | 21.4 (5.59) | 24.5 (5.14) | 26.5 (5.20) | 28.0 (3.42) | .56 | .54 | .57 | .45 | .51 |
| Letters | 16.3 (8.45) | 23.3 (11.28) | 30.9 (12.69) | 34.2 (12.26) | | .54 | .77 | .50 | .60 |
| Forms | 10.6 (3.83) | 13.1 (3.63) | 14.4 (3.77) | 16.4 (2.46) | | | .60 | .53 | .73 |
| Numbers | 21.2 (10.94) | 27.50 (11.53) | 33.0 (12.38) | 36.8 (11.08) | | | | .63 | .72 |
| Relational terms | 9.9 (3.06) | 11.4 (2.60) | 11.9 (2.71) | 13.1 (2.05) | | | | | .56 |
| Classification | 11.7 (4.81) | 14.8 (4.62) | 17.0 (4.62) | 18.8 (3.89) | | | | | |

matrices—such a sum is called a matrix *trace*. Proceed with the MANOVA if the $k + 1$ traces are also "in the same ballpark." Alternatively, if the $P$ value is $< .005$ and the logarithms and traces are clearly unequal, use statistical tests that do not require homogeneity. We discuss these tests below (A rationale for considering the determinants of the covariance matrices is that each may be viewed as a *generalized variance*. The trace of a covariance matrix, which is the sum of $p$ univariate variances, is also an index of generalized variance. The approximate equality of $k + 1$ generalized variances [determinant logarithms or traces] provides support for covariance matrix homogeneity.)

For our research example, the Box test resulted in significance ($F$ (63, 124701) = 2.039, $P \doteq .000$). This $P$ value would be considered "small" by most researchers, but the large $df_2$ value makes the test very powerful. As aforementioned, there is other numerical information (i.e., determinants and traces) that may be considered when making a judgment about the equal-covariance-matrix condition. For our data, there are four group covariance matrices to consider plus the error covariance matrix. The four group log determinants are 17.1, 16.8, 16.8, and 14.3, and the log determinant of the error covariance matrix is 16.8. The traces are, respectively, 269.6, 327.8, 384.4, 309.9, and 325.4. In our judgment, neither the five matrix log determinants nor the five matrix traces are "wildly different." Therefore, in this case, we judge that the equal-covariance-matrix condition is not violated. (Thus, it is reasonable to report the *error* correlation matrix as in Table 7.2.)

To statistically address the question of multivariate normality for our research example, we began with a graphical assessment by using quantile plots; SAS yields a plot for each group automatically in conjunction with the OUTLIER macro noted above (see Friendly, 1991, pp. 570–572). The plot for group 1 of our research example is given in Figure 7.2. This plot was the most deviant of the four. (The "2" in the upper-right position represents the outlier that we deleted.) Because the observed values did not deviate appreciably from the expected distribution (i.e., they generally fell on the expected line), we proceeded with our analysis under the assumption that the multivariate normal condition was plausible.

### 3. Test Statistics

When the $k$ covariance matrices are approximately equal, researchers can choose from at least four criteria (i.e., Bartlett-Pillai, Hotelling-Lawley, Roy, and Wilks) for testing the MANOVA and MANCOVA omnibus null hypothesis that the means of the $k$ groups on the $p$ outcome measures do not differ significantly. Detailed discussions of these criteria and their connections with MANOVA are available in Huberty (1994, pp. 183–189). The SAS CANDISC, SAS GLM, SPSS GLM, and SPSS MANOVA programs output the four criterion values and test statistic transformations of

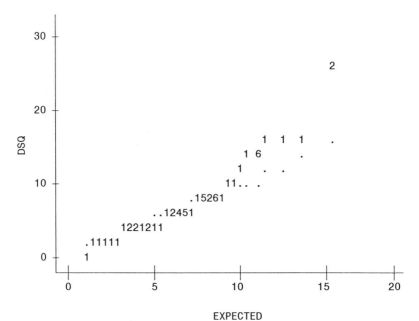

**FIGURE 7.2**   Chi-squared probability plot for group 1 (rarely watches). Note; 37 obs hidden. "." indicates expected value under normality.

all except Roy. There are some arguments for preferring one or more of these criteria in some research situations (see, e.g., Maxwell, 1992; Morrison, 1990, p. 256), but our discussion focuses on the Wilks's lambda criterion ($\Lambda$).

Applied researchers must use other criteria when they judge the $k$ covariance matrices to be clearly *unequal*. When $k = 2$, the Yao test provides a reasonable criterion (see Huberty, 1994, pp. 199–200, for a computer program for the Yao test); Johansen (1980) and James (1954) suggest criteria that are appropriate when $k \geq 3$ (Coombs, Algina, & Oltman, 1996).

### 4. Effect Size

The basis for rejecting the MANOVA and MANCOVA omnibus null hypothesis should be more than simply the $P$ value associated with the test statistic (e.g., an $F$ transformation of the Wilks's lambda). Stated equivalently, we must have more evidence than a "small" $P$ value. An *effect-size index* provides such additional evidence. A generalization of the univariate squared point-biserial correlation (for the two-group situation) and the univariate squared eta (in the $k$-group situation) is available in the multiple outcome variable context. A multivariate effect-size index that involves a single outcome variable composite is

$$\eta^2 = 1 - \Lambda, \tag{1}$$

which is applicable for $k \geq 2$. (This index is modified for multiple outcome variable composites, see Huberty, 1994, p. 194.) Just as in the univariate context, there is an adjusted index:

$$\eta^2_{\text{adj}} = \eta^2 - \frac{p^2 + q^2}{3N}(1 - \eta^2), \tag{2}$$

where $q$ is the hypothesis degrees-of-freedom value. Adjusting the $\eta^2$ value is highly desirable to reduce the bias in estimating the relation between the grouping variable and the leading composite of the outcome variables. If one opts for a MANOVA criterion other than the Wilks's lambda, related effect-size indexes are available (Huberty, 1994, p. 194). The SPSS MANOVA program outputs unadjusted effect-size index values (based on the maximum number of outcome variable composites) for the Bartlett-Pillai, Hotelling-Lawley, and Wilks's criteria; thus the researcher must calculate the adjusted values.

When covariance matrix *homogeneity* exists, researchers can conclude that real effects exist when the $P$ value is "small" and the effect-size value is "substantial." Guidance for assessing the magnitudes of the two values in a MANOVA context is very limited. Again, judgment calls are necessary. To our knowledge, no one has proposed an effect-size index for use when covariance *heterogeneity* exists. One possibility that shows some promise is based on the concept of group overlap and involves the use of predictive discriminant analysis results (see Huberty & Lowman, in press).

Using our research example, because we concluded that the conditions of covariance-matrix homogeneity and $p$-variate normality were satisfied, we conducted the four-group, six-variable (i.e., one-factor) MANOVA. Results of such an analysis are available from SAS CANDISC, SAS DIS-CRIM, SAS GLM, SAS STEPDISC, SPSS DISCRIMINANT, SPSS GLM, and SPSS MANOVA.

This first analysis yielded a Wilks's lambda ($\Lambda$) value of .6480, $F$ (18, 651.02) $\doteq$ 5.995, $P \doteq .000$. Also, $\eta^2 \doteq 1 - .6480 = .3520$, and $\eta^2_{\text{adj}} \doteq .3520 - [(6^2 - 18^2)/(3)(239)](1 - .3520) \doteq .311$. Considering the $P$ value and $\eta^2_{\text{adj}}$ value jointly, we concluded that frequency of viewing *Sesame Street* had a "real" effect on knowledge of the six preschool-related skills. The outcome variable means in Table 7.2 suggest that a mean increase in knowledge and skills accompanied increases in the frequency of viewing *Sesame Street* (from group 1 to group 4).

### 5. Variable Ordering

Researchers often want to know about the relative "importance" of the $p$ outcome variables in distinguishing among the groups. Answering the relative importance question involves assessing group separation when a variable is deleted. If the deletion of a variable decreases the group separation, then the variable is considered "important." Therefore, conceptually the researcher must conduct $p$ MANOVAs, each with $p - 1$ variables, and determine the relative decreases in separation attributable to each variable. The variable that when deleted is associated with the largest decrease in overall group separation (i.e., increase in Wilks's lambda) is considered the most important.

It turns out, however, that for a one-factor MANOVA only a single analysis, usually termed a "stepwise discriminant analysis," (output via SAS STEPDISC or SPSS DISCRIMINANT) is necessary. Examination of the $F$-to-remove values in the summary step will reveal the rank order of the variables' relative importance. Assign tie ranks to $F$ values that are "close" in the *summary* step (see Huberty & Wisenbaker, 1992). It is important that researchers not become confused and use the multiple-step information from a stepwise analysis to determine variable rank ordering (Huberty, 1989).

Using our research example, we conducted six MANOVAs, each using five outcome variables, to assess the relative contributions of the six outcome variables to group differentiation. We concluded that the six outcome variables were equally important with respect to contribution to overall group separation (see Table 7.3). (Huberty and Wisenbaker, 1992, discuss a more formal approach—as opposed to eye-balling the lambda values—to determining variable rank ordering.)

### 6. Descriptive Discriminant Analysis

To further interpret these omnibus results, applied researchers should use some descriptive discriminant analysis (DDA) techniques (see Brown & Wicker, chapter 8, this volume). Two types of interpretations that may be considered (see Huberty, 1994, chap. XV, for details) are (a) structure, that is, the meaning of the constructs underlying the group differences defined by the linear discriminant functions (LDFs); and (b) the typology described graphically by plotting the four group centroids in the space of the LDFs. It is in the context of a DDA that Thomas (1997) suggests an index for

**TABLE 7.3** Wilks's Lambda Values Associated with Deleted Outcome Variables

| Outcome variable deleted | Wilks's lambda | Rank |
|---|---|---|
| Letters | .681 | 3.5 |
| Forms | .667 | 3.5 |
| Classification | .657 | 3.5 |
| Body parts | .654 | 3.5 |
| Numbers | .654 | 3.5 |
| Relational terms | .654 | 3.5 |
| (None) | .648 | |

outcome variable ordering that is an alternative to the above-noted all-but-one-variable approach.

## C. Additional Analyses

### 1. Variable Selection

A stepwise analysis is commonly used first to *order* the outcome variables and then, inappropriately, to *delete* some of the outcome variables. The researcher should have devoted considerable thought to the relevant substantive theory, beliefs, and related research and should have selected variables to measure some specific underlying (latent) construct(s). Therefore, the finding that one or more variables do not contribute to construct definition is, in itself, informative. Furthermore, construct definition is dependent on the collection of variables being studied. The bottom line from our perspective is that use of MANOVA implies that the initial choice of outcome variables was made on some theoretical basis, and outcome variable selection in a MANOVA context is inappropriate (see Section III. B. Response Variables and Design).

### 2. Contrast Analyses under Covariance-Matrix-Homogeneity

Applied researchers who perform a MANOVA and MANCOVA with three or more groups may want to study overall group differences by conducting an *omnibus* test as noted above. The group typology can then be examined by plotting the group centroids (i.e., group mean vectors) in the orthogonal geometric space formed by the linear composites of the outcome variables (see Donnay & Borgen, 1996, and Walsh et al., 1996, for examples). However, in many instances the investigator may be interested in more focused questions than the omnibus question. Researchers who want to study a more specific set of questions should conduct one or more *contrast* analyses rather than an omnibus analysis. A mean vector contrast is simply a linear combination of the $k$ outcome variable mean vectors (such that the sum of the combination weights is 0).

In our research example a comparison of only two groups (e.g., rarely watches versus watches more than five times per week) might be more interesting to some researchers than an omnibus comparison. Another investigator might want to compare the two groups who watch rarely and sometimes with those who watch more than five times per week. These comparisons would require a *pairwise contrast* analysis and a *complex contrast* analysis, respectively.

Our earlier discussion of the test criterion (e.g., Wilks's lambda), effect size, linear composite, covariance matrix equality, and variable ordering in the context of an omnibus analysis applies with equal force to contrast analyses, but the procedure for interpreting the $P$ value for a single contrast test differs. To get the $P$ value for each test, multiply the test tail-area probability by the number of contrast tests conducted—this is an application of the Bonferroni adjustment idea. Researchers should note that performing an omnibus analysis is *not* a prerequisite for performing a set of contrast analyses.

With the current example, at least three *pairwise* contrasts (represented as $C_1 = \underline{\mu}_1 - \underline{\mu}_4$, $C_2 = \underline{\mu}_2 - \underline{\mu}_4$, and $C_3 = \underline{\mu}_3 - \underline{\mu}_4$) may be of interest to address the question, Are there any differences between each of the first three *Sesame Street* viewing frequencies and the fourth frequency? Alternatively, an investigator might want to address the more *complex* question, How does average or less frequent viewing compare with very frequent viewing (i.e., $C_4 = \underline{\mu}_1 + \underline{\mu}_2 + \underline{\mu}_3 - 3\underline{\mu}_4$)? Contrast analyses can be accomplished via either SAS GLM, SPSS MANOVA, or SPSS GLM.

The top part of Table 7.4 notes results of the four contrast analyses under covariance matrix homogeneity. Two of the pairwise contrasts ($G_1$ vs. $G_4$ and $G_2$ vs $G_4$) and the complex contrast ($G_1$, $G_2$, $G_3$ vs $G_4$) yielded substantial effects. That is, the differences between the two lowest frequencies and most frequent *Sesame Street* viewing had an effect on a linear composite (not defined in this chapter) of the six knowledge variables. The conclusions are similar with regard to the complex contrast.

For the first two pairwise contrasts and the complex contrast it may be of substantive interest to determine an ordering of the six outcome variables. The all-but-one-variable analyses indicated, in our judgment, that the six variables are equally important contributors to each of the three effects.

Finally, typically it would be of interest to determine the single substantive construct that underlies a significant contrast effect. This may readily be accomplished by examining the structure $r$,s for the first and only LDF.

### 3. Contrast Analyses under Covariance Matrix Heterogeneity

For purposes of illustration, suppose that the assumption of covariance matrix equality is *not* tenable for our research example. The usual contrast analyses conducted via the major statistical packages are inappropriate because they assume homogeneity. To conduct a pairwise contrast analyses under heterogeneity one can use a program written by Y. Yao, which is on a diskette that accompanies Huberty (1994). We analyzed the three above-noted pairwise contrasts ($C_1$, $C_2$, $C_3$) via the Yao program and presented the results in the bottom part of Table 7.4. The conclusions regarding these contrast effects would be the same as those obtained under the covariance homogeneity assumption (The *I* index value in Table 7.4 reflects the amount of group overlap; see Huberty & Lowman, in press.)

### 4. Multiple Analyses

Some investigators may have a theoretical interest in studying multiple systems of outcome variables. This is likely to occur when the researcher

**TABLE 7.4   Wilks's Lambda F, P, and Effect Size Values for Contrast Analysis (N = 239)[a]**

| Contrast | Lambda | $F(6, 230)$ | P | $\eta^2_{adj}$ |
|---|---|---|---|---|
| Homogeneous covariance matrices | | | | |
| $G_1$ vs $G_4$ | .690 | 17.184 | .000 | .274 |
| $G_2$ vs $G_4$ | .860 | 6.217 | .000 | .095 |
| $G_3$ vs $G_4$ | .947 | 2.131 | .204 | .004 |
| $G_1$, $G_2$, $G_3$ vs $G_4$ | .783 | 10.610 | .000 | .176 |

| Contrast | $T^2$ | $F(,)$ | P | I |
|---|---|---|---|---|
| Heterogeneous covariance matrices | | | | |
| $G_1$ vs $G_4$ | 123.59 | 19.656 (6, 104.3) | .000 | .670 |
| $G_2$ vs $G_4$ | 40.85 | 6.497 (6, 104.3) | .000 | .459 |
| $G_3$ vs $G_4$ | 16.62 | 2.628 (6, 92.8) | .064 | .381 |

[a] $G_1$ = rarely watches, $G_2$ = sometimes watches, $G_3$ = often watches, $G_4$ = watches > 5 times per week. Computer-generated tail areas multiplied by 3 to obtain the contrast $P$ values.

is interested in two or more outcome variable systems, and it would make no theoretical sense to combine them in an effort to identify construct(s) that would "cut across" the systems. In this case, *multiple* independent MANOVAs (or MANCOVAs) would be conducted.

## D. Two-Factor Multivariate Analysis of Variance

Approximately 64% of the studies found in The Review used a multiple-grouping-variable design. We illustrate this design with a 2 (setting)-by-4 (viewing frequency) analysis with the same six outcome variables as for the one-factor MANOVA example. We judged two of the original 240 children to be outliers, thus we proceeded by analyzing the 238 × 6 data matrix. To conserve space, we do not report in this chapter the numerical descriptives for the eight cells. Examination of the eight quantile plots yielded via the SAS OUTLIER macro indicated that the multivariate normality condition was not seriously violated for any of the eight groups, and we judged that the $P$ values for the Box $M$ test ($P \doteq .000$) in conjunction with the log determinants and the traces of the nine covariance matrices (eight cells plus error) did not provide enough evidence to conclude covariance heterogeneity. We therefore conducted the two-factor MANOVA.

First we tested for a setting-by-viewing frequency interaction and judged that the results (Wilks's lambda ($\Lambda$) $\doteq .8772$, $F(18, 639.9) \doteq 1.677$, $P \doteq .039$, and $\eta^2_{adj} \doteq .075$) did not support the conclusion of real interaction effects. Next, we investigated the possibility of significant main effects for setting and viewing frequency. [If we had judged interaction effects to be real, then we would have assessed viewing frequency omnibus or contrast effects for *each* setting level.] We judged the setting effect to be unimportant ($\Lambda \doteq .9369$, $F(6, 225) \doteq 2.523$, $P \doteq .022$, $\eta^2_{adj} \doteq .012$), while the omnibus viewing frequency analysis ($\Lambda \doteq .6185$, $F(18, 636.9) \doteq 6.551$, $P \doteq .000$, and $\eta^2_{adj} \doteq .348$) revealed an important effect. One way to view these effects is to examine a plot of the four VIEWCAT group centroids in the orthogonal space of the first two LDFs. A next step for the three sets of LDFs for which the effects are considered real would be to examine sets of LDF structure $r$'s for substantive constructs underlying the effects.

## E. Multivariate Analysis of Covariance

Multivariate analysis of covariance (MANCOVA) is more involved than MANOVA from three standpoints: substantive theory, study design, and data analysis. Deciding on whether to remove, or partial out, the effects of one or more concomitant variables from a set of outcome variables is a serious decision that takes considerable thought, study, substantive knowledge, and judgment. There are no general rules for making such a decision. With respect to design, the measures on the covariates are typically collected *prior to* collecting data on the outcome variables. An important design issue is that the covariates must not be numerically affected by the grouping variable(s) (see Rencher, 1998, pp. 178–179, for a little more detail). In terms of analysis, the regression homogeneity condition analog in MANCOVA is considerably more involved than the condition in MANOVA. In MANCOVA we are concerned with the parallelism of the $k$ multiple regression hyperplanes; that is, the lack of interaction between the set of covariates and the grouping variable. The condition of simple regression (or correlation) homogeneity across the $k$ groups must be checked to ensure the equality of slopes of the $k$ regression hyperplanes and to establish the lack of interaction between the covariate and the grouping variable. (We will abbreviate discussion of results similar to those in the MANOVA context.)

The research example we use to illustrate MANCOVA is an extension of that used for MANOVA example. The grouping variable is viewing frequency, and the six outcome variable measures are the same as for the MANOVA example. The covariate measures are the six corresponding pretest measures. A 239 × 12 data matrix now replaces the previous 239 × 6 matrix. None of the scores on the six pretests were aberrant. We identified one additional outlying child (via the OUTLIER macro); thus, the $N$ for our MANCOVA is 238. A reporting of numerical descriptives such as those in Tables 7.1 and 7.2 would include the six covariates.

MANCOVA essentially analyzes the effects of the grouping variable on the six outcome variables after these have been adjusted for the six covariates. An examination of the five adjusted covariance matrices—four group matrices and one error matrix—indicated that the four adjusted covariance matrices are not appreciably different. (Even though the $P$ value for the test is .000, the $F[234, 113038] \doteq 1.371$ is not "large," and the five determinant matrix logarithms were very close in numerical value, as were the five covariance matrix traces.)

Including the set of covariates in a study may not be too meaningful unless they have some relation with the set of outcome variables. It is also advisable to supply evidence that the collection of covariates is correlated with the collection of outcome variables. We recommend assessing that correlation via perusal of the leading *error* canonical correlation between the set of outcome variables and the set of covariates. SPSS MANOVA automatically provides this information and labels it "canon cor." under "Eigenvalues and canonical correlations" associated with the "EFFECT … WITHIN CELLS Regression" results. To obtain this information via SAS, the researcher must first obtain the MANCOVA error correlation or covariance matrix and then subject this matrix to a canonical correlation analysis via PROC CANCORR. For the current data set, the leading *error* (i.e., pooled across the four groups) canonical correlation is .714, which we judged to be appreciable.

The SPSS results and the SAS results regarding assumptions are somewhat conflicting because SPSS assesses the interaction between the *set* of covariates and the grouping variable, whereas the SAS GLM procedure outputs tests of interactions between *each* covariate and the grouping variable. The SPSS results indicated that it may not be reasonable to assume that the four regression hyperplanes are approximately "parallel" because the single multivariate test result was $\Lambda \doteq .5316$, $F(108, 1182.04) \doteq 1.276$, $P \doteq .035$, and the six SAS $P$ values range from .056 to .700 with a median of approximately .200; the six $\eta^2_{adj}$ values range from .008 to .073 with a median of approximately .050. We will proceed with our analysis example under the assumption that the parallelism condition is met.

The Wilks's lambda for adjusted mean vector equality is .7356 with $F(18, 631.22) \doteq 4.021$, $P \doteq .000$, and $\eta^2_{adj} \doteq .218$. Thus, we concluded that the four adjusted mean vectors (see Table 7.5) are not equal; that is, viewing frequency had an effect on the six outcome variables when adjusted for the six covariates. That information, in and of itself, may not be too helpful, although a plot of the adjusted mean vectors in the two-dimensional space of the LDFs may very well reveal an interesting typology of the four groups. Contrast analyses that answer focused research questions involving subsets of the groups may be of greater interest to the researcher.

Whether one is interested in MANCOVA omnibus or contrast effects, it would be of substantive interest to examine the structure associated with the resulting effects. Just as in a MANOVA context, one may accomplish this by examining constructs underlying each significant MANCOVA effect—via LDF structure $r$'s.

**TABLE 7.5** Adjusted Means on the Outcome Variables for the Four Viewer Frequency Groups ($N = 238$)[a]

| Outcome variable | Group | | | |
|---|---|---|---|---|
| | 1 | 2 | 3 | 4 |
| Body parts | 23.00 | 24.87 | 25.83 | 26.67 |
| Letters | 19.31 | 23.81 | 29.86 | 31.21 |
| Forms | 11.62 | 13.29 | 13.96 | 15.57 |
| Numbers | 24.80 | 28.41 | 31.62 | 33.40 |
| Relational terms | 10.45 | 11.58 | 11.70 | 12.66 |
| Classification | 13.08 | 15.23 | 16.40 | 17.53 |

[a] Group 1 = rarely watches, Group 2 = sometimes watches, Group 3 = often watches, Group 4 = watches > 5 times per week.

Four unresolved issues regarding the analysis of data from a group-comparison design involving multiple covariates require brief mention. One issue is the testing of the parallel hyperplane condition via the SAS and SPSS procedures. It appears that SPSS MANOVA conducts one test on the set of covariate-by-treatment interactions—is it the correct one? On the other hand, SAS GLM conducts $q$ (the number of covariates) interaction tests. Do all of these test results need to be "nonsignificant" to conclude the parallelism condition is satisfied? The second issue pertains to the next analysis step to take if one concludes that the parallel hyperplane condition is not met—a Johnson-Neyman analog. We are not aware of any statistical software that can test equality of adjusted mean vectors with nonparallel hyperplanes. Similarly, the third issue pertains to testing the equality of adjusted mean vectors when the group-adjusted covariance matrices are clearly not equal—a Yao analog; again, we know of no adjusted mean analog. The fourth issue pertains to calculating an effect-size index value when either the parallel-hyperplane-condition or the equal-adjusted-covariance-matrix condition is clearly violated.

## V. RECOMMENDED PRACTICES

According to The Review (Keselman et al., 1998), the reporting of information from a MANOVA or MANCOVA generally was very limited and, in some cases, misleading. In this section we address issues pertaining to the design, analysis, interpretation, and description of a MANOVA or MANCOVA study.

## A. Design

The Review revealed that applied researchers seldom consider the relation among sample size, number of outcome variables, and statistical power in a MANOVA context. Only three of the 208 studies explicitly addressed the statistical power issue. Only six other articles noted sample size, by itself, as a potential concern. We recognize that researchers may have a strong desire to study a particular collection of outcome variables despite limited resources. Applied researchers should, in such cases, explicitly be cognizant of the statistical power–sample size issue.

We cannot overemphasize the importance of the initial choice of outcome variables. Researchers should make clear the theories and beliefs underlying the rationale for including the outcome variables in a study.

## B. Analysis

Approximately 84% ($N = 175$) of the 208 MANOVA and MANCOVA applications in The Review reported and focused on the results of multiple univariate analyses. This reflects the widespread misconception that conducting a multivariate analysis prior to conducting multiple univariate analyses "protects" one from a compounding of a Type I error probability. We do *not* buy that argument! If the research interest focuses on the multiple univariate results, then the researcher should go directly to them and use a Bonferroni adjustment; a MANOVA/MANCOVA is *not* a necessary preanalysis! As we mentioned earlier, researchers should conduct a multivariate analysis only when they are interested in *multivariate* outcomes. A multivariate outcome pertains to some linear composite(s) of the outcome variables (or of "adjusted" composites that take into consideration one or more concomitant variables).

It is unfortunate that the statistical packages lead some researchers to think that a stepwise analysis is appropriate for a MANOVA. Some researchers seem to consider it routine to order and to delete variables via stepwise programs. A number of methodologists (e.g., Huberty, 1989; Thompson, 1995) have condemned such practices. The advice given here is simple: Don't do it!

Finally, the reporting of some results related to the conduct of MANOVA and MANCOVA has been somewhat scanty. The Review revealed very rare reporting of covariance matrix equality, any kind of variable intercorrelations, and effect-size index values, although reporting of variable means was almost routine. Researchers often do *not* indicate the statistical packages they use. As pointed out repeatedly in this chapter and by many others in this book, alternative analysis procedures may be appropriate, but they call for rationales or at least some reference(s) for support. Many published works do *not* provide adequate references, and reference to a statistical package is *not* adequate to support an analysis approach.

## C. Interpretation

Determining that two or more vectors of response variable means are statistically significantly different is not too informative in and of itself. It is a rare occurrence when one does not get a "small" *P* value when conducting a MANOVA or MANCOVA, and with that information one can conclude that there is a significant relation between the grouping variable and something, or that the grouping variable has a significant effect on something; however, it is the "something" that is of basic concern when interpreting significant results. Researchers often fail to address two issues of the number of dimensions in the data that may be interpreted and their structure (i.e., the constructs identified with the linear composites that are the basis for a MANOVA and MANCOVA).

## D. Description

Throughout the chapter we emphasize thoroughness of descriptions, and The Review revealed a lack of this in much research. For example, one can find in many writings a blurring of the purpose of MANOVA and the purpose of the generic "discriminant analysis," and, as we pointed out, these two sets of analyses differ considerably in purpose. The purpose of MANOVA and MANCOVA is to assess grouping variable effects, whereas (descriptive) discriminant analysis describes the resultant MANOVA and MANCOVA effects.

With regard to design and analysis, we encourage researchers to provide complete descriptions of initial choice and assignment of analysis units, variables, and variable measures. Descriptions should be so thorough that the reader can mentally formulate the associated data matrix.

Some considerations in-between design and analysis also require description for consumers of research. Provide a description of the procedures considered in handling outliers and missing data. Cognizance of these two problems should be evident, even when these problems do not exist for a given data set. Additional preanalysis considerations pertain to the assessment of MANOVA data conditions and to quantitative descriptive information. The main purpose of describing such information is to give a clear picture of the data set being analyzed.

When it comes to the actual MANOVA, describe the MANOVA test criterion considered, the test statistic value, and the *P* value, along with an effect-size index value. The research questions investigated must be crystal clear—that is, note the omnibus question or the contrast questions. The analysis description will be a little more involved when conducting a two-factor MANOVA or a MANCOVA.

## REFERENCES

Algina, J., & Keselman, H. J. (1997). Detecting repeated measures effects with univariate and multivariate statistics. *Psychological Methods, 2,* 208–218.

Coombs, W. T., Algina, J., & Oltman, D. O. (1996). Univariate and multivariate omnibus hypothesis tests selected to control Type I error rates when population variances are not equal. *Review of Educational Research, 66,* 137–179.

Crowder, M. J., & Hand, D. J. (1990). *Analysis of repeated measures.* London: Chapman and Hall.

Donnay, D. A., & Borgen, F. H. (1996). Validity, structure, and content of the 1994 Strong Interest Inventory. *Journal of Counseling Psychology, 43,* 275–291.

Friendly, M. (1991). *SAS system for statistical graphics.* Cary, NC: SAS Institute Inc.

Harris, R. J. (1985). *A primer of multivariate statistics.* New York: Academic Press.

Harrist, A. W., Zaia, A. F., Bates, J. E., Dodge, K. A., & Pettit, G. S. (1997). Subtypes of social withdrawal in early childhood: Sociometric status and social-cognitive differences across four years. *Child Development, 68,* 278–294.

Huberty, C. J (1989). Problems with stepwise methods: Better alternatives. In B. Thompson (Ed.), *Advances in social science methodology* (Vol. 1, pp. 43–70). Greenwich, CT: JAI Press.

Huberty, C. J (1994). *Applied discriminant analysis.* New York: Wiley.

Huberty, C. J, & Lowman, L. L. (1996, June). *Use of multivariate analysis of variance in behavioral research.* Paper presented at the annual meeting of the Psychometric Society, Banff, Canada.

Huberty, C. J, & Lowman, L. L. (in press). Group overlap as an effect-size basis. *Educational and Psychological Measurement.*

Huberty, C. J, & Wisenbaker, J. M. (1992). Variable importance in multivariate group comparisons. *Journal of Educational Statistics, 17,* 75–91.

Huberty, C. J, Wisenbaker, J. M., Smith, J. D., & Smith, J. C. (1986). Using categorical variables in discriminant analysis. *Multivariate Behavioral Research, 21,* 479–496.

James, G. S. (1954). Tests of linear hypotheses in univariate and multivariate analysis when the ratios of the population variances are unknown. *Biometrika, 41,* 19–43.

Johansen, S. (1980). The Welch-James approximation to the distribution of the residual sum of squares in a weighted linear regression. *Biometrika, 67,* 85–92.

Keselman, H. J., & Algina, J. (1996). The analysis of higher-order repeated measures designs. In B. Thompson (Ed.), *Advances in social science methodology* (Vol. 4, pp. 45–70). Greenwich, CT: JAI Press.

Keselman, H. J., Huberty, C. J, Lix, L. M., Olejnik, S., Cribbie, R. A., Donahue, B., Korvalchuk, R. K., Lowman, L. L., Petoskey, M. D., Keselman, J. C., & Levin, J. R. (1998). Statistical practices of educational researchers. *Review of Educational Research, 68,* 350–386.

Keselman, H. J., & Keselman, J. C. (1993). The analysis of repeated measurements. In L. Edwards (Ed.) *Applied analysis of variance in the behavioral sciences,* (pp. 105–145). New York: Marcel Dekker.

Kres, H. (1983). *Statistical tables for multivariate analysis: A handbook with references to applications.* New York: Springer-Verlag.

Little, R. J. A., & Schenker, N. (1995). Missing data. In G. Arminger, C. C. Clogg, & M. E. Sobel (Eds.), *Handbook of statistical modeling for the social and behavioral sciences* (pp. 39–75). New York: Plenum Press.

Looney, S. W., & Stanley, W. B. (1989). Exploratory repeated measures analysis for two or more groups. *The American Statistician, 43,* 220–225.

Maxwell, S. E. (1992). Recent developments in MANOVA applications. In B. Thompson (Ed.), *Advances in social science methodology,* Vol. 2 (pp. 137–168). Greenwich, CT: JAI Press.

Morrison, D. F. (1990). *Multivariate statistical methods.* New York: McGraw-Hill.

O'Brien, R. G., & Kaiser, M. K. (1985). MANOVA method for analyzing repeated measures designs: An extensive primer. *Psychological Bulletin, 97,* 316–333.

Porter, A. C., & Raudenbush, S. W. (1987). Analyses of covariance: Its model and use in psychological research. *Journal of Counseling Psychology, 34,* 383–392.

Rencher, A. C. (1995). *Methods of multivariate analysis.* New York: Wiley.

Rencher, A. C. (1998). *Multivariate statistical inference and applications.* New York: Wiley.

Roth, P. L. (1994). Missing data: A conceptual review for applied psychologists. *Personnel Psychology, 47,* 537–560.

Schafer, J. L. (1997). *Analysis of incomplete multivariate data.* London: Chapman and Hall.

Simon, S. J., & Werner, J. M. (1996). Computer training through behavior modeling, self-paced, and instructional approaches: A field experiment. *Journal of Applied Psychology, 81,* 648–659.

Stevens, J. P. (1990). *Intermediate statistics: A modern approach.* Hillsdale, NJ: Erlbaum.

Stevens, J. P. (1996). *Applied multivariate statistics for the social sciences* (3rd ed). Mahnah, NJ: Erlbaum.

Thomas, D. R. (1997). A note on Huberty and Wisenbaker's "Views of variable importance." *Journal of Educational and Behavioral Statistics, 22,* 309–322.

Thompson, B. (1995). Stepwise regression and stepwise discriminant analysis need not apply here: A guidelines editorial. *Educational and Psychological Measurement, 55,* 525–534.

Walsh, B. D., Vacha-Haase, T., Kapes, J. T., Dresden, J. H., Thomson, W. A., & Ochoa-Shargey, B. (1996). The Values Scale: Differences across grade levels for ethnic minority students. *Educational and Psychological Measurement, 56,* 263–275.

# 8

# DISCRIMINANT ANALYSIS

**MICHAEL T. BROWN AND LORI R. WICKER**

*Graduate School of Education, University of California, Santa Barbara,*
*Santa Barbara, California*

## I. INTRODUCTION

Discriminant analysis (also known as discriminant function analysis) is a powerful descriptive and classificatory technique developed by R. A. Fisher in 1936 to (a) describe characteristics that are specific to distinct groups (called descriptive discriminant analysis); and (b) classify cases (i.e., individuals, subjects, participants) into pre-existing groups based on similarities between that case and the other cases belonging to the groups (sometimes called predictive discriminant analysis). The mathematical objective of discriminant analysis is to weight and linearly combine information from a set of $p$-dependent variables in a manner that forces the $k$ groups to be as distinct as possible. A discriminant analysis can be performed using several different statistical programs including DISCRIMINANT (SPSS; Norusis & Norusis, 1998), DISCRIM (SAS; Shin, 1996), and stepwise discriminant analysis (BMDP; Dixon, 1992).

Specific descriptive questions that can be answered through discriminant analysis include the following: (a) In what ways do various groups in a study differ? (b) What differences exist between and among the number of groups on a specific set of variables? (c) Which continuous variables

*Handbook of Applied Multivariate Statistics and Mathematical Modeling*
**209**

best characterize each group, or, which continuous variables are not characteristic of the individual groups? (d) Given the results of a multivariate analysis of variance (MANOVA) indicating that group differences exist in the data, what specific variables best account for these differences? Predictive discriminant analysis addresses the question, Given the characteristics of a case as measured by a set of variables, to what predefined group does the case belong? Predictive discriminant analysis requires the use of classification rules (namely, discriminant functions) derived from a previous descriptive discriminant analysis on a data set for which the group membership of the cases is known. Those classification rules (i.e., derived weights and linear combination resulting from the descriptive discriminant analysis) are used in the classificatory analysis to assign new cases to the predefined groups.

In this chapter we give a thorough and complete discussion of what investigators need to know and do to use discriminant analysis properly. We begin with a brief layout of the specific steps and procedures necessary to conduct a descriptive discriminant analysis. This is followed by a more detailed discussion of each step, and of the information investigators need to complete the steps. Information about how to properly interpret the results of a descriptive discriminant analysis is provided, followed by a discussion of predictive discriminant analysis. Finally, we describe reporting requirements for publishing results of either a descriptive or predictive discriminant analysis. Throughout the discussion we integrate examples of the specific procedures using data drawn from a published discriminant analysis study.

## II. ILLUSTRATIVE EXAMPLE

The illustrative study we will use is "Cranial shape in fruit, nectar, and exudate feeders: Implications for interpreting the fossil record" (Dumont, 1997). Dumont was interested in examining whether the morphological characteristics of primate cranial and dentary fossils could be used to distinguish among fruit feeders, nectar feeders, and exudate feeders (primates that gouge bark to feed on the sap). If successful in that effort, the researcher wanted to know in what ways the three types of feeders differed with respect to cranial and dentary morphology. In addition, Dumont wanted to derive classification rules to use in categorizing unclassified primate fossils. With a sample size ($n$) of 131 fossils and 22 ($p$) discriminator variables, Dumont concluded that, indeed, it was possible to discriminate among these 3 ($k$) groups of primates based on cranial and dentary characteristics. Dumont made available to us the complete data for 28 cases to use for illustrative purposes throughout this chapter. We conducted a descriptive discriminant analysis on the three groups (fruit, nectar, and exudate feeders) using three of the eight discriminator variables that evidenced strength in discriminating between the groups: total skull length, minimum skull width, and dentary depth at canine. We were unable to obtain a data set of cases (similar to the Dumont, 1997, data set) for which group membership was unknown.

Our chosen illustrative study will demonstrate discriminant analysis relevance and use with data most relevant to anthropology and the biological and medical sciences. The interested reader is referred to Betz (1987), Brown and Tinsley (1983), Klecka (1975), and Tankard and Harris (1980) for case illustrations relevant to other areas of the social sciences and humanities (i.e., discrimination among college freshmen who differ in intent to continue mathematics coursework; participants in different leisure activities; political factions in the British Parliament, and television viewers and nonviewers).

## III. DESCRIPTIVE DISCRIMINANT ANALYSIS

We recommend the following steps in performing a descriptive discriminant analysis:

1. Determine if discriminant analysis will provide statistical results that answer your research questions (i.e., is descriptive discriminant analysis suitable for answering your question?).
2. Determine the appropriateness of the data set for discriminant analysis.
3. Define the groups that will be used in the analysis.
4. Select the variables that will be used in the analysis.
5. Test the data to assure that the assumptions of discriminant analysis are met. If some assumptions are not met, determine whether discriminant analysis is robust for those assumptions.
6. Perform the analysis.
7. Interpret the results.

### A. Data Requirements for Discriminant Analysis

We have already determined that discriminant analysis will help answer Dumont's (1997) question of whether fossil morphological characterics can be used to distinguish among primate feeder groups. The next step is to determine if the Dumont data meet the requirements for performing a discriminant analysis.

### 1. Groups and Discriminant Variables

Discriminant analysis requires a data set that contains two or more mutually exclusive groups and scores on two or more variables for each case in the group. The groups may be constructed on the basis of demographic characteristics (e.g., sex, ethnicity, and socioeconomic status), personal attributes (e.g., height, psychological disorder, and blood type), or past or present behavior (e.g., whether one watches television, hallucinates, or smokes cigarettes). The groups should be constructed on a dimension that the investigator deems critical to understanding the differences that exist between the groups. In our example, we have three mutually exclusive groups of primates: fruit, nectar, and exudate feeders.

Variables should be chosen that are believed to represent dimensions on which the groups are expected to differ. Labeled discriminator or discriminant variables in discriminant analysis, they are called dependent variables in analyses such as analysis of variance (ANOVA) and MANOVA or predictor variables in analyses such as Multiple Linear Regression (cf. Brown & Tinsley, 1983). Discriminant analysis evaluates the degree to which such variables differentiate the groups, hence the name "discriminator variables." The effectiveness of the discriminant analysis depends on the extent to which the groups differ significantly on these variables. Therefore, the decision to select certain variables as potential discriminator variables is critical to the success of discriminant analysis.

Several guidelines can be followed in selecting discriminator variables. First, relevant theory should be consulted to determine which variables have the greatest potential for distinguishing among the groups of interest. Second, investigators should consider variables that have been shown to be relevant to group discrimination by previous research or theory (Brown & Tinsley, 1983). ANOVA results indicating group differences on single variables and factor analysis results indicating underlying factors among sets of variables are additional kinds of empirical evidence that can be consulted in selecting discriminator variables. Of course, an investigator's professional opinion also can be relied upon when selecting potential discriminator variables. These methods can be used singularly, but we recommend that investigators use them conjointly in choosing the discriminator variables. In our illustrative study, Dumont (1997) recognized that dietary habits are known to distinguish among live primates, and those habits result from or in differences in cranial and dentary morphology. She reasoned, therefore, that the differences in total skull length, minimum skull width, and dentary depth at the canine tooth would be useful in describing and classifying primate fossils.

Investigators should take care to select discriminator variables that are not highly intercorrelated. Intercorrelated variables causes difficulty in accurately determining the precise loadings (viz., partial correlations) of the individual variables on the discriminant functions (a problem discussed later). In short, if the variables are highly correlated with each other, they will likely load on the same function and, thus, not contribute in a unique (nonredundant) way to group discrimination. We recommend that investigators consult published correlation matrices (or compute the complete correlations among the variables if data are available) and select variables that are not correlated significantly to use as discriminator variables. In the illustrative example, the correlations among three discriminator variables ($r$ length-width $= -.21$, length-depth $= -.17$, and width-depth $= .02$) were not statistically significant (see Table 8.1).

As in the illustrative study, the selected variables should satisfy the requirements for ordinal level measurement, at minimum. Ordinality makes it possible to describe the relations among the groups along the continuum underlying the variables of interest. Specifically, it permits the development of continuous scales on which the differences among the groups on the variables can be denoted. Nondichotomous nominal variables can be used, but they must be dummy coded into dichotomous categories (see Morrison, 1974). Doing so may have the effect of greatly increasing the number of variables used in the analysis. In the illustrative case, the cranial and dentary measurements were made to the nearest 0.1 mm using digital calipers and with a high degree of reliability (see Dumont, 1997).

Although two or more variables are required for the analysis, the number of ($p$) variables should not exceed the number of ($k$) groups. It is important to note that the difficulty of interpreting the results of the analysis increases with the number of discriminator variables selected for analysis (Brown & Tinsley, 1983). This is due to the increasing complexity of interrelations among the variables in accounting for group differences and the increasing subtlety in differences among the groups. In addition, the probability that significant results will be spurious increases as the ratio of discriminator variables to individual cases exceeds $10:1$, as discussed in the next section. For these reasons, we suggest that researchers restrict the discriminator variables to those that have major theoretical and empirical relevance.

In summary, the data set should consist of two or more groups having at least ordinal scores on two or more discriminator variables. The discriminator variables should measure dimensions that the investigator views as important to understanding the differences that exist among groups. Importance can be determined on the basis of theory, past research, or another compelling rationale. The discriminator variables should not be intercorrelated with each other. The number of variables should not exceed the number of groups.

## 2. Sample Size

The investigator needs to determine the sample size needed to perform a discriminant analysis. We suggest that investigators avoid very large sample sizes because statistical tests are more likely to yield significant results for trivial differences under such circumstances. As sample size increases, the variability in the sampling distributions (i.e., the denominator in most correlational statistics) decreases, assuming all other factors are held constant. This will result in a reduction of the overlap between sampling distributions under the null and alternative hypotheses, leading to a greater probability of finding statistical significance (viz., rejecting the null hypothesis). However, small sample sizes are not recommended because the idiosyncrasies in the sample will unduly influence the statistical results, thereby lessening the stability and generalizability of the results (Huberty, 1975, Morrison, 1974).

Small sample sizes also should be avoided because of the need to determine the reliability of the results of a discriminant analysis by confirming their applicability to other samples; this is known as cross-validation. Some cross-validation procedures involve using some part of the total sample to derive the discriminant functions (called a "developmental sample") and evaluating the performance of those functions to correctly classify the remaining part of the sample (called a "cross-validation sample") (see Tinsley & Brown, chapter 1, this volume).

Brown and Tinsley (1983) recommended that the total sample size should be at least ten times the number of discriminator variables. Stevens (1996) argued that the ratio of cases to variables should be more on the order of 20 to 1. We recommend the total sample size should lie within these two recommendations, with care given to ensuring that the sample size for the developmental sample meets the requirements for the number of cases in each group.

Ideally, the cases in the sample would be distributed evenly across the groups (Huberty, 1975; Tatsuoka, 1970), but often this is not possible nor reasonable. When the groups of interest differ in the number of cases, Brown and Tinsley (1983) recommend that the number of cases in each group be at least equal to the number of variables. A more conservative criterion (Huberty, 1975) is that the number of cases in the smallest group be at least three times the number of variables, allowing one-third of the cases to be used for cross-validation purposes. The criteria of Brown and Tinsley (1983) and Huberty (1975) are not as inconsistent as they may appear. The difference in their recommendations is that Brown and Tinsley did not assume, as does Huberty, that the cross-validation and developmental samples would be drawn from one sample.

Descriptive discriminant analysis requires that there are complete data on the discriminator variables and that group membership is known. Those cases that do not have scores on all of the discriminator variables automatically are excluded from the analysis, possibly affecting the adequacy of the data in satisfying sample and group size requirements adversely. Furthermore, cases having missing data may be systematically different from those cases containing complete data, so use of those with incomplete data could result in generalizability problems (Norusis, 1990). We recommend that, upon completion of the analysis, the researcher examine the analysis output and determine whether the number of cases included in the analysis still meets the sample and group size requirements. If sample size requirements are not met, acknowledge the limitations associated with low sample and group sizes when reporting and interpreting the results. Furthermore, applied researchers should compare the cases with missing data to those with complete data to see if they differ systematically. If they differ, generalizability problems are evident and should be discussed in reporting the results.

Our illustrative data set consisting of three discriminator variables and 28 cases does not satisfy the minimum sample size criterion ($N = 30$) of having ten cases per variable recommended by Brown and Tinsley (1983). In situations such as this, researchers must use caution in interpreting the results. In our data set, the fruit, nectar, and exudate groups have 12, 11, and 5 cases, respectively, thereby satisfying the Brown and Tinsley requirement that the number of cases in each group be equal to or exceed the number of variables, but not Huberty's (1975) more conservative criterion of at least 3 cases for every variable in each group. Group 3 does not meet the latter requirement. This is another reason to use caution when interpreting the results. Lastly, it is important to note that there are complete data on the discriminators and group membership variables on all 28 cases.

### 3. Assumptions

Because the data used in discriminant analysis involve multiple variables, the assumptions required for a discriminant analysis are the same as those required for other multivariate analyses (i.e., MANOVA). Violations of the following assumptions are expected to inflate the Type 1 error rate: (a) independence of observations (independent measures); (b) multivariate normality (the observations based on the discriminator variables are normally distributed), and (c) homogeneity of covariance matrices (the population covariance matrices based on the discriminator variables are equal). (Huberty and Petoskey discuss these assumptions in detail in chapter 7, this volume). Current evidence suggests that discriminant analysis is robust with respect to violation of assumptions of multivariate normality and of homogeneity of covariance matrices (Stevens, 1996). Nevertheless, conscientious researchers should examine and report whether these assumptions hold.

The assumption of independent observations is critical; the likelihood of making a Type I error increases markedly if this assumption is violated. To assure that the scores for each of the variables (observations) are independent (that is, not highly correlated), the researcher needs to examine the correlation matrix for the discriminator variables. In general, correlations having an absolute value of less than .3 have no interpretive value (Tabachnick & Fidell, 1989) and do not violate the assumption. Correlations having an absolute value equal to or greater than .3 indicate violation of the assumption.

We would like to note some ideas advanced by Stevens (1996) that appear to have merit, but we have found no corroborating empirical or theoretical evidence to warrant an opinion on our part; we offer them to the reader in order to be complete in our treatment of discriminant analysis and because they are interesting. Stevens (1996) argues that correlations greater than .3 are not problematic if they are not statistically significant. In the event that they are, Stevens contends that the researcher could alter the alpha level to a more conservative one and see if they are still significantly correlated. Stevens (1996) also contends that significant correlations within each group are not as problematic to the validity of a discriminant analysis as those that might occur in the pooled within-groups matrix. Therefore, researchers should examine the pooled within-groups correlation matrix and those within each group for the discriminator variables to determine whether the observations (measures) are correlated across groups or only within groups.

No statistical test for multivariate normality is available on SPSS, SAS, or BMDP. However, Stevens (1996) suggests that the data follow a multivariate normal distribution if the distributions of scores on the discriminator variables are normal. Investigators should examine these distributions for indications that the assumption of multivariate normality is likely to have been violated (Norusis, 1990). Keep in mind that this simple alternative is no guarantee that the multivariate distributions do not depart from normality.

Discriminant analysis is especially robust to violations of the assumption of homogeneity of the covariance matrices if the ratio of the largest group $n$ divided by the smallest group $n$ is less than 1.5 (Stevens, 1996). Investigators can test for violations of the homogeneity of covariance assumption using the Box's M statistic in SAS and SPSS. When violated, Klecka (1975) notes that the worst consequence is that cases are more likely to be classified into the group with the greatest dispersion. When the assumption has been violated, some statistical packages permit investigators the option of proceeding with the analysis using separate covariance matrices for each group instead of the pooled within-group covariance matrix (cf. Cocozzelli, 1988). In short, the available literature indicates that violation of the homogeneity of covariances assumption is not of major importance to conducting a valid discriminant analysis and, even when violated, a number of effective options are available to the researcher.

### 4. Using Prior Probabilities

When calculating the classification function, coefficients used to assign preclassified cases to groups, it is ordinarily assumed that a case has an equal probability of being a member of any of the groups; statistical packages use this assumption by default. Sometimes investigators may be tempted to use unequal prior probabilities (e.g., the proportion of cases in each group);

the commercially available statistics packages allow the investigator to input other probabilities. However, we caution that the use of different prior probabilities can cause the function coefficients to differ radically from those derived using equal prior probabilities. We agree with Stevens (1996) that the decision to assume other than equal prior probabilities is justifiable only when the investigator has a firm theoretical or empirical basis for believing that the differences in group size reflect actual differences in the size of the respective populations, as in the case of racial and ethnic groups.

### 5. The Statistical Assumptions and Our Illustrative Case

In our example, the assumption, independent observations, was tested by calculating a pooled within-groups correlation matrix. As previously presented, the three discriminator variables were minimally correlated. The assumption of multivariate normality was assessed by inspecting the distribution of scores on the three discriminator variables (see Figure 8.1). These figures show a normal distribution of scores on each group of the three variables. The third assumption of equal population covariance matrices for the dependent variables was tested using Box's M statistic. The statistically significant $F$ ratio ($F(12, 744) = 2.53, p \leq .003$) suggested that the population covariance matrices were not equal. Nevertheless, we concluded that this violation was not very problematic for discriminant analysis (cf. Klecka, 1975), and we recognized that we should be able to evaluate the likelihood of errors by examining the determinants of the pooled within-groups covariance. Lastly, in the absence of any research documenting that the differences in group sizes are reflective of actual population sizes, we assumed that cases have an equal probability of being a member of any of the groups.

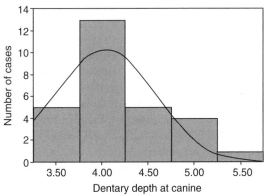

**FIGURE 8.1**    Distribution of the dentary depth at canine for the sample cases. $SD = .54$; mean $= 4.04$; $N = 28.00$.

## B. Components of Discriminant Analysis Results

A proper interpretation of discriminant analysis requires an accurate understanding of four important concepts: discriminant coefficient, discriminant function, discriminant scores, and group centroids.

### 1. Discriminant Coefficients

Discriminant analysis calculates mathematical weights for scores on each discriminator variable that reflect the degree to which scores on that variable differ among the groups being discriminated. Thus, the discriminator variables on which more groups differ and on which the groups differ most receive the most weight; these weights are referred to as discriminant coefficients.

Most computer programs used to perform discriminant analysis allow investigators to consider both unstandardized and standardized discriminant function coefficients. Unstandardized coefficients, calculated on raw scores for each variable, are of most use when the investigator seeks to cross-validate or replicate the results of a discriminant analysis or to assign previously unclassified subjects or elements to a group. However, the unstandardized coefficients cannot be used to compare variables or to determine what variables play the greatest role in group discrimination because the scaling for each of the discriminator variables (i.e., their means and standard deviations) usually differ. In our illustrative study, for example, the means and standard deviations of the three discriminator variables differ across the three primate feeder groups, making direct comparisons of the variables difficult: total skull length ($m = 4.04$, $sd = .54$), minimum skull width ($m = 1.06$, $sd = .32$), and dentary depth at canine ($m = .29$, $sd = .06$). A comparison of unstandardized coefficients for the discriminator variables will distort the apparent contributions of the variables to the function because of the scaling differences.

Standardized discriminant coefficients are used to determine the comparative relations of discriminator variables to the functions. Standardized discriminant coefficients have been converted to $z$ scores (i.e., $m = 0$, $sd = 1$) to eliminate scaling differences among the discriminator variables. As a result, investigators can determine the degree to which each discriminator variable is associated with differences among the groups by examining the absolute magnitude of the standardized discriminant coefficients and the relative importance of each discriminator variable to group discrimination.

Table 8.1 presents the standardized and unstandardized discriminant function coefficients for our illustrative study. If we attempted to interpret the unstandardized functions, we would conclude that total skull length and dentary depth at canine were equally important in defining Function 1. However, inspection of the standardized function coefficients reveals that Function 1 is defined primarily by total skull length. The unstandardized

**TABLE 8.1    Discriminant Function Coefficients**

| | Standardized function | | Unstandardized function | |
|---|---|---|---|---|
| | **1** | **2** | **1** | **2** |
| Total skull length | 1.0 | 0.16 | 3.14 | 0.49 |
| Minimum skull width | 0.05 | 0.95 | 0.15 | 3.21 |
| Dentary depth at canine | 0.18 | −0.37 | 2.93 | −5.99 |
| Constant | | | −13.73 | −3.64 |

coefficients suggest that dentary depth at canine receives twice the weight of minimum skull width on Function 2, but the standardized coefficients reveal that the latter actually should be given three times the weight of the former.

## 2. Discriminant Functions and Discriminant Scores

Discriminant analysis forms one or more weighted linear combinations of discriminator variables called discriminant functions. Each discriminant function has the general form:

$$D = a + b_1 x_1 + b_2 x_2 + \cdots + b_p x_p, \tag{1}$$

where $D$ is the discriminant score, $a$ is the Y-intercept of the regression line, $b$ is the discriminant function coefficient, $x$ is the discriminator variable raw score, and $p$ is the number of discriminator variables. The Y-intercept is a constant that adjusts the function to account for the scaling differences in the discriminator variables that are present when raw (unadjusted) scores are used for the analysis; this adjustment is unnecessary when standard scores are analyzed. A discriminant score for each function is calculated for each case in a sample by multiplying each case's discriminator variable raw score by the associated weight (viz., discriminant coefficient). Then, the discriminator scores are used to calculate the average discriminant score of the cases belonging to a group (i.e., group centroid) for each discriminant function. In descriptive discriminant analysis, these scores are used to assign each case to a group. Discriminant scores can be requested as output from most statistical packages. The maximum number of functions that can be produced in a discriminant analysis is equal to the lesser of the number of discriminator variables ($p$) or number of groups minus one ($k - 1$).

In the illustrative study, discriminant analysis produced two unstandardized discriminant functions. Function 1 had unstandardized weights ($b$) of 3.14, .15, and 2.94 and an intercept ($a$) of −13.73 (see Table 8.1). Function 2 had unstandardized weights of .49, 3.21, and −5.99 and an intercept of −3.64. There were three primate feeder groups in our example

**TABLE 8.2   Evaluating Discriminant Functions**

| Function | Eigenvalue | Relative percent | Canonical correlation | Wilks's lambda | Chi-square | df | p |
|----------|-----------|------------------|-----------------------|----------------|------------|-----|-----|
| 1 | 2.00 | 87.0 | .82 | .26 | 32.56 | 6 | .00 |
| 2 | .30 | 13.0 | .48 | .77 | 6.24 | 2 | .04 |

study and three discriminator variables, so our discriminant analysis could produce up to two discriminant functions ($k - 1 = 3 - 1 = 2$) (see Table 8.2), maximally.

The discriminant functions produced by an analysis will differ in the amount of group discrimination for which they can account and the particular groups they best differentiate. These points will be elaborated upon in our later discussion of the interpretation of discriminant functions.

## 3. Group Centroids

In discriminant analysis, the group centroids represent the mean discriminant score of the members of a group on a given discriminant function. For classification and prediction purposes, the discriminant score of each group case (e.g., each individual) is compared to each group centroid, and the probability of group membership is calculated. The closer a score is to a group centroid, the greater the probability the case belongs to that group. This will be discussed in greater detail in the predictive analysis section.

Group centroids reveal how much and in what ways the groups are differentiated on each function. The absolute magnitude of the group centroids indicates the degree to which a group is differentiated on a function, and the sign of the centroid indicates the direction of the differentiation. Table 8.3 shows that Function 1 discriminates nectar feeders from fruit and exudate feeders. Nectar feeders scored at the positive end on the bipolar function and exudate feeders at the negative end of the function. Function 2 discriminated exudate feeders from fruit and nectar feeders,

**TABLE 8.3   Group Centroids**

| Group | Function 1 | Function 2 |
|-------|-----------|-----------|
| Fruit feeders | −1.03 | −.44 |
| Nectar feeders | 1.66 | .02 |
| Exudate feeders | −1.17 | 1.01 |

with exudate feeders at the positive end of the function and fruit and nectar feeders at the negative end of the function.

## C. Interpreting Discriminant Functions

Interpreting the results of a discriminant analysis depends, in large part, on the interpretation of the discriminant functions. The function is defined by the discriminant coefficients that are used to weight a case's scores on the discriminator variables. The task facing the researcher is to examine the standardized discriminant functions to identify the conceptual dimension (or construct) underlying each function.

First, researchers should identify the discriminator variables that have the highest and lowest weights on a function. The size of the coefficient tells the investigator how much a discriminator variable contributes to group discrimination on a function, and the sign tells the investigator the direction of the relation. For example, in our illustrative study, all of the discriminator variables on Function 1 are positively related to the function, but only total skull length contributes meaningfully to group discrimination on the function (see Table 8.1). Thus, high discriminant scores on Function 1 are associated with greater total skull length. On Function 2, minimum skull width contributes most to group discrimination on the function and in the positive direction; dentary depth at canine is moderately and negatively related to group discrimination on the function. Consequently, high scores on Function 2 are associated with greater minimum skull width and, to a lesser extent, with decreased dentary depth at canine.

Also, researchers should examine the structure matrix coefficients, which show the correlation of discriminator variables with the function. In a manner similar to interpreting factors in factor analysis (Cudeck, chapter 10, this volume), investigators can identify the construct underlying each function by studying the discriminators on a given function. The absolute magnitude indicates the strength of the relation between each variable and the function and the sign indicating the direction of the relation. For example, the structure matrix indicated that total skull length ($r = .98$) is most highly correlated with Function 1, and the other discriminator variables contributed little to group discrimination ($r$'s for minimum skull width and dentary depth at canine were $-.16$ and $.01$, respectively). Minimum skull width ($r = .91$) is most highly correlated with Function 2, but dentary depth at canine also contributed meaningfully, albeit negatively, to group discrimination ($r = -.37$; $r$ for total skull length was $.03$). Structure matrix correlation coefficients less than .30 typically are not interpreted because the square of the structure matrix coefficient reveals that such discriminators account for less than 10% of variability in the function. Though labeling and describing the functions includes an element of subjectivity, it is an essential aspect of achieving a parsimonious description of group differences. This procedure is consistent with the rationale underlying discriminant analysis, that information from many variables considered together is more useful in differentiating groups than information on singular variables in isolation.

In addition, researchers should determine the meaning associated with high or low scores on a function. In our example, we conclude that high scores on Function 1 are associated with higher scores on total skull length. Higher scores on Function 2 are associated with high scores on minimum skull width and moderately lower scores on dentary depth at canine.

Finally, researchers should examine the group centroids for each discriminant function and identify the groups having the highest and lowest scores. Those groups are differentiated the best on that function (see Table 8.3), and the group centroids provide some insight into the unique attributes of each group.

## D. Rotating Discriminant Functions

Much like in the case of factor analysis, rotating significant discriminant functions can facilitate the interpretation of the functions. The investigator is cautioned to keep in mind that when rotation (varimax) is performed, the relative standing of the functions is altered so that the first function may not necessarily account for the maximum amount of group differences on the discriminator variables (cf. Stevens, 1996). With regards to the illustrative example, we did not rotate the discriminant functions because the structure matrix provided sufficient information to assist in the interpretation of the functions.

## E. Evaluating the Importance of Functions

### 1. Primacy

In a manner analogous to factor analysis, the first function generated in a discriminant analysis accounts for the greatest proportion of differences among the groups. Each subsequent function that is generated accounts for a decreasing proportion of the differences among the groups. In our example, two functions were generated (see Table 8.2), with Function 1 accounting for more group differences than Function 2. Furthermore, the proportion of group differences accounted for by each function is statistically independent of that accounted for by the other functions. Discriminant analysis will generate the maximum possible number of discriminant functions, in decreasing order of importance in accounting for group differences, but it is possible that the functions do not differ significantly (either statistically or practically) in the proportion of group differentiation for which they account.

## 2. Wilks's Lambda and Chi-Square

Wilks's lambda, the ratio of within-group variability to total variability on the discriminator variables, is an inverse measure of the importance of the functions (Betz, 1987; Brown & Tinsley, 1983; Huberty, 1975; Tatsuoka, 1970). Values close to 1 indicate that almost all of the variability in the discriminator variables is due to within-group differences (differences between cases in each group); values close to 0 indicate that almost all of the variability in the discriminator variables is due to group differences. A chi-square test based on lambda indicates whether the variability that is systematically related to group differences is statistically significant. The test is performed over and over again after each function is extracted until a nonsignificant result is obtained or all of the functions have been generated. In our example, the Wilks's lambda value for Function 1 is .26 (see Table 8.2), signifying that most of the variability captured by Function 1 can be attributed to group differences. The Wilks's lambda value for Function 2 is .77, suggesting that little of the variability captured by Function 2 is attributed to between-group differences. Because the purpose of a discriminant analysis is to detect group differences, these lambda statistics suggest that Function 2 is of little importance.

The chi-square values show that, prior to extraction, the variability related to group differences is statistically significant at the .05 level. After extracting Function 1, the variance remaining in the discriminator variables is still significantly related to the group differences. Therefore, a second function was extracted. A reasonable conclusion for the investigator based on Wilks' lambda and the chi-square test is that, although only a small amount of group differences is accounted for by Function 2, the proportion of group differences is statistically significant and worthy of attention.

## 3. Eigenvalues

Eigenvalues indicate the ratio of between-groups variability to within-groups variability for a function. The larger the eigenvalue, the better at accounting for the group differences are the discriminator variables loading on the function. In our illustrative case, the eigenvalue for Function 1 (2.00) is much larger than the eigenvalue of Function 2 (.30) (see Table 8.2). This is consistent with the Wilks' lambda statistic, which indicated that the ratio of between-group variability to within-group variability detected by Function 1 is much larger than that detected by Function 2. Given our earlier interpretation of the functions and study of the group centroids, we can conclude that total skull length, as opposed to minimum skull width and dentary depth and canine, plays an almost exclusive role in group discrimination on Function 1 and that nectar feeders are the most unique of the three groups with respect to this variable.

## 4. Relative and Absolute Percent

Unlike Wilks's lambda and the associated chi-square test, the relative and absolute percent statistics are indices of the practical rather than the statistical significance of the functions for group discrimination. Not all functions that account for a statistically significant amount of group variability on the discriminator variables are of conceptual or practical value in distinguishing among the groups. The relative percent statistic, also known as the percent of variance accounted for and percent of variance explained, is obtained by dividing each eigenvalue by the sum of the eigenvalues and is a direct index of the relative importance of each function in accounting for between-group differences. The relative percent shows the proportion of the total amount of between-group variance that is attributable to a particular function. In our example, the relative percent for Functions 1 and 2 are 87% and 13%, respectively (see Table 8.2).

The absolute percent, calculated by hand by dividing each eigenvalue by the number of discriminator variables used in the analysis, indicates the magnitude of between-group variability explained by each function relative to the amount of between-group variation. Thus, with our example, the absolute percent for Function 1 is 2.00/3, which equals 66.67%. The absolute percent for Function 2 is .30/3, which equals 10%. The total percent of between-group variability explained by the functions is 76.67% (i.e., 66.67% + 10%) or the sum of the eigenvalues divided by the number of discriminator variables (i.e., [2.00 + ... 30]/3 = .77).

The absolute percent is useful in gauging how effective the discriminator variables were in distinguishing among the groups. It is in this context that the relative percent is interpreted. It is possible for a statistically significant function with a large relative percent to account for little real variability among the groups when the total amount of the variability in the groups that is accounted for by discriminators is small. In such a case, the amount of variance in the discriminator variables that is systematically related to group differentiation is actually less than it appears from an observation of the relative percent alone.

## 5. Canonical Correlation

Some of the variability represented by a function is unrelated to group differences. This variability may be related to within-group differences, or to various types of errors that occured in the data collection or data entry process. The canonical correlation, characteristically generated by discriminant analysis, indicates the relation between scores on the function and group differences, and, therefore, provides yet another indication of the practical value of the functions. The canonical correlation indicates the amount of group variability captured by each function, and the canonical correlation squared indicates the proportion of variance in a function that is related to group differences. As with the eigenvalues, larger canonical correlations indicate more important functions. In our example, the canonical correlations for Functions 1 and 2, respectively, are .82 and .48 (see Table 8.2). The squares of these correlations are .67 and .23, respectively. These values indicate a strong relation between scores on Function 1 and group differences, and a weak relation between scores on Function 2 and group differences.

## 6. Integrating Information about the Functions

Integrating the information yielded by these statistics requires a solid grasp of the research and theory relevant to the area under investigation. All of the statistics will yield the same rank order of importance of the functions, but they give different clues as to the degree of importance of a function. The most common strategy (cf. Betz, 1987) is to use the chi-square test of significance to determine the maximum number of functions that may be important. Then examine the Wilks' lambda value, the canonical correlations, and the absolute percent values to determine the adequacy of the significant functions in explaining the variability among the groups. Finally, examine the relative percent values to decide the importance of each function.

With respect to our illustrative study, the chi-square test indicates that there are two statistically significant functions. The Wilks' lambda statistic indicates that a large portion of between-group differences is captured by Function 1, and a much smaller portion is captured by Function 2. The canonical correlations indicate a strong relation between scores on Function 1 and group differences, and a weak relation between scores on Function 2 and group differences. Lastly, the relative percents and absolute percents confirm that there is a substantially greater portion of between-group variability captured by Function 1 as compared to Function 2.

## F. Evaluating Classification Accuracy

The final determination of the usefulness of a set of functions rests on the ability of the functions to accurately classify members to groups (Betz, 1987). An evaluative classification can be performed as part of the descriptive discriminant analysis to determine the accuracy with which the discriminant functions assign individuals to groups. This evaluative classification includes multiple procedures for determining accuracy (hit) rates and cross-validation.

## 1. Classification Accuracy

Evaluating classification accuracy (i.e., hit rates) is an important means of determining the statistical and the practical usefulness of a set of computed functions. Investigators often intend to use the functions generated in a discriminant analysis to classify cases that have not yet been classified or that have been classified by some alternative and possibly inferior method. Before attempting to classify new cases, however, the investigator needs to have some indication of the accuracy of the functions in classifying cases.

The actual procedure involves classifying the original cases in the sample using the functions and evaluating the accuracy of these classifications. When performing a descriptive discriminant analysis, the investigator begins with a sample of cases whose group membership is known. After the discriminant analysis has been performed and the discriminant functions generated and evaluated, discriminant scores are calculated on each selected function, and group centroids are calculated for each function. Next, the group to which each case would be assigned on the basis of their scores on the discriminator variables is determined. Finally, the group to which each case would be assigned is compared to the group to which the case actually belongs, and the percentage of correct assignments is calculated. The procedure results in a percentage indicating the proportion of cases assigned to the groups correctly. Norusis (1990) provides a useful discussion of the mathematical computations involved in determining group assignments.

Table 8.4 presents the classification resutls for our illustrative study. It shows that the percentage of primate fossils correctly classified using the functions generated in the analysis were 83.3% for the fruit feeder group,

**TABLE 8.4** Percent of Correct Classifications in the Developmental and Cross-Validation Samples

| Actual group membership | Predicted group membership | | | |
| --- | --- | --- | --- | --- |
| | Fruit feeders | Nectar feeders | Exudate feeders | Total |
| Developmental sample: Classification count | | | | |
| Fruit feeders | 10 | 0 | 2 | 12 |
| Nectar feeders | 1 | 10 | 0 | 11 |
| Exudate feeders | 1 | 0 | 4 | 5 |
| Developmental sample: Classification percent | | | | |
| Fruit feeders | 83.3 | .0 | 16.7 | 100 |
| Nectar feeders | 9.1 | 90.9 | .0 | 100 |
| Exudate feeders | 1.0 | .0 | 80.0 | 100 |
| Cross-validation sample: Classification count | | | | |
| Fruit feeders | 9 | 1 | 2 | 12 |
| Nectar feeders | 1 | 10 | 0 | 11 |
| Exudate feeders | 1 | 0 | 4 | 5 |
| Cross-validation sample: Classification percent | | | | |
| Fruit feeders | 75.0 | 8.3 | 16.7 | 100 |
| Nectar feeders | 9.1 | 90.9 | .0 | 100 |
| Exudate feeders | 20.0 | .0 | 80.0 | 100 |

90.9% for the nectar feeder group, and 80% for the exudate feeder group. The overall hit rate (i.e., percent of correct predictions) for the analysis was 85.7%.

## 2. Cross-Validation

It is critical to cross-validate the results of a discriminant analysis, especially if the researcher intends to classify other samples into the groups of interest. As stated previously, discriminant analysis is a maximizing procedure in which the discriminant coefficients are derived mathematically to maximize the distinctions among the groups. In doing so, discriminant analysis capitalizes on chance differences among the sample groups that occur because of idiosyncratic characteristics of the sample. Because these idiosyncratic characteristics differ from sample to sample, the usefulness of the classification functions (i.e., discriminant functions) is uncertain. Generalizability is particularly uncertain for the lesser functions generated in an analysis. Thus, the investigator needs to determine how much shrinkage in the hit rate can be expected when classifying cases that were not used to derive the functions.

Methods of cross-validation abound (see Tinsley & Brown, chapter 1, this volume); we will illustrate two that are among the more feasible. In the jackknife method (also referred to as the "leave-one-out" method and the U-method), one sample case is systematically held out and the discriminant analysis is performed on the remaining sample. Then, the excluded observation is classified into one of the groups using the discriminant functions generated by the analysis. This procedure is repeated until each member of the sample has been held out and classified. The information resulting from this procedure can be used to determine the stability of various statistics produced by the discriminant analysis. The proportion of cases correctly classified can be compared to that expected on the basis of chance using the $z$ test of proportions that will be discussed later in the chapter. The jackknife procedure is available on BMDP, SAS, and the newer versions of SPSS.

Cross-validation using the hold-out method involves splitting the sample, randomly, into two parts with two-thirds of the sample belonging to a "developmental" sample and one-third being allocated to a "cross-validation" sample. The discriminant analysis is performed on the developmental sample, and the functions derived are used to classify the members of the smaller cross-validation sample. The accuracy of classification for the smaller sample indicates the hit rate investigators can expect to achieve for newer samples.

Examination of Table 8.4 reveals that the analysis correctly classified 85.7% of the 28 cases in the developmental sample using the derived functions. We performed the jackknife cross-validation procedure to determine shrinkage and found that 82.1% of cross-validated classifications were correct. This value is lower than the hit rate achieved for the derivation sample, which will almost always be the case. Thus, the use of the discriminant functions derived from the developmental sample to classify independent samples can be expected to result in approximately 82% of the cases being correctly classified. Next, we need to determine the significance of this hit rate.

## 3. Chance Proportions and Hit Rates

Computer programs do not provide a significance test of the proportion of cases correctly classified in a discriminant analysis. Researchers can evaluate classification accuracy by comparing the proportion of cases correctly classified to the base or chance rate using a $z$ test of proportions (cf. Betz, 1987; Brown & Tinsley, 1983; see Formula 5). The base rate is the proportion of cases expected to be correctly classified on the basis of the best available alternative strategy (i.e., results from current classification practices including the use of previously derived discriminant functions). The base rate is the preferred standard against which to evaluate classification accuracy, but it is often the case in research that no alternate strategy exists. In such circumstances, investigators can test the significance of their classification results by comparing the proportion of correct predictions to the proportion expected on the basis of chance.

There are a number different ways of determining the chance rate (Brown & Tinsley, 1983; Morrison, 1974). When the groups to which cases were classified are of equal size, the proportion of correct classifications expected to occur on the basis of chance alone is

$$1/k, \tag{2}$$

where $k$ equals the number of groups into which members are being classified. This assumes that each case had an equal chance of being classified into any one of the groups. In our three-group example, Formula 2 yields a chance rate of 1/3 or 33%, signifying that approximately nine cases of the 28 would be expected to be classified correctly on the basis of chance.

Notice, however, that the groups in our example are not of equal size. Therefore, the chances of being classified into each group are not equal. Two strategies for calculating chance proportions exist when the group sizes are unequal. Sometimes, investigators are interested only in maximizing the overall classification accuracy (i.e., they do not care about the particular groups into which cases are being classified nor about differences in the types of classification errors being made). In this situation, classification accuracy would be maximized by classifying all members into the largest group and the chance proportion is calculated using the formula

$$n/N, \tag{3}$$

where $n$ is the number of persons in the largest group and $N$ is the total sample size (i.e., the sum of all members of all groups). The three groups in our example, consisted of 12, 11, and 5 cases, so 12/28 (43%) would be classified correctly by chance if the investigator was only interested in maximizing correct classifications.

Typically, researchers are interested in more than simply maximizing the number of correct classifications. In many situations, the different types of assignment errors differ in their theoretical or practical significance; investigators might be interested in correctly classifying cases into all of the groups. Formulas 2 and 3 can grossly overestimate the chance probability of correct predictions in these circumstances. The proper formula for calculating chance probability in such cases is one that considers the proportion of the total sample in each group:

$$p_1 a_1 + p_2 a_2 + \cdots + p_k a_k, \tag{4}$$

where $p$ is the proportion of the total sample actually belonging to a group, $a$ is the actual proportion of cases classified by discriminant analysis into a particular group, and $k$ is the number of the groups (Betz, 1987; Brown & Tinsley, 1983). In our example, $p_1$, $p_2$, and $p_3$ equal .43, .39, and .18, respectively, $a_1$, $a_2$, and $a_3$ equal .43, .38, .21, respectively, and the chance rate equals .43(.43) + .39(.38) + .18(.21) = .19 + .15 + .04 = .38. Thus, if the researcher was interested in classifying the cases into even the smallest group, the likelihood of doing so correctly by chance would be 38%.

Formulas 2, 3, and 4 will yield identical results when the number of cases in each group are equal, but they diverge as a function of the extent to which the group sizes differ. Even small differences in the calculated chance rate can influence the outcome of the $z$ test of proportions (Brown & Tinsley, 1983, p. 304). In our example, Formulas 1, 2, and 3 yield different chance rates of 33%, 43%, and 38%. Given the actual hit rate observed in a given study, such differences in the chance rate can be of enormous significance.

The chance proportion can be compared to the proportion correctly classified using the $z$ test of proportions. This test must be performed by hand because it is not provided by the commercially available statistical packages. The formula for the $z$ test of proportions is

$$z = (N p_a - N p_c)/(N p_c (1 - p_c)), \tag{5}$$

where $N$ = the total sample size, $p_a$ = the proportion of cases correctly classified using discriminant analysis, and $p_c$ = the proportion of cases expected to be classified correctly on the basis of chance. The obtained $z$ value then can be compared to the $z$ value for a one-tailed test from the $z$ table found in the appendices of most statistics books. For example, if alpha for the test is chosen to be .05, the critical value of $z$ is 1.65. In our illustrative study, $N = 28$, $p_a = .82$, and $p_c = .38$. Applying Formula 5,

$z = (28(.86) - 28(.38)/(28(.38) (1-.38)) = 1.88$, which exceeds the critical value of 1.65. Therefore, we conclude that the classification accuracy of our analysis, even after cross-validation, exceeds that expected on the basis of chance at the .05 level of significance.

## IV. PREDICTIVE DISCRIMINANT ANALYSIS

Up to this point, our discussion has focused primarily on the descriptive uses of discriminant analysis. There are occassions in which professionals encounter cases for which group membership is not known, but it is important or desirable to determine its category membership. It may be important to know whether a newly discovered poem or painting belongs to the body of work of an artist of historical significance. It may be a newly discovered fossil whose phylum is debated by paleontologists. It might be to determine whether an individual who possesses a given set of characteristics is likely to be successful or unsuccessful in an educational or medical/psychological treatment program. Descriptive discriminant analysis will identify the distinct groups that exist within the sample, establish classification rules (viz., discriminant functions) based on the characteristics of the cases whose group membership is known, and assign new cases to these groups using the established classification rules. In this analysis, however, discriminant analysis is classifying cases for which group membership is known. Predictive discriminant analysis involves the use of derived discriminant functions to classify new cases for which group membership is not known.

In order to perform a predictive discriminant analysis, researchers must have the classification rules (viz., discriminant functions) from a previous descriptive discriminant analysis or researchers must perform such an analysis to obtain those rules. The investigator can perform a descriptive discriminant analysis on the cases of known membership and then use the derived discriminant functions to classify cases for which group membership is not known. When the cases of unknown group membership are contained in a sample of cases for which group membership is known, some statistical packages allow the investigator to perform a descriptive discriminant analysis on the cases of known membership and, as an option, use the discriminant functions that were derived in that analysis to classify cases in the data set for which group membership is not known.

Another method of performing classifications is to use the discriminant functions derived from one sample of cases in a descriptive analysis, which have been preassigned to a set of groups, to classify unassigned cases from other samples to those groups. This method of classifying cases makes use of the set of coefficients, called Fisher's linear discriminant functions or classification function coefficients, which are produced for each group by discriminant analysis. These functions can be obtained from one or two sources: an investigator's earlier descriptive discriminant analysis or that performed by earlier investigators. Fisher's linear discriminant function coefficients for our illustrative analysis are presented in Table 8.5.

It is important to note that classification functions differ from the discriminant functions discussed earlier in that classification functions are produced for each group for which classification is sought. The discriminant functions discussed earlier were not group specific; a case would receive a score on a disriminant function and be assigned to the group for which discriminant analysis determines that the statistical probability is highest for group membership. Using classification functions, new cases will receive a discriminant score on each group-specific function. The classification function coefficients are unstandardized and are used to multiply the scores of a case on the appropriate discriminator variables. Next, the products are summed and the constant (a) is added. This yields a discriminant score for each classification function. If we had a case of unknown group membership and were interested in classifying it, the case would have received three discriminant scores based on the three classification functions presented in Table 8.5. The case should be assigned to the group for which it has the largest discriminant score.

It is important to reemphasize that researchers cannot use the standardized discriminant functions and their coefficients in these computations because the means and standard deviations used to develop the standardized coefficients in the descriptive discriminant analysis sample (also called the derivation or developmental sample) most likely will differ from the means and standard deviations of the sample to-be-classified (i.e., the cross-validation, replication, and classification samples).

We had no data for primate fossils for which group membership was unknown. Therefore, consider the following fabricated situation where an anthropologist discovers a primate fossil whose total skull length, minimum skull width, and dentary depth at canine measured 4.59 mm, .99 mm, and .30 mm, respectively. The anthropologist's objective is to determine the primate classification of the fossil. With the classification functions in Table 8.5, the fossil obtains the following discriminant scores: fruit feeders, 142.44 $[(42.50 \times 4.59 \text{ mm}) + (20.53 \times .99 \text{ mm}) + (115.67 \times .30 \text{ mm}) + (-107.66) =$

**TABLE 8.5**  **Fisher's Linear Discriminant Functions**

|  | Fruit feeders | Nectar feeders | Exudate feeders |
|---|---|---|---|
| Total skull length | 42.50 | 51.19 | 42.80 |
| Minimum skull width | 20.53 | 22.43 | 25.17 |
| Dentary depth at canine | 115.67 | 120.80 | 106.60 |
| Constant | −107.66 | −147.05 | −111.69 |

142.44); nectar feeders, 146.36 ((51.192 × 4.59 mm) + (22.43 × .99 mm) + (120.80 × .30 mm) + (−147.05) = 146.36); and exudate feeders, 141.66 ((42.80 × 4.59 mm) + (25.17 × .99 mm) + (106.60 × .30 mm) + (−111.69) = 141.66)]. The fossil obtained the largest discriminant score for the nectar feeders primate group and would be assigned to that group.

## V. OTHER ISSUES AND CONCERNS

### A. Following up Multivariate Analysis of Variance

A common but somewhat unfortunate practice in the social sciences is to follow MANOVA by a series of univariate ANOVAs or $t$-tests. With few exceptions, we argue against such uses of ANOVA or $t$ tests. We join a number of researchers who argue, instead, for the use of discriminant analysis (cf. Betz, 1987; Borgen & Seling, 1978; Bray & Maxwell, 1982; Brown & Tinsley, 1983; Huberty, 1975; Tatsuoka, 1970). MANOVA is used to determine whether groups formed on the basis of one or more independent variables differ on two or more dependent variables (see Huberty & Petoskey, chapter 7, this volume). MANOVA controls for the inflation of the Type 1 error rate associated with performing MANOVAs or multiple $t$-tests on correlated dimensions. A statistically significant MANOVA indicates the presence of group differences on the dependent variables or combinations of the dependent variables (viz., multivariates). Typically, subsequent analysis is necessary to determine what groups differ on the dependent variables and how they differ.

Discriminant analysis allows the investigator to examine group distinctiveness on the multivariates present in the set of variables; ANOVA and $t$-tests do not. In discriminant analysis, the variables relate to similar qualities and quantities that distinguish among the groups' "load" on the same functions, thereby revealing the latent dimensions underlying the dependent variables related to the group differences.

It is somewhat paradoxical that this advantage of discriminant analysis, identifying the multidimensionality underlying correlated dependent measures, is also a potential disadvantage. The interpretability of discriminant functions, specifically of the discriminant coefficients, is problematic when dependent variables (viz., discriminator variables) are highly correlated (Betz, 1987; Bray & Maxwell, 1982; Brown & Tinsley, 1983). The discriminatory power of the analysis is unaffected, but the weights assigned to the correlated variables by the discriminant analysis are reduced because they provide statistically redundant information. Though this information is redundant for predictive purposes, it may be essential for interpretive purposes. Thus, the relation of a particular discriminator variable to a particular function may be obscured if that discriminator variable is related to the other discriminator variables.

The disadvantage of discriminant analysis as a follow-up to MANOVA can be nullified by selecting uncorrelated dependent (discriminant) variables or by factor analyzing the variables to obtain orthogonal factors that can be used as the discriminator variables (cf. Brown & Tinsley, 1983; Sanathanan, 1975). An alternate strategy is to examine the correlations (i.e., the structure matrix coefficients) between scores on each discriminator variable and each function to discover whether relations exist that were obscured in the calculation of the discriminant function coefficients.

## B. Stepwise Discriminant Analysis

Researchers often use stepwise discriminant analysis in an effort to discover the "best" subset of discriminator variables to use in discriminating groups. Space limitations prohibit a full treatment of how to perform and interpret such an analysis; interested readers are referred to Stevens (1996) and Huberty (1994). However, we caution readers that stepwise discriminant analysis is far less likely to yield replicable results than descriptive discriminant analysis because of its "outrageous" capitalization on chance and the nuances of the particular sample (Thompson, 1995). Using large-sample sizes and a small number of predictor variables reduces this problem only slightly (cf. Thompson, 1995). Furthermore, the significance tests used with stepwise discriminant analysis are positively biased and erroneous (cf. Rencher & Larson, 1980; Stevens, 1996; Thompson, 1995). Therefore, we agree with Thompson (1995) that stepwise discriminant analysis should be eschewed for publication purposes. We recommend instead that investigators interested in the relative importance of variables to discrimination inspect the standardized discriminant function coefficients and the structure matrix coefficients. The reader is also referred to Huberty (1994) for other strategies.

## C. Reporting the Results of a Discriminant Analysis

Consumers of research in which discriminant analysis was performed need specific information to understand and evaluate the results. Researchers should report the following:

1. The mean and standard deviation of the sample on the study variables and the correlations among the variables
2. The Wilks' lambda and the chi-square statistics, the relative percents, and the canonical correlation
3. The proportion of cases correctly classified, the chance or base rate, the procedure used in calculating the chance or base rate, and the $z$ test of proportions comparing the proportion of expected and observed classifications

4. The standardized discriminant function coefficients
5. The unstandardized classification functions for each group
6. The unstandardized function coefficients and the group centroids
7. The results of the cross-validation analysis

## VI. CONCLUSION

Discriminant analysis is a powerful tool for analyzing and describing group differences and for classifying cases into groups formed on the basis of their similarities and differences on multiple variables. We used an anthropological example to illustrate the applicability of this technique in the physical and biomedical sciences, in addition to the social sciences, where its use is greater. It should be obvious that discriminant analysis also can be useful in the arts and humanities. Discriminant analysis has great utility as a tool for applied researchers confronted with the problems of selection, placement, and taxonomy, regardless of discipline. However, we can not overemphasize the importance of cross-validation and of testing classification hit rates against the base or chance rate.

## REFERENCES

Betz, N. E. (1987). Use of discriminant analysis in counseling psychology research. *Journal of Counseling Psychology, 34*(4), 393–403.

Borgen, F. H., & Seling, M. J. (1978). Uses of discriminant analysis following MANOVA: Multivariate statistics for multivariate purposes. *Journal of Applied Psychology, 63,* 689–697.

Bray, J. H., & Maxwell, S. E. (1982). Analyzing and interpreting significant MANOVA's. *Review of Educational Research, 52,* 340–367.

Brown, M. T., & Tinsley, H. E. A. (1983). Discriminant analysis. *Journal of Leisure Research, 15*(4), 290–310.

Cocozzelli, C. (1988). Understanding canonical discriminant function analysis: Testing typological hypotheses. *Journal of Social Service Research, 11*(2/3), 93–117.

Dixon, W. J. (Ed.). (1992). *BMDP statistical software manual: The data manager/to accompany release 7/version 7.0, Vol. 3.* Berkeley: Unversity of California Press.

Dumont, E. R. (1997). Cranial shape in fruit, nectar, and exudate feeders: Implications for interpreting the fossil record. *American Journal of Anthropology, 102,* 187–202.

Huberty, C. J. (1994). Why multivariable analyses? *Educational and Psychological Measurement, 54* (3), 620–627.

Huberty, C. J. (1975). Discriminant analysis. *Review of Educational Research, 45,* 543–598.

Klecka, W. R. (1975). Discriminant analysis. In N. H. Nie, C. H. Hull, J. G. Jenkins, K. Steinbrenner, & D. H. Bent (Eds.), *Statistical package for the social sciences* (2nd ed., pp. 434–467). New York: McGraw-Hill.

Morrison, D. G. (1974). Discriminant analysis. In R. Ferber (Ed.), *Handbook of marketing research.* New York: McGraw-Hill.

Norusis, M. J. (1990). *SPSS advanced statistics user's guide.* Chicago, IL: SPSS.

Norusis, M. J., & Norusis, M. J. (1998). *SPSS 8.0 guide to data analysis.* Chicago: SPSS.

Rencher, A. C., & Larson, S. F. (1980). Bias in Wilks' lambda in stepwise discriminant analysis. *Technometrics; 22,* 349–356.

Sanathanan, L. (1975). Discriminant analysis. In D. Amick & H. Walberg (Eds.), *Introductory multivariate analysis* (pp. 236–256). Berkeley, CA.: McCutchan Publishing Corporation.

Shin, K. (1996). *SAS guide: For DOS version 5.0 and Windows version 6.10.* Cary, NC: SAS Institute.

Stevens, J. (1996). *Applied multivariate statistics for the social sciences* (3rd ed.). Mahwah, NJ: Lawrence Erlbaum.

Tabachnick, B. G., & Fidell, L. S. (1989). *Using multivariate statistics* (2nd ed.) New York: Harper and Row.

Tankard, J. W. & Harris, M. C. (1980). A discriminant analysis of television viewers and nonviewers. *Journal of Broadcasting, 24* (3), 399–409.

Tatsuoka, M. (1970). *Discriminant analysis: The study of group differences.* Champaign, IL: The Institute for Personality and Ability Testing.

Thompson, B. (1995). Stepwise regression and stepwise discriminant analysis need not apply here: A guidelines editorial. *Educational and Psychological Measurement, 55* (4), 525–534.

# CANONICAL CORRELATION ANALYSIS

**ROBERT M. THORNDIKE**

*Department of Psychology, Western Washington University, Bellingham, Washington*

Canonical correlation analysis and the closely related method Cohen (1982) calls set correlation can be viewed as the most general of the traditional least-squares methods for the analysis of data structures. In the opening section of this chapter I describe set and canonical correlation analysis as a widely applicable taxonomy for data analysis, show how they are related, and explain how all other least-squares procedures can be derived as conceptual special cases. Then I develop the traditional canonical analysis model, show its relations to principal components analysis, and review significance tests and other interpretive aids such as redundancy and the rotation of canonical components. I conclude with a discussion of methodological issues in the use of canonical analysis, including stepwise procedures and the need for cross-validation. I provide an example, using one of the commercially available statistics packages, to illustrate the elements of a canonical analysis.

## I. APPROPRIATE RESEARCH SETTINGS

A search of Psychological Abstracts from 1971 to the present produced 568 references to canonical analysis and related techniques. Almost 500 of these were applications of canonical analysis methodology to substantive questions. The most common application is one in which the relations between two sets of personality scales are examined (e.g., the Sixteen Personality Factor Questionnaire and the NEO Personality Inventory [Gerbing & Tuley, 1991]). However, the uses to which canonical analysis have been put are quite varied. The range of topics includes relations between interview reports of television viewing patterns and motives for watching religious television programs (Abelman, 1987), perception/preference for shoreline scenes and demographic variables (Nasar, 1987), affective variables and competitive performance in Olympic marathon runners (Silva & Hardy, 1986), attitudes toward gambling and gambling habits (Furnham, 1985), and speech patterns and schizophrenic symptoms (Alpert, Pouget, Welkowitz, & Cohen, 1993). In many of these studies, the methodology left much to be desired, but the list illustrates the range of topics and variables that can fruitfully be analyzed using canonical correlation.

## A. An Example

In this section, I present a set of data from a study by Kleinknecht, Kleinknecht, and Thorndike (1997) in which 679 respondents completed two measures of the emotion of disgust. The Disgust Scale (DS) contained eight subscales, whereas the Disgust Emotion Scale (DES) contained five subscales (see Table 9.1). The original study did not use canonical analysis, but the data present questions for which canonical analysis is appropriate and has often been used (i.e., do these two instruments measure underlying constructs that are substantially related, and along what dimensions?).

### 1. Development Sample Preliminary Results

To allow a cross-validation of the results (see Tinsley & Brown, this volume, chapter 1, and below), the sample was divided into 470 individuals for development of the canonical variates and 209 for cross-validation. It is essential to provide the bivariate correlations among the study variables when reporting a canonical analysis (see Table 9.1). As we might expect for scales like these, all of the correlations are positive, suggesting a general dimension of susceptibility to experience disgust emotions. The raw data were submitted to the canonical analysis program included in the SYSTAT statistical package (SYSTAT, 1997).

## II. GENERAL TAXONOMY OF RELATIONSHIP STATISTICS

Consider the case where we have four variables, for example, general cognitive ability $(G)$, scholastic achievement $(A)$, parents' cognitive ability $(F)$, and social status $(S)$. We are interested in the degree of association

**TABLE 9.1  Bivariate Correlations between the Subscales of the Disgust Scale and the Disgust Emotion Scale for the Development Sample**

|  | $X_1$ | $X_2$ | $X_3$ | $X_4$ | $X_5$ | $X_6$ | $X_7$ | $X_8$ | $Y_1$ | $Y_2$ | $Y_3$ | $Y_4$ | $Y_5$ |
|---|---|---|---|---|---|---|---|---|---|---|---|---|---|
| **Disgust Scale ($X$s)** | | | | | | | | | | | | | |
| Food | 1.00 | | | | | | | | | | | | |
| Animals | .46 | 1.00 | | | | | | | | | | | |
| Body products | .46 | .56 | 1.00 | | | | | | | | | | |
| Sex | .30 | .30 | .31 | 1.00 | | | | | | | | | |
| Body envelope violations | .39 | .52 | .39 | .28 | 1.00 | | | | | | | | |
| Death | .39 | .48 | .44 | .31 | .55 | 1.00 | | | | | | | |
| Hygiene | .36 | .42 | .42 | .38 | .29 | .40 | 1.00 | | | | | | |
| Magical thinking | .40 | .47 | .45 | .29 | .39 | .51 | .41 | 1.00 | | | | | |
| **Disgust Emotion Scale ($Y$s)** | | | | | | | | | | | | | |
| Rotting food | .49 | .40 | .41 | .19 | .30 | .34 | .34 | .38 | 1.00 | | | | |
| Odors | .44 | .49 | .60 | .21 | .34 | .39 | .45 | .41 | .65 | 1.00 | | | |
| Animals | .41 | .59 | .41 | .21 | .39 | .46 | .41 | .42 | .56 | .64 | 1.00 | | |
| Blood, injury, injections | .28 | .28 | .21 | .14 | .41 | .40 | .23 | .26 | .28 | .33 | .43 | 1.00 | |
| Mutilated bodies | .42 | .53 | .45 | .22 | .64 | .58 | .34 | .43 | .47 | .59 | .60 | .53 | 1.00 |

between cognitive ability $(G)$ and scholastic achievement $(A)$, $r_{GA}^2$. (Squared correlations are used because they are proportions of variance and because once we begin to consider more than two variables the sign of the coefficient ceases to be meaningful.) $r_{GA}^2$ is the ordinary squared product moment correlation between the two variables, called the *whole correlation* or the *zero-order correlation*. Suppose further that we believe parental ability $(F)$ influences $G$. We might then be interested in the relation between $G$ and $A$ with the effect of $F$ removed or statistically controlled in $G$. This relation is called a *semipartial correlation* (or sometimes a *part correlation*), which we symbolize as $r_{(G \cdot F)A}^2$; the symbol $(G \cdot F)$ indicates that $F$ has been partialed out of $G$ but not $A$. (A second semipartial correlation is possible for this situation, the one formed when a variable, such as social status, is removed from achievement. Call this correlation $r_{G(A \cdot S)}^2$.)

Two additional types of relations are possible. We could remove the effect of $F$ (or $S$ or both) from both $G$ *and* $A$. If one variable, say $F$, is removed from both $G$ and $A$, the result is called a first-order *partial correlation*, symbolized $r_{GA \cdot F}^2$. When two or more variables are partialed out of the relation, we have a higher-order partial correlation, for example, $r_{GA \cdot FS}^2$. Alternatively, we could ask what the relation is between that part of $G$ which is independent of $F$ $(G \cdot F)$ and that part of $A$ which is independent of $S$ $(A \cdot S)$. The resulting statistic, $r_{(G \cdot F)(A \cdot S)}^2$, is known as a *bipartial* correlation.

Cohen (1982) expanded this conceptual scheme to the situation where all of the single variables are considered to be sets of predictor variables $(\mathbf{X})$ and criterior variables $(\mathbf{Y})$. The $\mathbf{X}$-set can be composed of some variables to be used as predictors, $\mathbf{P}$, and some variables to be partialed out, $\mathbf{Q}$. Likewise, the set $\mathbf{Y}$ can be seen as containing some criterion variables, $\mathbf{C}$, and some covariates, $\mathbf{D}$. Using this notation, the five types of relations between the sets $\mathbf{P}$ and $\mathbf{C}$ are as follows:

| | | | |
|---|---|---|---|
| Whole | $\mathbf{R_{PC}^2}$ | | |
| Predictor semipartial | $\mathbf{R_{(P \cdot Q)C}^2}$ | Criterion semipartial | $\mathbf{R_{P(C \cdot D)}^2}$ |
| Partial | $\mathbf{R_{PC \cdot Q}^2}$ or $\mathbf{R_{PC \cdot D}^2}$ or $\mathbf{R_{PC \cdot QD}^2}$ | | |
| Bipartial | $\mathbf{R_{(P \cdot Q)(C \cdot D)}^2}$ | | |

It is well known (e.g., Cohen & Cohen, 1983) that analysis of variance (ANOVA) is a special case of multiple regression/correlation in which the independent variables (set $\mathbf{P}$) are categorical and the single dependent variable $(C)$ is continuous. When $\mathbf{C}$ consists of a set $(>1)$ of dependent variables, the general result is a canonical analysis. Multivariate analysis of variance (MANOVA) is the special case of canonical analysis in which the predictors are categorical. The reverse case, categorical criterion variables and continuous predictors, yields discriminant function analysis. Analysis of covariance (ANCOVA) and multivariate analysis of covariance (MANCOVA) are special cases of multivariate semipartial correlation (Timm & Carlson, 1976), where a set of one or more continuous variables **D** is partialed out of the **C** set ($R^2_{P(C \cdot D)}$). A less common analysis, but one that Cohen (1982) illustrates, is the semipartial case where a set of **Q** variables is partialed out of the **P** set ($R^2_{(P \cdot Q)C}$). Finally, in multivariate bipartial correlation, also called bipartial canonical correlation (Timm & Carlson, 1976), the relation is found between a set of variables **P,** from which a set of variables **Q** has been partialed, and a set of variables **C,** from which a set of variables **D** has been partialed. If the sets **Q** and **D** are identical, the result is a multivariate partial correlation analysis. Cohen (1988) shows that this scheme also applies to contingency tables, that is, data sets where all variables are categorical. The presentation of canonical analysis that follows is in terms of whole correlations, but the general principles apply equally if one or both sets represent residual variables from which other sets of variables have been partialed.

## III. HOW RELATIONS ARE EXPRESSED

There are two ways to view the relations between sets of variables. The first is a generalization of a Venn diagram, in which the correlation or covariance between sets can be viewed as the degree to which the spaces defined by the two sets overlap. This view results in Cohen's (1982) set correlation, a single statistic that expresses the generalized relation between the sets. Although related to canonical analysis, the set correlation approach is different in that (a) it produces a single value for the degree of association; (b) that value is symmetric; and (c) the method does not yield weighted composites of the variables. Cohen argues that judicious use of partialing in set correlation yields more interpretable results than does canonical analysis.

The alternative way to view relations is to see a set of variables as producing weighted linear composites that are interpreted as dimensions of the variable space similar to the components of principal component analysis. This approach led Bartlett (1948) to describe canonical analysis as *external factor analysis* and factor or component analysis as *internal factor analysis*. The difference between the two is the criterion by which the composites are determined.

In principal, component analysis all of the variables are viewed as belonging to the same set. Successive components are determined such that each component in turn has the largest possible sum of squared correlations with the observed variables within that set, subject to the restriction that each component be uncorrelated with (orthogonal to) those that precede it. The result is a set of dimensions for the variable space, such that each successive dimension accounts for the greatest possible amount of variance in the observed variables. The location of the component in the variable space is determined solely by the properties of the variables in the set, hence the name internal factor analysis.

In canonical analysis, the criterion for locating a composite is that it have the maximum correlation with a composite of the other set. This is the same criterion that applies in multiple correlation. That is, in multiple correlation a composite of the $p$ predictors is developed of the general form

$$\hat{Z}_{X_i} = \beta_{X_1} Z_{X_{1_i}} + \beta_{X_2} Z_{X_{2_i}} + \cdots + \beta_{X_p} Z_{X_{p_i}}, \tag{1}$$

subject to the restriction that $\hat{Z}_X$ have maximum correlation with the criterion variable $Z_Y$. The difference is that in canonical analysis a second composite is also developed,

$$\hat{Z}_{Y_i} = \gamma_{Y_1} Z_{Y_{1_i}} + \gamma_{Y_2} Z_{Y_{2_i}} + \cdots + \gamma_{Y_c} Z_{Y_{c_i}}, \tag{2}$$

from the $c$ criteria, subject to the restriction that the two composites have the highest possible correlation. Thus, a canonical correlation (often given the symbol $R_c$) is the product–moment correlation between the two weighted composites, $r_{\hat{Z}_X \hat{Z}_Y}$, and the composites of a pair are defined so that they maximize their canonical correlation. The composites defined by these sets of weights are called **canonical variates.** The sets of weights, $\beta$ and $\gamma$, define the contents of the two variates and are often used for interpretation (Harris, 1985, 1989). However, some investigators (e.g., Meredith, 1964; Thorndike & Weiss, 1968) have argued that **canonical component loadings** (the correlations between the observed variables and the composites), which are equivalent to principal component loadings, are preferable for interpretation.

A single pair of composites cannot extract all of the variance that is common to the two sets whenever the smaller set has more than one variable. Completely explaining the relations between the sets requires as many dimensions as there are variables in the smaller set. If $c \leq p$, a complete canonical analysis requires $c$ pairs of composites. Each succeeding pair must satisfy two conditions: (a) both new composites be uncorrelated with all previous composites, and (b) the pair be as highly correlated as possible within the limits imposed by condition (a). Under these conditions, the $c$ composites of the criterion set will exhaust the variance of that set. If $p > c$, there will be variance in the predictor set that is not included in the analysis, but that variance is necessarily uncorrelated with the criterion set.

### A. Results from the Kleinknecht et al. (1997) Study

The initial results for our example are presented in Table 9.2. There are five variables in the smaller set, so the analysis produced five pairs of

**TABLE 9.2    Canonical Weights for the Disgust Scale (X-Set) and Disgust Emotion Scale (Y-Set)**

| Variables | X-set canonical variates ($\beta$s) | | | | | | Y-set canonical variates ($\gamma$s) | | | | |
|---|---|---|---|---|---|---|---|---|---|---|---|
| | 1 | 2 | 3 | 4 | 5 | | 1 | 2 | 3 | 4 | 5 |
| Food | .18 | .15 | .00 | 1.08 | .14 | Rotting food | .11 | −.06 | .01 | 1.30 | .40 |
| Animals | .26 | .11 | −1.12 | −.49 | .46 | Odors | .20 | −1.09 | −.74 | −.62 | −.40 |
| Body products | .18 | .68 | .92 | −.46 | .27 | Animals | .24 | −.08 | 1.37 | −.33 | −.07 |
| Sex | .11 | −.04 | −.07 | .05 | .12 | Blood, | .07 | .26 | −.19 | .35 | −1.10 |
| Body envelope | .31 | −.75 | .57 | −.14 | .23 | injury, and | | | | | |
| violations | | | | | | injections | | | | | |
| Death | .26 | −.41 | −.07 | −.05 | −.70 | Mutilated | .57 | .88 | −.45 | −.33 | .80 |
| Hygiene | .13 | .37 | −.08 | −.09 | −.90 | bodies | | | | | |
| Canonical | .76 | .51 | .35 | .28 | .10 | | | | | | |
| correlations | | | | | | | | | | | |

variates and their associated statistics. The $\beta$-weights for the eight predictors are given first, followed by the $\gamma$-weights for the five criteria. The last row of the table contains the canonical correlations. We will discuss the interpretation of these weights shortly. For now it is sufficient to note that the weights are all positive for the first pair of variates and that succeeding pairs have about half positive and half negative weights. The positive weighting of all variables on the first variates is consistent with our earlier observation that the positive bivariate correlations might indicate that a general dimension of proneness to disgust emotions exists in both scales. Bipolar variates are also the norm for the later dimensions. Note that some of the weights exceed 1.0, a point to which we return shortly.

## IV. TESTS OF SIGNIFICANCE

Bartlett (1941) provided a test for the statistical significance of canonical correlations. The null hypothesis tested is that none of the canonical correlations differ from zero by more than would be expected by chance from a population in which $\mathbf{R}_{cp}$ is a null matrix, that is, that there are only random associations between the variables in the two sets. The test is distributed approximately as $\chi^2$ when sample size is reasonably large and the population is multivariate normal (see Cooley & Lohnes, 1971, or Thorndike, 1978, for the test statistics).

A statistically significant value for this $\chi^2$ implies that at least one of the canonical correlations, the first one, is larger than would be expected by chance, and hence is considered to be statistically significant. Tests for the second and succeeding canonical correlations follow a sequential procedure in which the last correlation tested is omitted and a new $\chi^2$ is

**TABLE 9.3  Summary of Canonical Correlations and Significance Tests**

| Root | $R_c$ | $\chi^2$ | df | p |
|---|---|---|---|---|
| 1 | .76 | 640.41 | $8 \times 5 = 40$ | .000 |
| 2 | .51 | 242.11 | $7 \times 4 = 28$ | .000 |
| 3 | .35 | 101.69 | $6 \times 3 = 18$ | .000 |
| 4 | .28 | 42.21 | $5 \times 2 = 10$ | .000 |
| 5 | .10 | 4.68 | $4 \times 1 = 4$ | .320 |

computed from the remaining correlations. The process is repeated, each time omitting the next largest correlation and reducing the degrees of freedom until a nonsignificant $\chi^2$ is found, which indicates that there are no more significant correlations between the two sets. Ordinarily, interpretation of the canonical correlations and composites is limited to those that are statistically significant.

Table 9.3 presents the significance tests for the example data. The entry for root refers to the ordinal position of the canonical correlation. The canonical correlations ($R_c$), significance tests ($\chi^2$), degrees of freedom (df), and the associated probabilities (p) are given. The first four correlations are statistically significant.

## V. VARIANCE ACCOUNTED FOR—REDUNDANCY

The canonical correlation ($R_{c_j}$) is the correlation between the jth pair of linear composites; its square ($R_{c_j}^2$) expresses the proportion of variance in each composite that is related to the other composite of the pair. Thorndike and Weiss (1968) pointed out 30 years ago that $R_c^2$ cannot be interpreted as the degree of relation between the sets of variables. That is, an $R_c^2$ of .5 means that the pair of *variates* share 50% of their variance, *not that the sets of variables share 50% of their variance.* (See Wakefield & Carlson, 1975, and comments by Pielstick & Thorndike, 1976, for the consequences of failing to appreciate this point.) Stewart and Love (1968) proposed an index that would express the relation between the sets in terms of the variance accounted for, which they called *redundancy.*

### A. Vocabulary for Canonical Analysis

How to express the amount of variance that one set accounts for in the other has produced a vocabulary problem for canonical analysis. The basic analysis produces pairs of composites that are defined by sets of weights. The weights are equivalent to multiple regression beta-weights; that is, they reflect the *independent* contribution of each variable to its composite. However, the variables are themselves correlated, so the weights do not reveal how much of a variable's variance is accounted for by the composite. The correlation between the observed variable and the composite (or, more accurately, its square) provides this information.

We have used the generic term *composite* to refer to the combinations of variables formed in a canonical analysis, but now we must differentiate two expressions for these linear combinations. I will use the term *canonical variate* to refer to the composite as it is expressed by the weights of equations (1) and (2), and the weights in those equations will be called *canonical weights*. Some authors (e.g., Thompson, 1984) refer to these as functions and function weights.

When referring to the composites as they are described by the correlations between the canonical variates and the observed variables, we will use the term *canonical component* because the composites expressed in this way are similar to the components of principal component analysis; only the criterion by which they are located in the variable space is different. The component —observed-variable correlations will be called *component loadings* because they are very similar to the loadings of principal component analysis. (Some writers prefer the term *structure coefficient* to *component loading* because this is the terminology used in internal factor analysis. I prefer component loading because canonical analysis has parallels with principal component analysis, but does not have similar parallels to common factor analysis. The latter uses reduced correlation matrices.)

The canonical weights relate the observed variables to the canonical variates of one set, but it is possible and often useful to consider the relations of the observed variables in one set with the canonical variates of either set. To do this we must have the loadings. We can use the correlations of the observed variables in set **P** with the components of set **P** (which I call *intraset loadings*,) or the correlations of the observed variables in set **P** on the components of set **C** (called *interset loadings*), and vice versa (Thorndike & Weiss, 1968, 1970).

### 1. How Weights and Loadings are Related

There is a straightforward relation between the weights and the loadings. Each canonical variate is defined by a set of weights arrayed in a vector ($\boldsymbol{\beta}$ for the **P** set, $\boldsymbol{\gamma}$ for the **C** set). For example, $\boldsymbol{\beta}_1$ refers to the set or vector of weights defining the first canonical variate of the **P** set, and $\boldsymbol{\gamma}_3$ symbolizes the vector of weights defining the third canonical variate of the **C** set. The vector of intraset loadings is computed from the weight vector and the matrix of correlations between the variables of that set.

Using the symbol $s_{Pp_1}$ to represent the vector of intraset loadings for the **P** set on its first component, we obtain

$$s_{Pp_1} = \mathbf{R}_{pp}\boldsymbol{\beta}_1, \tag{3}$$

where $\mathbf{R}_{pp}$ is the matrix of correlations among the $P$-set variables. The vector of intraset loadings of the **C** variables on the third component of their set is $s_{Cc_3} = \mathbf{R}_{cc}\gamma_3$.

The interset loadings can be computed in a similar way. The difference is that we use the matrix of interset correlations, $\mathbf{R}_{pc}$ or $\mathbf{R}_{cp}$. For example, the vector of correlations of the **C**-set variables with the first component of the **P**-set is given by

$$s_{Cp_1} = \mathbf{R}_{cp}\boldsymbol{\beta}_1. \tag{4}$$

The vector of loadings of the **P**-set variables on the third component of the **C** set is given by $s_{Pc_3} = \mathbf{R}_{pc}\gamma_3$. Alternatively, the loading of any individual variable on a component of the other set can be found by multiplying the variable's intraset loading by the canonical correlation between the components. That is, the interset loading of the $i^{th}$ variable ($p_i$) in set **P** on $j^{th}$ component of set $\mathbf{C}(c_j)$ can be found by

$$s_{p_i c_j} = s_{p_i p_j} R_j. \tag{5}$$

Because some computer programs do not compute any component loadings, and many (including SYSTAT) do not compute the interset loadings, you may have to apply equations (3) through (5) to the canonical weights yourself if you wish to use the loadings for interpretation. An alternative is to calculate scores for the individual participants on the canonical variates using equations (1) and (2). Then an ordinary correlation program can be used to obtain the intraset and interset loadings because scores on the variates are available.

## 2. Loadings for the Kleinknecht et al. Study

The intraset loadings for both sets of variables from the study of the two disgust scales are presented in Table 9.4. Note that all variables in both sets have high positive loadings on the first component except Disgust with Sex ($X_4$). This is consistent with the pattern of positive bivariate correlations observed in Table 9.1. The bottom row contains the column sums-of-squares (e.g., $.66^2 + .80^2 + \cdots + .64^2 = 3.64$) or amount of variance accounted for by each component. The last column contains the intraset canonical communalities, which we will discuss shortly. The communalities are all (within rounding) 1.0 for the DES because the five components completely account for the five variables, but there is variance in each of the DS subscales that is not accounted for by its five components.

**TABLE 9.4** Intraset Loadings for the Subscales of the Disgust Scale and the Disgust Emotion Scale for the Development Sample

| Disgust Scale | X-set canonical variates | | | | | |
|---|---|---|---|---|---|---|
| | 1 | 2 | 3 | 4 | 5 | h² |
| Food | .66 | .20 | .00 | .67 | .12 | .94 |
| Animals | .80 | .14 | −.45 | −.23 | .26 | .98 |
| Body products | .71 | .50 | .37 | −.20 | .17 | .96 |
| Sex | .33 | .07 | −.04 | .08 | −.02 | .12 |
| Body envelope violations | .78 | −.48 | .22 | −.09 | .12 | .91 |
| Death | .77 | −.26 | −.01 | −.01 | −.35 | .78 |
| Hygiene | .58 | .38 | −.10 | −.02 | −.59 | .84 |
| Magical thinking | .64 | .10 | −.10 | .18 | .09 | .47 |
| Sums of squares | 3.64 | .77 | .41 | .59 | .60 | |
| Proportion of variance | .46 | .10 | .05 | .07 | .08 | |

| Disgust Emotion Scale | Y-set canonical variates | | | | | |
|---|---|---|---|---|---|---|
| | 1 | 2 | 3 | 4 | 5 | h² |
| Rotting food | .67 | −.32 | .03 | .65 | .17 | 1.00 |
| Odors | .79 | −.57 | −.19 | −.07 | −.07 | 1.00 |
| Animals | .80 | −.16 | .56 | −.05 | −.10 | 1.00 |
| Bood, injury, and injections | .57 | .32 | −.08 | .19 | −.73 | 1.00 |
| Mutilated bodies | .93 | .30 | −.16 | −.10 | .12 | 1.00 |
| Sums of squares | 2.90 | .64 | .38 | .48 | .59 | |
| Proportion of variance | .58 | .13 | .08 | .10 | .12 | |

The interset loadings are given in Table 9.5. Few programs compute the interset loadings, so the interset loadings in Table 9.5 were computed using equation (5) and a calculator. It is easy to see that these values drop off quickly for the later, less highly correlated components. The values in the final row are the column sums-of-squares or proportions of variance of this set of variables that are related to the components of the other set. They are also the redundancy values (see below). The values in the final column are the interset communalities (see below) or proportion of variance of each variable that is accounted for by the variables of the other set.

## B. Computing Redundancy

Once the loadings have been computed, it is a simple step to obtain a measure of the association between the two sets of variables. Stewart and Love (1968) approached the problem from the perspective of the intraset loadings. Drawing on principal component thinking, they noted that the sum of squared intraset loadings of the variables on a component is the amount of variance of the set that is accounted for by a component of that set (the sum-of-squares row in Table 9.4), which is similar to the eigenvalues of principal components analysis. This quantity, divided by the number of variables, produces the proportion of variance in the set that is accounted for by the component (last row in Table 9.4). For the $j^{th}$ component of set $\mathbf{P}$

$$V_{p_j} = \frac{1}{p} \mathbf{s}_j' \mathbf{s}_j = \frac{1}{p} \sum_{i=1}^{p} s_{p_i p_j}^2, \tag{6}$$

where $\mathbf{s}_j$ is the vector of component loadings for the variables of set $\mathbf{P}$ on that component.

The squared canonical correlation is the proportion of a component's variance accounted for by the paired component in the other set. Therefore, the proportion of variance in set $\mathbf{P}$ accounted for by the $j^{th}$ component of set $\mathbf{C}$ is

$$\text{Red}_{\mathbf{P}_{C_j}} = V_{p_j} R_{c_j}^2, \tag{7}$$

and the total redundancy of set $\mathbf{P}$ with set $\mathbf{C}$ is the sum of the individual redundancies.

Thorndike and Weiss (1968, 1970) suggested determining the degree of association between sets of variables directly from the interset loadings. The squared interset loadings give the proportion of each variable's variance that is accounted for by a component of the other set. Therefore, the mean of squared interset loadings for a given component is its redundancy. That

**TABLE 9.5** Interset Loadings for the Subscales of the Disgust Scale and the Disgust Emotion Scale for the Development Sample

| Disgust Scale | X-set components | | | | | |
| --- | --- | --- | --- | --- | --- | --- |
| | 1 | 2 | 3 | 4 | 5 | $h^2$ |
| Food | .50 | .10 | .00 | .19 | .01 | .30 |
| Animals | .61 | .07 | −.16 | −.06 | .03 | .41 |
| Body products | .54 | .26 | .13 | −.05 | .02 | .38 |
| Sex | .25 | .04 | −.01 | .02 | −.00 | .06 |
| Body envelope violations | .59 | −.24 | .08 | −.03 | .01 | .41 |
| Death | .59 | −.13 | −.00 | −.00 | −.04 | .37 |
| Hygiene | .44 | .19 | .04 | −.01 | .06 | .24 |
| Magical thinking | .49 | .05 | .04 | .05 | .01 | .25 |
| Sums of squares | 2.11 | .20 | .05 | .05 | .01 | 2.42 |
| Redundancy | .26 | .03 | .01 | .01 | .00 | .30 |

| Disgust Emotion Scale | Y-set components | | | | | |
| --- | --- | --- | --- | --- | --- | --- |
| | 1 | 2 | 3 | 4 | 5 | $h^2$ |
| Rotting food | .51 | −.16 | .01 | .18 | .02 | .32 |
| Odors | .60 | −.29 | −.07 | −.02 | −.01 | .45 |
| Animals | .61 | −.08 | .20 | −.01 | −.01 | .42 |
| Blood, injury, and injections | .43 | .16 | −.03 | .05 | −.07 | .22 |
| Mutilated bodies | .71 | .15 | −.06 | −.03 | .01 | .53 |
| Sums of squares | 1.68 | .16 | .05 | .04 | .01 | 1.94 |
| Redundancy | .34 | .03 | .01 | .01 | .00 | .39 |

is, the proportion of variance in set **P** that is related to the $i^{th}$ component of set **C** is

$$\text{Red}_{\mathbf{P}_{c_j}} = \mathbf{s}'_{\mathbf{P}_{c_j}}\,\mathbf{s}_{\mathbf{P}_{c_j}} = \frac{1}{p}\sum_{i=1}^{p} s^2_{p_ic_j}, \tag{8}$$

where $\mathbf{s}_{\mathbf{P}_{c_j}}$ is the vector of interset loadings of the variables in the **P** set with the $i^{th}$ component of the **C** set. The total redundancy of one set given the other is the sum of the redundancies of the individual components. If you apply equation (7) to the values in the last row of Table 4, you will get the values in the last row of Table 5, which are the redundancies of the components of each set given the other.

These data illustrate the asymmetry of the redundancy index. Thirty percent of the variance in the DS is accounted for by the DES, but 39% of the variance in the DES is related to the DS. In both cases, the first component accounts for the vast majority of the redundancy.

## C. Redundancy as the Average Squared Multiple Correlation

The redundancy of each individual variable with the other set, its interset communality (see below), is the proportion of that variable's variance accounted for by the other set of variables (Thorndike, 1976). But the squared multiple correlation (*SMC*) of a variable with a set of predictors is also the proportion of variance that the predictors account for. Therefore, an individual variable's redundancy or interset communality is equal to its *SMC* with the variables of the other set. Because total set redundancy is the average of the individual variable redundancies, total set redundancy is the average *SMC* of the variables in one set with the variables of the other set. Note, however, that this equality will not hold if we are interested only in the statistically significant canonical relations. Several computer packages, including SPSS and SYSTAT, provide the *SMC*s of the variables in one set with the variables of the other set.

There is no necessary relation between the proportion of variance that the first component of set **P** accounts for in its set and the proportion of variance that the first component of set **C** accounts for in its set. Nor is it always the case that earlier components account for more variance in their own sets. In Table 9.4 some of the later components account for more of the variance in their own sets than do some earlier ones. Each canonical variate is located to be maximally correlated with its paired variate without regard to intraset variance. Therefore, although the canonical correlations reveal a symmetric level of association between the variates, the redundancies will almost always be unequal. This is true even in the special case where $p = c$, and the analysis will account for all of the variance in each set. Redundancy will be high when the early, highly correlated components account for a large proportion of the variance in their sets, and it will be low when the later components account for most of the variance.

## VI. INTERPRETING THE COMPONENTS OR VARIATES

Some authors (e.g., Meredith, 1964; Thorndike & Weiss, 1973) have argued that interpretation of the pairs of composites should be based on the canonical component loadings, whereas Harris (1985, 1989) has presented examples where interpreting the canonical variates in terms of the weights provides a more reasonable result. We now examine some of the issues.

### A. Differences between Weights and Loadings for Interpretation

Canonical weights are generalized regression weights and may be expressed either in standardized form (comparable to beta weights in multiple regression) when correlation matrices are analyzed, or in unstandardized form (comparable to $b$ weights) when the analysis is performed on covariance matrices. Because differences in the metrics of variables are seldom meaningful in behavioral and social science research, the analysis is usually carried out on the standardized variables and the interpretation is of the standardized weights. (The SYSTAT program used to perform the analyses for this chapter does not have an option for unstandardized weights. The SPSS canonical analysis package produces both standardized and unstandardized weights whenever raw data are used as input.) The standardized weights reflect each variable's relative *independent* contributions to the variates, and this fact has some important consequences.

Unlike loadings, the weights, even in standard score form, are not bounded by zero and one. Several weights in Table 9.2 are greater than 1.0. Because the weights are unbounded, their relative magnitudes *for a particular variate* are informative, but weights cannot be compared between variates. If variables $X_1$ and $X_2$ have weights of .5 and .8 on variate 1, we can conclude that $X_2$ makes a larger contribution to that variate than does $X_1$. However, if $X_1$ has weights of .5 and .8 on variates 1 and 2 respectively, we do not know whether $X_1$ makes a larger contribution to the first or second variate. For the later variates in analyses in which $p$ and $c$ are large, weights greatly exceeding 100 can be encountered. This is most likely to occur when there are moderate to large correlations between variables within a set. In contrast, the loadings are bounded by plus and minus 1 and are, in effect, standardized *across* components. Therefore, the magnitude of a loading has meaning both within a single component and across components.

The second consequence of using weights for interpretation is that correlations among the observed variables affect the weights. A variable may receive a small weight simply because it is highly correlated with another variable in its set, even though both variables have high correlations with the variate. This makes the weights sensitive to the selection of variables, a fact that may be particularly important if stepwise procedures are used because the weight for a particular variable may change dramatically when another variable is added to or dropped from the set. The loadings appear to be less subject to such fluctuations.

Finally, Darlington, Weinberg, and Walberg (1973) suggested that regression weights tend to have larger standard errors, and therefore be less stable than correlations. That means we can expect the weights to show greater variability between samples than would the loadings if an analysis is replicated or cross-validated (Thorndike & Weiss, 1973). However, a Monte Carlo study by Thompson (1991) found that the sampling error of the loadings was essentially equal to that of the weights. He concluded that "in general, the results do not indicate that either type of coefficient is inherently less sensitive to sampling error" (p. 380). Thompson's investigation did not explore the possible effects of variable selection on the stability of the weights and loadings, nor the relative interpretability of the weights and loadings. Therefore, the majority of opinion and practice still favors loadings over weights for interpretation.

## 1. Intraset and Interset Canonical Communalities

Another feature of the loadings gives them an interpretive advantage. The loadings are used to compute redundancies, which are important for determining the amount of variance two sets of variables share. Thorndike (1976) pointed out an additional feature of the components and their loadings. When computed by the procedures developed by Stewart and Love (1968), redundancies are analogous to factor variances. Like factor variances, they are the column sums of squares of the loadings. However, the row sums of squares can be interpreted in the same manner as communalities in factor analysis. The row sums of squared *intraset* loadings, which Thorndike (1976) dubbed *intraset communalities,* reveal the proportion of an observed variable's variance included in the shared variance of a canonical analysis. Likewise, the row sums of squared *interset* loadings, which yield the *interset communalities,* reflect the proportion of an individual variable's variance that is related to the other set of variables.

In a complete canonical analysis, a variable's interset communality will be equal to its *SMC* with the variables of the other set. When a reduced set of components (e.g., those that have statistically significant canonical correlations) is retained for interpretation, the interset communalities reveal the proportion of each variable's variance explained by the retained components. The intraset and interset communalities provide interpretive information that is not available with the weights.

## B. Rotating Canonical Components

The similarities between canonical analysis and principal component analysis have led investigators to apply factor analytic concepts such as rotation to the canonical components as an aid to interpretation. The within-set and between-set variances accounted for (the intraset and interset communalities) will remain constant, and total redundancy will not be altered by a rotation because once a basis space is determined for a set of variables, rotation of the axes of that space cannot add or remove variance, only redistribute it (see Thorndike, 1976). Communalities are invariant under rotations (even oblique rotations) of the axes and the projection of each variable in a component space is defined by its communality.

Cliff and Krus (1976) performed a Varimax rotation on the weights of the combined set of variables (i.e., on the $p + c$ by $k$ matrix of canonical weights [where $k$ is the number of pairs of variates retained for rotation, usually the number of significant $R_c$s]) as though the variables were a single set. They showed that the same transformation applied to either the loadings or the weights would produce the same result. Also, for a given transformation matrix **T,** the correlations between the components after rotation can be obtained by

$$L = T' \Lambda T, \tag{9}$$

where $\Lambda$ is the matrix of correlations between components before rotation. $\Lambda$ is diagonal, but **L** will not be. That is, after rotation a pair of components need not be orthogonal to those pairs preceding or succeeding it. A point Cliff and Krus did not address, however, is that the transformation chosen (the elements of **T**) probably would be different if simple structure were sought for the loadings rather than the weights. The decision to rotate toward simple structure does not imply a preference for interpreting the weights or the loadings, but the resulting solution may be different, depending on whether the weights or loadings are used.

## C. Rotated Components in the Kleinknecht et al. (1998) Study

SYSTAT has an option to apply Varimax rotation to the number of components specified by the user. The paired components of the full loading matrix are rotated after the fashion of Cliff and Krus (1976). Table 9.6 gives the rotated components and the correlations between components after rotation. As is often the case when the number of components subjected to rotation approaches the number of variables, the components of the DES become associated with single subscales, except for the Blood/Injection scale, which, not too surprisingly, is associated with the Mutilated Bodies component. This component is paired with a component defined

**TABLE 9.6 Component Loadings after Rotation and Correlations between Components**

| Disgust Scale | X-set canonical variables | | | | Disgust Emotion Scale | Y-set canonical variates | | | |
|---|---|---|---|---|---|---|---|---|---|
| | 1 | 2 | 3 | 4 | | 1 | 2 | 3 | 4 |
| Food | .18 | .18 | −.14 | .92 | Rotting food | .25 | −.30 | .23 | .87 |
| Animals | .32 | .24 | −.85 | .20 | Odors | .22 | −.88 | .28 | .32 |
| Body products | .22 | .89 | −.23 | .22 | Animals | .29 | −.28 | .88 | .24 |
| Sex | .14 | .13 | −.18 | .23 | Blood, injury, | .63 | −.03 | .12 | .23 |
| Body envelope | .91 | .16 | −.13 | .16 | and injections | | | | |
| violations | | | | | Mutilated bodies | .88 | −.36 | .27 | .07 |
| Death | .68 | .15 | −.33 | .28 | | | | | |
| Hygiene | .08 | .43 | −.44 | .32 | | | | | |
| Magical thinking | .28 | .21 | −.35 | .46 | | | | | |
| Sums of squares | 1.58 | 1.18 | 1.27 | 1.41 | | 1.37 | 1.07 | .99 | .98 |

**Correlations between rotated components**

| Disgust Scale components | Disgust Emotion Scale components | | | |
|---|---|---|---|---|
| | $Y_1$ | $Y_2$ | $Y_3$ | $Y_4$ |
| $X_1$ | .66 | −.10 | −.16 | −.32 |
| $X_2$ | .02 | .38 | −.33 | .09 |
| $X_3$ | .09 | .22 | .25 | −.01 |
| $X_4$ | .12 | −.05 | .02 | .25 |

by the Body Envelope Violations and Death subscales of the Disgust Scale, and the correlation between them is .66 $(r_{X_1 Y_1})$. The correlations between paired components are given in the diagonal of the matrix. The fact that there are nonzero values off the diagonal shows that the rotation process has sacrificed the orthogonal property of the original canonical analysis. Indeed, components 3 and 4 of the DES correlate more highly (−.33 and −.32) with components 2 and 1, respectively, of the DS than they do with their paired components. The variance that is accounted for in each set, the total canonical variances and redundancies, remain unchanged by the rotation, but the pattern of relations has become complex. In attempting to simplify the interpretation of the individual components by rotation we have greatly complicated the interpretation of the interset relations.

The column sums of squares are shown at the bottom of the table of rotated loadings. These values represent the amounts of variance accounted for by each component for the original orthogonal components. The sums of squares after rotation have a similar interpretation, but cannot strictly be interpreted as component variances because the components are no longer orthogonal. These sums-of-squares illustrate how the Varimax rotation has spread the variance out across the components, just as it does in factor analysis.

How many components to rotate is a problem in canonical analysis, just as it is in factor analysis, and there is no definitive answer in either case. We used the results of the significance test in this example, but this almost certainly retained too many components because of the large sample size ($N > 30[p + c]$). Very little work has been done on rotation in canonical analysis, so it is unclear whether any of the guidelines for the number of factors used in factor analysis can be applied. One certainly should not rotate components for which $R_c^2$ is not statistically significant. Perhaps some guideline based on the scree test applied to the $R_c s$ might be appropriate. In the present case, the scree test would indicate one, or perhaps two, substantive components. An alternative would be to apply a criterion such as Kaiser's eigenvalue $> 1.0$ rule to the component variances, but given that the components are not ordered by the variance accounted for, this might imply retaining some later components and rejecting earlier ones. At this stage, the best guide probably is the informed judgment of the investigator. This is a problem desperately in need of further investigation.

### 1. Alternative Rotation Strategies

Existing studies and programs (for example, SYSTAT, 1997) have generally used the procedure described by Cliff and Krus (1976), which ignores the boundary between sets and rotates the combined matrix toward simple structure. I do not advocate this procedure. Investigators interested in the structure of the combined sets probably should perform a traditional factor analysis. If you regard the sets as logically distinct, ignoring that distinction in the rotation step makes little conceptual sense. Thorndike (1976) described two other strategies that might guide the rotation of the components.

The first alternative is to rotate each set independently within its space, subject to the constraint that the interset communalities be invariant (Thorndike, 1976). Independent rotations would preserve the separate identities of the sets, but would probably produce complex relations between the components that would be difficult to interpret, such as those in Table 9.6.

A second strategy involves rotating the components of one set to a simple structure or other desired location (for example, by a Procrustes transformation) and then determining the locations of the components of the other set such that maximum correlations would be obtained (Thorndike, 1976). Assuming that we rotate the $\mathbf{C}$ set first, the vector of rotated weights, $\hat{\gamma}_j$, for variate $\mathbf{c}_j$ can be used as known values in the equation

$$(\mathbf{R}_{cc}^{-1}\mathbf{R}_{cp}\mathbf{R}_{pp}^{-1}\mathbf{R}_{pc} - \hat{\lambda}_j)\hat{\gamma}_j = 0 \tag{10}$$

to solve for $\hat{\lambda}_j$ ($\lambda = R_c^2$). Then $\hat{\gamma}_j$ and $\hat{\lambda}_j$ can be used in the equation

$$\hat{\beta}_j = (\mathbf{R}_{cc}^{-1}\mathbf{R}_{cp}\hat{\gamma}_j)\hat{\lambda}_j^{-1/2} \tag{11}$$

to obtain the rotated weights for variate $\mathbf{p}_j$. There are no examples in the literature where this rotation strategy has been used, probably because no program has implemented it, but Thorndike (1977) employed a stepwise procedure to achieve a similar result (see below). This rotation strategy holds promise for cases where primary interest is in the structure of one set of variables relative to a defined structure of another set. For example, the Verbal-Performance distinction between scales of the Wechsler test series is well documented. An investigator might wish to determine how a new instrument relates to the Wechsler Adult Intelligence Scale-III (WAIS-III). By defining the composites of the WAIS-III to correspond to the usual V-P alignment, the components of the new instrument that have maximum relation with this structure could be determined. It is hoped that standard computer packages will implement this alternative soon. Until then investigators will have to write their own routines or use a less desirable alternative, such as the Cliff and Krus (1976) method.

## VII. METHODOLOGICAL ISSUES IN CANONICAL ANALYSIS

### A. Stepwise Procedures

Stepwise procedures involve the addition or deletion of variables from one or both sets of variables either to achieve optimal prediction with the smallest number of variables or to clarify or modify the nature of the canonical variates. There are two reasons an investigator might wish to employ a stepwise procedure. First, the stability of any least-squares procedure is affected by the number of variables involved. The larger the set, the greater the opportunity to capitalize on sample-specific covariation, and hence the less stable the resulting parameter estimates. Thus, an investigator might seek the smallest set of variables that will do the job of prediction in order to minimize this instability. Second, the composition of the composites depends on which variables are included in the analysis. Rotation is one way to change the composition of the composites to make them more interpretable, but the weights can be affected by variable selection. Stepwise procedures can guide variable selection so the resulting composite has a desired composition.

Thorndike and Weiss (1969, 1983) introduced stepwise (actually, variable deletion) procedures in canonical analysis (technically, a full stepwise procedure considers both addition and deletion at each step). They listed several criteria that can be used in choosing variables to be added to or deleted from the equation, including a variable's interset *SMC,* the intraset or interset component loadings, and the weights. Additionally, an investigator could try each variable in turn and add the one that produces the largest change in the canonical correlation. There are no significance tests for the canonical weights or for the difference between canonical correlations, so choosing variables based on the change in a significance test (as is done in many multiple regression computer programs) is not available (except that one might use a test based on the multiple partial correlation of the variable in question with the variables of the other set).

We noted earlier that the canonical weights are influenced by the selection of variables, and that this might be a reason not to use them for rotation. This property of the weights can be used to guide variable selection to modify the nature of the variates in a desired direction (Thorndike, 1977). When an investigator wishes to define or control the structure of a complex criterion, I recommend inspecting the canonical weights from an initial solution and adding or deleting variables to alter the canonical variate until it has the desired composition. This procedure should assure that a criterion composite includes the appropriate dimensions, but the results should always be cross-validated because stepwise procedures maximize the opportunity to capitalize on chance relations in the data.

## B. Estimating Bias in a Canonical Correlation

Canonical analysis, like multiple regression, is a least-squares procedure that will capitalize on any characteristics in the data that lead to a larger correlation. As an index of relation the canonical correlation is positively biased. Two ways to estimate the degree of bias in a canonical correlation are cross-validation and formula estimation, which involves an estimate based on a formula incorporating the number of variables and the sample size.

### 1. Cross-Validation

Any sample will differ from its population in unpredictable random ways. For a canonical analysis of a combined set of $p + c$ variables there is a theoretical population correlation supermatrix **P** with elements $\rho_{ik}$ arranged in the usual two within-set and two between-set matrices. The data being analyzed represent a sample from this population whose elements $r_{ik}$ differ from those of **P** in random ways and by amounts that depend on sample size, $N$. Each sample element is an estimate of its population element. About half of the estimates will be too large, and about half will be too small, but the least-squares nature of canonical analysis assures that the weights will take full advantage of overestimates and ignore underestimates. This property of capitalizing on sampling errors is what causes the positive bias in multivariable least-squares procedures.

Cross-validation (i.e., applying the equation developed in one sample to a second sample, see Tinsley & Brown, chapter 1, this volume) is based on the assumption that two independent samples from a given population cannot share sample-specific covariation, they can only share true population covariation. Therefore, a regression equation developed in one sample will fit the data from the second sample only to the extent that both samples reflect the same underlying population covariation. A cross-validation is carried out by solving for the weights in one sample and then applying those weights as fixed values to the data from the second sample. The degree of association found in the second sample is an estimate of how well the regression equation would work in the population (or other samples from the population).

Thorndike, Weiss, and Dawis (1968) investigated the influence of sample-specific covariation on the results of a canonical analysis using double cross-validation on a large set of data (41 variables and a total sample of 505) and concluded that "cross-validation ... is required before any attempt at interpretation of canonical variates can be made." (p. 494). Despite their initial warning and subsequent cautions by Thompson (1984) and others, most uses of canonical analysis in applied situations have not been accompanied by a cross-validation.

Two recent studies of cross-validation in canonical analysis have used bootstrapping methodology (Dalgleish & Chant, 1995; Thompson, 1995). Bootstrapping differs from conventional cross-validation in that samples of the chosen size are taken from the main data set *with replacement,* which means that (a) the bootstrap sample is a subset of the development sample, and (b) the same individual's data may enter the bootstrap sample more than once. Lederer (1996) compared bootstrapping, conventional cross-validation, and various formula estimates in the multiple regression case and concluded that the bootstrap method exhibited a positive bias in estimating the multiple correlation unless the total sample size was quite large ($N > 10p$) and the bootstrap samples were quite small ($n < .10N$). This suggests that bootstrapping should not be used in canonical analysis until we learn more about its properties.

## 2. Cross-Validation of the Kleinknecht et al. Results

The question of whether the relations found in the Kleinknecht et al. (1997) analysis can be generalized across new samples is addressed by cross-validation. Recall that 209 individuals from the original sample were held out for cross-validation. Cross-validation consists of calculating scores on the canonical variates for the holdout sample using the canonical weights from the development sample in equations 1 and 2. The correlations between variate scores in the holdout sample are the cross-validation coefficients. The standard-score weights computed from the development sample are used, so the cross-validation sample data were also converted to standard score form. Each individual's $Z$-score on each variable was multiplied by the weight for that variable on a variate, and the products were summed to yield a score for the variate. For example, to compute the score for an individual on the first component of the DES, the person's standard score on the Rotting Food scale was multiplied by .11 (see Table 9.2), their score on Odors was multiplied by .20, and so forth; then the products were summed. When the process has been completed for the desired variates, the bivariate correlations between these weighted composites are calculated. These are the cross-validation correlations (see Table 9.7).

The values in Table 9.7 reveal two things. First, compare the correlations between $X$–$Y$ pairs of variates in the diagonal (e.g., $X_1$, $Y_1$ = .68) to the $R_c$s at the bottom of Table 9.2. There is very little shrinkage on cross-validation, indicating that the relations we observed in the development sample are robust. The largest loss is 0.08, and the correlation between the third pair of variates has actually increased by 0.01. All of these pairwise correlations are also statistically significant at $p < .01$, but they may not be of practical importance because of the relatively large sample size. As I and others have warned, excessive power has often led investigators to interpret trivial correlations (see, for example, Thorndike, 1972).

Second, the orthogonality of the variates is specific to the development sample. That is, there are small correlations between variates (none statistically significant) within each set and there are also nonzero off-diagonal correlations between sets. The departure from orthogonality is not serious here, but investigators should compute these correlations to determine whether a substantial deviation from orthogonality exists. A significant departure from orthogonality might imply a change in structure from the development to the cross-validation sample.

### 3. Formula Estimation

Several formulas have been offered for estimating the bias in multiple correlation coefficients and at least two studies have explored the use of one of these formula estimates in canonical analysis. The formulas fall into two broad categories, depending on whether they estimate the actual degree of association in the population, $\rho^2$, or how well the sample equation will

**TABLE 9.7   Correlations between Cross-Validated Canonical Variates**[a]

|       | $Y_1$ | $Y_2$ | $Y_3$ | $Y_4$ | $X_1$ | $X_2$ | $X_3$ | $X_4$ |
|-------|-------|-------|-------|-------|-------|-------|-------|-------|
| $Y_1$ | 1.00  |       |       |       |       |       |       |       |
| $Y_2$ | −.02  | 1.00  |       |       |       |       |       |       |
| $Y_3$ | −.12  | −.00  | 1.00  |       |       |       |       |       |
| $Y_4$ | .02   | −.02  | −.10  | 1.00  |       |       |       |       |
| $X_1$ | **.68** | .03 | .02   | −.02  | 1.00  |       |       |       |
| $X_2$ | −.08  | **.46** | −.01 | −.11  | −.00  | 1.00  |       |       |
| $X_3$ | .04   | .05   | **.36** | .07 | .05   | .00   | 1.00  |       |
| $X_4$ | −.04  | .03   | −.16  | **.23** | −.02 | .08   | .09   | 1.00  |

[a] Cross-validated canonical correlations in boldface.

work in future samples. A formula proposed by Fisher (1924) estimates $\rho^2$ from $R^2$, the sample multiple correlation.

$$\hat{\rho}_{FW}^2 = 1 - \{(1 - R^2)([N - 1]/[N - p])\}, \tag{12}$$

where $N$ is sample size, and $p$ is the number of predictor variables. Monte Carlo studies by Dawson-Saunders (1982) and Thompson (1990) concluded that this formula provides a reasonable estimate of the population canonical correlation when sample size is reasonably large ($N \geq 10p$). Clearly, the major factor in this correction is the relation between the $p$ and $N$, and the empirical studies have confirmed that bias will be small when $N$ is large relative to $p$.

A separate question, which is equivalent to cross-validation, is how well the regression equation will work in other samples from the population. This is the same as asking how well the sample equation would work if applied to the whole population because this would be the average result obtained from samples. Several formulas have been offered to answer this question. The most comprehensive (and potentially useful because it was developed for small samples) formula was derived by Browne (1975).

$$\hat{\rho}_B^2 = \frac{(N - p - 3)\hat{\rho}^4 + \hat{\rho}^2}{(N - 2p - 2)\hat{\rho}^2 + p}, \tag{13}$$

where $\hat{\rho}^2 = \max(\hat{\rho}_{FW}^2, 0)$

and $\hat{\rho}^4 = \max\{(\hat{\rho}^2)^2 - [2p(1 - \hat{\rho}^2)^2/(N - 1)(N - p + 1)], 0\}$.

Constraining $\hat{\rho}^2$ and $\hat{\rho}^4$ to be positive produces a small positive bias in the statistic when the population correlation is small, but otherwise, the formula works well (Lederer, 1996). The results obtained by Thompson (1990) and Dawson-Saunders (1982) suggest that this formula might provide a good estimate of the cross-validation correlations if $p + c$ is substituted for $p$. The drawback of any formula estimate is that it cannot provide a check on the correlations between variates that were not paired in the development analysis.

## VIII. CONCLUDING REMARKS

One factor that contributes to poor methodology in canonical analysis is that there are no fully adequate computer programs generally available to researchers. Each major statistics package can perform a canonical analysis, but none of them include all of the desirable features. I used the Set and Canonical Correlations routine newly provided in SYSTAT 7.0 for the example in this chapter. This is the only program that provides rotation of the components. However, SYSTAT provides no interset information other than the redundancies and does not output the communalities, the component variances, or the weights after rotation. The latter are essential for cross-validating the rotated solution. It also has no option for analyzing covariances to obtain the deviation-score weights that are more appropriate for cross-group comparisons.

You can obtain interset loadings from SPSS (which is much more difficult to use than SYSTAT), but the output is hard to follow and the program provides neither rotation nor a simple way to output the weight or loading matrices for input to a rotation program. I find it easier to apply equation (7) to the output from SYSTAT than to work with SPSS. SAS, with its matrix manipulation features, can produce most of the results a sophisticated user would want, but there the user must almost write the program.

In this chapter I have tried to provide all the equations an investigator would need to do the job of canonical analysis correctly and completely. I hope applied researchers will bring pressure to bear on statistical software providers to upgrade the options their programs offer and to make it easier to do methodologically sophisticated canonical analyses. In an ideal world, the program should perform a random split of the input sample, save the two subsamples separately (or create a new variable indicating development or cross-validation sample membership) and do a cross-validation automatically, thereby fostering perhaps the most important aspect of sound methodology. The program would also output intra- and interset loadings and communalities, have options for the three rotation strategies described above, and provide the weights following rotation. Cross-validation could be performed either before or after rotation. It would not be hard to incorporate these features into programs already available, but programmers are most likely to undertake the upgrade in response to demands from users.

## REFERENCES

Abelman, R. (1987). Religious television uses and gratifications. *Journal of Broadcasting and Electronic Media, 31,* 293–307.

Alpert, M., Pouget, E. R., Welkowitz, J., & Cohen, J. (1993). Mapping schizophrenic negative symptoms onto measures of a patient's speech: Set correlation analysis. *Psychiatry Research, 48,* 181–190.

Bartlett, M. S. (1941). The statistical significance of canonical correlations. *Biometrika, 32,* 29–38.

Bartlett, M. S. (1948). Internal and external factor analysis. *British Journal of Psychology (Statistical Section), 1,* 73–81.

Browne, M. W. (1975). Predictive validity of a linear regression equation. *British Journal of Mathematical and Statistical Psychology, 28,* 79–87.

Cliff, N., & Krus, D. J. (1976). Interpretation of canonical analysis: Rotated vs. unrotated solutions. *Psychometrika, 41,* 35–42.

Cohen, J. (1982). Set correlation as a general multivariate data-analytic method. *Multivariate Behavioral Research, 17,* 301–341.

Cohen, J. (1988). Set correlation and contingency tables. *Applied Psychological Measurement, 12,* 425–434.

Cohen, J., & Cohen, P. (1983). *Applied multiple regression/correlation analysis for the behavioral sciences,* (2nd ed.) Hillsdale, NJ: Erlbaum.

Cooley, W. W., & Lohnes, P. R. (1971). *Multivariate data analysis.* New York: Wiley.

Dalgleish, L. I., & Chant, D. (1995). A SAS macro for bootstrapping the results of discriminant analysis. *Educational and Psychological Measurement, 55,* 613–624.

Darlington, R. B., Weinberg, S. L., & Walberg, H. H. (1973). Canonical variate analysis and related techniques. *Review of Educational Research, 43,* 433–454.

Dawson-Saunders, B. K. (1982). Correcting for bias in the canonical redundancy statistic. *Educational and Psychological Measurement, 42,* 131–143.

Fisher, R. A. (1924). The influence of rainfall on the yield of wheat at Rothamstead. *Philosophical Transactions of the Royal Society of London, Series B, 213,* 89–142.

Furnham, A. (1985) Attitudes to, and habits of, gambling in Britain. *Personality and Individual Differences, 6,* 493–502.

Gerbing, D. W., & Tuley, M. R. (1991). The 16PF related to the five-factor model of personality: Multiple-indicator measurement versus the a priori scales. *Multivariate Behavioral Research, 26,* 271–289.

Harris, R. J. (1985). *A primer of multivariate statistics,* (2nd ed.). Orlando, FL: Academic Press.

Harris, R. J. (1989) A canonical cautionary. *Multivariate Behavioral Research, 24,* 17–39.

Kleinknecht, R. A., Kleinknecht, E., & Thorndike, R. M. (1997). The role of disgust and fear in blood and injection-related fainting symptoms: A structural equation model. *Behavior Research and Therapy, 35,* 1–13.

Lederer, M. (1996). *Cross-validation, formula estimation, and a bootstrap approach to estimating the population cross-validity of multiple regression equations.* Unpublished master's thesis, Western Washington University, Bellingham, WA.

Meredith, W. (1964). Canonical correlations with fallible data. *Psychometrika, 29,* 55–65.

Nasar, J. L. (1987). Physical correlates of perceived quality in lakeshore development. *Leisure Sciences, 9,* 259–279.

Pielstick, N. L., & Thorndike, R. M. (1976). Canonical analysis of the WISC and ITPA: A reanalysis of the Wakefield and Carlson data. *Psychology in the Schools, 13,* 302–304.

Silva, J. M., & Hardy, C. J. (1986). Discriminating contestants at the United States Olympic marathon trials as a function of precompetitive affect. *International Journal of Sport Psychology, 17,* 100–109.

Stewart, D., & Love, W. (1968). A general canonical correlation index. *Psychological Bulletin, 70,* 160–163.

SYSTAT. (1997). *New statistics for SYSTAT 7.0.* Chicago: SPSS.

Thompson, B. (1984). Canonical correlation analysis. *Sage University Paper Series on Quantitative Applications in the Social Sciences,* 07–047.

Thompson, B. (1990). Finding a correction for the sampling error in multivariate measures of relationship: A Monte Carlo study. *Educational and Psychological Measurement, 50,* 15–31.

Thompson, B. (1991). Invariance of multivariate results: A Monte Carlo study of canonical function and structure coefficients. *Journal of Experimental Education, 59,* 367–382.

Thompson, B. (1995). Exploring the replicability of a study's results: Bootstrap statistics for the multivariate case. *Educational and Psychological Measurement, 55,* 84–94.

Thorndike, R. M. (1972). On scale correction in personality measurement. *Measurement and Evaluation in Guidance, 4,* 238–241.

Thorndike, R. M. (1976). *Strategies for rotating canonical components.* Paper presented to the Annual Meeting of the American Educational Research Association, San Francisco. (ERIC Document Reproduction Service, ED 123 259).

Thorndike, R. M. (1977). Canonical analysis and predictor selection. *Multivariate Behavioral Research, 12,* 75–87.

Thorndike, R. M., (1978). *Correlational procedures for research.* New York: Gardner Press.

Thorndike, R. M., & Weiss, D. J. (1968, May). *Multivariate relationships between the Kuder Occupational Interest Survey and the Minnesota Importance Questionnaire.* Paper presented at the Annual Meeting of the Midwestern Psychological Association, Chicago.

Thorndike, R. M., & Weiss, D. J. (1969). An empirical investigation of stepwise canonical correlation. Work Adjustment Project Research Report No. 27. Minneapolis, MN.

Thorndike, R. M., & Weiss, D. J. (1970, September). *Stability of canonical components.* Paper presented at the annual meeting of the American Psychological Association, Miami, FL.

Thorndike, R. M., & Weiss, D. J. (1973). A study of the stability of canonical correlations and canonical components. *Educational and Psychological Measurement, 33,* 123–134.

Thorndike, R. M., & Weiss, D. J. (1983). An empirical investigation of step-down canonical correlation with cross-validation. *Multivariate Behavioral Research, 18,* 183–196.

Thorndike, R. M., Weiss, D. J., & Dawis, R. V. (1968). Multivariate relationships between a measure of vocational interests and a measure of vocational needs. *Journal of Applied Psychology, 52,* 491–496.

Timm, N. H., & Carlson, J. E. (1976). Part and bipartial canonical correlation analysis. *Psychometrika, 41,* 159–176.

Wakefield, J. A., & Carlson, R. E. (1975). Canonical analysis of the WISC and the ITPA. *Psychology in the Schools, 12,* 18–20.

# 10

# EXPLORATORY FACTOR ANALYSIS

**ROBERT CUDECK**

*Department of Psychology, University of Minnesota, Minneapolis, Minnesota*

## I. EXPLORATORY FACTOR ANALYSIS

Factor analysis is a collection of methods for explaining the correlations among variables in terms of more fundamental entities called factors. It grew out of the observation that variables from a carefully formulated domain, such as tests of human ability or measures of interpersonal functioning, are often correlated with each other. This means that scores on each variable share information contained in the others. The scientific objective is to understand why this is so. According to the factor analytic perspective, variables correlate because they are determined in part by common but unobserved influences. These influences must be superordinate to the variables that are actually measured because they account for the individual differences in the tests. The goals of factor analysis are to determine the number of fundamental influences underlying a domain of variables, to quantify the extent to which each variable is associated with the factors, and to obtain information about their nature from observing which factors contribute to performance on which variables. The historical background of the method in the study of tests of ability is fascinating and relevant to understanding objectives and rationale. Brief reviews of this

*Handbook of Applied Multivariate Statistics and Mathematical Modeling*

history are summarized by Bartholomew (1987), Maxwell (1977), and Mulaik (1972). Several historical articles were published in the November, 1995, issue of the *British Journal of Mathematical and Statistical Psychology*. Both Bartholomew (1995) and Lovie and Lovie (1993) give interesting overviews.

Specialized versions of the factor analysis model exist for different types of data (e.g., categorical variables) and for specific research designs, such as longitudinal studies or multiple batteries of tests. The most popular version is appropriate for continuous variables that are normally distributed or at least roughly symmetrical and where the factors describe tests from a single battery in a linear fashion.

## A. The Regression Connection

It is useful to illustrate the basic ideas of the method with an example from a related problem. Figure 10.1 shows a path diagram representing the associations between exam scores on three school subjects with an explanatory variable, number of hours of homework (HW), shown as the single predictor. The correlations between the exam scores vary in magnitude, but typically are all positive. Generally speaking, such a pattern of correlations means that a student who performs well on one of the tests tends to perform well on the others, and conversely, that students who do poorly tend to do poorly on them all. It is natural to attempt to understand how the correlations arise. Teachers and parents know (and students privately concede) that HW is the main component responsible for exam performance. To improve performance on achievement tests, increase time on homework.

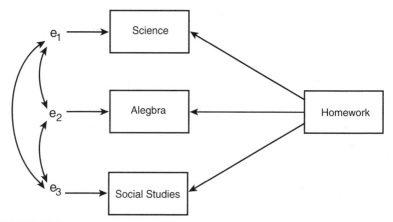

**FIGURE 10.1** Path diagram for the multivariate regression of scores on three exams on hours of homework. The residual correlations between the exam scores with homework partialed out are small but not zero in magnitude.

In relating this idea to factor analysis, it is important to emphasize that the manner in which an independent variable "accounts for" scores on the dependent variables is in the partial correlation sense. That is, when HW is used to predict each of the variables, the residual correlations among the three exams after the contribution of HW is partialed out are small. If this is so, then it is said that the correlations between the exam scores are "explained" by HW, or that hours spent on HW accounts for the observed correlations among the three test scores.

This goal of explaining the *correlations* among the exam scores in terms of the influence of HW is distinct from a related but conceptually quite different perspective of accounting for the *variance* of the former by the latter. The correlations between exam scores are accounted for in terms of HW if the residual correlations after HW is partialed out are near zero, irrespective of whether the residual variances are large or small.

### B. Factor Analysis and Partial Correlation

Factor analysis resembles the partial correlation problem in several ways. The path diagram in Figure 10.2 shows the relationships between a single factor and three measures of personality. The variables are Perseverance, Industriousness, and Perfectionism. It is assumed that the reason the tests correlate is because they are each predicted from a "common cause," represented in the figure by the circled variable, Need for Achievement (*n*Ach). *n*Ach, the *factor* in this example, is not a variable that is actually measured. Rather it is thought to be superordinate to the particular collection of personality markers being used, more than the sum of parts of the

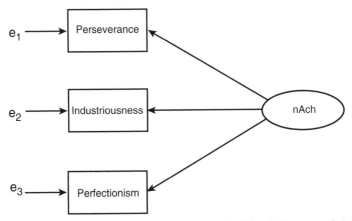

**FIGURE 10.2**   Path diagram for a single factor of Need for Achievement (nAch) that accounts for correlations between three personality variables. The residual correlations between the observed variables with the factor partialed out are zero.

variables measured to study it. Investigators often assume that the factor is an operationalization of a basic human characteristic or trait that functions as a hypothetical construct for their measurement domain.

The three personality measures in the example are dependent variables in the usual regression sense, whereas $n$Ach is the independent variable. It is assumed that all information the tests share in common is due to the contribution of the factor. One of the major conceptual differences between factor analysis and multivariate regression is the assumption that the factor accounts for the original correlations in the population. Consequently, the covariance between all pairs of tests with the factor partialed out is zero. This is shown in Figure 10.2 where the residual variables, as contrasted with those in Figure 10.1, are uncorrelated.

## C. What Is a Factor Analysis?

A fundamental question in applications of mathematical models concerns the existence of unknown variables used in the representation. Much of the utility of factor analysis is the supposition that the method operationalizes basic human traits. The extent to which this is actually accomplished is open to debate, probably on a case-by-case basis. It is useful in any event to state two main schools of thought regarding the nature of factors.

### 1. Classification Scheme

Factor analysis is often used as a method for grouping variables according to a similar correlation pattern. In this capacity, factor analysis is just a way to sort objects, to arrange like items with like. This goal has a strong similarity to cluster analysis and even to principal components analysis. There is no trait or latent variable theory underlying this approach. The objective of the analysis is to partition the variables into subsets that are hopefully distinct from those of other groups and homogeneous within a group. The rationale for the classification in terms of the underlying mechanism explaining it is of secondary interest, if it is relevant at all.

### 2. Latent Variable Model

The more frequent, and traditional, interpretation of a factor is as a fundamental causal variable postulated to explain the covariance among variables that are dependent on it. Within this tradition, two different viewpoints of a factor are encountered. The first, which might be termed the *domain score* definition, holds that a factor is a hypothetical and unobserved, but still empirically grounded, measure. A factor is the combination of all relevant variables in a particular measurement domain. For example, one might think of the verbal factor assumed to underlie performance on tests dealing with written and spoken communication as the weighted sum of the many verbal tests used in the study of individual differences. This definition is concrete, tied to an actual combination of test scores that could conceivably be obtained, even if it rarely is. The second perspective is to suppose that a factor is a *latent variable*. In this view, a factor is a basic trait, ability, or human characteristic that is essentially outside of measurement. A factor is not a consequence of tests in a domain. Verbal ability, for example, is a feature of human linguistic behavior. It is not associated with any actual measure of it. One might call this a predisposition or capacity that is inherent in individuals.

## D. The Factor Analysis Model

The model represented in the path diagram of Figure 10.2 is a regression structure, with many similarities to classical regression. To be concrete, the ideas of the diagram should be framed algebraically. Let $y_{i1}, y_{i2}, \ldots, y_{ip}$ be scores on $p$ tests or variables for individual $i$, and let $z_{i1}, z_{i2}, \ldots, z_{im}$ be the scores on $m$ unobserved factors. Figure 10.2 shows the simplest possible case, with $p = 3$ and $m = 1$. It is convenient to suppose that both the variables and factor are in standard score form with means all equal to zero and variances equal to unity. There are two main parts of the model, the first being the collection of regression relationships between variables and factors. The second part connects the correlations between the variables with the correlations between the factors.

The regression coefficient for predicting variable $y_j$ from factor $z_t$ is $f_{jt}$. The collection of coefficients is assembled in the matrix $\mathbf{F}$. For example, when there are $m = 3$ factors, the regression model for individual $i$ on the $j$th variable is

$$y_{ij} = f_{j1}z_{i1} + f_{j2}z_{i2} + f_{j3}z_{i3} + e_{ij} \tag{1}$$

where $e_{ij}$ is the residual of the $j$-th variable. It is assumed that the residuals are uncorrelated with each of the factors, and that the residuals are mutually uncorrelated among themselves. There are three sets of variables in (1), namely the observed scores, the factors, and the residuals. Although this regression relationship is the beginning point of the model, its actual use requires that factor scores, $z_t$, be available. In practice, this is seldom possible.

The second part of the model arises algebraically from the regression model. It specifies that the correlation between any two observed variables is equal to the product of the regression coefficients and factor correlations. The correlations among observed scores will be denoted as $\mathbf{R}_y$, the correlations between factors as $\mathbf{R}_z$, and the covariance matrix between the residuals as $\mathbf{C}_e$. Due to the assumption that the regression residuals are uncorrelated, $\mathbf{C}_e$ is a diagonal matrix. All off-diagonal elements are equal to zero, while the $j$-th diagonal element is equal to the variance of $e_j$. With these assumptions, the model for the correlations among the observed variables is

$$[\mathbf{R}_y]_{jk} = \sum_{t=1}^{m} \sum_{u=1}^{m} f_{jt} f_{ku} [\mathbf{R}_z]_{tu} \qquad j \ne k \tag{2a}$$

The notation $[\mathbf{R}_y]_{jk}$ denotes the $j$, $k$-th element of $\mathbf{R}_y$ and, similarly, $[\mathbf{R}_z]_{tu}$ is the $t$, $u$-th element of $\mathbf{R}_z$. The magnitude of the correlations in $\mathbf{R}_y$ depends on the magnitude of the correlations among the factors in $\mathbf{R}_z$ and also on the strength of the regression relationships between variables and the factors in $\mathbf{F}$. The representation of the variances of the variables is

$$var(y_j) = \sum_{t=1}^{m} \sum_{u=1}^{m} f_{jt} f_{ju} [\mathbf{R}_z]_{tu} + [\mathbf{C}_e]_{jj} \tag{2b}$$

Because the variance of a standardized variable is unity, the sum in this expression equals 1.

The entire cast of variables and parameter matrices from (1), (2a), and (2b) consists of the seven members summarized below.

| Variables | | Parameter matrices | |
|---|---|---|---|
| $\mathbf{y}_i$ | Observed scores | $\mathbf{R}_y$ | Correlations between variables |
| $\mathbf{z}_i$ | Factors | $\mathbf{R}_z$ | Correlations between factors |
| $\mathbf{e}_i$ | Residuals | $\mathbf{C}_e$ | Covariance matrix of residuals |
| | | $\mathbf{F}$ | Regression coefficients |

Neither $\mathbf{z}_i$ nor $\mathbf{e}_i$ actually appear in the correlation structure. The parameter matrices of the model in expressions (2a) and (2b), which will be estimated with sample data, are $\mathbf{R}_z$, $\mathbf{C}_e$ and $\mathbf{F}$. The population correlation matrix among $\mathbf{y}$, $\mathbf{R}_y$, is viewed as a consequence of the model. It is computed from (2a) from the other parameter matrices.

In the argot of factor analysis, the regression coefficients are called *factor loadings*. The matrix $\mathbf{F}$ is the *matrix of factor loadings* or the *factor pattern matrix*. $\mathbf{R}_z$ is the *factor correlation matrix*. Elements in $\mathbf{e}_i$ are called *unique variables*. They contain information in $\mathbf{y}_i$ that is not predictable from the factors. The variance of the $j$-th unique variable, $[\mathbf{C}_e]_{jj}$, is known as the *unique variable variance* or *uniqueness* of $j$. The difference

$$h_j = 1 - [\mathbf{C}_e]_{jj} \tag{3}$$

is called the *communality* of $y_j$. It is the squared multiple correlation predicting $y_j$ from the factors, or the proportion of variance of $y_j$ predicted by $\mathbf{z}$.

One other matrix not included among the parameters of the model above should be identified. It is the $p \times m$ matrix $\mathbf{R}_{yz}$ of correlations between $y_j$ and $z_t$, known as the *structure matrix*. The factor structure and factor pattern are easy to confuse. Elements of $\mathbf{R}_{yz}$ are correlations that lie within the range of $-1$ to $1$. Elements of $\mathbf{F}$ are regression coefficients.

In many cases, elements of $\mathbf{F}$ are <1 in absolute value; however, because they are regression coefficients, they may be greater than unity. A further element of confusion is that if $[\mathbf{F}]_{jt} \simeq 0$, then $z_t$ has no direct influence on $y_j$. It may still be the case, however, that the correlation between $y_j$ and $z_t$ is sizable; $[\mathbf{R}_{yz}]_{jt}$ is not necessarily equal to zero even when $[\mathbf{F}]_{jt}$ is. In the special case in which the factors are uncorrelated so that $\mathbf{R}_z = \mathbf{I}$, then $\mathbf{R}_{yz} = \mathbf{F}$.

## E. Example

Suppose a small collection of tests includes four different measures of Verbal Achievement in English and two other tests of Foreign Language Aptitude (cf. Carroll, 1993, chapter 5). The correlation matrix among these six variables in the population appears in Table 10.1. The correlations were actually generated by a model with two factors that have the following values

$$\mathbf{F} = \begin{pmatrix} .7 & .0 \\ .6 & .2 \\ .8 & .1 \\ .6 & -.1 \\ .1 & .8 \\ 0 & .9 \end{pmatrix} \qquad \mathbf{R}_z = \begin{pmatrix} 1 & \\ .4 & 1 \end{pmatrix}$$

$$\mathbf{C}_e = \begin{pmatrix} .510 & & & & & \\ & .504 & & & & \\ & & .286 & & & \\ & & & .678 & & \\ & & & & .286 & \\ & & & & & .190 \end{pmatrix}$$

The pattern of the regression coefficients in $\mathbf{F}$ reflects the fact that the first factor, $z_1$, strongly influences performance on the Verbal tests, whereas

**TABLE 10.1  Population Correlation among Six Tests**

|       | $V_1$ | $V_2$ | $V_3$ | $V_4$ | $FL_1$ | $FL_2$ |
|-------|------|------|------|------|------|------|
| $V_1$    | 1.00 |      |      |      |      |      |
| $V_2$    | .476 | 1.00 |      |      |      |      |
| $V_3$    | .588 | .588 | 1.00 |      |      |      |
| $V_4$    | .392 | .364 | .462 | 1.00 |      |      |
| $FL_1$   | .294 | .420 | .420 | .168 | 1.00 |      |
| $FL_2$   | .252 | .378 | .378 | .126 | .756 | 1.00 |

the second factor, $z_2$, accounts to a large extent for the Foreign Language Aptitude tests. General ability on verbal tasks and general ability in foreign language are positively associated with $corr(z_1, z_2) = .4$. In this artificial example, the regression coefficients and factor correlation account completely for the correlations between all pairs of tests. For example, using (2a), the correlation between $V_1$ and $V_3$ is

$$[\mathbf{R}_y]_{31} = (.7)(.8) + 0(.1) + (.7)(.1)(.4) + 0(.8)(.4)$$
$$= .588$$

Although the factors account for the correlations among tests completely, they do not fully describe the variances of the tests. The unique variable variance for $V_2$ is $[\mathbf{C}_e]_{22} = .504$. That is, approximately half the variance of $V_2$ is not associated with $z_1$ or $z_2$. From (3), the squared multiple correlation predicting $V_2$ from the factors is the difference

$$h_{V_2} = 1 - .504 = .496$$

## F. The Rotation Problem

Although $\mathbf{F}$, $\mathbf{R}_z$, and $\mathbf{C}_e$ generate the correlation matrix in Table 10.1, they are not the only parameter matrices that do so. One of the main conceptual difficulties of factor analysis is that for any given correlation matrix, where $m \geq 2$ factors exist, there is an infinite number of sets of matrices that perform equally well in describing the data. Denote the matrices comprising any one of these alternative solutions as $\mathbf{F}^*$, $\mathbf{R}_z^*$, and $\mathbf{C}_e^*$ to distinguish them from the true parameters $\mathbf{F}$, $\mathbf{R}_z$, and $\mathbf{C}_e$. All the alternative solutions have the same values for unique variances in $\mathbf{C}_e$ as are obtained above, so that $\mathbf{C}_e^* = \mathbf{C}_e$ in each case. It is the factor loadings and factor correlations that differ. Two of the many possible alternatives for $\mathbf{F}^*$ and $\mathbf{R}_z^*$ are given in Table 10.2.

**TABLE 10.2   Two Equivalent Solutions for the Data in Table 10.1**

$$
\mathbf{F}^*: \quad
\begin{pmatrix}
.662 & .117 \\
.567 & .300 \\
.757 & .233 \\
.567 & 0 \\
.095 & .817 \\
0 & .900
\end{pmatrix}
\qquad
\begin{pmatrix}
.696 & -.076 \\
.696 & .108 \\
.845 & 0 \\
.547 & -.152 \\
.497 & .683 \\
.447 & .781
\end{pmatrix}
$$

$$
\mathbf{R}_z^*: \quad
\begin{pmatrix}
1 & \\
.247 & 1
\end{pmatrix}
\qquad
\begin{pmatrix}
1 & \\
0 & 1
\end{pmatrix}
$$

Both of these possibilities account for the data in Table 10.1 the same as the original matrices $\mathbf{F}$, $\mathbf{R}_z$, and $\mathbf{C}_e$. For example, using the coefficients from the left-hand side of Table 10.2, the correlation between $V_1$ and $V_3$ can be obtained equivalently from

$$[\mathbf{R}_y]_{31} = (.662)(.757) + (.117)(.233) + (.662)(.233)(.247) + (.117)(.757)(.247)$$

$$\simeq .588$$

The upshot of the rotation problem is that there is no way to choose among the many possibilities on the basis of the data. Every different factor pattern with its associated factor correlation matrix fit $\mathbf{R}_y$ the same as the others. Sometimes the parameters from two possible solutions are so different that two researchers can reach opinions that are nearly contradictory about the nature of the factors. In this illustration, the correlation between factors in the original matrix $\mathbf{R}_z$ was .4, a reasonable value given the nature of the variables. In the left-hand solution of Table 10.2, the factor correlation is .247, a substantially lower figure, whereas in the solution on the right-hand side the factors are uncorrelated. When the range of values of $corr(z_1, z_2)$ is zero to .4, with all values equally plausible in terms of the data, then the nature of the factors is nearly impossible to understand.

Thurstone (1945) is credited with circumventing the rotation problem. He did not "solve" the problem, and the extent to which his argument is reasonable is open to debate. The approach is based on an appeal to parsimony: If there are underlying factors that determine performance on tests, then not every factor is expected to have an influence on every test. Although many sets of factor matrices account equally well for the correlations among tests, the best of these is the one that has the simplest pattern of effects on the variables. It is especially helpful when a factor does *not* directly influence test performance, for then inferring the nature of the factors from the content of the variables is easier to achieve. A configuration of factor loadings in which some factors affect only particular tests is said to display *simple structure*. Considering again the hypothetical example of this section, if one were to choose only among the three alternatives presented above, the first is preferable to the others because its pattern of effects is simplest.

The ideal of simple structure has enormous practical consequences for applications of the model. (Indeed, Adkins, 1964, p. 14 judged the simple structure concept to be "Thurstone's most noteworthy single contribution to factor analysis" (p. 14)). It effectively dispenses with the rotation problem so that estimation can be carried out in a meaningful way. It implies that a particular structure exists in the population, and the structure has an interpretable form. The concept is evoked again in estimation when the transformation of estimated loadings is considered, and where a single, sample-based version of the model is required.

This completes the main features of the population model. The model resolves correlations among the variables in terms of a small number of factors. The parameters are the factor loadings, factor correlations, and unique variable variances. The influence of the factors on the variables is presumed to display simple structure. This assumption addresses the rotation problem, clarifies the nature of the factors, and makes estimation feasible by restricting attention to a single set of parameters.

## II. FACTOR ANALYSIS AND PRINCIPAL COMPONENTS

Principal component analysis (PCA) is a technique for summarizing the information contained in several variables into a small number of weighted composites. PCA is a data analysis method focused on a particular collection of variables. It is often incorrectly used as a kind of factor analysis, and many published studies erroneously present PCA results as a variety of factor analysis. To make maters worse, commercial statistics packages sometimes produce PCA as the default method of their factor analysis programs. The two methods are not part of the same analytic approach, they have different scientific objectives, and the algebraic forms are distinct.

Consider the form of the first principal component. The score for individual $i$ on component, $c_{i1}$, uses weights $w_{11}, \ldots, w_{p1}$ in the linear combination

$$c_{i1} = y_{i1}w_{11} + y_{i2}w_{22} + \cdots + y_{ip}w_{p1}$$

This linear combination is chosen so that the sum of squares of $c_1$ is as large as possible subject to the condition that $w_{11}^2 + \cdots + w_{p1}^2 = 1$. The second principal component is another linear combination of $y_j$

$$c_{i2} = y_{i1}w_{12} + y_{i2}w_{22} + \cdots + y_{ip}w_{p2}$$

where the variance of $c_2$ is maximal, subject to the conditions that $corr(c_1, c_2) = 0$ and that $w_{12}^2 + \cdots + w_{p2}^2 = 1$. The criterion of summarizing the information in $p$ variables by a few components is valuable as a means of reducing the number of variables needed in an analysis.

Major differences in the two models are too often disregarded. Because the $y_{ij}$ are observed scores, the principal components are observed scores also. This is certainly different from a factor that is an unobserved latent variable. Components are directly connected with a particular set of observed $y$-variates. If a variable is added or deleted from the collection, the principal components change accordingly. Because a factor is superordinate to any specific group of variables, its nature is unchanged when variables are added or deleted. Principal components can always be computed from any set of variables. Factor analysis, in contrast, is a mathematical model that sometimes fits a domain of variables satisfactorily, but in other applications does not. Some writers have observed that the PCA weights obtained in an analysis are sometimes similar to the regression coefficients of factor analysis estimated with the same data. It is argued that because of this similarity, the two methods are the same. This numerical coincidence does not occur with every collection of variables, however (see Hubbard & Allen, 1987, and Fabrigar, Wegener, MacCallum & Strahan, 1999, for examples), so it is difficult to evaluate the claim in general.

A key practical difference is that the principal components are not designed to account for the correlations among the observed scores, but instead are constructed to maximally summarize the information among variables in a data set. It is specifically the summary of correlations that factor analysis is designed to accomplish. As a case in point, the correlation matrix in Table 10.1 can be completely summarized by two factors, but not even five principal components account for correlations to the same extent.

In summary, there are substantial differences between principal components and factor analysis in detail and orientation. The two have distinct statistical objectives and reflect different theoretical rationales. Morrison (1990) gives a lucid review of the unique features of each. There are frequent uses in applied research for the summarization provided by PCA, but the method in many ways is unlike the descriptive and conceptual framework of factor analysis.

## III. ESTIMATING THE PARAMETERS

Everything to this point pertains to the population model. The goal in estimation is to obtain values of $\mathbf{F}$, $\mathbf{R}_z$, and $\mathbf{C}_e$ from a sample of $N$ individuals. The objective is not to actually estimate the factors, but rather to estimate the parameter matrices. To do this, the sample correlation matrix for the $N$ subjects, denoted $\mathbf{R}_S$, is computed, and matrices $\hat{\mathbf{F}}$, $\hat{\mathbf{R}}_z$, and $\mathbf{C}_e$ are found by fitting the model to the data. In the theoretical model in the population of subjects of equation (2), it is assumed that the factors account for the population correlations exactly. In a sample of subjects, it is not the case that the factors account for the correlations in $\mathbf{R}_s$ exactly, although if the population model is reasonable for the variables in the population, it typically performs well in a sample also.

There are three general issues related to estimation that must be addressed in every analysis. These are the method of parameter estimation, the decision regarding the number of factors, and the choice of rotation.

### A. Method of Estimation

Choosing from among a long list of different estimation methods is not always straightforward. Most computer programs prespecify a default method. This can make the decision seem nonexistent, but there are practical differences between estimation methods that should be understood by a user of the model. Extended treatments of the major alternatives are reviewed in texts on factor analysis (e.g., Harman, 1976) and are available in major software packages. Some methods that were popular a few years ago were attractive because calculations are easy, not because the estimators have useful statistical properties. Fortunately, only a few of the alternatives are really relevant for general practice. The following three options have proven to be especially valuable in many different contexts.

### 1. Ordinary Least Squares

OLS is perhaps the most popular method. It also known as Principle Factors with Iteration and sometimes also called the MINRES method. This approach is so named because the objective is to find parameter estimates that minimize the squared differences between $R_S$, the sample correlation matrix, and the reproduced correlation matrix based on parameter estimates.

### 2. Maximum Likelihood

Maximum Likelihood (ML) is attractive if it is reasonable to assume that the distribution of variables is normal or at least roughly symmetric (Lawley & Maxwell, 1971, chapter 4; Morrison, 1990, chapter 9.3). ML estimates are the values that are most consistent with the sample data under the assumed normal distribution. Estimates from this method have several desirable statistical properties, and it is the basis for most new developments in factor analysis and related methods of the past 20 years.

### 3. Noniterative Estimators

Noniterative Estimators have been proposed that are simple to compute and therefore especially valuable with large numbers of variables (Hägglund, 1982; Jöreskog, 1983; Kano, 1990). An approach called PACE is very efficient computationally (Cudeck, 1991). Although this method is not yet used frequently, it generally performs well and is so fast that many analyses can be conducted quickly.

As a rule of thumb, ML is the preferred approach if the assumption of normality is plausible and the number of variables is 50 or fewer. If the distribution of the variables is decidedly nonnormal, then least squares may be most appropriate for general conditions. PACE and estimators similar to it are attractive when the number of variables is moderate to large. PACE is also useful as a way to produce starting values for either ML or Least Squares. If the model is appropriate for the variables being studied, then estimates produced by any one of these methods tend to agree with the others.

## B. Number of Factors

The decision about the number of factors is perhaps the most difficult part of an analysis. Every aspect of the results is affected by this choice. Even if parameter estimates from two different methods of estimation are comparable so that estimator differences can be ignored, the choice of number of factors can be difficult to make with assurance. This is especially so in exploratory studies where the measurement structure of a test battery is being investigated with little previous work to guide the analysis.

Although there are well-established and popular criteria for making a decision, no definitive answer to the number of factors question exists. Experienced researchers working with the same data may not agree about what is most appropriate. The best that can be done is to decide the matter pragmatically, being explicit about the criteria used in the process. Replicating the analysis on fresh data, perhaps by splitting a large sample into two subsamples and working with each independently, can verify that the decision reached in one analysis is justifiable in light of a second (Cudeck & Browne, 1983).

The decision must balance two features of an analysis. It is important that the model account for data to a reasonable degree. This means, all things equal, that more factors are preferable to fewer factors, because larger numbers of factors describe the information in a sample better than a few. On the other hand, although a complicated model with many factors may be understandable, it is not always the case that larger numbers of factors are easier to interpret than fewer. Typically, in fact, the converse is true. If the analysis is to be helpful, it must be interpretable in light of what is known about the variables and the wider substantive domain of which the study is a part. Both aspects of the problem—the degree to which the model describes data and the interpretability of the solution—are important. A model that fits well but makes little theoretical sense may have no practical significance to researchers who wish to use it. Conversely, a model that seems understandable but fits data poorly usually means the analysis is incomplete or that a different kind of summary may be more satisfactory than factor analysis.

Complicating the process is the fact that for the population correlation matrix there is no definite number of factors in operation. No matter how many factors might seem to be appropriate, the model is an *approximation* for the population correlations, not an exact summary (Cudeck & Henly, 1991). Consequently, in actual fact there is no "correct" number of factors to identify. The objective of an analysis instead is to develop a model that makes substantive sense and that describes the data to a reasonable extent. Even though there is no particular form of the model in operation, it is still valuable to summarize the correlations among many variables by a simple and psychologically meaningful factor structure. The idea that latent variables determine or account for complex facets of behavior seems almost inescapable in the study of human development and individual functioning. When a researcher settles on a particular form of the model, it does not mean the model is true, but rather that it is scientifically useful to view the behavior from this perspective.

There are both formal and informal criteria used to decide the number of factors. The edited book by Bollen and Long (1993) is an excellent survey of ideas with many references to a wider literature. The most popular informal methods are the Eigenvalues-Greater-than-Unity rule and the so-called Scree Test. The Test of Exact Fit is the classic formal method for determining $m$. Another promising formal method is an index called the Root Mean Square Error of Approximation (RMSEA). All of these procedures are used to suggest a possible number of factors. The choice ultimately involves a subjective evaluation on the part of the investigator as to a particular model's suitability. Personal opinion is not only unavoidable in this decision, it is also desirable. No procedure should be blindly applied in an automatic way.

### 1. Eigenvalues Greater Than Unity

Eigenvalues are mathematical summaries used extensively in engineering and the sciences. The most common use of eigenvalues in statistics is as generalized measures of variance contained in a set of variables (Green & Carroll, 1976). For $p$ standardized variables with no linear dependencies, there are $p$ eigenvalues, each of which are greater than zero and which collectively, sum to $p$. An influential rule based on eigenvalues was initially justified by Guttman (1954) for the particular and artificial case where a factor analysis model holds exactly in the population. He showed in that case that a lower bound to the correct number of factors equals the number of eigenvalues of the population correlation matrix that are greater than unity. It was later argued by Kaiser (1960) that this rule be applied quite broadly to situations where the model does not hold in the population exactly. Taking $m$, the number of factors equal to the number of eigenvalues greater than unity, is popular because it is easy to apply; however, this rule is not infallible and should not be used as an automatic procedure. It can be helpful as an initial guide.

### 2. Scree Test

Many researchers, especially Cattell (1966), have suggested that rather than determining $m$ by the number of eigenvalues greater than unity, that the number of factors should be decided by the number of eigenvalues that are of appreciable size compared to the others in the distribution. To make this judgement, a plot of the ordered eigenvalues is constructed. One scans the graph *from the smaller to the larger coefficients,* checking for a break in magnitude. The number of eigenvalues that are especially large supposedly corresponds to the number of factors in the analysis. Picking $m$ according to this procedure is also subjective, although again it can be valuable as a way to focus attention on a provisional number of factors.

## 3. Test of Exact Fit

The test of exact fit is the classical hypothesis test used especially with maximum likelihood to evaluate that a factor analysis model with $m$ factors applies to the population correlation matrix. The alternative hypothesis is that the population correlation matrix has some other unstructured form. The test statistic and degrees of freedom for evaluating this hypothesis are denoted $T_m$ and $df_m$, respectively. They are commonly printed out as part of the results in computer programs. If $T_m$ is small compared to $df_m$, then the model with $m$ factors cannot be rejected. If $T_m$ is large compared to $df_m$, then the hypothesis that exactly $m$ factors operate in the population is rejected. An automatic procedure is to evaluate a sequence of these hypotheses for $m = 0, 1, \ldots,$ and select $m$ as the value for which the associated hypothesis is not rejected. The practical problem with this approach is that it evaluates a proposition that is untrue because, as noted above, no version of the model actually operates in the population. When a researcher settles on a particular value for $m$, it means only that the summary is sensible and effective empirically, not that the null hypothesis is true. The major objection to the test of exact fit is that its premise is implausible, a priori. As with the two preceding guidelines, this test is sometimes helpful as a guideline but seldom ever as an automatic decision rule.

## 4. Root Mean Square Error of Approximation

The RMSEA (Steiger & Lind, 1980; Steiger, 1990) was developed for use with ML estimation. It can be computed from $T_m$, $df_m$, and sample size, $N$. For cases in which $T_m > df_m$, the statistic is

$$RMSEA = \sqrt{\frac{T_m - df_m}{(N-1)df_m}}$$

and when $T_m \leq df_m$ then one would set $RMSEA = 0$. Large values of $RMSEA$ indicate poor performance for a particular number of factors, so it is actually a "badness of fit" measure that incorporates information about the number of parameters being used in the model. $RMSEA$ usually decreases in magnitude for increasing numbers of factors. Based on experience, Browne and Cudeck (1993) suggested that values of $RMSEA > .10$ usually indicate models that fit unacceptably, $RMSEA \leq .05$ corresponds to models that fit well, and models producing values in between are intermediate.

## C. Example

Janssen, De Boeck, and Vander Steene (1996) studied verbal skills required for a task in which subjects generate synonyms to stimulus words. The eight measures of verbal ability and fluid intelligence were (a) Raven's Matrices, (b) Cattell's Culture Fair Intelligence test, (c) a multiple-choice vocabulary test, (d) the vocabulary subtest in free response form of the Verbal-Numerical Intelligence Test, (e) a 40-item verbal classification test, (f) a 40-item verbal analogies test, (g) the Controlled Associations Test, and (h) the Opposites Test, the latter two from the Kit of Factor-Referenced Cognitive Tests. These tests were administered to $N = 227$ grade school children as part of a battery of background variables. The original data are listed in Janssen et al. (1996, p. 297). Using restricted factor analysis, the authors suggested that three factors satisfactorily account for the correlations among these variables. This analysis applies the exploratory methods reviewed in this chapter.

Many variables used in the study of human abilities have distributions that are roughly symmetrical. For this reason, ML does not seem inappropriate here. The eigenvalues of the sample correlation matrix are the following:

$$3.64 \quad 1.14 \quad .84 \quad .56 \quad .52 \quad .47 \quad .43 \quad .40$$

According to the Eigenvalues-Greater-than-Unity criterion, $m = 2$ factors might be entertained initially. The upper section of Figure 10.3 shows the relative magnitudes more clearly. To get an impression of the number of relatively large and small coefficients, one fits a line to the smaller eigenvalues in the attempt to identify where a break in size occurs. This break seems to take place between the third and fourth eigenvalues. The straight line through the smaller terms in the figure is a least squares fit applied to the last five points. As generally happens in most data sets, the first eigenvalue is much larger than the others, making it difficult to appreciate differences in size among the coefficients at the break. Consequently, in the lower section of Figure 10.3 the graph has been redrawn with the first eigenvalue excluded. This gives more clarity at the point where the smaller and larger values occur. According to these results, $m = 3$ factors seem plausible.

Table 10.3 shows measures of the Test of Exact Fit and RMSEA. Using the former, the hypothesis of $m = 0$ and $m = 1$ would be rejected, but the test of $m = 2$ would not. Using RMSEA would suggest that also $m = 1$ is a poor fit, but $m = 2$ is not.

It is always a good idea to examine the residual correlations after a tentative number of factors has been decided. Although it is not essential that the model account for the sample correlations exactly, the residuals should be small for the most part. It is especially satisfactory when the great majority of residuals are $< |.10|$. If residuals are much larger than

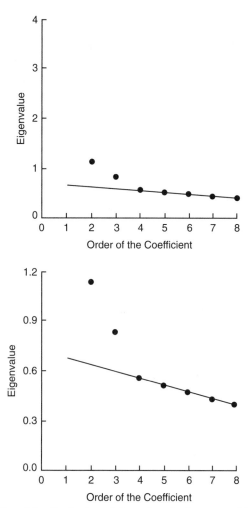

**FIGURE 10.3** Plot of the distribution of eigenvalues for Janssen et al.'s (1996) data. The bottom graph is drawn excluding the large first eigenvalue in order to show more detail at the point of the break.

**TABLE 10.3  Fit Measures for Eight Variables**

| $m$ | $T_m$ | $df_m$ | $RMSEA$[a] |
|---|---|---|---|
| 0 | 563.6 | 28 | .29 |
| 1 | 70.8 | 20 | .11 |
| 2 | 21.7 | 13 | .05 |
| 3 | 1.4 | 7 | 0 |

[a] $RMSEA$, root mean square error of approximation.

**TABLE 10.4   Residual Correlations for $m = 2^a$**

| 1 | 2 | 3 | 4 | 5 | 6 | 7 | 8 |
|---|---|---|---|---|---|---|---|
| 00 | | | | | | | |
| 00 | 0 | | | | | | |
| 00 | 00 | 00 | | | | | |
| 01 | 00 | 01 | 00 | | | | |
| −02 | 01 | 03 | 00 | 00 | | | |
| 02 | 00 | 01 | −01 | 00 | 00 | | |
| 01 | 00 | −04 | 00 | −04 | 00 | 00 | |
| 01 | −01 | −05 | −01 | −02 | −01 | 18 | 00 |

[a] Decimal points omitted.

.10, there could be other factors to consider in the data. Table 10.4 shows the residuals for $m = 2$ factors. These are all small, with the exception of the residual between variables $y_7$ and $y_8$ of .18. This is large enough to suggest that a third factor might be more appropriate than the simpler two-factor model.

True to life then, even in this small example, the decision about the number of factors is not completely clear. In general, $m = 2$ performs well in terms of several criteria. Based on the consistency of these measures of fit, two factors could be justified. However, because of the large residual and also because Janssen et al. argued in favor of $m = 3$ factors on substantive grounds, a decision to use three factors does not seem unwarranted either. The issue will be revisited in the section below after the matter of factor interpretation is reviewed.

The measures of fit in Table 3 are based on maximum likelihood. It was observed earlier that when the model gives a satisfactory summary, the parameter estimates from different estimation methods usually agree. To give an indication of this correspondence, the estimates of the unique variable variances, $\hat{\mathbf{C}}_e$, are shown for maximum likelihood, LS and PACE. Except for the estimate from PACE of the second unique variable variance, the three sets are very similar.

| | | | | | | | |
|---|---|---|---|---|---|---|---|
| ML: | .71 | .15 | .37 | .49 | .49 | .46 | .46 | .55 |
| LS: | .70 | .17 | .37 | .49 | .49 | .46 | .46 | .55 |
| PACE: | .67 | .27 | .38 | .49 | .50 | .45 | .49 | .52 |

## D. Analytic Rotation of Factors

The key assumption of simple structure is that each variable has some loadings that are near zero in magnitude with others that are nonzero but with unspecified magnitude. That a particular structure operates is an important theoretical idea of the population model. In an actual study, the goal is to estimate the parameters from a sample. Five conditions regarding the pattern of elements in **F** were proposed by Thurstone (1945) for simple structure. McDonald (1985) summarizes these along with interesting commentary. However, simple structure is not a precise concept, nor is it a condition that can be confidently claimed to occur. (There is no statistical test of a pattern of factor loadings that allows one to claim, "Simple structure holds, $p < .05!$") Experienced researchers sometimes differ in their opinion as to whether a satisfactory pattern has been achieved. What appears to be simple to one investigator may seem complicated to another.

In the years before 1950, the estimation of the model required considerable subjective ingenuity and individual effort. It was recognized from the beginning that this involved as much artistic sense as scientific expertise. Even when the same investigator repeated the task there could be differences in the solutions. So it was a significant accomplishment that starting in the 1950s procedures were developed that implemented algebraic criteria to approximate simple structure. Several of these procedures work well with many types of data.

These algorithms are called *factor rotation methods*. The reason for the name has to do with the geometric basis of factor analysis for which Harman (1976) gives an especially thorough coverage. It was noted earlier that for a specific number of factors there are an infinite number of sets of factor pattern and factor correlation matrices that fit an associated matrix of correlations equally well. All of the many sets of factor loadings and factor correlations that perform identically for a particular correlation matrix make up what is called the *equivalence class of solutions*. The rotation methods are designed to select one of these from the many possibilities. The final result hopefully displays a simple pattern, to the extent that the data under the rotation algorithm can manage it.

To arrive at the estimated model requires two steps. For a fixed value of *m*, an estimation method such as ML or least squares arbitrarily but unambiguously identifies one member of the equivalence class. This is usually labeled the *initial* or *unrotated solution* in output from a computer program. Denote the matrix of initial factor loadings as $F_0$. The unrotated solution is seldom of substantive interest in itself. Each rotation method then picks from the equivalence class defined by *m* and the estimation method one other solution that is the final form. This is called the *rotated solution*. It consists of the estimated factor pattern, $\hat{F}$, the factor correlation matrix, $\hat{R}_z$, and the unique variable variances, $\hat{C}_e$.

There are perhaps half a dozen popular rotation methods. Alternatives appear periodically in the literature. Although results from two or more of them sometimes appear similar, they are not the same procedures, and estimates from different approaches can differ in substantively meaningful ways. Some personal favorites are described briefly below. It should be noted that if ML or least squares estimation is used, for example, then any of the different results are correctly referred to as "ML estimates" or "least squares estimates." To be unambiguous, one should report both the method of estimation and the rotation option.

Rotation methods are distinguished between those that restrict the factors to be uncorrelated and those that do not. The former are known as *orthogonal* rotation methods, the latter as *oblique*. In the first case, all off-diagonal elements of $\mathbf{R}_z$ are zero as an assumption, whereas in the second they are estimated from the data. Effective orthogonal algorithms were developed a decade before successful oblique methods. Consequently, orthogonal rotation is the default method in many computer programs. In spite of the popularity of orthogonal rotation, it would be preferable if correlated factors were the usual approach. Here are the reasons why:

1. A satisfactory approximation to simple structure is more likely to be achieved when the factors are correlated than when they are not. Many examples exist where the factor pattern with orthogonal factors is complex but which is simple with oblique factors. If simple structure is a reasonable objective, then oblique rotation methods are more likely to satisfy it.

2. Oblique rotation produces estimates of the factor correlations given the data. If correlations of zero are most consistent with the data, then an oblique rotation gives values of zero as estimates. If nonzero correlations are more suitable, then oblique rotation can estimate these correlations accurately. Oblique rotation does not restrict the correlations to be specific values.

3. It is possible to test whether one or more factor correlations are zero, just as it is possible to test whether any other conventional set of correlations among variables are zero (see Cudeck & O'Dell, 1994, for an example). Correlations of zero are a proposition to evaluate, not a predetermined fact to accept uncritically.

4. It is not in general possible to identify uncorrelated factors in a population of subjects that are also uncorrelated in subpopulations selected from the original population (Meredith, 1993). Thus if a set of factors are orthogonal in a combined population of men and women, the same factors typically are correlated in both constituent subpopulations. The converse is not in general true.

Which rotation to choose? In a comprehensive study, Hakstian and Abell (1974) presented some Monte Carlo data regarding the accuracy of oblique rotation algorithms. Harman (1976) presents much valuable information on options available before 1970, but gives little critical evaluation of the possibilities. McDonald (1985) reviews nine rotation options. He recommends three specific methods and some members of the Harris and Kaiser (1964) family (a general rotation scheme whose members vary according to the value of a weighting parameter). In the limited space of this chapter, it is not possible to review either the computational philosophy or the empirical behavior of the various rotation options. A personal opinion is that the method known as Direct Oblimin with parameter zero, or Direct Quartimin (Jennrich & Sampson, 1966), is best in a wide variety of circumstances on both algebraic as well as practical grounds. Of the orthogonal methods, Varimax (Kaiser, 1958) performs well.

## E. Example

A model with $m = 3$ factors was suggested for the eight-variable data from Janssen et al. (1996). Table 10.5 gives the estimates of the factor loadings, **F** for ML with Direct Quartimin rotation. The estimates of unique variable variances, $\hat{\mathbf{C}}_e$, are repeated from above. The last group of three columns is the structure matrix, $\hat{\mathbf{R}}_{yz}$. These results are quite similar to results given by Janssen et al. To understand the factors, one studies the variables that have large loadings on a factor and attempts to interpret the characteristics that the variables have in common. The common psychological or behavioral features are attributed to the underlying factor. The first factor is exclusively associated with the two general intelligence tests. Following Carroll's (1993) empirical taxonomy, Janssen et al. labeled this factor Induction or Fluid Intelligence. Four tests with high loadings on factor 2 all relate to understanding verbal material, and so the label Verbal Comprehension seems appropriate. The last two tests require that subjects write lists of synonyms or antonyms in response to several stimulus words in a fixed time limit. The common factor responsible for their correlation seems to be the production ability of Verbal Fluency.

**TABLE 10.5  Maximum Likelihood Estimates with Direct Quartimin Rotation**[a]

| | $\hat{\mathbf{F}}$ | | | | | $\hat{\mathbf{R}}_{yz}$ | | |
|---|---|---|---|---|---|---|---|---|
| | **1** | **2** | **3** | $\hat{\mathbf{C}}_e$ | $\hat{h}_j$ | **1** | **2** | **3** |
| 1. Raven's Matrices | 47 | 12 | 03 | 71 | 29 | 52 | 34 | 31 |
| 2. Cattell's Intelligence Test | 94 | −07 | 02 | 15 | 85 | 92 | 35 | 37 |
| 3. Vocabulary, multiple-choice | 02 | 84 | −07 | 37 | 63 | 35 | 79 | 53 |
| 4. Vocabulary, free response | −14 | 68 | 11 | 49 | 51 | 20 | 69 | 52 |
| 5. Classification | 10 | 67 | −01 | 49 | 51 | 38 | 71 | 51 |
| 6. Analogies | 12 | 59 | 12 | 46 | 54 | 42 | 72 | 59 |
| 7. Associations | 05 | 00 | 71 | 46 | 54 | 36 | 53 | 73 |
| 8. Opposites | −02 | 01 | 67 | 55 | 45 | 27 | 48 | 67 |

[a] Decimal points omitted. $\hat{\mathbf{F}}$, factor loadings; $\hat{\mathbf{C}}_e$, unique variable variances; $\hat{h}_j$, communality; $\hat{\mathbf{R}}_{yz}$, factor structure.

The communalities are the squared multiple correlations for predicting $y_j$ from the factors. Raven Matrices, $y_1$, is predicted most weakly from these factors, whereas Cattell's Test, $y_2$, is predicted best. The estimates from (3) are

$$\hat{h}_1 = 1 - .72 = .29$$
$$\hat{h}_2 = 1 - .15 = .85$$

The factor loadings in $\hat{\mathbf{F}}$ describe the direct influence of a factor on a variable. The associated factor correlation matrix is

$$\hat{\mathbf{R}}_z = \begin{pmatrix} 1 & & \\ .43 & 1 & \\ .43 & .71 & 1 \end{pmatrix}$$

Even though, for example, $z_2$ and $z_3$ have a negligible direct influence on $y_2$ (the estimated loadings are $-.07$ and $.02$), there is a moderate correlation between the Raven's and the factors. Corresponding elements from the structure matrix are $.35$ and $.37$. The regression coefficients between variables and factors in $\hat{\mathbf{F}}$, and the estimated correlations between $\mathbf{y}$ and $\mathbf{z}$ in $\hat{\mathbf{R}}_{yz}$, are often confused. The reason for the confusion is generally because elements of $\hat{\mathbf{F}}$ are mistakenly interpreted as correlations. Elements in $\hat{\mathbf{F}}$ are regression coefficients, not correlations. The regression coefficient for a dependent variable predicted from an independent variable can be small, whereas the associated correlation can be large, because regression coefficients are computed from the partial correlation of the dependent variables on the predictors. Most textbooks on regression discuss this issue of interpretation between regression coefficients and correlations. Cliff (1987) gives an especially clear summary.

## IV. STANDARD ERRORS FOR PARAMETER ESTIMATES

Practical methods have been developed for large-sample estimates of the sampling variances and covariances of parameter estimates of the model. Work by Jennrich (1973, 1974) is especially important in this research. A general overview of the area, including many examples, is given by Cudeck and O'Dell (1994). A comprehensive technical approach has been described by Browne and Cudeck (1997). The main conceptual issues are as follows.

If simple structure holds in a particular situation, then for most variables there should be coefficients in the pattern matrix that are small while others are large. The inferential question concerns whether certain coefficients in the population are small. Accordingly, one might wish to test whether a particular set of parameters is zero in the population. Alternatively, there might be interest in constructing interval estimates of factor loadings, or there may be a question as to whether the difference between two parameters exceeds a user-specified minimum value. Lacking proper statistical tools, these questions can only be handled informally.

Estimates of standard errors are most useful when judging whether a sample factor loading is so small as to indicate that the associated population loading is essentially zero. Some researchers interpret only the largest loading in a row of $\hat{\mathbf{F}}$ as having a nonzero population loading, treating all other values for the same variable as zero. This is obviously simplistic, unless, as in Table 10.5, the structure is very clean. An even more popular approach is to pick an arbitrary cutoff value, such as .30, and use it as the critical point for every loading. This practice is more defensible, but does not take into account the fact that different parameter estimates have different sampling variability. Information about the standard errors is invaluable because it incorporates all sources of variability that are relevant to each estimate.

To see how this information can be used, the estimates for the three-factor model with Janssen et al.'s data are repeated in Table 10.6, but this time standard errors are included in parentheses. This example is a complicated multivariate problem involving 24 factor loadings and 3 factor correlations. The overall estimation picture is predictably a complicated one. Variables $y_3$, $y_4$, $y_5$, and $y_6$ have salient loadings on Factor 2. The standard errors for these loadings are relatively small and of similar magnitude. This would be interpreted to mean that the parameters are close in value to the estimates. By contrast, there is greater uncertainty associated with the estimate $\hat{f}_{21} = .94$ than with $\hat{f}_{11} = .47$ where $se(\hat{f}_{21}) = .23$ and $se(\hat{f}_{11}) = .15$. A few of the variances are notably large, especially for $y_7$.

**TABLE 10.6** **Parameter Estimates with Standard Errors**[a]

|  | 1 | 2 | 3 |
|---|---|---|---|
| Factor loadings |  |  |  |
| 1. Raven's Matrices | 47(15) | 12(10) | 03(09) |
| 2. Cattell's Intelligence Test | 94(23) | −07(07) | 02(10) |
| 3. Vocabulary, multiple-choice | 02(05) | 84(08) | −07(07) |
| 4. Vocabulary, free response | −14(06) | 68(09) | 11(10) |
| 5. Classification | 10(06) | 67(10) | −01(09) |
| 6. Analogies | 12(07) | 59(10) | 12(11) |
| 7. Associations | 05(09) | 00(21) | 71(32) |
| 8. Opposites | −02(05) | 01(18) | 67(27) |
| Factor correlations |  |  |  |
| Factor 1 | 100 |  |  |
| Factor 2 | 43(07) | 100 |  |
| Factor 3 | 43(10) | 71(07) | 100 |

[a] Decimal points omitted.

Nominal 95% interval estimates for the parameters are formed by using the estimates plus or minus 1.96 times the standard error. The point estimates for $y_7$ are $\hat{f}_{71} = .05$, $\hat{f}_{72} = 0.0$, and $\hat{f}_{73} = .71$. It is a mistake to treat these as exactly the same as the parameters. The uncertainty in the estimates is essential in interpreting the results. To illustrate, 95% interval estimates are

$$
\begin{array}{ccc}
f_{71} & f_{72} & f_{73} \\
(-.12 \text{ to } .23) & (-.41 \text{ to } .41) & (.08 \text{ to } 1.34)
\end{array}
$$

Not only are these intervals of different widths, but the range of possible values is striking. Although both $f_{71}$ and $f_{72}$ might be zero in the population, this fact is of secondary importance to the ranges. And although $f_{73}$ may not be zero, there is considerable doubt what the parameter might be. Standard errors for the factor correlations are small. It does not seem likely that the factors are uncorrelated in the population.

The magnitude of estimated standard errors is a complicated function of sample size, number of factors, estimation method, rotation option, and parameter values. Changing any of these affects both the parameter estimates and standard error estimates. The influence of the population parameters may seem commonsensical, but it is actually a subtle source of the overall complexity in estimates. In particular, the value of the factor correlations affects both the factor loading estimates and their standard errors. In large samples with good simple structure and several variables per factor, the standard errors may be uniformly small. Interpretation is then relatively straightforward because uncertainty due to variability of estimation is minimal. In other cases where sample size is small, where the number of factors is relatively large, or the structure is less clear—in other words, precisely in the situations where exploratory factor analysis is often applied—there can be considerable uncertainty about population parameters. Sampling variability of the estimates often is large when either sample size is small or when too many factors are studied for a particular collection of variables than can be accurately estimated. An easy, but not completely satisfactory, way to address the problem of uncertainty in the estimates is to reanalyze and consider a model with fewer factors. Standard errors for simpler models are usually smaller than those for more complex models that utilize greater numbers of factors.

## V. TARGET ROTATION

The goal of analytic rotation methods is to find a transformation of the factors that gives a simple pattern of regression coefficients. Rotation procedures have the advantage of being objective and reliable and often work well in approximating the simple structure ideal. There is no guarantee that the clearest solution is the one they might produce, however. After all, rotation algorithms are quantitative procedures that do not reflect subjective information about how interpretable a solution is. For this reason, a class of transformation methods has been developed that allows an investigator to explore other structures that are substantively plausible for the data. With these methods, a researcher specifies a *target matrix* of factor loadings that reflects a hypothesis or conjecture about the structure. The objective is to rotate the initial matrix of loadings to be as close as possible to specified elements of the target. The resulting matrix of loadings performs the same as all other matrices in the equivalence class, but hopefully more nearly approximates a researcher's subjective opinion about the factors. The most general procedure for target rotation was presented by Browne (1972a, 1972b).

To illustrate target rotation, data will be used from a study by Shutz (reported in Carroll, 1993) based on $p = 9$ variables and $N = 2562$ observations. Four factors fitted with least squares work well in this case, the largest residual correlation being only .013. In the left-hand side of Table 10.7 are estimated loadings with Varimax rotation, $\hat{\mathbf{F}}_V$. Taking into account the content of the variables, Carroll suggested that these factors are Verbal Ability, Spatial Ability, Numerical Ability, and General Reasoning Ability. Accepting the interpretation, it can be seen that this solution is not completely satisfactory. Most factors have large loadings for tests that are not

**TABLE 10.7** Varimax Rotation (Left) Estimates after Oblique Rotation to a Target (Right)[a]

| | $\hat{\mathbf{F}}_V$ | | | | $\hat{\mathbf{C}}_e$ | $\hat{\mathbf{F}}_T$ | | | |
|---|---|---|---|---|---|---|---|---|---|
| Word meaning | 76 | 14 | 20 | 17 | 34 | 77* | 00 | −01 | 06 |
| Odd words | 94 | 15 | 23 | 13 | 02 | 101* | 01 | 01 | −04 |
| Boots | 15 | 75 | 18 | 16 | 35 | −01 | 77* | 00 | 05 |
| Hatchets | 16 | 84 | 18 | 12 | 22 | 02 | 90* | 00 | −04 |
| Mixed arithmetic | 28 | 15 | 75 | 18 | 30 | 04 | −03 | 77* | 09 |
| Remainders | 21 | 18 | 88 | 10 | 13 | −03 | 04 | 97* | −06 |
| Mixed series | 33 | 29 | 35 | 53 | 41 | −01 | 02 | 11 | 69* |
| Figure changes | 29 | 31 | 20 | 59 | 44 | −05 | 04 | −09 | 81* |
| Teams | 34 | 19 | 24 | 44 | 60 | 10 | −04 | 02 | 57* |

| | | | | $\hat{\mathbf{R}}_z$ | | |
|---|---|---|---|---|---|---|
| | | | 100 | | | |
| | | | 38 | 100 | | |
| | | | 52 | 43 | 199 | |
| | | | 67 | 61 | 63 | 100 |

[a] Decimal points omitted. Starred elements are unconstrained; others were fit to the target with elements of zero.

thought to be strongly associated with them, and some variables, in particular the last three, have large loadings on two or more factors.

In an attempt to "clean up" the Varimax solution, the following target was defined for a target rotation. The values of zero in the matrix are target elements for the corresponding entry in **F**, whereas "?" means that no constraint is imposed for the specified element.

$$
\mathbf{F}_T = \begin{pmatrix}
? & 0 & 0 & 0 \\
? & 0 & 0 & 0 \\
0 & ? & 0 & 0 \\
0 & ? & 0 & 0 \\
0 & 0 & ? & 0 \\
0 & 0 & ? & 0 \\
0 & 0 & 0 & ? \\
0 & 0 & 0 & ? \\
0 & 0 & 0 & ?
\end{pmatrix}
$$

This estimates $\hat{\mathbf{F}}_T$ after rotating as closely as possible to the target matrix is in the right-hand side of Table 10.7. The solution is simple and readily interpretable in light of the content of the variables. The factor correlations are of moderate size. Estimates of unique variable variances are unaffected by any rotation method, and so are the same for both the original Varimax solution as well as the subsequent target rotation.

## VI. CASE STUDY

In order to illustrate some of the ideas presented earlier, an analysis of a larger and more realistic problem is presented in this section. Since the 1930s, exploratory factor analysis has been the tool of choice in the investigation of basic features of personality and temperament. A good deal of recent activity has focused on the classification of natural-language personality descriptors and the major personality factor dimensions embedded in these descriptors. Goldberg (1990, 1993) in particular has conducted a comprehensive program of the structure of personality terms selected from the lexicon. Complementing and extending the earlier work of many others, Goldberg's empirical analyses have converged on five comprehensive dimensions of personality functioning. These are the well-known Big Five factors.

There are two central issues in all these investigations. Obviously, the first objective is to ascertain whether the number of factors needed to explain correlations among everyday descriptors of personality, across different subsets of variables and different samples of subjects, is really five. This *dimensional* consistency is needed to support the contention that the

Big Five are replicable and consistent. The second objective is to explore the structure, and verify that over a broad sampling of items and subjects these five factors should correspond to the expected dimensions of Extraversion, Agreeableness, Conscientiousness, Emotional Stability, and Intellect or Openness. This *structural* requirement is needed to provide assurance that the actual content and nature of the factors is broadly generalizable.

Saucier (1994) recently developed an efficient 40-item adjective set designed to measure the Big Five. He showed that these "Mini-markers" have adequate reliability and satisfactory correspondence with longer inventories. In order to study further the structure of this collection of adjectives, the instrument was administered to a sample of 705 undergraduate university students who rated themselves with these items using a four-point scale. (I am grateful to Auke Tellegen for making these data available.)

Preliminary information about the number of factors is available from the distribution of the eigenvalues of the 40 variable correlation matrix shown in Figure 10.4. The number of eigenvalues greater than unity is nine, which is probably too many for this case. In terms of relative size, there is a clear break between the fifth and sixth eigenvalues, which supports the expected five-factor structure. However, a case can be made for considering six or even more factors, and it seems unwarranted to be dogmatic about just five. Values of RMSEA for $m = 5$, 6 and 7 are .08, .07, and .06, respectively. Again, larger numbers of factors obviously perform better

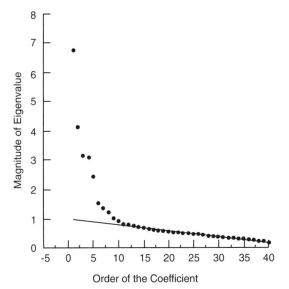

**FIGURE 10.4**    Distribution of eigenvalues for 40 Mini-markers of the Big Five. The straight line is the least squares fit through the smallest 30 eigenvalues.

**TABLE 10.8  Mini-Marker Scales of the Big-Five Personality Dimensions Ordinary Least Squares Estimates with Oblique Rotation[a]**

|  | Variable | 1 | 2 | 3 | 4 | 5 |
|---|---|---|---|---|---|---|
| 32 | Talkative | 73 | 14 | −06 | −14 | −01 |
| 13 | Extraverted | 65 | 04 | −07 | −08 | −02 |
| 2 | Bold | 60 | −19 | 04 | 02 | 22 |
| 11 | Energetic | 54 | 17 | 16 | −05 | 07 |
| 28 | Shy | −80 | 12 | 02 | −08 | −01 |
| 25 | Quiet | −85 | 03 | 10 | 02 | 07 |
| 1 | Bashful | −72 | 13 | −02 | −19 | −01 |
| 40 | Withdrawn | −61 | −16 | −12 | −14 | 10 |
| 30 | Sympathetic | 00 | 68 | −01 | −12 | 09 |
| 39 | Warm | 08 | 77 | 01 | −07 | 09 |
| 20 | Kind | −01 | 69 | 03 | −04 | 14 |
| 6 | Cooperative | 02 | 49 | 11 | 01 | 00 |
| 4 | Cold | −12 | −57 | 00 | −11 | 02 |
| 38 | Unsympathetic | −04 | −55 | −04 | 08 | −20 |
| 27 | Rude | 09 | −52 | −15 | −15 | −02 |
| 15 | Harsh | 07 | −56 | −01 | −26 | 11 |
| 22 | Organized | −03 | −02 | 83 | −11 | −04 |
| 10 | Efficient | 10 | −06 | 70 | 00 | 06 |
| 31 | Systematic | −08 | −09 | 56 | −13 | 03 |
| 24 | Practical | −04 | 10 | 40 | 02 | 00 |
| 9 | Disorganized | 01 | 03 | −75 | 00 | 14 |
| 29 | Sloppy | 00 | −05 | −60 | −04 | 02 |
| 17 | Inefficient | −12 | −05 | −64 | −07 | 03 |
| 3 | Careless | 03 | −15 | −50 | −09 | −03 |
| 36 | Unenvious | −02 | −14 | −04 | 57 | 07 |
| 26 | Relaxed | −05 | 21 | −03 | 29 | −02 |
| 21 | Moody | −11 | −18 | −01 | −53 | 16 |
| 19 | Jealous | −04 | 10 | −10 | −73 | −17 |
| 33 | Temperamental | 04 | −28 | 09 | −54 | 18 |
| 12 | Envious | 01 | 15 | −01 | −72 | −14 |
| 34 | Touchy | −02 | −17 | 06 | −45 | 09 |
| 14 | Fretful | −17 | −02 | −15 | −46 | −03 |
| 7 | Creative | 13 | 07 | −06 | 08 | 66 |
| 16 | Imaginative | 11 | 11 | −06 | 03 | 61 |
| 23 | Philosophical | −10 | 03 | −09 | 01 | 58 |
| 18 | Intellectual | 00 | −02 | 19 | 10 | 46 |
| 5 | Complex | −05 | −09 | −09 | −12 | 46 |
| 8 | Deep | −02 | 11 | 00 | −09 | 52 |
| 35 | Uncreative | −11 | −10 | 00 | −11 | −58 |
| 37 | Unintellectual | 01 | −09 | −16 | −14 | −42 |
| Factor correlations |  | — |  |  |  |  |
|  |  | 12 | — |  |  |  |
|  |  | 08 | 24 | — |  |  |
|  |  | 17 | 26 | 14 | — |  |
|  |  | 15 | 10 | 08 | 01 | — |

[a] Decimal points omitted. Item numbers correspond to order in the original inventory. Factor 1, Extraversion; Factor 2, Agreeableness; Factor 3, Conscientious; Factor 4, Emotional Stability; Factor 5, Intellect or Openness.

than fewer. Still there is no compelling reason to consider more than five factors on the basis of overall fit alone. In light of this information, and considering that the inventory was designed to reflect five factors, the solution with $m = 5$ was provisionally accepted.

OLS estimation was used because these data were scored on a 1 to 4 scale. OLS does not make a specific assumption about the distribution of the variables, and seems more appropriate than ML for data of this kind. Estimates for $m = 5$ factors with direct quartimin rotation are presented in Table 10.8. To make the presentation of this large table of coefficients as clear as possible, the variables and factors were rearranged to correspond to the order presented in Saucier (1994), decimal points are omitted, and the loadings on each item's hypothesized factors is highlighted. With this large sample, the standard errors of estimate for all the loadings are small. None is greater than .03. Sampling variability can be safely ignored here. The structure is unusually clear. It is also notable that the factor correlations are generally small. The largest correlations are between Factors 2 and 3 and Factors 2 and 4. Although .24 and .26 are not zero, the factors are not strongly associated.

This does not mean that only a five-factor model is defensible. The model with six factors naturally fits better than five. Moreover, the results are equally interpretable. The main difference between the five- and six-factor solutions are that items associated with Factor 5 in Table 10.8 split into two different factors in the $m = 6$ solution. One cluster has large loadings for items Intellectual, Unintellectual, Philosophical, Complex and Deep, whereas the second includes the markers Creative, Uncreative, and Imaginative. The factor correlation for the pair is $-.29$. This is an obvious differentiation between ability-type and creativity-type items, which is consistent with analyses based on larger collections of ability variables. This second set of results reinforces the fact that more than one interpretable solution can often be identified in a single analysis. The decision to retain one or the other requires knowledge of the domain under investigation. Ideally, it will lead to further investigation to refine the instruments and provide more information about the structure of constructs.

## VII. SUMMARY

The exploratory factor analysis model is a statistical method for investigating common but unobserved sources of influence in a collection of variables. The empirical basis for this technique is the observation that variables from a carefully chosen domain are often intercorrelated. In attempting to understand why this is so, it is natural to hypothesize that the variables all reflect a more fundamental influence that contributes to individual differences on each measure. The model explicitly breaks down the variability of variables into a part attributable to the factors and shared with other variables, and a second part that is specific to a particular variable but unrelated to the factors. The major theoretical appeal of the method is that it provides a way to investigate constructs and traits.

Although the method has an important theoretical basis, the majority of uses are descriptive in nature. The goal is to summarize complicated patterns of correlations between variables into a simpler explanatory framework. In this capacity, the methodology is typically referred to as exploratory factor analysis. There is, however, a strong confirmatory tradition as well. For example, a confirmatory use of the method can be carried out to test whether a particular number of factors is adequate for a domain. Other more specific structural hypotheses can be investigated with target rotations. The availability of standard errors of estimate of the parameters allows considerable flexibility in setting up and testing hypotheses about the magnitude of loadings or factor correlations. An even greater degree of generality in confirmatory analyses can be accomplished with methods of structural equation models (SEM) (Bollen, 1989); however, classical factor analysis plays an important role in the theory of SEM, so the two approaches are complementary rather than competitive.

There are many decision points in a factor analysis. Some of these, such as the method of estimation and rotation option, are of obvious importance and should be not only carefully considered but also completely described in research publications. A variety of other practical matters must be decided as well. These include sample size considerations and information about relevant populations, scale of the variables and type of correlation index, the number of variables per factor, approaches to interpretation, and the particular software package used. A practical review of reporting standards for factor analysis is given by Ford, MacCallum and Tate (1986) that includes recommendations about which aspects should be presented in an analysis.

Factor analysis is an area of active methodological research. Elaborations of the model, improvements in computational methods, and extensions to specific research designs and populations occur regularly. Many of the methods reviewed in this chapter are not included in popular statistical software. The computer program used for these examples is a special-purpose package called CEFA (Comprehensive Exploratory Factor Analysis—Browne, Cudeck, Tateneni, & Mels, 1998).[1]

## ACKNOWLEDGMENTS

This research was partially supported by National Institute of Mental Health grant MH5-4576.

[1] Available on the web site of Dr. Browne at Ohio State University: quantrm2.psy.ohio-state.edu:80/browne

# REFERENCES

Adkins, D. C. (1964). Louis Leon Thurstone: Creative thinker, dedicated teacher, eminent psychologist. In N. Frederiksen & H. Gulliksen (Eds.), *Contributions to mathematical psychology* (pp. 1–39). New York: Holt, Rinehart & Winston.

Bartholomew, D. J. (1987). *Latent variable models and factor analysis.* London: Charles Griffin.

Bartholomew, D. J. (1995). Spearman and the origin and development of factor analysis. *British Journal of Mathematical and Statistical Psychology, 48,* 211–220.

Bollen, K. A. (1989). *Structural equations with latent variables.* New York: Wiley.

Bollen, K. A., & Long, J. S. (1993). *Testing structural equation models.* Newbury Park, CA: Sage.

Browne, M. W. (1972a). Orthogonal rotation to a partially specified target. *British Journal of Mathematical and Statistical Psychology, 25,* 115–120.

Browne, M. W. (1972b). Oblique rotation to a partially specified target. *British Journal of Mathematical and Statistical Psychology, 25,* 207–212.

Browne, M. W., & Cudeck, R. (1993). Alternative ways of assesing model fit. In K. A. Bollen, & J. S. Long (1993). *Testing structural equation models* (pp. 136–162). Newbury Park, CA: Sage.

Browne, M. W., & Cudeck, R. (1997, July). *A general approach to standard errors for rotated factor loadings.* Paper presented at the 10th European Meeting of the Psychometric Society, Santiago de Compostela, Spain.

Browne, M. W., Cudeck, R., Tateneni, K. & Mels, G. (1998). *CEFA: Comprehensive Exploratory Factor Analysis.* Unpublished Manuscript, Ohio State University, Columbus, OH.

Carroll, J. B. (1993). *Human cognitive abilities.* Cambridge, UK: Cambridge University Press.

Cattell, R. B. (1966). The scree test for the number of factors. *Multivariate Behavioral Research, 1,* 245–276.

Cliff, N. (1987). *Analyzing multivariate data.* New York: Harcourt, Brace, Jovanovich.

Cudeck, R. (1991). Noniterative factor analysis estimators, with algorithms for subset and instrumental variable selection. *Journal of Educational Statistics, 16,* 35–52.

Cudeck, R., & Browne, M. W. (1983). Cross-validation of covariance structures. *Multivariate Behavioral Research, 18,* 147–167.

Cudeck, R., & Henly, S. J. (1991). Model selection in covariance structures analysis and the "problem" of sample size: A clarification. *Psychological Bulletin, 109,* 512–519.

Cudeck, R., & O'Dell, L. L. (1994). Applications of standard error estimates in unrestricted factor analysis: Significance tests for factor loadings and correlations. *Psychological Bulletin, 115,* 475–487.

Fabrigar, L. R., Wegener, D. T., MacCallum, R. C., & Strahan, E. J. (1999). Evaluating the use of the exploratory factor analysis in psychological research. *Psychological Methods, 4,* 272–299.

Ford, J. K., MacCallum, R. C., & Tate, M. (1986). The application of exploratory factor analysis in applied psychology: A critical review and analysis. *Personnel Psychology, 39,* 291–314.

Goldberg, L. R. (1990). An alternative "Description of Personality:" The Big-Five factor structure. *Journal of Personality and Social Psychology, 59,* 1216–1229.

Goldberg, L. R. (1993). The structure of phenotypic personality traits. *American Psychologist, 48,* 26–34.

Green, P. E., & Carroll, J. D. (1976). *Mathematical tools for applied multivariate analysis.* New York: Academic.

Guttman, L. (1954). Some necessary conditions for common-factor analysis. *Psychometrika, 19,* 149–161.

Hakstian, A. R., & Abell, R. A. (1974). A further comparison of oblique factor transformation methods. *Psychometrika, 39,* 429–444.

Harman, H. H. (1976). *Modern factor analysis.* Chicago: University of Chicago Press.

Harris, C. W., & Kaiser, H. F. (1964). Oblique factor analytic solutions by orthogonal transformations. *Psychometrika, 29,* 347–362.

Hägglund, G. (1982). Factor analysis by instrumental variables methods. *Psychometrika, 47,* 209–222.

Hubbard, R., & Allen, S. J. (1987). A cautionary note on the use of principal components analysis: Supportive empirical evidence. *Sociological Methods and Research, 16,* 301–308.

Janssen, R., De Boeck, P., & Vander Steene, G. (1996). Verbal fluency and verbal comprehension abilities in synonym tasks. *Intelligence, 22,* 291–310.

Jennrich, R. I. (1973). Standard errors for obliquely rotated factor loadings. *Psychometrika, 38,* 593–604.

Jennrich, R. I. (1974). Simplified formulae for standard errors in maximum-likelihood factor analysis. *British Journal of Mathematical and Statistical Psychology, 27,* 122–131.

Jennrich, R. I., & Sampson, P. F. (1966). Rotation for simple loadings. *Psychometrika, 31,* 313–323.

Jöreskog, K. G. (1983). Factor analysis as an errors-in-variables model. In H. Wainer & S. Messick (Eds.), *Principals of modern psychological measurement* (pp. 185–196). Hillsdale, NJ: Erlbaum.

Kaiser, H. F. (1958). The varimax criterion for analytic rotation in factor analysis. *Psychometrika, 23,* 187–200.

Kaiser, H. F. (1960). The application of electronic computers to factor analysis. *Educational and Psychological Measurement, 20,* 141–151.

Kano, Y. (1990). Noniterative estimation and the choice of the number of factors in exploratory factor analysis. *Psychometrika, 55,* 277–291.

Lawley, D. N., & Maxwell, A. E. (1971). *Factor analysis as a statistical method.* New York: American Elsevier.

Lovie, A. D., & Lovie, P. (1993). Charles Spearman, Cyril Burt and the origins of factor analysis. *Journal of the History of the Behavioral Sciences, 29,* 308–321.

Maxwell, A. E. (1977). *Multivariate analysis in behavioral research.* London: Chapman & Hall.

McDonald, R. P. (1985). *Factor analysis and related methods.* Hillsdale, NJ: Erlbaum.

Meredith, W. (1993). Measurement invariance, factor analysis and factorial invariance. *Psychometrika, 58,* 525–543.

Morrison, D. F. (1990). *Multivariate statistical methods* (3rd ed.). New York: McGraw-Hill.

Mulaik, S. A. (1972). *The foundations of factor analysis.* New York: McGraw-Hill.

Saucier, G. (1994). Mini-markers: A brief version of Goldberg's unipolar Big-five markers. *Journal of Personality Assessment, 63,* 506–516.

Steiger, J. H. (1990). Structural model evaluation and modification: An interval estimation approach. *Multivariate Behavioral Research, 25,* 173–180.

Steiger, J. H., & Lind, J. M. (1980). *Statistically based tests for the number of common factors.* Paper presented at the annual meeting of the Psychometric Society, Iowa, IA.

Thurstone, L. L. (1945). *Multiple factor analysis.* Chicago: University of Chicago Press.

# 11

# CLUSTER ANALYSIS

**PAUL A. GORE, JR.**

*Department of Psychology, Southern Illinois University, Carbondale, Illinois*

Linnaeus, whose system of biological taxonomy survives in modified form to this day, believed that all real knowledge depends on our capacity to distinguish the similar from the dissimilar. The classification of objects and experiences is a fundamental human activity. Perhaps in an attempt to simplify a complex world, we tend to organize aspects of our environment into meaningful units such as gender, political party, or species—sorting objects and events into categories based on their characteristics or uses. During early childhood for example, we learn that a chair and a table are both pieces of furniture. The very process of classification may be a necessary prerequisite for the acquisition of language. Furthermore, the human capacity to learn over the life span is greatly enhanced as a result of our ability to assimilate new information into existing cognitive categories or schemas.

Classification is also fundamental to the development of science (Gould, 1989; Medin, 1989). Early attempts at categorizing celestial bodies, for example, gave rise to modern-day astronomy. And, although scientists no longer rely on measurement of the four humors to understand the nature and behavior of physical substances, this early Greek classification system spawned inquiry into physics, chemistry, biology, and philosophy. Ra-

*Handbook of Applied Multivariate Statistics and Mathematical Modeling*
Copyright © 2000 by Academic Press. All rights of reproduction in any form reserved.

tionally based classification gave way to more sophisticated techniques when Tryon (1939) and Cattell (1944) introduced mathematical procedures for organizing objects based on observed similarity. It was not until Sokal and Sneath's (1963) publication of *Principles of Numerical Taxonomy*, however, that clustering methods gained widespread acceptance in the sciences. Today, the literatures of biology, zoology, chemistry, earth sciences, medicine, engineering, business and economics, and the social sciences are replete with cluster analysis studies.

Although the use of cluster techniques is now quite commonplace, there are valid reasons why many investigators continue to be ambivalent about using the procedures in their research. For example, prior to beginning a cluster analysis, researchers must make several critical methodological decisions with little or no guidance. Further, attempts to familiarize oneself with cluster methodology may necessitate an interest in soil science or icosahedral particle orientations as methodological advances are frequently embedded within content-specific empirical reports. The purpose of this chapter, therefore, is to (a) provide an overview of the uses of cluster methods, (b) describe the procedures involved in conducting cluster analyses, and (c) provide some general recommendations to the researcher interested in using cluster procedures.

## I. GENERAL OVERVIEW

Cluster analysis is a term used to describe a family of statistical procedures specifically designed to discover classifications within complex data sets. The objective of cluster analysis is to group objects into clusters such that objects within one cluster share more in common with one another than they do with the objects of other clusters. Thus, the purpose of the analysis is to arrange objects into relatively homogeneous groups based on multivariate observations. Although investigators in the social and behavioral sciences are often interested in clustering people, clustering nonhuman objects is common in other disciplines. For example, a marketing researcher might be interested in clustering metropolitan areas into test-market subsets based on observations of population demographics, economic trends, and retail sales data. Alternatively, researchers in the physical sciences might be interested in clustering proteins or bacteria based on relevant observations of these objects.

It is helpful to consider the relation of cluster analysis to other multivariate procedures. Although cluster and discriminant analyses are both concerned with the characteristics of groups of objects, there is an important conceptual difference between the two procedures. Discriminant analysis is used to identify an optimal subset of variables that is capable of distinguishing among discrete predetermined groups (see M. Brown & Wicker, chapter 8, this volume). In contrast, cluster analysis begins with undifferentiated groups and attempts to create clusters of objects based on the similarities observed among a set of variables. Cluster analysis is also frequently compared to exploratory factor analysis (see Cudeck, chapter 10, this volume). In fact, some social scientists advocate the use of "inverse factor analysis" or Q-analysis as a method of clustering objects (Overall & Klett, 1972; Skinner, 1979). Both cluster and factor analyses attempt to account for similarities among a set of observations by grouping those observations together. Through an inspection of item covariability, factor analysis creates groups of variables. In contrast, cluster methods are used to group people together based on their scores across a set of variables. These contrasting approaches to observing similarities have been likened to the contrast between the Aristotelian and the Galilean approach to scientific investigation (Cattell, Coulter, & Tsujioka, 1966; Filsinger, 1990). Filsinger notes that although the dominant Galilean approach permits accurate prediction of one variable given knowledge of another, the typological Aristotelian approach characterized by cluster analysis allows us to better understand the unique combination of characteristics that an organism possesses.

Most cluster analyses share a similar process. A representative sample must be identified and variables selected for use in the cluster method. Samples and variables should be carefully selected so as to be both representative of the population in question and relevant to the investigator's purpose for clustering. The researcher must decide whether to standardize the data, which similarity measure to use, and which clustering algorithm to select. The researcher's conception of what constitutes a cluster and his or her research question can provide some guidance in making these decisions. The final stages of cluster analysis involve interpreting and testing the resultant clusters, and replicating the cluster structure on an independent sample.

Morris, Blashfield, and Satz (1981), among others, have summarized some generic problems inherent in using cluster analysis and interpreting the findings from cluster studies. Differences exist within and across disciplines with respect to the terms used to describe cluster methods. A single cluster algorithm, for example, might be called complete linkage, maximum method, space-distorting method, space-dilating method, furthest neighbor method, or diameter analysis. Furthermore, as is often the case with multivariate methods, a set of procedural decisions must be made prior to conducting a cluster analysis. The investigator is faced with choosing from among hundreds of possible analytic combinations—some of which have been described but rarely used. Further complicating these decisions is the fact that statistical packages often include only a subset of these procedures, and some procedures are available only through the use of privately distributed software programs. Emerging research emphasizes the importance of these decisions in reporting that different classification procedures are often best suited to different classification questions or to the nature of the data being analyzed. Finally, researchers often fail to replicate and validate their clusters because of the difficulty involved. Nonreplicable clusters, and clusters that fail to relate to variables outside of the cluster solution are of limited practical utility. In this volume, Tinsley and S. Brown (chapter 1), Dawis (chapter 3), M. Brown and Wicker (chapter 8), and Thorndike (chapter 9) describe cross-validation, bootstrapping, and jackknife procedures for evaluating the generalizability of multivariate analyses.

## II. USES FOR CLUSTER ANALYSIS

Cluster methods lend themselves to use by investigators considering a wide range of empirical questions. Investigators in the life sciences, for example, are often interested in creating classifications for life forms, chemicals, or cells. They may be interested in developing complete taxonomies or in delimiting classifications based on their particular research interests. Medical scientists rely on clinical diagnoses and may use cluster methods to identify groups of people who share common symptoms or disease processes. The use of cluster methods in the behavioral sciences is as varied as the fields that constitute this branch of inquiry. A psychologist might be interested in exploring the possible relations among types of counseling interventions. In contrast, the economist may be charged with identifying economic similarities among developing countries. Clustering methods are useful whenever the researcher is interested in grouping together objects based on multivariate similarity.

Cluster analysis can be employed as a data exploration tool as well as a hypothesis testing and confirmation tool. The most frequent use of cluster analysis is in the development of a typology or classification system where one does not already exist. For example, Heppner and his colleagues (Heppner et al., 1994) identified nine unique clusters of persons presenting for treatment at a university counseling center. Persons in different clusters differed both in the constellation and magnitude of their presenting concerns and on external characteristics (e.g., ethnicity) and family of origin variables (e.g., family history of alcoholism). Alone, these findings might be of use to practitioners working in similar settings. It might be helpful to know, for example, that clients resembling members of clusters 1 and 2 (the severe and high generalized distress clusters) are less likely to come from alcoholic parents compared to clients resembling members of cluster 6 (severe somatic concerns). It is far more likely, however, that findings such as these will lead the researcher to generate new hypotheses that might help to explain the obtained cluster solution.

Cluster algorithms might assemble observations into groups in ways that researchers would not have considered. The result of such an unexpected grouping should challenge the investigator to develop and further explore hypotheses that might account for the cluster solution. Similarly, applied researchers might use information obtained from a cluster solution to develop hypotheses regarding treatment or intervention. Larson and Majors (1998) recently provided an example of this type of study. These investigators identified four clusters of individuals based on their reports of subjective career-related distress and problem-solving efficacy. Their report outlines recommended strategies for working with career clients who may resemble members of these four clusters.

Investigators might also use clustering methods to test a priori assumptions or hypotheses or to confirm previously established cluster solutions (Borgen & Barnett, 1987). Chartrand and her colleagues (Chartrand et al., 1994), for example, used cluster analysis to compare a continuum versus an interactional view of career indecision, and McLaughlin, Carnevale, and Lim (1991) used cluster methods to compare competing models of organizational mediation. As researchers become more familiar with clustering procedures, it is likely they will develop new and innovative uses for this methodology. Several recent studies (Gati, Osipow, & Fassa, 1994; Sireci & Geisinger, 1992) apply cluster analysis to the psychometric evaluation of measurement instruments, and the authors make cogent arguments for using cluster techniques to complement existing methods of psychometric evaluation.

## III. CLUSTER ANALYSIS METHODS

### A. Theory Formulation

The first stage of a framework for classification, according to Skinner (1981), is theory formulation (see Hetherington, chapter 2, this volume). It is during this stage, and prior to collecting data, that the researcher should focus on grounding the proposed analysis in theory, specifying the purpose of the study, and determining the population and variables to be used. Ideally, a researcher should describe the nomological network guiding the study by providing precise definitions of expected classifications and the theoretical relations among them. In reality, however, social scientists rarely specify the conceptual framework guiding their classification research. This may stem in part from the common perception of cluster analysis as an exploratory technique. Other exploratory procedures such as exploratory factor analysis (see Cudeck, chapter 10, this volume) and multiple regression (see Venter & Maxwell, chapter 6, this volume) are routinely used in the absence of explicit theoretical assumptions. One of the primary arguments for beginning with theory when using cluster analysis is the fact that cluster analyses will cluster any data (even random data). When guided by theory, the researcher will have some mechanism for evaluating the meaning of, and potential uses for, the resulting clusters.

Speece (1995) encourages researchers to consider the purpose for their classification during this stage of the study. Cluster analysis may be used to develop a typology or classification system, as a test of existing classification systems, or simply to explore possible undiscovered patterns and similarities among objects. Speece notes that classification systems may be used either to promote communication with practitioners or to enhance prediction. As Blashfield and Draguns (1976) have observed, these two goals are frequently in direct opposition to one another. Speece encourages researchers to be explicit about the purpose for conducting their cluster analyses and to avoid altering their purpose midstream.

Similar to other analytic procedures, cluster analysis requires investigators to select variables for study and to define a population of interest. Although the selected variables represent only one particular subset of measurements available to the investigator, they determine how the clusters will be formed. Because cluster analysis will always produce a classification scheme, regardless of the data available, it is essential that the investigator select variables that relate adequately to the classification problem at hand. A biologist attempting to establish a classification scheme for living organisms, for example, would generate very different clusters using embryonic morphology as opposed to adult features. Psychologists can imagine how different their primary classification tool (*DSM-IV*) would look if it had been established using etiological variables instead of current symptomatology. Everitt (1986) points out that the initial choice of which variables to include in the analysis is itself an initial categorization of the data and one for which the investigator receives no statistical guidance.

A related issue involves determining the number of variables to include in a cluster analysis. Although there is no clear-cut rule of thumb, researchers whose studies are guided by theory will have an advantage in specifying which variables are most likely to contribute to a meaningful cluster solution. Research exploring the issue of how many variables to include in an analysis is conflicting. For example, Hands and Everitt (1987) found that increasing the number of variables used in a cluster analysis resulted in better identification of a known cluster structure. Price (1993), on the other hand, found that increasing the number of variables used in the analysis resulted in poorer cluster identification. Further, Milligan (1980) cautions against the use of irrelevant variables. Researchers are encouraged to select variables based on sound theoretical grounds, to select variables that will maximally discriminate among objects, and to avoid the indiscriminant inclusion of variables.

Selection of a representative population also needs to be considered during this stage of a cluster study. As is true with any statistical procedure, inferences drawn from a sample will generalize most adequately to the population from which the sample was drawn. Although there are no clear-cut recommendations with respect to how many objects to include in an analysis, data from several studies (Hands & Everitt, 1987; Schweizer, Braun, & Boller, 1994) suggest that increasing sample sizes result in an increase in cluster reliability.

## B. Measures of Association

Clustering objects based on their similarity to one another requires that you determine the degree of similarity among the objects. Generally, the matrices used in cluster analyses are referred to as either similarity (proximity) or distance matrices. Interpretation of the magnitude of the values depends on the type of matrix generated. High levels of similarity among objects are indicated by large values in a similarity or proximity matrices and small value in a distance matrix.

### 1. Similarity Matrices

There are several commonly used measures of similarity. The product–moment correlation coefficient is often used with continuous data. Contrary to its more common usage as an index of the association between two variables, however, the correlation coefficient used in similarity matrices describes the relations between two objects (e.g., people) on a set of variables (i.e., Cattell's Q correlation; Cattell, 1988). This is accomplished by inverting the Person (rows) × Variable (columns) matrix so that it becomes a Variable (rows) × Person (columns) matrix. Correlations calculated using this inverted matrix represent the degree of similarity between two objects (persons) on the defined set of variables. Table 11.1 provides an example of a raw data matrix in the classical (i.e., Person × Variable) and inverted (i.e., Variable × Person) formats. Table 11.2 provides a similarity matrix (product–moment correlations) for that raw data matrix.

When continuous data are not available, other procedures are used to calculate the similarity matrix. The contingency table can be used in combination with one of several different formulae to describe the similarity between two objects when only binary variables are available. For example, a contingency table could be calculated for the binary data in Table 11.1 and used to establish the degree of similarity between two persons on five separate binary variables (see Table 11.2). Once such a table has been established, the investigator has a large number of similarity coefficients from which to choose. Many of the coefficients differentially weigh values that signify matching pairs, mismatches, or instances where both objects lack some attribute (negative matches; for example see Anderberg, 1973, and Romesburg, 1984). Imrey (chapter 14, this volume) provides a more extensive coverage of multivariate procedures for analyzing binary observations.

**TABLE 11.1    Raw Data Matrices**

**Classical format**

|              | Variable |     |     |     |     |
|              | **1** | **2** | **3** | **4** | **5** |
|--------------|-----|-----|-----|-----|-----|
| Participant 1 | 1.0 | 5.0 | 7.0 | 7.0 | 5.0 |
| Participant 2 | 2.0 | 6.0 | 2.0 | 8.0 | 6.0 |
| Participant 3 | 4.0 | 4.0 | 5.0 | 5.0 | 4.0 |
| Participant 4 | 5.0 | 1.0 | 4.0 | 4.0 | 3.0 |
| Participant 5 | 7.0 | 2.0 | 3.0 | 9.0 | 2.0 |

**Inverted ("Q") format**

|      | Participant |     |     |     |     |
|      | **1** | **2** | **3** | **4** | **5** |
|------|-----|-----|-----|-----|-----|
| Var1 | 1.0 | 2.0 | 4.0 | 5.0 | 7.0 |
| Var2 | 5.0 | 6.0 | 4.0 | 1.0 | 2.0 |
| Var3 | 7.0 | 2.0 | 5.0 | 4.0 | 9.0 |
| Var4 | 7.0 | 8.0 | 5.0 | 4.0 | 9.0 |
| Var5 | 5.0 | 6.0 | 4.0 | 3.0 | 2.0 |

**Classical format for binary observations**

|              | Variable |     |     |     |     |
|              | **1** | **2** | **3** | **4** | **5** |
|--------------|-----|-----|-----|-----|-----|
| Participant 1 | 1 | 0 | 0 | 1 | 1 |
| Participant 2 | 0 | 0 | 1 | 1 | 0 |

**Classical format for use in centroid cluster method**

|               | Variable |     |     |     |     |
|               | **1** | **2** | **3** | **4** | **5** |
|---------------|-----|-----|-----|-----|-----|
| Participant 1 | 1.0 | 5.0 | 7.0 | 7.0 | 5.0 |
| Participant 2 | 2.0 | 6.0 | 2.0 | 8.0 | 6.0 |
| Participant (3,4) | 4.5 | 2.5 | 4.5 | 4.5 | 3.5 |
| Participant 5 | 7.0 | 2.0 | 3.0 | 9.0 | 2.0 |

Nominal measures having greater than two levels and ordinal observations can be dealt with using a variety of methods. For example, Cohen's (1960) Kappa Coefficient ($P_o - P_c / 1 - P_c$, where $P_o$ = Probability of Observed Agreement and $P_C$ = Probability of Chance Agreement) is often used with nominal data and has the added advantage of correcting similarity values for the occurrence of chance. Tinsley and Weiss (chapter 4, this volume) review the use of the Kappa Coefficient in evaluating interrater agreement. Kendall's Tau Coefficient (Kendall, 1963) is a method available to researchers when dealing with observations that are ranked.

Two of the most common concerns among investigators using binary or polychotomous variables are how to treat absences and how to treat negative matches. Aldenderfer and Blashfield (1984) provided an excellent example of the difficulty of dealing with absences in anthropology. They noted that failing to discover a particular type of artifact at one site does not necessarily reflect on the behaviors of the deceased inhabitants. It may, for example, reflect decay patterns idiosyncratic to a particular site. Dealing with negative matches is often dependent on the nature of the variables used and the investigator's research question. In some instances it may be inappropriate to consider two objects similar based on the fact that they both lack some characteristic. To consider two people similar because neither of them lives on the East Coast, for example, may not be appropriate. On the other hand, we might be more justified in assuming some similarity among two individuals because they both lack a high school education. The Rand coefficient $[(a + d)/a + b + c + d]$ counts negative

**TABLE 11.2   Distance and Similarity Matrices for Use in Cluster Algorithms**

| | Similarity and distance matrices corresponding to raw data matrix | | | | | | | | | | |
|---|---|---|---|---|---|---|---|---|---|---|---|
| | Product-moment correlation matrix | | | | | Squared Euclidean distance matrix | | | | | |
| | Object | | | | | | Object | | | | |
| Object | 1 | 2 | 3 | 4 | 5 | Object | 1 | 2 | 3 | 4 | 5 |
| 1 | 0 | | | | | 1 | 0 | | | | |
| 2 | .46 | 0 | | | | 2 | 29 | 0 | | | |
| 3 | .75 | .07 | 0 | | | 3 | 19 | 30 | 0 | | |
| 4 | −.27 | −.47 | .36 | 0 | | 4 | 54 | 63 | 13 | 0 | |
| 5 | −.13 | .16 | .40 | .66 | 0 | 5 | 74 | 59 | 37 | 32 | 0 |

Contingency table used to calculate similarity values for binary observations

|  |  | Person 1 | | |
|---|---|---|---|---|
|  |  | 1 | 0 | |
| Person 2 | 1 | $1^a$ | $1^b$ | 2 |
|  | 0 | $2^c$ | $1^d$ | 3 |
|  |  | 3 | 2 | $5^p$ |

matches as bona fide matches, whereas the Jaccard coefficient [a/a + b + c] ignores negative matches entirely. Consideration of negative matches becomes more important when dealing with polychotomous data (Blong, 1973; Romesburg, 1984).

## 2. Distance Matrices

In contrast to similarity measures, distance (or dissimilarity) measures emphasize the differences between two objects on a set of observations. Large values in a distance matrix reflect dissimilarity. The squared Euclidean distance measure has received the most widespread use among researchers in the social and behavioral sciences. Represented as $D^2$, values are calculated for each object pair by summing the squared differences between observations. The distance between participant 1 and participant 2 in Table 11.1 (see inverted matrix) can be described by the following equation $(1 - 2)^2 + (5 - 6)^2 + (7 - 2)^2 + (7 - 8)^2 + (5 - 6)^2$ or 29. A squared Euclidean distance matrix (see Table 11.2) is made up of all pairwise comparisons among objects, in this case participants. Using this distance measure has its drawbacks, however, in that changes in the scale of measurement can result in quite different pairwise rankings. For this reason, most researchers choose to standardize their data prior to using the squared distance measure. Another potentially useful distance measure is Mahalanobis's (1936) distance, which is the distance of a case from the centroid of the remaining cases. Conceptually, this measure of distance accounts for possible correlations among variables—something that is not taken into consideration when Euclidean distances are calculated.

Related to the use of distance matrices is the issue of variable standardization. Although considerable debate has surrounded the issue of whether or not to standardize variables, most authors currently agree that standardization of variables is advisable when differences between variables may be an artifact of different measurement units or scales. However, several recent studies also suggest that standardization may not influence the outcome of a cluster analysis as much as was once thought. For example, Punj and Stewart (1983) reviewed 12 cluster studies and concluded that standardizing variables appears to have little effect on the clustering solution. More recent studies echo Punj and Stewart's conclusions (Milligan & Cooper, 1988; Schaffer & Green, 1996).

There are numerous problems associated with the process of selecting an appropriate similarity or distance measure. One of the most frequently discussed issues involves how similarity among objects is captured by distance versus similarity measures. Cronbach and Gleser (1953) described how the similarity between profiles could be decomposed into shape, scatter, and elevation information. Shape describes the pattern of rises and falls across variables. A typical Minnesota Multiphasic Personality Inventory (MMPI) profile, for example, will contain patterns of high and low scores on 10 clinical and 3 validity scales. Scatter describes the distribution of scores around a central average. Psychologists occasionally discuss the scatter of subtest scores relative to the mean performance level of some construct. Finally, elevation represents the degree of rise or fall across variables and describes a profile's absolute magnitude. Objects that are deemed similar on one of these three characteristics are not necessarily similar on another characteristic. For example, MMPI profiles from two clients might be similar with respect to shape (e.g., rising scores for depression and anxiety) but dramatically different with respect to elevation (e.g., client 1 may be in the "clinically significant" range, whereas the scores for client 2 fall well within the average range).

The product–moment correlation coefficient is often criticized for discarding information about elevation and scatter—a by-product of standardizing the variables and subtracting the mean from each score. Two variables will correlate highly regardless of the magnitude or variability observed among scores, provided the scores rise or fall in parallel. In situations where nonlinear relations exist, the product–moment correlation coefficient also fails to adequately capture shape characteristics. Distance matrices calculated on standardized variables will result in the similar loss of elevation and scatter information. Distance matrices calculated on nonstandardized data, on the other hand, retain shape, elevation, and scatter information. Therefore, clusters resulting from a nonstandardized distance matrix may be based on any combination of shape, elevation, and scatter. An investigator hoping to capture severity (elevation) in the formation of clusters may in fact be forming clusters on shape or scatter alone.

Although the debate over the superiority of using either the distance or similarity method appears to have subsided somewhat in recent years, studies investigating the benefits or drawbacks of both methods continue to appear in the literature. Punj and Stewart (1983) concluded from their review of cluster studies, that the choice of similarity or distance measures does not appear to be a critical factor in identifying clusters. In contrast, Scheibler and Schneider (1985) reported striking differences between the Pearson correlation and $D^2$ in a recent Monte Carlo study. Overall, Gibson, and Novy (1993) echoed these findings more recently.

For the applied researcher, the question of whether to use a similarity or distance matrix often depends upon the nature of the classification question. When classification based on elevation or scatter is desirable, then researchers are encouraged to consider using the squared Euclidean distance measure on nonstandardized raw data. In contrast, when a researcher is primarily concerned with establishing clusters based on the shape of relations among observations, then the similarity matrix is the method of choice. Researchers who conduct their cluster analyses using both procedures and obtain similar outcomes are assured their cluster structure is not an artifact of the method used.

## C. Clustering Algorithms

During the next stage of cluster analysis research, the investigator must select a clustering procedure. Whereas the similarity or distance measures provide an index of the similarity among objects, the cluster algorithm operationlizes a specific criterion for grouping objects together (Speece, 1995). Although researchers often disagree on the most appropriate classification scheme for cluster procedures (Lorr, 1994; Milligan & Cooper, 1987), cluster methods are frequently classified into the following four general categories: hierarchical methods, partitioning methods, overlapping cluster procedures, and ordination techniques. The first two categories have received widespread use and support among researchers in the social and behavioral sciences and will thus be described in more detail.

## D. Hierarchical Methods

Some of the most widely used procedures for conducting cluster analyses are referred to as hierarchical methods or sequential agglomerative hierarchical methods. In general, these procedures begin by assuming that each entity in the similarity or distance matrix is an individual cluster. At each successive stage of the procedure, clusters are combined according to some algorithm. This process continues until all objects have been combined to form one cluster. Hierarchical methods generate strictly nested structures and are often presented graphically using a dendrogram similar to the one shown in Figure 11.1. In this example, objects 3 and 4 are clustered together during the first step of the procedure because that pair of objects have the highest similarity index or the smallest distance value. During the second, third, and fourth steps of this example, objects 1, 2, and 5 are added to the cluster, respectively. Since the cluster method proceeds until only one cluster remains, it is up to the researcher to decide how many clusters to interpret.

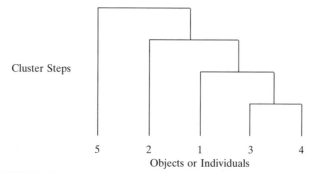

Cluster Steps

| 5 | 2 | 1 | 3 | 4 |

Objects or Individuals

**FIGURE 11.1**   Steps of cluster formation portrayed using a dendrogram.

## 1. Single Linkage Procedure

Four different agglomerative methods are found frequently in the literature. These methods differ primarily in how elements are compared to one another at each successive stage of clustering. Sneath (1957) and McQuitty (1957) proposed the *single linkage procedure* in which an object is included in a cluster if it shares a sufficient degree of similarity with at least one member of a cluster. This method also has been referred to as the *elementary linkage, minimum method,* and *nearest neighbor cluster analysis* (Johnson, 1967; Lance & Williams, 1967). The squared Euclidean distance matrix in Table 11.2 can be used as a starting point to demonstrate this relatively simple algorithm. The first step in establishing clusters is to join the pair of objects with the smallest distance value (or largest similarity value). Thus, in this example, objects 3 and 4 ($D^2 = 13$) are joined to form a cluster. Then, the distance matrix is recalculated to reflect the smallest distance between either 3 or 4 and the remaining objects, thereby reducing the $5 \times 5$ matrix to a $4 \times 4$ matrix. For example, the distance between cluster 1 and 3 (19) is compared to that between cluster 1 and 4 (54) so the value of 19 is used in the recalculated matrix to signify the distance between clusters 1 and 3–4. This recalculated matrix is presented in Table 11.3. During the next step of the procedure, cluster 1 is added to cluster (3,4). This process continues until all objects are clustered together. Single linkage procedures can use distance or similarity matrices, but notice that a different cluster solution would emerge if the similarity matrix in Table 11.1 was used.

## 2. Maximum Linkage Procedure

A second agglomerative method, referred to as *complete linkage analysis, maximum method,* or *furthest neighbor analysis* (Horn, 1943), is the opposite of the single linkage method in that the distance between clusters is defined as the distance between the most remote pair of cluster members. The distance matrix in Table 11.2 can again serve to illustrate this procedure. The first step of the complete linkage analysis is the same as for the single linkage method. Following the initial clustering of objects 3 and 4, however, the distance matrix is recalculated to reflect the maximum distance between the element within a cluster and the comparison cluster. The recalculated matrix is shown in Table 11.3. The larger of the distances between either 3 or 4 and 1 is included (i.e., 54) in this recalculated matrix. The next step in the procedure involves the formation of a second cluster containing objects 1 and 2. Table 11.2 illustrates how the same principle can be applied when comparing two clusters that contain multiple elements. The larger distance between 1 or 2 and either 3 or 4 is used for subsequent clustering (i.e., 63). As can be seen from this example, the complete method yielded a pattern after two stages of the analysis that was different from that generated by the single linkage method. These two methods often yield

**TABLE 11.3** Recalculated Distance Matrices Obtained Using Four Cluster Algorithms

Single linkage method squared Euclidean distance matrices

| Object | Object | | | | Object | Object | | |
|---|---|---|---|---|---|---|---|---|
| | 1 | 2 | (3,4) | 5 | | (1,3,4) | 2 | 5 |
| 1 | 0 | | | | (1,3,4) | 0 | | |
| 2 | 29 | 0 | | | 2 | 29 | 0 | |
| (3,4) | 19 | 30 | 0 | | 5 | 32 | 59 | 0 |
| 5 | 74 | 59 | 32 | 0 | | | | |

Maximum linkage method squared Euclidean distance matrices

| Object | Object | | | | Object | Object | | |
|---|---|---|---|---|---|---|---|---|
| | 1 | 2 | (3,4) | 5 | | (1,2) | (3,4) | 5 |
| 1 | 0 | | | | (1,2) | 0 | | |
| 2 | 29 | 0 | | | (3,4) | 63 | 0 | |
| (3,4) | 54 | 63 | 0 | | 5 | 74 | 37 | 0 |
| 5 | 74 | 59 | 37 | 0 | | | | |

Unweighted pair-group method squared Euclidean distance matrices

| Object | Object | | | | Object | Object | | |
|---|---|---|---|---|---|---|---|---|
| | 1 | 2 | (3,4) | 5 | | (1,2) | (3,4) | 5 |
| 1 | 0 | | | | (1,2) | 0 | | |
| 2 | 29 | 0 | | | (3,4) | 41.5 | 0 | |
| (3,4) | 36.5 | 46.5 | 0 | | 5 | 66.5 | 34.5 | 0 |
| 5 | 74 | 59 | 34.5 | 0 | | | | |

Centroid method squared Euclidean distance matrix

| Object | Object | | | |
|---|---|---|---|---|
| | 1 | 2 | (3,4) | 5 |
| 1 | 0 | | | |
| 2 | 29 | 0 | | |
| (3,4) | 33.3 | 43.3 | 0 | |
| 5 | 74 | 59 | 31.2 | 0 |

quite different cluster patterns. In general, the maximum linkage method tends to generate clusters that are compact and spherical.

### 3. Average Linkage Method

The average linkage cluster method (Sokal & Michener, 1958) joins elements to existing clusters based on the average degree of similarity between the element and the existing members of the cluster. The average linkage method actually represents a set of related methods that differ with respect to how the "averaging" is conducted. Romesburg (1984) describes the *unweighted pair-group method* that uses arithmetic averages of pairs to recalculate the distance matrix. Table 11.3 illustrates distance matrices derived from the squared Euclidean distance matrix using the unweighted pair-group method. Note that the distance between 1 and cluster (3,4) is the average of the distances between 1 and 3 (19) and 1 and 4 (54). Similarly, the distance between cluster (1,2) and cluster (3,4) is represented by the equation (19 + 54 + 30 + 63)/4.

### 4. Centroid Method

Somewhat related to the unweighted pair-group method is the centroid method of cluster analysis. In this method, distance is defined as the distance between group centroids. Whereas the arithmetic average procedure recalculates distance matrices by averaging previously determined distance values, the recalculation of a distance matrix in the centroid method is preceded by updates to the raw data matrix. Table 11.1. shows an updated raw data matrix reflecting the combination of objects 3 and 4 during the first stage of a centroid cluster analysis. Note that cluster (3,4) now has raw scores that are the arithmetic average of the scores previously belonging to participant 3 and participant 4. Table 11.3 also shows the updated distance matrix resulting from these raw scores. Note that this method generates a distance matrix slightly different from that generated by the averaging method described above.

### 5. Minimum Variance Method

A final group of agglomerative methods frequently used by researchers are *sum-of-square methods* or *minimum variance methods.* Ward's (1963) is probably the most widely used minimum variance method, although other sum-of-squares procedures are available to the researcher (Friedman & Rubin, 1967; Wishart, 1969). The relative proximity of a set of objects can be described using the concept of sum of squares (the squared sum of the distances of each object from the mean value of the cluster). Using Ward's method, the cluster that results in the smallest increase in the sum of squares is formed during each step. Every possible combination of cluster formations is considered at each subsequent step.

The raw data matrix in inverted (Q) format shown in Table 11.1 can be used to illustrate this procedure. For the purposes of brevity, let us consider only participants 1–3 and variables 1 and 2. In this reduced data set, there are three possible initial cluster formations (1,2), (1,3), and (2,3). The following equations are analyzed to determine the lowest sum of squares values:

Cluster (1,2): $(1-1.5)^2 + (5-5.5)^2 + (2-1.5)^2 + (6-5.5)^2 + (4-4)^2 + (4-4)^2 = 1.00$

Cluster (1,3): $(1-2.5)^2 + (5-4.5)^2 + (4-2.5)^2 + (4-4.5)^2 + (2-2)^2 + (6-6)^2 = 5.00$

Cluster (2,3): $(2-3)^2 + (6-5)^2 + (4-3)^2 + (4-5)^2 + (1-1)^2 + (5-5)^2 \quad = 4.00$

Note that the sum of squares for clusters that contain only one object always sum to zero. Each of these formulae captures the average overall sum of squares of each object from its group (or potential "cluster") mean on each variable measured. The lowest value (1.0) determines that objects 1 and 2 will be combined to form the first cluster.

## 6. Divisive Methods

Whereas hierarchical agglomerative methods begin by treating every object as a unique cluster, hierarchical divisive methods begin by assuming that all objects are members of a single cluster and proceed by establishing clusters with successively smaller membership. Monothetic divisive methods are typically used with binary data. Objects are initially separated based on whether or not they posses some specific attribute. Lambert and Williams (1962, 1966) and MacNaughton-Smith (1965) described related procedures for determining which variable to select for clustering at each step of the analysis. These procedures often rely on chi-square pair-wise comparisons, among variables to determine the cluster criterion.

## 7. Polythetic Hierarchical Methods

In contrast to monothetic methods, *polythetic hierarchical methods* can accommodate clusters of objects that have been measured on continuous scales. MacNaughton-Smith, Williams, Dale, and Mockett (1964) described the most commonly used procedure. An initial cluster consisting of one object is formed by determining which object has the largest average distance from other objects. In the distance matrix in Table 11.2, object 1 differs from objects 2 through 5 by an average of 44 units (e.g., [29 + 19 + 54 + 74]/4). This value is larger than the average distance among other comparisons and thus object 1 is clustered out. During subsequent stages the average distances of each object from other objects in the main cluster (e.g., "within-group" distance) are calculated, and the between-group distance is subtracted from the within-group distance (see Table 11.4). The object having the highest positive value (i.e., object 2 in Table

**TABLE 11.4   Calculations Demonstrating the Use of the Polythetic Cluster Method**

|  | Average distance to group (1) (a) | Average distance to main group (b) | Difference (b)–(a) |
|---|---|---|---|
| Participant 2 | 29 | 50.6 | 21.6 |
| Participant 3 | 19 | 26.3 | 7.3 |
| Participant 4 | 54 | 36.0 | −18.0 |
| Participant 5 | 74 | 42.6 | −31.0 |

11.4) is incorporated into the newly formed cluster 1, and the process is repeated. The interested reader is referred to Everitt (1986) for a more comprehensive example.

## E. Partitioning Methods

Iterative partitioning methods, also frequently referred to as two-stage cluster analysis or k-means partitioning, were developed, in part, in response to one of the major shortcomings of the hierarchical methods. Once an object is clustered in hierarchical methods, it cannot be reassigned to a "better fitting" cluster at some subsequent stage of the process. Iterative partitioning methods simultaneously minimize within-cluster variability and maximize between-cluster variability. A typical partitioning method begins by partitioning a set of data into a specific number of clusters. Objects are then evaluated for potential membership in each of the formed clusters, and the process continues until no further cluster assignment or reassignment occurs among objects or until a predetermined number of iterations has been reached. Although the most efficient means of partitioning the data into clusters would be to generate all possible partitions, this method is computationally impractical given the size of most data sets.

Partitioning procedures differ with respect to the methods used to determine the initial partition of the data, how assignments are made during each pass or iteration, and the clustering criterion used. Research shows that initial partitions based on random assignment results in poor cluster recovery (Milligan, 1980; Milligan & Sokal, 1980; Scheibler & Schneider, 1985), leading investigators to propose methods for generating initial cluster partitions (Milligan & Sokal, 1980; Punj & Stewart, 1983). The most frequently used method assigns objects to the clusters having the nearest centroid. This procedure creates initial partitions based on the results from preliminary hierarchical cluster procedures such as the average linkage method or Ward's method, a procedure that resulted in partitioning methods being referred to as two-stage cluster analysis. Some partitioning methods use multiple passes during which cluster centroids are recalculated and objects are re-evaluated, whereas other methods use a single-pass procedure. Partitioning methods also differ with respect to how they evaluate an object's distance from cluster centroids; some procedures use simple distance and others use more complex multivariate matrix criteria. Finally, most partitioning methods require that the user specify a priori how many clusters will be formed. As such, these methods prove most useful when the researcher has well-formulated hypotheses about the number of clusters to expect.

### F. Overlapping and Ordination Methods

Overlapping cluster methods or "clumping techniques" permit clustered objects to belong to more than one cluster and are thus useful when an investigator has sound reason to believe there to be overlap among hypothesized clusters. These are methodologically complex procedures that have not been widely used by social and behavioral scientists. Ordination clustering algorithms are often referred to as inverse factor analysis or Q-type factoring because they essentially involve factor analyzing an inverted raw data matrix ("Q" matrix) like that shown in Table 11.1. Several authors have discussed the potential problems of applying factor analytic technology to object clustering problems (Everitt, 1986; Fleiss, Lawlor, Platman, & Fieve, 1971; Gorsuch, 1983). Among the charges levied against Q-type factoring is the fact that these procedures may be in violation of assumptions of the general linear model.

### G. Empirical Investigations of Cluster Methods

Studies evaluating cluster methods typically use the recovery of a known cluster structure as a benchmark. Many of these studies adopt Monte Carlo techniques and vary one or more characteristics (e.g., cluster structure, number of objects, number of variables, or type of algorithm). Although the literature examining the performance of cluster methods is voluminous, the vast majority of studies have focused on evaluating the most widely used hierarchical methods. Punj and Stewart (1983) reviewed 12 validity studies conducted between 1972 and 1980. The reviewed studies used a wide range of hierarchical methods including single, complete, centroid, median, and average linkage, and Ward's method, in addition to the k-means partitioning method. In general, these authors concluded that Ward's method and the average linkage method outperformed all other hierarchical methods. The k-means partitioning method appears to provide recovery that rivals that of the better hierarchical methods, but only when a nonrandom starting point is used. Punj and Stewart also pointed out the deleterious effects of adding spurious variables to the analysis and of including all objects in the final solution. Cluster solutions that incorporate most or all of the available objects tend to include more outliers (see Venter and Maxwell, chapter 6, and Huberty and Petoskey, chapter 7, this volume, for a discussion of procedures for detecting and dealing with outliers.) This issue relates directly to an investigator's determination of how many clusters to interpret and will be addressed in more detail below.

In comparing nine hierarchical and four nonhierarchical cluster methods, Scheibler and Schneider (1985) found hierarchical procedures to be quite robust at all but the most extreme levels of coverage (e.g., those situations where all objects must be clustered). Ward's and the average methods of clustering were the most accurate of the methods explored. Consistent with previous findings, partitioning methods performed best when nonrandom starting partitions were used. Similar findings were reported by Overall et al. (1993) and Milligan (1981a).

The interested reader is referred to Milligan and Cooper (1987) for a comprehensive review of cluster algorithm performance. In general, however, researchers using hierarchical methods should consider using either Ward's or the average linkage method. These methods perform well under a range of circumstances (e.g., in the presence of outliers and overlapping cluster structures). Some data suggest that Ward's method is best used with distance matrices, whereas the average linkage method works best with similarity matrices. I recommend that researchers consider using both the average linkage method and Ward's method with similarity and distance matrices. That will allow you to rule out method artifact if consistent cluster structures are obtained across the analyses. Investigators using partitioning methods should establish initial partitions based on preliminary hierarchical analyses.

### H. Deciding on the Number of Clusters to Interpret

Once a cluster algorithm has been executed, the researcher is faced with the task of deciding how many clusters to interpret. By definition most clustering procedures will continue unabated until all objects in the data set have been assigned to one or more clusters. In the case of hierarchical cluster algorithms, all objects will ultimately exist within a single all-inclusive cluster—a solution that holds no interest for applied researchers. Fortunately, a number of external and internal criteria exist to assist the researcher in determining the best number of clusters to interpret.

External criteria use information outside of the cluster solution to evaluate the data partition. A large number of external criteria have been developed including the Rand (Rand, 1971), the Jaccard (Anderberg, 1973), and the kappa statistics (Blashfield, 1976). Monte Carlo studies suggest that some external criteria are better than others. For example, Milligan and Cooper (1986) found the Hubert and Arabie (1985) adjusted Rand index to be superior to four other external criteria across various combinations of cluster number and cluster algorithm. Unfortunately, external criteria are rarely of much use to the applied researcher since the true cluster structure is almost never known a priori.

Internal criteria, or stopping rules, use information inherent in the cluster solution to determine the degree of fit between the data partition and the data used to create the partitions. The large number of internal criteria precludes a detailed description of them here, but the results from a number of empirical studies reveal that the criteria vary in validity. For example, Milligan (1981b) compared thirty different internal criteria to two external criterion indices and concluded that the Baker and Hubert (1975) Gamma and the Hubert and Levin (1976) C-index were the most successful at identifying the correct number of clusters in a data set. Overall and his colleagues (Atlas & Overall, 1994; Overall & Magee, 1992) advocate cross-validating the cluster structure (partition replication) as a means of determining the correct number of clusters in a set of data. Although the procedure described by Overall will result in a rigorous test, it is quite cumbersome and involves multiple hierarchical and higher order cluster procedures on subsets of data. McIntyre and Blashfield (1980) described a somewhat less cumbersome replication process.

The applied researcher may find it difficult to implement many of the stopping rules advocated above. They are often statistically complex and not readily available in popular statistical software (e.g., SAS and SPSS). SPSS currently includes a "coefficient" as part of the default agglomerative schedule printout. Coefficient values represent the square Euclidean distance between the two objects (or clusters) being joined. As such, small coefficients indicate that fairly homogeneous clusters are being joined, whereas larger values indicate that dissimilar clusters or objects are being joined. Many researchers rely on these values to suggest the number of interpretable clusters in much the same way as factor analysis researchers rely on the scree plot to determine how many factors to extract. SAS includes the cubic clustering criterion—a method that was favorably reviewed by Milligan and Cooper (1985). Atlas and Overall (1994) described a method of calculating ANOVAs at each hierarchical level that is relatively simple to perform and provides a more empirical approach to determining the number of clusters to interpret. Until more internal stopping rules are integrated into the popular computer statistics programs, however, researchers will need to rely on theoretical rationale, subjective inspection, or additional time-consuming statistical computations to determine the best number of clusters to interpret.

## I. External Validity

The final stage of developing a classification system involves establishing the external validity of the cluster structure. Issues that might be addressed in such studies include exploring the cluster structure's predictive and descriptive utility and its consistency across samples.

At the very least, researchers are encouraged to replicate their cluster solutions on independent samples from the same population. If the researcher wishes to make inferences beyond the population used in the study, then the cluster solution should be validated on diverse populations. As Skinner (1981) points out, the inferences that we draw regarding the meaning of our classification scheme rely on these generalizability studies. Cross-validation samples from the same population are often obtained by randomly dividing an existing sample into two or more subsamples prior to conducting a cluster analysis (see Tinsley & S. Brown, chapter 1, this volume).

In addition to investigating the generalizability of their results to other samples, researchers should also attend to the generalizability of their classification structures to alternative measures of the same constructs. A cluster solution gains strength when an investigator is able to replicate that solution using different observations. When the cluster study is well grounded in theory, the researcher is able to evaluate the validity of the data partition by comparing the classification obtained to theoretical postulates about the relations among relevant constructs.

Finally, the researcher's ultimate objective is often to use the obtained cluster solution to predict other phenomenon. Thus, the concurrent and predictive utility of the cluster solution must be evaluated carefully and separately across all samples to which the researcher wishes to generalize. For example, it is one thing to establish a highly reliable classification of political and economic stability among emerging nations, it is somewhat more impressive, however, if that classification system is capable of predicting economic growth within the United States resulting from increased foreign trade with those nations. It is not sufficient merely to establish a classification system; a critical analysis of the ability of that system to provide information relevant to real-world questions and problems is necessary to establish the system's usefulness.

## J. Presenting Cluster Analysis Results

Aldenderfer and Blashfield (1984), among others (e.g., Romesburg, 1984), provide investigators with guidelines for reporting cluster analysis results. Researchers should begin by explaining and describing the theoretical or empirical framework used to guide the cluster study and how objects and observations were selected for inclusion in the study. They should provide unambiguous descriptions of the cluster and similarity or distance methods used in a study so that researchers from different content areas will be able to identify their method from among the myriad of procedures. Because the choice of similarity or distance measures can affect the outcome of a cluster solution, it is vital that researchers report which procedure was used in their study. Also identify the computer program used to establish the cluster solution because various computer programs will occasionally generate different cluster solutions using the same data and cluster method (Blashfield, 1977). Finally, researchers must provide a cogent description of how clusters were selected and evidence for the validity of their cluster structure.

## IV. CONCLUSION

The use of cluster methods has increased dramatically in the last 30 years, but many researchers still fail to use the procedure when applicable or they use it improperly. Because cluster analysis is not a single standardized procedure, and there are pitfalls associated with its improper use, care is required in its application. Nevertheless, the perceived difficulties involved in conducting a cluster analysis study are greatly outweighed by the potential usefulness and flexibility of these procedures. Investigators should use theory to guide their research questions and to identify the populations and variables of interest; it is imperative to make this step explicit. Also, use theory to guide your choice of a measure of association (i.e., a distance or similarity measure) and clustering algorithm (e.g., hierarchical or partitioning method). Conduct your analyses using more than one method to increase your confidence in your findings (and indirectly to contribute to cluster methods research). Although a number of statistical options exist for determining the best number of clusters to interpret, cross-validation remains one of the best ways of demonstrating the internal validity of a cluster solution. Consider the cluster study as a first step and not as an end in itself whenever you are interested in the how clusters relate to other phenomenon. Finally, thoroughly describe your procedure when preparing a written report.

## REFERENCES

Aldenderfer, M. S., & Blashfield, R. K. (1984). *Cluster analysis.* Thousand Oaks, CA: Sage Publications.

Anderberg, M. R. (1973). *Cluster analysis for applications.* New York: Academic Press.

Atlas, R. S., & Overall, J. E. (1994). Comparative evaluation of two superior stopping rules for hierarchical cluster analysis. *Psychometrika, 59,* 581–591.

Baker, F. B., & Hubert, L. J. (1975). Measuring the power of hierarchical cluster analysis. *Journal of the American Statistical Association, 70,* 31–38.

Blashfield, R. K. (1976). Mixture model tests of cluster analysis: Accuracy of four agglomerative hierarchical methods. *Psychological Bulletin, 83,* 377–388.

Blashfield, R. K. (1977). On the equivalence of four software programs for performing hierarchical cluster analysis. *Psychometrika, 42,* 429–431.

Blashfield, R. K., & Draguns, J. G. (1976). Evaluative criterion for psychiatric classification. *Journal of Abnormal Psychology, 85,* 140–150.

Blong, R. J. (1973). A numerical classification of selected landslides of the debris slide-avalanche-flow type. *Engineering Geology, 7,* 99–114.

Borgen, F. H., & Barnett, D. C. (1987). Applying cluster analysis in counseling psychology research. *Journal of Counseling Psychology, 34,* 456–468.

Cattell, R. B. (1944). A note on correlation clusters and cluster search methods. *Psychometrica, 9,* 169–184.

Cattell, R. B., Coulter, M. A., & Tsujioka, B. (1966). The taxonometric recognition of types and functional emergents. In R. B. Cattell (Ed.), *Handbook of multivariate experimental psychology.* Chicago: Rand McNally.

Cattell, R. B. (1988). The data box: Its ordering of total resources in terms of possible relational systems. In: J. R. Nesselroade & R. B. Cattell (Eds.), *Handbook of multivariate experimental psychology* (2nd ed., pp. 69–130). New York: Plenum Press.

Chartrand, J. M., Martin, W. F., Robbins, S. B., McAuliffe, G. J., Pickering, J. W., & Calliotte, J. A. (1994). Testing a level versus an interactional view of career indecision. *Journal of Career Assessment, 2,* 55–69.

Cohen, J. (1960). A coefficient of agreement for nominal scales. *Educational and Psychological Measurement, 20,* 37–46.

Cronbach, L. J., & Gleser, G. C. (1953). Assessing similarity between profiles. *Psychological Bulletin, 50,* 456–473.

Everitt, B. (1986). *Cluster analysis* (2nd ed.). New York: John Wiley & Sons.

Filsinger, E. E. (1990). Empirical typology, cluster analysis, and family-level measurement. In T. W. Draper, A. C. Marcos (Eds.), *Family variables: Conceptualization, measurement, and use.* Thousand Oaks, CA: Sage Publications.

Fleiss, J. L., Lawlor, W., Platman, S. R., & Fieve, R. R. (1971). On the use of inverted factor analysis for generating typologies. *Journal of Abnormal Psychology, 77,* 127–132.

Friedman, H. P., & Rubin, J. (1967). On some invariant criteria for grouping data. *Journal of the American Statistical Association, 62,* 1159–1178.

Gati, I., Osipow, S. H., & Fassa, N. (1994). The scale structure of multi-scale measures: Application of the split-scale method to the Task Specific Occupational Self-Efficacy Scale and the Career Decision Making Self-Efficacy Scale. *Journal of Career Assessment, 2,* 384–397.

Gorsuch, R. L. (1983). *Factor analysis* (2nd ed.). Hillsdale, NJ: Lawrence Erlbaum Associates.

Gould, S. J. (1989). *Wonderful life: The Burgess Shale and the nature of history.* New York: Norton.

Hands, S., & Everitt, B. (1987). A Monte Carlo study of the recover of cluster structure in binary data by hierarchical clustering techniques. *Multivariate Behavioral Research, 22,* 235–243.

Heppner, P. P., Kivlighan, D. M. Jr., Good, G. E., Roehlke, H. J., Hills, H. I., & Ashby, J. S. (1994). Presenting problems of university counseling center clients: A snapshot and multivariate classification scheme. *Journal of Counseling Psychology, 41,* 315–324.

Horn, D. (1943). A study of personality syndromes. *Character and Personality, 12,* 257–274.

Hubert, L. J., & Arabie, P. (1985). Comparing partitions. *Journal of Classification, 2,* 193–218.

Hubert, L. J., & Levin, J. R. (1976). A general statistical framework for assessing categorical clustering in free recall. *Psychological Bulletin, 83,* 1072–1080.

Johnson, S. C. (1967). Hierarchical clustering schemes. *Psychometrika, 32,* 241–254.

Kendall, M. G. (1963). *Rank correlation methods* (3rd ed.). London: Griffin.

Lambert, J. M., & Williams, W. T. (1962). Multivariate methods in plant ecology, IV. Nodal analysis. *Journal of Ecology, 50,* 775–802.

Lambert, J. M., & Williams, W. T. (1966). Multivariate methods in plant ecology, IV. Comparison of informational analysis and association analysis. *Journal of Ecology, 54,* 635–664.

Lance, G. N., & Williams, W. T. (1967). A general theory of classificatory sorting strategies: 1 hierarchical systems. *Computer Journal, 9,* 373–380.

Larson, L. M., & Majors, M. S. (1998). Applications of the Coping with Career Indecision instrument with adolescents. *Journal of Career Assessment, 6,* 163–179.

Lorr, M. (1994). Cluster analysis: Aims, methods, and problems. In: S. Strack & M. Lorr (Eds.), *Differentiating normal and abnormal personality* (pp. 179–195). New York: Springer Publishing.

MacNaughton-Smith, P. (1965). Some statistical and other numerical techniques for classifying individuals. *Home office research unit report No. 6.* London: H.M.S.O.

MacNaughton-Smith, P., Williams, W. T., Dale, M. B., & Mockett, L. G. (1964). Dissimilarity analysis. *Nature, 202,* 1034–1035.

Mahalanobis, P. C. (1936). On the generalized distance in statistics. *Proceedings of the National Institute of Science, Calcutta, 12,* 49–55.

McIntyre, R. M., & Blashfield, R. K. (1980). A nearest centroid technique for evaluating the minimum-variance clustering procedure. *Multivariate Behavioral Research, 15,* 225–238.

McLaughlin, M. E., Carnevale, P., & Lim, R. G. (1991). Professional mediators' judgements of mediational tactics: Multidimensional scaling and cluster analysis. *Journal of Applied Psychology, 76,* 465–472.

McQuitty, L. L. (1957). Elementary linkage analysis for isolating orthogonal and oblique types and typal relevancies. *Educational and Psychological Measurement, 17,* 207–229

Medin, D. L. (1989). Concepts and conceptual structure. *American Psychologist, 44,* 1469–1481.

Milligan, G. W. (1980). An examination of the effect of six types of error perturbation on fifteen clustering algorithms. *Psychometrika, 45,* 325–342.

Milligan, G. W. (1981a). A review of Monte Carlo tests of cluster analysis. *Multivariate Behavioral Research, 16,* 379–407.

Milligan, G. W. (1981b). A Monte Carlo test of thirty internal criterion measures for cluster analysis. *Psychometrika, 46,* 187–199.

Milligan, G. W., & Cooper, M. C. (1985). An examination of procedures for determining the number of clusters in a data set. *Psychometrika, 50,* 159–179.

Milligan, G. W., & Cooper, M. C. (1986). A study of the comparability of external criteria for hierarchical cluster analysis. *Multivariate Behavioral Research, 21,* 441–458.

Milligan, G. W., & Cooper, M. C. (1987). Methodology review: Clustering methods. *Applied Psychological Measurement, 11,* 329–354.

Milligan, G. W., & Cooper, M. C. (1988). A study of standardization of variables in cluster analysis. *Journal of Classification, 5,* 181–204.

Milligan, G. W., & Sokal, L. M. (1980). A two-stage clustering algorithm with robust recovery characteristics. *Educational and Psychological Measurement, 40,* 755–759.

Morris, L. C., Blashfield, R. K., & Satz, P. (1981). Neuropsychology and cluster analysis. *Journal of Clinical Neuropsychology, 3,* 79–99.

Overall, J. E., Gibson, J. M., & Novy, D. M. (1993). Population recovery capabilities of 35 cluster analysis methods. *Journal of Clinical Psychology, 49,* 459–470.

Overall, J. E., & Klett, C. (1972). *Applied multivariate analysis.* New York: McGraw Hill.

Overall, J. E., & Magee, K. N. (1992). Replication as a rule for determining the number of clusters in hierarchical cluster analysis. *Applied Psychological Measurement, 16,* 119–128.

Price, L. J. (1993). Identifying cluster overlap with NORMIX population membership probabilities. *Multivariate Behavioral Research, 28,* 235–262.

Punj, G., & Stewart, D. W. (1983). Cluster analysis in marketing research: Review and suggestions for application. *Journal of Marketing Research, 20,* 134–148.

Rand, W. M. (1971). Objective criteria for the evaluation of clustering methods. *Journal of the American Statistical Association, 66,* 846–850.

Romesburg, H. C. (1984). *Cluster analysis for researchers.* Belmont, CA: Lifetime Learning Publications.

Schaffer, C. M., & Green, P. E. (1996). An empirical comparison of standardization methods in cluster analysis. *Multivariate Behavioral Research, 31,* 149–167.

Scheibler, D., & Schneider, W. (1985). Monte Carlo tests of the accuracy of cluster analysis algorithms—A comparison of hierarchical and nonhierarchical methods. *Multivariate Behavioral Research, 20,* 293–304.

Schweizer, K., Braun, G., & Boller, E. (1994). Validity and stability of partitions with different sample sizes and classification methods: An empirical study. *Diagnostica, 40,* 305–319.

Sireci, S. G., & Geisinger, K. F. (1992). Analyzing test content using cluster analysis and multidimensional scaling. *Applied Psychological Measurement, 16,* 17–31.

Skinner, H. A. (1979). Dimensions and clusters: A hybrid approach to classification. *Applied Psychological Measurement, 3,* 327–341.

Skinner, H. A. (1981). Toward the integration of classification theory and methods. *Journal of Abnormal Psychology, 20,* 68–87.

Sneath, P. H. A. (1957). The application of computers to taxonomy. *Journal of General Microbiology, 17,* 201–226.

Sokal, R., & Sneath, P. (1963). *Principles of numeric taxonomy.* San Francisco: W. H. Freeman.

Sokal, R., & Michener, C. D. (1958). A statistical method for evaluating systematic relationships. *University of Kansas Science Bulletin, 38,* 1409–1438.

Speece, D. L. (1995). Cluster analysis in perspective. *Exceptionality, 5,* 31–44.

Tryon, R. (1939). *Cluster analysis.* New York: McGraw Hill.

Ward, J. H. (1963). Hierarchical grouping to optimize an objective function. *Journal of the American Statistical Association, 58,* 236–244.

Wishart, D. (1969). An algorithm for hierarchical classifications. *Biometrics, 25,* 165–170.

# MULTIDIMENSIONAL SCALING

**MARK L. DAVISON**

*Department of Educational Psychology, University of Minnesota, Minneapolis, Minnesota*

**STEPHEN G. SIRECI**

*Department of Educational Policy, Research, and Administration, University of Massachusetts, Amherst, Massachusetts*

Multidimensional scaling (MDS) is a versatile technique for understanding and displaying the structure of multivariate data. This technique has seen wide application in the behavioral sciences and has led to increased understanding of complex psychological phenomena. MDS has been used, for example, to assess cognitive developmental theories (Howard & Howard, 1977; Miller & Gelman, 1983), study interracial relations among children (Collins, 1987), determine consumer preferences (Hoffman & Perreault, 1987), and evaluate the dimensional structure and content validity of tests and questionnaires (Napior, 1972; Sireci, Rogers, Swaminathan, Meara, & Robin, 1997). MDS uses observed proximity data defined over all possible pairs of stimuli to derive a representation of stimulus structure. In this chapter, we limit the discussion to MDS applications founded on distance assumptions; that is, the assumption that the essential features of the proximity data can be represented as a stochastic function of symmetric distances among stimuli in a multidimensional space spanned by continuous, latent dimensions.

A very simple example of MDS is presented in Table 12.1, Table 12.2, and Figure 12.1. In this example, the stimuli are the six Basic Theme Scales from the Strong-Campbell Interest Inventory (Campbell & Hansen, 1985):

*Handbook of Applied Multivariate Statistics and Mathematical Modeling*

**TABLE 12.1 Intercorrelations of Six Holland Basic Theme Scales from the Strong Interest Inventory**

| Scale | R | I | A | S | E | C |
|-------|-----|-----|------|------|------|---|
| Realistic (R) | 1 | | | | | |
| Investigative (I) | 0.52 | 1 | | | | |
| Artistic (A) | 0.06 | 0.33 | 1 | | | |
| Social (S) | 0.18 | 0.32 | 0.26 | 1 | | |
| Enterprising (E) | 0.3 | 0.13 | 0.01 | 0.38 | 1 | |
| Conventional (C) | 0.37 | 0.37 | −0.09 | 0.34 | 0.46 | 1 |

the Realistic (R), Investigative (I), Artistic (A), Social (S), Enterprising (E), and Conventional (C) interest scales. The proximity data are the correlations among the six stimulus scales shown in Table 12.1. This proximity data matrix was submitted to a nonmetric scaling analysis in two dimensions, an analysis which attempts to represent the structure of the six stimuli as coordinates in such a way that stimulus scales that are the most highly correlated are represented as being near each other, and stimulus scales that do not correlate highly are represented as far apart. The first two columns of Table 12.2 show the empirical coordinate estimates, and these estimates have been plotted in Figure 1. Note, for instance, that the smallest correlation in Table 12.1 is between the Artistic and Conventional scales ($r = -.09$), and these two scales are quite far apart in Figure 12.1. The highest correlation in Table 12.1 equals .52 for the Realistic and Investigative stimulus scales; these two scales are adjacent in Figure 1. The higher the correlation between two scales in Table 12.1, the smaller the distance between them in Figure 12.1. MDS methods based on distance assumptions are designed to reproduce a proximity matrix with a dimensional coordinate structure such that the distances between stimuli in that structure reproduce

**TABLE 12.2 Empirical and Theoretical Configurations for the Six Holland Interest Scales**

| Scale | Empirical configuration | | Theoretical configuration | |
|-------|-------------|--------------|-------------|--------------|
| | **Dimension I** | **Dimension II** | **Dimension I** | **Dimension II** |
| Realistic | −0.36 | 0.73 | −0.50 | 0.87 |
| Investigative | 0.35 | 0.66 | 0.50 | 0.87 |
| Artistic | 1.53 | −0.12 | 1.00 | 0.00 |
| Social | 0.15 | −0.75 | 0.50 | −0.87 |
| Enterprising | −0.70 | −0.64 | −0.50 | −0.87 |
| Conventional | −0.97 | 0.12 | −1.00 | 0.00 |

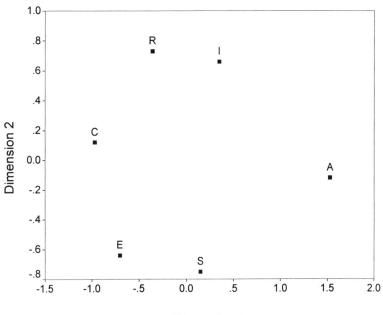

**FIGURE 12.1** Two-dimensional scale values based on an analysis of the Basic Theme Scales from the Strong Vocational Interest Inventory and the proximity data in Table 12.1. $R$ = Realistic, $I$ = Investigative, $A$ = Artistic, $S$ = Social, $E$ = Enterprising, $C$ = Conventional.

the proximity data; stimuli that are more proximal or similar in the data are represented as closer to each other in the structure.

The most commonly applied MDS methods are exploratory, rather than formal hypothesis testing or confirmatory techniques, although the past two decades have seen an increase in what are called "constrained" or "confirmatory" methods that are designed for investigation of explicit hypotheses. Constrained approaches will be briefly described below.

In this chapter, we begin by describing the types of data to which MDS is applied and then MDS itself. Illustrative applications to direct and derived proximity data are then presented. This is followed by a discussion of the relationships between MDS and other multivariate techniques discussed in this volume. Throughout, we have tried to weave references to applications in various fields. Finally, we consider needed further improvements in the technique itself and in the application of the method.

## I. PROXIMITY DATA

A *proximity datum,* $\delta_{ij}$, is an index that quantifies the degree of (dis)similarity or association between a pair of objects or stimuli. Proximity data come in two varieties: *direct* and *derived.* If the respondents are asked to examine pairs of objects and then to quantify explicitly the (dis)similarity of each pair, then the data are said to be direct proximity judgments. For instance, if the objects were three American universities (Northwestern University, University of Massachusetts, and University of Minnesota), respondents might be asked to consider each pair of universities in turn and to rate the pair on a four point scale (1 = Highly Similar, 4 = Highly Dissimilar). Such data are called direct proximity data, because respondents are asked to report directly their perception of the proximity or similarity between the universities.

Derived proximities are so named because the proximity measures are computed (derived) from a persons-by-objects (or stimulus) data matrix. Table 12.1 is just such a derived proximity matrix, since the correlations were computed from a persons-by-scales data matrix containing a profile of scores for each person on six scales. Whether derived or direct, the proximity data are measures of the similarity or association between the pairs of stimuli.

## II. MODEL AND ANALYSIS

The statistical aspects of MDS are often difficult to grasp because several features critical to understanding MDS (e.g., estimating functions rather than parameters) are not typically covered in traditional statistics courses. In this section, we will briefly introduce the critical MDS statistical models, and then illustrate them with actual applications in subsequent sections. Several excellent textbooks are available for more in-depth study (e.g., Borg & Groenen, 1997; Davison, 1991). MDS algorithms can be found in most statistical software packages, including SAS, SPSS, SYSTAT, and STATISTICA. Other computer programs include the Guttman--Lingoes series of nonmetric programs (e.g., Roskam & Lingoes, 1981), KYST (Kruskal, Young, & Seery, 1978), MULTISCALE-II (Ramsay, 1991), and PROXSCAL (Commandeur & Heiser, 1993). Borg and Groenen (1997) have provided an excellent comparison of various programs.

As indicated earlier, the purpose of MDS is to find the most parsimonious, but comprehensive, set of continuous latent dimensions that can account for the proximity data. This is accomplished (as will be illustrated subsequently) by using the proximity data to find a set of coordinate points (one point for each stimulus) so that the distances among stimulus points accurately represent the proximities among stimuli. For example, stimulus pairs judged most similar are closer together in the multidimensional space; stimulus pairs judged most dissimilar are farther apart. We now describe this distance principle more technically.

According to one of the more common expressions of the MDS model for a single data matrix,

$$\tau(\delta_{ij}) = d_{ij} + e_{ij} \tag{1a}$$
$$d_{ij} = [\Sigma_k(x_{ik} - x_{jk})^p]^{1/p}, \tag{1b}$$

where $\tau$ is a function of the observed data $\delta_{ij}$, $d_{ij}$ is the distance between stimuli $i$ and $j$ in the space defined by the $K$ dimensions, $x_{ik}$ and $x_{jk}$ are the coordinates for stimuli $i$ and $j$ respectively along dimension $k$ ($k = 1, \ldots, K$), and $p$ is a Minkowski exponent specified by the user. Usually, $p = 2$, giving the well-known Euclidean distance function. For a specified value of $K$ and $p$, the analysis yields least squares (or maximum likelihood) estimates of coordinates ($x_{ik}, x_{jk}$) so that the stimuli, $i$ and $j$, can be plotted in multidimensional space and the distances between them ($d_{ij}$) can then be computed. It also yields an estimate of the function $\tau$, which can then be displayed graphically.

If the data are thought to be at the interval level, then the function $\tau$ is a linear function, estimating $\tau$ involves estimating the slope and intercept constants in the linear function $\tau$, and the analysis is said to be *metric*. If the data are thought to be ordinal, then the analysis is said to be *nonmetric*, and $\tau$ is taken to be an unknown, monotonic function such that for any two data points $\delta_{ij} > \delta_{i'j'}$ then $\tau(\delta_{ij}) \geq \tau(\delta_{i'j'})$. The optimally rescaled data points, $\tau(\delta_{ij})$, are variously known as disparities, rank images of the data, or dhats ($\hat{d}_{ij}$). Here, however, we will refer to the values $\tau(\delta_{ij})$ as the transformed data. In a nonmetric analysis, the solution consists of the coordinate estimates, $x_{ik}, x_{jk}$, and the estimate of the monotone function $\tau$.

One of the most puzzling features about nonmetric MDS is the transformed data, $\tau(\delta_{ij})$. To understand its importance, consider the following three ordinal proximity data points from a hypothetical study with three stimuli and therefore three stimulus pairs: $\delta_{12} = 4$, $\delta_{23} = 4.1$, $\delta_{13} = 21$. If these data are ordinal, then the essential feature of the data is the ordering of the three stimulus pairs. Any alternative numerical assignment is equally good if it orders the stimulus pairs in the same way as the original proximity data. Here is one alternative numerical assignment: $\tau(\delta_{12}) = 1$, $\tau(\delta_{23}) = 2$, and $\tau(\delta_{13}) = 3$. Here is another: $\tau(\delta_{12}) = 2$, $\tau(\delta_{23}) = 21$, and $\tau(\delta_{23}) = 22$. There is an infinite number of these alternative, equally good numerical assignments for the rank order of the data. Not all of these numerical assignments can be represented equally well by a spatial distance model. For instance, in our little example, the second numerical representation $\tau(\delta_{12}) = 1$, $\tau(\delta_{23}) = 2$, and $\tau(\delta_{13}) = 3$, can be represented perfectly by three points along a straight line (a unidimensional solution) with coordinates $x_1 = -1$, $x_2 = 0$, and $x_3 = 1$. Neither of the other two representations (including the original data itself) can be represented perfectly by distances between three points in one dimension. From the possible, alternative representations for the data, nonmetric MDS seeks to find the one, $\tau(\delta_{ij})$, which can best be represented by distances, and it finds the coordinate estimates $(x_{ik}, x_{jk})$ corresponding to those distances. The fit is then measured in terms of the mismatch between the transformed data, $\tau(\delta_{ij})$ and the distances.

For the solution based on the proximity data in Table 12.1, the coordinate estimates are displayed in Table 12.2 and Figure 12.1. The estimated monotone function, shown in Figure 12.2, is monotonically decreasing (smaller correlations are associated with larger distances), and displays clear nonlinearities tending to confirm the choice of a nonmetric analysis.

## A. Weighted Euclidean Model

Not all applications of MDS involve a single proximity matrix. More specifically, the proximity data matrix can be replicated with data entries $\delta_{ij}^r$ where $\delta_{ij}^r$ is the $r$th replication of the proximity measure for stimulus pair $(i, j)$. Note that henceforth $r$ is not an exponent, but rather a superscript designating a replication. When the replications correspond to people, then the model of Equation 1 can be extended to accommodate replications and individual differences as follows:

$$\tau^r(\delta_{ij}^r) = d_{ij}^r + e_{ij}^r, \tag{2a}$$

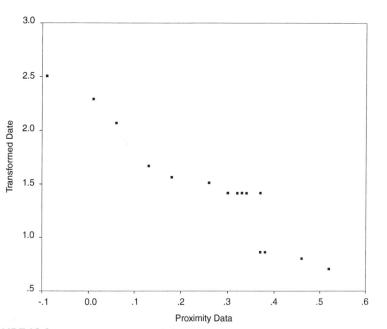

**FIGURE 12.2**    Data transformation $\tau(\delta_{ij})$ for the two-dimensional analysis of the proximity data in Table 1.

where

$$d_{ij}^r = [\Sigma_k (y_{ik}^r - y_{jk}^r)^2]^{1/2} \tag{2b}$$
$$= [\Sigma_k w_k^r (x_{ik} - x_{jk})^2]^{1/2}. \tag{2c}$$

According to the weighted Euclidean model (WEM) model, each respondent's data satisfies the assumptions of Equation 1 in the Euclidean metric $p = 2$, with an optimal rescaling function unique to respondent $r$, $\tau^r$, and with distances expressed with respect to coordinates idiosyncratic to respondent $r$, $(y_{ik}^r, y_{jk}^r)$. However, the idiosyncratic coordinates are assumed proportional to coordinates in a common space, such that $y_{ik}^r = (w_k^r)^{1/2} x_{ik}$, where $w_k^r$ refers to a weight for respondent $r$ along dimension $k$. In its most general form, an analysis based on Equation 2 will yield estimates of the subject-specific optimal rescaling function, $\tau^r$, the stimulus coordinates ($x_{ik}$, $x_{jk}$) said to constitute the stimulus space, and the individual differences parameters $w_k^r$ said to constitute the subject space. Roughly speaking, the larger the subject weight $w_k^r$, the larger the variance accounted for by dimension $k$ in the proximity data of respondent $r$. Because it contains both person and stimulus parameters, an analysis based on the WEM leads to a description of the stimuli, the stimulus coordinate space, ($x_{ik}$, $x_{jk}$), and persons, the subject weight space $w_k^r$.

## B. Generalized Euclidean Model

There is an even more general model (Tucker, 1972), sometimes called the generalized Euclidean model (GEM). It has seldom been used because it is complex and difficult to interpret. It differs from the WEM by including an additional individual differences parameter to characterize the interaction of perceptual dimensions. For instance, in research on the size–weight illusion, one might posit an interaction between the perceived volume and perceived weight of an object such that the perceived weight of an object increases as its volume increases. Furthermore, the size of this interaction might vary across people. As applied to judgments about the dissimilarity of objects varying in volume and weight, the GEM can be formulated to yield (a) a representation of stimulus variation presumably along dimensions of perceived volume and weight ($x_{ik}$, $x_{jk}$), (b) a representation of individual differences in the salience or importance attached to the dimensions $w_k^r$, and (c) a representation of individual differences in the sizes of the interaction, $c_{kk'}^r$.

## III. CONDUCTING A MULTIDIMENSIONAL SCALING

In conducting an MDS study, the researcher will face a number of issues: some are matters of measurement, some involve research design, and others involve statistical questions.

## A. Selection of Stimuli and Participants

As in any investigation, the research questions primarily determine the stimuli (objects) to be analyzed. When the number of stimuli to be analyzed is relatively small, such as when an experimenter is interested in studying attitudes toward specific ethnic groups, selecting the stimuli to be scaled is straightforward (i.e., include all ethnic groups of interest). However, if the stimuli scaled in an MDS experiment represent some larger domain, or are too numerous to analyze together, selection of the stimuli to be scaled from the larger domain or stimulus set directly affects the external validity of the experiment. For example, suppose an investigator wanted to discover the dimensional structure underlying the nebulous field known as multivariate statistics. A simple (and defensible) procedure would be to use the topics covered in this book as the stimuli to be scaled. However, another textbook may include very different topics that would represent the field quite differently, yielding a different representation. Thus, selection of the stimuli to be scaled is important when considering the likely generalizability of the results from an MDS experiment. The goal is selection of a subset of stimuli to be scaled that adequately represents the larger domain of interest.

When direct proximity data are gathered, another critical external validity issue is selection of the participants. For example, in the hypothetical study of the domain of multivariate statistics, different results may be obtained if only teachers or only practitioners were used to gather the proximity data. Similarly, one researcher may sample participants from the membership directory of the American Statistical Association, and another may select participants from the American Sociological Association. An attractive feature of individual differences MDS models is that proximity data can be gathered from many different groups of participants and, in addition to focusing on the structure of the stimuli rated, variations in the structure among the different groups can also be investigated.

## B. Selecting a Proximity Measure

In selecting a proximity measure, the researcher must choose between a direct or derived proximity measure. That is, should respondents be asked directly to report the (dis)similarity of stimulus pairs, or should the researcher "infer" the similarity by computing a derived proximity measure based on respondent reactions to the stimuli as captured in a subjects-by-stimulus matrix? Direct proximity judgments are a tool for studying perceptions of stimuli, but studies of perception sometimes employ derived proximity measures computed from unidimensional perceptual judgments. In studies of perception, the direct proximity judgment approach seems best when (a) the researcher prefers to let respondents determine the stimulus dimensions on which judgments will be based, rather than specifying those dimensions for respondents in a series of unidimensional rating scales, or (b) there is doubt that respondents can perceptually (or cognitively) parse separate stimulus dimensions in unidimensional tasks, so that a global or holistic dissimilarity judgment is more consistent with respondents' cognitive and perceptual capacities.

Outside of perception (broadly defined), derived proximity measures are used. There are many such measures from which to choose. For interval data, Pearson correlations among variables are often employed as proximities; although the squared Euclidean distance function can also be a good choice (Davison, Kuang, & Kim, 1999). For dichotomous data, tetrachoric correlations and binary measures of association can be used. For ordinal data, there are rank order correlations. Proximity measures for nominal data have also been proposed (e.g. Goodman-Kruskal tau). Many of these options can be seen in the proximity modules of packages such as SPSS and SYSTAT.

## C. Gathering Direct Proximity Data

Two common approaches are *paired comparisons* and *sorting* procedures.

### 1. Paired Comparisons Judgments

The paired comparison procedure involves presenting respondents with pairs of stimuli and asking them to rate the similarity of the stimuli comprising each pair. For example, Sireci et al. (1997) used a paired comparisons procedure to evaluate the structure of the 1996 National Assessment of Educational Progress (NAEP) Science Assessment. Ten science teachers who were experienced in state and national assessment efforts rated pairs of NAEP Science items in terms of their similarity to one another with respect to the science knowledge and skills measured. Item similarity rating booklets were prepared for each teacher. Each page of the booklet contained a pair of test items with an eight-point similarity rating scale printed at the bottom of the page. The lower end point of the scale was labeled "highly similar" and the upper end point of the scale was labeled "very different." The teachers completed their similarity ratings by reviewing the item pair on each page and circling their similarity rating on the bottom of each page.

The paired comparisons procedure provides a comprehensive set of similarities data for a given set of stimuli. By comparing each stimulus with all others, $n - 1$ ratings are obtained for each of the $n$ stimuli. Although this procedure provides multiple ratings per stimulus and a complete set of proximities for all possible stimulus pairings, it can be burdensome for the respondents. For $n$ stimuli, $n(n - 1)/2$ similarity ratings are required (e.g., 20 stimuli require 190 stimulus pairs). Thus, as $n$ increases, factors such as fatigue, attentiveness, and motivation can seriously affect the similarity ratings. The 45 items rated in the Sireci et al. study involved 900 ratings.

If data are to be gathered using a paired comparison procedure, there are several factors to control. First are unwanted interdependencies in the data that can arise from improper ordering of stimulus pairs, or from respondents' unfamiliarity with the rating task. For example, it is possible for respondents to rate stimulus similarities very liberally at first, and then after becoming more familiar with the stimuli and rating tasks, gradually become more conservative in their ratings (or vice versa). For this reason, we recommend that respondents be given adequate time to review all stimuli before beginning the similarity ratings and practice with similarity ratings before the primary data of interest are collected.

Interdependencies among stimulus pairings can also occur when the order of stimulus pair presentation is not carefully designed. To avoid these problems, the stimulus pairs should be ordered in random or systematic fashion, so appearances of any one stimulus are evenly spaced throughout the paired comparisons. Each stimulus should also appear as the first stimulus in a pair about the same number of times as it appears as the second stimulus in a pair. These two recommendations aim towards counterbalancing the ordering of stimuli so that order effects (i.e., space and time effects, Davison, 1991) do not contaminate respondents' perceived similarity ratings. Ross (1934) provided an algorithm, computerized by Cohen and Davison (1973), for counterbalancing stimuli to be used in paired comparison rating tasks. Another important issue is whether a common ordering should be used across all respondents. Some designs randomly order stimulus pairs across respondents. These stimulus ordering issues are important for alleviating potential threats to the validity of MDS results. However, the presence or magnitude of these potential threats has not been extensively studied.

Incomplete paired comparisons designs have been proposed to maximize the information provided by participants in a MDS study, while simultaneously minimizing the burden on them. There are two general types of incomplete paired comparison designs. The first type limits the number of interstimulus ratings, resulting in an incomplete matrix of interstimulus similarities. In these designs, numerous ratings are gathered for each stimulus, but not all stimuli are compared with one another. The goal of this design is to gather enough ratings on the similarity of specific pairings so that the missing similarity ratings can be estimated accurately. Consider, for example, four stimuli, A, B, C, and D. There are six possible stimulus pairings of these four stimuli. After gathering five of them, for example, AB, AC, BC, BD, and CD, a fairly good estimate of how the remaining pairing, AD, would be rated could be derived. Obviously, such estimates get increasingly complex as $n$ increases, and so the reader is referred to the work of Spence (1982, 1983) in this area.

The other option for reducing the number of paired comparisons per respondent is to require that all interstimulus comparisons be made by a subset of respondents, rather than by all respondents. This strategy was used by Sireci et al. (1997). Across the 10 respondents, 7 rated each stimulus pairing. This design reduced the number of similarity ratings required from each respondent by 200 (i.e., 700 similarity ratings rather than 900), but still provided multiple ratings for all possible stimulus pairs. Thus, the similarity matrix for each respondent was incomplete, but a complete interitem proximity matrix could be derived across the respondents.

## 2. Sorting Tasks

Another way to reduce the burden on participants is to use a sorting procedure rather than a paired comparison procedure. Sorting procedures typically require respondents to sort stimuli into "piles" representing highly similar groupings. For example, rather than rate the similarity of all possible pairings of test questions, teachers could be asked to group the questions into subsets that are homogeneous with respect to the knowledge and skills measured. Typical instructions for participants are to place the stimuli into mutually exclusive and exhaustive categories so that stimuli in the same category are more similar to one another than they are to stimuli in other categories.

Sorting procedures should be iterative (Rosenberg & Kim, 1975). For instance, the respondents might be asked first to sort the stimuli into three piles. Then, they might be asked to sort each of the original three piles into subpiles. Then they might be asked to break each subpile into subsubpiles, and so on. The number of piles may be specified by the investigator or left up to the respondents. A common practice is to provide a lower and upper limit to the number of piles (e.g., 3 to 10 piles), and let each respondent choose the number of piles she or he believes is required to adequately represent homogeneous subsets of stimuli.

A dichotomous matrix of similarity ratings is then derived for each participant. For example, zeros could be used to indicate items that are placed together in the same pile, and ones could be used to represent items that are placed in different piles. By summing the cells of these matrices across judges, a matrix of stimulus proximities can be computed, with the elements of the matrix ranging from zero to $k$, where $k$ represents the number of respondents.

It should be evident that information is lost when a sorting procedure is used in lieu of a paired comparisons procedure. A direct, multicategory similarity rating of stimulus $i$ to stimulus $j$ is more informative than a dichotomous rating dependent in part on the other stimuli grouped together with each stimulus. However, sorting procedures greatly reduce the amount of time required to gather direct proximity data.

## D. Choosing an Appropriate Scaling Model: Weighted versus Unweighted Models

Having gathered the data, the researcher faces various data analysis decisions. If direct proximity data are gathered from more than two individuals or groups of individuals, or if multiple derived proximity matrices are available, researchers have a choice between fitting unweighted or weighted MDS models. Such data are often called *three-way data,* because the data matrix has three factors: stimuli × stimuli × replications. If the research questions pertain only to the structure of the stimuli, then an unweighted MDS model should suffice. However, if differences between individuals, groups of individuals, or other entities that the multiple proximity matrices may represent are also of interest, then a weighted MDS model should be used. As described earlier, these models configure the stimuli as points in a multidimensional "group space" and the individuals as vectors in a "subject space" of the same dimensionality. These two spaces are related to one another, as in equation 2 $[y_{ik}^r = (w_k^r)^{1/2} x_{ik}]$. The vector of dimension weights derived for each matrix describes the relative weighting of the dimensions from the group space needed to best account for the (transformed) proximity data $[\tau^r(\delta_{ij}^r)]$ of each matrix designated by the superscript $r$ (for replication). Examples of a common stimulus space, and subject (weight) space are provided below.

Although WEM MDS models are popular for analyzing three-way data (i.e., multiple matrices of proximity data), it should be noted that the mere presence of multiple matrices does not necessitate a weighted analysis. Weighted models should only be used if individual differences are of interest, or if a fixed orientation (i.e., nonrotatable solution) is desired. Other options for analyzing multiple proximity matrices using unweighted MDS are to average the $\delta_{ij}^r$ over replication matrices to compute a single, average matrix, or to fit the matrices in a single analysis using the classical (unweighted) MDS model. This latter option is called *replicated MDS* (Schiffman, Reynolds, & Young, 1981), which derives a single stimulus configuration, but allows for a separate transformation of the data from each replication matrix in deriving the stimulus space. Thus, in replicated MDS, individual differences (i.e., personal spaces) cannot be derived from the stimulus space. However, individual transformations [i.e., separate $\tau^r(\delta_{ij}^r)$] are applied to each proximity matrix ($r$) to improve the fit of the solution. Allowing for separate transformations $\tau^r(\delta_{ij}^r)$ is useful in situations where subjects may have used the similarity rating scale in different ways.

Once the data are gathered and selection of the MDS model is completed, two other issues must be resolved. The first issue pertains to how "ties in the data" are handled. This issue is important in nonmetric MDS because the input data are usually in the form of short, ordinal scales. If the *secondary approach* to ties is used, the transformed data must satisfy the constraint $\delta_{ij} = \delta_{i'j'}$ implies $\tau(\delta_{ij}) = \tau(\delta_{i'j'})$; that is, the transformed data must mirror equalities in the proximity data. However, if the *primary approach* is used, then ties in the data can be untied in the transformed data: i.e., $\delta_{ij} = \delta_{i'j'}$ implies nothing about the relationship of $\tau(\delta_{ij})$ and $\tau(\delta_{i'j'})$. By employing the primary approach (i.e., relaxing the equality constraint of the secondary approach), a better fit between the distances and transformed data $\tau(\delta_{ij})$ can be obtained. This more liberal primary approach is often, but not always, justified because the original proximities are ordinal and are not error-free. Thus, most nonmetric applications employ the primary approach.

### E. Determining Dimensionality and Interpreting the Solution

An MDS analysis typically involves solving for solutions in several dimensionalities and then selecting the one that appears most appropriate. There are three general criteria for selecting dimensionality: model-data fit, interpretability of the solution, and reproducibility (replication). Because interpretability of the solution is one criterion for selecting dimensionality, the tasks of determining the best dimensional representation of the data, and then interpreting it, go hand-in-hand.

Numerical fit indexes are commonly used as a first criterion for selecting dimensionality. In general, these indexes describe either goodness of fit to the data in terms of percentage of variance in the transformed proximities $[\tau(\delta_{ij})]$ accounted for by the MDS-distances, or badness of fit in terms of differences between the scaled distances $(d_{ij})$ and transformed proximities. For nonmetric MDS the STRESS fit measure proposed by Kruskal (1964) is most common. STRESS is a normalized, least-squares index of the fit between the distance estimates and the transformed proximities, $\tau(\delta_{ij})$:

$$\text{STRESS} = \sqrt{\frac{\sum_{i,j>i}[\tau(\delta_{ij}) - d_{ij}]^2}{\sum_{i,j>i} d_{ij}^2}} \tag{3}$$

Conceptually, STRESS is an index of the mismatch between the distances $d_{ij}$ and the transformed data $\tau(\delta_{ij})$. It ranges between zero and one, with zero indicating a perfect match between the distances and transformed data. If STRESS is low (close to zero), then similar stimulus pairs (according to the data) are generally close together in the spatial representation, and dissimilar points (according to the data) are far apart in the spatial representation. On the other hand if STRESS is high, then similar points (according to the data) are not necessarily close together in the spatial representation; dissimilar points (according to the data) may not be far apart in the spatial representation.

Another descriptive fit index provided by many MDS programs is the squared multiple correlation between the transformed proximities and the distances, R-squared (RSQ). RSQ is interpretable as the proportion of variance in the transformed proximities accounted for by the distances. Thus, higher values of RSQ indicate better fit.

STRESS, RSQ, and most other MDS fit indexes are descriptive, and so they do not provide an absolute criterion for selecting the "best" dimensionality. Furthermore, increasing the dimensionality of the MDS model almost always leads to a reduction in STRESS and an increase in RSQ. Therefore, the change in fit across increasing dimensional solutions, rather than the value of a fit index for a particular solution, is most useful for selecting the most appropriate MDS solution. If very little improvement in fit occurs when adding an additional dimension to the model, it is unlikely that the additional dimension is necessary. Kruskal and Wish (1978) and others suggest plotting the fit values as a function of dimensionality, similar to an eigenvalue plot in factor analysis, to determine at which dimensionality improvement in fit levels off. Figure 12.3 shows the plot for the proximity data in Table 12.1, which clearly suggests a two-dimensional solution. It is rare for such a clear "elbow" to appear in these plots, and for the plot to unambiguously answer the question of dimensionality.

Although fit indexes are seldom "decisive" for determining dimensionality, they are generally helpful for narrowing down various dimensional solutions to those that are the best candidates for interpretation. For non-

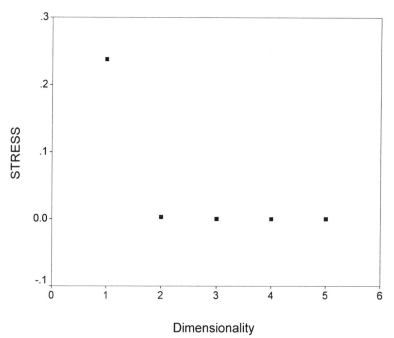

**FIGURE 12.3**   Scree-plot of STRESS by dimensions for the proximity data in Table 12.1.

metric MDS of a single proximity matrix, we suggest guidelines based on Kruskal and Wish (1978) and various simulation studies (e.g., Dong, 1985; MacCallum, 1981). STRESS values less than .15 obtained from a one-dimensional MDS solution are suggestive of unidimensionality. Accepting multidimensional solutions with STRESS values above .10 is generally not recommended, unless the number of stimuli is large or the proximity data contain large amounts of measurement error. In the context of maximum likelihood solutions, formal hypothesis testing approaches to dimensionality have been proposed (Ramsay, 1977, 1981, 1982, 1991; Takane, 1981), although the performance of these formal hypothesis tests with real data remains largely unexplored.

## F. Interpreting Multidimensional Scaling Solutions

After considering fit indexes, an MDS researcher is typically faced with trying to interpret two or more competing MDS solutions. In general, lower-dimensional solutions are preferable to higher-dimensional solutions for reasons of parsimony and ease of communicating the results. However, if additional dimensions lead to substantive understanding of the structure of the data, the increased information provided by the higher-dimensional solution cannot be ignored. Thus, interpretation of a configuration is a key factor in selecting dimensionality. The task for the investigator is to determine the lowest dimensionality in which all meaningful aspects of the structure of the data appear.

There are both qualitative and quantitative methods for interpreting MDS solutions. Both methods rely on external information about the stimuli scaled (and on the participants if three-way data are employed). Starting with the stimulus configuration, the most intuitive qualitative method of interpretation is visual. A common practice is to look at the data configured in two dimensions, which may involve sequentially evaluating several two-dimensional subspaces in high-dimensional solutions. Each dimension can be interpreted by evaluating whether the dimension is ordering the stimuli according to some continuous stimulus characteristic, or grouping stimuli according to a discrete characteristc. Rotation of the configuration, which we describe below, may be helpful when interpreting a configuration visually. The task for the interpreter is to consider known characteristics of the stimuli and identify a characteristic that explains the ordering of the stimuli along the dimension, or describes a common feature of stimuli grouped together. For example, Sireci and Geisinger (1992, 1995) used the known content specifications of test items to help interpret the MDS configurations derived from respondents' similarity ratings of these items. In both studies, they discovered groupings of items from the same content area that were clustered in multidimensional space. Thus, the external information known about the items (i.e., the content areas they were constructed to measure) was used to interpret the dimensions.

A straightforward way to interpret a dimension visually is to compare stimuli located at opposite ends of the dimension. If there is some conspicuous difference between stimuli "pulled" in opposite directions, the dimension probably reflects the attribute on which the stimuli differ. This evaluation can be done visually by examining the stimuli in two-dimensional space and focusing on one dimension at a time, or by inspecting the stimulus coordinates for each dimension, looking for relatively large positive and negative coordinates.

Similarly, a MDS configuration can be inspected to determine if the stimulus structure conforms to expected hypotheses about the data. For example, MDS analysis of the six *Vocational Preference Inventory* scales by Rounds, Davison, and Dawis (1979) confirmed the hexagonal structure of these scales predicted by Holland's (1973) theory of career choice.

Interpreting MDS solutions visually is extremely difficult in greater than three dimensions. A useful empirical aid in such cases is to cluster-analyze the stimulus coordinates. The cluster solution provides a means for looking at the item groupings simultaneously across all dimensions. For example, Figure 12.4 presents a two-dimensional subspace portraying forty items from the auditing section of the 1990 Uniform Certified Public Accountants' Examination (Sireci & Geisinger, 1995). The ellipses encircling

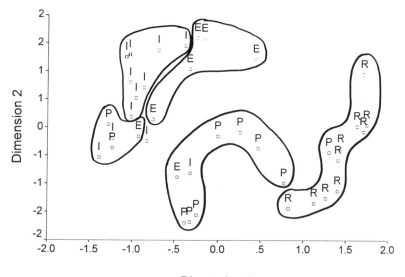

**FIGURE 12.4**   Stimulus space of 40 auditing items. R = Reporting, E = Evidence and Procedures, I = Internal Control, P = Professional Responsibilities.

the items are based on groupings obtained from a hierarchical cluster analysis of the stimulus coordinates of these items from a six-dimensional solution. The clustering results are strongly related to the general content structure of the exam, with three of the four content areas emerging as clusters.

Other empirical methods for facilitating interpretation of MDS stimulus spaces use traditional univariate and multivariate statistical techniques to relate external data to the MDS configuration. External data refer to any information about the stimuli in addition to the proximity data analyzed. Various authors have used external data along with multiple regression (Schiffman et al., 1981; Sireci & Geisinger, 1995; Young & Hamer, 1987), canonical correlation (DeVille, 1996), nonmetric regression (Hoffman & Perrault, 1987), and analysis of variance (Sireci et al., 1997).

Although we described both objective and subjective methods for interpreting MDS solutions, selection and interpretation of the final solution is always primarily subjective. Using external data to help interpret the solution is usually very helpful, but when direct correspondence is not found between external variables and the dimensions, the subjective interpretive skills of the investigator become paramount. For this reason, MDS configurations are often interpreted by several individuals who are familiar with the attributes of the stimuli.

## G. Rotation

The distances among stimuli in an unweighted MDS solution are unique; however, the orientation of the configuration is arbitrary. Orthogonal rotation of the dimensions from an unweighted model does not affect the interstimulus distances, and so rotation can be conducted with no loss of fit to the data. Such rotation is often desirable because it can greatly facilitate interpretation. The simplest method for rotating a solution is visual. This is usually accomplished by inspecting the configuration and orienting the axes so that the configuration makes more intuitive sense. For example, the dimensions can be rotated so that certain stimuli are more closely aligned along a particular dimension. Reorienting the solution is not restricted to rotation of the dimensions. The dimensions can be permuted or dilated, the origin of the space can be shifted, and the dimensions can be reflected. For an unweighted MDS solution, these adjustments preserve the distances among the stimuli calculated by the model.

The stimulus space for a weighted MDS solution is generally considered nonrotatable. This claim stems from the unique orientation of the space in order to best fit the disparity data for each matrix. However, when the weights for two or more dimensions are proportional to one another, or show little variation, partial rotation can be accomplished without loss of fit to the data (Davison, 1991). Partial rotation involves rotation of a two-dimensional plane within a higher-dimensional solution. Thus, when inter-preting a weighted MDS solution, the proportionality of the matrix weights should be considered.

### H. Interpreting the Subject Space in Weighted Euclidean Multidimensional Scaling

As described previously, weighted MDS models portray the differences among the multiple proximity matrices in terms of differences in the weights used to relate the "personal spaces" (configuration of stimuli most appro-priate for each matrix $r$) to the group space (configuration of stimuli most appropriate for the entire group of matrices). The relationship between these two spaces is as in equation 2 $[y_{ik}^{r} = (w_k^r)^{1/2}x_{ik}]$. Thus, the differences among matrices, which may be individuals or groups of individuals, are represented as differences in the set of weights $(w_k^r)$ for each matrix. The differences among the matrices can be interpreted by visual inspection of the weight space. There are two factors that determine the location of the weight vector for each matrix: length of the vector and the orientation with respect to the dimensions.

Consider the weight space presented below for a social studies test (Figure 12.6). Teacher "8" provides an example of someone who essentially used both of the dimensions equally. The two-dimensional subset of weights for this subject (.40 and .36) portray her vector along a 45° angle between these two dimensions. The two-dimensional weight vector for subject "13" on the other hand, is aligned much more closely with Dimension 1, reflecting a relatively strong emphasis on the first (geography) dimension.

The lengths of the weight vectors reflect the proportion of variance of the transformed proximity data accounted for by the stimulus coordinates of the personal space for a given matrix. Distances between the endpoints of the vectors in the weight space cannot be interpreted as distances between points in the space. It is the direction and length of the vectors in the space that describes the variation among matrices. The variance of the transformed proximities accounted for by the personal space for a matrix is given by the square root of the sum of the squared weights (Davison, 1991; Young & Harris, 1993). Referring back to Figure 12.6, the difference in the length of the weight vectors for teachers 8 and 5 indicate the large difference in the percentage of variance of these two teachers' transformed proximity data accounted for by the two-dimensional subspace (30% versus 3%).

### I. Concluding Remarks in Selecting Dimensionality and Interpreting the Solution

Interpreting and selecting an MDS solution involves several tasks: evalua-tion of dimensionality and fit, interpretation of the stimulus space, interpre-tation of the weight space (in weighted MDS solutions), and evaluating the stability of the solution. This last criterion can be an important aid in evaluating the other criteria. If the fit indexes and interpretations of the stimulus and weight spaces for a particular solution change over repeated samples, acceptance of the solution may not be defensible. On the other hand, consistency in fit and interpretations over repeated sampling may help indicate appropriate dimensionality. Therefore, replicating the MDS analysis (using a new sample or splitting the data) is recommended whenever possible.

## IV. EXAMPLES

### A. Example 1: Direct Proximity Data

Sireci and Geisinger (1995) used a paired comparisons procedure to evaluate the content structure of a nationally standardized social studies achievement test for junior high school students. Fifteen social studies teachers rated all possible pairs of items in terms of similarity. Thus, the experimental task yielded 15 proximity matrices that were submitted to an analysis based on the WEM of Eq. (2). Figure 12.5 shows the two-dimensional

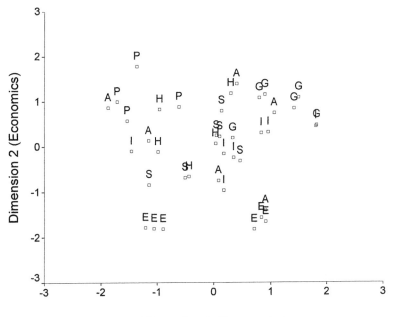

**FIGURE 12.5**  Two-dimensional stimulus space for social studies items. G = Geography, E = Economics, H = History, P = Politics, S = Sociology/Anthropology, I = Interrelated, A = Applied Social Studies.

stimulus space (e.g., the estimates of stimulus coordinates $x_{ik}$ in Eq. 2) obtained from a metric analysis based on the WEM. The items reflected seven content areas shown in the legend of Figure 12.5.

As can be seen in Figure 12.5, the horizontal dimension tends to separate the six items measuring geography from the other items, and the vertical dimension tends to separate the six economics items from the others. Thus, the MDS solution suggests that the teachers attended to the geography and economics content characteristics of these items when making their judgments.

Figure 12.6 displays the "subject space" for the two dimensions displayed in Figure 12.5. The vectors portrayed in Figure 12.6 represent the relative differences among the teachers in their use of the "Geography" and "Economics" dimensions. The tip of each vector is a point whose coordinates equal the teacher's weights along the two dimensions (i.e., the estimates of the weights $w^r_k$ in Eq. 2). The closeness of teacher "13" to the horizontal dimension indicates that this teacher had a substantial weight on the Geography dimension and a near zero weight on the Economics dimension. The weights indicate that this teacher seemed to ignore the Economics dimension and attend heavily to the Geography dimension in

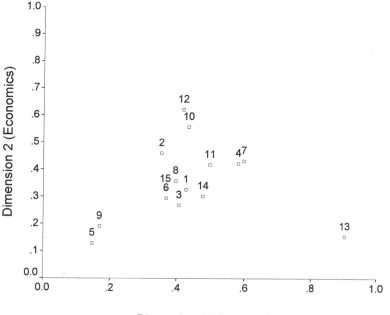

**FIGURE 12.6** Two-dimensional subject space for 15 teachers based on the multidimensional scaling of social studies items.

making judgments about item pairs. Teacher "5" has a near zero weight on both dimensions and did not seem to emphasize either dimension in making her similarity ratings. Follow-up analysis revealed that this teacher primarily rated the item similarities based on cognitive, not content, considerations.

## B. Example 2: Derived Proximity Data

The two-dimensional configuration in Figure 12.1 is based on an analysis of the derived proximity data in Table 12.1. This analysis is based on the nonmetric model of Eq. 1. Such an analysis yields estimates of the stimulus coordinates, $x_{ik}$, which are shown in Figure 12.1 and the first two columns of Table 12.2. However, such an analysis also yields estimates of the monotone function, $\tau$ in Figure 12.2. This mapping shows the relationship between the original data, $\delta_{ij}$, and the transformed data $\tau(\delta_{ij})$ to which the distances have been fitted in a least-squares fashion. In this case, the function is monotonically decreasing and shows definite nonlinearities. For the solution composed of the coordinates in Figure 12.1 and the transformation in Figure 12.2, STRESS = .0004, suggesting a near perfect least-squares fit between the transformed data $\tau(\delta_{ij})$ and the distances computed from the empirical coordinates in Table 12.2.

## C. Constrained Multidimensional Scaling

The data in Table 12.1 provide a convenient illustration of constrained MDS. In a constrained MDS analysis, the coordinates, $x_{ik}$, and/or the monotone transformation $\tau$ are constrained to satisfy some hypothesis, and the fit of the model is examined when these constraints are applied. A constrained analysis differs from a confirmatory analysis in that, in a constrained analysis, the fit measure has no known distribution under the null hypothesis. Therefore, the hypothesis cannot be formally accepted or rejected at some specified alpha level.

Holland (1973) proposed a two-dimensional, hexagonal model of these data. In the strictest form of the model, a form more strict than Holland probably intended, the six scales represent evenly spaced points along a circle in the order R-I-A-S-E-C. In Figure 12.1, the points fall in the order R-I-A-S-E-C along a two-dimensional contour, but the contour is hardly a perfect circle, and the points are definitely not evenly spaced. Would the stronger form of the model fit the data equally well? The last two columns of Table 12.2 give the coordinates of six points evenly spaced along a perfect circle. Using SYSTAT (Wilkinson & Hill, 1994) the nonmetric model of Eq. 1 was fitted to the data holding the coordinates at the values specified in the last two columns of Table 12.2 while estimating the transformation $\tau$ freely. The resulting fit statistic was STRESS = .1480, as compared to

STRESS = .0004 when the coordinates and the transformation were both estimated freely. In our judgment, relaxing the constraints leads to a marked improvement in fit. This in turn leads to the conclusion that the Holland model is supported in broad outline—points arrayed along a contour in two-dimensions in the order R-I-A-S-E-C, but it does not support the particulars of the model in its strongest form—points evenly spaced around a perfect circle.

## V. MULTIDIMENSIONAL SCALING AND OTHER MULTIVARIATE TECHNIQUES

Comparing multivariate techniques is complicated by the fact that the boundaries between them are often unclear. For instance, cluster analysis, factor analysis, and MDS can all be defined so broadly that any one encompasses the other two. For our purposes, the defining features of MDS are distance assumptions and continuous dimensions. That is, MDS attempts to represent proximity data in terms of distances between stimulus points in a continuous space. Our goal is to compare and contrast the kinds of research questions addressed by MDS.

### A. Cluster Analysis

MDS and cluster analysis are both methods applied to direct or derived proximity data defined over pairs of objects for the purpose of studying the structure of those objects. As we have defined the term here, MDS represents the structure in terms of continuous dimensions, whereas cluster analysis portrays the structure in terms of discrete stimulus groupings.

At times, cluster analysis and MDS have been viewed as competing methods. Often, however, the two methods have been used together in a complementary fashion, as in Figure 12.4, because the stimuli vary both in terms of discrete (or qualitative) features and continuous dimensions. For instance, in studies of kinship relationship terms (son, daughter, grandparent, etc.), the terms vary along a continuous dimension of age (son vs. father) and a qualitative gender feature (son vs. daughter).

In analyzing proximity data, either direct or derived, cluster analysis is more appropriate than MDS when the primary research purpose is to discover qualitatively different subgroups of stimuli (e.g., Sireci, Robin & Patelis, 1999). On the other hand, MDS is more appropriate when the purpose is to discover continuous dimensions on which the stimuli can be scaled. In many cases, however, the two procedures are used in a complementary fashion to represent continuous and discrete attributes along which stimuli vary.

## B. Exploratory Factor Analysis

Both MDS and factor analysis portray structure in terms of coordinates along continuous axes, either dimensions or factors. In factor analysis, however, the coordinates are assumed to combine multiplicatively in a scalar products fashion [e.g., $\delta_{ij} = f(\Sigma_k x_{ik} x_{jk})$] to reproduce the proximity data, whereas in MDS, the coordinates are assumed to combine according to some distance function, such as Eq. 1.

In recent years, MDS (and cluster analysis) has been heavily favored over factor analysis as a method for analyzing direct proximity data. Factor analysis, particularly variants of components analysis, were used in early studies of ratio-scaled similarity judgments (e.g., Ekman & Sjoberg, 1965; Micko, 1970). Possibly because the ratio-scaled judgments for which these models were designed are employed less commonly in psychological research, the more factor-like scalar product models have received less attention. The MDS distance model became the dominant method for studying direct proximity judgments. However, a recent analysis of direct proximity data using a principal components analysis may reawaken interest in analyses based on assumptions other than distance assumptions (Dix & Savickas, 1995).

In practice, components analysis tends to yield a somewhat more complex solution (more components than dimensions; Davison, 1985), and the choice between the two methods can depend, in part, on the interpretability of any extra components. Furthermore, MDS dimensions often have a contrastive character not found in components (or factors) after some rotation to simple structure. For instance, in their study of career development tasks, Dix and Savickas (1995) found a "networking with colleagues" and a "keeping current in the field" category, each appearing as a separate, rotated component. Had these same data been analyzed with MDS, the solution may have contained a dimension contrasting "networking with colleagues" at the positive end of the dimension versus "keeping current in the field" at the negative end, suggestive of the trade-off that workers must make between types of tasks when time or other resources are limited. In our view, this contrastive character of MDS dimensions can make them a more suitable vehicle for studying some types of questions, such as those involving trade-offs.

Whereas MDS has been the dominant method for studying direct proximity judgments, factor analysis has been dominant in the study of derived proximity measures, at least when the derived proximity measures are correlation coefficients and covariances. Because the MDS model contains no counterpart to the uniquenesses of factor analysis, comparative studies have often examined MDS and components analysis. Some studies have attempted to relate the factors of factor analysis to the dimensions of MDS (Davison, 1985; Silverstein, 1987); others have related the factors to regions or clusters in the MDS space, rather than to the dimensions per se (e.g., Guttman & Levy, 1991). Comparisons of components and MDS representations for the same data have led to three conclusions.

First, because it assumes a distance function, rather than scalar products, even metric MDS tends toward simpler representations of the structure (i.e., fewer dimensions than factors) (Davison, 1985; Silverstein, 1987), and nonmetric MDS tends toward even simpler representations yet. Second, in many (but not all) data sets, the unrotated components analysis tends to contain a general factor that is missing from the MDS representation of the data. Unrotated components beyond the first correspond to MDS dimensions (Davison, 1985; Silverstein, 1987). Finally, an MDS analysis rests on fundamentally different assumptions about the original subjects-by-variables data matrix (Davison & Skay, 1991). As a result, Davison argues (Davison, Gasser, & Ding, 1996; Davison, et al., 1999) that MDS and factor analysis address different questions. MDS is a method for studying the major profile patterns in a population; factor analysis is a method for studying latent constructs.

## C. Item Response Theory

Like MDS, item response theory (IRT) is a scaling model designed to locate stimuli (usually test items) and respondents (usually test takers) in a common metric space. Most IRT applications posit a unidimensional space, although multidimensional IRT models also exist (Hambleton, Robin, & Xing, chapter 19, this volume). Whereas MDS has been used to study data structures in a variety of domains, IRT is used primarily in psychological and educational testing for studying items and individual differences. There are relatively few studies relating the two scaling methods, although both Fitzpatrick (1989) and Chen and Davison (1996) found that, when applied to the same data, MDS stimulus coordinates and IRT location parameters can be virtually identical (after linear transformation).

In IRT, one must first address the question of dimensionality; that is, the number of abilities or traits underlying responses to items. MDS has been suggested as one method of addressing the item dimensionality issue (Chen and Davison, 1996; Li, 1992; Reckase, 1982). To use MDS in this fashion, one must first choose a derived proximity measure and apply MDS to the resulting items-by-items proximity matrix. For the most part, the proximity measures have been chosen in an ad hoc fashion. Future research requires more careful justification of the proximity measure either through an explicit model for the item data (Chen & Davison, 1996) or based on simulation research (e.g., De Ayala & Hertzog, 1991; Li, 1992).

## D. Circumplexes

MDS has been used to study hypothesized circumplex structures, at least in the early phases of a research program (Gurtman, 1994; Rounds et al., 1979). As an exploratory approach it seems better suited to the study of circumplex structures than factor analysis, because exploratory factor analyses often reveal the circumplex only when applied in an unorthodox fashion that involves throwing out one factor (often the one that accounts for the largest amount of variance) and that avoids any traditional rotation to simple structure. In the early stages of a research program, MDS has the advantage that it tends to yield a two-dimensional, roughly circular structure for data that satisfy any of the various models leading to a circumplex. The analysis can be successful in revealing the circumplex even if the researcher has little understanding of the underlying process. As the research program proceeds, however, often a more confirmatory approach becomes useful as the researcher begins testing more specific hypotheses about the exact nature of the underlying structure leading to the circumplex. Such a progression from exploratory MDS to confirmatory structural equations modeling can be seen in Round's work on the circumplex structure of vocational interests (i.e., from Rounds et al., 1979, to Tracey & Rounds, 1993), and is described in Tracey (chapter 22, this volume).

## VI. CONCLUDING REMARKS

### A. Statistical Developments

The past decade has seen continuing development of constrained MDS approaches, analyses that fit the data with coordinates constrained to satisfy a specified hypothesis (Heiser & Groenen, 1997; Heiser & Meulman, 1983; Borg & Groenen, 1997). We look forward to continued development of such constrained approaches. More importantly, however, we hope the algorithms for constrained MDS become more widely available and begin to see wider use.

Most MDS analyses do not include standard errors on parameter estimates. The development of such estimates would greatly aid researchers in interpreting the precision of their solutions. Although the earliest attempts at estimation of standard errors followed a maximum likelihood approach (e.g., Ramsay, 1977), later authors (Ding et al., 1997; Heiser & Meulman, 1983; Weinberg, Carroll, & Cohen, 1984) have been exploring methods based on the jackknife and bootstrap methods. These seem particularly promising because they require no distributional assumptions, are readily adapted to metric or nonmetric algorithms, and avoid asymptotics which seem to fail in individual differences applications of MDS (Weinberg et al., 1984). The computational intensity of the bootstrap and jackknife approaches will pose less and less of a problem as computing power increases.

The fundamental nonmetric assumption of MDS—data monotonically related to distances except for error—seems minimal. Yet there are theoretical arguments to suggest that it is too strict. For instance, Heffner (cited in Cohen, 1973; Zinnes & McKay, 1983) presented a very plausible error model for direct proximity data whose nonmetric extension would be $\tau(\delta_{ij}) = [\Sigma_k(x_{ik} - x_{jk})^2 + \sigma_i^2 + \sigma_j^2 + e_{ij}]$, where $\sigma_i^2$ and $\sigma_j^2$ are error variances associated with the locations of stimuli i and j respectively. Relaxing the unreasonable assumption of equal error variances across tests in Davison and Skay's (1991) vector model leads to a similar form. Even if $\tau$ is a monotone function, the data are no longer monotonically related to the distances, and hence the fundamental assumption of nonmetric MDS is violated.

## B. Applications

Multiattribute scaling is a common way of developing a multidimensional representation of stimulus perceptions. In this method, the researcher specifies dimensions along which respondents will rate the stimuli, and then the respondent rates the stimuli along one dimension at a time. This approach makes two assumptions. First, it assumes the researcher (rather than respondents) can and should specify the dimensions along which stimuli are judged. Such an assumption is unwise when the researcher's purpose is to study the dimensions salient to the respondent and individual differences in dimension saliences. Second, it assumes that, based on their experience and global perception of each stimulus, respondents can abstract each dimension specified by the researcher and judge that dimension independently of all other dimensions. At times, this assumption may be doubtful (Bock & Jones, 1968; Johnson, 1987).

In making direct proximity judgments, the respondent makes a global judgment over dimensions that are respondent selected. Therefore, direct proximity judgments are ideal for studying the perceptual dimensions salient to the respondent even when the respondent may not be able to cleanly abstract each dimension from others (Johnson, 1987). Consequently, MDS based on direct proximity judgments will remain a major tool in various types of perception research: psychophysics, social perception, and so on. Its potential for the study of perceptual interactions between dimensions has yet to be fully exploited.

The major obstacle to wider application of MDS with direct proximity data is the large number of judgments required by paired comparisons tasks. Sorting tasks reduce respondent labor, but yield only a dichotomous raw proximity matrix, a particularly severe limitation for individual differences applications of MDS.

MDS research with derived proximity data will continue to be so varied that it is difficult to characterize. MDS based on derived proximity data has been used as a general data reduction technique, as a method of examining formal structural hypotheses, and as a tool for studying the dimensionality of a data structure. Where possible, future applications to derived proximity data need to better justify the choice of proximity measure using previous simulation literature or an explicit model of the original subjects-by-variables matrix.

MDS will continue to be a major tool of perception research. With exploratory factor analysis and cluster analysis as alternatives, MDS will also remain a significant method for studying derived proximity measures. If the past few years are any indication, statistical developments will continue on methods for constrained analyses, methods for estimating the precision (standard errors) of parameter estimates, explicit models for derived proximity matrices, methods for reducing the respondent labor in studies using direct proximity judgments, and methods for estimating scale values in models, such as Hefner's (cited in Zinnes & McKay, 1983), in which the data are nonmonotonically related to distances.

## REFERENCES

Bock, R. D., & Jones, L. V. (1968). *The measurement and prediction of judgment and choice.* San Francisco, CA: Holden-Day.

Borg, I., & Groenen, P. (1997). *Modern multidimensional scaling: Theory and applications.* New York: Springer-Verlag.

Campbell, D. P., & Hansen, J. C. (1985). *Strong-Campbell Interest Inventory.* Palo Alto, CA: Stanford University Press.

Chen, T., & Davison, M. L. (1996). A multidimensional scaling, paired comparisons approach to assessing unidimensionality in the Rasch model. In G. Engelhard & M. Wilson (Eds.), *Objective measurement: Theory into practice* (Vol. 3, 309–334). Norwood, NJ: Ablex.

Cohen, H. S. (1973). *A multidimensional analogy to Thurstone's law of comparative judgment.* Unpublished doctoral dissertation, University of Illinois at Urbana Champaign, Champaign, IL.

Cohen, H. S., & Davison, M. L. (1973). Jiffy-scale: A FORTRAN IV program for generating Ross-ordered paired comparisons. *Behavioral Science, 18,* 76.

Collins, L. M. (1987). Deriving sociograms via asymmetric multidimensional scaling. In F. W. Young & R. M. Hamer (Eds.), *Multidimensional scaling: History, theory and applications,* (pp. 179–196). Princeton, NJ: Lawrence Erlbaum.

Commandeur, J. J. F., & Heiser, W. J. (1993). *Mathematical derivations in the proximity scaling (PROXSCAL) of symmetric data matrices* (Tech. Rep. No. RR-93-03). Leiden, The Netherlands: Department of Data Theory, University of Leiden.

Davison, M. L. (1985). Multidimensional scaling versus components analysis of test intercorrelations. *Psychological Bulletin, 97,* 94–105.

Davison, M. L. (1991). *Multidimensional scaling.* Malabar, FL: Krieger.

Davison, M. L., Gasser, M., & Ding, S. (1996). Identifying major profile patterns in a population: An exploratory study of WAIS and GATB patterns. *Psychological Assessment, 8,* 26–31.

Davison, M. L., Kuang, H. & Kim, S. (1999). The structure of ability profile patterns: A multidimensional scaling perspective on the structure of intellect. In P. L. Ackerman, P. C. Kyllonen, & R. D. Roberts (Eds.), *Learning and individual differences: Process, trait, and content determinants*. Washington, DC: APA Books.

Davison, M. L., & Skay, C. L. (1991). Multidimensional scaling and factor models of test and item responses. *Psychological Bulletin, 110*, 551–556.

De Ayala, R. J., & Hertzog. M. A. (1991). The assessment of dimensionality for use in item response theory. *Multivariate Behavioral Research, 26*, 765–792.

Deville, C. W. (1996). An empirical link of content and construct validity evidence. *Applied Psychological Measurement, 20*, 127–139.

Ding, S., Bielinski, J., Davison, M. L., Davenport, E. C., Jr., Kuang, H., & Kim, S. (1997). *SPSS[R] and SAS[R] Macros for Estimating Standard Errors of Scale Values in Nonmetric Multidimensional Scaling based on Euclidean Distance Proximities from a Persons by Variables Matrix* (Tech. Rep. R999B40010.B). Minneapolis, MN: Department of Educational Psychology, University of Minnesota.

Dix, J. E., & Savickas, M. L. (1995). Establishing a career: Developmental tasks and coping responses. *Journal of Vocational Behavior, 47*, 93–107.

Dong, H. (1985). Chance baselines for INDSCAL's goodness of fit index. *Applied Psychological Measurement, 9*, 27–30.

Ekman, G., & Sjoberg, L. (1965). Scaling. *Annual Review of Psychology, 16*, 451–474.

Fitzpatrick, S. J. (1989). *Rasch model parameter estimation: A comparison of IRT and nonmetric multidimensional scaling methods.* Paper presented at the annual meeting of the American Educational Research Association, San Francisco.

Gurtman, M. B. (1994). The circumplex as a tool for studying normal and abnormal personality: A methodological primer. In S. Strack & M. Lorr (Eds.), *Differentiating normal and abnormal personality* (243–263). New York: Springer.

Guttman, L., & Levy, L. (1991). Two structural laws for intelligence tests. *Intelligence, 15*(1), 79–104.

Heiser, W. J., & Groenen, P. J. F. (1997). Cluster differences scaling with a within-cluster loss component and a fuzzy successive approximation strategy to avoid local minima. *Psychometrika, 62*, 63–83.

Heiser, W. J., & Meulman, J. J. (1983). Constrained multidimensional scaling including confirmation. *Applied Psychological Measurement, 7*, 381–404.

Hoffman, D. L., & Perreault, W. D. (1987). Consumer preference and perception. In F. W. Young & R. M. Hamer (Eds.), *Multidimensional scaling: History, theory and applications* (pp. 199–218). Princeton, NJ: Lawrence Erlbaum.

Holland, J. L. (1973). *Making vocational choices: A theory of careers.* Englewood Cliffs, NJ: Prentice Hall.

Howard, D. V., & Howard, J. H. (1977). A multidimensional scaling analysis of the development of animal names. *Developmental Psychology, 13*, 108–113.

Johnson, L. J. (1987). *Multidimensional scaling of speech naturalness in stutterers.* Unpublished doctoral dissertation, University of Minnesota, Minneapolis, MN.

Kruskal, J. B. (1964). Multidimensional scaling by optimizing goodness of fit to a nonmetric hypothesis. *Psychometrika, 29*, 1–27.

Kruskal, J. B., & Wish, M. (1978). *Multidimensional scaling.* Beverly Hills, CA: Sage.

Kruskal, J. B., Young, F. W., & Seery, J. B. (1978). How to use KYST-2, a very flexible program to do multidimensional scaling and unfolding (Tech. Rep.). Murray Hill, NJ: Bell Labs.

Li, X. (1992). *An investigation of proximity measures with dichotomous item response data in nonmetric multidimensional scaling.* Unpublished doctoral dissertation, University of Minnesota.

MacCallum, R. (1981). Evaluating goodness of fit in nonmetric multidimensional scaling by ALSCAL. *Applied Psychological Measurement, 5*, 377–382.

Micko, H. C. (1970). A "halo-" model for multidimensional ratio scaling. *Psychometrika, 35,* 199–227.

Miller, K., & Gelman, R. (1983). The child's representation of number: A multidimensional scaling analysis. *Child Development, 54,* 1470–1479.

Napior, D. (1972). Nonmetric multidimensional techniques for summated ratings. In R. N. Shepard, A. K. Romney, & S. B. Nerlove (Eds.), *Multidimensional scaling: Volume 1: Theory* (pp. 157–178). New York: Seminar Press.

Ramsay, J. O. (1977). Maximum likelihood estimation in multidimensional scaling. *Psychometrika, 42,* 241–266.

Ramsay, J. O. (1981). MULTISCALE. In S. S. Schiffman, M. L. Reynolds, & F. W. Young (Eds.), *Introduction to multidimensional scaling* (pp. 389–405). New York: Academic Press.

Ramsay, J. O. (1982). Some statistical approaches to multidimensional scaling data. *Journal of the Royal Statistical Society, 145,* 285–312.

Ramsay, J. O. (1991). *MULTISCALE-II manual: extended version.* Montreal, Quebec: McGill University.

Reckase, M. D. (1982). *The use of nonmetric multidimensional scaling with dichotomous test data.* Paper presented at the annual meeting of the American Educational Research Association, New York.

Rosenberg, S., & Kim, M. P. (1975). The method of sorting as a data-gathering procedure in multivariate research. *Multivariate Behavioral Research, 10,* 489–502.

Roskam, E. E., & Lingoes, J. C. (1981). Minissa. In S. S. Schiffman, M. L. Reynolds, & F. W. Young (Eds.), *Introduction to multidimensional scaling* (pp. 362–371). New York: Academic Press.

Ross, R. T. (1934). Optimum order for presentation of pairs in paired comparisons. *Journal of Educational Psychology, 25,* 375–382.

Rounds, J. B., Jr., Davison, M. L., & Dawis, R. V. (1979). The fit between Strong-Campbell Interest Inventory general occupation themes and Holland's hexagonal model. *Journal of Vocational Behavior, 15,* 303–315.

Schiffman, S. S., Reynolds, M. L., & Young, F. W. (1981). *Introduction to multidimensional scaling.* New York: Academic Press.

Silverstein, A. B. (1987). Multidimensional scaling vs. factor analysis of Wechsler's intelligence scales. *Journal of Clinical Psychology, 43,* 381–386.

Sireci, S. G., & Geisinger, K. F. (1992). Analyzing test content using cluster analysis and multidimensional scaling. *Applied Psychological Measurement, 16,* 17–31.

Sireci, S. G., & Geisinger, K. F. (1995). Using subject matter experts to assess content representation: A MDS analysis. *Applied Psychological Measurement, 19,* 241–255.

Sireci, S. G., Robin, F., & Patelis, T. (1999). Using cluster analysis to facilitate standard setting. *Applied Measurement in Education, 12,* 301–325.

Sireci, S. G., Rogers, H. J., Swaminathan, H., Meara, K., & Robin, F. (1997). *Evaluating the content representation and dimensionality of the 1996 Grade 8 NAEP Science Assessment.* Commissioned paper by the National Academy of Sciences/National Research Council's Committee on the Evaluation of National and State Assessments of Educational Progress, Washington, DC: National Research Council.

Spence, I. (1982). Incomplete experimental designs for multidimensional scaling. In R. G. Goledge & J. N. Rayner (Eds), *Proximity and preference: Problems in the multidimensional analysis of large data sets.* Minneapolis, MN: University of Minnesota Press.

Spence, I. (1983). Monte Carlo simulation studies. *Applied Psychological Measurement, 7,* 405–426.

Takane, Y. (1981). Multidimensional successive-categories scaling: A maximum likelihood method. *Psychometrika, 46,* 9–28.

Takane, Y., Young, F. W., & DeLeeuw, J. (1977). Nonmetric individual differences multidimensional scaling: An alternating least-squares method with optimal scaling features. *Psychometrika, 42,* 7–67.

Tracey, T. J., & Rounds, J. B. (1993). Evaluating Holland and Gati's vocational interest models: A structural meta-analysis. *Psychological Bulletin, 113,* 229–246.

Tucker, L. R. (1972). Relations between multidimensional scaling and three-mode factor analysis. *Psychometrika, 37,* 3–27.

Weinberg, S. L., Carroll, J. D., & Cohen, H. S. (1984). A quasi-nonmetric method for multidimensional scaling via an extended Euclidean model. *Psychometrika, 48,* 575–595.

Wilkinson, L., & Hill, M. (1994). SYSTAT: The system for statistics. Evanston, IL: SYSTAT, Inc.

Young, F. W., & Hamer, R. M. (1987). (Eds.). *Multidimensional scaling: History, theory and applications.* Princeton, NJ: Lawrence Erlbaum.

Young, F. W., & Harris, D. F. (1993). Multidimensional scaling. In M. J. Norusis (Ed.). *SPSS for windows: Professional statistics* (computer manual, version 6.0) (pp. 155–222). Chicago, IL: SPSS, Inc.

Zinnes, J. L., & McKay, (1983). Probabilistic multidimensional scaling: Complete and incomplete data. *Psychometrika, 48,* 27–48.

# 13

# TIME-SERIES DESIGNS
# AND ANALYSES

**MELVIN M. MARK**

*Department of Psychology, Pennsylvania State University,*
*University Park, Pennsylvania*

**CHARLES S. REICHARDT**

*Department of Psychology, University of Denver, Denver, Colorado*

**LAWRENCE J. SANNA**

*Department of Psychology, Washington State University, Pullman, Washington*

*And time, that takes survey of all the world*
—William Shakespeare (*Henry IV, Part 1, Act 5, Scene 4*)

Whether time takes survey of all the world, as Shakespeare claims, we mortals may never know. Nonetheless, time is an important element in research, basic and applied, as well as in everyday ways of coming to know about the world. For example, the fundamental concepts of change and causality presume processes that play out and can be observed, at least indirectly through their consequences, over time.

One very important, though generally underused, way to assess change and causality is with time-series data. A *time series* is simply a set of repeated observations of a variable on some entity or unit, where the number of repetitions is relatively large. Generally the observations are repeated at a constant interval (such as yearly, monthly, weekly, or daily), but this is not always necessary. Examples of time-series data include the monthly birth rate for the state of Michigan, the annual Gross National Product of the United States, the weekly number of production errors in a factory, and the number of events of a specified type entered daily into a diary.

## I. ALTERNATIVE PURPOSES OF TIME-SERIES STUDIES

Time-series data can be used for a variety of purposes. In this section we briefly describe four different types of studies that motivate the use of time-series data.

First, time series can be used simply to *assess the pattern of change over time*. For example, Sanna and Parks (1997) were interested in whether, and to what extent, there had been a change in the level of research on groups in social and organizational psychology (see also Moreland, Hogg, & Hains, 1994). Toward this end, Sanna and Parks tallied on an annual basis the number and proportion of published articles that studied groups and the number and proportion of pages allocated to research on groups in a set of academic journals. Their primary question was whether the level of published research on groups had changed over the time period they observed. In another example, Hofmann, Jacobs, and Geras (1992) examined individual differences in the pattern of performance across time. They assessed whether the performance of some workers (in this case, baseball players) tended to improve over time while the performance of others declined. Although both Sanna and Parks (1997) and Hofmann et al. (1992) focused on *trend* (i.e., on the overall upward or downward slope of the data in a time series), a time-series assessment of change can alternatively focus on *seasonality*. Seasonality, also known as periodicity, involves the degree to which there are regular cycles of ups and downs associated with conventional time periods. For example, Hopkins (1994) assessed whether there are weekly patterns in self-reported anxiety among individuals who suffered from Generalized Anxiety Disorder.

A second purpose of time-series studies is *forecasting,* that is, predicting future values of a variable that has been measured in time-series form. For example, predictions about an economic variable such as unemployment are often based on time-series data about unemployment and perhaps other indicators (e.g., Montgomery, Zarnowitz, Tsay, & Tiao, 1998). Although quite common in economics, finance, and related business fields, forecasting is not as common in most other social sciences.

A third purpose of time-series studies is to *assess the effect of a treatment or intervention.* Such studies take the form of an interrupted time-series (ITS) design (Campbell & Stanley, 1966; Cook & Campbell, 1979), wherein a time series of observations is collected over a period of time, a treatment is introduced, and the series of observations is continued. In a frequently cited example of the ITS design (e.g., McCain & McCleary, 1979; McCleary & Hay, 1980), McSweeney (1978) examined the effect of introducing a 20-cent fee for calls to local directory assistance in Cincinnati on the number of such calls made. Figure 13.1 presents a time-series plot of the average daily number of calls to local directory assistance each month in Cincinnati. The first observation is for the month of July 1967, when

52,100 calls were placed to directory assistance each day on average, and the last observation in the time series is for December 1976, when 24,100 calls were placed to directory assistance each day on average. Note that we have omitted the first third of the observations from McSweeney (1978), making a "cleaner" example (but with no change in substantive conclusions). Starting in March 1974, Cincinnati Bell charged 20 cents per call placed to directory assistance where previously these calls had been free. The sudden "interruption" in the time series of observations, occurring when the 20-cent charge for calls (i.e., the treatment) was introduced, suggests that the fee had an abrupt downward effect on the number of calls placed to directory assistance.

A fourth purpose of time-series studies is to *assess the relationship between two (or more) time-series variables.* For example, past studies have investigated the relationship, over time, between advertising expenditures and sales of a product (e.g., Helmer & Johannson, 1977) and between unemployment and suicide rate (Brenner, 1973). In many cases, the covariation between two time series is studied in an attempt to determine the causal relationship between the two processes. For example, Brenner (1973) was interested not simply in whether unemployment and suicide rates covaried over time, but in whether changes in the unemployment rate caused changes in the rate of suicide. Similarly, time-series research on advertising and sales is often motivated by an interest in assessing the effect of advertising expenditures on sales. However, as we shall see, considerable caution is generally called for when causal inferences are based on findings about covariation between two (or more) time-series variables.

In the remainder of the present chapter, we describe statistical methods appropriate for time-series data. To provide a concrete context, we will focus on ITS designs, such as the McSweeney (1978) study illustrated in Figure 13.1. Although we focus on ITS designs, the issues of statistical analysis that we discuss are central to time-series studies, regardless of purpose.

Before describing statistical analyses, we should note that there are two general approaches to time-series analysis. In the present chapter, we are concerned only with approaches that analyze time-series data in the "time domain." Such approaches examine the relationship between the value of a variable at one point in time and the value of that variable at later points in time. The alternative is spectral analysis, which analyzes time-series data in the "frequency domain." That is, spectral analysis examines the extent to which a time-series variable exhibits oscillations (i.e., regular, wave-like behavior) across time at different frequencies. Spectral analysis is far more mathematically complex than the approaches we will consider in the present chapter and is far less frequently used in the social sciences. In particular, spectral analysis appears to be both most useful and most frequently used in substantive areas in which the time series of interest

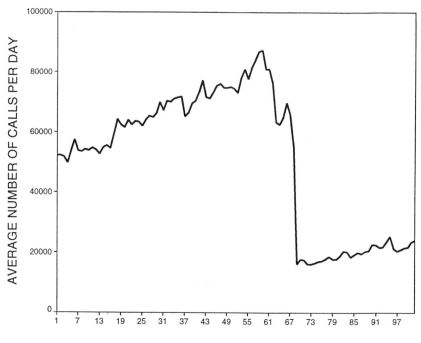

MONTH (July 1967 - December 1976)

**FIGURE 13.1** Average daily calls to local directory assistance in Cincinnati (based on partial data from McSweeney, 1978; obtained from McCleary & Hay, 1980).

includes strong, regular cyclical patterns (Kendall & Ord, 1990). For example, spectral analyses might be usefully applied to certain psychophysiological data, where the question of interest often involves the strength of certain cycles (e.g., Boutcher, Nugent, McLaren, & Weltman, 1998).

## II. THE REGRESSION APPROACH TO FITTING TRENDS

As noted above, the purpose of ITS designs is to *estimate the effect of a treatment* (such as a new fee for directory assistance) on an outcome measured in time-series form (such as the number of calls made to directory assistance). The logic of the analysis of data from a simple ITS design, such as the McSweeney (1978) study, is straightforward. In essence, the trend in the pretreatment observations is estimated and projected ahead in time to provide an indication of how the posttreatment data would have appeared if there had been no treatment. The estimate of the treatment effect is then derived from the difference, if any, between the projected and the actual posttreatment trends. Returning to the McSweeney study, the trend in the pretreatment observations appears to be close to a straight line (i.e., linear). Although more complex analyses are usually required, as we shall see in the next section, one might begin by modeling this trend with a simple linear regression equation, such as the following:

$$Y_t = \alpha + \beta_1 t + \varepsilon_t \tag{1}$$

In this equation, $Y_t$ represents the time-series observations, such as the average daily number of directory assistance calls per month, with the $t$ subscript denoting time. For example, given that the first observation in the graph in Figure 13.1 is for the month of July 1967, during which time there were 52,100 calls to directory assistance per day on average, $Y_1$ here equals 52,100. The "$\alpha + \beta_1 t$" portion of Equation 1 represents the straight line that is being fit to the pretreatment observations. This straight-line fit to the pretreatment observations is shown in Figure 13.2, which reproduces the data in Figure 13.1. The value of $\alpha$ in Equation 1 is the intercept of the line on the vertical axis (that is, the height of the line when it intersects the vertical axis), which is about 53,500. The value of $\beta_1$ is the slope of the straight line. In equation (1), $\beta_1$ is multiplied by $t$ where, as noted above,

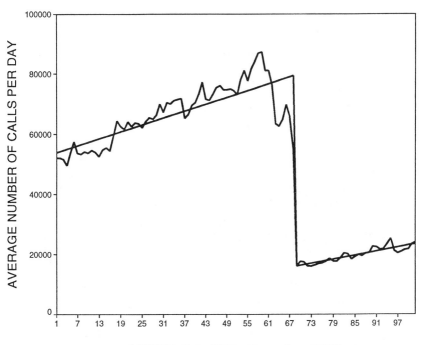

MONTH (July 1967 - December 1976)

**FIGURE 13.2**   Average daily calls, with linear trend lines fit for pretreatment and posttreatment periods.

$t$ represents "time." That is, $t$ is a variable where the value of 1 is assigned to the first observation in the time series, 2 is assigned to the second observation, and so on (in a subsequent section we discuss alternatives to such line fitting based on $t$). Finally, the $\varepsilon_t$ in Equation 1 represents the residual, disturbance, or error term. That is, for each time point $t$, $\varepsilon$ is the discrepancy between the straight line in Figure 13.2 and the actual observation.

As Figure 13.2 shows, a straight (rather than curved) line appears to fit the trend in these pretreatment observations fairly well. However, in other cases the trend in the pretreatment observations may be curvilinear rather than linear. In such instances, one could transform the data to try to make the trend linear. Alternatively, one could fit a model to the untransformed data that included quadratic, cubic, or additional polynomial components as appropriate.

So far we have been concerned with fitting the trend in the pretreatment observations. In ITS studies the primary focus, however, is on the effect of the treatment on posttreatment observations. Assume for the moment that the slope of the trend in the posttreatment observations is the same as that in the pretreatment observations. In such cases, the treatment simply lowers (or raises) the level of the observations but otherwise leaves the trend in the observations unchanged. Under these circumstances, the effect of the treatment can be estimated with the following regression model:

$$Y_t = \alpha + \beta_1 t + \beta_2 X_{1t} + \varepsilon_t. \tag{2}$$

Equation 2 is identical to Equation 1 except for the addition of the term $\beta_2 X_{1t}$. In this term, $X_{1t}$ is a variable that takes on only one of two values, either zero or one. For each *pre*treatment observation, $X_1$ is set to zero, while for each *post*treatment observation, $X_1$ is set to one. In essence, equation (2) is a regression equation where the observations ($Y_t$) are regressed onto the two variables, $t$ and $X_{1t}$. If you performed such a regression, you would obtain an estimate for $\beta_2$, which would be an estimate of the size of the treatment effect. This estimate would reveal how much the level of the trend in the time series was altered by the treatment. For example, the pretreatment observations in Figure 13.2 dropped from a level of about 79,000 just before the treatment to a level of about 16,000 after the treatment. The difference between these, $-63,000$, would be the approximate size of the value estimated by $\beta_2$ in Equation 2. This drop in the level of the time series in Figure 13.2 is represented by the abrupt downward line.

As we discuss in more detail later, a treatment might alter not just the level but also (or only) the trend of the observations. This appears perhaps to be the case in Figure 13.2, where the trend in the posttreatment observations is not quite as steep as that in the pretreatment. This could be an effect of the treatment, whereby the new fee not only reduced the absolute number of calls to directory assistance, but also reduced the rate of growth in these calls. Equation 3 allows for the treatment to bring about both a shift in level and a shift in trend:

$$Y_t = \alpha + \beta_1 t + \beta_2 X_{1t} + \beta_3 X_{2t} + \varepsilon_t. \tag{3}$$

Equation 3 is identical to Equation 2 except for the additional term $\beta_3 X_{2t}$. In this term, $X_{2t}$ is set equal to zero for each observation before the treatment is introduced; for the first observation after the treatment is introduced, the value of $X_{2t}$ is set equal to 1; for the second observation after the treatment is introduced, $X_{2t}$ is set equal to 2, and so on. In essence, Equation 3 is a regression equation where the outcome variable ($Y$) is regressed onto the variables $t$, $X_{1t}$, and $X_{2t}$. If you performed such a regression, you would obtain an estimate for $\beta_3$, which would be an estimate of the *change* in the slope of the trend in the observations from before the treatment was introduced to after the treatment was introduced. For example, if the slope in the posttreatment observations were the same as the slope in the pretreatment observations, then $\beta_3$ would equal zero. If the slope in the posttreatment observations were greater than the slope in the pretreatment observations, then $\beta_3$ would be positive and equal to the difference in the two slopes. Conversely, a negative value of $\beta_3$ would indicate that the posttreatment slope is less than the pretreatment slope. In Figure 13.2, the slope in the pretreatment observations (the solid line) is estimated to be 380. The slope in the posttreatment observations (the dashed line) is estimated to be 223. Accordingly, $\beta_3$ is estimated to be $-157$, which is the difference between 223 and 380.

## III. THE PROBLEM OF AUTOCORRELATION

In the preceding section, we discussed fitting time-series data using regression models. The standard approach to fitting regression models to data uses ordinary least squares (OLS) regression. Time-series data commonly have a characteristic that can make OLS regression inappropriate. As a consequence, more complex, and seemingly more arcane statistical analyses are often required.

Ordinary least squares regression is based on the assumption that the residuals ($\varepsilon_t$ in the preceding equations) are uncorrelated with each other. This is often a reasonable assumption when regression analysis is applied to data where each data point comes from a different individual. However, in classic time series, the data usually come from the same individual (or social aggregate) over time. As a result, the residuals or disturbances, which are represented by the $\varepsilon_t$ terms, usually persist across time periods and so are usually correlated among themselves. In particular, the residuals from data points that are close together in time are likely to be more similar than the residuals from data points that are far apart in time. For example, the residuals from the number of calls to directory assistance in July is likely to be similar to the residuals from the number of calls in June and August, and more similar with these adjacent residuals than with the residuals from the number of calls in January or December. This pattern of correlation between adjacent residuals is called *autocorrelation* or *serial dependency.*

Such autocorrelation violates the assumption of uncorrelated residuals in OLS regression. Violations of this assumption do not bias the estimates of treatment effects in OLS regression. However, autocorrelation does bias estimates of the standard errors of these estimates, which can seriously invalidate hypothesis tests and confidence intervals. In principle, the bias can be either positive or negative, so that a statistical inference can be either too liberal or too conservative. In practice, however, autocorrelation usually causes test statistics to be artificially inflated. Indeed, McCain and McCleary (1979) suggest that the magnitude of autocorrelation observed in practice can inflate a test statistic by a factor of three or even more. Thus, autocorrelation can lead researchers to reject the null hypothesis far more often than they should, and it can similarly lead to narrower confidence intervals than are appropriate. As a result, time-series analysis requires techniques, unlike OLS regression, that can account for the effects of autocorrelation.

## IV. THE AUTOREGRESSIVE INTEGRATED MOVING AVERAGE APPROACH TO MODELING AUTOCORRELATION

Autoregressive Integrated Moving Average (ARIMA) models are commonly used in the social sciences to take account of the effects of autocorrelation in time-series analysis. Huitema and McKean (1998) note that ARIMA models are popular at least in part because they have been strongly advocated by influential writers (e.g., Box, Jenkins, & Reinsel, 1994; Cook & Campbell, 1979; Glass, Willson, & Gottman, 1975; McCain & McCleary, 1979), and because they are available in popular software packages. Nonetheless, ARIMA models are relatively unfamiliar to many researchers in the social sciences, presumably because time-series analysis has been relatively infrequently used, at least until recently.

### A. Autoregressive Integrated Moving Average Models

The term *ARIMA* stands for a family of models, where each member of the family represents a different structure for the autocorrelation in a time series. The *AR* in ARIMA stands for autoregressive, and denotes a model in which each residual is directly dependent on previous residuals. A simple, or "first-order" AR structure is commonly noted as AR(1). When applied to the error term (or residual) of a model (such as in Equations 1–3), AR(1) fits each residual ($\varepsilon_t$) as a function of the previous residual ($\varepsilon_{t-1}$) plus an independent disturbance or "random shock," ($U_t$). In algebraic form, this model can be written as

$$\varepsilon_t = \phi_1 \varepsilon_{t-1} + U_t, \tag{4}$$

where $\phi_1$ is in essence the coefficient from regressing $\varepsilon_t$ onto $\varepsilon_{t-1}$.

The disturbances (i.e., the $U_t$s) in an AR model are persistent (i.e., they continue to have an effect over time), although that effect diminishes as the length of time increases, with the speed of the drop-off being an exponential function of $\phi_1$. As an example, imagine a time series consisting of the end-of-month checking account balances of a graduate student whose basic income and expenses are fairly steady over time. This balance will fluctuate somewhat, based on the happenstance of a particular month. For example, an unexpected big check from Grandma will lead to an unusually large balance at the end of that month. Moreover, the effect of this unusually high "disturbance" will tend to persist over time, having some effect the following month, less effect the next month, and so on. In at least some cases when time series follow AR models, it is possible to think of actual physical mechanisms through which the effect of a disturbance persists over time. In AR models, however, the storage mechanism is imperfect or "leaky," so that a disturbance is not completely carried over indefinitely (as we expect anyone who has gotten a big unexpected check has found to be the case months later).

For higher-order AR models, the equations include terms that represent the linkage between the current residual and even earlier ones. For example, a second-order AR model, represented as AR(2), can be written as

$$\varepsilon_t = \phi_1 \varepsilon_{t-1} + \phi_2 \varepsilon_{t-2} + U_t. \tag{5}$$

In this model, the current residual ($\varepsilon_t$) is taken to be a function of the previous residual ($\varepsilon_{t-1}$) and of the residual before that ($\varepsilon_{t-2}$). The size of the coefficients ($\phi_1$ and $\phi_2$) represent the magnitude of the effect of previous residuals on current ones. When AR(2) models occur, our experience is that the $\phi_1$ coefficient is usually larger and positive, whereas the $\phi_2$ coefficient is smaller and negative. In at least some contexts, this is sensible, in that the negative $\phi_2 \varepsilon_{t-2}$ term can arise from a compensatory or homeostatic mechanism. For example, in the case of a graduate student's checking account, low balances one month may lead subsequently to some belt-tightening whereas, in contrast, an unusually high balance may stimulate some discretionary spending. Higher-order AR models, such as AR(3) and AR(4), would include additional autoregressive terms but are rare in practice (McCleary & Hay, 1980).

Equations (4) and (5) are written in terms of $\varepsilon_t$. Much of the writing about ARIMA models instead specifies the equations in terms of $Y_t$, the time-series observations themselves (e.g., calls to directory assistance). In this case, equation (4) would be presented as

$$Y_t = \phi_1 Y_{t-1} + \varepsilon_t. \qquad (6)$$

However, autocorrelation is a problem for standard statistical analyses, not when it occurs in the actual time series, but when it occurs in the error term. If, for example, the regression model in equation (3) is applied to the time-series data (e.g., calls to directory assistance), and the residuals are not autocorrelated, then ARIMA modeling is not necessary. This has long been realized but, as Huitema and McKean (1998) point out, many advocates of ARIMA modeling write as though autocorrelation in $Y_t$ were the problem and even check for autocorrelation in $Y_t$ rather than in $\varepsilon_t$. However, note that if one's model—unlike the models in equations 1–3—includes only one parameter for the mean, these two formulations are equivalent.

In AR models, the residuals in the data are influenced by one or more previous residuals. In contrast, in moving average (MA) models the residuals in the data are influenced by one or more previous random shocks. For example, the equation for a first-order MA model, noted as MA(1), can be written as

$$\varepsilon_t = U_t - \theta_1 U_{t-1}, \qquad (7)$$

where $U_t$ is a new random shock, $U_{t-1}$ is the random shock from the previous observation, and $\theta_1$ is essentially the regression coefficient when $\varepsilon_t$ is regressed on $U_{t-1}$. In parallel fashion, a second-order MA model, MA(2), would have an additional MA term (and would be written as $\varepsilon_t = U_t - \theta_1 U_{t-1} - \theta_2 U_{t-2}$), and so on for higher-order models.

For AR models, as we saw previously, the disturbances (i.e., the $U_t$s) have effects that persist over time (though with increasingly smaller influence). For MA models, in contrast, the effect of a disturbance is short-lived. In the MA(1) model, a disturbance at one time is felt only at that time point and at the very next time point. Moreover, the lagged effect of the disturbance is only partial, depending on the size of $\theta_1$.

The I or integration component of ARIMA models (which would better be termed "difference") represents the number of times a series is differenced. Differencing refers to subtracting the previous value of the time-series observations from the current value and can be represented as:

$$\nabla Y_t = Y_t - Y_{t-1}. \qquad (8)$$

Many approaches to time-series analysis, including ARIMA modeling, require that the time-series data that are to be analyzed be (or be made to be) "stationary." Stationarity means that characteristics (e.g., the mean, variance, autocorrelation) of a time series are stable (in expected value) over time. When a time series has a long-term upward or downward trend in level (as in Figure 13.1), for example, the time series is nonstationary, and stationarity must be induced. Within the ARIMA approach, differencing is often used instead of regression equations (e.g., equation 3) to remove trends in level. However, differences of opinion exist about when it is best to model trends in level using differencing and when it is best to model trends in level using regression models.

Regression models are generally to be preferred to differencing when the trend is "deterministic" and differencing when it is "stochastic." In essence, a trend is stochastic if it is the direct result of a random process, such as a so-called random-walk process. To understand a random-walk process, imagine a person who spends each day in a casino gambling on a modified game of roulette, where the object is to bet on red or black, with half of the slots being each color. Imagine further that the game is fair and that the person always places the same number of bets, each for the same amount. The amount of money won or lost each day would then be purely random, and most days the outcome would be about zero, whereas some days would see a large loss or gain. Finally, imagine that each day's winnings or losses go into or are drawn from an account (with no other withdrawals or deposits). The time series for this account would be a random walk. Over the long haul, this time series of observations would center around a common mean, because the expectation would be of zero winnings. However, it might wander far from that mean for an extended period of time. More importantly, a particular stretch of the time series could appear to have a linear (or other) trend. When a trend may plausibly exist because we are examining a limited section of a random-walk process, differencing is appropriate. The differenced series, as shown in Equation 8, consists of change scores so that, in our hypothetical roulette example, $\nabla Y_t$ is the difference between yesterday's balance and today's, which is equal to today's winnings or losses.

When, instead, a trend is likely to be the result of a nonrandom factor, such as population growth that would increase calls to directory assistance, regression models are to be preferred. The variable causing the trend (e.g., population growth) or a proxy can often be used as a predictor. When these are unavailable in time-series form, then linear or simple curvilinear models might be fit in their stead. Some caution should be exercised, however, as it is possible to fit curves where there is no theoretical reason for doing so. Another cautionary note is that some time-series computer programs appear to fit AR and MA parameters directly to the observed time series, $Y_t$, and do not allow these parameters to be fit to the error term from a regression model. This makes the regression approach to trend infeasible with that software. Users can check for this possibility by comparing the size of the treatment effect estimate from a standard regression analysis with that from the specialized time-series software. Assuming that no differencing has been applied, if the coefficients diverge greatly across the two analyses, the software is probably fitting the ARIMA model to the observations rather than to the errors.

When there is any question about whether to difference a time series or to model it with regression, the conservative approach is to do both and compare the conclusions that result from each (see the discussion of "bracketing" below). It is also possible to use both differencing and regression models combined.

The different combinations of autoregressive, integrated, and moving average components comprise the family of ARIMA models. The extent to which each component is included in the ARIMA model for a given set of data is generally represented by the notation ARIMA($p,d,q$), where $p$, $d$, and $q$ refer, respectively, to the order of the AR, I, and MA elements. Thus, for example, an ARIMA(1,0,0) is a first-order autoregressive model with no differencing and no moving average components. ARIMA models can also be expanded to account for seasonality (see Box, Jenkins, & Reinsel, 1994; Kendall & Ord, 1990). The notation ARIMA($p,d,q$)($P,D,Q$)$_s$ is typically used to denote models that include seasonal components, with $p$, $d$, and $q$ retaining the same meaning as before, whereas $P$, $D$, and $Q$ indicate, respectively, the order of the AR, I, and MA components of the seasonal model, and with $s$ specifying the length of the seasonal period. For instance, an ARIMA(1,0,0)(1,0,0)$_{12}$ model has one regular autoregressive term and one seasonal autoregressive term, where the latter has a period length of 12, as would be expected with monthly data. Although there are some added complexities, for the most part it suffices to conceptualize the seasonal components as seasonally lagged counterparts of regular autoregressive, integration, and moving average models.

## B. The Iterative Autoregressive Integrated Moving Average Modeling Process

ARIMA modeling is conducted in a multistage process involving (a) identification of a tentative model, (b) estimation of the model, (c) diagnosis of the model's adequacy, and, we advocate, (d) bracketing of parameter estimates. The process is usually iterative, in that the results of one stage can lead the analyst to cycle back through earlier stages. For example, if tests of the model at the diagnostic stage indicate an unsatisfactory fit, the analyst would return to the identification stage and select an alternative or additional model. In this section, we describe the ARIMA modeling process and illustrate it with data from the McSweeney (1978) study.

Before addressing the steps in the process, it is important to be clear about what precisely is being subjected to the iterative process. If a regression model is being used to remove trends, the estimated residuals from the model are used in the iterative ARIMA modeling process. For purposes of identifying and diagnosing the ARIMA model, the residuals could be obtained from either an OLS regression analysis or a transfer function model (Box, et al. 1994), where a transfer function analysis is essentially a regression model capable of modeling different patterns of the treatment effect over time, and which is fit using a specialized ARIMA computer program (as discussed further below). However, once the ARIMA structure of the model has been identified and diagnosed, the final estimation of model parameters (including the estimation of the treatment effect and any other regression parameters) must be performed in a unified step with the regression and ARIMA models combined. This can be done using generalized least squares analysis with a classic OLS regression software package (Hibbs, 1974; Johnston, 1972), but is probably best and most easily performed using a transfer function analysis with a specialized time-series computer program.

If differencing is carried out, differencing becomes a part of the iterative ARIMA modeling process. That is, subsequent steps in the ARIMA modeling process are carried out using the differenced data as the input. This is simple to achieve in ARIMA software programs, which provide for identification, estimation, and diagnosis. If neither differencing nor a regression model is used to remove trends, the ARIMA modeling process is applied to the $Y_t$ scores (the raw data).

Whether the ARIMA modeling is applied to $\varepsilon_t$, $Y_t$, or $\nabla Y_t$, the treatment effect can obscure the ARIMA structure in the identification phase of modeling. Consider, for example, how the large drop after the intervention in Figure 13.1 would alter the pattern of correlation between adjacent observations, relative to the autocorrelation that occurs in the pretreatment period only. Several approaches can be employed to avoid such distortion. If the pretreatment series is long enough, the autocorrelation structure can be identified from it alone (following the identification process, described next). Alternatively, the autocorrelation structure can be identified from analyses that include a regression term or a transfer function for the intervention (for other approaches, see Kendall & Ord, 1990, p. 223).

During the *identification* stage, the analyst tentatively selects an ARIMA model that appears to best fit the autocorrelation structure in the data. For example, the analyst might consider whether the time series appears to fit an ARIMA (1,0,0) or an ARIMA (0,1,1) model. The principle tools in the identification stage are the autocorrelation function (ACF) and the partial autocorrelation function (PACF). The ACF, which is usually displayed graphically, presents the correlation between the time series and itself lagged different time periods. The lag-one correlation of the ACF, for example, is the correlation between the time series and the time series shifted backward one time point (so that, for example, in monthly data, January and February are paired, February and March are paired, and so on). The lag-two correlation similarly is the correlation between the series and itself shifted backward two time points. The PACF is similar to the ACF, except the correlations are effectively partial correlations, where the variables being partialed are the ones for the intermediate time lags. Thus, the lag-two partial correlation is the autocorrelation at lag two (i.e., between the variable and itself shifted two time units backward) partialling out the lag-one variable. The lag-three partial correlation is the autocorrelation at lag three (i.e., between the variable and itself shifted three time units backward) partially out both the lag-one and lag-two variables. Note that the lag-one value of the PACF equals the lag-one value of the ACF, because there is no intermediate lag to partial out. Another identification tool, the inverse autocorrelation function (IACF), is provided by some time-series computer programs. Its role is similar to that of the PACF, although some time-series analysts contend that it works better for identifying seasonal autocorrelation.

The patterns of the correlations in the ACF and PACF (and, perhaps, the IACF) are used to determine which ARIMA model underlies the autocorrelation structure in the data. The relationship is summarized in Table 13.1. As indicated in that table, for autoregressive models the ACF will decay exponentially and may appear like a sine wave that dies off rapidly. The PACF for autoregressive models will show spikes at the beginning, with the number of spikes equal to $p$, the order of the AR process (an example follows shortly). For moving average models, the picture is reversed. The ACF will have spikes, with the number of spikes equal to $q$, the order of the MA process, whereas the PACF will decay exponentially or like a damped sine wave. When differencing is needed, the ACF will

**TABLE 13.1    Summary of the Behavior of the Autocorrelation and Partial Autocorrelation Functions for Various Autoregressive Integrated Moving Average Models**[a]

|  | *ACF* | *PACF* |
|---|---|---|
| ARIMA (0,0,0) | (Near) zero at all lags | (Near) zero at all lags |
| ARIMA (1,0,0) | Decays exponentially (with alternating sign if $\phi$ is negative) | Spike at lag 1 |
| ARIMA ($p$,0,0) | Decays exponentially, may show damped oscillations | Spikes at lags $1 - p$ |
| ARIMA (0,0,1) | Spike at lag 1 | Decays exponentially (with alternating sign if $\theta$ is positive) |
| ARIMA (0,0,$q$) | Spikes at lags $1 - q$ | Decays exponentially, may show damped oscillations |
| ARIMA (0,1,0) | Decays slowly, nearly linearly | Large spike at lag 1 |
| ARIMA ($p$,0,$q$) | Decays exponentially, may show damped oscillations | Decays exponentially, may show damped oscillations |

[a] ACF, autocorrelation function; PACF, partial autocorrelation function; ARIMA, autoregressive integrated moving average.

decay slowly, and nearly linearly, whereas the PACF will show one spike, usually very large.

Seasonal ARIMA models are also indicated by patterns in the ACF and PACF (and, if available, the IACF). For seasonal models, the spikes and decays will occur at lags corresponding to the seasonal period (e.g., at lags 12, 24, and 36 for annual cycles in monthly data; see Box, et al., 1994; Kendall & Ord, 1990). Whether or not there appears to be seasonal autocorrelation, the analyst tentatively selects an ARIMA model based on the ACF and PACF that are observed for the time series being analyzed.

As an example of the identification stage, Figures 13.3 and 13.4 display the ACF and PACF for the McSweeney data presented in Figure 13.1. For illustrative reasons only, Figure 13.3. shows the ACF and PACF for the original pretreatment observations themselves (i.e., calls to local directory assistance). The dropoff in the ACF is slow and nearly linear. The PACF has a single, large spike at lag 1. This pattern suggest that, if the trend were not modeled otherwise, the series would need to be differenced before proceeding. If we were to apply differencing to this time series, we would then proceed to carry out the identification phase on the differenced series.

Instead, we are interested in assessing the autocorrelation in the McSweeney data when trend is accounted for by a regression term, as in equations (1)–(3). As previously noted, if it is not accounted for, the treatment effect can obscure the pattern of autocorrelation in a time series. One way to avoid this is to perform identification with the pretreatment observations only. Figure 13.4 presents the ACF and PACF from the residuals of a linear regression of the pretreatment data (see equation(1)). The ACF decays like a damped sine wave. The PACF has a single significant spike (confidence intervals are shown on the figure). Looking at Table 13.1, we can see that this pattern corresponds to an autoregressive model, specifically AR(1). Accordingly, we would tentatively identify an AR-IMA(1,0,0) model and proceed to the estimation stage. This decision is bolstered by the fact that if we carry out identification on the residuals from applying Equation 3 to the combined pre- and posttreatment data, the ACF and PACF look similar to those in Figure 13.4, again suggesting an ARIMA (1,0,0).

At the estimation stage, the parameters of the ARIMA models are estimated. Box and Jenkins (1970) developed a nonlinear least-squares estimation procedure, and exact likelihood procedures have been developed since then (see Kendall & Ord, 1990, for details and for alternative estimation procedures). Most time series analysts rely on specialized commercial time-series computer programs to derive parameter estimates.

For the analysis of the McSweeney data, if we fit Equation 3 with an ARIMA (1,0,0) models of the error term, the following parameter estimates result. For $\alpha$, the intercept, the estimate is 53,525. $\beta_1$, the index of linear trend in the pretreatment observations, is estimated at 380. $\beta_2$, the interven-

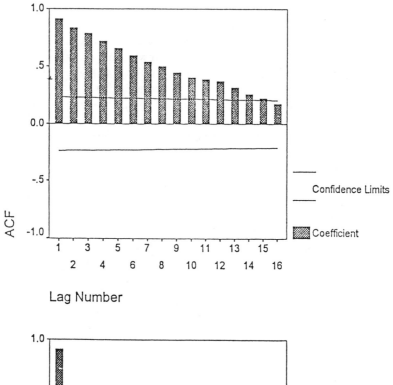

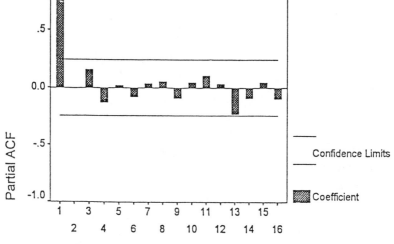

**FIGURE 13.3** Autocorrelation function (ACF) and partial autocorrelation function (PACF) for pretreatment observations, with trend not removed.

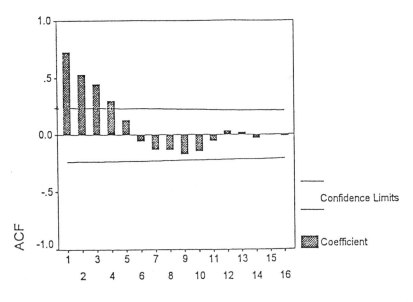

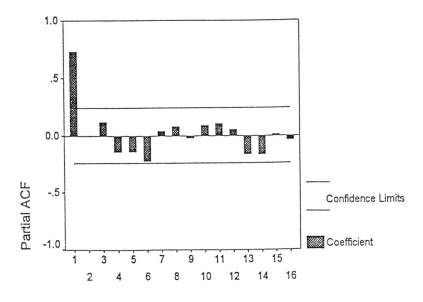

**FIGURE 13.4** Autocorrelation function (ACF) and partial autocorrelation function (PACF) for unstandardized residuals from a regression modeling linear trend in the pretreatment observations.

tion effect, is estimated as $-63,360$, representing the drop in the average daily calls to directory assistance. $\beta_3$, the possible change in slope following the intervention, is estimated at $-159$. All are significant ($p < .0001$), with the exception of $\beta_3$, for which $p < .10$.

In the diagnosis phase, checks are made on the adequacy of the estimated model (Kendall & Ord, 1990). First, it is suggested that the residuals from the ARIMA model be plotted against time, and examined both for unusual values and apparent increases or decreases in variability, which could suggest the need for further detrending or a transformation of the data. No apparent problems were revealed by inspection of the residuals from the McSweeney data as fit by a regression model (Equation 3) with an error structure of ARIMA(1,0,0) (the plot of these is not shown here). Second, the analyst examines the results of a statistical significance test (or confidence interval; cf. Reichardt & Gollob, 1997) for estimates of the ARIMA parameter, and changes are made accordingly. For example, if an ARIMA (2,0,0) were fit, and the estimate of the second AR term were small and not statistically significant, the model might be re-estimated as an ARIMA (1,0,0). For the McSweeney data, the autoregessive parameter AR(1) was significant, with a $t$ of 10.345.

Third, the ACF and the PACF of the residuals from the ARIMA model are examined for evidence that additional terms are needed in the ARIMA model. For example, significant spikes at values corresponding to possible seasonal cycles may suggest the need to add a seasonal term. However, a single, statistically significant, spike can occur by chance, or a set of individually nonsignificant elevated values might mistakenly be overlooked. The "Box-Pierce-Ljung statistic," commonly denoted by Q, can be used to assess the statistical significance of the values of the ACF or PACF over a specified set of lags. Essentially, the Q statistic is a chi-square test of whether a set of ACF (or PACF) values differ from what would be expected from a time series with no autocorrelation (or partial autocorrelation). The Q statistic is typically evaluated across several different sets of time lags (e.g., Q may be computed for lags 1–10, 1–20, and so on). Figure 13.5 presents the ACF and PACF of the residuals that result when equation (3) is fit to the McSweeney data with an ARIMA (1,0,0) error model. Contrasting this picture with Figure 13.4, we can see that the autocorrelation has generally been removed. Although several spikes approach the confidence interval, none exceed it. Some concern might be raised by the small spike at lag 12 in the ACF and PACF, accompanied by a smaller, clearly nonsignificant spike in the ACF at lag 24. This might suggest the need for a seasonal autoregressive parameter (more on this in the discussion of bracketing).

A fourth type of model check in the diagnostic phase involves verifying that the model coefficients fall within acceptable bounds. There are limits to the values that the $\phi$ and $\theta$ parameters can coherently take. For $\phi$, these are called the bounds of stationarity. For example, for an AR(1) model,

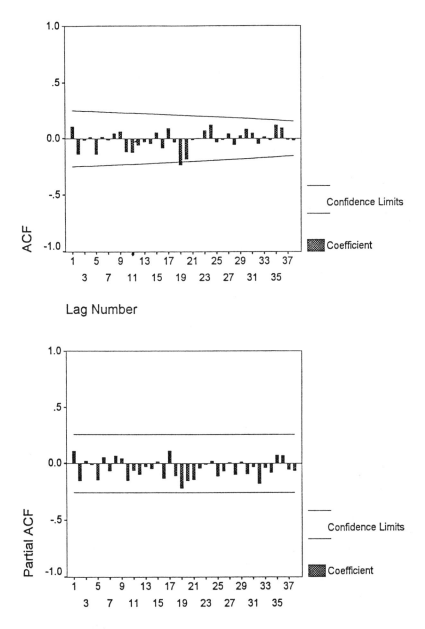

**FIGURE 13.5**  Autocorrelation function (ACF) and partial autocorrelation function (PACF) for residuals from a regression model (see Equation 3) with an ARIMA(1,0,0) error model.

the absolute value of $\phi_1$ must be less than 1. Were the estimate of $\phi_1$ to exceed these bounds, the estimated model would not be stationary. Similarly, in an MA(1) model, the so-called bounds of invertibility restrict $\theta_1$ to have an absolute value less than one. More complex bounds of stationarity and invertibility apply to higher order models (Box et al., 1994; Kendall & Ord, 1990). In the case of the McSweeney data, the $\phi_1$ parameter was estimated as .72, well within the bounds of stationarity.

In addition, the overall "predictability" of the model can be assessed. As Kendall and Ord (1990) note, this is done less often in ARIMA modeling than in standard regression analyses. Measures of $R^2$ and adjusted $R^2$ are available. However, as Harvey (1984) notes, these measures can be misleading when a series has been differenced. Harvey (1984) provides an alternative for this case. Kendall and Ord (1990) examine both an adjusted $R^2$ and Harvey's alternative. In the case of the McSweeney data, the model had an $R^2$ of .95. This unusually high value occurs because the model predicted well both the large linear trend in the data and the large treatment effect.

Finally, although not traditionally considered a part of the formal iterative analysis process, a bracketing step might usefully be added at the end. Too often time-series (and other) analysts simply report results from a single model even when the final choice of that model is uncertain. Doubt about the correct model can arise because, as Orwin (1997) points out, although explicit decision rules exist about the iterative model-building process, "in practice, analysts must frequently make decisions while 'in the cracks' between one rule or the other, making replicability an issue" (p. 449). The best solution is to assess the robustness of the substantive conclusions empirically and to report the results from multiple models so as to "bracket" estimates of the size of parameters within plausible ranges (Reichardt & Gollob, 1987). That is, when there is doubt as to the correct ARIMA model, one should estimate the treatment effects using a range of plausible ARIMA model specifications.

For example, in an ITS investigation of the effect of a community action group's protest, Steiner and Mark (1985) reported that although an ARIMA (1,0,0) provided the best fit, some analysts might have chosen an ARIMA (2,0,0) or (3,0,0) model instead. As a result, Steiner and Mark reported the findings for all three ARIMA models (although, as we expect will often occur in journal articles, results for the latter two models were only very briefly summarized). Alternatively, one could assess the replicability of the decisions made during the analysis by having two or more analysts work independently, as the U.S. General Accounting Office does in its studies using ARIMA modeling (Orwin, 1997). Either of these approaches is consistent with the frequent calls elsewhere in this *Handbook* for procedures that replicate and assess the robustness of findings. In the case of the McSweeney data, one question is whether a seasonal model might fit. Accordingly, we added a seasonal autoregressive parameter to the error term model. It did not achieve significance and the other estimates, including the treatment effect, were virtually unchanged.

## C. More on Transfer Function Modeling

In addition to modeling trends in the pretreatment and posttreatment observations, a transfer function can also be used to model the pattern the treatment effect takes over time (Box et al., 1994; Box & Tiao, 1975, Kendall & Ord, 1990). The temporal pattern of the treatment's effect can be quite elaborate, and one of the great advantages of a time-series design is that it can allow you to assess this pattern. With time series data, the researcher can see, for example, whether the effect is abrupt and persistent, as in the McSweeney (1978) case, whether it starts out large and then decays over time, or whether it starts out small and increases. Without repeated observations over time, such differences are impossible to assess.

A few of the possible patterns of the time course of a treatment effect are presented in Figure 13.6. In each of the idealized time series in Figure 13.6, the vertical line indicates when the treatment was implemented. As

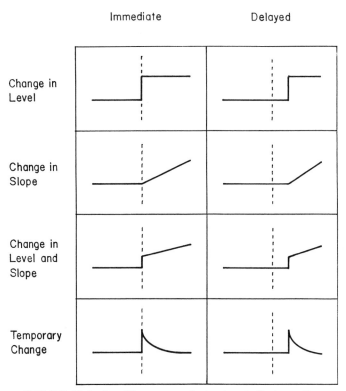

**FIGURE 13.6**    Possible patterns of a treatment effect over time.

shown in Figure 13.6, a treatment might change an outcome variable's level, slope, or both. In addition, a treatment effect can be either immediate or delayed, and it can be either permanent or temporary. Although Figure 13.6 illustrates many patterns, it certainly does not exhaust the variety of patterns that effects may take over time and that can be assessed given multiple posttreatment observations.

Such different temporal patterns of treatment effects can be modeled using a transfer function. In this approach, the treatment is represented in the analysis in one of two ways, either as (a) a "pulse," where the treatment variable is coded as 1 at the time of the intervention and 0 otherwise, or (b) a "step function," where the treatment variable is coded as 0 before the intervention and 1 afterwards (as was done with the $X_1$ variable previously). In general, use of a pulse function is preferred if the treatment effect is expected to be temporary, whereas a step function is preferred if the treatment effect is expected to be enduring.

Transfer functions can employ a number of parameters to model a wide range of possible treatment effect patterns. In practice, however, many if not most patterns of interest can be modeled with one or two parameters, as summarized in Figure 13.7, where $B$ is the "backward shift operator," which serves to move a term backward in time. For example, if we expect a permanent abrupt shift, we would represent the treatment with a step function (noted as $S_t$ in Figure 13.7), and include one parameter, $\omega$. This

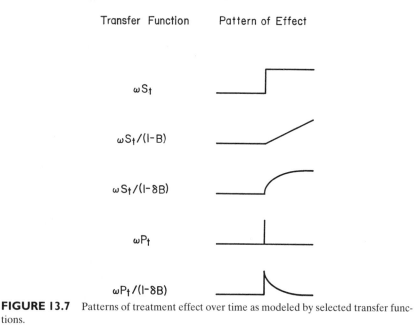

Transfer Function      Pattern of Effect

$\omega S_t$

$\omega S_t/(1-B)$

$\omega S_t/(1-\delta B)$

$\omega P_t$

$\omega P_t/(1-\delta B)$

**FIGURE 13.7** Patterns of treatment effect over time as modeled by selected transfer functions.

simple transfer function is equivalent to the step-function regression term described in Equation 2. However, addition of another parameter allows us to model a gradually building permanent effect. In this case, $\omega$ represents the asymptotic effect of the treatment and $\delta$ represents the rate of change (as the treatment effect builds up). Examples of different transfer functions, and the pattern of effects to which they correspond, are given in Figure 13.7. As shown there, if a pulse function $(P_t)$ is used to represent the treatment, the same set of parameters results in corresponding temporary effects. If they use a transfer function approach, time-series analysts should use theory to guide the choice of a transfer function model. To fit a wide array of transfer function models is to engage in exploratory research.

Transfer functions also allow the inclusion of additional time-series variables as covariates in the analysis, and these variables can be either lagged or contemporaneous (see Orwin et al., 1984, Orwin, 1997, for an example). In this way, transfer functions can be used to model the covariation between two ongoing time series for assessing causal relationships or to improve forecasts (Box et al., 1994; Kendall & Ord, 1990).

## D. Some Anticipated Advances

Time-series analysis is an area of active development within statistics and econometrics. There are (at least) two relatively recent advances that have not yet influenced applied time-series analysis, but that may in the near future. One is fractional autoregressive integrated moving-average (FARIMA) models. In standard ARIMA models, only whole integers can be used to represent the order of a component. FARIMA models relax this restriction, in particular for differencing (see, e.g., Sowell, 1992).

Second, autoregressive conditional heteroskedasticity (ARCH) and generalized autoregressive conditional heteroskedasticity (GARCH) models have been developed to address another problem that, like autocorrelation, can plague the error term of time-series data: heteroskedasticity (e.g., Li & Mak, 1994). Heteroskedasticity occurs in time-series data when the variance changes over time. For example, a time series might be marked by big movements up and down at the beginning and by relative stability later (as Figure 13.2 suggests, this occurs subtly in the McSweeney, 1978, data). Within ARIMA modeling, this form of nonstationarity has been addressed by transforming the original time series, usually by taking the logarithm. ARCH and GARCH models assume that a time series can be modeled by an autoregressive process, but allow the variance to change over time, conditional on past information (with the predictors of this change being slightly different in ARCH and GARCH).

Today's applied time-series analyst need not be conversant with FARIMA or ARCH/GARCH models. But the same may not be true for readers of the second edition of this *Handbook*.

## V. MULTIPLE CASES

So far we have considered the analysis of time-series data where observations are collected on a single unit or case (e.g., the city of Cincinnati or a single individual). In the present section, we consider analysis options when separate time series are collected on each of a number of units or cases (e.g., separate time series for each of the census tracks within Cincinnati or for each of several individuals). The number of cases will be denoted by $N$ and will refer to the number of individuals or groups of individuals for which there are separate time series of observations (e.g., if distinct time series of observations are collected and analyzed for each of 20 individuals, there are 20 cases, whereas if a single time series of observations is collected for a city, there is only one case). In addition, $K$ will denote the number of repeated observations over time (such as the number of months of monthly data) for each of the $N$ separate time series. Time-series designs require multiple time points (i.e., require that $K > 1$) but can have any number of cases (i.e., can have $N = 1$ or $N > 1$). Designs with multiple cases (i.e., $N > 1$) and multiple time points ($K > 1$) are sometimes called pooled cross-sectional and time-series designs (Dielman, 1983, 1989).

There are two general approaches to the analysis of time-series designs with multiple cases (i.e., $N > 1$). The first approach is to analyze each time series of observations individually. For example, a researcher could apply ARIMA modeling separately to each of the distinct time series of observations. The second approach is to model the multiple time series jointly. For example, a researcher could calculate correlations between variables across the cases and analyze the resulting correlation (or variance–covariance) matrix within a single structural equation model.

The advantage of the "separate" analysis approach is that a completely different model can be fit to each time series. The drawback to the separate approach is that one generally needs many time points (i.e., $K$ must be large). For example, it is often said that ARIMA modeling requires a bare minimum of 40–50 time points, and even more time points are to be preferred (e.g., Box et al., 1994; Box & Tiao, 1975; Glass et al., 1975; McCain & McCleary, 1979; McCleary & Hay, 1980). The advantages and disadvantages of the "joint" analysis approach are the converse. With the joint analysis approach the number of time points ($K$) can be small. But to make up for the small number of time points, the number of cases ($N$) must often be substantial. For example, some joint approaches require that the number of cases ($N$) be at least as large as the number of time points ($K$), and the number of cases may have to be substantially larger than the number of time points, if good estimates of model parameters are to be obtained. In addition, the joint analysis approaches tend to be less flexible than the separate approach in that some features of the fitted models (e.g., the model for the pattern of autocorrelation) may have to be assumed to be the same across the time series of individual cases. Nonetheless, the joint modeling approaches are rapidly increasing in popularity, and we shall say a few more words about them next.

There have been three generations of joint analysis approaches. The first generation of joint approaches tend to fall into two categories: the multivariate general linear model (Algina & Swaminathan, 1979; Swaminathan & Algina, 1977) and generalized least squares regression (Hibbs, 1974; Johnston, 1972; Orwin, 1997; Simonton, 1977, 1979). These approaches are both based on regression models such as in equation 3, wherein the values of the regression coefficients (i.e., the $\beta$s) are assumed to be the same for all cases (i.e., for each of the $N$ individual time series). In other words, these classes of models estimate average treatment effects (e.g., average changes in levels or average changes in slopes due to the treatment) across cases, but do not estimate individual differences in treatment effects across cases. The advantage of these approaches, however, is that they can be very flexible in allowing for patterns of autocorrelation. For example, the multivariate general linear model can take into account any pattern of autocorrelation without imposing the restrictions, including the stationarity assumptions, of ARIMA models; nor does this approach require that a specific model for the pattern of autocorrelation be identified or diagnosed.

The second generation of joint analysis approaches grew out of multilevel analysis models that were initially developed for dealing with nested data (Bryk & Raudenbush, 1987, 1992; Burstein, 1980; Raudenbush & Bryk, 1986; Raudenbush, 1988). These models spawned a number of special purpose computer programs such as HLM, MIXREG, and MLwinN (de Leeuw & Kreft, 1999; Kreft, de Leeuw, & van der Leeden, 1994). These models and computer programs provide two noteworthy advances compared to the first-generation approaches. First, the new models fit the same regression equations as in Equation 3 but allow for individual differences in the regression intercepts, slopes, and treatment effects. That is, these models allow the researcher to assess differences in the regression coefficients in the time-series models across individual cases. In addition, the researcher can assess the degree to which individual differences in the regression coefficients of the time-series models are related to other individual characteristics. For example, if McSweeney (1978) had collected separate time series for each census tract in Cincinnati, and if he had obtained information on individual census tract characteristics such as average socioeconomic status, then multilevel models could have been used to assess the relationship, across census tracks, between the size of the treatment effect and average socioeconomic status. Second, the special purpose multilevel models are typically designed for use even if observations for different cases are collected at different time points, or if some cases have different numbers of observations (e.g., due to missing data). One disadvantage is that not all of these multilevel computer programs, at least in their most current versions, allow for autocorrelation among the observations.

The third generation of joint analysis approaches evolved from the realization that general purpose, structural equation modeling (SEM) programs (such as AMOS, EQS, LISREL, Mplus, Mx, and RAMONA) could be used to mirror either of the first two generations of analyses, even though SEM computer programs had not been initially designed with that purpose in mind (McArdle, 1998; McArdle & Epstein, 1987; Meredith & Tisak, 1990; Muthén & Curran, 1997). In classic SEM conceptualizations, the values of the $\beta$s in models such as equation 3 are regression coefficients that are to be estimated, and the $t$ and $X$ scores represent manifest variables. However, the roles of these parameters and variables can be switched in time-series analyses. For example, in a simple time-series model, the $t$ and $X$ variables become regression parameters that are fixed at the same value for each individual, whereas the $\beta$s become latent variables. With this reparameterization, SEM can impose the same assumptions as in the multivariate generalized linear model, for example, by fixing the values of the $\beta$s to be constant across cases and letting the residuals correlate freely. Or SEM can perform multilevel analyses by letting the values of the $\beta$s vary across individuals. For example, the size of the treatment effect can be allowed to vary across individuals and, as a result, the relationship between the size of the treatment effect and other individual characteristics can be assessed. In addition, if an appropriate ARIMA structure can be identified, it could be added to the residuals in a multilevel SEM model so that the pattern of autocorrelation can be taken into account; however, this additional step has often been overlooked in practice, to the potential detriment of the analysis. (Perhaps as awareness of the use of ARIMA modeling as applied to $N$-of-1 studies increases, researchers will become increasing aware of the potential value of ARIMA modeling in $N$-greater-than-1 studies as well.) The relative drawback to the SEM approach is that SEM computer programs tend not to be nearly as user-friendly as the special purpose multilevel analysis programs when observations in the individual time series are collected at different points in time or when there are missing data (for details, see McArdle, 1998; McArdle & Hamagami, 1991, 1992). In addition, the most popular SEM computer programs require that $N$ be at least as large as $K$ when performing multilevel analyses, although it is possible for $N$ to be less than $K$ with the specialized multilevel programs.

By using the generational nomenclature, we do not mean to imply that analysis strategies from later generations should necessarily replace those from previous generations. The approaches from each generation have their own unique niches. Nonetheless, the use of SEM with pooled cross-sectional and time-series data does appear to be receiving a particularly enthusiastic reception. The use of SEM to model pooled cross-sectional and time-series data has come to be called latent growth modeling or latent growth curve analysis (among other variants), and is discussed by DiLalla and by Willett and Keiley (chapters 15 and 23, this volume) in further detail. Interestingly, SEM computer programs are even being used to model $N$-of-1 time series using block Toeplitz matrices (Hershberger, Molenaar, & Corneal, 1996), and this approach can be generalized to $N$-greater-than-1 analyses (Jones, 1991). But at its present stage of development, the Toeplitz approach is relatively untested and, at this point in time, appears to us to be less attractive than either the more specialized ARIMA $N$-of-1 time-series programs or the latent growth curve approach.

## VI. THREATS TO INTERNAL VALIDITY IN THE SIMPLE INTERRUPTED TIME-SERIES DESIGN: OLD AND NEW CONSIDERATIONS

Regardless of the number of units $(N)$, a study in which a single treatment is introduced and in which all the units receive that treatment at a single point in time is called a "simple" ITS design. For example, the McSweeney (1978) study is a simple ITS design. Although the present chapter focuses on statistical analysis, it is important for time-series researchers not to inflate the importance of analysis relative to design issues, which includes thoughtful consideration of threats to validity.

Campbell and Stanley (1966), Cook and Campbell (1979), and others (e.g., Reichardt & Mark, 1998) have discussed the validity of ITS designs. As these authors point out, the addition of time-series data has advantages for internal validity, relative to a simple "pretest–posttest" design. That is, the simple ITS design increases the confidence one can have that the treatment, and not some other factor, is responsible for the observed change in the outcome variable. Specifically, the addition of time-series data allows researchers to assess and to rule out the threats to internal validity of maturation and statistical regression to the mean. For example, most maturational patterns would be apparent across the time series, rather than leading to an abrupt interruption at the time of the treatment. Nonetheless, the simple ITS design is susceptible to other threats to internal validity, in particular, history, instrumentation, and attrition, as Cook and Campbell (1979) and others have noted.

Another threat to internal validity in the simple ITS design, which has not generally been recognized in the literature, can be labeled "anticipation." This threat to validity arises if the treatment is implemented in expectation of a likely change in the outcome variable or its correlates. For example, imagine that a researcher started collecting weekly time-series data in June on the number of passersby outside a faculty member's office. Also imagine that the researcher attempted to assess, using an ITS design, whether the faculty member's preparation of a course syllabus influenced the number of passersby. For the typical faculty member, the researcher would probably detect an apparent treatment effect. But it would not be because the syllabus preparation caused more people to pass by. Instead, the faculty member's syllabus preparation simply anticipated the beginning of the Fall semester and the attendant upswing of people on campus. The threat of anticipation can be controlled if the change in the outcome variable is a regular seasonal process that can be modeled (e.g., if the increase in the number of passersby at the start of the fall semester can be modeled using data from previous years). However, it is probably best to try to avoid the threat by having the timing of the treatment implementation under the researcher's control or a result of an arbitrary administrative decision that does not correspond to likely changes in the outcome variable or its correlates (for an example, see Hennigan et al., 1982).

Another, related threat to validity that is not widely recognized in the ITS literature is an instance of the classic third-variable causation problem. This threat arises when both the initiation of the treatment and the subsequent change in the outcome variable are manifestations (or consequences) of the same underlying processes. We call this threat "co-indicators," because the "treatment" and the "outcome variable" both arise from the same shared processes. For example, getting a driver's licence does not cause graduation from high school. Instead, both are "differentially lagged" indicators or manifestations of the ongoing developmental processes of aging. The fact that getting a driver's license typically precedes high school graduation does not make the relationship between these two causal. Our sense is that the problem of co-indicators arises infrequently in ITS designs, but is far more common in time-series investigations in which causality is inferred from the covariation between two time-series variables.

## VII. MORE ELABORATE INTERRUPTED TIME-SERIES DESIGNS

The plausibility of a potential threat can sometimes be assessed using extant records (e.g., records about attrition, if attrition is a threat). More commonly, plausibility can be assessed using knowledge and evidence about the particular processes being investigated. In the McSweeney (1978) study, for example, even without time-series data it would seem implausible that a maturational process would lead to such a large, rapid decline in the use of local directory assistance. In addition, many threats to validity can be rendered implausible with more elaborate ITS designs.

The simple ITS design can be extended in a variety of ways. First, a control group can be added. For example, McSweeney (1978) could have examined time-series data on local directory assistance calls in another city where no fee was introduced. To the extent that the control time series is affected by everything that affects the experimental time series, except for the treatment, the control group's pre-posttreatment change should reflect the effects of validity threats, such as history. An analogous extension of the simple ITS design is to add, not a control group, but a "nonequivalent dependent variable." That is, in addition to the dependent variable that should be affected by the treatment, a time-series variable is also examined that should not be affected by the treatment but should be affected by plausible threats to validity. In fact, McSweeney employed this design feature, in that he examined calls in Cincinnati to directory assistance for *long distance* (as opposed to local) calls, for which no fee was introduced.

Analytically, two general approaches are available when either a control group or nonequivalent dependent variable is added (and when $K$ is large and $N = 1$ for each group). The standard approach is to conduct a separate ARIMA analysis on the control time series. Alternatively, the control series could be used as a covariate in the analysis of the outcome time series, to see if its inclusion reduces the size and significance of the treatment effect. Analogous approaches are available when $N > 1$.

A further elaboration of the simple ITS design is what Cook and Campbell (1979) call the "switching replications" ITS design. In this design, two groups receive the same treatment, but at different times so that each group's time series serves alternatively, at different points in time, as the experimental series and as the control series. Imagine, for example, that McSweeney (1978) had found a control site, where a fee for local directory assistance calls was implemented at a different point in time than in Cincinnati.

When the effects of a treatment are transitory, in that they disappear shortly after the treatment is removed, causal inference can be strengthened by repeatedly introducing and removing the treatment. Although it would probably not be feasible in a study of fees for directory assistance, this approach has been used successfully in other time-series studies. For instance, Schnelle et al. (1978) examined the effect of police-helicopter surveillance on the frequency of home burglaries. Helicopter surveillance was either present as an adjunct to patrol-car surveillance, or patrol-car surveillance was employed alone. Helicopter surveillance was added and removed several times, and the frequency of burglaries decreased and increased, respectively.

## VIII. ELABORATION IN TIME-SERIES STUDIES OF COVARIATION

Our focus in this chapter has been on ITS studies, which estimate the effect of a treatment. Nevertheless, it is worth noting that the same procedures through which the simple ITS design can be elaborated, particularly the use of control groups and nonequivalent dependent variables, can also be valuable when covariation between two time-series variables is examined in an attempt to infer causality.

Efforts to infer causality from the covariation between time series are often based on the definition of causality developed by Granger (1969) for time-series investigations. According to this definition, variable $X$ is a cause of variable $Y$ if present values of $Y$ can be better predicted using past values of $Y$ plus past or present values of $X$, than using past values of $Y$ alone. Although widely used, this conception of cause is generally inadequate. One potentially serious problem is the threat to validity of co-indicators. According to Granger's definition, if changes in $X$ occur before changes in $Y$, $X$ could be deemed the cause of $Y$ even if the relationship between the two series is solely the result of a common underlying process that causes both $X$ and $Y$.

Another shortcoming of Granger's (1969) definition of causality is the threat to validity of anticipation. Consider an example concerning time-series research on the causal relationship between advertising and sales. Advertising is at least sometimes timed to precede and overlap with peak periods of sales. For example, in some college towns, certain stores advertise most heavily at the beginning of the new school year, before and concurrent with home football games, and at holidays such as Christmas when higher volumes of sales are anticipated. Moreover, the advertising time series $(X)$ may allow us to predict sales better than we can with past sales data $(Y)$ alone (particularly because the start of the semester and home football games do not fall on the same date every year, so analyses for seasonal patterns would not adequately predict them). Thus, even if the advertising were ineffective (e.g., a sales increase resulted solely from increased foot traffic past the store), we would find evidence consistent with Granger's definition of causality—simply because of anticipation effects in the sales data.

Adding control time series can help eliminate these types of alternative explanations. Consider, for example, a study in which the radio advertising expenditures of a shoe store is compared with sales data. Adding a control time series of sales from another shoe store could aid in drawing causal inferences. That is, if advertising causes sales, advertising for the Shoes-R-Us store should predict sales at that store better than it predicts sales at another shoe store down the street. However, if the covariation is due to an anticipation effect only, then Shoes-R-Us' advertising should more comparably predict covariation for both stores' sales. Alternatively, a non-equivalent dependent variable could be added. If the ads were for women's shoes, for example, then advertising expenditures should predict sales of women's shoes better than it predicts sales of men's shoes.

## IX. DESIGN AND IMPLEMENTATION ISSUES

A number of issues arise in designing a time-series study regardless of the purpose. In the present section, we briefly address four common issues.

## A. Time Interval between Observations

At what time intervals should observations be collected? Is it better to collect data at annual, quarterly, monthly, weekly, daily, or hourly intervals? At least three considerations apply. First, choices are sometimes constrained by the nature of data archives. Second, expectations or initial assumptions about the stability of observations should influence the choice of the time interval. For example, in the Sanna and Parks (1997) study, monthly data on the proportion of journal pages allocated to research on groups would probably be characterized by wide swings, because each article could greatly influence the outcome variable. However, when data collection is under the researcher's control, measurement in smaller time intervals is probably preferable, to the extent resource constraints allow, because the data can always be aggregated to longer time intervals if the data from shorter intervals are unstable. Third and most important, the time interval should be selected so as to match the nature of the research question. If a researcher is interested in seasonality, for example, the nature of the expected seasonality should determine the time interval. For example, Hopkins's (1994) interest in possible daily and weekly cycles in anxiety dictated that measures be taken on multiple occasions during each day. Similarly, in causality-probing time-series studies, the nature of the hypothesized causal mechanism should influence the selection of the time interval between observations. For example, to increase power, one would likely want to match the interval for observations with the causal lag that is hypothesized to exhibit the effect at its largest.

## B. Number of Time Points

The number of time points $(K)$ for which observations should be collected depends on many factors including (a) the type of analysis to be conducted, (b) the number of cases $(N)$, (c) the nature of the time series, and (d) issues of power and precision. For example, although ARIMA modeling of a single case is often said to demand a minimum of 40–50 time points, analyses involving multiple cases can require only a few time points. In general, there is a trade-off between the number of time points and the number of cases. A study with only a few time points can often compensate by having a large number of cases. Conversely, a study with only a few cases can often compensate by having a large number of times points.

The number of time points needed will also depend on the extent to which the time-series data are "well-behaved," that is, conform to the model being applied and contain little residual error variance. For instance, ARIMA analysts find that some series can readily be modeled with 40 observations (i.e., time points), whereas other series are difficult to model with even two or three times as many observations.

## C. Number of Cases

As just mentioned, the number of cases ($N$) that should be observed is at least partly linked to the number of time points that are observed ($K$), and both are linked, at least partly, to the choice of statistical analysis. In many instances, such as in the Sanna and Parks's (1997) investigation of the prevalence of research on groups, a single "case" is used that is actually an aggregate. That is, Sanna and Parks examined the number and proportion of journal articles allocated to research on groups, and their single case (i.e., their single time series) was created by combining data from several journals. On the other hand, in their study of performance patterns over time, Hofmann et al. (1992) examined individual-level data from 128 cases. The numbers of cases differed in these two studies at least in part because Sanna and Parks were interested in the overall representation of research on groups in organizational psychology journals collectively, whereas Hofmann et al. were interested in whether individuals differ in their patterns of performance over time. Obviously, individual differences can be assessed only when there are multiple cases.

## D. Archives as a Source of Data and Mediating Processes

As Cook and Campbell (1979) noted, time-series investigations on many topics often rely heavily on archival data, rather than on "primary" data collected by the researcher. Indeed, which variables happen to exist in time-series form in a data archive may come to define the research question for a time-series researcher. This flies in the face of the conventional wisdom that methods and measurements should be tailored to fit the research question, rather than the reverse. But variables often exist in time-series form precisely because they have been identified as important enough to warrant ongoing measurement.

Cook and Campbell (1979) also noted another significant manner in which archives may constrain time-series studies; namely, measures of outcomes are far more likely to be collected and stored than are measures of mediating processes. Measures of mediating processes may not be necessary when estimating the size of causal relationships, but they are critical if researchers are to understand *why* the relationship exists (or why it has a given size). To take the classic British Breathalyzer example of Ross, Campbell, and Glass (1970), which examined the effect of a "Breathalyzer" crackdown on drunken driving, we may be convinced that the Breathalyzer caused a decrease in fatalities of a certain size, but wonder why—was it, for example, that potential drinkers thought it was more likely that they would be apprehended? Or did the publicity surrounding the Breathalyzer increase their awareness of the penalties they would face if apprehended? Or was it a combination of both?

That time-series studies, especially those that draw their data from archives, commonly leave questions of mediating process unaddressed was documented by Mark, Sanna, and Shotland (1992). In a *PsycLIT* search of time-series articles abstracted between January, 1983, and March, 1990, only 1 of 302 articles explicitly addressed the question of mediators in the relationship between a treatment and an outcome. Although researchers may have no recourse when data are drawn from archives, mediating processes can be investigated far more often than is the case at present when investigators generate their own data, as is typically the case in time-series studies in areas such as educational psychology, experimental psychology, and clinical-counseling psychology (Mark et al., 1992). We recommend that investigators collecting time-series data consider including measures not only of possible mediating processes, but also of possible moderators of effects. This is because examining the moderators of a treatment effect can often provide insight into mediating processes: Different mediating mechanisms often result in different patterns of moderation, that is, about the conditions under which the effect will hold or be larger or smaller (Mark, 1990; Mark, Hofmann, & Reichardt, 1992).

## X. SUMMARY AND CONCLUSIONS

Time-series data can be applied to an array of different research purposes and analyzed with a variety of statistical techniques. In the present chapter, we have focused primarily on the interrupted time-series design and on ARIMA statistical analyses.

In the interrupted time-series design, a series of observations is collected over time, both before and after a treatment is implemented. The pattern in the pretreatment observations provides a comparison with which to assess a possible treatment effect in the posttreatment observations. Advantages of interrupted time-series designs include that they can be used to (a) estimate the effects of a treatment when only a single individual, or on a single aggregated unit such as a city, is available for study, (b) reveal the pattern of the treatment effect over time, and (c) provide an estimate of the treatment effect without withholding the treatment from anyone who is eligible to receive it (Reichardt & Mark, 1998). Various design adjuncts can often strengthen the credibility of inferences about treatment effect in ITS designs, including control groups, nonequivalent dependent variables, switching replications, and removed and repeated treatment. Especially when elaborated in such ways, interrupted time-series designs can be among the most credible quasi-experimental designs (Cook & Campbell 1979; Marcantonio & Cook, 1994; Reichardt & Mark, 1998). Nonetheless, researchers need to be careful that the somewhat arcane nature of the statistical analysis of time-series data not lead to an overemphasis on statistical technique and correspondingly to a failure to recognize threats to validity. Although they need to be thoughtful about issues of statistical analysis, which can indeed be complex, researchers using ITS designs need to be just as thoughtful in their assessment of the plausibility of threats to validity, including threats not generally recognized, such as anticipation and co-indicators.

In addition to ITS designs, the present chapter has also noted other uses of time-series data, including assessing the degree and nature of change, forecasting, and assessing covariation (and perhaps) causal linkages between two or more time-series variables. Again, researchers need to focus just as much attention on threats to validity as on sophisticated analysis techniques such as ARIMA and transfer function modeling. In the case of studies assessing covariation between two time series, the threats of anticipation and co-indicators are generally sufficiently plausible that we encourage great caution when attempting to draw causal inferences. However, the same logic of elaborating a design that is often used in ITS studies can also be applied to designs assessing covariation, and in so doing one can increase confidence in the tentative causal inferences derived from them (Reichardt & Mark, 1998).

In the language of Shakespeare, time takes "survey of all the world" (*Henry IV, Part 1*). Humans, in attempting to read time's survey, must pay attention to threats to validity as well as to the technical sophistication of statistical analyses. For, as always, warranted inferences depend on the considered design of research and on the critical analyses carried out "in the quick forge and working-house of thought" (Shakespeare, *Henry V*).

## REFERENCES

Algina, J., & Keselman, H. J. (1997). Detecting repeated measures effects with univariate and multivariate statistics. *Psychological Methods, 2,* 208–218.

Algina, J., & Swaminathan, H. (1979). Alternatives to Simonton's analyses of the interrupted and multiple-group time-series designs. *Psychological Bulletin, 86,* 919–926.

Bentler, P. M. (1992). *EQS Structural Equations Program Manual.* Los Angeles, CA: BMDP Statistical Software.

Boutcher, S. H., Nugent, F. W., McLaren, P. F., & Weltman, A. L. (1988). Heart period variability of trained and untrained men at rest and during mental challenge. *Psychophysiology, 35,* 16–22.

Box, G. E. P., & Jenkins, G. M. (1970). *Time-series analysis: Forecasting and control.* San Francisco, CA: Holden-Day.

Box, G. E. P., Jenkins, G. M., & Reinsel, G. C. (1994). *Time-series analysis: Forecasting and control* (3rd ed.). Englewood Cliffs, NJ: Prentice Hall.

Box, G. E. P., & Tiao, G. C. (1975). Intervention analysis with applications to economic and environmental problems. *Journal of the American Statistical Association, 70,* 70–92.

Brenner, M. H. (1973). *Mental illness and the economy.* Cambridge, MA: Harvard.

Bryk, A. S., & Raudenbush, S. W. (1987). Application of hierarchical linear models to assessing change. *Psychological Bulletin, 101,* 147–158.

Bryk, A. S., & Raudenbush, S. W. (1992). *Hierarchical linear models: Applications and data analysis methods.* Newbury Park, CA: Sage.

Burstein, L. (1980). The analysis of multi-level data in educational research and evaluation. *Review of Research in Education, 8,* 158–233.

Campbell, D. T., & Stanley, J. C. (1966). *Experimental and quasi-experimental designs for research.* Chicago: Rand McNally.

Cook, T. D., & Campbell, D. T. (1979). *Quasi-experimentation: Design and analysis issues for field settings.* Chicago: Rand McNally.

de Leeuw, J., & Kreft, I. G. G. (1999). Software for multilevel analysis. UCLA statistics preprints #239. Available: http://www.stat.ucla.edu/papers/preprints/239/.

Dielman, T. E. (1983). Pooled cross-sectional and time series data: A survey of current statistical methodology. *American Statistician, 37,* 111–122.

Dielman, T. E. (1989). *Pooled cross-sectional and time series data analysis.* New York: Marcel Dekker.

Glass, G. V, Willson, V. L., & Gottman, J. M. (1975). *Design and analysis of time-series experiments.* Boulder, CO: Colorado Associated University Press.

Granger, C. W. J. (1969). Investigating causal relations by econometric models and cross-spectral methods. *Econometrica, 37,* 424–438.

Greenhouse, S. W., & Geisser, S. (1959). On methods in the analysis of profile data. *Psychometrika, 77,* 95–112.

Harvey, A. C. (1984). A unified view of statistical forecasting procedures. *Journal of Forecasting, 3,* 245–275.

Helmer, R. M., & Johnsson, J. K. (1977). An exposition of the Box-Jenkins transfer function analysis with an application to the advertising-sales relationship. *Journal of Marketing Research, 14,* 227–239.

Hennigan, K. M., DelRosario, M. L., Heath, L., Cook, T. D, Wharton, J. D., & Calder, B. G. (1982). Impact of the introduction of television on crime: Empirical findings and theoretical implications. *Journal of Personality and Social Psychology, 42,* 461–477.

Hershberger, S. L., Molenaar, P. C. M., & Corneal, S. E. (1996). A hierarchy of univariate and multivariate structural times series models. In G. A. Marcoulides & R. E. Shoemacker (Eds.), *Advanced structural equation modeling: Issues and techniques* (pp. 159–194). Mahwah, NJ: Erlbaum.

Hibbs, D. A., Jr. (1974). Problems of statistical estimation and causal inference in time-series regression models. In H. L. Costner (Ed.), *Sociological methodology 1973–1974.* San Francisco: Jossey-Bass.

Hofmann, D. A., Jacobs, R. R., & Geras, S, J. (1992). Mapping individual performance over time. *Journal of Applied Psychology, 77,* 185–195.

Hopkins, M. B. (1994). *Rhythms of reported anxiety on persons experiencing Generalized Anxiety Disorder.* Unpublished doctoral dissertation Pennsylvania State University, University Park, Pennsylvania.

Huitema, B. E., & McKean, J. W. (1998). Irrelevant autocorrelation in least-squares intervention models. *Psychological Methods, 3,* 104–116.

Johnston, J. (1972). *Econometric methods* (2nd ed.). New York: McGraw-Hill.

Jones, K. (1991). The application of time series methods to moderate span longitudinal data. In L. M. Collins & Horn, J. L. (Eds.), *Best methods for the analysis of change: Recent advances, unanswered questions, future directions* (pp. 75–87). Washington, DC: American Psychological Association.

Kendall, M., & Ord, J. K. (1990). *Time series* (3rd ed.). London: Edward Arnold.

Kreft, I. G. G., de Leeuw, J., & van der Leeden, R. (1994). Review of five multilevel analysis programs: BMDP-5V, GENMOD, HLM, ML3, and VARCL. *The American Statistician, 48,* 324–335.

Li, W. K., & Mak, T. K. (1994) On the squared residual autocorrelations in nonlinear time series with conditional heteroskedasticity. *Journal of Time Series Analysis, 15,* 627–636.

Marcantonio, R. J., & Cook, T. D. (1994). Convincing quasi-experiments: The interrupted time series and regression-discontinuity designs. In J. S. Wholey, H. P. Hatry, & K. E. Newcomer (Eds.), *Handbook of practical program evaluation* (pp. 133–154). San Francisco: Jossey Bass.

Mark, M. M. (1990). From program theory to test of program theory. In L. Bickman (Ed.), *Program theory in program evaluation* (pp. 37–51). San Francisco: Jossey-Bass.

Mark, M. M., Hofmann, D. A., & Reichardt, C. S. (1992). Testing theories in theory-driven evaluations: (Tests of) Moderation in all things. In H. T. Chen & P. H. Rossi (Eds.), *Using theory to improve program and policy evaluations* (pp. 71–84). New York: PSO-Greenwood Press.

Mark, M. M., Sanna, L. J., & Shotland, R. L. (1992). Time series methods in applied social research. In F. B. Bryant, J. Edwards, R. S. Tindale, E. J. Posavac, L. Heath, E. Henderson, & Y. Suarez-Balcazar (Eds.), *Methodological issues in applied social research* (Social Psychological Applications to Social Issues, Vol. 2.) (pp. 111–133). New York: Plenum.

McArdle, J. J. (1998). Modeling longitudinal data by latent growth curve methods. In G. A. Marcoulides (Ed.), *Modern methods for business research* (pp. 359–406). Mahwah, NJ: Erlbaum.

McArdle, J. J., & Epstein, D. (1987). Latent growth curves within developmental structural equation models. *Child Development, 58,* 110–133.

McArdle, J. J., & Hamagami, F. (1991). Modeling incomplete longitudinal and cross-sectional data using latent growth structural models. In L. M. Collins & J. L. Horn. (Eds.), *Best methods for the analysis of change: Recent advances, unanswered questions, future directions* (pp. 276–304). Washington, DC: American Psychological Association.

McArdle, J. J., & Hamagami, F. (1992). Modeling incomplete longitudinal and cross-sectional data using latent growth structural models. *Experimental Aging Research, 18,* 145–166.

McCain, L. J., & McCleary, R. (1979). The statistical analysis of simple interrupted time-series quasi-experiments. In T. D. Cook & D. T. Campbell, *Quasi-experimentation: Design and analysis issues for field settings* (pp. 233–293). Chicago: Rand McNally.

McCleary, R., & Hay, R. A., Jr. (1980). *Applied time series analysis for the social sciences.* Newbury Park, CA: Sage.

McSweeney, A. J. (1978). Effects of response cost on the behavior of a million persons: Charging for directory assistance in Cincinnati. *Journal of Applied Behavior Analysis, 11,* 47–51.

Meredith, W., & Tisak, J. (1990). Latent curve analysis. *Psychometrika, 55,* 107–122.

Montgomery, A. L., Zarnowitz, V., Tsay, R. S., & Tiao, G. C. (1998). Forecasting the U.S. unemployment rate. *Journal of the American Statistical Association, 93,* 478–493.

Moreland, R. L., Hogg, M. A., & Hains, S. C. (1994). Back to the future: Social psychological research on groups. *Journal of Experimental Social Psychology, 30,* 527–555.

Muthén, B. O., & Curran, P. J. (1997). General longitudinal modeling of individual differences in experimental designs: A latent variable framework for analysis and power estimation. *Psychological Methods, 2,* 371–402.

Orwin, R. G. (1997). Twenty-one years old and counting: The interrupted time series comes of age. In E. Chelimsky & W. R. Shadish (Eds.), *Evaluation for the 21st century: A handbook* (pp. 443–465). Thousand Oaks, CA: Sage.

Orwin, R. G., Shucker, R. E., & Stokes, R. C. (1984). Evaluating the life cycle of a product warning: Saccharin and diet soft drink sales. *Evaluation Review, 8,* 801–822.

Raudenbush, S., & Bryk, A. S. (1986). A hierarchical model for studying school effects. *Sociology of Education, 59,* 1–17.

Raudenbush, S. W. (1988). Educational applications of hierarchical linear models: A review. *Journal of Educational Statistics, 13,* 85–116.

Reichardt, C. S., & Gollob, H. F. (1987). Taking uncertainty into account when estimating effects. In M. M. Mark & R. L. Shotland (Eds.), *Multiple methods for program evaluation. New directions for program evaluation, No. 35* (pp. 7–22). San Francisco: Jossey-Bass.

Reichardt, C. S., & Gollob, H. F. (1997). When confidence intervals should be used instead of statistical tests, and vice versa. In L. L., Harlow, S. A. Mulaik, & J. H. Steiger (Eds.). *What if there were no significance tests?* (pp. 259–284). Hillsdale, NJ: Lawrence Erlbaum.

Reichardt, C. S., & Mark, M. M. (1998) Quasi-experimentation. In L. Bickman & D. J. Rog (Eds.), *Handbook of applied social research methods* (pp. 193–228). Thousand Oaks, CA: Sage.

Ross, H. L., Campbell, D. T., & Glass, G. V (1970). Determining the social effects of a legal reform. The British 'Breathalyser' crackdown of 1967. *American Behavioral Scientist, 13,* 493–509.

Sanna, L. J., & Parks, C. D. (1997). Group research trends in social and organizational psychology: Whatever happened to intragroup research? *Psychological Science, 8,* 261–267.

Schnelle, J. F., Kirchner, R. E., Macrae, J. W., McNees, M. P., Eck, R. H., Snodgrass, S., Casey, J. D., & Uselton, P. H., Jr. (1978). Police evaluation research: An experimental and cost-benefit analysis of a helicopter patrol in a high-crime area. *Journal of Applied Behavior Analysis, 11,* 11–21.

Simonton, D. K. (1977). Cross-sectional time-series experiments: Some suggested statistical analyses. *Psychological Bulletin, 84,* 489–502.

Simonton, D. K. (1979). Reply to Algina and Swaminathan. *Psychological Bulletin, 86,* 927–928.

Sowell, F. (1992). Maximum likelihood estimation of stationary univariate fractionally integrated time series models. *Journal of Econometircs, 53,* 165–188.

Steiner, D., & Mark, M. M. (1985). The impact of a community action group: An illustration of the potential of time series analysis for the study of community groups. *American Journal of Community Psychology, 13,* 13–30.

Stoolmiller, M. (1995). Using latent growth curve models to study development processes. In J. M. Gotman (Ed.), *The analysis of change* (pp. 104–138). Mahweh, NJ: Erlbaum.

Swaminathan, H., & Algina, J. (1977). Analysis of quasi-experimental time-series data. *Multivariate Behavioral Research, 12,* 111–131.

# POISSON REGRESSION, LOGISTIC REGRESSION, AND LOGLINEAR MODELS FOR RANDOM COUNTS

## PETER B. IMREY

*Departments of Statistics and Medical Information Science,*
*University of Illinois at Urbana-Champaign, Champaign, Illinois*

## I. PRELIMINARIES

### A. Introduction

A categorical variable is observed whenever an individual, event, or other unit is identified and placed into one of a finite collection of categories. Gender, ethnicity, socioeconomic class, clinical diagnosis, and cause of death are examples. Each possible category is a *level* of the variable that may assume it; for example, female and male are the levels of gender. Categorical variables may be qualitative (nominal), or quantitative (ordinal, interval, or ratio scaled), depending on the nature of their levels. This chapter considers statistical models for counts of independently occurring random events, and counts at different levels of one or more categorical *outcomes* (i.e., responses, dependent variables). Attention is focused primarily on qualitative categorizations, but the reader also will find here a basis for understanding models for quantitative categorical variables that are described in several references.

Categorical variables also form the primary subject matter of Huberty and Petoskey (chapter 7, this volume), where levels are employed as fixed predictors (i.e., independent variables) that partially explain differences between observations of continuous dependent random variables. Counts appear there only as fixed sample sizes. In contrast, we consider categorical dependent variables whose associated *random counts* require explanation by other predictors. The predictors may be categorical, continuous, or both. We assume the reader is familiar with regression and analysis of variance (ANOVA) for continuous observations, hopefully in the context of the general linear model. Presentation of theoretical material is primarily expository, but mathematical detail is included where essential.

Researchers require tools for working with dependent counts and categorizations that fulfill the same scientific purposes as the classical general linear model analyses for continuous data: regression, correlation, ANOVA and analysis of covariance (ANCOVA). This chapter gives a unified discussion of Poisson regression, logistic regression, and loglinear modeling of contingency tables. These are three special cases of the general *loglinear* model, wherein expected category counts are *products* of effects of independent variables. This contrasts with the general *linear* model, in which expected means of continuous measurements are *sums* of such effects. Poisson and logistic regression each provide regression, ANOVA, and ANCOVA-like analyses for response counts with, respectively, one and two levels. Poisson regression is most commonly used to analyze rates, whereas logistic regression is used to analyze proportions, a distinction to be further clarified below. Loglinear modeling of contingency tables is a correlation analysis for multiple categorical responses that can also be used, in a manner akin to ANOVA, to model a dichotomous or polytomous dependent variable as a function of multiple independent categorical variables. Readers desiring a more complete discussion of the material of this chapter may refer to Agresti (1996), an excellent, readable text.

## B. Scope of Application

We illustrate the similarities and differences of the methods with three data sets. Table 14.1 summarizes the frequencies of respiratory illnesses, from late August 1991 to late February 1992, reported by a stratified random sample of students from a very large college population. These frequencies are cross-tabulated by three factors plausibly related to exposure or susceptibility: current cigarette smoking, bedroom occupancy (single or shared), and gender. Each cell of Table 14.1 corresponds to a combination of these three dichotomous predictors, for which the number of students surveyed (in parentheses) and their collective total of respiratory illnesses are shown. Note that many students reported more than one illness, and each separate illness was counted.

Poisson regression is often the method of choice for such problems, which require analysis of random counts of one class of event (here, respiratory illnesses) observed in multiple places, times, or under varying condi-

**TABLE 14.1  Respiratory Illnesses among 116 College Students during a 6-Month Period, by Gender, Smoking, and Single versus Shared Bedroom: Illnesses (Students)[a]**

| Gender | Own bedroom | | Shared bedroom | | |
| --- | --- | --- | --- | --- | --- |
| | Smoker | Nonsmoker | Smoker | Nonsmoker | Total |
| Women | 9 (7) | 44 (26) | 14 (6) | 34 (17) | 101 (56) |
| Men | 7 (5) | 27 (24) | 7 (5) | 33 (26) | 74 (60) |
| Total | 16 (12) | 71 (50) | 21 (11) | 67 (43) | 175 (116) |

[a] Unpublished data from the control group of a matched case-control study reported by Imrey et al. (1996). The stratified sampling induced by the matching procedure is ignored for the illustrative purposes of this example.

tions. The observed respiratory illness rate in each cell is the random illness count divided by the product of the known number of students in the cell and the fixed 6-month observation period, expressed on an appropriate scale, such as illnesses/100 persons/year. The objective is to use the observed random counts to model the dependence of expected rates on the three susceptibility factors. More generally, the purpose of Poisson regression is to relate variation among a collection of statistically independent random counts, or among rates for which they serve as numerators, to categorical and/or quantitative descriptors of the domains from which the counts are observed.

Table 14.2 gives results of an experiment to study effects of dietary fat source and level on the development of breast tumors in a rat cancer model, and to examine whether these effects depend on the hormonal status of the animal. Rats were randomized to one of three hormonal groups: removal of ovaries (OVX), removal of ovaries with subsequent estrogen replacement by continuous infusion from a silastic implant (OVX/E), or sham surgery (OVX/S). The latter consisted of anesthesia and opening of the abdominal cavity, without removal of the ovaries. After dosing with a chemical carcinogen, 32 members of each hormonal group were then randomly assigned to each of four diets containing high or low levels (40% or 10% of total calories) of either beef tallow or corn oil. After premature death or eventual sacrifice at 70 weeks of age, the rats were classified by presence or absence of mammary adenocarcinomas at dissection.

The proportions of rats with adenocarcinoma varied dramatically across groups, from a low of 6% to a high of 84%. The analytic objective is to study how the three experimental factors contribute, individually and jointly, to this variation. Logistic regression is useful here and, more generally, when a response variable is dichotomous and one wishes to determine how qualitative and/or quantitative characteristics (e.g., surgical treatment, dietary fat percentages) of an observational unit (e.g., rat) determine its chances

**TABLE 14.2  Mammary Adenocarcinoma Prevalence at Necropsy, by Treatment and Dietary Fat Source and Level[a]**

| Treatment | Fat source | Fat level (%) | Adenocarcinoma Present | Adenocarcinoma Absent |
|-----------|-----------|---------------|---------|--------|
| OVX/S | Beef tallow | 10 | 19 | 13 |
| OVX/S | Beef tallow | 40 | 20 | 12 |
| OVX/S | Corn oil | 10 | 13 | 19 |
| OVX/S | Corn oil | 40 | 22 | 10 |
| OVX | Beef tallow | 10 | 3 | 29 |
| OVX | Beef tallow | 40 | 5 | 27 |
| OVX | Corn oil | 10 | 2 | 30 |
| OVX | Corn oil | 40 | 5 | 27 |
| OVX/E | Beef tallow | 10 | 15 | 17 |
| OVX/E | Beef tallow | 40 | 27 | 5 |
| OVX/E | Corn oil | 10 | 15 | 17 |
| OVX/E | Corn oil | 40 | 24 | 8 |
| All | Both | Both | 170 | 214 |

[a] Data from Clinton, Li, Mulloy, Imrey, Nandkumar, & Visek, 1995, used by permission of the American Society for Nutritional Sciences.

of falling into one response category or the other (e.g., adenocarcinoma present or absent).

The data in Table 14.3, a $2^4$ contingency table arranged as a $4 \times 4$ table to save space, were extracted from a case-control study of risk factors for nasopharyngeal cancer (NPC) in Malaysian Chinese. Detailed occupational

**TABLE 14.3  Occupational Exposures among Malaysian Chinese[a]**

| | | Total smokes Yes | | Total smokes No | |
|---|---|---|---|---|---|
| | | Gasoline fumes | | Gasoline fumes | |
| Wood dusts | Paint fumes | Yes | No | Yes | No |
| Yes | Yes | 16 | 4 | 11 | 6 |
| Yes | No | 19 | 16 | 5 | 27 |
| No | Yes | 10 | 4 | 15 | 6 |
| No | No | 40 | 104 | 23 | 258 |

[a] Unpublished data from a study reported by Armstrong et al. (1999). The author is grateful to R.W. Armstrong for permission to use these data for this example.

histories were taken from all participants. Subjects were then classified into categories depending upon whether or not they had been occupationally exposed to each of a variety of inhalants and other environmental conditions. Table 14.3 cross-classifies all subjects in the study by exposure to four types of inhalants: wood dusts, paint fumes, gasoline fumes, and smoke of any origin.

Interest here is in understanding the patterns of occupational exposures that individuals experienced over their working lifetimes, by describing the statistical relationships among these four exposures. The categorical outcome variable has 16 levels, one corresponding to each possible pattern of presence or absence of the four exposures; there are no predictors *per se.* If the exposures had been expressed quantitatively as a roughly multivariate normal vector, then the six pairwise correlations would have characterized the entire association structure—that is, all statistical dependence among the exposures. But no comparably parsimonious summarization is routinely available for four-dimensional categorical data such as shown in Table 14.3. Loglinear modeling may be used here to find simpler representations of statistical dependencies among the four occupational exposure dichotomies than the counts themselves provide. More generally, loglinear models are used for two purposes: (a) to represent associations in contingency tables cross-classifying an arbitrary number of variables with any numbers of levels, and/or (b) to model one polytomous categorical variable (e.g., first choice among five candidates for political office) as a function of others (e.g., urban, suburban, or rural location; gender; and socioeconomic status category).

### C. Why Not Linear Models?

Prior to the development of specialized methods for modeling categorical data, counts were often analyzed using heuristic adaptations of classical linear modeling. For example, individual dichotomous observations might be represented by values of 1 or 0 and these analyzed by ordinary least-squares (OLS) regression or ANOVA on a set of predictors. This seems reasonable because the sample mean of these values is the proportion $p$ falling in the first category, whereas its expectation, the population mean, is the associated probability $\pi$. Alternatively, observed proportions $p$ at different levels of predictor variables, or transformations of them (e.g., arcsin $p$), might be similarly modeled.

Such methods continue to see some use on the supposition that their results, although not optimal, will be satisfactory approximations for practical purposes. This is frequently the case, but under more narrow conditions than is widely appreciated. We review limitations of such heuristic methods.

The primacy of the classical linear model for means owes much to how easily it represents situations where predictors have stable effects in varying circumstances. Simple linear regression portrays relationships in which a change of magnitude $\delta$ in a continuous predictor has constant additive impact on the mean response, whatever the predictor's initial value. Similarly, main effect models in regression or ANOVA portray relationships in which a change in any predictor has constant additive impact on the mean response, whatever the initial values or levels of other predictors.

However, when proportions are represented by modeling of 0–1 indicator variables, such "strictly linear" models may easily yield nonsense predictions—probabilities above 1 or below 0 for predictor combinations within or near the region of the data. Such models may also be unrealistic even when predictions are admissible. Increasing a probability $\pi$ by 0.3 from 0.1 by 0.4 is additively the same as increasing $\pi$ from 0.65 to 0.95, but rarely would one expect precisely the same stimulus to produce both these effects. This suggests that stability of effect is not best represented on the linear scale. Similar considerations also apply to the analysis of rates rather than proportions.

Aldrich and Nelson (1984) provide an excellent illustration of the conceptual failure of a strictly linear model: modeling the probability of home ownership as a function of family income. It is inconceivable that a fixed increase in income, say by $5,000, would increase the probability of home ownership equally across the entire income distribution. Such an increase could have little or no effect upon the poorest or richest among us, but must have a discernible effect upon those in an intermediate range, where income level is the primary determinant of the ability to obtain a mortgage. Thus, the probability of home ownership must follow an $S$-shaped or "sigmoid" curve, near zero and rising very slowly at low incomes, rising more rapidly in an intermediate range, then leveling off and creeping upwards towards one at high incomes. The cumulative logistic distribution is one such curve, and it forms the basis for logistic regression.

Even where the strictly linear model is reasonable, OLS regression and ANOVA may be poor approaches to statistical inference. Inferential methods for the classical linear model are predicated on the assumptions that individual observations are statistically independent, share a common variance, and have at least approximately normal (i.e., Gaussian) distributions. Categorical data analyses may often be formulated so that the assumptions of statistical independence and approximate normality are satisfied. However, the assumption of a common variance is essentially always violated for analyses of counts. This is so because, in contrast to Gaussian observations, the variability of either a single dichotomous outcome or a category count is inherently linked to the corresponding mean. Consequently, predictors that influence means of proportions or rates inevitably change the variability of these proportions or rates as well. For analyses involving small sample sizes, approximate normality may also be unrealistic.

Although OLS and ANOVA parameter estimates remain unbiased

under such violations, they lose precision relative to optimal methods. Worse, variances of these estimates are mis-estimated by OLS, and the assumed distributional shapes of test statistics do not apply in small samples. Consequently, the desired coverage properties of confidence intervals and the intended Type I error rates ($\alpha$'s) of hypothesis tests may be lost.

When modeling proportions, these difficulties are most pronounced when the true variances of observations are highly discrepant and sample sizes are small. A common rule of thumb is that OLS regression and ANOVA are unlikely to greatly mislead if sample sizes are at least moderate and the range of true probabilities underlying the data is from 20–80%. Within this range, the standard deviations of individual dichotomous observations can vary by no more than 25%. Though apparently based only on this latter fact and the rough intuition of knowledgeable practitioners, the guideline is probably fairly reliable and gives some assurance that conclusions from many past analyses are valid.

However, it is currently very difficult to justify the use of OLS regression or ANOVA even for analyses that fall within this guideline. Strictly linear models are often substantively inappropriate and, when they are appropriate, generalized linear modeling (see below and McCullagh & Nelder, 1989) and weighted least-squares (Goldberger, 1964; Grizzle, Starmer, & Koch, 1969; Koch, Imrey, Singer, Atkinson, & Stokes, 1985) provide superior methods for fitting and drawing statistical inferences from them. Our purpose in this chapter, therefore, is to introduce and illustrate the methods for nonlinear modeling of counts that form the foundation of contemporary categorical data methodology.

## II. MEASURING ASSOCIATION FOR COUNTS AND RATES

The parameters of linear models for continuous data represent differences in means, or functions of such differences. The parameters of Poisson, logistic regression, and loglinear models represent measures of association used to describe count data. Specifically, parameters of Poisson models reflect differences of logarithms of rates. These differences are, equivalently, logarithms of *rate ratios*. Similarly, parameters of logistic regression and loglinear models reflect differences of logarithms of odds, which are equivalently logarithms of *odds ratios*. In some circumstances a rate ratio or *odds ratio* may be used to approximate a related quantity, the *risk ratio*. This section provides a basic overview of these measures of association and their use.

### A. Rate Ratios

A *rate r* measures change over time and is the quotient of a count of events and a measure, in space–time or person–time units, of the extent of the observational domain from whence the count was obtained. Such quotients are often reexpressed in conventional units. Thus, a rate of bald eagle hatchings might be expressed as 1 hatching per 1000-square miles per year, a rate of meteorites passing through a portion of the upper atmosphere as 10 meteorites per 1000 cubic miles per day, and a rate of new colds as 2 colds per person-year. A person-year represents one person observed for a year, 12 observed for 1 month each, one observed for 3 and another for 9 months, or any other such combination. A more common example is the annual death rate in a *dynamic* community, that is, one with continuously changing population such as any city. This is expressed as deaths per 100,000 person-years, where person-years for the computation are estimated by the midyear population, under the assumption that population change is relatively steady throughout the year. We may think of an observed rate as estimating the long-run rate parameter $\lambda$ underlying a process that generates new events. One use of Poisson regression models is to provide smoothed, statistically stable estimates of such rate parameters for fixed, observed values of independent variables, and possibly for interpolations to intermediate unobserved values as well.

Rates may be estimated for events from an entire domain of observation, or rates from subdomains may be calculated and compared. Thus, the overall or *crude* rate of respiratory illnesses in Table 14.1 is $r_T = 175/(0.5 \times 116) = 3.02$ colds per person-year, whereas the rates for female and male nonsmokers with their own bedrooms are, respectively, $r_{FNO} = 44/(0.5 \times 26) = 3.38$, and $r_{MNO} = 27/(0.5 \times 24) = 2.25$ colds per year. The terms *conditional* or *specific* are used to describe these latter rates, because each is calculated under conditions that specify a portion of the observational domain (e.g., female nonsmokers with their own bedrooms). The rates for all females, all males, and all nonsmokers with their own bedrooms are respectively $r_F = 101/(0.5 \times 56) = 3.60$, $r_M = 74/(0.5 \times 60) = 2.47$, and $r_{NO} = 71/(0.5 \times 50) = 2.84$ colds per year. These are called *marginal* rates because, inasmuch as they condition on some variables and collapse over categories of others, they may be calculated from the right row margin and bottom column margin of Table 14.1. The concept of obtaining marginal measures by jointly conditioning and collapsing on complementary sets of variables is, however, completely independent of the format in which a particular table is displayed.

A natural measure of the association of gender with illness is the rate ratio, the quotient of a rate among women to a rate among men. This may be calculated marginally, by ignoring smoking and bedroom occupancy, as $r_W/r_M = 3.60/2.47 = 1.46$ estimating $\lambda_W/\lambda_M$, or fully conditionally as $r_{WNO}/r_{MNO} = 3.38/2.25 = 1.50$ estimating $\lambda_{WNO}/\lambda_{MNO}$. In this instance the marginal and conditional odds ratios are quite similar, but in general they need not be. Marginal and conditional associations of illness with each of the other variables may be calculated similarly. In Poisson regression models, the natural logarithms of conditional rate ratios play a role analogous to those of slopes in linear regression and of differences between cell means in ANOVA. Under such models estimated conditional rate ratios are functions of, and sometimes exactly reproduce, their marginal counterparts.

## B. Probability (Risk) Ratios

The probability of an adverse outcome is sometimes referred to as *risk*. Such a probability $\pi$ is estimated from a sample by its corresponding observed proportion $p$. The terminology of conditional and marginal rates applies straightforwardly to probabilities and their estimates. Thus, from Table 14.2, the overall proportion of rats with adenocarcinoma is $p_{+++} = 170/384 = 44.3\%$, where the three subscripts denote treatment, fat source, and fat level, and a subscripted $+$ is used to indicate pooling over all levels of the corresponding variable. Similarly, the marginal proportions of rats with adenocarcinoma among those with OVX/S, OVX, and OVX/E are respectively $p_{1++} = 74/128 = 57.8\%$, $p_{2++} = 15/128 = 11.7\%$, and $p_{3++} = 81/128 = 63.2\%$, whereas the corresponding conditional proportions among rats fed 10% beef tallow are $p_{111} = 19/32 = 59.4\%$, $p_{211} = 3/32 = 9.4\%$, and $p_{311} = 15/32 = 46.7\%$.

In this context, risk ratios that are analogous to rate ratios measure association between the response variable, adenocarcinoma, and any of the three predictors. Thus, the marginal association between adenocarcinoma occurrence and surgery may be measured by the marginal risk ratios $p_{1++}/p_{2++} = 4.93$ estimating $\pi_{1++}/\pi_{2++}$ for OVX/S relative to OVX, and $p_{3++}/p_{2++} = 5.40$ estimating $\pi_{3++}/\pi_{2++}$ for OVX/E relative to OVX. The corresponding conditional risk ratios among rats fed 10% beef tallow are $p_{111}/p_{211} = 6.33$ and $p_{311}/p_{211} = 5.00$, respectively.

When an event occurs at constant rate $r$ over time, the probability of at least one occurrence during a period of length $t$ may be shown to be $\pi = 1 - e^{-rt}$. So the risk ratio comparing two groups observed for the same period is $\pi_1/\pi_2 = (1 - e^{-r_1 t})/(1 - e^{-r_2 t})$. When $r_1 t$ and $r_2 t$ are small, this closely approximates the rate ratio $r_1/r_2$; however, $\pi_1/\pi_2$ must approach 1 as $t$ increases. Hence, the risk ratio is used as an estimate of the rate ratio for events that are uncommon during the time period observed, but otherwise tends to be nearer to 1 than the rate ratio. Because of the close relationship of the two measures for uncommon events, the term *relative risk* and the abbreviation *RR* are often used for either rate ratio or risk ratio, depending upon the context.

## C. Odds Ratios

A probability $\pi$ may instead be represented as the corresponding odds $\omega = \pi/(1 - \pi)$, with inverse transformation $\pi = \omega/(1 + \omega)$. Although probabilities take values between 0 and 1, odds take values between 0 and $\infty$. The probability and odds uniquely determine one another. The odds ratio $\Omega = \omega_1/\omega_2 = [\pi_1/(1 - \pi_1)]/[\pi_2/(1 - \pi_2)]$ is the measure of association on the odds scale analogous to the risk ratio on the probability scale. When $\pi_1$ and $\pi_2$ are both small, $1 - \pi_1$ and $1 - \pi_2$ are both close to 1 and $\Omega$ closely approximates both risk ratio and rate ratio. Otherwise, $\Omega$ may differ from them greatly; if $\pi_1 = 3/4$ and $\pi_2 = 1/4$, then the risk ratio is 3 but $\Omega = 9$. Population odds $\omega$ and odds ratios $\Omega$ are readily estimated from samples by replacing each $\pi$ by the corresponding proportion $p$ in the above expressions; thus, $o = p/(1 - p)$ estimates $\omega$ and $OR = o_1/o_2$ estimates $\Omega$. Crude, conditional and marginal odds and odds ratios are defined analogously to their counterparts for rates and probabilities.

Now consider the $2 \times 2$ subtable at the bottom of Table 14.2, relating adenocarcinoma to fat level among OVX/E rats fed corn oil. Letting $y_{ij}$ represent the count in the cell at the intersection of the $i^{th}$ row and $j^{th}$ column of the subtable, then the ratio of adenocarcinoma odds for 10% versus 40% fat is $OR = y_{11}y_{22}/y_{12}y_{21} = (15 \times 8)/(17 \times 24) = 5/17$. Note that the result of this computation remains the same if we interchange row and column dimensions by flipping the table over its main diagonal. Nor does the formula change even if fat level instead of adenocarcinoma is (mistakenly) treated as random, and one calculates the ratio of conditional odds of being fed 10% fat among adenocarcinoma victims to those conditional odds among rats without adenocarcinoma. Thus, the "risk" OR, from (a) a "cohort" study that observes the development of an outcome in groups defined by some preexisting characteristic, is calculated identically as the "exposure" OR, from either (b) a "case-control" study that compares the frequency of that characteristic in groups having experienced or escaped the outcome event, or (c) a "cross-sectional" study of a random sample from the same population. One may show also that the ORs from all three types of studies estimate the same parameter $\Omega$. Hence, data from case-control and cross-sectional studies may be used to estimate relative odds of an outcome, and thereby relative risk (rate ratio or risk ratio) of an *uncommon* outcome, without directly observing the period during which the outcome develops—a point of great scientific utility.

In logistic regression models, conditional odds ratios play the same role as do conditional rate ratios in Poisson regression, and their logarithms play a role analogous to those of slopes in linear regression and differences in cell means in ANOVA. Under a logistic regression model, estimated conditional odds ratios are functions of and sometimes exactly reproduce their marginal counterparts.

The concepts of odds, odds ratios, and estimated odds ratios OR may be readily generalized to arbitrary $I \times J$ contingency tables by applying the formulae for a $2 \times 2$ table to all $2 \times 2$ subtables formed by intersecting any two rows of the $I \times J$ table with any two of its columns. The OR extracted above from the $12 \times 2$ Table 14.2 is of this character. In contingency tables of arbitrary dimension, $2 \times 2$ subtables may be extracted from each two-dimensional "slice" obtained by fixing levels of all but two dimensions. For instance, one may also think of Table 14.2 as a $3 \times 2 \times 2 \times 2$ table from which, conditioning on fat source = corn oil, fat level = 40%, and treatment = OVX/S or treatment = OVX, one might calculate the generalized odds ratio relating adenocarcinoma to treatment as OR = $(22 \times 27)/(10 \times 5) = 11.88$. These conditional generalized ORs play the same role in loglinear models as do conditional odds ratios in logistic regression models.

## III. GENERALIZED LINEAR MODELS

Poisson regression, logistic regression, and loglinear models are special cases of the *generalized linear model* (GLM). Consequently, they use intimately related probability distributions to portray random variation, employ structural models of similar form, apply the same maximum likelihood-based toolkit for statistical inference, and share computational algorithms. In this section we describe the one-parameter generalized linear model. Acquaintance with the framework it provides is essential to an integrated rather than fragmented understanding of models for count data. Note that the classical general linear model is a special case of the (two-parameter) generalized linear model, so that the reader will find some familiar concepts here, embedded in a richer context.

### A. Exponential Family, Link Function, and Linear Predictor

Consider a sequence of random observations $y_1, y_2, \ldots, y_I$, and write $\mu_i = E(y_i)$. The structure of any GLM takes the form $g(\mu_i) = \Sigma_{\tau=1}^{T} x_{i\tau} \beta_\tau$ or, in equivalent vector form, $g(\boldsymbol{\mu}) = X\boldsymbol{\beta}$, where $X = (x_1', x_2', \ldots, x_I')'$ is a known *model matrix*, $\boldsymbol{\beta}$ is a $T$-vector of unknown parameters, and $g(\cdot)$ is an invertible and differentiable *link function*. When the link function $g(\cdot)$ is the identity transformation, this is the structure of the general linear model, which further assumes that the $y_i$ arise independently from Gaussian distributions sharing a common unknown variance $\sigma^2$.

However, GLMs allow the use of other *exponential family* distributions. Univariate one-parameter exponential families will be sufficient for our purposes. These distributions have probability density or mass functions $f(y)$ that may be written as $f(y) = Ae^B$, wherein each of $A$ and $B$ can be factored into a pair of terms, with the random observation $y$ and the single unknown parameter, say $\lambda$, appearing only in different members of each pair. This separability condition is satisfied by the Poisson and binomial mass functions, as well as the densities of geometric, logarithmic, and gamma distributions, and of the Gaussian (normal) and inverse Gaussian families for any fixed variance.

A GLM is then defined by three components: (a) the choice of probability law from within the exponential family, (b) the link function $g(\cdot)$, and (c) the model matrix $\mathbf{X}$ that specifies the *linear predictor* $\mathbf{X}\boldsymbol{\beta}$. Of these, the two former choices represent extensions of the usual fixed effects linear models for continuous data. The model matrix $\mathbf{X}$ can and usually does take the same forms in generalized as in classical linear models, although the interpretations of resulting parameters differ depending on the underlying distribution and choice of link. The link $g(\cdot)$ suggests, but must be clearly distinguished from, the data transformations often used with classical linear models to stabilize variance and/or to bring an empirical distribution into better conformity with the Gaussian assumption. The link $g(\cdot)$ does not transform the data; ideally, in a generalized linear model the data are never transformed. Rather, a distribution is chosen that is inherently justifiable, and the link function acts solely as a choice of scale on which a linear parametric model is reasonable.

Due to the special structure of exponential family distributions, likelihood-based inference may be dealt with conveniently and in a unified fashion for all GLMs. Specifically, the equations for maximum likelihood estimation may be solved using Fisher scoring, a well-known iterative computational algorithm. Additional simplification occurs when the link function is matched to the exponential family by choosing as $g(\cdot)$ the factor of the exponent $B$ (of $f(y) = Ae^{B}$) that contains the parameter $\lambda$. This factor is called the *canonical link* for the distribution $f(y)$. The three types of models in this chapter employ Poisson and binomial distributions and their canonical links. We now briefly review classical likelihood theory as used with GLMs.

## B. Likelihood-Based Estimation and Testing

Maximum likelihood-based methods that generalize OLS are used to estimate and test hypotheses about the parameter vector $\boldsymbol{\beta}$ of any GLM. Consider an observation vector $\mathbf{y}$ of length $n$ with associated loglikelihood function $l(3\boldsymbol{\beta};\mathbf{y})$; that is, $l(\boldsymbol{\beta}; \mathbf{y})$ is the natural logarithm of the probability mass or density function of $\mathbf{y}$, considered as a function of the parameters in $\boldsymbol{\beta}$ for the fixed observed data in $\mathbf{y}$. The *likelihood score* $\mathbf{S}(\boldsymbol{\beta})$ for $\boldsymbol{\beta}$ is the vector of derivatives, that is, slopes, of $l(\boldsymbol{\beta}; \mathbf{y})$ with respect to the parameters $\beta_i$, evaluated at $\boldsymbol{\beta}$. Similarly, the *information matrix* $\mathbf{I}(\boldsymbol{\beta})$ for $\boldsymbol{\beta}$ is the matrix of expectations (i.e. average values over the distribution of $\mathbf{y}$) of the slopes of the elements of $\mathbf{S}(\boldsymbol{\beta})$ with respect to the $\beta_i$, also evaluated at $\boldsymbol{\beta}$.

Then the approximate distribution of the score vector $\mathbf{S}(\boldsymbol{\beta})$ in large-samples is $\mathrm{MN}(\mathbf{0}, \mathbf{I}(\boldsymbol{\beta}))$, that is, multivariate normal with mean $\mathbf{0}$ and covariance matrix $I(\beta)$. When $l(\beta; y)$ is a smooth function near $\beta$, the *likelihood equations* $S(\beta) = 0$ have a solution $\hat{\beta}$ that maximizes $l(\beta; y)$ (and hence the likelihood itself) and that converges to $\beta$ with increasing $n$. This is the *maximum likelihood estimate (mle)* of $\beta$. The approximate distribution of $\hat{\beta}$ in large samples is $MN(\beta, I(\beta)^{-1})$. The mle may be determined using Fisher scoring.

If $\beta = (\beta_1', \beta_2')'$, then three methods are commonly employed to test linear hypotheses of the form $H_0: \beta_2 = 0$. Let $\hat{\beta}^0 = (\hat{\beta}_1^{0\prime}, 0')'$ be the mle of $\beta$ under $H_0$. The above distributions of $\hat{\beta}$ and $S(\hat{\beta})$ imply, when $H_0$ is true and the sample is sufficiently large, that $\hat{\beta}_2$ and $S(\hat{\beta}^0)$ should both be close to $0$, and $l(\hat{\beta}^0; y)$ should be close to $l(\hat{\beta}; y)$. $H_0$ may thus be tested using either (a) the likelihood ratio statistic $Q_L = -2\log\Lambda = 2[\log(l(\hat{\beta}^0; y) - \log(l(\hat{\beta}; y))]$ from the difference of these two loglikelihoods, (b) the quadratic form *Wald statistic* $Q_W$, which measures the distance of $\hat{\beta}_2$ from $0$ against its variability estimated from $I(\hat{\beta})^{-1}$, or (c) the quadratic form *score statistic* $Q_S$, which measures the distance from $0$ of the subvector of scores corresponding to elements of $\beta_2$ (evaluated at $\hat{\beta}^0$), against its variability estimated from $I(\hat{\beta}^0)$.

For the general linear model the three statistics are identical. For other GLMs, $Q_W$ and $Q_S$ may be interpreted as approximations to $Q_L$. Under $H_0$, $Q_L$, $Q_W$, and $Q_S$ all have the same large-sample chi-square distribution with degrees of freedom (df) equal to the length of $\beta_2$. Each may be extended to test general linear hypotheses of form $C\beta = 0$ for arbitrary matrix $C$. Note that the Pearson chi-square test for independence in a two-way contingency table is a score statistic, and many other classical categorical data tests use score and Wald statistics.

With GLMs, these chi-square statistics are used for two distinct purposes: checking goodness-of-fit and comparing alternative models. When $\beta$ contains as many elements as the number of observed patterns of values of predictor variables, in which case the observed means for each pattern may be precisely predicted, the model is said to be *saturated*. For instance, a factorial model including all main effects and interactions among its dimensions is saturated. In this context, tests of $H_0$ are "goodness-of-fit" statistics that compare model-based predictions directly to the observed means. For a GLM, this likelihood ratio goodness-of-fit statistic is known as the *deviance* of the model.

When $\beta$ does not saturate the data, tests of $H_0$ examine the statistical significance of one or more parameters in the context of a "reduced" model from which some (presumably random) variation in the data has already been smoothed away. By definition of the loglikelihood ratio, $Q_L$ for such tests may always be calculated as the difference in deviance statistics for the models corresponding to $\beta$ and one of its subvectors. These tests are undertaken in confirmatory analyses because $H_0$ suitably expresses a scientific question of interest, or in exploratory analyses to entertain either simplification of an acceptable model or expansion of a possibly oversimplified model. In comparison to likelihood ratio tests, Wald tests are computationally efficient and commonly employed for purposes of model simplification, whereas score tests play a similar role in model expansion. Frequently, data analysis of GLMs will involve examination and testing of one or more series of models that are "nested," in the sense that each involves an additional expansion (forward selection) or simplification (backward elimination) of the model from the prior step.

## C. Hierarchical Models, Sufficient Statistics, and Marginal Distributions

Models for counts numerically preserve certain features of the data, while smoothing other features away. To use these models properly, one needs to appreciate how the choice of a model determines which features of the data are preserved and which are removed. Below we define hierarchical GLMs, and note that this class of models preserves particularly simple, natural data features. This will be elaborated further in the context of each type of GLM.

*Hierarchical models* refer here to polynomial regression models that are complete up to a given order (this excludes, for instance, the model $E(y_i) = \alpha + \beta x_i^2$), and ANOVAs or ANCOVAs that incorporate the main effects and other lower order relatives of any included interactions (this excludes, for instance, the model $E(y_{ij}) = \mu + \alpha_i + \gamma_{ij}$). Most generalized linear models used in practice are hierarchical in this sense, because nonhierarchical models are only occasionally scientifically sensible.

For all GLMs with canonical links, the likelihood equations take the remarkably simple form $\sum_{i=1}^{n} x_{i\tau}\mu_i = \sum_{i=1}^{n} x_{i\tau}y_i$, $\tau = 1, \ldots, T$. These are solved by reexpressing the $\mu_i$ in terms of $\boldsymbol{\beta}$, and then applying Fisher's method of scoring. The solution $\hat{\boldsymbol{\beta}}$ can thus depend on the observed data $\boldsymbol{y}$ only through the $T$ linear combinations $\sum_{i=1}^{n} x_{i\tau}y_i$, $\tau = 1, \ldots, T$. These linear combinations are known as *sufficient statistics* for $\boldsymbol{\beta}$, because knowledge of them conveys all the information in the data needed for finding the mle $\hat{\boldsymbol{\beta}}$. Thus, all data sets that share the same values of these sufficient statistics will yield the same $\hat{\boldsymbol{\beta}}$ and $\hat{\mu}_i$, no matter how they may otherwise differ.

However, the likelihood equations state also that the fitted means $\hat{\mu}_i$, $i = 1, \ldots, n$, precisely preserve the sufficient statistics. For hierarchical models, the preserved statistics are characteristics of the marginal distributions obtained by collapsing over some, perhaps most, of the predictor variables. The particular forms of these statistics are simple: sums of counts, overall sums of quantitative predictor values, sums of such values within levels of qualitative predictors, and cross-product sums. In general, these statistics will determine certain marginal measures of association (e.g., rate or odds ratios) that are then reflected in the fitted GLM. Each such fitted model is thus a synthesis of specific marginal properties of the observed data with a conditional structure imposed by the model. Indeed, the conditional structure dictates precisely which marginal properties of the data will be preserved, and vice versa.

As noted earlier, the marginal associations that are preserved may differ from the conditional associations implied by the model. Sometimes, however, the model structure will require that these marginal and conditional associations be identical. In such cases the data may be "collapsed" over certain variables—that is, those variables may be ignored—without loss or distortion of information about other variables. Collapsing tables of counts over some variables is a ubiquitous method of data summarization. The structures of fitted GLMs provide guidance as to when collapsing may be inappropriate. We will consider this in greater detail later.

## IV. POISSON REGRESSION MODELS

Poisson regression models are used to describe variation in frequencies or rates of phenomena that occur randomly and independently from place to place and/or time to time, at uniform rates throughout homogeneous domains of observation, and singly rather than in concurrent clusters. The domains may be areas, spatial regions, or fixed populations of individuals. Domains may also be aggregated collections of experience, such as person-years observed or passenger miles traveled, for either a static group of individuals or a population experiencing births, deaths, immigration, and emigration. Studies of disease incidence or mortality rates are examples in which such domains of experience are typically used. By definition, the expected number of events occurring in any observational domain is the product of the extent of the domain with the rate of events within it.

The assumption that each event occurs independently of events elsewhere in the domain, or in the past, excludes situations with a fixed limitation on the number of observable events. Thus, Poisson regression is ordinarily not suitable for studies of events that can occur only once to each member of a group of fixed, known size. Rather, members of such a group are categorized into those who have and those who have not experienced the event and are analyzed by logistic regression.

However, a notable exception in practice arises in the study of quite rare events, when the expected number of events in each fixed-size group is very small relative to the size of the group. In that case the theoretical limitation on events has negligible impact. Poisson regression models of incidence and mortality rates from rare chronic diseases, such as specific cancers, are examples. More commonly, Poisson regression for such studies is justified because the observational domains are aggregated person-years from dynamic populations, rather than a group of $n$ fixed individuals with each observed for a full year. In addition, incidence studies of many diseases include invididuals who may experience more than one case.

Poisson regression uses the Poisson probability law to represent random variation, with the link function $g(\mu) = \log(\mu)$. Here, as with all GLMs, log denotes natural logarithms for mathematical simplicity. This link makes the parameters logarithms of rate ratios, or linear functions of logarithms of rate ratios. For categorical independent variables, the rate ratios compare predicted rates at individual levels to the predicted rate at a baseline level, or to the geometric mean of rates at all levels, while holding other independent variables constant. For continuous independent variables, the rate ratios compare the predicted rate for a given value of the independent variable to the rate associated with a one-unit reduction in that value, also holding all other independent variables constant. The term *Poisson regression* incorporates models that would be described as regression, ANOVA, and ANCOVA if applied to Gaussian observations. Although at this writing most current software packages do not include programs specifically for Poisson regression, these models are readily fit using general-purpose GLM routines such as GLIM©, SAS© PROC GENMOD, and the S-PLUS© glm function.

We begin with a basic description of the Poisson distribution, and of how fixed denominators of rates are incorporated into Poisson regression models. We then illustrate with analyses of two data sets.

## A. Technical Overview

### 1. Poisson Distributions

The Poisson law Poi($\lambda$) with rate parameter $\lambda$ gives the probability distribution of a count of events observed in a region of unit extent. It is presumed that events occur homogeneously, randomly, and independently, and that the probability of events occurring together in clumps is vanishingly small. If $y$ is the random count, the mathematically implied form of the Poisson probability mass function is $Pr\{y\} = e^{-\lambda}\lambda^y/y!$ for all non-negative integer $y$, and 0 elsewhere. The mean of this distribution is $\lambda$, hence the term "rate parameter," but $\lambda$ is the variance as well. Thus, counts of Poisson events that occur frequently must vary more than counts of less common events—a variance heterogeneity that propogates through all of categorical data analysis. Poisson distributions are discrete, taking values only on positive integers, and for small $\lambda$ have long right tails. Poisson distributions with larger $\lambda$ have Gaussian shapes, and cumulative probabilities that may be approximated by Gaussian distributions with $\mu = \sigma^2 = \lambda$.

By the homogeneity and independence conditions stated above, expanding or shrinking the domain of observation simply multiplies the parameter, hence the mean and variance, by the expansion or shrinkage factor; for a domain of extent $t$, $Pr\{y\} = e^{-(\lambda t)}(\lambda t)^y/y!$.

## 2. Implementation

Since GLMs model a collection $\{y_i\}$ of independent counts, the underlying probability law for Poisson GLMs is the product of $I$ Poisson distributions with respective parameters $\lambda_i t_i$. The log link specifies the Poisson regression model log $y = X\beta$, with $X$ determining the linear predictor and specific parameterization.

Consider initially the case of all $t_i = 1$, so that all counts arise from observing domains of the same extent, and the rate parameters are expressed in that unit. This would apply, for instance, to the modeling of cold rates per 6-month period from the data summarized in Table 14.1, had those data been reported individually for each of the 116 students, rather than aggregated by combinations of levels of Gender, Smoking, and Bedroom Sharing. The likelihood equations are $X'\lambda = X'y$ and the information matrix for $\beta$ is $I(\lambda) = X'D_\lambda X$, where in general $D_z$ is the diagonal matrix with main diagonal $z$. The large-sample covariance matrix of $\hat{\beta}$ may thus be estimated by $\text{Cov}(\hat{\beta}) = I^{-1}(\hat{\lambda}) = (X'D_\lambda X)^{-1}$.

Data with varying $t_i$ (e.g., Table 14.1 as presented) are readily handled by slight alteration of the canonical link. Since $\lambda_i = \mu_i/t_i$, the model log $\lambda = X\beta$ is equivalent to log $\mu - \log t = X\beta$, where $t$ is the vector of extents $t_i$. The link function for this model is $g(\mu_i) = \log \mu_i - \log t_i$, and we have log $\mu_i = x_i'\beta + \log t_i$, where $x_i'$ is the $i^{th}$ row of $X$. The linear predictor $X\beta$ is thereby offset by log $t$ due to the varying $t_i$. The previous paragraph remains correct with $\lambda$ replaced by $\mu$. To accomplish this adjustment, the "offset vector" log $t$ must, however, be supplied explicitly to computing software.

## B. Respiratory Illness Rates among College Students

We illustrate with analyses of two data sets. Consider first Table 14.1. We use log (# students) as offset for each group. The deviance of the model $X = (1, 1, \ldots, 1)'$ with only an intercept parameter is the likelihood ratio statistic for testing equality of respiratory illness rates across all eight groups, analogous to the omnibus one-way ANOVA F-statistic. This takes the value 9.28, for which the asymptotic $\chi_7^2$ distribution yields $p = .23$. Thus, from a purely exploratory standpoint, there is insufficient evidence to conclude that any of gender, smoking, or bedroom sharing influence the respiratory illness rate.

### 1. Factorial Analysis

However, since the data have a factorial structure, a more aggressive approach is to analyze Table 14.1 by modeling akin to a factorial ANOVA. Writing $\lambda_{ijk}$ for the respiratory illness rate per 6 months among students in the $i^{th}$ level of Gender, $j^{th}$ level of Smoking, and $k^{th}$ level of Bedroom Sharing, a full factorial model is log $\beta_{ijk} = \beta + \beta_i^G + \beta_j^B + \beta_k^S + \beta_{ij}^{GB} + \beta_{ik}^{GS} + \beta_{jk}^{BS} + \beta_{ijk}^{GBS}$, where $G$, $B$ and $S$ represent Gender, Bedroom Sharing, and Smoking, respectively, and $i,j,k = 1,2$. As in ANOVA, this model is overparameterized. Constraints must be applied that simultaneously identify the parameters, and allow their estimation by determining a full rank model matrix $X$. Several choices for these are available, and the default implementation varies across available software. We apply the classical ANOVA "deviation" constraints $\Sigma_{i=1}^2 \beta_i^G = \Sigma_{j=1}^2 \beta_j^B = \Sigma_{k=1}^2 \beta_k^S = \Sigma_{i=1}^2 \beta_{ij}^{GB} = \Sigma_{j=1}^2 \beta_{ij}^{GB} = \Sigma_{i=1}^2 \beta_{ik}^{GS} = \Sigma_{k=1}^2 \beta_{ik}^{GS} = \Sigma_{j=1}^2 \beta_{jk}^{BS} = \Sigma_{k=1}^2 \beta_{jk}^{BS} = \Sigma_{i=1}^2 \beta_{ijk}^{GBS} = \Sigma_{j=1}^2 \beta_{ijk}^{GBS} = \Sigma_{k=1}^2 \beta_{ijk}^{GBS} = 0$. These imply that

$$
\log \lambda = X_S \beta = \begin{bmatrix}
1 & 1 & 1 & 1 & 1 & 1 & 1 & 1 \\
1 & 1 & 1 & -1 & 1 & -1 & -1 & -1 \\
1 & 1 & -1 & 1 & -1 & 1 & -1 & -1 \\
1 & 1 & -1 & -1 & -1 & -1 & 1 & 1 \\
1 & -1 & 1 & 1 & -1 & -1 & 1 & -1 \\
1 & -1 & 1 & -1 & -1 & 1 & -1 & 1 \\
1 & -1 & -1 & 1 & 1 & -1 & -1 & 1 \\
1 & -1 & -1 & -1 & 1 & 1 & 1 & -1
\end{bmatrix} \beta,
$$

with $\beta = [\beta, \beta_1^G, \beta_1^B, \beta_1^S, \beta_{11}^{GB}, \beta_{11}^{GS}, \beta_{11}^{BS}, \beta_{111}^{GBS}]$.

$X_S$ may be written much more compactly using the Kronecker product $\otimes$ and $\bar{\mathbf{1}} = [1 \ -1]'$. Thus, $X_S = [\mathbf{1}_2 \otimes \mathbf{1}_2 \otimes \mathbf{1}_2, \bar{\mathbf{1}} \otimes \mathbf{1}_2 \otimes \mathbf{1}_2, \mathbf{1}_2 \otimes \bar{\mathbf{1}} \otimes \mathbf{1}_2, \mathbf{1}_2 \otimes \mathbf{1}_2 \otimes \bar{\mathbf{1}}, \bar{\mathbf{1}} \otimes \bar{\mathbf{1}} \otimes \mathbf{1}_2, \bar{\mathbf{1}} \otimes \mathbf{1}_2 \otimes \bar{\mathbf{1}}, \mathbf{1}_2 \otimes \bar{\mathbf{1}} \otimes \bar{\mathbf{1}}, \bar{\mathbf{1}} \otimes \bar{\mathbf{1}} \otimes \bar{\mathbf{1}}]$, where $u \otimes v$ is a block matrix with $(i, j)^{th}$ block $u_{ij} v$. For brevity, we will occasionally use such notation.

Solving for $\beta$ yields a multiplicative decomposition of respiratory disease rates structurally identical to the additive decomposition of means given by factorial ANOVA. In particular, the exponentiated parameters are interpretable in terms of the fitted respiratory disease rates as follows: $e^\beta$ = geometric mean of respiratory disease rates across all eight groups; $e^{2\beta_1^G}$ = geometric mean, averaged over the four combinations of Bedroom Sharing and Smoking levels, of the respiratory disease rate ratios for women relative to men; $e^{2\beta_1^B}$, $e^{2\beta_1^S}$: main effects of Bedroom Sharing and Smoking, analogous to $e^{2\beta_1^G}$; $e^{4\beta_{11}^{GB}}$: geometric mean, over the levels of Smoking, of the quotient of the female versus male rate ratio among students with their own bedrooms to the same rate ratio among students sharing bedrooms (in other words, a geometric mean ratio of rate ratios); $e^{4\beta_{11}^{GS}}$, $e^{4\beta_{11}^{BS}}$: respectively, interactions of Gender with Smoking and Bedroom Sharing with Smoking on respiratory disease rate, analogous to $e^{4\beta_{11}^{GB}}$; $e^{8\beta_{111}^{GBS}}$: Bedroom Sharing by Gender ratio of rate ratios for smokers, divided by that for non-smokers. As a consequence of the imposed constraints, the above "log ratio of rate ratios" interpretations of two-variable interaction terms are equivalent, within the model, to their alternatives formed by exchanging the roles of the interacting predictors. The analogous statement also applies to the interpretation of $\beta_{111}^{GBS}$.

These data yield $\hat{\boldsymbol{\beta}} = [.418, .161, -.111, -.080, .025, -.055, -.039, .069]'$. In the orthogonal saturated model, each parameter has the same estimated standard error (se) of .096, obtainable as the square root of any diagonal element of the estimated information matrix. For simplicity, letting $l$ index the eight elements of $\boldsymbol{\beta}$, the Wald test of $H_0$: $\beta_l = 0$ is $Q_W = (\hat{\beta}_l/se(\hat{\beta}_l))^2$, with large-sample $\chi_1^2$ distribution. Wald statistics are frequently used for multiple significance tests of individual parameters or groups of parameters within the same model, because their computation does not require fitting an alternative model for each test, in contrast to the likelihood ratio and score statistic counterparts. Using Wald tests, no parameters are statistically significant at $\alpha = .05$, but the test for Gender main effect is suggestive (p = .093).

The tests for each parameter are adjusted for all other parameters in the model. In a balanced ANOVA all estimates would be independent; such adjustment would have no impact, and model reduction would leave remaining estimated parameters unchanged. In the categorical data setting, the inherent variance heterogeneity and the imbalance due to differing offsets produce correlation between the estimated parameters. Thus, each main effect test may be misleading, as it is adjusted for interactions that contain and are correlated with the main effect. Since the $p$-values for all individual interactions exceed 40%, it is reasonable to examine the main effects model $X_M$, incorporating only the first four columns of $X_S$.

The deviance of the main effects model, $Q_L^2 = 1.39$, yields $p = .85$ when evaluated as $\chi_4^2$, an apparently adequate fit. Within the main effects model, the likelihood ratio test of the joint significance of the three predictors may be obtained by direct comparison of loglikelihoods, or as the difference between the deviance of the main effect model and that of the intercept-only model. The latter approach gives a likelihood ratio chi-square for significance of the main effects model of $(9.28 - 1.39) = 7.89$ with $df = 3, p = .048$, providing modest evidence for some effect. The estimated effects (se's) are $\hat{\beta}_1^G = 0.199(0.077)$, $\hat{\beta}_1^B = 0.096(0.077)$, and $\hat{\beta}_1^S = 0.018(.093)$. The corresponding Wald and likelihood ratio statistics of $Q_W = 6.64$ ($p = .010$), $Q_L = 6.73$ ($p = .010$) for Gender, $Q_W = Q_L = 1.58$ ($p = .21$) for Bedroom Sharing, and $Q_W = Q_L = 0.04$ ($p = .85$) for Smoking show clearly that the association of interest is with Gender. The model estimates that women contract new respiratory illness at roughly $e^{2 \times .199} = 1.49$ times the rate for men at the same levels of the other two factors. Using the approximate normal distribution of $\hat{\beta}_1^G$, a 95% confidence interval for the true rate ratio may be obtained as $e^{2\hat{\beta}_1^G \pm 1.96 \times (2 \times se(\hat{\beta}_1^G))} = (1.10, 2.01)$.

Note that the estimated rate ratio for gender is close but not equal to the marginal rate ratio of 1.46, which is also the estimated rate ratio obtained from a model excluding effects of bedroom sharing and smoking entirely. The discrepancy, in a no interaction model that implies a constant gender rate ratio, is due to adjustment for bedroom sharing and smoking when their associated extent measures—person-years in this instance—are not distributed proportionately (i.e. balanced) across genders. With balanced data, the parameter estimates for the main effects model would have precisely reproduced the marginal rate ratios: conditional and marginal rate ratios would have been equal for all three variables.

To confirm the fit of the model, we may examine the departures ($y_{ijk}$ − $\hat{\mu}_{ijk}$) of the counts in each group from their predicted counts based on $\hat{\boldsymbol{\beta}}$. The variances of these residuals differ, and they are somewhat correlated through their joint dependence on $\hat{\boldsymbol{\beta}}$. Computer software may offer *standardized residuals* and/or *adjusted residuals*. Standardized residuals account for the differing variances of the $y_{ijk}$ through division by $\hat{\mu}_{ijk}$, the estimated standard deviation of $y_{ijk}$. Squares of these standardized residuals $e_{ijk}$ = ($y_{ijk}$ − $\hat{\mu}_{ijk})/\hat{\mu}_{ijk}$ sum to the Pearson chi-square, or score statistic $Q_S$, for fit of the model. However, this standardization neglects the variability in the $\hat{\mu}_{ijk}$, as well as the differing influences of the $y_{ijk}$ on the fitted model, and hence on their associated residuals. As a consequence, although standardized residuals do tend to be normally distributed in large samples, they are underdispersed relative to the standard Gaussian law. The adjusted residuals $e'_{ijk} = e_{ijk}/\sqrt{1-\text{h}_{ijk}}$ are standard Gaussian in large samples, where $\text{h}_{ijk}$ depends on the information matrix and is a generalization to GLMs of the leverage of $y_{ijk}$ (see Venter & Maxwell, chapter 6, this volume). The adjusted residuals for the main effects model for respiratory infection counts range from −1.12 to 0.62 with no striking pattern, giving further assurance that the model is adequate.

## 2. Sample Size Considerations

Repeated reference has been made to a "large sample" requirement and large-sample distributions, and this issue now requires some discussion. Although the 116 students from whom the eight counts in Table 14.1 were assembled is not a small number, as few as five students contributed to some individual cells, producing counts as low as seven. Are the methods used above justified for these data?

The validities of tests and confidence intervals under a given model $X_1$ depend on the accuracies of (a) the large sample Gaussian approximation to the distribution of $\hat{\boldsymbol{\beta}}$, and (b) the large sample chi-square approximations to the distributions of likelihood ratio, Wald, and score statistics. One may show that the former depends on the closeness of $X'_1 y$ to its limiting Gaussian distribution and on closeness to linearity, near $X'_1 \hat{\mu}$, of $\hat{\boldsymbol{\beta}}$ expressed as a function of $X'_1 y$. The deviance and alternative fit statistics $Q_W$ and $Q_S$ may be approximated by quadratic functions of the asymptotically Gaussian residuals ($y_{ijk} - \hat{\mu}_{ijk}$), and hence of $y$. The limiting chi-square distributions of these test statistics stem from such quadratic approximations. The accuracy in practice of these distributions for fit statistics therefore depends on the closeness to normality of $y$ as well as the accuracy of the quadratic approximations to the test statistics. For entertaining the reduction of $X_1$ to a simpler, nested model $X_2$, the same considerations apply, where $X_2'y$ and $\hat{\mu}_1 = E(y)$ under the model $X_1$ take the roles of $X_1'y$ and $y$, respectively. These considerations make general recommendations hard to come by, as the sample size requirements may vary depending on the choice of GLM, the overall frequency of the response variable, the choices of $X_1$ and $X_2$, and whether model fit or inferences about parameter values are at issue.

Most practice is justified heuristically by extrapolation from an extensive literature on two-way contingency tables arising from product-binomial, multinomial, and product-multinomial distributions. These distributions are formed by conditioning on Poisson distributions, and are discussed below in the technical overviews to logistic regression and loglinear models. Due to the close relationships of the Poisson, binomial, and multinomial distributions, these results provide some guidance for all the models of this chapter. A classical rule of thumb requires expected counts no less than 5 to validate the nominal distribution of Pearson's chi-square statistic (here $Q_S$ for the main effects ANOVA model). The Gaussian laws that approximate distributions for such counts include at least 97.5% of their areas within the admissible nonnegative range. This criterion has repeatedly been shown to be more conservative than necessary for a satisfactory approximation to the distribution of Pearson's chi-square statistic. Adjustments to chi-square critical points have been developed that allow use of Pearson's chi-square with sets of quite low counts. However, when the various tests are to be used with nominal critical points, it is not clear how far the classical criterion can be relaxed. Though probably adequate for most situations when the classical rule holds, in the presence of small expectations the nominal chi-square distributions seem less dependable for likelihood ratio and Wald test statistics, with Wald tests somewhat more vulnerable. Broadly speaking, it is generally accepted that inferences are unlikely to go far astray when based on sufficient statistics for which approximate normality is clearly justifiable, from models for total counts of at least 30, with few expected values much below 5 and none below 1. In other circumstances consultation with a specialist may be advisable, as "exact" procedures that bypass concerns about large-sample approximations are now available for many analytic situations.

Under the main effects model for Table 14.1, the sufficient statistics are 175 total illnesses, and excesses of 27 total illnesses among women over men, one illness among students sharing bedrooms over students with their own bedrooms, and 101 illness among nonsmokers over smokers. By simple linear transformation, these are equivalent to fixing the total numbers of illnesses among women and men at 101 and 74, among students not sharing and sharing bedrooms at 87 and 88, and among smokers and nonsmokers at 37 and 138. Under the main effects model, each of these totals is the sum of four Poisson counts of apparently moderate size, justifying Gaussian approximations. Further, the smallest expected counts from the fitted model are 5.7 and 6.9, with the rest exceeding 10, whereas all $y_{ijk}$ are at least 7, and most exceed 25. So the main effects analysis seems on reasonably firm ground.

However, it is informative to consider the same analysis applied to the respiratory illness experiences of the 116 individual students, rather than to the pooled experiences of the eight groups of students. The relevant likelihood is then product-Poisson with 116 rather than eight terms. All properties of the fitted main effects model are unchanged with this approach except the deviance, which now assesses the model's fit to *each individual's illness experience* rather than to the aggregated experiences within each of the eight groups. The new deviance is 154.44, with $df = 112$, $p = .005$, indicating significant lack of fit. One possibility is that the students are not homogeneous within groups, so that the dispersion across students within the same group is greater than the Poisson distribution explains. The groups may be internally heterogeneous with respect to some omitted factor that affects respiratory illness rate, or there may be positive association among illnesses, with any respiratory illness weakening a student's resistance to another. Either of these situations produces *overdispersion,* that is, greater variability than explained by Poisson distributions. On the other hand, the expected counts for individual students under the saturated model average $175/116 = 1.51$, and do not justify use of the nominal chi-square distribution to evaluate the deviance. The apparent departures from expectations on a per student basis may thus be due to chance. Nor can residuals from individual students be evaluated using nominal criteria based on an approximate Gaussian distribution, because the Gaussian distribution does not adequately describe the behavior of Poisson residuals for counts with such low expectations.

Many if not most Poisson and logistic regression models utilize continuous predictors, so that a natural pooling analogous to that of Table 14.1 is not available. For such data, the situation is as just above: the deviance and related fit tests, as well as the adjusted residuals, cannot be evaluated using nominal asymptotics. Either the fit of some simplified model must be assumed *a priori,* or an artificial pooling must be introduced to test model fit.

When substantial inherent overdispersion is present, Poisson regression and indeed all GLMs are overly liberal; that is, they yield incorrectly low $p$-values and unrealistically narrow confidence intervals. See remarks on this under Exceptions at the end of the chapter.

## C. Firearm Suicide Rates

Poisson modeling with a regression flavor is illustrated by analysis of Table 14.4, showing 1990 Illinois deaths from firearm suicides by age and gender, for 15 age groups. The July 1990 census count for the population in each age by gender group is used as an estimate of person-years accumulated by that group during calendar year 1990, and thus as the denominator for the tabulated firearm suicide rates.

The age groups being of equal widths, these may be coded $-7$ to 7 for analysis; centering this variable reduces collinearity of its powers in a polynomial regression model. Gender may be coded as 1 for women and $-1$ for men. Trends of higher than cubic order being difficult to interpret, one might commence an analysis by fitting a full cubic interaction model, or equivalently a nested model fitting separate cubics with age for each gender. This model includes four parameters for each gender and yields deviance and Pearson lack of fit chi-squares of 19.73 and 18.30, respectively, with $df = 30 - 8 = 22$. Estimated expected counts for the four oldest female age groups are below 2, and for the the two oldest are 0.57 and 0.33. Several other counts for female groups are between 3 and 5. This suggests the fit statistics cannot be relied upon. However, both are well below their expected values and just over half the nominal critical value

**TABLE 14.4  1990 Illinois Deaths by Firearm Suicide, by Age and Gender, with Person-Years Exposure and Suicides per 100,000 Person-Years**[a]

|  | Men | | | Women | | |
|---|---|---|---|---|---|---|
| Age group | Deaths | Person-years | Rate | Deaths | Person-years | Rate |
| 15–19 | 45 | 422033 | 10.66 | 3 | 392807 | 0.76 |
| 20–24 | 55 | 440328 | 12.49 | 5 | 426036 | 1.17 |
| 25–29 | 49 | 491778 | 9.96 | 7 | 491964 | 1.42 |
| 30–34 | 40 | 503265 | 7.95 | 11 | 510897 | 2.15 |
| 35–39 | 41 | 450205 | 9.11 | 5 | 460796 | 1.09 |
| 40–44 | 33 | 394499 | 8.37 | 9 | 409567 | 2.20 |
| 45–49 | 28 | 309738 | 9.04 | 8 | 326291 | 2.45 |
| 50–54 | 28 | 257738 | 10.86 | 2 | 274311 | 0.73 |
| 55–59 | 14 | 231693 | 6.04 | 4 | 252452 | 1.58 |
| 60–64 | 36 | 230167 | 15.64 | 4 | 260443 | 1.54 |
| 65–69 | 28 | 201609 | 13.89 | 5 | 251533 | 1.99 |
| 70–74 | 29 | 152861 | 18.97 | 1 | 213986 | 0.47 |
| 75–79 | 26 | 108373 | 23.99 | 1 | 176676 | 0.57 |
| 80–84 | 22 | 60717 | 36.23 | 1 | 122050 | 0.82 |
| 85–89 | 15 | 38713 | 38.75 | 0 | 107554 | 0.00 |

[a] Suicide counts from 1995 State of Illinois data and person-years exposure approximated using 1990 U.S. Census Bureau tabulations by age and gender, downloaded from the U.S. Centers for Disease Control (CDC) Wonder website.

of 33.92 at $\alpha$ = .05. No adjusted residuals exceed 2, though the 14 observed deaths among 55–59-year-old men are quite low relative to counts in neighboring age categories and to the model prediction of 24. Overall, this model appears an adequate starting point. Simplification to a quadratic model yields a deviance of 19.77, virtually identical to the deviance from the cubic model. The likelihood ratio test of (19.77-19.73) = 0.04 with $df$ = 2 indicates no evidence whatsoever for a cubic trend.

In a quadratic interaction model, although the linear and quadratic age main effects show small Wald statistics, the main effect of gender and its interactions with both linear and quadratic age terms each show $Q_w > 12$ with $df$ = 1, $p$ < .0005. The significance of the quadratic trends may be tested separately in each gender using likelihood ratio or Wald statistics for the sum and difference of the quadratic age main effect and its interaction with gender. These yield 35.85 and 38.34 for males ($p$ < .0001) and 6.72 ($p$ = .010) and 5.74 ($p$ = .017) for females, so no further model reduction is reasonable. The sufficient statistics for this model are the total deaths for each gender and the linear and quadratic contrasts in deaths with age within each gender. Rough normality of these as well as the cubic trends seems reasonable, so statistics comparing the quadratic to cubic model and the Wald statistics within the quadratic model may be evaluated

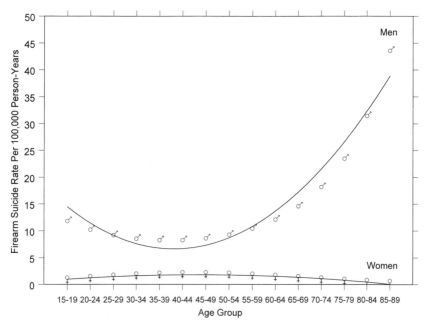

**FIGURE 14.1** Observed and fitted 1990 United States firearm suicide rates, by age and gender.

using their nominal distributions—although all are so definitive that this is not a major concern.

Recasting as a model nesting Age within Gender yields predicted firearm suicide rates of $\exp(-10.909 - .074(Code) - .025(Code)^2)$ for women and of $\exp(-9.246 - .092(Code) + .018(Code)^2)$ for men, in terms of the integer age group codes from $-7$ to $7$ with $0 = 50$–$54$. These are plotted, with corresponding observed rates, in Figure 14.1. The rate ratio for men relative to women, conditional on age group, is $\exp(1.663 + 0.166 \times Code + 0.043 \times Code^2)$. This decreases between 15–19 and 40–44, but increases very rapidly thereafter. The rate ratio associated with an increase of one age category from the starting category $Code$ is $\exp(-0.99 - 0.50 \times Code)$ for women, which decreases with increasing age, and $\exp(-0.74 + 0.36 \times Code)$ for men, which increases as age rises.

## V. LOGISTIC REGRESSION ANALYSIS

Logistic regression models are used to describe counts $y_{ij}$, $i = 1, \ldots, I$, $j = 1, 2$ from classifying individual observational units according to a dichotomous criterion, say "success" or "failure." The $y_{ij}$ may each be 1 or 0 corresponding to unique individuals or, as in Table 14.2, pooled counts for groups distinguished by patterns of predictor variables. These latter are analogous to the counts from Tables 14.1 and 14.4 discussed in connection with Poisson regression, but with defined failures observed along with successes, and fixed total numbers of events observed in each group. Since in this context the success count determines the failure count by subtraction, one need be concerned only with predicting or explaining the success frequencies $y_{i1}$.

The objects of inference are probabilities $\pi_i$ with which successes occur for the different individuals or groups indexed by $i$. Logistic regression provides models for variation in these $\pi_i$, with values of continuous and/ or categorical predictor variables. The parameters are logarithms of odds ratios of predicted probabilities, or logarithms of multiplicative functions of these odds ratios. Thus, in Table 14.2, the intent is to model the probability of finding at least one mammary tumor at autopsy as a function of the combination of surgical treatment, dietary fat source, and dietary fat level.

For logistic as contrasted to Poisson analyses, each observation on an individual must be 0 or 1. Hence, no count from a group may exceed the group size. The approximation of person-years by census counts, as in Table 14.4, sometimes leads to confusion in the choice between Poisson and logistic regression. If counts of both successes and failures are known, and reported counts are in principle limited to known maxima, then the data are more suited to logistic than to Poisson regression. On the other hand, in observing microbial colonies on a culture plate or nests of eagles sighted in a forest preserve, there are no "failures" to count. Similarly, in rates using person-year denominators, the units of the numerator (events) and denominator (person-years) are incommensurable: events are not person-years, or passenger-miles, and cannot be transformed into them. Logistic regression cannot be used in such cases, but Poisson regression is often reasonable.

Technically, logistic regression models are GLMs using the binomial distribution and the *logit* link. We summarize the GLM components, and then illustrate by modeling the probabilities of tumor occurrence for Table 14.2. Multiple mammary tumors in a rat are ignored in this analysis. To model the total numbers of mammary tumors found in each rat or experimental group of rats, Poisson regression would be used instead.

## A. Technical Overview

### 1. Binomial Distributions

A success count $y$ from a homogeneous stratum of independent observational units is conceptually equivalent to the count of heads in repeated, statistically independent flips of a weighted coin. This has a binomial distribution with the coin's "success" probability $\pi$ and stratum sample size $N$ as parameters. "Binomial regression" is sometimes used synonymously with "logistic regression."

The binomial probability of $y$ successes may be determined in two distinct ways. Directly, one may note that the probability of any sequence of $N$ independent observations, of which $y$ are successes and $N - y$ are failures, is $\pi^y(1 - \pi)^{N-y}$, and then multiply this by the number of distinct sequences $N!/y!(N - y)!$ of size $N$ for which this occurs. This yields $Pr\{y|N, \pi\} = [N!/y!(N - y)!] \pi^y(1 - \pi)^{N-y}$. The collection of these probabilities for all possible success counts $y = 0, \ldots N$ is the binomial distribution $Bin(N, \pi)$. Binomial distributions have mean $N\pi$, variance $N\pi(1 - \pi)$, and unimodal shapes ranging from long-tailed to the right when $N\pi$ is small to long-tailed to the left when $N(1 - \pi)$ is small. Binomial distributions with large $N$ and small $\pi$ are very close to Poisson distributions. Binomial distributions with both $N\pi$ and $N(1 - \pi)$ at least moderate are fairly symmetrical, and are approximated well by Gaussian distributions with $\mu = N\pi$ and $\sigma^2 = N\pi(1 - \pi)$.

### 2. The Logit Link Function

The *logit* corresponding to a probability is the log of its odds: $logit(\pi) = \log \omega = \log(\pi/(1 - \pi))$. The logit is the canonical link function for the binomial distribution. More substantive considerations also support its use to model counts of dichotomous events.

Logistic regression for grouped data evolved from "logit analysis" of quantal bioassays. These experiments measure the potency of drugs from responses of test animals to varying doses, when only a dichotomous response (e.g., death) can be reliably or conveniently measured. The *logistic distribution* is a continuous probability law with density function much like the Gaussian bell-shaped curve, but a bit shorter in the middle and a bit higher on the tails. Define the *tolerance* of an animal as the minimal dose required to elicit the response from that animal. If the distribution of tolerances across animals follows a logistic distribution, as it often does approximately when dose is expressed on a proper (e.g., logarithmic) scale, then the logit of the response probability is a linear function of the scaled dose. Logistic regression then yields the natural covariance analysis with which to compare potencies of different preparations. "Logit analysis" was developed just prior to midcentury, but is conceptually and for most practical purposes virtually identical to psychophysical and related psychometric models based on Gaussian laws, such as "probit analysis," that have roots a century earlier. Note that both the logistic and Gaussian laws have the proper sigmoid shape for the home ownership example discussed earlier.

As observed previously, parameters of logistic regression models are log odds ratios and functions of them. The invariance of odds ratios to dimensional exchange, coupled with Bayes' Theorem on conditional probabilities, can be used to show that data on suspected predictors collected retrospectively from independent samples of "successes" and "failures" (e.g., of sick and healthy) may be used to estimate odds ratios by logistic regression analysis. Numerically, this involves pretending that the data on predictors had been collected initially and the success/failure dichotomy observed subsequently as a random outcome. The intercept in such "case-control" models is meaningless, but all other analytic results remain available. Such analyses are enormously useful in epidemiology and other fields where rare events are studied or prospective observation is otherwise often infeasible. When the outcome category of interest is relatively rare, as for example is the development of most cancers, then logistic regression parameters may be used to estimate prospective relative risks from such case-control data.

Lastly, the logistic regression model arises naturally in discriminant analysis. Under classical assumptions, where two groups share a set of multivariate Gaussian discriminators with common within-group covariance matrix, Bayes' Theorem implies a logistic regression model for the probability of either group conditional on the predictors. Under added assumptions, this extends to dichotomous predictors as well.

### 3. The Formal Generalized Linear Model

Logistic regression analysis models the probabilities $\pi_i$ associated with $I$ counts $y_{i1}$ having independent $Bin(N_i, \pi_i)$ distributions, $i = 1, \ldots, I$. Accordingly, the underlying probability law for statistical inference is the product–binomial distribution $\Pi_{i=1}^{I} [(N_i!/y_{i1}(N_i - y_{i1})!) \pi_i^{y_{i1}} (1 - \pi_i)^{N_i - y_{i1}}]$.

The GLM is $\text{logit}(\pi_i) = \log[\pi_i/(1 - \pi_i)] = x_i'\beta$ or, in full matrix representation, $\text{logit}(\pi) = X\beta$. If $y$ consists solely of zeroes and ones reflecting individual Bernoulli trials, the likelihood equations are $X'\pi = X'y$. The information matrix is $I(\pi) = X'D_\pi (I - D_\pi)X$, and the large-sample covariance matrix is estimated by $Cov_A(\hat{\beta})_{\hat{\pi}}I^{-1}(\hat{\pi}) = (X'D_{\hat{\pi}} (I - D_{\hat{\pi}})X)^{-1}$, where $I$ is an identity matrix.

On the other hand, suppose observations are grouped so that each $y_i$ is the count of successes from a collection of units with the same combination of predictor values. Then the above must be modified by replacing $\pi$ by $m = E(y)$ at appropriate points, where the expectation $m_i$ is the product of the number of units with the $i^{th}$ combination of predictors and the success probability $\pi_i$ for that combination. The likelihood equations and information matrix become $X'\hat{m} = X'y$ with $I(\pi) = X'D_m(I - D_\pi)X$, yielding $Cov_A(\hat{\beta}) = I^{-1}(\hat{\pi}) = X'D_{\hat{m}}(I - D_{\hat{\pi}})X)^{-1}$. Standardized and adjusted Pearson residuals for logistic regression models are obtained from those for Poisson regression by replacement of $\hat{\mu}$ in their denominators by $\hat{\mu}(1 - \hat{\pi})$, viz. $e = (y - \hat{\mu})/\sqrt{\hat{\mu}(1 - \hat{\pi})}$ and $e' = e/\sqrt{1 - h}$.

Sample size issues for logistic regression are essentially those for Poisson regression as discussed earlier, but apply not only to the success counts explicitly modeled but to the failure counts as well. The asymptotic results are compromised when expected counts of either successes or failures are too small.

### B. Mammary Carcinogenesis in Rats

As illustration, we show for Table 14.2 that the probabilities of mammary carcinoma at necropsy depend on surgical treatment and dietary fat in a simple manner. For examination of a full $3 \times 2 \times 2$ factorial model for the twelve observed tumor proportions, surgical treatment is coded using dummy variables with OVX/E as the reference cell; fat source and fat level are coded as 1, $-1$ variables. Main effects of surgical treatment and fat level appear enormously significant (adjusted likelihood ratio and Wald statistics of 94.62 and 63.63 with $df = 2$ for surgical treatment, 14.59 and 13.73 with $df = 1$ for fat level), while none of the interaction terms approach statistical significance and most are below their standard errors. Consequently, a main effects model $X_M = [\mathbf{1}_{12}, [1\ 0\ -1]' \otimes \mathbf{1}_4, [0\ 1\ -1]' \otimes \mathbf{1}_4, \mathbf{1}_3 \otimes \bar{\mathbf{1}} \otimes \mathbf{1}_2, \mathbf{1}_6 \otimes \bar{\mathbf{1}}]$ may be entertained. This shows a deviance of 5.37 with $df = 7$. Although the nominal $p = 0.61$ is imprecise in view of small expected values for the OVX rats fed low fat, this imprecision is of no concern, since the deviance is so low.

The surgical treatment factor incorporates two biologically separable effects: absence versus presence of estrogen [OVX vs. (OVX/S or OVX/E)] and artificial versus normal hormonal milieu (OVX/S vs. OVX/E). The estimate from $X_M$ of the OVX/S versus OVX/E contrast, $2\hat{\beta}_2 + \hat{\beta}_3$, is below its estimated standard error with $p = 0.36$ by Wald test. The test of fat source effect yields a similar result. An aggressive analyst might remove fat source from the model and replace the two treatment parameters in $X_M$ by a single parameter, distinguishing OVX from OVX/S and OVX/E. Fitting the reduced model using dummy variables with reference categories OVX and 10% dietary fat, $X_R = [\mathbf{1}_{12}, [1\ 0\ 1]' \otimes \mathbf{1}_4, \mathbf{1}_6 \otimes [0\ 1]']$, yields an increase in deviance of 1.76 with $df = 2$, to a total deviance of 7.14 with $df = 7$. One may expect the distribution of the difference in deviances to be closer to its limiting distribution than is the distribution of the total deviance. In this case, neither statistic causes any concern about model fit.

Logistic regression diagnostics for individual counts are also widely implemented in software. These extend linear regression diagnostics in Venter and Maxwell (chapter 6, this volume) to GLM's. They include, as well as the residuals mentioned earlier, leverage values and exact or approximate case deletion statistics. The latter measure sensitivities of lack-of-fit statistics, point estimates of parameters, and confidence regions for parameter sets to removal of individual observations from the data set. Such diagnostics are useful for grouped data in this experimental setting, as well as for the individual Bernoulli observations typical in analyses of observational data with continuous predictors. Adjusted residuals reveal that $X_R$ describes the OVX/S beef tallow-fed rats and the OVX/E 40% beef tallow-fed rats less well than the nine other groups: observed counts for the former do not reflect a fat level effect, whereas the latter count appears to exaggerate that effect. The data for OVX rats fed 40% fat also have higher leverage than the other groups. Although the three poorer fitting observations individually have substantial influences, none by itself suggests more than random chance at work. Since their impacts on parameter estimates are also not consistent with one another and together have only moderate net effect, collectively they cast no particular doubt on the proposed model.

The estimated parameters of 2.57 for surgical treatment and 1.00 for fat level are log ORs that associate adenocarcinoma occurrence with the presence of estrogen and high fat. Variance heterogeneity from the wide range of proportions in the experiment induces some positive correlation between these estimates, approximately 0.18 in $\text{Cov}_A(\hat{\boldsymbol{\beta}})$. For practical purposes this is small, and each parameter may be confidently interpreted quite separately. The estimated conditional ORs of 13.04 (95% CI 7.06–24.08) for either estrogen group versus OVX and of 2.71 (95% CI 1.68–4.35) for 40% versus 10% dietary fat show that both variables had substantial impact on tumor occurrence. The lack of evidence for interaction allows us to treat each conditional OR as constant across levels of the other predictor. The odds of tumor for OVX/S or OVX/E rats fed 40% dietary fat are thus estimated as $13.04 \times 2.71 = 35.34$ times the odds of tumor for OVX rats fed 10% dietary fat. Removal of estrogen and dietary fat reduced the odds of tumor by 34.34/35.34 = 97.2%. However, as Table 14.2 makes clear, this is not the same as reducing the risk (probability) of tumor by 97%; the odds ratios cannot approximate risk or rate ratios when the outcome is, as here, rather common.

From the form of $X_R$, the sufficient statistics for this reduced model are the counts of rats with adenocarcinoma in (a) the full experiment (b) the combined OVX/S and OVX/E groups, and (c) the combined 40% fat level groups. With the known total number of rats in each group, these specify the marginal proportions of rats with and without adenocarcinomas in the 10% fat, 40% fat, OVX, and combined OVX/S-OVX/E subgroups of the experiment. These marginal proportions, hence the ORs from them, are precisely reproduced by the margins of fitted counts from $X_R$. Moreover, because of the simple structure of the model, the fitted conditional ORs are precisely equal to their marginal counterparts. We later consider more complex loglinear and logistic regression models, for which conditional and marginal ORs differ.

## C. Variable Selection

Due to its suitability for use with both case-control and cohort observational studies, logistic regression is commonly employed with large numbers of potential predictor variables and little *a priori* information on their relative importance. For instance, the nasopharyngeal cancer study that produced Table 14.3 used logistic regression to develop a dietary predictive index from raw data on 110 five-level categorical variables. Such exploratory analysis cannot approach the scientific rigor of hypothesis-driven analysis of a well-designed experiment. However, highly multivariate observational research is often the only feasible approach to examine issues of critical importance to society. Numerous automated methods have been developed to aid variable selection in these situations, and we discuss several in the context of logistic regression. Each attempts to find a relatively small set of variables that jointly capture the predictive capacity of a larger set of candidate variables.

### 1. Forward Selection

Candidate predictors essentially compete in tournaments for entry into the model. In round 1, the score statistic for every single-variable model is calculated. Among models with resulting $p$-values below a preset criterion for entry, the "winner" with the lowest $p$-value is selected. In round 2, score statistics are separately calculated for adding each remaining variable to the model from round 1. The winning variable is determined similarly, and added to the model. Competition continues in this way until no remaining variable produces an adjusted $p$-value below the entry criterion. In principle, likelihood ratio or Wald statistics might also be used for this process. However, the score statistics for all candidate variables in any round may be calculated after fitting only the model including winners of previous rounds. This offers large computational savings.

Forward selection risks stopping too early, when linear combinations of remaining candidate variables retain predictive power that the individual variables do not. Forward selection also retains variables admitted early that lose their predictive power to combinations of variables admitted later.

## 2. Backwards Elimination

To avoid omitting important variables, backwards elimination starts with all variables included. This may be infeasible for very large numbers of candidate variables, as numerical instabilities may arise. (Indeed, a common rule of thumb is that the ratio of subjects to predictors in a final logistic regression model should be at least twenty. The origins of this rule are obscure, and many sophisticated practitioners compromise by maintaining a ratio of 10 or more.)

Backwards elimination, when feasible, tests each candidate variable for removal using the Wald statistic within the model containing all candidates, and permanently eliminates the variable with highest $p$-value among those above a prespecified retention criterion. The process is then repeated with the reduced model, until the Wald tests for each remaining predictor give $p$-values below the retention criterion.

## 3. Stepwise Selection

This adds steps to forward selection to allow removal of redundant predictors. After entry of a new variable into the model, Wald statistics for all predictors in the new model are calculated. The predictor with highest $p$-value above a specified retention criterion is then removed from the model, which is then refit to the remaining variables. The process is iterated until the $p$-value for each retained variable is below the retention criterion. Entry steps and removal steps alternate until no remaining candidate variables meet the criterion for entry, and all included candidates meet the criterion for retention. Note that unless the retention criterion is somewhat higher than the entry criterion, a variable may cycle in and out of the model endlessly. Indeed, if entry and retention criteria are similar, random discrepancies between score and Wald statistics will sometimes cause this. Consequently, most software halts the process when any variable enters and then leaves the model on consecutive steps. In principle, score or likelihood ratio statistics might be used to screen for variable retention. The Wald statistic offers computational economy, since Wald tests for retention of each parameter in a model may be calculated noniteratively from the model-based estimated covariance matrix. In contrast, score and likelihood ratio tests require fitting a separate reduced model to test each parameter.

Stepwise selection cleans out variables that become redundant for prediction once other variables are added. If the effects of several predictors are confounded, stepwise regression may be better than forward selection at distinguishing variables whose effects are most likely to be meaningful. As with forward selection, however, stepwise selection may stop too early.

### 4. Best Scoring Models

Additional methods analogous to those of linear regression examine (a) score statistics of all possible models, (b) the $k$ $l$-variable models with highest score statistics or lowest score statistic $p$-values for different values of $l$, or (c) models selected by comparing the score statistics obtainable for $l-$ and $(l - 1)$-variable models. Score statistics are used because they do not require fitting a new model for each test; hence, very large numbers of models may be screened with reasonable computing resources.

### 5. Cautionary Notes

Though extremely useful as exploratory tools, these techniques are easily abused. Literatures of many fields are replete with uncritical applications.

By default, current software implementations omit interactions and curvilinear regression terms. Although the analyst may include them, when using fully automated variable selection and removal it is typically difficult to distinguish interactions from main effects, higher from lower order polynomial terms, or otherwise incorporate precedence relations important to orderly model construction. Thus, including interaction and/or curvature frequently results in automated selection of an uninterpretable nonhierarchical model.

The variable $C$ "confounds" the relationship between $A$ and $B$ if the marginal association of $A$ and $B$, ignoring $C$, differs noticeably from the conditional association between $A$ and $B$, given $C$. This can occur when $C$ is associated with both $A$ and $B$ in the data set, and distortion can be substantial even if these associations are not statistically significant. The above methods do not check for statistically nonsignificant confounding variables, and may well exclude confounders of substantive importance.

When many intercorrelated variables are used as predictors, the variable selection/exclusion sequences may be sensitive to small fluctuations in the data. Analyses of slightly different data sets may select quite disparate sets of predictors, even while closely agreeing on fitted probabilities for individuals or groups. Conventional $p$-values in models obtained by automated variable selection take no account of the multiplicity of tests that created the model, and hence may grossly exaggerate statistical significance in this context. Infatuation with a precise set of selected variables, and exaggeration of their statistical significance, lead to many failures of replication in observational research.

Nowhere is the need for cross-validation more evident than when assessing small models obtained from large numbers of variables. Full replication is most desirable. Withholding a portion of the sample for validation purposes is helpful but inefficient. Available computationally intensive methods include, for instance, assessing how well each observation is predicted by a model determined by the other observations. We emphasize that, ingenuity of the analyst notwithstanding, the information contained in observational data sets of modest size is rarely sufficient to solve large, complex problems. The naive user of automated variable selection is fated to learn this the hard way.

## D. Conditional Logistic Regression

Logistic regression as outlined above is more precisely termed "unconditional logistic regression." The method can fail badly, as with a seemingly innocuous binary analog of the paired $t$-test. Recall that the paired $t$-test is equivalent to the $F$-test of treatment effect in a randomized block design, with each pair of observations forming a block. Now consider paired binary observations, as in a pre–post design for individuals, or a matched-pairs study with subjects in each homogeneous pair randomized to treatments $A$ and $B$. This yields a stratified $2 \times 2$ contingency table, with each matched pair a stratum and precisely one member of the pair in each row of its $2 \times 2$ subtable. Adding a block effect for each pair to a logistic regression model with a single dichotomous predictor gives a natural binary analog to the paired $t$-test.

However, expected counts under this model are all below 1 and do not increase with sample size. Since each additional matched pair requires that a parameter for the new block be added to the model, the information in the data does not increase on a per parameter basis. Conventional large-sample results do not apply to such situations. Distressingly, $\hat{\beta}$ for the treatment effect converges in large samples to $2\beta$ rather than $\beta$, exaggerating in this instance any real effect. Other distortions arise from the same cause, for any predictor(s), when data occur in small matched sets whose number increases with sample size.

A solution is to base analysis on the likelihood obtained by simultaneously conditioning on the observed marginal distributions of the binary outcome and on the combinations of predictor variables within each matched set. The randomness remaining pertains to which levels and/or values of predictor variables accompany a given binary outcome (e.g., survival or death) in matched sets where both outcomes occur. The conditional likelihood is based on the possible assignments of predictor variable combinations to individuals within matched sets. Inferential techniques described earlier are then applied to this permutation likelihood.

There is little software designed specifically for conditional logistic regression models. The conditional likelihood is not in the exponential family, so these models are not GLMs. For matched pairs studies, unconditional logistic regression software may be manipulated to fit conditional models. More generally, the conditional logistic regression likelihood is formally identical to the *partial likelihood* used with semiparametric proportional hazards models (i.e., *Cox models*) for survival data. Documentation of software for proportional hazards survival analysis typically describes its use to fit conditional logistic regression models as well. Happily, after the change to the conditional likelihood, both descriptive and inferential interpretation of conditional logistic regression models are virtually the same as for unconditional models. However, the conditional models contain no intercept, and the GLM-based approach to residual analysis must be replaced by different methods based on the partial likelihood.

In unconditional or conditional logistic regression modeling of sparse data, further conditioning on sufficient statistics produces likelihoods for "exact" small sample inference. The resulting inferential procedures are extensions of Fisher's exact test for $2 \times 2$ tables, and are computationally intensive in the context of today's computing power. Where computing allows, they are useful both for small sample problems that clearly require them, and as a check on equivocal asymptotic results for somewhat larger data sets (Cytel Software Corporation, 1999).

## VI. LOGLINEAR MODELS FOR NOMINAL VARIABLES

Loglinear models represent association structures in multidimensional contingency tables, and serve the same function for qualitative variables as do correlation matrices for multivariate Gaussian data. The pairwise correlations in a correlation matrix completely specify the joint distribution of such Gaussian variables, and hence all association amongst them. Specification of a full association structure for qualitative variables in a contingency table is potentially much more complex, but can be greatly simplified if (a) some sets of variables can be treated as independent of others within slices of the table, and/or (b) within-slice generalized population odds ratios $\Omega$, or functions of them among some sets of variables, can be treated as the same from slice to slice. Looking within slices of a table means conditioning on values of some other variables. Thus, the above means (a) treating some sets of variables as conditionally independent of other sets, and (b) treating conditional dependence among some sets of variables as stable, or homogeneous, across levels of other variables.

When subsets of variables in a table reflect approximate conditional independence and/or conditional homogeneity, loglinear modeling allows representation of association in the full table by a smaller set of marginal and/or conditional tables. The simplest models reflect joint or conditional independence among sets of variables or, for variables that are related, uniformity of conditional generalized population ratios across levels of other variables. The parameters of these models are logarithms of conditional generalized odds ratios or differences of these, and usually have straightforward substantive interpretations.

The number of parameters in a saturated loglinear model for nominal variables is one less than the number of cells in the multidimensional contingency table. This can be quite large, and the objective of a loglinear model analysis is to find a far simpler, parsimonious representation of the association structure. Usually this is done by successive classwise elimination of groups of higher-order association terms representing complex heterogeneous patterns of association, much as high-order interaction terms are often eliminated from ANOVA models. Although each parameter of a loglinear model has a clear interpretation, the number and complexity of such parameters often make these awkward to use directly. Guided inspection of fitted marginal or conditional counts or probabilities is often more useful. The function of analytic modeling is to direct choice, and guide interpretation, of a sensible and parsimonious set of marginal and/or conditional tables with which to summarize the association structure of the full table.

Loglinear models are based on multinomial or product-multinomial distributions. However, for computational purposes they may be formally treated as Poisson GLMs with factorial ANOVA-like structures. We review the technical basis for this and then explore the concepts particular to loglinear modeling in the context of three-dimensional contingency tables. A graphical representation of loglinear models is presented as an aid to discovering when a contingency table may be safely collapsed over certain of its variables. These concepts are illustrated through analysis of data on possible cancer risk factors. We then touch briefly on the relationship of loglinear to logit models, and the loglinear modeling of ordinal variables.

## A. Technical Overview

### 1. Multinomial Distributions

Multinomial distributions generalize binomial distributions from dichotomies to polytomies. The multinomial distribution that describes classification of $N$ individuals into $J$ categories with respective probabilities $\pi_j$, $j = 1, \ldots, J$, is $Pr\{\{y_j\}|N, \pi\} = Mn(N, \pi_1, \ldots \pi_j) = N!\Pi_{j=1}^{J} \pi_j^{y_j}/y_j!$, with $\Sigma_{j=1}^{J} \pi_j = 1$. This may be derived by the same methods as the binomial. For multinomial distributions, $y_j$ has respective mean and variance $m_j = N\pi_j$ and $N\pi_j(1-\pi_j) = m_j(N - m_j)/N$ and $Cov(y_j, y_{j'}) = -m_j m_{j'}/N$. Thus, variability of each count $y_j$ still depends on its mean $m_j$ but, as with the binomial, increases not directly with $m_j$ but rather with $\min(m_j, N - m_j)$, the closeness of $m_j$ to an end of its possible range.

Let the count vector $y$ be composed of subvectors, $y = [y_1', \ldots, y_I']'$, with the subvectors $y_i = [y_{i1}, \ldots, y_{iJ}]'$ distributed independently $Mn(N_i, \pi_i) = Mn(N_i, \pi_{i1}, \ldots \pi_{iJ_i})$. Then the distribution of $y$ is said to be product-multinomial with sample size parameter vector $N = [N_1, \ldots, N_I]'$ and probability parameter vector $\pi = [\pi_1', \ldots, \pi_I']'$. The subscripts $i$ and $j$ often represent vectors $i = [i_1, \ldots, i_u]'$ and $j = [j_1, \ldots, j_v]'$, in which case the multinomial populations themselves and their response categories are cross-classifications respectively of $u$ and $v$ variables. Table 14.2 may be so described with $u = 3$ indexing the three experimental variables, with 3,2 and 2 levels respectively and $v = 1$ with 2 levels of the adenocarcinoma response.

The multinomial distribution is also a conditional distribution of independent Poisson counts, conditioning on the total count of $N$, and this perspective is critical to loglinear model fitting and interpretation. Thus, suppose that the $J$ counts $y_j$ have independent $Poi(\lambda_j)$ distributions over a single region of extent $t$, and that the relative numbers of events of different types are of interest but the total number of events $N = \Sigma_{j=1}^{J} y_j$ is not. Then the conditional distribution $Pr\{\{y_j\}|N\}$ is most relevant to statistical analysis. Using the fact that $N$ is $Poi(\Sigma_{j=1}^{J} \lambda_j)$, this conditional distribution is readily seen to be $Mn(N, \pi_1, \ldots, \pi_J)$, with $\pi_j = \lambda_j/\Sigma_{u=1}^{J} \lambda_u$. Because of this, it may be shown that a Poisson regression model with $N$ as a sufficient statistic yields the same mle of $\beta$ as would a multinomial distribution with $N$ fixed. In essence, cross-classified counts with one or more fixed marginal totals, for which the appropriate probability model is multinomial or product-multinomial, may be computationally treated as Poisson regression models for the purposes of likelihood-based analyses. This allows us to view loglinear models as GLMs, and to use any computer program designed for GLMs to fit and interpret loglinear models.

## 2. The Formal Model

The loglinear model is $\pi_i = \exp(X_i\beta)/[1_J'\exp(X_i\beta)]$ for $i = 1, \ldots, I$, with the columns of $X_i$ linearly independent of $1_J$. Taking logarithms of both sides yields $\log \pi_{ij} + C_i = \Sigma_{k=1}^{K} x_{ijk}\beta_k$, where $C_i = \log[1_J' \exp(X_i\beta)]$. This looks like a Poisson regression model with offset $C_i$; here $C_i$ just embodies the probability constraint $\Sigma_{j=1}^{J} \pi_{ij} = 1$. Using $\pi = (\pi_1', \ldots, \pi_I')'$, $X = [X_1', \ldots, X_I']'$ and $L = I_I \otimes 1_J$, we may write this compactly as $\pi = D_\eta^{-1} \exp(X\beta)$, with $\eta = L' \exp(X\beta)$ and $\log \eta$ incorporating the $C_i$ above, $i = 1, \ldots, I$.

The product-multinomial likelihood equations for the loglinear model are thus found to be the equations $X'm = X'y$ for the Poisson regression model designated by $X$, augmented by the constraints $L'm = L'y = N$, where $N$ is the vector of sample sizes for the component multinomials. The two sets of equations may be written together as $X^{*'}m = X^{*'}y$, where $X^* = [X, L]$. These are precisely the likelihood equations for the Poisson regression model $\log m = X^*\beta$. The product-multinomial information matrix $I(\pi) = X'(D_m - m\pi')X$ yields the estimated covariance matrix $\text{Cov}_A(\hat{\beta}) = I^{-1}(\hat{\pi}) = [X'(D_{\hat{m}} - \hat{m}\hat{\pi}')X]^{-1}$. Standardized (Pearson) residuals are as with Poisson regression, and denominators for adjusted residuals may be obtained using $\text{Cov}_A(\hat{\beta})$. In principle, $X$ may contain categorical and/or continuous predictors. We will focus on ANOVA-like models, as these have been most extensively developed and applied.

## B. Hierarchical Analysis of Variance Models for Three-Way Tables

ANOVA models for contingency tables employ matrices $X*$ identical to those for ANOVA models of continuous data. Independence, conditional independence, and homogeneous dependence are the basic simplified representations used by these models. We review terminology. Members of a set of variables $\mathscr{S}$ are *mutually independent* when the probability of any combination of their levels is the product of the marginal probabilities for each member. The variables in $\mathscr{S}$ are *conditionally independent* given the variables in another set $\mathscr{T}$ if the above holds at all combinations of levels of the variables in $\mathscr{T}$. They are *jointly independent* of the variables in $\mathscr{T}$ if the joint probability of any combination of levels of all variables is the product of the marginal joint probability for the variables in $\mathscr{S}$ with the marginal joint probability for the variables in $\mathscr{T}$. Note also that if $D$ and $E$ are independent, the generalized logits of $E$ must be the same within all levels of $D$; when $D$ and $E$ are dependent, the differences in generalized logits of $E$, between levels of $D$, characterize the dependence. Such differences are logarithms of population generalized odds ratios (i.e., odds ratios $\Omega$ constructed from $2 \times 2$ tables formed by intersecting a pair of levels of $E$ with a pair of levels of $D$). The association between $D$ and $E$ is *homogeneous over* $\mathscr{T}$ when the generalized odds ratios relating $D$ and $E$ are the same within all combinations of categories of variables in $\mathscr{T}$.

The pivotal concepts are well captured by the three-way layout for a single multinomial table, where $N$ observations are cross-classified by row, column, and layer response variables. To simplify terminology, henceforth we use "odds ratios" to include generalized odds ratios for polytomies as well as conventional odds ratios for dichotomies. Below we catalog the possible combinations of main effects and interactions that may be used to hierarchically model such three-way contingency tables. The parameters of these loglinear models are logarithms of generalized odds ratios, or functions of them. As with continuous responses, some models may also be reparameterized to represent nested effects. The dimensions of the table are referred to as $A$, $B$, and $C$, respectively with $I$, $J$, and $K$ categories. Notation of the form $\beta$, $\beta_i^A$, $\beta_{ij}^{XY}$, and $\beta_{ijk}^{XYZ}$ represents the intercept, main effects (level $i$ of $A$), two-way interactions (level $i$ of $A$ with level $j$ of $B$), and three-way interactions (level $i$ of $A$ with levels $j$ of $B$ and $k$ of $C$).

Each model below exemplifies a class of models, obtainable by permuting dimensions of the table. With each model is a brief description. As with ANOVA, subscripted "+" represents summation over all levels of a subscript. Most models for three-way tables have closed-form solutions for estimated expected values $\hat{m}_{ijk}$, and these are provided. Model classes (1–3) impose uniform distributions on some variables, and are rarely used in practice.

1. $m_{ijk} = \beta$. Mutual independence, with uniform distribution of each variable: $\hat{m}_{ijk} = N/IJK$.

2. $m_{ijk} = \beta + \beta_i^A$. Mutual independence. Combinations of $B$ and $C$ are uniformly distributed within each level of $A$, and all conditional distributions of $A$ are equal to its marginal distribution: $\hat{m}_{ijk} = n_{i++}/JK$.

3. $m_{ijk} = \beta + \beta_i^A + \beta_j^B$. Mutual independence. $C$ is uniform within each combination of $A$ and $B$, and all conditional distributions of $A$ and $B$ are equal to their marginal distributions: $\hat{m}_{ijk} = n_i + + n_{+j+}/NK$.

4. $m_{ijk} = \beta + \beta_i^A + \beta_j^B + \beta_k^C$. Mutual independence. All conditional distributions of any variable are equal to their marginal distributions: $\hat{m}_{ijk} = n_{i++}n_{+j+}n_{++k}/N^2$.

5. $m_{ijk} = \beta + \beta_i^A + \beta_j^B + \beta_k^C + \beta_{ik}^{AC}$. Joint independence of $A$ and $C$ from $B$. At each level of $B$, the joint distribution of $A$ and $C$ is the same as their marginal joint distribution. At each combination of levels of $A$ and $C$, the distribution of $B$ is the same as its marginal distribution: $\hat{m}_{ijk} = n_{i+k}n_{+j+}/N$.

6. $m_{ijk} = \beta + \beta_i^A + \beta_j^B + \beta_k^C + \beta_{ik}^{AC} + \beta_{jk}^{BC}$. Conditional independence of $A$ and $B$, given $C$. At each level of $C$, $A$ and $B$ are independent. The joint distribution of $A$ and $C$ at each level of $B$ is the same as the joint marginal distribution, and the joint distribution of $B$ and $C$ at each level of $A$ is the same as the joint marginal distribution: $\hat{m}_{ijk} = n_{i+k}n_{+jk}/n_{++k}$.

7. $m_{ijk} = \beta + \beta_i^A + \beta_j^B + \beta_k^C + \beta_{ij}^{AB} + \beta_{ik}^{AC} + \beta_{jk}^{BC}$. Homogeneous association. No variables are independent or conditionally independent, but the conditional odds ratios for each pair of variables are constant across levels of the third variable. The relationship between each pair of variables is confounded by the third variable, and no conditional odds ratios necessarily equal their marginal counterparts, even approximately. There is no closed form expression for $\hat{m}_{ijk}$.

8. $m_{ijk} = \beta + \beta_i^A + \beta_j^B + \beta_k^C + \beta_{ij}^{AB} + \beta_{ik}^{AC} + \beta_{jk}^{BC} + \beta_{ijk}^{ABC}$. Saturated model with no data reduction. Conditional odds ratios between any two variables may vary across levels of the remaining variable: $\hat{m}_{ijk} = y_{ijk}$.

The equation for a hierarchical ANOVA loglinear model is a cumbersome means of expression. It is easier to designate a model by a parenthesized list of its sufficient statistics, which are the marginal distributions corresponding to the highest order terms containing each variable in the model. If "A," "AB," and "ABC" are used to represent, respectively, the marginal distribution of A, the marginal joint distribution of A and B, and the joint distribution of A, B, and C, then models (1–8) above may be represented, respectively, as (), (A), (A, B), (A, B, C), (AC, B), (AC, BC), (AB, AC, BC), and (ABC).

## C. Collapsibility and Association Graphs

Models (5–8) incorporate dependence between $A$ and $C$. Under models (5) and (6), all conditional odds ratios relating $A$ to $C$ at any specific level of $B$ are equal to their counterparts at other levels of $B$, and in the marginal $A \times C$ table that ignores $B$ by collapsing over its levels. Under these models the observed marginal $A \times C$ ORs are the mle's of their expected values. Hence, the marginal $A \times C$ table collapsed over $B$ may be used to describe the conditional association of $A$ and $C$ within levels of $B$. For this purpose, the table is *collapsible* over $B$. Models (6) and (7) do not share this property, as the conditional odds ratios differ from their marginal counterparts due to confounding. Collapsing over $B$ is not possible without distortion of the $A \times C$ relationship.

Because marginal tables are easier to interpret than a full table or slices, it is useful to know when a table is collapsible without bias. The conditional independence model (6) is the prototype. If a table is well described by this model, the $A \times C$ conditional association may be observed after collapsing over $B$, and the $B \times C$ conditional association may be observed by collapsing over $A$. Collapsing over $C$, however, distorts the $A \times B$ conditional odds ratio. Consider circumstances in which $A$, $B$, and $C$ are each composite variables representing cross-classifications respectively of $A_1, \dots, A_a, B_1, \dots, B_b$, and $C_1, \dots, C_c$. Then applying these results from model (6) to the composite variables implies that collapsing over any subset of the $A$'s will not alter associations among the $B$s or between them and the $C$s, and that collapsing over any subset of the $B$s will not alter associations among the $A$s or between them and the $C$s.

As a high-dimensional table may often be partitioned in different ways, a graphical method helps to recognize the collapsibility suggested by a specific loglinear model. Represent each variable by a vertex (equivalently, node) of a graph, and connect two vertices with an edge exactly when the corresponding pair of variables occurs together in some term of the model. Then any partitioning of the vertices into subsets $A_1, \dots, A_a, B_1, \dots, B_b$, and $C_1, \dots, C_c$, such that a path from any of the $A$s to any of the $B$s must pass first through at least one of the $C$s, represents the conditions of the last paragraph and allows easier recognition of how the table may be collapsed without distortion from confounding.

Note that such association graphs do not uniquely characterize a model, because they represent only pairwise associations. Thus, models (7) and (8) of the previous section are both represented by a triangle, as they differ only in the presence or absence of a three-factor term. However, the graph indicates correctly that, under either model, collapsing over any variable distorts the relationship between the other two.

## D. Occupational Associations among Malaysian Chinese

We illustrate by examining the association structure of Table 14.3, abbreviating the four variables by $W$, $P$, $S$, and $G$, and coding each as 1 or $-1$ as the history of occupational exposure was present or absent. The deviance for the model that excludes only the four-way interaction is $Q_L = 0.09$, $df = 1$, $p = .76$, suggesting adequate fit. Wald and likelihood ratio tests of three-way interaction terms within this model yield p-values for three terms that do not approach statistical significance, but a marginally significant $p = .04$ for the $P \times S \times G$ term. In the model retaining only this three-way interaction with all two-way interactions and main effects, $W \times P$, $W \times S$, and $W \times G$ are each statistically significant, so that no further reduction is reasonable. Thus Model I = $(PSG, WP, WS, WG)$, for which no adjusted residuals are large, is a possible model for these data.

However, inasmuch as this model reduction process gave five opportunities to make a Type I error at the conventional $\alpha = .05$ in testing each three or four-way interaction, Model I may overfit the data. Simultaneous removal of all four three-way interactions from the no four-way interaction model yields a difference in deviance statistic of 7.52, $df = 4$, $p = .11$, with an overall deviance of 7.61, $df = 5$, $p = .18$. So it is also reasonable to omit all three-way interactions and examine the homogeneous association Model II = $(WP, WS, WG, PS, PG, SG)$. Within Model II, $p$ for the $P \times S$ term just slightly exceeds .10. Removal of $PS$ yields Model III, $(WP, WS, WG, PG, SG)$, with deviance 10.29, $df = 6$, $p = .11$. Within Model III, $p = .06$ for the $W \times S$ term, removal of which yields Model IV = $(WP, WG, PG, SG)$, with deviance 13.77, $df = 7$, $p = .06$. Remaining interactions within Model IV are all significant at the 1% level, so no further simplification of this highly smoothed model is appropriate.

The choice among models I–IV is not clear, and depends largely on the modeling philosophy of the analyst. Table 14.5 gives ORs between exposure pairs for each fitted model. For the heterogeneous association Model I, the fitted ORs among pairs from Gasoline or Paint Fumes and Total Smokes differ across values of the third member of the triad. For Model I only, the effects in Table 14.5 must be divided by 1.87 to give the fitted OR for the exposure pair in the presence of the third exposure in the triad, and multiplied by 1.87 in its absence.

All models show very strong positive association of $G$ with $P$, fairly strong positive associations of $G$ with $S$ and of $P$ with $W$, and moderate positive association of $G$ with $W$. Models I–III add a weaker association of $S$ with $W$. Model II adds a negative association of smaller but similar magnitude between $P$ and $S$, and Model I allows this association to vary

**TABLE 14.5** Estimated Conditional Odds Ratios from Loglinear Models for Occupational Exposures in Malaysian Chinese

| Exposure association | Model I | Model II | Model III | Model IV |
|---|---|---|---|---|
| Gas fumes × Paint fumes | $9.61^a$ | 10.85 | 9.27 | 9.27 |
| Gas fumes × Total smokes | $2.53^b$ | 3.84 | 3.31 | 3.65 |
| Paint fumes × Total smokes | $0.73^c$ | 0.61 | — | — |
| Gas fumes × Wood dusts | 2.11 | 2.10 | 2.20 | 2.49 |
| Paint fumes × Wood dusts | 4.52 | 4.52 | 4.25 | 4.25 |
| Total smokes × Wood dusts | 1.74 | 1.72 | 1.55 | — |

[a] Geometric mean of odds ratios with and without Total Smoke exposure.
[b] Geometric mean of odds ratios with and without Paint Fume exposure.
[c] Geometric mean of odds ratios with and without Gasoline Fumes exposure.

across levels of $G$. Association is more strongly negative with $G$ present, but weakly positive when $G$ is absent. In Model I, fitted ORs relating each pair of the $G$, $P$, $S$ triad in the absence of the third exposure are 3.5 times their values when the third exposure is present.

Examination of fitted counts and residuals indicates how each of Models II–IV sacrifice closeness of fit for greater simplicity. Model II overestimates (10.4 vs. 6) the number of subjects whose only exposure is $P$. Model III does better (9.2 vs. 6) in this cell, but underpredicts the counts exposed to both $P$ and $G$ but not to $S$ or $W$ (9.4 vs. 15) and those exposed to all but $P$ (14.2 vs. 19). Model IV underpredicts both these counts more severely (with respective estimated expectations of 8.6 and 12.9), and underprediction of subjects exposed to all factors (250.9 vs. 258) becomes noticeable by adjusted residual.

Figure 14.2 shows association graphs for Models I–IV. Models I and II share the same totally connected graph, so no collapsing is possible under either model without distortion. Under Model III, $P$ and $S$ are separated by the set $\{G, W\}$, reflecting the conditional independence of $P$ and $S$ given $G$ and $W$. Thus, the table may be collapsed over $P$ without distorting relationships among $S$, $G$, and $W$, or collapsed over $S$ without distorting relationships among $P$, $G$, and $W$. Under Model IV, $S$ is connected to $P$ and $W$ only through $G$, reflecting the joint independence of $P$ and $W$ from $S$, given $G$. The table may be collapsed as above, and also collapsed over $W$ or both $P$ and $W$, without distorting the association between $G$ and $S$.

A conservative analyst would likely summarize these data using Model I, recognizing the possibility of overfitting. A very aggressive analyst might choose Model IV, recognizing the possibility of oversmoothing the table. Models II and III are reasonable compromises. An alternate and possibly superior approach to choosing a single model is to present the most prominent structural features of the data using Model IV, and then point out the

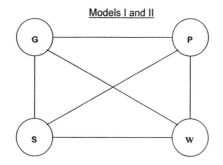

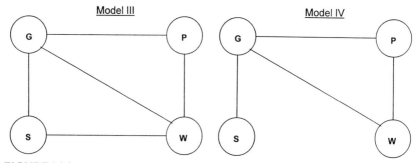

**FIGURE 14.2** Association graphs of loglinear models for occupational exposures in Malaysian Chinese. G, gas fumes; P, paint fumes; S, total smokes; W, wood dusts.

possible additional complexities as reflected in Models I–III. One need not force a choice, and rigidity in such matters makes for bad science.

### E. Logit Models as Loglinear Models

Generalized logits are linear functions of logarithms of cell counts. Hence, every loglinear model for expected cell counts implicitly incorporates linear models for the generalized logits of each variable. These are found by simply subtracting the model expressions for the log numerator and denominator of the logit. Terms identical for the numerator and denominator counts vanish, leaving only the modeled variable and its interactions with remaining variables, which now play the role of predictors. Different loglinear models may therefore imply the same equation for a particular logit, for example, $(AB, AC, BC)$ and $(AC, BC)$ yield the same model for any generalized logit from $C$, but different models for those from $A$ or $B$. The common model for a generalized logit from $C$ is a main effects model with $A$ and $B$ as independent variables.

Although they imply an identical parametric structure for a generalized logit from $C$, the loglinear models $(AB, AC, BC)$ and $(AC, BC)$ will generally yield different predicted logits and estimated probabilities for the levels of $C$. In the latter, conditional population odds ratios relating $C$ to $A$ will reproduce the marginal $A \times C$ population odds ratios whereas in the former they generally will not. And, assuming for simplicity that $C$ is dichotomous, only the former model yields the same estimated logits as the logistic regression model for $C$ with main effects $A$ and $B$. More generally, if several loglinear models share identical embedded models for a selected "outcome" variable, which—if any—yields the same fitted counts as the corresponding logistic regression model?

The answer is implied by the product-binomial GLM for logistic regression, in which sample sizes for each observed combination of predictors are fixed parameters. Predicted counts from logistic regression thus always conform to the distribution of predictors in the sample. For an hierarchical loglinear model to similarly conform, this distribution must be among the model's sufficient statistics, which requires that the model include the interaction term involving all predictors. Of the main effects models that regress $C$ on $A$ and $B$, only $(AB, AC, BC)$ includes $AB$, and hence will reproduce the expected counts of the logistic regression. The parameters of the logistic regression and loglinear models may be recovered from one another by direct transformation, or through use of the fitted counts. As illustration, note that (a) Model I for the occupational exposures is equivalent to a logistic regression main effects model for $W$ as a function of $G$, $P$, and $S$; and (b) the no $k$-way interaction hierarchical loglinear model for a $2^k$ table embodies, simultaneously, the no $k-1$-way interaction logistic regression models for each dimension as a function of the others.

### F. Simplified Structures for Ordinal Variables

Quantitative information on ordinal or higher-scaled variables may also be used in loglinear modeling by placing category scores into columns of $X$, producing models with direct connections to continuous analyses of covariance and correlation (see, e.g., Agresti (1984)). The effect is to impose symmetries determined by the scores on odds and odds ratios. Two brief examples for the $I \times J$ table give an impression of these ordinal association models.

Suppose the column categories are ordered with known scores $v_j$, $j = 1, \dots, J$, and consider the loglinear model $\log m_{ij} = \beta + \beta_i^A + \beta_j^B + v_j \mu_i$, with the $\beta$s and $\mu$s unknown parameters to be estimated. Under this model, the generalized log odds ratio in the $2 \times 2$ table formed by arbitrary pairs of rows and columns is found to be $(\mu_i - \mu_{i'})(v_j - v_{j'})$. Thus, for any rows $i$ and $i'$, the odds of row $i$ versus row $i'$ changes multiplicatively across columns by a row-dependent power of $\exp(v_j - v_{j'})$. If the $v_j$ are equidistant by $c_0$, then this base $\exp(v_j - v_{j'})$ is simply $c^{(j-j')}$, with $c = e^{c_0}$.

Suppose the rows are also ordered with known scores $u_i$, $i = 1, \dots I$, and consider the *linear by linear association model* $\log m_{ij} = \beta + \beta_i^A + \beta_j^B + u_i v_j \beta^{AB}$. The log odds ratios then take the form $(u_i - u_{i'})(v_j - v_{j'})\beta^{AB}$ for the known sets of scores. If rows and columns are equidistant and the scores are standardized to unit separation of adjacent levels, then all $2 \times 2$ tables formed from adjacent rows and columns have the same odds ratio $\exp(\beta^{AB})$. The single parameter $\beta^{AB}$ of this *uniform association model* thus describes the full association structure of the $I \times J$ table. Significance tests of $\hat{\beta}^{AB}$ are tests of the broad hypothesis of independence, with power narrowly focused on "trend" alternatives for which relative frequencies in rows with higher scores tend to steadily increase or steadily decrease with increasing column score.

Specifically, in the linear-by-linear association model, $\sum_{i=1}^{I} \sum_{j=1}^{J} u_i v_j n_{ij}$ is the sufficient statistic generated by the last column of $X$. The full set of sufficient statistics determines the Pearson product–moment correlation; tests of $\hat{\beta}_{AB}$ test statistical significance of the Pearson correlation of scores. Among loglinear models for contingency tables, this is the closest analog to simple linear regression and correlation analyses.

## VII. EXCEPTIONS

We remark on two circumstances requiring alterations or alternatives to the methods of this chapter.

### A. Overdispersion and Clustered Observations in Generalized Linear Models

Since the Poisson distribution depends on a single parameter, variance under this distribution or any of its conditional descendants is a known function of the mean. We may examine whether stipulated Poisson, binomial, or multinomial distributions hold by checking whether estimates of means and variances from the data conform to the implied functional relationship. In a GLM the mean depends on predictors in a manner that must be estimated, so such an examination becomes model dependent. In practice, variances are often too large. Such apparent overdispersion may be due to random chance, poor choice of $X$, or intrinsic heterogeneity of underlying parameters among observations that share the same constellation of predictor values. In the presence of intrinsic overdispersion due to such heterogeneity, the inferential methods of this chapter are inadequate, and their failures are anticonservative. Tests exceed their nominal levels, and confidence intervals fall short of their nominal coverage probabilities, sometimes very substantially.

One indicator of intrinsic overdispersion is a ratio of deviance to its degrees of freedom that departs from unity beyond what chance can explain, in a model that should a *priori* fit well. However, such a departure can indicate unexpected but real lack of fit, and assessment of chance effects is difficult in nonasymptotic situations that commonly arise with Poisson and logistic regression. Although apparent overdispersion is common in practice, it is frequently difficult to isolate an explanation.

Clustered data, as arise in studies of individuals within families, rodents within litters, classrooms within schools, or teeth within individuals, generate overdispersion because each cluster may reflect a random effect unaccounted for by the model. Such random effects generate intraclass correlation which, in conjunction with cluster sizes, determines the impact of the random effects on inference. With intraclass correlation coefficient $\rho$ and clusters of $q$ elements, standard variances are multiplied by the *variance-inflation factor (VIF)* $[1 + (q - 1)\rho]$.

For these reasons, the methods of this chapter should not be applied directly to make formal inferences about clustered data, although the modeling procedures can be used for descriptive purposes. For inference, methods that take specific account of the clustering and intraclass correlation should be used. When a good estimate of the VIF is available, one may adjust the usual inferences by multiplying reported variances and dividing reported chi-square statistics by the VIF. Sometimes the VIF is estimated by the ratio of a deviance to its degrees of freedom for a saturated or other "most complex" model. Such adjustment is generally available as an option and sometimes a default in GLM software. It may be justified both heuristically and by *quasi-likelihood* considerations beyond the scope of this review (McCullagh & Nelder, 1989). But the decision to adjust should be taken carefully, since correction for spurious overdispersion sacrifices power and misleadingly decreases apparent precision of estimates.

### B. Population-Averaged Models and Symmetry Considerations

As illustrated earlier in the discussion of conditional logistic regression, analyses involving comparisons within matched clusters may also be problematic for the models of this chapter. Other important issues arise with repeated measures and split-plot types of categorical data. Consider, for example, a square contingency table in which rows and columns both correspond to preferences of potential voters among political candidates, respectively before and after a televised debate. Interest is in characterizing any net changes in candidate preferences. Shifts in preferences are directly reflected in the difference between the marginal column distribution and the marginal row distribution. For the purpose of estimating such shifts, the odds ratios within the table, on which loglinear models have been based, are of little or no interest.

In contrast, consider a similar table of higher dimension reflecting responses of patients with a chronic disease to different medications, e.g., responses of migraine sufferers to several pain relievers. A physician treating a newly diagnosed patient should refer to the marginal distributions for each drug, from which the most effective drug overall might be chosen. On the other hand, a physician treating a patient on whom one or more drugs had failed should refer to the conditional odds of success of each remaining drug, among patients who experienced the same drug failures. A loglinear model is of little interest for treating the former patient, but may be of considerable help for treating the latter.

Statistical models that compare marginal distributions are sometimes called *population-averaged models,* as they summarize overall effects on a population. In this terminology, models that examine effects conditional on experiences at other times or under other conditions are called *subject-specific.* Both population-averaged and subject-specific modeling frequently deal with symmetries in a contingency table, but the symmetries for population-averaged models involve marginal distributions and for subject-specific models involve joint distributions. In most situations, the scientific question and/or the mode of data collection dictates the type of model that is most appropriate. Sometimes both a population-averaged and a subject-specific odds ratio may be estimated for the same scientific contrast. Because the marginal odds ratio estimated by a population-averaged model and the conditional odds ratio estimated by a subject-specific model are different parameters with different interpretations, these estimates do not duplicate one another. Typically, subject-specific effects are attentuated in population-averaged analyses; hence, subject-specific models may be advantageous for the purpose of discovering relationships.

However, where population-averaged models are dictated by the form of data or purpose of a study, the methods of this chapter are rarely the most natural or useful. For such studies, as well as other research for which the modeling of functions of multiple counts takes precedence over the modeling of individual counts themselves, alternative methods based on weighted least-squares and possibly generalized estimating equations are usually to be preferred (Koch et al., 1985; Diggle, Liang, & Zeger, 1994).

## REFERENCES

Agresti, A. A. (1984). *Analysis of ordinal categorical data.* New York: John Wiley and Sons.

Agresti, A. A. (1996). *An introduction to categorical data analysis.* New York: John Wiley and Sons.

Aldrich, J. S., & Nelson, F. D. (1984). *Linear probability, logit and probit models.* Beverly Hills, CA: Sage Publications.

Armstrong, R. W., Imrey, P. B., Lye, M. S., Armstrong, M. J., Yu, M. C., & Sani, S. *Nasopharyngeal cancer in Malaysian Chinese: Occupational exposures to particulates, formaldehyde, and heat.* Unpublished manuscript.

Clinton, S. C., Li, SP-S., Mulloy, A. L., Imrey, P. B., Nandkumar, S., & Visek, W. J. (1995).

The combined effects of dietary fat and estrogen on survival, 7,12-dimethylhydrazine-induced breast cancer, and prolactin metabolism in rats. *Journal of Nutrition 125,* 1192–1204.

Cytel Software Corporation (1999). *LogXact 4 for Windows user manual.* Cambridge, MA: Cytel Software Corporation.

Diggle, P. J., Liang, K.-Y., & Zeger, S. (1994). *Analysis of longitudinal data.* Oxford: Clarendon Press.

Goldberger, A. S. (1964). *Econometric theory.* New York: John Wiley & Sons.

Grizzle, J. E., Starmer, C. F., & Koch, G. G. (1969). Analysis of categorical data by linear models. *Biometrics 25,* 489–504.

Hosmer, D. W., & Lemeshow, S. (1989). *Applied logistic regression.* New York: John Wiley and Sons.

Imrey, P. B., Jackson, L. A., Ludwinski, P. H., England, A. E., Fella, G. A., Fox, B. C., Isdale, L. B., Reeves, M. W., & Wenger, J. D. (1996). An outbreak of serogroup C meningococcal disease associated with campus bar patronage. *American Journal of Epidemiology, 143,* 624–630.

Koch, G. G., Imrey, P. B., Singer, J. M., Atkinson, S. S., & Stokes, M. E. (1985). *Analysis of Categorical Data.* Montreal: University of Montreal Press.

McCullagh, P., & Nelder, J. A. (1989). *Generalized Linear Models* (2nd ed.). London: Chapman and Hall.

# EVALUATION OF
# MATHEMATICAL MODELS

# 15

# STRUCTURAL EQUATION MODELING: USES AND ISSUES

**LISABETH F. DILALLA**

*School of Medicine, Southern Illinois University, Carbondale, Ilinois*

Structural equation modeling (SEM) has become a popular research tool in the social sciences, including psychology, management, economics, sociology, political science, marketing, and education, over the past two to three decades. Its strengths include simultaneous assessment of various types of relations among variables and the ability to rigorously examine and compare similarities among and differences between two or more groups of study participants. However, one of its major limitations is the ease with which researchers can misinterpret their results when anxious to "prove" the validity of a model or to attempt to assess causality in the relation between two or more variables when the research design does not allow for such conclusions. These will be elaborated below. First, however, it is essential to understand exactly what SEM is in order to better understand when it is most useful to apply it as a statistical tool. The purpose of this chapter is to provide an overview of some of the possible uses of this technique and to discuss some of the important assumptions and necessary conditions for using SEM and for interpreting the results.

*Handbook of Applied Multivariate Statistics and Mathematical Modeling*
Copyright © 2000 by Academic Press. All rights of reproduction in any form reserved.

## I. DEFINING STRUCTURAL EQUATION MODELING

Latent variable analysis (Bentler, 1980) involves the study of "hidden" variables that are not measured directly but that are estimated by variables that can be measured. Latent variable analysis includes such techniques as factor analysis, path analysis, and SEM. Factor analysis (see also Cudeck, chapter 10, and Hoyle, chapter 16, this volume) involves assessing a latent factor that is operationalized by measured variables. The latent factor is identified by the variance that is shared among the measured variables; it is the "true" variable that affects the measured variables. Path analysis, first described by Sewall Wright (1934, 1960), determines the causal relations between a series of independent and dependent variables. SEM encompasses both of these and is a method for testing carefully delineated models based on hypotheses about how observed and latent variables are interrelated (Hoyle, 1995b) in order to meaningfully explain the observed relations among the variables in the most parsimonious way (MacCallum, 1995).

The strength of SEM lies in its ability to rigorously test a hypothesized model of relations among manifest and latent variables. The key to this is that the model must be specified *a priori* and be theoretically based. SEM can then provide a series of indices that indicate the extent to which the specified model appears to fit the observed data. The results cannot be used to assert that a given model with a good fit must therefore precisely identify the mechanisms through which the variables are intercorrelated because, although that one model may fit, there will also be countless others that might fit equally well or even better. Therefore, as long as the model is theoretically strong, a good-fitting model can be said to be supported by the data and to be sensible, but, as with any other statistical technique, SEM cannot be used to "prove" the model.

Clearly, careful forethought is essential in developing a model. The first step is to develop a theoretical model delineated by the latent variables (drawn in ovals), which are the constructs that are of interest theoretically (see Figure 15.1). This part of the model is the structural model and is composed only of the latent, unmeasured variables. Then the variables that are actually measured (drawn in rectangles) and are used as indicators of the latent variables can be added to the model (see Figure 15.1). This part of the model that specifies the relations between the latent and manifest variables is called the measurement model. The final model, including both the structural and the measurement parts of the model, has the advantage of allowing the researcher to explore the relations between the latent variables that are of interest theoretically by including the operationalized versions of these variables as manifest variables. Thus, for instance, home environment and aggression can be operationalized using several measures of environment to represent two "home environment" factors ("Home"

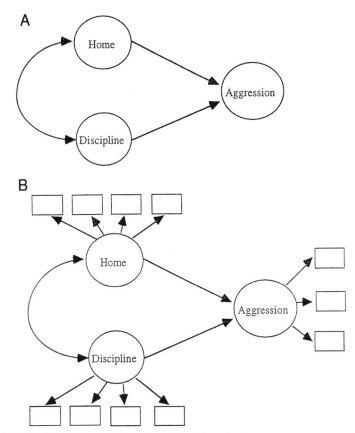

**FIGURE 15.1** Structural model and its elaboration into a measurement model.

and "Discipline" in Figure 15.1) and several measures of aggression to represent an "aggression" factor, and then the causal relation between the two latent variables can be assessed using these measures.

## II. COMMON USES OF STRUCTURAL EQUATION MODELING

A path diagram can be created to test a set of relations among variables simultaneously. Thus, a variable can be regressed on several other variables and can in turn simultaneously predict another outcome variable. This set of relations cannot be tested using standard regression analysis because of the complex set of simultaneous relations. Another particularly useful application of SEM is to test factor analysis models, both in an exploratory framework as well as a confirmatory one. SEM also can be used to assess the applicability of a given model on more than one group, to make direct comparisons between groups, to run growth curve models, and to compare nested models, to name some of the more common uses.

The most important rule to bear in mind is that a strong theoretical model must be posited prior to model testing. Within this theoretical framework, a series of nested models can be compared to come up with a parsimonious explanation for the relations among the variables. Of course, exploratory analyses can be run, but then the structure of the relations among the variables is being assessed for that particular data set, and replication using an independently assessed sample is essential in order to draw conclusions with confidence. As with any statistical analysis, analyses that capitalize on the chance relations among variables in a given data set are not useful for advancing our understanding of any of the sciences. This would result if analyses are conducted without a theoretical framework to guide the model fitting.

## A. Exploratory Factor Analyses

One of the typical uses of SEM is to conduct exploratory factor analyses. In exploratory factor analysis, no pre-existing model is hypothesized. Rather, the relations among the variables are explored by testing a series of factors that account for shared variance among the variables. This is feasible with SEM because there is a very specific type of model included in the exploration. Only models with no causal links between factors and with paths only between the factors and the manifest variables are tested (Loehlin, 1992). Therefore, there are only a specific number of models tested. This method is useful in allowing the researcher to determine the simplest factor model (the one with the fewest number of factors and the fewest nonzero paths in the pattern matrix) that will adequately explain the observed intercorrelations among the various manifest variables (Loehlin, 1992). In an SEM factor analysis, the latent variables are the factors, the manifest variables are the variables that make up the factors, the loadings between the manifest and latent variables form the factor pattern matrix, and the residuals on the manifest variables are the specific or unique factors. Thus, SEM can be used to explore the latent factor structure of a data set by considering successively increasing or decreasing numbers of factors and comparing each new model to the preceding one to determine the optimal and most parsimonious model that best accounts for the manifest variable interrelations.

## B. Confirmatory Factor Analyses

A related use of SEM is to conduct confirmatory factor analyses. Exploratory and confirmatory factor analyses are not necessarily independent (Nesselroade, 1994). The researcher must have some theoretical starting point for the exploratory analysis, even in that a certain set of manifest variables are assumed to be related sufficiently to submit to factor analysis. Additionally, confirmatory factor analysis may still involve a small amount of tweaking of the model to improve its fit. With confirmatory factor analysis (see also Hoyle, chapter 16, this volume), a path model is designed that describes the interrelations among the manifest variables by hypothesizing a specific set of latent factors that account for those relations. These factors are based in theory, as are the items that load on them. This last fact is essential because it both justifies the model and requires that changes not be made to the model without reconsidering the theory behind the analysis. A common example of misuse of SEM is to modify the factor structure of the model based on modification indices without regard to the underlying theory on which the model was based. Assuming the theory is sound, it probably is wiser to maintain the original items as long as there are enough items with high loadings to define the latent factor.

Confirmatory factor analyses can be conducted using SEM. The model is based on theory; that is, a hypothesized structural model is fit to the data to determine how well the interrelations among the variables are accounted for by the *a priori* model. Thus, the important difference between confirmatory and exploratory factor analysis with SEM is that the factors that account for the variable intercorrelations are specified up front with confirmatory analyses rather than becoming revealed through exploration of the variance accounted for with a different number of factors, as with exploratory analyses.

The measurement model of an SEM is a confirmatory factor analysis because it reflects the theoretically designed configuration of manifest variables as they load on the latent factors. For this reason, it is important that changes not be made to the measurement model lightly when attempting to achieve a parsimonious and significant fit. It is prudent to maintain the original measurement model even when some of the manifest variables load more highly on a different latent factor than the one originally specified or when a manifest variable does not add significantly to the definition of a latent factor because the model supposedly was based on sound theoretical underpinnings. Using SEM in an exploratory fashion is acceptable and can be enormously useful as one explores the ways in which certain variables relate to each other, but it is essential that at the end of this exploratory foray the researcher is aware of the exploratory nature of the analyses and describes them as such. Under these circumstances, the confirmatory aspect of the modeling is lost and further confirmation with another data set is necessary to determine that the final model is not simply a reflection of an unusual data set.

## C. Regression Analyses

SEM is also useful when a series of simultaneous regressions are needed, or when the variables in the analyses are based on factors rather than single variables. For example, if we hypothesize that socioeconomic status (SES) predicts the home environment, which in turn predicts children's play behaviors with peers, then a model can be designed with paths from SES to home environment and from home environment to play behaviors, and these paths can be tested simultaneously. This analysis cannot be conducted with a standard regression analysis. Additionally, if we want to measure each of these variables with several items, then with SEM we can form three factors—SES, home environment, and play behaviors—and a number of items that load on each of the factors to define them. So for instance, parental education and occupation and family income can load on SES, a set of items assessing parental discipline and sociability can load on home environment, and items based on videotaped analyses of children's play behaviors can load on the play behaviors factor. Then each of these three factors can be regressed according to the theory that was originally specified, and these interrelations can be assessed directly and simultaneously.

### D. Moderator–Mediator Analyses

The use of moderator and mediator variables also can be tested using SEM. For instance, with the above example, we can ask whether home environment mediates the relation between SES and children's play behaviors or whether SES has a direct influence on play behaviors even after the effects of the home environment are included in the model. Thus, we can test the model specified above and then add the direct path from SES to play behaviors to determine whether this path significantly improves the overall model fit. The use of moderator variables is more complex, but these variables also can be tested using SEM. Methods for assessing linear, quadratic, and stepwise effects of the moderating variable on the dependent variable are described in Baron and Kenny (1986) and can be applied to SEM.

### III. PLANNING A STRUCTURAL EQUATION MODELING ANALYSIS

Clearly, SEM has a number of unique methodological advantages, such as using multiple measures as both independent and dependent variables. However, one distinct disadvantage, as with many of the procedures described in this volume, is that it has become so easy to use many of the SEM programs that a user can run analyses without being aware of some of the basic assumptions that are necessary for conducting the analyses correctly. Incorrect results can be reported if the user does not read and input data correctly or read the output (including error messages) accurately. Therefore, the remainder of this chapter focuses on fundamental necessary steps for understanding the basics of performing and interpreting an SEM analysis.

Prior to setting up an SEM analysis, the variables must be examined to determine their appropriateness for the analysis. Certain assumptions must be met. Sample size is important because SEM requires larger samples than most other statistical procedures. The format of data input also must be considered. Estimation procedures must be chosen based on the types of variables in the model. Each of these issues is considered in the following sections.

## IV. DATA REQUIREMENTS

Basic assumptions common to all regression-type analyses include data multivariate normality and a sufficiently large sample size. SEM also assumes continuous data, although there are ways to model ordinal data by using tetrachoric or polychoric correlations, and ways to model categorical data as well (Mislevy, 1986; Muthen, 1984; Parry & McArdle, 1991). A number of bootstrapping and Monte Carlo techniques have been used in an attempt to determine the ramifications of violations of these basic assumptions. There is not a clear consensus on the seriousness of relaxing these constraints, but parameter estimates appear to be fairly robust even when fit statistics may be severely compromised by this relaxation (Loehlin, 1992).

### A. Multivariate Normality

Multivariate normality of the variables in the model is necessary for a well-behaved analysis. Multivariate normality is simply a generalization of the bivariate normal situation. With a multivariate normal distribution, each variable is normally distributed when holding all other variables constant, each pair of variables is bivariate normal holding all other variables constant, et cetera, and the relation between any pair of variables is linear. In order to test for multivariate normality, the third and fourth order moments of the variables must be examined for multivariate skewness and multivariate kurtosis. Mardia (1970) has devised measures to assess these that are available in the EQS (Bentler, 1992) and PRELIS (Joreskog & Sorbom, 1993b) computer software packages for data preparation.

It is important to assess the multivariate normality of the data, because in its absence model fit and standard errors may be biased or irregular (Jaccard & Wan, 1996; West, Finch, & Curran, 1995). However, there are no definitive studies detailing at exactly what degree of nonnormality errors will begin to be detectable. Jaccard and Wan (1996) outline four options for minimizing nonnormality when it is present in the data set. First, outliers can be located and eliminated from the data. However, it is imperative that this be done carefully and with a great deal of thought by the researcher. If the outliers are truly errors, for instance, if someone were tested under improper conditions, then the outlier can be eliminated and the resulting sample is still the one that was originally expected. However, there are times when an outlier occurs under reasonable circumstances, such as when a person behaves very differently from expectation and from how others in the sample behave, but there are no noticeable differences in test administration, and the person did not suffer from a known condition that would affect test performance. In these cases, it is not acceptable to eliminate this person from the total sample because he or she is a randomly ascertained member of the population. In other words, it is not acceptable to drop respondents simply because the researcher is not happy with the results that were obtained. If such atypical participants are dropped from the sample, then the researcher must be clear that results of the SEM may not generalize to extreme members of the population.

A second acceptable procedure for dealing with nonnormality is to transform the variable that is causing the problem. Several transformation options exist, and it depends on the type of variable and the degree of skewness or kurtosis present as to which transformation is preferred. For instance, a linear transformation will not affect the distribution of the variable, whereas a nonlinear transformation may alter the variable's distribution and its interactions and curvilinear effects (West et al., 1995). Censored variables (variables that have "ceiling" or "floor" effects such that a large majority of the respondents receive the top or bottom possible score, and only a small percent of the respondents score along the continuum of possible values) also may bias parameter estimation (van den Oord & Rowe, 1997).

It is important after transforming a variable to reassess the skewness or kurtosis to determine whether this has improved. It also is important to be aware that the interpretation of the variable following a transformation cannot be the same as prior to the transformation. Transformation causes a loss of metric so the variable cannot be easily compared across studies or to other, comparable variables (West et al., 1995). Examples of typical and appropriate transformations are provided in Cohen and Cohen (1975) and West et al. (1995).

A third approach to the violation of multivariate normality is to use a routine other than maximum likelihood (e.g., weighted least squares) to estimate the model parameters. Maximum likelihood estimation (MLE) is the least tolerant of nonnormality in the data. However, most other methods require much larger sample sizes, which may not be reasonable for all research projects. Finally, the last approach is to use bootstrapping to estimate the standard errors. This entails repeatedly sampling with replacement from the data to derive a large number of samples, and then running the model on each of these samples to ascertain the stability of the fit index across the samples. This allows an estimation of whether the normality violation is a problem for the particular model. The new version of LISREL-8 makes this method fairly simple.

## B. Sample Size

The difficult part of choosing an appropriate sample size is that there is no clear-cut rule to follow. Different researchers, experimenting with different types of data and models, have found varying results in terms of the necessary sample size for obtaining accurate solutions to model-fitting analyses (Bentler & Chou, 1987; Guadagnoli & Velicer, 1988; Hu & Bentler, 1995; Loehlin, 1992). One consideration is the number of manifest variables used to measure the latent factors. The larger this is, and the larger the loadings are for the indicators, then the smaller the necessary sample can be. For instance, Guadagnoli and Velicer (1988) found that as long as the factors were sufficiently saturated with loadings of at least .80, the total number of variables was not important in determining the necessary sample size, and the sample size could possibly go as low as 50. (With factor loadings less than .80, however, the sample size requirements were more stringent.)

Another important consideration is the multivariate normality of the measures. Smaller samples may be adequate if all measures are multivariate normal. Bentler and Chou (1987) suggested that if all variables are normally distributed, a 5:1 ratio of respondents to number of free parameters may be sufficient. Without multivariate normality, a sample size as large as 5000 may be necessary to obtain accurate results (Hu & Bentler, 1995). For most studies, Anderson and Gerbing (1988) suggest that sample sizes of at least 150 should be adequate.

All of these sample size suggestions assume continuous data sampled randomly from a population. It is possible that sample sizes need to be different when these assumptions are violated, but there are no definitive answers as yet. Of course, the basic concern with a small sample is how well the sample represents the population. Quirks specific to the sample may greatly affect the analysis, and that is more likely if a smaller sample happens to misrepresent the population. If the sample is truly randomly ascertained and is an accurate representation of the larger population of interest, then some of the concerns about small sample size may become less problematic (Loehlin, 1992). An excellent discussion of the important issues of sample size and the accompanying concern about power can be found in Jaccard and Wan (1996).

## V. PREPARING DATA FOR ANALYSIS

Once data are collected, the investigator must determine the type of matrix to use for data input and decide how to handle missing data.

## A. Input Data Matrix

Most scholars recommend the use of a covariance matrix for analysis because the methods that are most commonly used to solve structural equation models (i.e., maximum likelihood and generalized least squares) are based on theories that were derived using covariance matrices rather than correlation matrices (Loehlin, 1992). Also, when a correlation matrix is computed, the variables are standardized based on that sample. When those standardized scores are used for the SEM, the sample standard deviation must be used to calculate the standardized variable, resulting in the loss of one degree of freedom (Loehlin, 1992). The effects of this are most serious with small samples. It is imperative to use covariance matrices as input when a multiple-group analysis is conducted because otherwise variance differences across groups cannot be considered. Instead, the groups are treated as though they did not differ in variance, which may be quite misleading. Thus, covariance matrices are necessary for comparison across groups and across time, but correlation matrices may be used when the analysis focuses exclusively on within-group variations.

## B. Missing Data

A second issue concerns the handling of missing data. There are three main options for dealing with this problem. First, variables can be deleted in a listwise fashion, thereby excluding all participants who are missing any of the variables that are part of the analyses. If variables are missing randomly throughout the sample, however, this procedure often reduces the analysis sample to a size that is too small for reliable estimates to be produced. Furthermore, the smaller sample may be less representative of the general population that was originally sampled, and therefore issues of generalizability become a concern.

Second, variables can be deleted in a pairwise fashion, thereby only omitting participants for those elements in the covariance matrix for which they are missing variables. When pairwise deletion is used, the elements in the resulting covariance matrix are based on different participants and different sample sizes. This can result in impossible relations among variables and therefore in a matrix that is not positive definite (Jaccard & Wan, 1996; Kaplan, 1995). The matrix may then be uninvertable, or the parameter estimates may be theoretically impossible. Additionally, there is a difficulty in determining what sample size to claim for the analyses. Because of these shortcomings, I recommend against using pairwise deletion for SEM analyses (see also Schumacker & Lomax, 1996).

Finally, imputation methods may be used to estimate the values of the missing data. Although these methods maintain the original sample size, the values used where data are missing are rough estimates that may not be accurate. Even if the new values are reasonable, there may be problems with nonnormality and error variance heteroscedasticity (Jaccard & Wan, 1996). Problems also may arise if the imputation is based on the variables used in the model, because then the same variables are used for estimating another variable's values and for estimating the relations among the variables. This type of redundancy increases the probability of a Type I error. Rovine (1994) describes several methods of data estimation that may avoid some of these problems. These methods are too complex to describe here and may prove challenging for beginning users, but they may prove useful if missing data are a problem.

Despite its drawbacks, the safest method for dealing with missing data in SEM is to use listwise deletion. However, as long as the data are missing completely at random, the sample size remains large, and no more than 10% of the data are missing, there is not a large practical difference between the methods of dealing with missing data. If an imputation method is chosen, it is important to have a strong rationale for the type of imputation used and not to confound the imputation with the modeling itself. More research is necessary on the various ways to handle missing data before we can recommend one method over another unequivocally.

## C. Construction of Input Matrix

Once the appropriate type of input matrix has been chosen and the problem of missing data handled, the sample matrix must be constructed. If LISREL (Joreskog & Sorbom, 1993a) is the program of choice, a package called PRELIS (Joreskog & Sorbom, 1993b) can compute the input matrix in a form that is ready to be read by LISREL. This is deceptively simple; all that the user needs to do is provide a raw data file and specify the missing values and type of matrix to compute. As with all canned statistical packages, the user must be diligent in specifying the desired options. It is very simple to request the wrong type of matrix (e.g., a correlation matrix rather than a covariance matrix) or to misspecify the types of variables that have been measured (e.g., identifying a continuous variable as ordinal). PRELIS will compute the requested matrix, even if it is not necessarily sensible, LISREL will analyze that matrix, and the naive user may think that the model has been tested adequately. PRELIS is an excellent tool for preparing data for input to LISREL as long as the user correctly identifies the variables and the desired type of input matrix.

## D. Estimation Procedures

The purpose of the estimation procedure is to minimize the discrepancy between the estimated matrix of covariances among the variables, based on the model being tested, and the actual data covariance matrix. The choice of estimation procedure depends on the sample and the model being estimated.

The MLE procedure is the default option in LISREL and EQS. This method is based on the assumption of multivariate normality, and it requires a relatively large sample size to perform adequately. When these two assumptions are not met, it may be wise to consider a different estimation method. A second method is generalized least squares, which also assumes multivariate normality and zero kurtosis. A number of other methods have been developed, many of which are available through the standard SEM programs. In practice, there may be little difference in outcome between the various methods (Chou & Bentler, 1995), but when there is doubt about the appropriateness of a method, I recommend that the investigator use several methods and compare the results.


## VI. MULTIPLE GROUPS

One huge advantage of SEM analyses over other types of analyses is the ability to compare two or more independent groups simultaneously on the same model to determine whether the groups differ significantly on one or more parameters. There are a number of instances when it is useful to model more than one group simultaneously. A researcher may want to compare experimental groups to determine whether a treatment had an effect, or gender groups to determine whether the relations among a set of variables are comparable for boys and girls, or small versus large businesses to determine whether shareholder earnings affect worker productivity comparably. The basic methodology is to hold all or a subset of the parameters constant across groups and assess model fits for all groups simultaneously. Then these equal parameters can be freed and the models compared to determine whether holding the parameters equal across groups provides a significantly worse fit. If it does, then the parameters differ significantly across groups; if not, then the parameters do not differ significantly across groups and can be set equal across groups without loss of model fit.

Certain regulations must be observed when conducting multiple-group SEM. First, it is important to input covariance matrices rather than correlation matrices, as described earlier, because variances may not be comparable across groups. Second, the latent variables must be on a common scale across groups, which means that the same manifest variable(s) must be used to set the scale for the latent variable(s) in all groups. These two practices allow comparisons across groups as well as a test for whether the groups are equivalent in terms of item variances (see Joreskog & Sorbom, 1993a, for specific examples testing the equality of factor structures and factor correlation matrices).

## VII. ASSESSING MODEL FIT

Once the model has been specified properly and the data have been entered correctly, the fit of the data to the hypothesized model must be evaluated. A number of tests are used to evaluate how well the model describes the observed relations among the measured variables; different modeling programs provide different output. There is no consensus as to which one is "best" because each test statistic has advantages and disadvantages. Also, there is no consensus regarding the effect of factors such as sample size and normality violations on different fit indices (e.g., Hu & Bentler, 1995; Marsh, Balla, & McDonald, 1988).

It is imperative to examine several fit indices when evaluating a model and never rely solely on a single index (Hoyle, 1995b). Table 15.1 summarizes some of the tests and the situations under which most researchers agree that they are most and least useful. The following descriptions and recommendations are by no means definitive or exhaustive, but they incorporate the most recent suggestions in the literature. Hoyle (chapter 16, this volume) and Tracey (chapter 22, this volume) also provide discussions of fit indices.

### A. Absolute Fit Indices

These indices compare observed versus expected variances and covariances, thereby measuring absolute model fit (Jaccard & Wan, 1996). The earliest measure and one that is still frequently reported is the chi-square fit index. This index was designed to test whether the model fit is perfect in the population (Jaccard & Wan, 1996). It compares the observed covariance matrix with the expected covariance matrix given the relations among the variables specified by the model. The chi-square will be zero when there is no difference between the two matrices (i.e., there is perfect fit), and the chi-square index will increase as the difference between the matrices increases. A significant chi-square value signifies that the model predicts relations that are significantly different from the relations observed in the sample, and that the model should be rejected.

A problem with the chi-square statistic is that SEM requires the use of large samples, and under those conditions the chi-square test is powerful and rejects virtually all models. Also, the chi-square statistic may not be distributed as a chi-square when sample size is small or when the data are nonnormal, and under these conditions the significance test is not appropriate (Jaccard & Wan, 1996). However, the chi-square test is useful when testing nested models. Therefore, I recommend that this statistic be reported, but users should be mindful of its limitations.

Two related absolute fit indices are the goodness-of-fit index (GFI) and the adjusted goodness-of-fit index (AGFI), which is similar to the GFI

**TABLE 15.1  Summary of Some of the More Common Model Fit Indices**

| Test name | Ideal score[a] | Notes | Estimated well with nonnormality? | Consistent across sample size? | Assesses parsimony? |
|---|---|---|---|---|---|
| **Absolute fit indices** | | | | | |
| Chi-square statistic (Bollen, 1989b)[b] | $p > .05$ | Useful for comparing groups. | No | No | No |
| Goodness-of-fit index (GFI) (Tanaka & Huba, 1984) | >.90 | Behaves consistently across estimation methods. | Unknown | No | No |
| Adjusted goodness-of-fit index (AGFI) (Tanaka & Huba, 1984)[b] | >.90 | Adjusts GFI for degrees of freedom. Otherwise, same benefits and concerns as for GFI. | Unknown | No | Yes |
| Root mean square residual (RMR) (Joreskog & Sorbom, 1993a)[b] | <.05 | Can be used to compare the fit of two different models with the same data; easiest to interpret if all observed variables are standardized. | Unknown | Yes | No |
| Centrality index (CI) (McDonald, 1989) | >.90 | Small sample size becomes a problem if the latent variables are dependent. | Unknown | Yes, but see notes | No |
| Root mean square error of approximation (RMSEA) (Steiger & Lind, 1980)[b] | <.08 | Measures absolute fit but adds a penalty for lack of parsimony. | Unknown | No | Yes |
| **Comparative fit indices** | | | | | |
| Comparative fit index (CFI) (Goffin, 1993)[b] | >.90 | Accurate across estimation methods; useful for comparing nested models. | Modest underestimation | Yes | No |
| Normed Fit Index (NFI) (Bentler & Bonett, 1980) | >.90 | Of the comparative fit indices, most sensitive to violations of normality and to small sample size. | Severe underestimation with small $N$ | No | No |
| Tucker-Lewis Index (TLI or NNFI) (Tucker & Lewis, 1973) | >.90 | Performs best with maximum likelihood (ML) method; performs badly with Generalized Least Squares. Good for comparing nested models. | Modest underestimation | Unclear | No |
| Incremental fit index (IFI) (Bollen, 1989a)[b] | >.90 | Preferred over TLI/NNFI when using Generalized Least Squares. | Modest underestimation | Yes | No |

[a] "Ideal score" is the cutoff score used by most researchers. There is no empirical basis for these conventions. The appropriateness of the cutoff for each test depends on the model being tested, the sample size, and the normality of the data (Hu & Bentler, 1995), and the debate continues about the interpretation of fit indices.
[b] Recommended.

but which adjusts for the degrees of freedom in the model. A third index is the centrality index (CI; McDonald, 1989). Scores on all three indices can range from 0 to 1.0, with values closer to 1.0 being preferable. Many researchers have suggested that values greater than .90 on these indices can be interpreted as signifying acceptable model fit, but there is no empirical support for this. Although Gerbing and Anderson (1993) found the CI to be particularly robust with Monte Carlo simulations, it is not provided by LISREL and is used less frequently in general, making it less useful for comparing results across studies. The final two absolute fit indices are the standardized root mean square residual (RMR), which is the average discrepancy between the observed and the expected correlations across all parameter estimates, and the root mean square error of approximation (RMSEA; Steiger & Lind, 1980), which adjusts for parsimony in the model. A perfect fit will yield an RMR or RMSEA of zero; scores less than .08 are considered to be adequate, and scores of less than .05 are considered to be good (Jaccard & Wan, 1996). The RMR is a function of the metric used in measuring the variables and is most interpretable with standardized variables. The RMSEA is useful because it adds a penalty for including too many parameters in the model.

### B. Comparative Fit Indices

These indices compare the absolute fit of the model to an alternative model. The comparative fit index (CFI; Bentler, 1990), the DELTA2 or incremental fit index (IFI; Bollen, 1989a), the normed fit index (NFI; Bentler & Bonett, 1980), and the nonnormed fit index (NNFI; Bentler & Bonett, 1980), which is a generalized version of the Tucker and Lewis index (TLI; Tucker & Lewis, 1973), are the most widely used comparative fit indices (see Table 15.1). Each compares the fit of a target model to a baseline model.

The CFI compares the tested model to a null model having no paths that link the variables, therefore making the variables independent of each other. The CFI appears to be quite stable, especially with small samples. It can range from 0 to 1.0; scores less than .90 are considered to be unacceptable. The IFI also tends to be quite consistent even with small samples. The IFI typically ranges from 0 to 1.0, although values can exceed 1.0, which makes this more difficult to interpret than the CFI. Again, higher values indicate better model fit. There is some debate about the sensitivity of the TLI and the NNFI to sample size; Marsh et al. (1988) suggest that they are relatively independent of sample size, whereas Hu and Bentler (1995) suggest that their values may not stay within the easily interpreted 0 to 1 range if the sample size is small. The NFI clearly is more sensitive to sample size, does not perform as consistently with smaller samples, and tends to be underestimated (i.e., values may be unacceptably small for good-fitting models) when the data are not normally distributed.

## C. Summary

Many indices have been proposed for evaluating the fit of a model (see Byrne, 1995; Hu & Bentler, 1995; Jaccard & Wan, 1996), and many cutoff values have been suggested for interpreting these indices (e.g., Schumacker & Lomax, 1996). However, there is much discussion among SEM users as to whether these cutoffs are appropriate. No unambiguous interpretation exists whereby model fit can be described as "definitely good" or "definitely bad." Instead, fit indices are interpreted fairly subjectively, although the cutoff values suggested in Table 15.1 will be of some help. The best way to determine whether a model is acceptable is to use several of the indices listed in Table 15.1 and to look for agreement across indices. Confidence can be placed in the model if most or all indices are acceptable, but the model should be considered a poor fit to the data if several of the indices are unacceptable. A good general practice is to report the chi-square and AGFI statistics, but to rely more on the comparative fit indices for interpreting the model fit.

## VIII. CHECKING THE OUTPUT FOR PROBLEMS

I have insufficient space in this chapter to do more than highlight aspects of the SEM output that may cause readers some confusion or lead to erroneous interpretations if not examined carefully. Therefore, in this section I will focus on double-checking output, using standardized scores, interpreting parameter estimates, using model modification indices, and comparing nested models. An example of LISREL output is provided in Tables 15.2 and 15.4 through 15.6 to help clarify the points made below. This model hypothesizes that day care experience and child temperament are causes of childhood aggression (DiLalla, 1998; see Figure 15.2). SES and sex are the two exogenous variables, and all other variables are regressed on them. Aggression is measured by parent report and by lab ratings during a peer play encounter. Aggression of the peer also is rated in the lab. Thus, child lab aggression is correlated with parent-rated aggression and peer aggression, and is regressed on day care experience, child temperament, SES, and sex.

## A. Ensuring Accurate Input

It is obvious that errors in input will lead to inaccurate output. Therefore, prior to interpreting the analysis results, it is imperative that the user first examine the model parameters to ensure that the parameters being estimated and the parameters that are meant to be fixed were specified correctly. I cannot overstress how important it is to double check the model

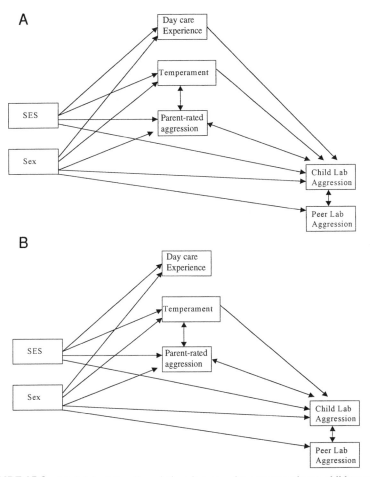

**FIGURE 15.2** Model depicting the relations between day care experience, child temperament, and child sociability (from DiLalla, 1998), and a nested version of the full model, with the day care experience parameter fixed to zero.

prior to examining the output for modeling results. An example of LISREL "Parameter Specification" output is provided in Table 15.2. All numbered parameters are supposed to be free and all zero parameters should be fixed. Note, for instance, that Beta (4,1) is free (parameter 3) because lab aggression is regressed on day care experience. Psi (4,1) is fixed (set at zero) because lab aggression and day care experience are not free to correlate.

Also, double check the input matrix before examining the model results. This requires that the user be familiar with the data set and understand how the input matrix should look (Hoyle, chapter 16, this volume). Always ensure that the entries in the matrix are accurate by comparing them to a separate print-out of the correlation or covariance matrix.

**TABLE 15.2 LISREL Output: Parameter Specifications for Model in Figure 15.2**

| | Beta matrix | | | | |
|---|---|---|---|---|---|
| | Day care experience | Tempera-ment | Parent-rated aggression | Child lab aggression | Peer lab aggression |
| Day care experience | 0 | 0 | 0 | 0 | 0 |
| Temperament | 1 | 0 | 0 | 0 | 0 |
| Parent-rated aggression | 2 | 0 | 0 | 0 | 0 |
| Child lab aggression | 3 | 4 | 0 | 0 | 0 |
| Peer lab aggression | 0 | 0 | 0 | 0 | 0 |

| | Gamma matrix | |
|---|---|---|
| | Socio-economic status | Sex of child |
| Day care experience | 5 | 6 |
| Temperament | 7 | 8 |
| Parent-rated aggression | 9 | 10 |
| Child lab aggression | 11 | 12 |
| Peer lab aggression | 0 | 13 |

| | Psi matrix | | | | |
|---|---|---|---|---|---|
| | Day care experience | Tempera-ment | Parent-rated aggression | Child lab aggression | Peer lab aggression |
| Day care experience | 14 | | | | |
| Temperament | 0 | 15 | | | |
| Parent-rated aggression | 0 | 16 | 17 | | |
| Child lab aggression | 0 | 0 | 18 | 19 | |
| Peer lab aggression | 0 | 0 | 0 | 20 | 21 |

## B. Check for Warnings and Errors

After ensuring the accuracy of the model and data, check the output for any warnings or errors. Fatal errors are easy to detect because they cause the program to crash and make it clear that something is wrong, but there are a number of other types of error messages that, if ignored, can lead to false confidence in the output. Table 15.3 lists some of the errors that beginning users tend to ignore or misunderstand the most frequently.

One of the most common mistakes made when specifying a model is inverting the direction of causality. It is necessary to completely understand the direction of causality of each path in the model and to specify each one correctly in the program. For instance, arrows point from the latent factor to the manifest variables used to approximate it. Thus, in LISREL notation, the lambda matrices are not symmetric and must be specified carefully so that the rows and columns are not reversed. Similarly, in EQS, equations are specified using regression semantics; the predicted variable (the arrowhead) is regressed on (equals) the predictor (the tail of the arrow).

A related error that may be more indicative of a misunderstanding of the entire process of SEM is the misinterpretation of causality. SEM analyses are based on the correlation or covariance between the variables, and it is axiomatic that causality cannot be inferred from correlations. A test of the models with the direction of causality reversed would yield the same fit index because the fit of the model depends only on how well the model recaptures the original covariance matrix. If two variables are correlated, it does not matter whether the first variable is regressed on the second, or vice versa; the resulting fit will be the same. As noted earlier, this underscores the importance of firmly grounding the model on theory and prior research.

Finally, poorly defined latent variables will cause the program to be underidentified, to fail to converge, or to yield a very poor fit value. The latent factor cannot be measured adequately unless the variables used to measure it are fairly highly intercorrelated. When the latent variables are

**TABLE 15.3    Selected LISREL Warnings or Errors**

| Warning/error message[a] | Interpretation | What to do |
|---|---|---|
| Solution written on "Dump" file | Solution was unable to converge; parameter estimates are saved in a file called "Dump"; fit indices and parameter estimates are provided because they may help locate the problem. These estimates should *not* be interpreted—they are *not* meaningful. | "Dump" values can be used as start values for the next run. |
| Inadmissibility error | Solution is nonadmissible. This happens if a Lambda matrix has a row of zeroes, meaning a latent variable is not defined by any manifest variables (e.g., often used in behavior genetic analyses). | Turn off the admissibility test if a row of zeroes is intended. |
| Sigma not positive definite | The covariance matrix based on the model (not the data) is not invertible; this may result if one or more parameters have been started at zero. | Change some of the start values and rerun the program. |

[a] Messages are not presented exactly as they appear on the output in order to facilitate generalizability.

poorly defined, the loadings of each of the manifest variables on the latent factor will be small, and the latent factor will not be well defined. Poorly defined factors are of little use in predicting another factor and the result will be a poor model fit, very small and nonsignificant parameter estimates, and a model that accounts for virtually no variance in the data. Thus, one model check is to ensure that the variables used to measure a factor are sufficiently intercorrelated that the factor will be defined.

## IX. INTERPRETING RESULTS

Only after the user is confident that there are no input errors should the output be examined to address the substantive issues guiding the research. Joreskog and Sorbom (1993a) describe a logical sequence to follow in examining the analysis output.

### A. Examine Parameter Estimates

First, examine the parameter estimates to ensure that they are in the right direction and of reasonable size. In many programs (e.g., LISREL), the estimates can be presented in standardized (see Table 15.4) or unstandardized form. Most users will find the standardized output to be more interpretable. For example, the partial regression of day care experience (.17) and lab aggression ($-.16$) on SES can be seen in the Gamma matrix of Table 15.4. As hypothesized, SES positively predicts day care quality and negatively predicts child aggression.

Also, the user should examine the standard errors of the parameters to ensure that they are not so large that the estimate is not reasonably determined. Additionally, the squared multiple correlations and the coefficients of determination indicate how well the latent variables are measured by the observed variables. These values should range from zero to one, with larger values indicating better fitting models.

### B. Examine Fit Indices

Second, examine the measures of overall fit (see Table 15.5). Remember to consider several indices, bearing in mind that interpretation of these indices is subjective and that you are looking for consistency across indices. For example, the fit indices in Table 15.5 yield conflicting results. The chi-square value is not significant ($\chi^2(4) = 7.40$, $p = .12$), which suggests a good fit between the model and the data. However, the RMR (.06) is marginal, suggesting that the residuals between the model estimates and the observed values are slightly larger than desirable. Also, the AGFI (.81) and the NNFI (.70) are low, and the RMSEA is high (.11). On the other

**TABLE 15.4** **LISREL Output: Standardized Solution for Model in Figure 15.2**

|  | Beta matrix | | | | |
| --- | --- | --- | --- | --- | --- |
|  | Day care experience | Tempera- ment | Parent-rated aggression | Child lab aggression | Peer lab aggression |
| Day care experience | — | — | — | — | — |
| Temperament | 0.03 | — | — | — | — |
| Parent-rated aggression | 0.19 | — | — | — | — |
| Child lab aggression | 0.17 | −0.11 | — | — | — |
| Peer lab aggression | — | — | — | — | — |

|  | Gamma matrix | |
| --- | --- | --- |
|  | Socio- economic status | Sex of child |
| Day care experience | 0.17 | 0.11 |
| Temperament | −0.33 | 0.12 |
| Parent-rated aggression | −0.21 | −0.01 |
| Child lab aggression | −0.16 | −0.52 |
| Peer lab aggression | — | −0.17 |

|  | Psi matrix | | | | |
| --- | --- | --- | --- | --- | --- |
|  | Day care experience | Tempera- ment | Parent-rated aggression | Child lab aggression | Peer lab aggression |
| Day care experience | 0.96 | | | | |
| Temperament | — | 0.87 | | | |
| Parent-rated aggression | — | 0.31 | 0.93 | | |
| Child lab aggression | — | — | −0.28 | 0.70 | |
| Peer lab aggression | — | — | — | 0.14 | 0.97 |

**TABLE 15.5** **LISREL Output: Partial List of Goodness-of-Fit Statistics for Model in Figure 15.2**

Chi-square with 4 degrees of freedom = 7.40 ($p$ = 0.12)
Root mean square error of approximation (RMSEA) = 0.11
Root mean square residual (RMR) = 0.057
Goodness-of-fit index (GFI) = 0.97
Adjusted goodness-of-fit index (AGFI) = 0.81
Normed fit index (NFI) = 0.91
Nonnormed fit index (NNFI) = 0.70
Comparative fit index (CFI) = 0.94
Incremental fit index (IFI) = 0.96

hand, the CFI (.94) and the IFI (.96) are in the acceptable range. There is not clear consensus among the fit indices, and therefore the most prudent interpretation is that the model requires further refinement.

## C. Examine Individual Aspects of the Model

The next step is to examine the standardized residuals and modification indices to determine what aspect or aspects of the model do not fit the data well. This step is important because the main function of SEM is to test a theoretical model and determine areas that require close scrutiny in future theory development and empirical research. Small standardized residuals indicate that the observed and the expected correlations are very similar and the model has done an adequate job of accounting for the data (Hu & Bentler, 1995). Modification indices assess the value of freeing parameters that are currently fixed or constrained (for instance, parameters that are forced to be equal to other parameters). For example, the correlation path between parent-rated aggression and peer aggression is fixed at zero in Table 15.2. Modification indices (Table 15.6) show that the largest change in chi-square would result from freeing this path (element 5,3 in the Psi matrix). The path can be freed if this makes sense theoretically, but it is essential to realize that such modifications result in *post hoc* analyses. They are useful in that they generate hypotheses for future research (Hoyle, 1995b), but until cross-validated on an independent sample there is no way to be certain that they are not capitalizing on chance associations in this particular data set (see Tinsley and Brown, chapter 1, this volume, for a discussion of cross-validation procedures). In the example presented earlier, it makes no sense theoretically to correlate a parent's rating of their own child's aggression with laboratory ratings of the aggression of an unfamiliar peer in the lab. Therefore, this path should not be freed in the model, even though it would improve the model's fit.

## D. Testing Nested Models

There are two ways to create nested models for determining the best fitting model for the data. One is to hypothesize *a priori* a full model and certain, more restricted models that are based in theory. For instance, it may be reasonable to form a model that describes the relations among day care experience, child temperament, and child aggression (i.e., the full model depicted in Figure 15.2), and then on theoretical grounds to postulate a nested model that includes only the path from temperament to sociability to determine whether day care experience was significant (the nested model in Figure 15.2). Alternatively, researchers can create an *a posteriori* nested model based on the modification indices derived from the analysis of the

**TABLE 15.6   LISREL Output: Modification Indices for Model in Figure 15.2**

| | Modification indices for beta | | | | |
|---|---|---|---|---|---|
| | Day care experience | Tempera-ment | Parent-rated aggression | Child lab aggression | Peer lab aggression |
| Day care experience | — | — | — | 0.04 | 0.04 |
| Temperament | — | — | — | 0.19 | 0.19 |
| Parent-rated aggression | — | — | — | 5.35 | 5.35 |
| Child lab aggression | — | — | — | — | — |
| Peer lab aggression | 0.15 | 0.00 | 4.24 | 4.90 | — |

| | Modification indices for gamma | |
|---|---|---|
| | Socio-economic status | Sex of child |
| Day care experience | — | — |
| Temperament | — | — |
| Parent-rated aggression | — | — |
| Child lab aggression | — | — |
| Peer lab aggression | 1.29 | — |

| | Modification indices for Psi | | | | |
|---|---|---|---|---|---|
| | Day care experience | Tempera-ment | Parent-rated aggression | Child lab aggression | Peer lab aggression |
| Day care experience | — | | | | |
| Temperament | — | — | | | |
| Parent-rated aggression | — | — | — | | |
| Child lab aggression | — | — | — | — | |
| Peer lab aggression | 0.04 | 0.19 | 5.35 | — | — |

Maximum modification index is 5.35 for element (5.3) of Psi.

full model. This nested model is equivalent to a *post hoc* test that requires cross-validation.

Regardless of how they are created, nested models can be statistically compared to determine whether dropping the paths from the full model resulted in a statistically significant decrease in model fit. The chi-square statistic is useful for comparing nested models. The difference between chi-squares (i.e., chi-square (Full) minus chi-square (Nested)) is distributed as a chi-square statistic, with degrees of freedom (df) equal to df (Full) minus df (Nested). A significant chi-square index of change means that the two models differ significantly and the more restricted model significantly worsened the model fit. A nonsignificant chi-square change value implies that the more restrictive model with the greater degrees of freedom can be accepted as the better (more parsimonious) model.

## X. CONCLUSION

SEM is a flexible analytic tool that can combine regression, correlation, and factor analyses simultaneously to address important issues in the social sciences, biological sciences, and humanities. Readers desiring a more detailed introduction to the topic will find useful treatments in Byrne (1994), Hayduk (1987), Hoyle (1995a), and Jaccard and Wan (1996). Related issues are addressed in the journal devoted to SEM entitled *Structural Equation Modeling: A Multidisciplinary Journal,* and advice about specific issues is available from SEMNET, the lively discussion forum that is available on an e-mail list server (listserv@ualvm.ua.edu).

There is still much to be learned about SEM, including appropriate uses of various fit indices and the interpretation of odd or impossible parameter estimates, but its value has been demonstrated in numerous studies. Furthermore, novel methods for addressing unique modeling issues continue to emerge. For instance, recent developments include the creation of phantom variables that enable one parameter estimate to be equal to a multiple of another or that allow a group of participants who are missing a variable to be included as a separate group rather than to be omitted from analysis (e.g., Loehlin, 1992; Rovine, 1994). As these innovations continue, the application of SEM to complex research questions will increase.

## REFERENCES

Anderson, J. C., & Gerbing, D. W. (1988). Structural equation modeling in practice: A review and recommended two-step approach. *Psychological Bulletin, 103,* 411–423.

Baron, R. M., & Kenny, D. A. (1986). The moderator-mediator variable distinction in social psychological research: Conceptual, strategic, and statistical consideration. *Journal of Personality and Social Psychology, 51,* 1173–1182.

Bentler, P. M. (1980). Multivariate analysis with latent variables: Causal modeling. *Annual Review of Psychology, 31,* 419–456.

Bentler, P. M. (1990). Comparative fit indices in structural models. *Psychological Bulletin, 107,* 238–246.

Bentler, P. M. (1992). *EQS structural equations program manual.* Los Angeles: BMDP Statistical Software.

Bentler, P. M., & Bonett, D. G. (1980). Significance tests and goodness-of-fit in the analysis of covariance structures. *Psychological Bulletin, 88,* 588–606.

Bentler, P. M., & Chou, C. (1987). Practical issues in structural modeling. *Sociological Methods and Research, 16,* 78–117.

Bollen, K. A. (1989a). A new incremental fit index for general structural equation models. *Sociological Methods and Research, 17,* 303–316.

Bollen, K. A. (1989b). *Structural equations with latent variables.* NY: Wiley.

Byrne, B. M. (1994). *Structural equation modeling with EQS and EQS/Windows: Basic concepts, applications, and programming.* Thousand Oaks, CA Sage Publications.

Byrne, B. M. (1995). One application of structural equation modeling from two perspectives: Exploring the EQS and LISREL strategies. In R. H. Hoyle (Ed.), *Structural equation modeling: Concepts, issues, and applications* (pp. 138–157). Thousand Oaks, CA: Sage Pub.

Chou, C.-P., & Bentler, P. M. (1995). Estimates and tests in structural equation modeling. In R. H. Hoyle (Ed.), *Structural equation modeling: Concepts, issues, and applications* (pp. 37–55). Thousand Oaks, CA: Sage Publications.

Cohen, J., & Cohen, P. (1975). *Applied multiple regression/correlation analysis for the behavioral sciences.* Hillsdale, NJ: Lawrence Erlbaum Associates, Pub.

DiLalla, L. F. (1998). Influences on preschoolers' social behaviors in a peer play setting. *Child Study Journal, 28,* 223–244.

Gerbing, D. W., & Anderson, J. C. (1993). Monte Carlo evaluations of goodness-of-fit indices for structural equation models. In K. A. Bollen & J. S. Long (Eds.), *Testing structural equation models* (pp. 40–65). Newbury Park, CA: Sage.

Goffin, R. D. (1993). A comparison of two new indices for the assessment of fit of structural equation models. *Multivariate Behavioral Research, 28,* 205–214.

Guadagnoli, E., & Velicer, W. F. (1988). Relation of sample size to the stability of component patterns. *Psychological Bulletin, 103,* 265–275.

Hayduk, L. A. (1987). *Structural equation modeling with LISREL: Essentials and advances.* Baltimore: Johns Hopkins University Press.

Hoyle, R. H. (1995a). (Ed.), *Structural equation modeling: Concepts, issues, and applications.* Thousand Oaks, CA: Sage Publications.

Hoyle, R. H. (1995b). The structural equation modeling approach: Basic concepts and fundamental issues. In R. H. Hoyle (Ed.), *Structural equation modeling: Concepts, issues, and applications* (pp. 1–15). Thousand Oaks, CA: Sage Publications.

Hu, L. T., & Bentler, P. (1995). Evaluating model fit. In R. H. Hoyle (Ed.), *Structural equation modeling: Concepts, issues, and applications* (pp. 76–99). Thousand Oaks, CA: Sage Publications.

Jaccard, J., & Wan, C. K. (1996). *LISREL approaches to interaction effects in multiple regression* (Sage University Paper series on Quantitative Applications in the Social Sciences, series no. 07-114). Thousand Oaks, CA: Sage.

Joreskog, K., & Sorbom, D. (1993a). *LISREL 8: Structural equation modeling with the SIMPLIS command language.* Hillsdale, NJ: Lawrence Erlbaum Associates Publishers.

Joreskog, K., & Sorbom, D. (1993b). *PRELIS 2: User's reference guide.* Hillsdale, NJ: Lawrence Erlbaum Associates Publishers.

Kaplan, D. (1995). Statistical power in structural equation modeling. In R. H. Hoyle (Ed.), *Structural equation modeling: Concepts, issues, and applications* (pp. 100–117). Thousand Oaks, CA: Sage Pub.

Loehlin, J. C. (1992). *Latent variable models: An introduction to factor, path, and structural analysis* (2nd ed). Hillsdale, NJ: Lawrence Erlbaum Associates, Pub.

MacCallum, R. C. (1995). Model specification: Procedures, strategies, and related issues. In R. H. Hoyle (Ed.), *Structural equation modeling: Concepts, issues, and applications* (pp. 16–36). Thousand Oaks, CA: Sage Pub.

Mardia, K. V. (1970). Measures of multivariate skewness and kurtosis with applications. *Biometrika, 57,* 519–530.

Marsh, H. W., Balla, J. R., & McDonald, R. P. (1988). Goodness-of-fit indexes in confirmatory factor analysis: The effect of sample size. *Psychological Bulletin, 103,* 391–410.

McDonald, R. P. (1989). An index of goodness-of-fit based on noncentrality. *Journal of Classification, 6,* 97–103.

Mislevy, R. J. (1986). Recent developments in the factor analysis of categorical variables. *Journal of Educational Statistics, 11,* 3–31.

Muthen, B. (1984). A general structural equation model with dichotomous, ordered categorical, and continuous latent variable indicators. *Psychometrika, 49,* 115–132.

Nesselroade, J. R. (1994). Exploratory factor analysis with latent variables and the study of processes of development and change. In A. von Eye & C. C. Clogg (Eds.), *Latent variables analysis: Applications for developmental research* (pp. 131–154). Thousand Oaks, CA: Sage Publishers.

Parry, C., & McArdle, J. J. (1991). An applied comparison of methods for least squares factor analysis of dichotomous variables. *Applied Psychological Measurement, 15,* 35–46.

Rovine, M. J. (1994). Latent variables models and missing data analysis. In A. von Eye & C. C. Clogg (Eds.), *Latent variables analysis: Applications for developmental research* (pp. 181–225). Thousand Oaks, CA: Sage Publishers.

Schumacker, R. E., & Lomax, R. G. (1996). A beginner's guide to structural equation modeling. Hillsdale, NJ: Lawrence Erlbaum Associates, Publishers.

Steiger, J. H., & Lind, J. C. (1980). Statistically-based tests for the number of common factors. Paper presented at the Annual Meeting of the Psychometric Society, Iowa City, IL.

Tanaka, J. S., & Huba, G. J. (1984). Confirmatory hierarchical factor analyses of psychological distress measures. *Journal of Personality and Social Psychology, 46,* 621–635.

Tucker, L. R., & Lewis, C. (1973). A reliability coefficient for maximum likelihood factor analysis. *Psychometrika, 38,* 1–10.

van den Oord, E. J. C. G., & Rowe, D. C. (1997). Effects of censored variables on family studies. *Behavior Genetics, 27,* 99–112.

West, S. G., Finch, J. F., & Curran, P. J. (1995). Structural equation models with nonnormal variables: Problems and remedies. In R. H. Hoyle (Ed.), *Structural equation modeling: Concepts, issues, and applications* (pp. 56–75). Thousand Oaks, CA: Sage Pub.

Wright, S. (1934). The method of path coefficients. *Annals of Mathematical Statistics, 5,* 161–215.

Wright, S. (1960). Path coefficients and path regressions: Alternative or complementary concepts? *Biometrics, 16,* 189–202.

# 16

# CONFIRMATORY FACTOR ANALYSIS

**RICK H. HOYLE**

*Department of Psychology, University of Kentucky, Lexington, Kentucky*

*Factor analysis* is a family of statistical strategies used to model unmeasured sources of variability in a set of scores. *Confirmatory factor analysis* (CFA), otherwise referred to as restricted factor analysis (Hattie & Fraser, 1988), structural factor analysis (McArdle, 1996), or the measurement model (Hoyle, 1991), typically is used in a deductive mode to test hypotheses regarding unmeasured sources of variability responsible for the commonality among a set of scores. It can be contrasted with *exploratory factor analysis* (EFA; see Cudeck, chapter 10, this volume), which addresses the same basic question but in an inductive, or discovery-oriented, mode.

Although CFA can be used as the sole statistical strategy for testing hypotheses about the relations among a set of variables, it is best understood as an instance of the general structural equation model (Bollen, 1989; Hoyle, 1995a; see Chapter 15). In that model, a useful distinction is made between the measurement model and the structural model (e.g., Anderson & Gerbing, 1988). The *measurement model* (i.e., CFA) concerns the relations between measures of constructs, indicators, and the constructs they were designed to measure (i.e., factors). The *structural model* concerns the directional relations between constructs. In a full application of structural equation modeling, the measurement model is used to model constructs, between which directional relations are modeled and tested in the structural model. Because CFA is a specific instance of the general structural equation model, many of the principles and strategies covered in this chapter apply to any instance of the structural equation model; however, there are issues and strategies specific to the measurement model, and these constitute the primary focus of the chapter.

## I. OVERVIEW

### A. Factors

The central concern of CFA is modeling factors, sometimes referred to as latent variables. *Factors* are influences that are not directly measured but account for commonality among a set of measurements. In Figure 16.1, the commonality among measures of a construct is depicted in two ways. On the left, a Venn diagram is used to illustrate a pattern of shared variance among scores on the three measures. Each circle represents the variance in one of the measures, $x_1$, $x_2$, and $x_3$. The overlap of the circles represents shared variance, or covariance. The shaded area, labeled "F," represents the area of overlap involving all three circles. It is this area that corresponds to a factor. Although it is simple and somewhat revealing, the Venn diagram

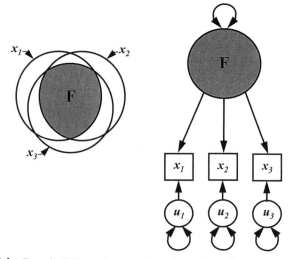

**FIGURE 16.1**  Two depictions of the commonality among three measures of a single construct. (Left: Venn diagram. Right: Path diagram.)

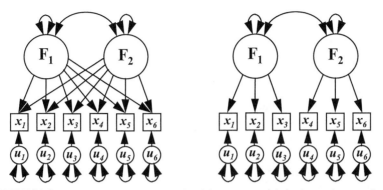

**FIGURE 16.2** Path diagrams of two correlated factors as modeled using exploratory factor analysis (EFA) (left) and confirmatory factor analysis (CFA) (right).

is but a graphical convenience; it does not imply a statistical means of modeling the factor. On the right in Figure 16.1, a *path diagram* is used to illustrate the same set of associations. In the path diagram, the rectangles represent the measured variables, generically referred to as *indicators*. Ellipses represent unmeasured variables. The large ellipse represents a factor, whereas the smaller ellipses represent errors of measurement or *uniquenesses*, which are unobserved sources of influence unique to each indicator. The single-headed arrows suggest causal influence, indicating specifically that each indicator is caused by two unmeasured influences: (a) a causal influence it shares with the other indicators; and (b) an additional causal influence not shared with the remaining indicators. The two-head curved arrows associated with the latent variables represent variances (see MacCallum, 1995; McArdle & Boker, 1990).

The path diagram can be translated directly into statistical form through a set of *measurement equations,* which specify the relations between factors and indicators. Consider the single equation, expressed in Bentler-Weeks (1980) notation (see Appendix for alternative notation),

$$x_i = *F_1 + \ldots + *F_j + u_i \tag{1}$$

According to this equation, variability in the $i$th indicator is an additive function of $j$ differentially weighted factors and the $i$th unique factor. Careful consideration of this equation illustrates a basic difference between CFA and EFA. In EFA, no restrictions are placed on the pattern of weights, denoted by "*," associated with the factors. As such, each indicator is, to some degree, influenced by each of the $j$ factors, and, in CFA language, is "free" to be estimated in the analysis. Thus, the EFA measurement equations for a two-factor model involving six indicators, displayed in Figure 16.2, would be,

$$x_1 = *F_1 + *F_2 + u_1 \tag{2}$$

$$x_2 = {}^*F_1 + {}^*F_2 + u_2 \tag{3}$$
$$x_3 = {}^*F_1 + {}^*F_2 + u_3 \tag{4}$$
$$x_4 = {}^*F_1 + {}^*F_2 + u_4 \tag{5}$$
$$x_5 = {}^*F_1 + {}^*F_2 + u_5 \tag{6}$$
$$x_6 = {}^*F_1 + {}^*F_2 + u_6 \tag{7}$$

In contrast, CFA requires the imposition of certain restrictions on the pattern of weights, or factor loadings, a seemingly limiting characteristic that, in fact, pays substantial inferential dividends. Equations 8 to 13 are the CFA measurement equations for a prototypic two-factor CFA model involving six indicators, shown in path diagram form in Figure 16.2.

$$x_1 = {}^*F_1 + 0F_2 + u_1 \tag{8}$$
$$x_2 = {}^*F_1 + 0F_2 + u_2 \tag{9}$$
$$x_3 = {}^*F_1 + 0F_2 + u_3 \tag{10}$$
$$x_4 = 0F_1 + {}^*F_2 + u_4 \tag{11}$$
$$x_5 = 0F_1 + {}^*F_2 + u_5 \tag{12}$$
$$x_6 = 0F_1 + {}^*F_2 + u_6 \tag{13}$$

Note that, wherereas in equations (2) to (7) each F was preceded by *, that is true but for a subset of the Fs in the CFA measurement equations. The zeroes indicate no influence of the factor on the indicator (in CFA language these parameters are "fixed" at zero). For instance, as can be seen in equations 8 to 13 or the path diagram on the right in Figure 16.2, $F_1$ influences $x_1$ but not $x_4$, and $F_2$ influences $x_6$ but not $x_3$.

This general pattern is an extreme example of what is referred to in EFA terms as *simple structure* (Thurstone, 1947). Whereas in EFA simple structure is a target of inductively oriented extraction and rotation algorithms, in CFA simple structure typically is assumed or imposed on the pattern of factor loadings.

## B. Hypothesis Testing

As noted earlier, the prototypic use of CFA is deductive, focusing on the correspondence between the pattern of associations in observed data and the pattern implied by a model specified apart from a knowledge of those data (i.e., hypothesized model). The statistical hypothesis is expressed as,

$$H_0 \colon \Sigma = \Sigma(\Theta) \tag{14}$$

In practice, **S**, the observed covariance matrix, is substituted for $\Sigma$, the unknown (and, typically, unknowable) population covariance matrix, and $\Sigma(\Theta)$ is expressed as $\Sigma(\hat{\Theta})$ or, more parsimoniously, as $\hat{\Sigma}$, the covariance matrix implied by the model given the data. In a departure from standard practice in the social and behavioral sciences, the desired outcome is a failure to reject $H_0$. This is because hypothesis testing in CFA focuses on *goodness of fit,* which is reflected in equivalence between S and $\hat{\Sigma}$ or, in population terms, $\Sigma$ and $\Sigma(\Theta)$.

## C. Confirmatory versus Exploratory Factor Analysis

CFA often is portrayed as different from, even superior to, EFA; however, in practice the distinction between CFA and EFA is substantially blurred. For instance, *post hoc* adjustments to misspecified CFA models is an exercise in induction, whereas EFA can be used in such a way that theory-based predictions are put to a convincing test (e.g., Fleming & Courtney, 1984). The two models, in prototypic form, might be more profitably construed as anchoring two ends of a continuum. In the prototypic application of EFA, no restrictions are imposed on the solution, so that the number of factors and pattern of loadings are ascertained after the analysis. In the prototypic application of CFA, the number of factors and pattern of loadings are hypothesized before the analysis by placing numerous restrictions on the solution (e.g., equations 8 to 13 hypothesize a two-factor structure with indicators 1, 2, and 3 loading on one factor and indicators 4, 5, and 6 loading on the other factor). As the actual implementation of either model departs from the prototype, the application shades toward the middle of the continuum. Indeed, in some circumstances, it is advantageous to use both models, thereby moving from back and forth along the continuum. In short, although the comparison between EFA and CFA is sometimes framed as "right vs. wrong," the two approaches each have strengths and weaknesses and, as routinely applied, are best viewed as complementary.

## II. APPLICATIONS OF CONFIRMATORY FACTOR ANALYSIS

### A. Basic Questions about Dimensionality

In the typical application, CFA is used to address questions about the dimensionality of an hypothetical construct assessed by multiple indicators. For instance, current conceptualizations of personality posit that people construe trait information along five broad dimensions (John, 1990). This fundamental hypothesis about the structure of personality information can be tested using CFA (e.g., Church & Burke, 1994; Panter, Tanaka, & Hoyle, 1994). In some domains, competing conceptualizations of a construct posit different numbers of dimensions, and, in some cases, the conceptualizations can be specified in such a way that CFA can be used to compare them statistically (e.g, Hoyle & Lennox, 1991). Finally, in instances where the number of factors that underlie a construct or measure has not been firmly established, CFA and EFA can be used in tandem to generate the factor model most consistent with a set of data.

An example of such an application was described by Tepper and Hoyle (1996). Their initial CFA of the Need for Uniqueness Scale (Snyder & Fromkin, 1980) revealed that it was not possible to modify the hypothesized three-factor model so as to adequately account for the commonality among the 32 items. Reasoning that a larger number of factors might be required, they conducted an EFA, which suggested four factors. Moving back to CFA, they evaluated the four-factor hypothesis by specifying an unrestricted model (Jöreskog, 1977), which permits a test of the number of factors without simultaneously evaluating a pattern of loadings. After obtaining support for this model, they used an empirical strategy to restrict nonsignificant loadings, thereby producing a four-factor model with a number of double loadings, a model that contrasted sharply with the hypothesized three-factor simple-structure model.

## B. Higher Order Models

Beyond the basic question of how many factors are implicated in the covariances among a set of scores, is the question of how the factors themselves are structured. If a model includes four or more factors that covary, then CFA can be used to test hypotheses about *higher order factors* that would account for commonality among first-order factors that are known to covary (e.g., Marsh & Hocevar, 1985; Rindskopf & Rose, 1988). For instance, one might ask whether the associations among the five factors that underlie personality description can be explained by a single second-order factor, or whether global self-esteem underlies domain-specific self-esteem (Hoyle, 1991).

## C. Measurement Invariance

An important but infrequent application of CFA is to questions about measurement invariance. *Measurement invariance* concerns the degree to which aspects of a CFA model (e.g., number of factors, pattern of loadings) are equivalent across groups or over time (Byrne, Shavelson, & Muthén, 1989; Hoyle & Smith, 1994). Measurement invariance is important because, if the structure of an hypothetical construct does not hold across groups or over time, then between-group or longitudinal comparisons of standing on the construct are not meaningful. Tests of measurement invariance often concern the comparability of a construct or measure across subsamples within a study (e.g., men, women) that are to be compared in terms of standing on the construct (e.g., Bollen & Hoyle, 1990; Byrne, 1989). For instance, Bollen and Hoyle (1990) demonstrated that the two-factor model hypothesized to underlie responses to their Perceived Cohesion Scale fit equally well for a sample of college students describing their experience as a member of the student body and a sample of adults living in a midsized city referring to their experience as a citizen of the city.

## D. Construct Validation

Once such basic questions as the number of factors and pattern of factor loadings have been addressed, CFA can be used to evaluate construct validity. *Construct validity* concerns the degree to which an hypothetical construct relates to other constructs in a theoretically meaningful pattern. One means of evaluating construct validity using CFA is to incorporate multiple constructs into a single model and evaluate the pattern of covariances among factors representing the constructs against a pattern predicted from theory or basic knowledge about the relations among the constructs (Hoyle, 1991). A more sophisticated approach, one that offers the added benefit of teasing apart construct and method variance is to apply CFA to a multitrait–multimethod matrix (MTMM) (for a review, see Marsh & Grayson, 1995). Specification of CFA models of MTMM data is challenging and estimation can be difficult, but the potential payoff is great. Creative approaches to modeling such data using CFA have been proposed (Kenny & Kashy, 1992; Marsh, 1989), and CFA remains a viable approach to evaluating the influence of method artifacts in measures of hypothetical constructs.

## E. Growth Models

Among the more sophisticated applications of CFA is the analysis of latent growth models. Such models introduce time into the analysis of an hypothetical construct, and, in so doing, provide a means of testing elegant and sophisticated hypotheses about the construct. Variants of the latent growth model can be used to evaluate the degree to which a construct is immutable (Kenny & Zautra, 1995), the point in a time sequence at which change in a time-varying construct takes places (Raykov, 1994), and the form and correlates of change in a construct over time (Willet & Sayer, 1994). For instance, Duncan, Duncan, and Stoolmiller (1994) showed that the trajectory of alcohol use from ages 14 to 18 is linear and positive. Moreover, the slope is steeper for adolescents who are not adequately monitored by their parents, and initial status (at age 14) is higher for adolescents who frequently engage in coercive interaction with their parents.

## III. DATA REQUIREMENTS

Any statistical method is valid only when certain minimal data requirements are met. For instance, a stringent assumption of inferential statistics is independence of observations. Violation of this assumption (e.g., data were gathered from members of intact groups but analyzed with individual as the unit of analysis) results in biased statistics and, possibly, errors of inference. A common criticism of CFA is that the data requirements are rarely met by most research efforts in the social and behavioral sciences. Increasingly, however, simulation research is demonstrating that CFA is robust to modest violations of theoretical assumptions. A brief consideration of the principal assumptions follows.

## A. Sample Size

An issue of considerable concern in the typical application of CFA is sample size. As mentioned earlier, applications of CFA require some parameters to be fixed and others to be free. The free parameters are estimated and statistical tests constructed to test model fit and whether parameter estimates, as hypothesized, differ from zero. All such tests constructed to date assume asymptotic, or large sample, properties of the data. Said differently, as sample size approaches infinity, estimators and statistical tests perform as statistical theory would prescribe. The large-sample assumption of CFA is evidenced by the fact that the theoretical distributions against which models $(z)$ and parameters $(\chi^2)$ are tested—unlike, for instance, the $t$ and $F$ distributions in mean comparisons—do not vary as a function of $N$. The question, then, is whether sample sizes characteristic of social and behavioral science research are sufficient to support CFA estimation and testing.

Unfortunately, there is no simple rule regarding the minimum or desirable number of observations for CFA. Indeed, any rule regarding sample size and CFA would need to include a variety of qualifications such as estimation strategy, model complexity, scale on which indicators are measured, and distributional properties of indicators. Tanaka (1987) proposed that, minimally, sample size determination should consider model complexity (cf. Marsh & Hau, 1999). He recommended a ratio of at least four subjects per *free parameter* in the model as a rule of thumb for determining minimum sample size. Although this rationale has merit, the sample size question is multifaceted and should consider additional features of the application. For instance, statistical power is a concern in any statistical analysis but particularly in CFA, wherein the desired statistical outcome is failure to reject the null hypothesis (eq. 14; MacCallum, Browne, & Sugawara, 1996). Reliability of model modifications is another (MacCallum, Roznowski, & Necowitz, 1992), and choice of estimator is still another (Hu, Bentler, & Kano, 1992). Although any recommendation is somewhat arbitrary, an $N$ of at least 200 seems advisable for anything beyond the simplest CFA model. An $N$ of at least 400 is preferable, because indexes of fit begin to evince their asymptotic properties at this number (Hu et al.,

1992). And, if an empirical strategy (e.g., LISREL's modification index) is used as a basis for respecifying a poor-fitting model, an $N$ of at least 800 is needed to protect against capitalizing on chance features of the data (MacCallum et al., 1992).

## B. Distributional Properties

CFA is a multivariate statistical model and, therefore, it is the multivariate distribution of the data that affects estimation and testing. With the exception of a small number of distribution-free estimators, estimators (e.g., maximum likelihood, generalized least squares) assume that the data to be analyzed are multinormally distributed (Bollen, 1989). Of course, actual data are never perfectly normally distributed, and, in many instances, are dramatically nonnormal (Micceri, 1989). The question then arises as to how robust estimators commonly used in CFA are to violations of the normality assumption? As with the sample size question, the answer seems to depend on a number of additional factors. If the model is reasonably accurate and the sample size reasonably large, then maximum likelihood, the standard estimation strategy in CFA, appears to perform reasonably well under most violations of multivariate normality (Chou & Bentler, 1995; Curran, West, & Finch, 1996; Hu et al., 1992; West, Finch, & Curran, 1995). In more extreme cases of nonnormality, either the asymptotic distribution free method can be used, although $N$ should be near 5,000, or, with more modest sample sizes, the Satorra-Bentler robust statistics (Satorra & Bentler, 1988) perform well.

## C. Scale of Measurement

Common estimators such as maximum likelihood assume that indicators are measured on a continuous scale (Jöreskog, 1994). Although many constructs in the social and behavioral sciences are, in theory, continuous, they typically are assessed using polychotomous scales (e.g., *strongly disagree, disagree, neither agree nor disagree, agree, strongly agree*). In such instances, indicators are said to be *coarsely categorized* (Bollen & Barb, 1981). Although estimation strategies are available for treating such indicators as ordered categories, they are not widely supported in software programs and, for models of typical size and complexity, require prohibitively large samples. Practical experience (Tepper & Hoyle, 1996) and some simulation work (Bollen & Barb, 1981) indicate that standard estimators perform well with indicators measured in five or more categories. Indicators measured with fewer categories result in nontrivially attenuated covariances and inflated fit statistics, so estimators designed for categorical data (e.g., Jöreskog, 1994; Muthén, 1984) are recommended in such instances.

## D. Type of Indicator

Traditionally, factor-analytic models have implicitly assumed that indicators are caused by factors (e.g., in Figures 16.1 and 16.2 the directional paths run *from* the factors *to* the indicators). Following from that assumption, indicators are assumed to reflect factors, whose presence is inferred from the pattern of covariation among the indicators. Bollen (1984) noted that, in some instances, indicators of a factor are uncorrelated and, rather than reflecting the factor, they cause it (i.e., the directional paths run *from* the indicator *to* the factor). Such indicators are termed *cause indicators* (Bollen, 1984) or *formative indicators* (Cohen, Cohen, Teresi, Marchi, & Velez, 1990) and can be contrasted with traditional *effect indicators* or *reflective indicators*. In practice, cause indicators are somewhat rare, perhaps because constructs typically are not conceptualized with reference to their causes. Because traditional psychometric criteria (e.g., internal consistency) assume, and common practice is dominated by, effect indicators (Bollen & Lennox, 1991), the present treatment does not consider formative or mixed-indicator models. Nevertheless, applied researchers are encouraged to consider indicators in light of this distinction in order to ensure that model specification maps onto the focal hypothetical construct as operationalized by the available indicators (illustrative examples are presented by Bollen & Lennox, 1991, and Cohen et al., 1990).

## IV. ELEMENTS OF A CONFIRMATORY FACTOR ANALYSIS

I now consider the primary elements of a typical application of CFA in the social and behavioral sciences (for greater detail, see Hoyle, 1995a, 1995b). These elements are presented and illustrated in the context of a prototypic application (Hoyle, Stephenson, Palmgreen, Lorch, & Donohew, 1998). In the context of a large-scale study of a mass media antidrug intervention, 764 adolescents completed a 20-item measure of sensation seeking. The scale was modeled after the widely used Form V of the Sensation Seeking Scale (Zuckerman, Eysenck, & Eysenck, 1978), which comprises four theory-based subscales: experience seeking, thrill and adventure seeking, disinhibition, and boredom susceptibility. The primary questions addressed in the analysis are (a) is there evidence of four related but distinct factors on which items load in a predicted pattern? and (b) are the individual items valid indicators of the factor they were designed to measure?

### A. Specification

Unlike EFA, which, in the typical application, is used to find the model that best explains the pattern of associations among a set of indicators,

CFA requires that one or more putative models be specified in advance. Specification of CFA models typically involves answering three questions about a set of indicators: (a) How many factors are present? (Competing theories or extant research might suggest more than one number.) (b) Which indicators are influenced by which factors? (Although indicators typically are specified as a function of one factor, this need not be the case.) (c) If there are two or more factors, how are they interrelated? (In EFA this question arises in the context of rotation, orthogonal or oblique, but in CFA it is handled more flexibly.) Once these questions are answered, the form of the model and its parameters can be specified in a path diagram, a set of measurement equations, or a set of matrices (see MacCallum, 1995).

## 1. Identification

A challenging and important aspect of specifying CFA models is establishing the identification of unknown parameters. Although unidentified parameters are rare in typical applications of CFA (i.e., at least three indicators, simple-structure specification, metric of latent variable(s) set, reasonably normal distributions), a basic understanding of identification is key to understanding CFA estimation and testing. In a CFA model, a parameter can be *identified* (i.e., it can take on a single value in the model) either by fixing it to a particular value in the specification (e.g., off loadings fixed at zero) or by solving for it as a function of parameters with fixed values (Bollen, 1989). If one or more free parameters is unidentified (i.e., they could take on multiple values given the pattern of fixed parameters in the model), then the model is said to be *underidentified,* and it cannot be satisfactorily estimated. If all free parameters are identified and their unique estimate can be obtained through only one manipulation of the fixed parameters, then the model is *just identified,* which means it can be estimated but not tested (i.e., it has no degrees of freedom). If one or more free parameters can be identified through multiple manipulations of the fixed parameters, then the model is *overidentified.* In overidentified models, unique but not exact estimates of free parameters can be derived. Any drawback associated with the lack of certainty regarding values of free parameters in overidentified models is compensated for by the capacity to test the overall fit of such models.

Bollen (1989) identifies four ways of evaluating the identification status of a CFA model:

1. The number of free parameters in the model must not exceed the number of elements in the covariance matrix. Although failure to satisfy this criterion ensures an underidentified model, meeting the criterion does not guarantee an identified model.

2. If each factor has at least three indicators and the loading of at least one indicator per factor is fixed at a nonzero value (usually 1.0) or the variance of each factor is fixed to a nonzero value (usually 1.0), then the model is identified.

3. A model that includes one or more factors, each having only two indicators, will be identified if there are at least two factors, the two-indicator factors are correlated with at least one other factor, and at least one indicator per factor or the factor variances is fixed to a nonzero value.

Rules 2 and 3 are no longer definitive when covariances between unique-nesses are permitted to covary. These identification rules are useful for standard applications of CFA but may not be sufficient for evaluating identification of complex models. In such cases, one must either trust the computer software to detect unidentified parameters, an appealing method that is not foolproof, or use the definitive strategy of algebraically expressing each free parameter as a function of the fixed parameters.

In the present example, the 20 items of the measure of sensation seeking are effect indicators of four factors. Each item is specified as loading on one factor. The four factors are specified as covarying but the uniquenesses are not. One loading on each factor is fixed at 1.0, which serves to set the metric of the factor and complete satisfaction of identification rule #2 detailed earlier. Identification rule #1 also is satisfied because there are 210 elements in the covariance matrix ($p = k[k + 1]/2 = 20[21]/2$) but only 46 free parameters (16 loadings + 20 uniqueness + 4 factor variances + 6 interfactor covariances).

A brief digression at this point will address a common question regarding specification of CFA models. The question concerns why and how the metric of the factors must be set. As is clear by now, factors are not directly measured and, therefore, have no scale of measurement. Factors are inferred on the basis of the association among indicators, and their influence on the indicators is ascertained by regressing each indicator onto the factors. The regression coefficients (or, in standardized form, factor loadings) cannot be derived apart from some assumption about the metric of the factors. By fixing a single loading to 1.0, the factor is ascribed the metric of the indicator whose loading is fixed. Alternatively, the variance of the factors can be fixed to 1.0, which puts them on a standard deviation metric. If the latter strategy is chosen, then covariances between factors are interpreted as correlation coefficients, a feature that is useful for testing hypotheses about whether certain factors are empirically distinct (i.e., are three factors any better than one?). The virtue of the former strategy is apparent when factors become outcome variables in SEMs. In such instances, the variances of the factors are a function of other parameters in the model, so fixing them to a particular value is not straightforward. Setting the metric by fixing a loading to 1.0 is no more difficult in this context than in the CFA context, making this strategy the more flexible of the two.

## B. A Word on Software

During the first decade following Jöreskog's (1967, 1969) pioneering developments in factor analysis, a significant hurdle for applied researchers interested in estimating CFA models was the cumbersome and nonintuitive computer software. As CFA has increased in popularity and prevalence, software programs have become more flexible and intuitive, and have, for the most part, moved from the mainframe to the personal computer environment (though relatively few programs are available for Macintosh users). Indeed, the 1990s has witnessed a proliferation of software programs that can be used to estimate CFA models. Among the widely available titles are AMOS, CALIS, COSAN, EQS, LISCOMP, LISREL, Mplus, Mx, RAMONA, RAMpath, and SEPATH. Programs may permit matrix, equation, or path diagram specification, and some allow the user to choose among the three. Although the software programs vary in comprehensiveness, any of the ones listed are satisfactory for most CFA applications.

Demonstration or student versions of some CFA software can be downloaded from the World Wide Web. Demo versions of EQS for Windows and Macintosh can be downloaded from Multivariate Software's site (http://www.mvsoft.com/). A free student version of AMOS is available for downloading from Smallwaters' site (http://www.smallwaters.com/amos/). COSAN is freeware, available for downloading at ftp://ftp.virginia.edu/public_access/jjm/cosan/. A student version of LISREL is available for a modest fee, though it is not available for downloading (details at http://www.ssicentral.com/lisrel/student.htm). A demo version of Mplus can be downloaded from the Web (http://www.stat model.com/mplus/demo.html). Mx is available free of charge either as a download (http://views.vcu.edu/mx/executables.html) or for interactive use on the Web (http://views.vcu.edu/mx/cgi-bin/mxi.cgi). RAMONA is part of the SYSTAT statistical software package; a demo version of SYSTAT can be downloaded from the web (http://www.spss.com/software/science/systat/down.html). Joel West, of University of California, Irvine, maintains a set of resource pages dedicated to structural equation modeling software on the Web (http://www.gsm.uci.edu/~joelwest/SEM/Software.html).

The results presented here were produced using version 5.6 of EQS for Windows.

## C. Estimation

Once a CFA model has been specified and a suitable covariance matrix obtained, values of free parameters can be estimated. Recall that, in the (typical) case of overidentified models, it is not possible to obtain exact values for the free parameters. Rather they must be estimated in such a way as to satisfy some alternative criterion. Technically, the criterion varies somewhat for different estimators. Yet, at a conceptual level, the criterion is the same for all estimators; namely, find the values of the free parameters that, when substituted into the equations, most precisely recover the observed variances and covariances. Numerous estimators are available in most CFA software packages, but relatively few are routinely used. Parameter estimates and fit statistics produced by the various estimators would be the same for normally distributed indicators assessed on very large samples fit to the correct model. Of course, in applied contexts, data rarely are normally distributed, samples typically are not large, and the correct model is not known. Thus, the decision regarding which estimator to use can be reframed as which estimator works best when data are modestly nonnormally distributed, sample size is around 200, and the correct model can only be approximated?

Maximum likelihood estimation has received the greatest attention and acceptance and tends to perform reasonably well in simulation studies that vary the degree to which the data and model fit the assumptions described earlier (e.g., Chou & Bentler, 1995; Hu et al., 1982). At present, the use of an estimator other than maximum likelihood requires explicit justification. Generalized least squares is an alternative normal theory estimator, but it has been less widely studied and, early research indicates, is more greatly affected by misspecification than maximum likelihood. This sensitivity to misspecification results in an upward bias in incremental fit indices based on comparisons with the null model (described later). The asymptotic distribution free estimator, which specifically incorporates information about the third and fourth moments of the multivariate distribution, would appear to be the ultimate all-purpose estimator; however, it does not evince theoretical properties until $N$ moves into the thousands (Hu et al., 1992; West et al., 1995). One of the most promising innovations is a *post hoc* correction to normal theory standard errors and fit statistics proposed by Satorra and Bentler (1988; see Hu et al., 1992, for results of simulation research). Estimation is a key area of research and development in CFA, and new techniques or adaptations of old one are proposed regularly (e.g., Bollen, 1996; Yuan & Bentler, 1995). Thus, the best advice is to make periodic forays into the technical literature in order to make an informed decision about method of estimation.

Returning to the example, the correlated four-factor model was estimated from the data displayed in Table 16.1 using the maximum likelihood estimator. As is to be expected, the data were not perfectly normally distributed. To evaluate the impact of nonnormality on fit indices and tests of parameter estimates, the Satorra-Bentler correction to the fit statistics and standard errors was applied to the maximum likelihood solution.

**TABLE 16.1** Covariance Matrix for Confirmatory Factor Analysis of Sensation-Seeking Items[a]

| Item | 1 | 2 | 3 | 4 | 5 | 6 | 7 | 8 | 9 | 10 | 11 | 12 | 13 | 14 | 15 | 16 | 17 | 18 | 19 | 20 |
|---|---|---|---|---|---|---|---|---|---|---|---|---|---|---|---|---|---|---|---|---|
| 1 | .941 | | | | | | | | | | | | | | | | | | | |
| 2 | .353 | 1.795 | | | | | | | | | | | | | | | | | | |
| 3 | .268 | .323 | 1.666 | | | | | | | | | | | | | | | | | |
| 4 | .029 | −.036 | .078 | 1.318 | | | | | | | | | | | | | | | | |
| 5 | .141 | .235 | −.023 | −.133 | 1.645 | | | | | | | | | | | | | | | |
| 6 | .266 | .453 | .583 | .088 | .007 | 1.580 | | | | | | | | | | | | | | |
| 7 | .141 | .206 | 1.003 | .043 | .117 | .583 | 1.503 | | | | | | | | | | | | | |
| 8 | .044 | .037 | .257 | .158 | −.061 | .157 | .226 | 1.466 | | | | | | | | | | | | |
| 9 | .054 | .148 | −.108 | −.001 | .272 | .039 | −.070 | −.052 | 1.422 | | | | | | | | | | | |
| 10 | .307 | .907 | .396 | .062 | .191 | .696 | .303 | .072 | .310 | 2.153 | | | | | | | | | | |
| 11 | .271 | .405 | .728 | .079 | .022 | .585 | .654 | .207 | −.002 | .648 | 1.548 | | | | | | | | | |
| 12 | .016 | .096 | .095 | .080 | .006 | .098 | .066 | .386 | .054 | .149 | .242 | 1.362 | | | | | | | | |
| 13 | .047 | .084 | .264 | −.029 | .093 | .253 | .241 | .069 | .114 | .136 | .250 | .031 | .919 | | | | | | | |
| 14 | .284 | .950 | .414 | −.009 | .126 | .702 | .363 | .088 | .145 | .955 | .627 | .147 | .121 | 2.253 | | | | | | |
| 15 | .089 | .185 | .601 | .100 | −.113 | .467 | .544 | .047 | −.086 | .398 | .500 | .124 | .122 | .433 | 1.776 | | | | | |
| 16 | .192 | .162 | .267 | −.089 | −.021 | .246 | .285 | .135 | −.004 | .200 | .331 | .254 | .114 | .149 | .243 | 1.320 | | | | |
| 17 | .350 | .387 | .511 | −.010 | .177 | .426 | .379 | .283 | .085 | .400 | .622 | .258 | .184 | .527 | .265 | .336 | 1.925 | | | |
| 18 | .285 | .843 | .507 | .137 | .160 | .717 | .409 | .177 | .163 | 1.450 | .579 | .103 | .167 | 1.039 | .359 | .250 | .504 | 2.221 | | |
| 19 | .235 | .256 | 1.044 | .088 | .048 | .659 | .910 | .294 | −.194 | .447 | .756 | .248 | .233 | .482 | .742 | .339 | .611 | .610 | 1.911 | |
| 20 | .253 | .227 | .553 | .100 | .090 | .405 | .579 | .232 | .018 | .317 | .553 | .169 | .230 | .243 | .286 | .264 | .449 | .390 | .632 | 1.207 |

[a] $N = 764$.

## 1. Iteration

Estimators such as maximum likelihood and generalized least squares are iterative. This means that they involve incremental attempts at optimizing parameter estimates, stopping when updated parameter estimates offer no appreciable improvement over prior estimates in terms of satisfying the estimation criterion (i.e., recovering the observed data). The degree to which the estimates satisfy this criterion is indexed by the *value of the fitting function,* which is zero for a set of estimates that perfectly recover the observed variances and covariances. Thus, an alternative characterization of the goal of iteration is that of minimizing the value of the fitting function. In order to fully grasp the process of iteration, familiarity with two additional concepts is required: start values and convergence. *Start values* are those values of the free parameters that initially are evaluated against the estimation criterion. Typically, start values are not optimal parameter estimates, which is evidenced by nontrivial decline in the value of the fitting function when the parameters are adjusted. When additional adjustments to the parameter estimates no longer result in meaningful change in the value of the fitting function, the iterative process is said to have *converged.* For various reasons, the iterative process might fail to converge after a very large number of iterations. Such failures to converge signal problems with either the start values, the data, the model, or some combination of these factors. The value of the fitting function at the point of convergence forms the basis for most indexes of model fit.

The iterative process for a portion of the sensation-seeking CFA is illustrated in Table 16.2. In the body of the table, each row corresponds

**TABLE 16.2   Change in Subset of Maximum Likelihood Parameter Estimates as Iterative Process Converges on a Minimum[a]**

| Iteration | Factor loadings | | | | | Variances | | | | | $F_1$ | $F_{ML}$ |
|---|---|---|---|---|---|---|---|---|---|---|---|---|
| | $x_1$[b] | $x_5$ | $x_9$ | $x_{13}$ | $x_{17}$ | $u_1$ | $u_5$ | $u_9$ | $u_{13}$ | $u_{17}$ | | |
| 0 | 1.000 | 1.000 | 1.000 | 1.000 | 1.000 | .941 | 1.645 | 1.422 | .919 | 1.925 | 1.000 | 4.26686 |
| 1 | 1.000 | 1.047 | .998 | .960 | 1.110 | .833 | 1.444 | 1.318 | .891 | 1.596 | .108 | 1.17209 |
| 2 | 1.000 | .493 | .452 | .505 | 1.792 | .764 | 1.576 | 1.364 | .850 | 1.542 | .177 | .93670 |
| 3 | 1.000 | .545 | .357 | .627 | 1.803 | .757 | 1.591 | 1.400 | .850 | 1.326 | .184 | .84648 |
| 4 | 1.000 | .481 | .312 | .656 | 1.862 | .763 | 1.605 | 1.406 | .842 | 1.305 | .178 | .83848 |
| 5 | 1.000 | .478 | .291 | .671 | 1.880 | .767 | 1.605 | 1.408 | .841 | 1.309 | .174 | .83798 |
| 6 | 1.000 | .467 | .281 | .678 | 1.889 | .768 | 1.607 | 1.409 | .839 | 1.308 | .173 | .83790 |
| 7 | 1.000 | .465 | .278 | .680 | 1.892 | .769 | 1.608 | 1.409 | .839 | 1.309 | .172 | .83789 |

[a] Estimates obtained using the maximum likelihood procedure in version 5.6 of EQS for Windows. $F_{ML}$ = maximum likelihood fitting function.
[b] Loading fixed to achieve identification.

**TABLE 16.3** Observed and Implied Covariance Matrices for the Indicators of Factor I in the Initially Specified Four-Factor Model

| Indicator | Observed (S) | | | | | Implied ($\hat{\Sigma}$) | | | | |
|---|---|---|---|---|---|---|---|---|---|---|
| | $x_1$ | $x_5$ | $x_9$ | $x_{13}$ | $x_{17}$ | $x_1$ | $x_5$ | $x_9$ | $x_{13}$ | $x_{17}$ |
| $x_1$ | .941 | | | | | .941 | | | | |
| $x_5$ | .141 | 1.645 | | | | .080 | 1.645 | | | |
| $x_9$ | .054 | .272 | 1.422 | | | .048 | .022 | 1.422 | | |
| $x_{13}$ | .047 | .093 | .114 | .919 | | .117 | .054 | .033 | .919 | |
| $x_{17}$ | .350 | .177 | .085 | .184 | 1.925 | .325 | .151 | .090 | .221 | 1.925 |

to an iteration (iteration 0 reflects start values). All but the final column includes information about parameters relevant to the first factor. The final column includes values of the maximum likelihood fitting function. The default start values in EQS are reflected in the first line: Factor loadings and the factor variance are 1.0, and values of the uniquenesses are the observed variances of the indicators (from the diagonal of the matrix in Table 16.1). Several features of the table are worth noting. First, notice that, after one iteration, some parameter estimates were adjusted upwards (e.g., factor loading $x_{17}$), whereas others were adjusted downward (e.g., the variance of $F_1$). Second, after only one iteration, the value of the fitting function dramatically declines, continues to decline for each iteration before leveling off after the sixth iteration, and converges after seven iterations (i.e., improvement in overall fit compared to the sixth iteration is trivial).

## 2. Implied Covariance Matrix

The final maximum likelihood estimates, when added to estimates of the other parameters constitute $\hat{\Theta}$, and are used to solve for $\hat{\Sigma}$, the implied covariance matrix. The portion of the observed and implied covariance matrices involving indicators of Factor 1 appear in Table 16.3. Note that values on the diagonal are identical. This is because the variances of the indicators are fully accounted for in the model as an additive function of one or more factors and a unique factor. On the other hand, the covariances in the two matrices do not match, and it is this discrepancy that reflects the degree to which the model falls short in recovering the data.

## 3. Residuals

If we subtract each element in the matrix on the right in Table 16.3 from its corresponding element in the matrix on the left, we are left with the *residual matrix*. For instance, looking to Table 16.3, if we compute the difference of element 2,1 in the two matrices, we obtain a residual of .061

($= .141 - .080$). This number reflects the amount of observed covariance not explained by the model. Note that, in some cases the covariance in $\hat{\Sigma}$ indicates that the model overestimated the association between two indicators (e.g., $x_1$ and $x_{13}$), whereas in other cases the model underestimated an association (e.g., $x_5$ and $x_9$). Beyond the obvious interpretation of over- or underestimation of particular associations, residuals are difficult to interpret. This is due to the fact that covariances reflect the metric of the indicators. Thus, for indicators assessed on a scale that ranges from 0 to 1, a residual of .25 would be large, whereas the same residual would be trivial for a pair of indicators assessed on a scale from 1 to 100. This problem is remedied by consulting a normalized (LISREL) or standardized (EQS) version of the residual matrix. The former construes residuals as $z$-scores, whereas the latter construes them as correlations.

## D. Fit

Examination of the residual matrix on an element-by-element basis provides insight into the adequacy of a model for explaining particular associations, but does not give a clear sense of the adequacy of the model as a whole. A single number that summarizes the correspondence between **S** and $\hat{\Sigma}$ is needed for such purposes. Unfortunately, there is no single number that leads to a definitive inference about the correspondence between **S** and $\hat{\Sigma}$ (Bollen & Long, 1993b; Tanaka, 1993). Rather, there are many such numbers, each imperfect in one way or another. The *fit indices* can be grouped into two categories: absolute and comparative.

### 1. Absolute Fit Indexes

Absolute fit indices reflect the degree to which $\hat{\Sigma}$ matches **S** without reference to a comparison model. The most prominent index in this class is $\chi^2$, which actually is an approximation that follows a $\chi^2$ distribution only in very limited (and unlikely) circumstances. The $\chi^2$ approximation is calculated as the product of the value of the fitting function and sample size minus one. Referring back to Tables 16.2 and 16.3, if estimation produces a set of parameter estimates that exactly reproduce the observed data, then the value of the fitting function is zero and, consequently, the value of $\chi^2$, regardless of sample size, is zero. As the value of the fitting function departs from zero, the $\chi^2$ approximation increases in value as a multiplicative function of sample size. The obtained valued is compared against a $\chi^2$ distribution based on degrees of freedom equal to the number of variances and covariances (210 in the present example) minus the number of free parameters (46 in the present example), or 164 for the hypothesized model. Looking to the bottom line in Table 16.2, the value of $\chi^2$ is 639.31 ($\approx .83789 \times 764$, within rounding error), which, evaluated on 164 degrees of freedom, is highly significant, indicating that the model does not adequately recover the observed data (i.e., $\mathbf{S} \neq \hat{\Sigma}$). (The Satorra-Bentler correction for nonnormality had little effect on the fit statistics and standard errors of the parameter estimates. Consequently, that information is not reported.)

Although the $\chi^2$ approximation maps nicely onto the null hypothesis evaluated in CFA (eq. 14), the assumptions that must be met in order for it to serve as a valid test are, in practice, virtually never met. As a result, other approaches to evaluating absolute fit have been proposed. Two that warrant consideration are the Goodness of Fit Index (GFI; Jöreskog & Sörbom, 1981) and the root mean square error of approximation (RMSEA; Steiger, 1990). GFI ranges from zero to 1.0. It indexes the relative amount of the observed variances and covariances accounted for by the model, making it analogous to the $R^2$ index associated with multiple regression analysis (Tanaka, 1993). Values greater than .90 are viewed as indicative of good fit. RMSEA indexes the degree of discrepancy between the observed and implied covariance matrices *per degree of freedom*. Thus, it is sensitive to model complexity in a manner similar to the adjusted $R^2$ in multiple regression (though it is not a proportion). An important virtue of RMSEA is that it follows a known distribution and, as such, is amenable to the construction of confidence intervals and the evaluation of the statistical power of the overall test of model fit (MacCallum et al., 1996). The minimum value of RMSEA is zero, a value it will take when a model exactly reproduces a set of observed data. Browne and Cudeck (1993) propose .05 as a value indicative of close fit, .08 as indicative of marginal fit, and .10 as indicative of poor fit of a model taking into account degrees of freedom of the model. For the four-factor model of sensation seeking, GFI = .92 and RMSEA = .06 with 90% confidence limits of .057 and .067.

## 2. Comparative Fit Indexes

A second approach to evaluating overall fit involves comparison of the fit of a proposed model with the fit of a strategically chosen baseline model (Bentler & Bonett, 1980). The standard baseline model is referred to as the *null model* or *independence model*. This model specifies no factors and no covariances between indicators—that is, no common variance among the indicators. Absolute fit information associated with the proposed and null models can be used to construct indices that reflect the degree to which the former improves over the latter in explaining common variance among the indicators. A plethora of these comparative fit indexes have been proposed and evaluated (for reviews, see Marsh, Balla, & Hau, 1996; Marsh, Balla, & McDonald, 1988; Mulaik et al., 1989). One that has demonstrated commendable performance in a variety of modeling circumstances is the comparative fit index (CFI; Bentler, 1990), which, like GFI, varies between zero and 1.0. CFI indexes the relative reduction in lack of fit of a proposed model over the null model. Values of .90 or higher are taken as indicative of acceptable fit (cf. Hu & Bentler, 1995). CFI for the four-factor model was .86.

## 3. Model Comparisons

As noted earlier, CFA can be used to select between two or more models of the same construct. These alternatives might be based on competing theoretical accounts of the construct, parsimony, or knowledge of the data. I considered two alternatives to the four-factor model. A one-factor model, though inconsistent with the construct definition the measure was designed to reflect, would provide a parsimonious account of the data and would point to a singular source of influence on responses to the 20 items. Maximum likelihood estimation of this model produced unfavorable fit statistics, $\chi^2(170, N = 764) = 1259.02, p < .001$, GFI $= .82$, RMSEA $= .092$, CFI $= .686$. Although the fit of this model clearly is poorer than the fit of the four-factor model, a statistical comparison of the two models would inspire confidence in that inference. Such a comparison is possible because the one-factor model is nested in the four-factor model. That is, the free parameters in the one-factor model are a subset of the free parameters in the four-factor model. Specifically, the one-factor model can be derived from the four-factor model by fixing the six interfactor correlations to 1.0. It is important that the *correlations,* not the covariances, be fixed at 1.0. This is accomplished by fixing factor variances at 1.0, which eliminates the need to fix a loading on each factor to 1.0.

Adding constraints to a model always results in an increase in $\chi^2$. At issue is whether the increase is significant relative to the change in degrees of freedom, an outcome that would indicate that the constraints are not statistically justified. $\chi^2$s are additive, so the difference between the two $\chi^2$s is itself distributed as a $\chi^2$ and can be evaluated with reference to the distribution defined by the difference in the degrees of freedom associated with the two models. In this instance, the comparison is highly significant $\Delta\chi^2(6, N = 764) = 619.92, p < .001$, indicating that the one-factor model provides a significantly poorer account of the data than the four-factor model.

A second alternative focuses on the nature of the associations among the four factors. Comparison of the one- versus four-factor model clearly supports the distinction among the four factors; however, it might be that, once distinguished, they reflect a single second-order factor. Such a model is shown in path diagram form in Figure 16.3. Note that the four factors that were specified as correlated in the hypothesized model now serve as indicators of a single factor, sensation seeking. And there now is a uniqueness factor associated with each of the four factors. This model is nested

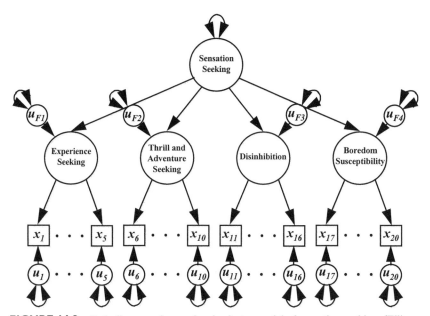

**FIGURE 16.3** Path diagram of second-order factor model of sensation seeking. (Ellipses indicate omitted indicators.)

in the correlated four factor model. The six covariances between the first-order factors provide the data to estimate that portion of the model, which includes four free parameters (three loadings and the variance of the second-order factor). Thus, the second-order model is tested on two more degrees of freedom than the four-factor model. Although the fit statistics for the second-order model are comparable to those for the four-factor model, $\chi^2(166, N = 764) = 662.71, p < .001$, GFI = .92, RMSEA = .063, CFI = .857, the two additional constraints in the second-order model result in a significant decline in overall fit compared to the four-factor model, $\Delta\chi^2(2, N = 764) = 23.61, p < .001$. Thus, we conclude that the four factors are best conceptualized as correlated but not in a pattern that reflects a single overarching factor.

### E. Respecification

Statistical support for the four-factor model is mixed. GFI exceeds .90 but CFI does not, and RMSEA is in the marginal range (.05–.08). Minor adjustments to the original specification would strengthen support for the model and perhaps provide additional insight into the performance of the measure. Indeed, most applications of CFA involve some manner of *respecification,* which, in the typical instance, involves either freeing fixed parameters (e.g., cross loadings or error covariances) or, less commonly, fixing free parameters (e.g., constraining interfactor covariances to zero) in the initially specified model (Bollen & Long, 1993a). With regard to inference, this process can be likened to *post hoc* comparisons in analysis of variance or model-trimming in multiple regression analysis: Unless the respecification is undertaken with proper attention to Type I error, there is a good chance that the respecified model, though adequate for the sample in hand, will not accurately characterize the process in the population represented by the sample (MacCallum et al., 1992). On the other hand, data suitable for CFA are obtained at a price, and responsible stewardship of such data does not permit passive acceptance of negative statistical results (Rosenthal, 1994). In short, cautious, well-informed respecification of theory-relevant CFA models is not to be discouraged and, as in the present case, may provide unexpected insight into the performance of a set of indicators (cf. Jöreskog, 1993). Of course, cross-validation in an independent sample increases confidence in the reliability of respecified models.

In CFA models, there are three primary sources of misspecification: (a) the wrong number of factors, (b) the wrong pattern of loadings, and (c) unmodeled subfactors. In the typical CFA specification, it is difficult to tease apart misspecification due to the wrong number of factors versus the wrong pattern of loadings. Because the number-of-factors question is the more basic of the two, a strategy that specifically addresses this question would be highly beneficial. Jöreskog (1977) proposed such a strategy in the *unrestricted model.* In the unrestricted model, a particular number of factors is specified, but only a single loading on each factor is fixed. In other words, with a single exception per factor, every indicator is permitted to load on every factor. So specified, the question of whether the pattern of loadings is correct is removed, and the analysis focuses specifically on whether the number of factors is correct. In order to estimate an unrestricted model, three problems must be solved. First, one or more plausible hypotheses about the number of factors must be developed. This can be addressed using EFA or by consulting prior research with the indicators. Second, the single best indicator of each factor, the *markers,* must be determined. This information can be gleaned from a rotated EFA solution. Finally, good start values must be provided; loadings from an EFA are a good place to start. Using conservative fit criteria (e.g., CFI > .95), the optimal number of factors can be ascertained by comparing different plausible numbers. Despite the obvious value of such a strategy, published examples of unrestricted models are rare (the application by Tepper & Hoyle, 1996, described earlier, is one example).

Because the present model is not dramatically misspecified and the number of factors is strongly justified on *a priori* grounds, we do not consider alternative numbers of factors. A second alternative, one that we do consider, concerns the pattern of loadings. The pattern of loadings can be misspecified in two ways. First, one or more indicators might have been assigned to the wrong factor. Second, one or more indicators might load on multiple factors. Misspecification of the first variety is unusual and highly unlikely in the present case. Misspecification of the second sort is likely and commonplace. As noted earlier, the typical CFA specification reflects a strong assumption of simple structure (i.e., that each indicator loads on a single factor; see Figure 16.2); however, this tendency is a norm not a rule (as we saw with the unrestricted model). Using an empirical specification-search strategy (e.g., modification index in LISREL, Lagrange multiplier test in EQS), we can ask whether freeing any of the factor loadings fixed at zero in the original specification would improve overall model fit. In the present example, such a search revealed that two items from the sensating seeking measure meaningfully reflect two factors. Freeing these two constraints results in dramatic improvement in the fit of the four-factor model, $\Delta\chi^2(2, N = 764) = 151.55, p < .001$; CFI rises to .91 and RMSEA drops to .051.

A third source of misspecification is the failure to model *subfactors,* subsets of indicators that continue to covary after commonality due to the specified factors has been accounted for. This idea is illustrated in the graphic to the left in Figure 16.1. Notice that most of the overlap between $x_1$ and $x_2$ is shared with $x_3$, but a small portion is not. In the model shown to the right in that figure, this variance is allocated to $u_1$ and $u_2$, which are modeled as uncorrelated. If the variance shared by $x_1$ and $x_2$ but not shared with $x_3$ were appreciable, then permitting $u_1$ and $u_2$ to covary would include that subfactor in the model. In the present example, the empirical specification search revealed that two pairs of uniqueness were correlated. Uniquenesses associated with the items, "I would like to try parachute jumping" and "I would like to try bungee jumping" covaried, as did uniqueness associated with the items, "I get bored seeing the same kids all the time" and "I get bored with people who always say the same things." Note how these pairs of items clearly tap similar preferences but ones that are not shared by other items in the scale. Relaxing the zero constraints on these two covariances produced an additional increment in fit over the model with two double loadings, $\Delta\chi^2(2, N = 764) = 86.31, p < .001$, leaving a model that fit well according to all fit criteria, $\chi^2(162, N = 764) = 441.40, p < .001$, GFI = .94, RMSEA = .048, CFI = .919.

### F. Parameter Estimates

Once the overall fit of a CFA model has been established, the focus shifts to the parameter estimates. Associated with all free parameters is a standard error, and the ratio of the unstandardized estimate to its standard error

**TABLE 16.4 Completely Standardized Maximum Likelihood Estimates for the Final Four-Factor Model[a]**

| Item | $F_1$ | $F_2$ | $F_3$ | $F_4$ | $u$ | $R^2$ |
|------|------|------|------|------|------|------|
| 1 | .416[a] | | | | .909 | .174 |
| 5 | .131** | | | | .991 | .018 |
| 9 | .090* | | | | .996 | .008 |
| 13 | .287*** | | | | .958 | .082 |
| 17 | .543*** | | | | .840 | .294 |
| 2 | | .642[a] | | | .767 | .412 |
| 6 | | .383*** | .338*** | | .796 | .366 |
| 10 | | .677*** | | | .736 | .458 |
| 14 | | .694*** | | | .720 | .482 |
| 18 | | .676*** | | | .737 | .457 |
| 3 | | | .790[b] | | .613 | .624 |
| 7 | | | .760*** | | .649 | .579 |
| 11 | .449*** | | .306*** | | .734 | .461 |
| 15 | | | .472*** | | .881 | .224 |
| 19 | | | .747*** | | .665 | .558 |
| 4 | | | | .073 | .997 | .006 |
| 8 | | | | .257*** | .966 | .067 |
| 12 | | | | .212*** | .977 | .045 |
| 16 | | | | .358*** | .934 | .128 |
| 20 | | | | .652[a] | .759 | .424 |

Interfactor correlations

| | | | | |
|---|---|---|---|---|
| $F_2$ | .703*** | | | |
| $F_3$ | .609*** | .409*** | | |
| $F_4$ | .886*** | .417*** | .789*** | |

[a] Two pairs of uniqueness were permitted to covary. The correlation between $u_8$ and $u_{12}$ was .232 and the correlation between $u_{10}$ and $u_{18}$ was .379.

[b] Unstandardized loading fixed at 1.0 to set metric of latent variable.

* $p < .05$; ** $p < .01$; *** $p < .001$.

provides a test of whether the estimate differs from zero. This *critical ratio* is distributed as a $z$, and is evaluated against the standard two-tailed criteria (i.e., 1.96, 2.58, and 3.33 for $p < .05$, .01, and .001 respectively).

Parameter estimates for the respecified model are presented in Table 16.4. Values in the table are completely standardized to facilitate interpretation. Thus, the variances are in standard deviation form, the interfactor associations are expressed as correlations, and the factor loadings reflect the number of standard units change in the indicators per standard unit change in the factor(s) on which they load. Reflecting the respecification of the four-factor model, items 6 and 11 have nonzero loadings on two factors. In general, the loadings are not large, an observation consistent with the broadband definition of the sensation-seeking construct. Items 4 and 9 are noticeably weak as indicators of their factor and probably should be rewritten.

In the final column of Table 16.4 are $R^2$ values for each item. These values were calculated as one minus the squared standardized uniqueness, and indicate the proportion of variance in each item attributable to the factor(s) on which it loads. These can be thought of as *validity coefficients,* indicative of the degree to which the items are indeed caused by one of the facets of sensation seeking. The relatively small value of $R^2$ for some items implicates unmodeled factors that contributed to variability in the responses.

## V. ADDITIONAL CONSIDERATIONS

### A. Conceptual Inference

When all fit statistics point to the adequacy of a particular model as an explanation for a set of data, one is tempted to infer that the model is correct or true; however, such an inference would be ill advised if it could be shown that, in addition to other models that could explain the data almost as well, there exist models that could explain the data *exactly* as well. Such models are referred to as *equivalent models,* models that portray the hypothetical construct differently but are statistically identical (MacCallum, Wegener, Uchino, & Fabrigar, 1993; Williams, Bozdogan, & Aiman-Smith, 1996). For instance, with only three first-order factors, a second-order model will produce precisely the same fit statistics (i.e., same $\chi^2$ and df) as a first-order model with correlated factors. At the conceptual level, this presents an obvious inferential conundrum. In some cases, this unfortunate state of affairs can be avoided. For instance, in the sensation-seeking analysis presented here, there were four first-order factors, which provided sufficient information at the factor level that the first- and second-order models could be directly compared. In other cases, equivalent models cannot be avoided (e.g., it is not possible to add a fourth factor to constructs that, by definition, comprise only three). In such cases, it is necessary to identify all equivalent models and make use of theory, logic, or results of extant research to argue for one over the others.

### B. Statistical Inference

An attractive feature of CFA is the capacity for statistical hypothesis testing. The null hypothesis, $\Sigma = \Sigma(\Theta)$, is straightforward, and would appear to be rather directly testable as $\mathbf{S} - \hat{\Sigma}$. Unfortunately, numerous factors conspire to cloud the inferential context within which this hypothesis is evaluated (Biddle & Marlin, 1987). The estimators used to derive $\hat{\Sigma}$ carry assumptions that are rarely, perhaps never, met in social and behavioral science research. Thus, the test statistic that summarizes $\mathbf{S} - \hat{\Sigma}$ does not exactly follow the $\chi^2$ distribution against which it is compared. An array of adjunctive indexes of fit have been proposed and evaluated, but, with few exceptions, these are descriptive statistics that are subject to biased interpretation. Moreover, because of the complexity typical of CFA models, respecification is prevalent, raising questions about the integrity of such models for explanation of the hypothetical construct in the population (Breckler, 1990). The best inferential strategy for CFA users at this time is twofold. First, investigators should consult a variety of fit indices that make differing assumptions and construe "fit" in different ways (Hoyle & Panter, 1995). Second, investigators need to replicate their models and demonstrate that they hold for different samples and, when applicable, alternative indicators.

## C. Statistical Power

In addition to the restrictive assumptions that govern the statistical test of $\mathbf{S} - \hat{\Sigma}$, there is the problem that, with enough statistical power, even trivial departures of $\hat{\Sigma}$ from $\mathbf{S}$ will result in rejection of the null hypothesis. On the other hand, insufficient statistical power will lead to a failure to detect nontrivial departures of $\hat{\Sigma}$ from $\mathbf{S}$. In short, as with all inferential statistics, knowledge about the statistical power of an analysis can provide important context for interpreting statistical outcomes. Determination of statistical power for CFA models is difficult, and, until recently, was generally ignored. Recent contributions to the technical literature on CFA simplify the determination of statistical power and promise to make statistical power a standard component of the inferential framework for CFA models.

Just as there are two levels of statistical testing involved in CFA (overall fit and parameter estimates), there are two sets of power considerations. On one level, there is the statistical power of the omnibus test of fit. MacCallum et al. (1996) have outlined a procedure and criteria for evaluating statistical power of the omnibus test of fit based on the RMSEA. A virtue of the strategy they propose is that it only requires knowledge of the number of variables, the degrees of freedom of a specified model, and some hypothetical sample size. Thus, power of the overall test of fit can be calculated *prior to* data collection. On another level, there is the statistical power of individual parameter estimates. At this level it is important to realize that statistical power is not the same for each and every free parameter in a particular model. Moreover, parameter estimates within a model are interdependent (Kaplan, 1995). Thus, it is difficult to ascertain power for particular parameters without data in hand. At this point, there are two strategies for evaluating statistical power of tests of parameter estimates. Saris and Satorra (1993) advocate a strategy that involves intentionally misspecifying a parameter of interest, then evaluating the degree to which the misspecification affects overall model fit. Kaplan (1995) has developed a strategy for using specification-search statistics as a means of indexing the statistical power of tests of particular parameters. In either case, overall fit or parameter estimates, knowledge about statistical power can provide important context for understanding the outcome of statistical tests in CFA.

## VI. CONCLUSIONS AND RECOMMENDATIONS

The study of hypothetical constructs is a fundamental endeavor in the social and behavioral sciences. Confirmatory factor analysis is a statistical strategy specifically designed to identify and explore hypothetical constructs as manifest in fallible indicators. The allure of CFA over other approaches to the study of hypothetical constructs is the capacity for testing detailed hypotheses in a deductive mode. Moreover, CFA models can be incorporated directly into general SEMs that include directional relations between hypothetical constructs. Recent developments in technical aspects of estimation and testing coupled with widely accessible and user-friendly software have rendered CFA more popular and appealing than ever.

With the increased accessibility of CFA comes a responsibility to understand the conditions under which CFA is appropriately applied and the factors relevant to interpretation of CFA results. On the basis of findings from a burgeoning technical literature, seven basic recommendations can be offered. First, users should aim for samples of at least 200 and, preferably, 400 cases. If more than minimal respecification of an hypothesized model is anticipated, then a sample of at least 800 cases is necessary. Second, the distributional properties of indicators should be well understood and corrective measures taken (e.g., transformations, parceling, scaled statistics) when distributions depart markedly from normality. Third, at least three and, preferably, four indicators of factors should be obtained. With at least three indicators, identification issues rarely arise. Fourth, simple structure should not be assumed in all models. With sufficient indicators per factor, cross-loadings are permissible and may be an important feature of a model. Fifth, the multifaceted nature of fit evaluation should be acknowledged by consulting two or more indicators of fit that rely on different computational logic. Sixth, whenever possible, multiple, nested models should be posited in order to rule out parsimonious or substantively interesting alternatives to the hypothesized model. Seventh, respecification is not to be eschewed, but it should be undertaken in a disciplined manner with due attention to the possibility of Type I errors. Substantially respecified models should be cross-validated in an independent sample. Beyond these basic considerations, applied researchers should routinely consult the latest literature on CFA, as new developments are frequent, particularly with regard to estimation and evaluation of fit.

## ACKNOWLEDGMENTS

During the preparation of this chapter, the author was funded by grants DA-05312 and DA-09569 from the National Institute on Drug Abuse and grant SP-07967 from the Center for Substance Abuse Prevention.

## REFERENCES

Anderson, J. C., & Gerbing, D. W. (1988). Structural equation modeling in practice: A review and recommended two-step approach. *Psychological Bulletin, 103,* 411–423.

Bentler, P. M. (1980). Multivariate analysis with latent variables: Causal modeling. *Annual Review of Psychology, 31,* 419–456.

Bentler, P. M. (1990). Comparative fit indexes in structural models. *Psychological Bulletin, 107,* 238–246.

Bentler, P. M., & Bonett, D. G. (1980). Significance tests and goodness-of-fit in the analysis of covariance structures. *Psychological Bulletin, 88,* 588–606.

Bentler, P. M., & Weeks, D. G. (1980). Linear structural equations with latent variables. *Psychometrika, 45,* 289–307.

Biddle, B. J., & Marlin, M. M. (1987). Causality, confirmation, credulity, and structural equation modeling. *Child Development, 58,* 4–17.

Bollen, K. A. (1984). Multiple indicators: Internal consistency or nor necessary relationship? *Quality and Quantity, 18,* 377–385.

Bollen, K. A. (1989). *Structural equations with latent variables.* New York: Wiley.

Bollen, K. A. (1996). An alternative two-stage least squares (2SLS) estimator for latent variable equations. *Psychometrika, 61,* 109–121.

Bollen, K. A., & Barb, K. H. (1981). Pearson's r and coarsely categorized measures. *American Sociological Review, 46,* 232–239.

Bollen, K. A., & Hoyle, R. H. (1990). Perceived cohesion: A conceptual and empirical examination. *Social Forces, 69,* 479–504.

Bollen, K. A., & Lennox, R. (1991). Conventional wisdom on measurement: A structural equation perspective. *Psychological Bulletin, 110,* 305–314.

Bollen, K. A., & Long, J. S. (1993a). Introduction. In K. A. Bollen & J. S. Long (Eds.), *Testing structural equation models* (pp. 1–9). Newbury Park, CA: Sage.

Bollen, K. A., & Long, J. S. (Eds.) (1993b). *Testing structural equation models.* Newbury Park, CA: Sage.

Breckler, S. J. (1990). Applications of covariance structure modeling in psychology: Cause for Concern? *Psychological Bulletin, 107,* 260–273.

Browne, M. W., & Cudeck, R. (1993). Alternative ways of assessing model fit. In K. A. Bollen & J. S. Long (Eds.), *Testing structural equation models* (pp. 136–162). Newbury Park, CA: Sage.

Byrne, B. M. (1989). Multigroup comparisons and the assumption of equivalent construct validity across groups: Methodological and substantive issues. *Multivariate Behavioral Research, 24,* 503–523.

Byrne, B. M., Shavelson, R. J., & Muthén, B. (1989). Testing for the equivalence of factor covariance and mean structures: The issue of partial measurement invariance. *Psychological Bulletin, 105,* 456–466.

Chou, C.-P., & Bentler, P. M. (1995). Estimates and tests in structural equation modeling. In R. H. Hoyle (Ed.), *Structural equation modeling: Concepts, issues, and applications* (pp. 37–55). Thousand Oaks, CA: Sage.

Church, A. T., & Burke, P. J. (1994). Exploratory and confirmatory tests of the big five and Tellegen's three- and four-dimensional models. *Journal of Personality and Social Psychology, 66,* 93–114.

Cohen, P., Cohen, J., Teresi, J., Marchi, M., & Velez, C. N. (1990). Problems in the measurement of latent variables in structural equations causal models. *Applied Psychological Measurement, 14,* 183–196.

Curran, P. J., West, S. G., & Finch, J. F. (1996). The robustness of test statistics to nonnormality and specification error in confirmatory factor analysis. *Psychological Methods, 1,* 16–29.

Duncan, T. E., Duncan, S. C., & Stoolmiller, M. (1994). Modeling developmental processes using latent growth structural equation methodology. *Applied Psychological Measurement, 18,* 343–354.

Fleming, J. S., & Courtney, B. E. (1984). The dimensionality of self-esteem: II. Hierarchical facet model for revised measurement scales. *Journal of Personality and Social Psychology, 46,* 404–421.

Hattie, J., & Fraser, C. (1988). The constraining of parameters in restricted factor analysis. *Applied Psychological Measurement, 12,* 155–162.

Hoyle, R. H. (1991). Evaluating measurement models in clinical research: Covariance structure analysis of latent variable models of self-conception. *Journal of Consulting and Clinical Psychology, 59,* 67–76.

Hoyle, R. H. (Ed.). (1995a). *Structural equation modeling: Concepts, issues, and applications.* Thousand Oaks, CA: Sage.

Hoyle, R. H. (1995b). The structural equation modeling approach: Basic concepts and fundamental issues. In R. H. Hoyle (Ed.), *Structural equation modeling: Concepts, issues, and applications* (pp. 1–15). Thousand Oaks, CA: Sage.

Hoyle, R. H., & Lennox, R. D. (1991). Latent structure of self-monitoring. *Multivariate Behavioral Research, 26,* 511–540.

Hoyle, R. H., & Panter, A. T. (1995). Writing about structural equation models. In R. H. Hoyle (Ed.), *Structural equation modeling: Concepts, issues, and applications* (pp. 158–176). Thousand Oaks, CA: Sage.

Hoyle, R. H., & Smith, G. T. (1994). Formulating clinical research hypotheses as structural equation models: A conceptual overview. *Journal of Consulting and Clinical Psychology, 62,* 429–440.

Hoyle, R. H., Stephenson, M. T., Palmgreen, P., Lorch, E. P., & Donohew, R. L. (1998). *Brief measure of sensation seeking for research on adolescents.* Manuscript submitted for publication.

Hu, L.-T., & Bentler, P. M. (1995). Evaluating model fit. In R. H. Hoyle (Ed.), *Structural equation modeling: Concepts, issues, and applications* (pp. 76–99). Thousand Oaks, CA: Sage.

Hu, L.-T., Bentler, P. M., & Kano, Y. (1992). Can test statistics in covariance structure analysis be trusted? *Psychological Bulletin, 112,* 351–362.

John, O. P. (1990). The "Big Five" factor taxonomy: Dimensions of personality in the natural language and in questionnaires. In L. A. Pervin (Ed.), *Handbook of personality theory and research* (pp. 66–100). New York: Guilford Press.

Jöreskog, K. G. (1967). Some contributions to maximum likelihood factor analysis. *Psychometrika, 32,* 443–482.

Jöreskog, K. G. (1969). A general approach to confirmatory maximum likelihood factor analysis. *Psychometrika, 34,* 183–202.

Jöreskog, K. G. (1973). A general method for estimating a linear structural equation system. In A. S. Goldberger & O. D. Duncan (Eds.), *Structural equation models in the social sciences* (pp. 85–112). New York: Academic Press.

Jöreskog, K. G. (1977). Author's addendum. In J. Magidson (Ed.), *Advances in factor analysis and structural equation models.* Cambridge, MA: Abt.

Jöreskog, K. G. (1993). Testing structural equation models. In K. A. Bollen & J. S. Long (Eds.), *Testing structural equation models* (pp. 294–316). Newbury Park, CA: Sage.

Jöreskog, K. G. (1994). On the estimation of polychoric correlations and their asymptotic covariance matrix. *Psychometrika, 59,* 381–389.

Jöreskog, K. G., & Sörbom, D. (1981). *LISREL V: Analysis of linear structural relationships by the method of maximum likelihood.* Chicago: National Educational Resources.

Kaplan, D. (1995). Statistical power in structural equation modeling. In R. H. Hoyle (Ed.), *Structural equation modeling: Concepts, issues, and applications* (pp. 100–117). Thousand Oaks, CA: Sage.

Keesling, J. W. (1972). *Maximum likelihood approaches to causal analysis.* Unpublished doctoral dissertation, University of Chicago.

Kenny, D. A., & Kashy, D. A. (1992). The analysis of the multitrait-multimethod matrix by confirmatory factor analysis. *Psychological Bulletin, 112,* 165–172.

Kenny, D. A., & Zautra, A. (1995). The trait-state-error model for multiwave data. *Journal of Consulting and Clinical Psychology, 63,* 52–59.

MacCallum, R. C. (1995). Model specification: Procedures, strategies, and related issues. In R. H. Hoyle (Ed.), *Structural equation modeling: Concepts, issues, and applications* (pp. 16–36). Thousand Oaks, CA: Sage.

MacCallum, R. C., Browne, M. W., & Sugawara, H. M. (1996). Power analysis and determination of sample size for covariance structure modeling. *Psychological Methods, 1,* 130–149.

MacCallum, R. C., Roznowski, M., & Necowitz, L. B. (1992). Model modifications in covariance structure analysis: The problem of capitalization on chance. *Psychological Bulletin, 111,* 490–504.

MacCallum, R. C., Wegener, D. T., Uchino, B. N., & Fabrigar, L. R. (1993). The problem of equivalent models in applications of covariance structure analysis. *Psychological Bulletin, 114,* 185–199.

Marsh, H. W. (1989). Confirmatory factor analyses of multitrait-multimethod data: Many problems and a few solutions. *Applied Psychological Measurement, 13,* 335–361.

Marsh, H. W., Balla, J. R., & Hau, K.-T. (1996). An evaluation of incremental fit indices: A clarification of mathematical and empirical properties. In G. A. Marcoulides & R. E. Schumacker (Eds.), *Advanced structural equation modeling: Issues and techniques* (pp. 315–353). Mahwah, NJ: Erlbaum.

Marsh, H. W., Balla, J. R., & McDonald, R. P. (1988). Goodness-of-fit indexes in confirmatory factor analysis: The effect of sample size. *Psychological Bulletin, 103,* 391–411.

Marsh, H. W., & Grayson, D. (1995). Latent variable models of multitrait-multimethod data. In R. H. Hoyle (Ed.), *Structural equation modeling: Concepts, issues, and applications* (pp. 177–198). Thousand Oaks, CA: Sage.

Marsh, H. W., & Hau, K.-T. (1991). Confirmatory factor analysis: Strategies for small sample sizes. In R. H. Hoyle (Ed.), *Statistical strategies for small sample research* (pp. 251–284). Thousand Oaks, CA: Sage.

Marsh, H. W., & Hocevar, D. (1985). Application of confirmatory factor analysis to the study of self-concept: First- and higher-order factor models and their invariance across groups. *Psychological Bulletin, 97,* 562–582.

McArdle, J. J. (1996). Current directions in structural factor analysis. *Current Directions in Psychological Science, 5,* 11–18.

McArdle, J. J., & Boker, S. M. (1990). *RAMpath.* Hillsdale, NJ: Erlbaum.

Micceri, T. (1989). The unicorn, the normal curve, and other improbable creatures. *Psychological Bulletin, 105,* 156–166.

Mulaik, S. A., James, L. R., Van Alstine, J., Bennett, N., Lind, S., & Stillwell, C. D. (1989). An evaluation of goodness-of-fit indices for structural equation models. *Psychological Bulletin, 105,* 430–445.

Muthén, B. (1984). A general structural equation model with dichotomous, ordered categorical, and continuous latent variable indicators. *Psychometrika, 49,* 115–132.

Panter, A. T., Tanaka, J. S., & Hoyle, R. H. (1994). Structural models for multimode designs in personality and temperament research. In C. F. Halverson, G. A. Kohnstamm, & R. P. Martin (Eds.), *The developing structure of temperament and personality from infancy to adulthood* (pp. 111–138). Hillsdale, NJ: Erlbaum.

Raykov, T. (1994). Studying correlates and predictors of longitudinal change using structural equation modeling. *Applied Psychological Measurement, 18,* 63–77.

Rindskopf, D., & Rose, T. (1988). Some theory and applications of confirmatory second-order factor analysis. *Multivariate Behavioral Research, 23,* 51–67.

Rosenthal, R. (1994). Science and ethics in conducting, analyzing, and reporting psychological research. *Psychological Science, 5,* 127–133.

Saris, W. E., & Satorra, A. (1993). Power evaluations in structural equation models. In K. A. Bollen & J. S. Long (Eds.), *Testing structural equation models* (pp. 181–204). Newbury Park, CA: Sage.

Satorra, A., & Bentler, P. M. (1988). *Scaling corrections for statistics in covariance structure analysis* (Rep. No. 2). Los Angeles: UCLA Statistics Series.

Snyder, C. R., & Fromkin, H. L. (1980). *Uniqueness: The human pursuit of difference.* New York: Plenum.

Steiger, J. H. (1990). Structural model evaluation and modification: An interval estimation approach. *Multivariate Behavioral Research, 25,* 173–180.

Tanaka, J. S. (1987). How big is big enough? Sample size and goodness-of-fit in structural equation models with latent variables. *Child Development, 58,* 134–146.

Tanaka, J. S. (1993). Multifaceted conceptions of fit in structural equation models. In K. A. Bollen & J. S. Long (Eds.), *Testing structural equation models* (pp. 10–39). Newbury Park, CA: Sage.

Tepper, K., & Hoyle, R. H. (1996). Latent variable models of need for uniqueness. *Multivariate Behavioral Research, 31,* 467–494.

Thurstone, L. L. (1947). *Multiple factor analysis.* Chicago: University of Chicago Press.

West, S. G., Finch, J. F., & Curran, P. J. (1995). Structural equation models with nonnormal variables. In R. H. Hoyle (Ed.), *Structural equation modeling: Concepts, issues, and applications* (pp. 56–75). Thousand Oaks, CA: Sage.

Wiley, D. E. (1973). The identification problem for structural equation models with unmeasured variables. In A. S. Goldberger & O. D. Duncan (Eds.), *Structural equation models in the social sciences* (pp. 69–83). New York: Academic Press.

Willet, J. B., & Sayer, A. G. (1994). Using covariance structure analysis to detect correlates and predictors of individual change over time. *Psychological Bulletin, 116,* 363–381.

Williams, L. J., Bozdogan, H., & Aiman-Smith, L. (1996). Inference problems with equivalent models. In G. A. Marcoulides & R. E. Schumacker (Eds.), *Advanced structural equation modeling: Issues and techniques* (pp. 279–314). Mahwah, NJ: Erlbaum.

Yuan, K.-H., & Bentler, P. M. (1995). *Some new test statistics for mean and covariance structure analysis with high dimensional data* (Rep. No. 195). Los Angeles: UCLA Statistics Series.

Zuckerman, M., Eysenck, S., & Eysenck, H. J. (1978). Sensation seeking in England and America: Cross-cultural age and sex comparisons. *Journal of Consulting and Clinical Psychology, 46,* 139–149.

## APPENDIX

Technical treatments of CFA typically are replete with Greek letters and matrix notation, and it is not uncommon to encounter such notation in research reports describing CFA results. Three pioneers in the development of CFA, Jöreskog (1973), Keesling (1972), and Wiley (1973), are credited with developing this relatively standard notational framework, and its use became widespread as a result of its prominent role in the influential LISREL software program. As a result, it is referred to as JKW or LISREL notation (Bentler, 1980).

As this chapter makes clear, it is not necessary to use JKW notation to explicate CFA; however, inasmuch as that notation dominates the technical literature on CFA and continues to appear in reports of CFA results, it is important that researchers acquaint themselves with basic aspects of JKW notation. A skeletal presentation of those aspects follows.

In JKW notation, a distinction is made between exogenous and endogenous factors. Exogenous factors are labeled $\xi$ and endogenous factors are labeled $\eta$. This distinction gives rise to two sets of measurement equations, those involving indicators of exogenous factors, labeled $x$, and those involving indicators of endogenous factors, labeled $y$. Factor loadings are labeled $\lambda_x$ and $\lambda_y$ for the exogenous and endogenous equations, respectively. Finally, the exogenous uniquenesses are labeled $\delta$, and the endogenous uniquenesses are labeled $\varepsilon$. The measurement equations are conveyed rather compactly in the two matrix equations,

$$x = \Lambda_x \xi + \delta \tag{A.1}$$
$$y = \Lambda_y \eta + \varepsilon \tag{A.2}$$

The dimensions of $\mathbf{x}$ and $\mathbf{y}$ are $q \times 1$ and $p \times 1$, respectively, where $q$ denotes the number of indicators of all exogenous factors and $p$ denotes the number of indicators of all endogenous factors. The dimensions of $\Lambda_x$ and $\Lambda_y$ are $q \times n$ and $p \times m$, respectively, where $q$ indexes the number of exogenous factors and $p$ the number of endogenous factors. In the typical case, $\delta$ and $\varepsilon$ are $q \times 1$ and $p \times 1$ vectors, respectively, though they become $q \times q$ and $p \times p$ lower-triangular matrices, and may be relabeled $\Theta_\delta$ and $\Theta_\varepsilon$, when uniquenesses are permitted to covary. To complete the measurement model, interfactor covariances are conveyed in $\Phi$ and $\Psi$, $q \times q$ and $p \times p$ lower-triangular matrices for the exogenous and endogenous latent variables, respectively.

Following these conventions, the JKW notation for the model shown on the right in Figure 16.2 would be as follows:

$$\begin{bmatrix} x_1 \\ x_2 \\ x_3 \\ x_4 \\ x_5 \\ x_6 \end{bmatrix} = \begin{bmatrix} 1 & 0 \\ \lambda_{21} & 0 \\ \lambda_{31} & 0 \\ 0 & 1 \\ 0 & \lambda_{52} \\ 0 & \lambda_{62} \end{bmatrix} \begin{bmatrix} \xi_1 \\ \xi_2 \end{bmatrix} + \begin{bmatrix} \delta_1 \\ \delta_2 \\ \delta_3 \\ \delta_4 \\ \delta_5 \\ \delta_6 \end{bmatrix} \tag{A.3}$$

The variances of $F_1$ and $F_2$ and the covariance between them are expressed in the $\phi$ matrix.

$$\Phi = \begin{bmatrix} \phi_{11} & \\ \phi_{21} & \phi_{22} \end{bmatrix} \tag{A.4}$$

Note that the metric of the factors, $\xi_1$ and $\xi_2$, is set by fixing $\lambda_{11}$ and $\lambda_{42}$ to 1. An alternative strategy would be to specify $\Phi$ as standardized (i.e., a correlation matrix) by fixing $\phi_{11}$ and $\phi_{22}$ to 1 and leaving $\lambda_{11}$ and $\lambda_{42}$ free to be estimated.

# 17

# MULTIVARIATE META-ANALYSIS

**BETSY J. BECKER**

*College of Education, Michigan State University, East Lansing, Michigan*

## I. WHAT IS META-ANALYSIS?

Meta-analysis is a term coined by Glass (1976) to mean the "analysis of analyses." Meta-analysis, or research synthesis, provides a way to examine results accumulated from a series of related studies, through statistical analyses of those results. Multivariate meta-analysis is the synthesis of studies that may report several related outcomes—a common occurrence in modern social-science research.

Glass developed the idea of meta-analysis in order to make sense of the results of studies of the effects of psychotherapy. An exhaustive search of the psychotherapy literature had uncovered hundreds of relevant studies. Glass (1976) used statistical analyses to synthesize the voluminous results, and called his approach "meta-analysis." Although the term was new, the idea of using quantitative combinations of research results dates to the early 20th century (e.g., Pearson, 1904; Tippett, 1931). Such summaries were more obvious in agriculture and medical research, where results have long been measured in terms of counts, weights, or other scales that do not vary across studies.

Glass was able to summarize treatment-effect data from studies of

*Handbook of Applied Multivariate Statistics and Mathematical Modeling*
Copyright © 2000 by Academic Press. All rights of reproduction in any form reserved.

psychotherapy thanks to his decision to express the various results in a standard metric, specifically as an "effect size" or standardized difference between group means. Thus although two studies might each use different scales to measure depression in psychotherapy recipients, if the construct underlying those scales represented the "same" view of depression, (e.g., the scales were linearly equatable), effect sizes could be compared across studies. Often, though, researchers examine treatment effects for several different outcomes (e.g., different constructs) within each study, leading to the potential for multivariate meta-analysis.

The purpose of this chapter is to review meta-analytic methods for synthesis of multivariate data. The approach to multivariate meta-analysis presented here can be applied regardless of the form of the effect of interest (be it an effect size, correlation, proportion, or some other index). In general the goals of a multivariate meta-analysis are the same as those of univariate syntheses: to estimate magnitudes of effect across studies, and to examine variation in patterns of outcomes. Often the reviewer hopes to explain between-study differences in effect magnitudes using explanatory variables.

In addition, in a multivariate synthesis the reviewer has powerful data concerning within-study differences between effects for different outcomes and differential relations of potential predictors with study outcomes. Using multivariate data to examine effects for different outcomes can reduce the confounding associated with between-study evidence that is common to univariate meta-analyses. For instance, suppose you wanted to compare coaching effects for math versus verbal outcomes, but no study had examined both. Given these circumstances, any observed math-versus-verbal differences could be confounded with other study characteristics that happened (by chance) to differ for the studies of math outcomes versus the studies of verbal outcomes. However, if all studies had measured both math and verbal outcomes, the confounding is removed because the subjects (and other study characteristics) would be identical for the two outcomes. The situation in a real synthesis is typically somewhere between these two extremes.

Differential relations (interactions) of potential predictors with multiple study outcomes also can be addressed directly in a multivariate synthesis. For example, a reviewer can model the effects of treatment duration on *any* outcome, or the effect of duration on *each* outcome separately, in one cohesive analysis. In cases where study features are likely to have different predictive power for different outcomes, it is critical to model their importance for each outcome. In a univariate synthesis researchers can address such questions only by creating distinct (and smaller) subsets of study results and examining different models for each subset.

The chapter begins with a brief description of how multivariate data arise in meta-analysis and the potential benefits and problems associated with several approaches to analyzing such data. I then discuss in detail how to conduct a meta-analysis in which one models the dependence in multivariate meta-analysis data. Effect-size and correlational data are given the most thorough treatment, as these are the two most common forms of multivariate meta-analysis data. Examples are drawn from two existing research syntheses.

## II. HOW MULTIVARIATE DATA ARISE IN META-ANALYSIS

Multivariate data arise in meta-analysis because of the nature of data in primary studies. Sometimes meta-analysis data are multivariate because the primary studies are multivariate—some or all studies measure multiple outcomes for each subject. For example, studies of the Scholastic Aptitude Test (SAT) typically report results for verbal and quantitative subtests. Multivariate meta-analysis data also can arise when primary research studies involve several comparisons among groups, based on a single outcome. For instance, two different treatments may be contrasted with a single control condition. In both cases the meta-analysis data are multivariate and involve dependence among study results.

Although all of these design features lead to dependence among study outcomes, the nature of the dependence depends on exactly what comparisons are computed and the metric(s) in which they are expressed. Many of the examples below concern data on means from treatment/control studies, but similar comments apply to studies of correlations, rates, and proportions.

### A. Multiple Outcomes

When a study examines a single outcome variable (say, $Y_1$) measured at one time point for one treatment (T) and one control (C) group, only one treatment versus control contrast can be computed in a particular metric (e.g., a single effect size or one odds ratio). However, study designs are usually not so simple. Rarely will a treatment affect only a single construct. Thus researchers may examine several constructs, and if they are correlated (which is typically true), treatment/control contrasts based on them will also be correlated. Similarly, if a single construct is measured at several times (such as pretreatment, posttreatment, and several months later), subjects' scores will likely be correlated, as will the treatment/control contrasts based on them.

Although it is common for some studies to report multiple outcomes, it is rare for all studies in a review to have the same outcomes. Even studies of the Scholastic Aptitude Test (SAT), an examination with a clear subtest structure and long history of use, do not all report results for the SAT's math and verbal subtests. Only half of the studies in Becker's (1990) review of SAT coaching gave results for both subtests. The numbers of studies with multiple outcomes vary even more when the outcome is less clearly defined. For example, only one study in Becker's (1988a) meta-analysis of gender differences in science achievement had multiple outcomes, whereas 29 of 43 studies in Chiu's (1997) synthesis of the effects of metacognitive instruction in reading reported multiple outcomes, with one study reporting twelve different outcomes.

When relations among variables are of interest in primary research, at least two variables must be measured so a measure of association can be computed. It is rare to find a study reporting only a single bivariate correlation (between a predictor $X$ and an outcome $Y$). But any set of correlations from one sample—for instance, the correlations between $X_1$ and $Y$ and between $X_2$ and $Y$—are dependent, leading to multivariate data in a synthesis examining several relations.

### B. Multiple Treatment Effects

In treatment/control research, researchers sometimes study variations in treatment leading to several treatment groups (e.g., $T_1$, $T_2$) or to several comparison groups (e.g., wait-list versus no-treatment controls). Two contrasts that compare results for groups $T_1$ and $T_2$ each to a single control group lead to dependent multivariate data in a meta-analysis. The more groups and contrasts involved, the more complex is the dependence in the multivariate data. For example, Ryan, Blakeslee, and Furst (1986) investigated the effects of practice on motor skill levels using a five-group design with three mental-practice groups, one physical-practice group, and one no-practice group. Comparisons of each of the four practice groups with the no-practice group would lead to four related effect sizes if all were included in a single meta-analysis.

### III. APPROACHES TO MULTIVARIATE DATA

Meta-analysts have treated multiple study outcomes in a variety of ways. Each approach has advantages and disadvantages, and each may be useful in certain situations. Your choice, as an analyst, will depend on the nature of the multivariate data in your synthesis. Below I use the terms *outcome* (or study outcome) and *effect* to refer to an index of study results, such as an effect size or correlation. A study reporting two effect sizes for one sample would have two dependent outcomes or effects.

Most procedures available for meta-analysis assume that the outcomes to be analyzed are independent (e.g., Hedges & Olkin, 1985, p. 108). Ad hoc approaches to multivariate meta-analysis may ignore this fact, whereas more sophisticated approaches deal directly with the dependence. Some problems arise when dependence in meta-analytic data is ignored or handled inadequately. Other difficulties arise when we try to model dependence fully. Problems specific to four main approaches to dependence are described below, but one common consequence of ignoring dependence in data (whether in meta-analysis or not) is a possible effect on Type I error level and the accuracy of probability statements made for the observed data. This problem has not been studied sufficiently for meta-analysis, but studies of other data-analytic techniques have shown that ignoring dependence affects Type I error levels in complicated ways, determined in part by the nature of the dependence in the data. This can be manifest via the values of estimates and standard errors; thus dependence can also affect bias and precision in estimation as well.

## A. Treating Multivariate Data as Independent

One approach to analyzing multivariate meta-analysis data is to ignore the dependence and treat outcomes as though they were independent. This can be done by creating a separate data record for each outcome and not linking those records. The multiple data records created from each study have many coded study features in common (e.g., type of treatment, length of follow-up interval, and theoretical orientation of therapist), but they may vary, for instance, in the type of variable (or construct) that was measured.

Generally speaking, I do not recommend this approach. A significant problem with this approach is that studies reporting more outcomes have more data records, and thus can unduly influence the combined results. This is particularly problematic if such studies are also unusual in ways that relate to treatment effectiveness (e.g., they use atypical samples or special treatment implementations). Some researchers reduce the influence of studies with many outcomes by weighting each outcome by the inverse of the count of outcomes from the study. This is similar to combining (averaging) across outcomes within each study (see below), but weighting by outcome count allows differences among multiple outcomes within studies to contribute to observed variation in outcomes, as is reasonable. This procedure reduces the influence of studies with more outcomes on the overall means, but it does not account for dependence among the outcomes.

When few studies report multiple outcomes the reviewer may find that doing more than simply ignoring the dependency requires significant extra effort and modifications to standard computational procedures. A simple approach that acknowledges the issue of dependence is to conduct sensitivity analyses (e.g., Greenhouse & Iyengar, 1994), essentially by conducting analyses with, and without, multiple outcomes per study. A reasonable approach is to first analyze an initial set of independent data consisting of only one outcome per study, then add the few "extra" outcomes to the data set and repeat the analyses. If the results of the two analyses appear similar, the dependence due to multiple outcomes can be overlooked, and the analyses of all outcomes reported. It is essential to report that sensitivity analyses were performed and that the reported results were similar to those based on independent data. More sophisticated approaches to dealing with the dependency are needed if the results for the full set of outcomes are not similar to those based on independent data.

## B. Combining across Different Outcomes

We can eliminate dependence in the data by first summarizing data within studies to produce one outcome (and data record) per study (e.g., Rosenthal & Rubin, 1986). Typical data-reduction techniques are often used, such as averaging across outcomes, using the median or midpoint of the outcomes from the study, or selecting a single best or most common outcome. This approach may be reasonable if the results represent measures of a single construct and show similar effects. Hedges and Olkin (1985, p. 221) also note that pooling estimators can lead to gains in efficiency of estimation.

However, problems may arise due to the information loss associated with data reduction. One value may not adequately represent the range of results within each study, particularly if the results differ systematically across meaningfully different outcomes (e.g., for math versus verbal scores). In addition, observed variation in outcomes is wrongly reduced when a central-tendency measure is used. Finally, the ability to examine differences in effects for different constructs is lost when outcomes are combined or when some constructs are omitted from the data set.

## C. Creating Independent Data Subsets

An easily implemented strategy for reducing dependency is to create independent subsets of data for analysis (Greenwald, Hedges, & Laine, 1996; Steinkamp & Maehr, 1984). When outcomes are found for several constructs (e.g., math and verbal effect sizes) or for several measurement occasions, separate analyses can be done for each construct or time point. However, the advantages of this approach are illusory. Comparisons of results across the data subsets (e.g., comparisons of mean math and verbal effects) are still influenced by the dependence that resides in the original data set, and therefore such comparisons are not valid under typical assumptions of independence. This approach may be reasonable if such comparisons are not of interest, or if the separate constructs have low intercorrelations within studies.

## D. Modeling Dependence

The most complete and accurate portrayal of the effects of dependence in multivariate meta-analysis data requires the use of multivariate techniques to model the dependence in study outcomes. Several approaches have been proposed (e.g., Becker, 1992, 1995; Gleser & Olkin, 1994; Hedges & Olkin, 1985, chapter 10; Kalaian & Raudenbush; 1996; Raudenbush, Becker, & Kalaian, 1988; Timm, 1999); all of these require that the reviewer incorporate information about the degree of correlation among outcomes into the analysis. To implement this approach the reviewer creates a separate data record for each outcome, but for analysis the records from each study are linked through a within-study covariance matrix among the outcomes.

Including separate records (and indices) for all reported outcomes enables the reviewer to ask whether effects are comparable across multiple outcomes (e.g., across measured constructs or time points), or whether potential explanatory variables (such as study features) relate differently to different outcomes. A fully multivariate approach should provide justifiable tests of significance for more complex questions than can be addressed using the ad hoc or univariate approaches described above, and more accurate probability statements for all tests conducted.

In theory multivariate approaches are applicable in any meta-analysis of multiple outcomes, but often the reviewer is faced with the practical problem of finding information about correlations among the outcomes. For instance, Becker (1990) found that 12 of 23 studies of coaching for the SAT had examined both SAT-Math and SAT-Verbal scores, but none reported the intercorrelation of math and verbal scores. Kim (1998) found relevant correlation values in four of nine multiple-outcome studies comparing paper-and-pencil versus computer-based testing, but Chiu (1997) found that fewer than 10% of multivariate studies of the impact of metacognitive strategy instruction in reading reported the necessary correlations.

In some cases reasonable correlational information may be available from study authors, test manuals, or test publishers, but published information may not be applicable to unusual or selective samples of subjects. When experimenter-made instruments are used, this information may not be available at all. When correlations are not available, various imputation schemes (Little & Rubin, 1987; Piggott, 1994) provide a way to insert estimates for missing correlations, or sensitivity analyses may be used to examine a range of correlation values (e.g., Chiu, 1997).

A final practical problem is that, to date, no standard commercial software package routinely produces analyses for multivariate meta-analysis data, though such analyses can be obtained using, for example, the matrix functions of SAS or SPSS. Also, the HLM program (Bryk, Raudenbush, & Congdon, 1996) for hierarchical linear modeling can be used for some analyses, with the stipulation that it be used for the case where each data point has known variance (the "V known" case). Kalaian (1989) has shown how to transform multivariate meta-analytic data for analysis using the standard implementation of HLM.

## IV. SPECIFIC DISTRIBUTIONAL RESULTS FOR COMMON OUTCOMES

In this section I give notation and distributional results for study outcomes expressed as effect sizes and Pearson product-moment correlations. The presentation on effect size is based directly on the work of Gleser and Olkin (1994). Methods for multivariate synthesis of other effect indices are less well developed, though exceptions include work on measures of variation by Raudenbush (1988) and Raudenbush and Bryk (1987). Multivariate methods for proportions (or log odds) or other forms of categorical data are not yet available.

### A. Effect Sizes

The effect size, or standardized mean difference, is probably the most common index of study results in social-science research syntheses. Though the term *effect size* has a more general use in mathematical statistics, Glass (1976) used it to refer to the difference between the means of two independent groups, divided by some measure of variation, typically a standard deviation. For simplicity, we consider a case where level of performance for a treated group (T) is compared to that of an independent comparison group (C). (The effect size for matched groups is considered by Dunlap, Cortina, Vaslow, & Burke, 1996). Probably the most common sources of multivariate effect-size data are study designs in which several distinct variables are measured for both groups, or one variable is measured repeatedly for the treatment and comparison groups.

### 1. Several Outcome Variables

Consider a series of $k$ studies, where the $i$th study examines scores $Y_{ijs}^T$, for subjects $s = 1$ to $n_i^T$, from a treated population, and $Y_{ijs}^C$, for $s = 1$ to $n_i^C$ from a comparison population, on the outcome variable $Y_j$. For simplicity we assume that all outcomes in study $i$ are measured for all subjects, and that individual scores on $Y_j$ are independent within and between populations. The $j$th effect size from study $i$ is

$$g_{ij} = \frac{\overline{Y}_{ij}^T - \overline{Y}_{ij}^C}{S_{ij}}, \tag{1}$$

where $\overline{Y}_{ij}^T$ and $\overline{Y}_{ij}^C$ are the sample means on $Y_j$ for treated and comparison subjects, respectively, and $S_{ij}^2$ is an estimate of the common variance $\sigma_{Yij}^2$ within study $i$. The effect size in the sample is an estimate of the population effect size $\delta_{ij} = (\mu_{ij}^T - \mu_{ij}^C)/\sigma_{Yij}$, where $\mu_{ij}^T$ and $\mu_{ij}^C$ are the population means of the treatment and comparison populations of study $i$, respectively, on $Y_j$.

The estimator $g_{ij}$ is biased, but Hedges's (1981) small-sample bias correction shrinks $g_{ij}$ to $d_{ij} = c(m_i) \, g_{ij}$, where $c(m) = 1 - 3/(4m - 1)$, and

$m$ is the degrees of freedom for the two-sample $t$-test, that is, $m_i = n_i^T + n_i^C - 2$. Hedges also showed that if the scores in study $i$ are normally distributed with possibly different means $\mu_{ij}^T$ and $\mu_{ij}^C$ and common variance $\sigma_{Yij}^2$ within each population, specifically, if $Y_{ijs}^T \sim N(\mu_{ij}^T, \sigma_{Yij}^2)$, and $Y_{ijs}^C \sim N(\mu_{ij}^C, \sigma_{Yij}^2)$, and if certain regularity conditions hold, then $d_{ij}$ has a large-sample normal distribution with mean $\delta_{ij}$ and variance

$$\sigma_{ij}^2(\delta_{ij}) = \text{Var}(d_{ij}) = \frac{(n_i^T + n_i^C)}{n_i^T n_i^C} + \frac{\delta_{ij}^2}{2(n_i^T + n_i^C)}. \tag{2}$$

If study $i$ has $p$ outcomes, then we may compute $p$ effect sizes $d_{i1}$ through $d_{ip}$ and $p$ variances for those effects. Gleser and Olkin (1994, formula 22–22, p. 349) showed that if $d_{ij}$ and $d_{ij*}$ are effect sizes for outcomes $j$ and $j^*$, and the within-study covariance between $Y_j$ and $Y_{j*}$ is the same for treated and comparison subjects (implying a common correlation $\rho_{ijj*}$ between the variables), then the covariance of $d_{ij}$ with $d_{ij*}$ is

$$\text{Cov}(d_{ij}, d_{ij*}) = \frac{(n_i^T + n_i^C)\rho_{ijj*}}{n_i^T n_i^C} + \frac{\delta_{ij}\delta_{ij*}\rho_{ijj*}^2}{2(n_i^T + n_i^C)}. \tag{3}$$

The large sample variances in (2) and covariances in (3) are estimated by substituting sample estimates of $\delta_{ij}$, $\delta_{ijj*}$, and $\rho_{ijj*}$ into (2) and (3) above.

As mentioned above, some or all estimates of $\rho_{ijj*}$ may not be reported, but values are needed in order to compute the covariances in (3). Various imputation schemes (often, substituting mean correlations from other studies) or sensitivity-analysis approaches may be used to obtain reasonable covariance values. If all $p$ variables are observed in study $i$, there are $p(p - 1)/2$ unique covariances among the effect sizes, and the covariance matrix has dimension $p \times p$. Gleser and Olkin (1994) provide other formulas for the effect sizes and their variance-covariance matrix (see, e.g., their formulas 22–23 through 22–26) for cases where the variance-covariance matrices cannot safely be assumed homogeneous within studies.

In practice, estimates of the variances and covariances can be computed by hand (though this is extremely tedious if many values are needed), by entering formulas into a spreadsheet program that supports matrix operations (e.g., PlanPerfect, Excel), or via the matrix language of a more sophisticated statistical package. I use PROC IML, the matrix language of the Statistical Analysis System (SAS) package. SPSS also supports matrix operations and can be used for both computation of variance–covariance matrices and further analyses.

## 2. Multiple Timepoints, Same Outcome Variables

The correlational structure of data from designs with $p$ measurements of the same construct is identical to that described above for $p$ correlated variables. A conceptual difference is that variable $Y_{ij}$ in study $i$ represents a measure taken at occasion $j$, rather than a measure of construct $j$. Similarly effect sizes $g_{ij}$ and $d_{ij}$ represent effects at occasion $j$, for $j = 1$ to $p$ occasions, and correlation $\rho_{ijj*}$ is the correlation between measures taken on occasions $j$ and $j*$ in study $i$. With this change in notation, formulas (1) through (3) can be applied in the new context.

A key consideration in this situation, however, is that outcomes from different studies may not have the same interpretation if the timing of the occasions of measurement differs across studies. For instance, if $p = 3$ effect sizes are recorded for each construct, as pretest ($j = 1$), posttest ($j = 2$), and follow-up ($j = 3$) measures, interpretations of $d_{i2}$ and $d_{i3}$ may depend on the length of treatment (the time between occasions 1 and 2) and the timing of the follow-up (e.g., was measure 3 taken a week, a month, or a year after the treatment concluded). Careful coding and analysis of the timing of measurements must be done to address this issue.

### 3. Multiple Treatment or Control Groups

Data from designs that use multiple treatment and control groups also can lead to correlated effect-size data (see Gleser & Olkin, 1994, pp. 341–346). One example is when two or more treatment means are compared with one control mean. Effect sizes for such contrasts share two terms, in that the same control group mean is subtracted in both numerators, and both may have the same denominator—the pooled standard deviation. These common terms lead to dependence between the effect sizes; the appropriate covariance is given by Gleser and Olkin (1994, p. 346).

### 4. Effect-Size Data: SAT Coaching

A number of studies have examined the effectiveness of coaching on the math and verbal subtests of the Scholastic Aptitude Test or SAT, the familiar college entrance examination published and administered by the Educational Testing Service. Over the years commercial coaching programs aimed at helping students to improve their SAT performance have been controversial (see, e.g., Federal Trade Commission, 1978), but they continue to offer their services to a wide cross-section of examinees. The data set includes seven studies, and was also used as an example by Gleser and Olkin (1994).

Table 17.1 shows unbiased effect sizes (standardized mean differences) for the seven comparisons of coached and control subjects on math and verbal SAT posttests, and their variances and covariances. Effect sizes were computed by subtracting each control-group mean from that of the coached group, and dividing by the pooled within-group standard deviation, then correcting for small sample bias using Hedges's (1981) correction. Covariances were computed using formula (3) with a math–verbal correlation of .66, reported in SAT technical manuals.

**TABLE 17.1   Results of Seven Studies of Scholastic Aptitude Test Coaching**[a]

| Sample sizes | | Effect sizes | | Duration | Variances and covariances among effects | | |
|---|---|---|---|---|---|---|---|
| $n^C$ | $n^T$ | $d_{math}$ | $d_{verbal}$ | (hours) | $Var(d_{math})$ | $Var(d_{verbal})$ | $Cov(d_{math}, d_{verbal})$ |
| 34 | 21 | 1.189 | 0.608 | 27 | 0.090 | 0.080 | 0.055 |
| 17 | 16 | 0.652 | −0.148 | 24 | 0.128 | 0.122 | 0.079 |
| 52 | 52 | −0.065 | 0.124 | 10 | 0.038 | 0.039 | 0.025 |
| 14 | 13 | −0.078 | 0.403 | 8.9 | 0.148 | 0.151 | 0.098 |
| 47 | 93 | 0.373 | −0.246 | 63 | 0.033 | 0.032 | 0.021 |
| 45 | 45 | 0.186 | 0.137 | 30 | 0.045 | 0.045 | 0.029 |
| 8 | 8 | −0.951 | 0.610 | 11 | 0.278 | 0.262 | 0.153 |

[a] The size of the uncoached (control) group is $n^C$ and the size of the coached (treatment) group is $n^T$. Positive effect sizes indicate coached students outscored controls.

## B. Correlations

Consider a series of $k$ studies, examining relations among a set of $q - 1$ predictors $Y_1, \ldots, Y_{q-1}$ and some outcome $Y_q$, using reasonable measures of those constructs. Assume that all $p = q(q - 1)/2$ correlations among the $q$ variables are reported and are based on complete data from the $n_i$ cases in study $i$. Missing correlations cause difficulties because the covariance formula between two $r$s (in 5 below) requires all other correlations that share subscripts with the two covarying correlations. In practice I have substituted average values (based on correlations reported in other studies) for missing correlations, checking whether the substituted values are consistent with the data reported in each study.

If $Y_{i1}, \ldots, Y_{iq}$ have the multivariate normal distribution, then let $r_{ist}$ and $\rho_{ist}$ be the sample and population correlations from study $i$ between $Y_s$ and $Y_t$. Olkin and Siotani (1976) showed that in large samples, $r_{ist}$ is approximately normally distributed with mean $\rho_{ist}$ and variance

$$\sigma_{ist} = Var(r_{ist}) \approx (1 - \rho_{ist}^2)^2/n_i, \tag{4}$$

and covariance $\sigma_{ist,uv} = Cov(r_{ist}, r_{iuv})$. A formula for $\sigma_{ist,uv}$ is given by Olkin and Siotani (1976, p. 238). The covariance can most easily be expressed by noting that if $r_{ist}$ is the correlation between the $s$th and $t$th variables in study $i$; $r_{iuv}$ is the correlation between the $u$th and $v$th variables, and $\rho_{ist}$ and $\rho_{iuv}$ are the corresponding population values, then

$$Cov(r_{ist}, r_{iuv}) = [0.5\rho_{ist}\rho_{iuv}(\rho_{isu}^2 + \rho_{isv}^2 + \rho_{itu}^2 + \rho_{itv}^2) + \rho_{isu}\rho_{itv} + \rho_{isv}\rho_{itu}$$
$$- (\rho_{ist}\rho_{isu}\rho_{isv} + \rho_{its}\rho_{itu}\rho_{itv} + \rho_{ius}\rho_{iut}\rho_{iuv} + \rho_{ivs}\rho_{ivt}\rho_{ivu})]/n_i. \tag{5}$$

In large samples the variances and covariances may be estimated by substituting corresponding sample estimates for the parameters in (4) and (5). However, simulations of mean correlations weighted using these variance-covariance estimators for the case of $q = 3$ variables (Becker & Fahrbach, 1994) have shown severe overestimation of means with moderate samples within studies ($n_i < 100$).

An alternative approach to analyzing correlations is to first transform them using Fisher's (1921) $Z$ transformation, $Z = 0.5 \log (1 + r)/(1 - r)$. Analyses are completed using the transformed values, then estimates corresponding to mean correlations and endpoints of interval estimates are returned to the correlation scale via the inverse transformation $r = [\exp(2Z) - 1]/[\exp(2Z) + 1]$. Variances and standard errors in the $Z$ scale cannot be returned to the correlation scale accurately.

Some researchers (e.g., James, Demaree, & Mulaik, 1986) strongly recommend the use of Fisher's transformation, whereas Thomas (1989, p. 37) argues that application of the transformation to $r = (\rho + e)$ is not justified because the variance of $r$ cannot be related to the variances of the population correlation $\rho$ and the error $e$. One key justification for using the $Z$ transformation is that it removes the factor involving $\rho$ from the variance of each correlation. Because the stabilized variance of $Z$ is approximately $1/(n - 3)$ in large samples, the weights used in averages and homogeneity tests do not depend on sample data. Also the transformation normalizes the (skewed) distribution of the correlation. Under this approach, the data for analysis would be the transformed correlations, with variances defined as

$$\mathrm{Var}(z_{ist}) \approx 1/(n_i - 3), \tag{6}$$

for all $p$ correlations in study $i$. Covariances among $z$s are more complicated, and are computed as the covariance in (5) divided by an added product term

$$\mathrm{Cov}(z_{ist}, z_{iuv}) = \sigma_{ist,uv}/[(1 - \rho_{ist}^2)(1 - \rho_{iuv}^2)]. \tag{7}$$

Simulations (Becker & Fahrbach, 1994) have shown that though this covariance formula is more complex than (5), mean correlations based on transformed correlations have minimal bias (less than .03 in the correlation metric) and good levels of precision for within-study sample sizes as small as 20 and $q = 3$ variables. It remains to be seen if this advantage holds for more than three variables (when the number of covariances can greatly exceed the number of variances), but our research to date supports the use of Fisher's $Z$ transformation, rather than $r$, in meta-analysis.

### 1. Correlational Data: Child Outcomes of Divorce

Whiteside and Becker (2000) examine the literature on social, emotional, and cognitive factors contributing to outcomes for young children in divorcing families. They consider two key variables in a synthesis of twelve studies of parental characteristics and co-parenting variables: the quality of the father's relationship with the child and the frequency of father visitation. Father visitation is one of the few aspects of the child's postdivorce life that is accessible to modification by the courts.

Whiteside's conceptual model of major factors in children's postdivorce adjustment guided the synthesis, and led to consideration of 17 variables investigated in between one and nine studies each. In this illustration I focus on six predictive factors (i.e., father–child relationship quality, frequency of father visitation, preseparation father involvement, maternal warmth, maternal restrictiveness, and coparent cooperation) and one outcome (internalizing symptoms, including anxiety, nervousness, and similar behaviors). These seven variables were examined in 15 samples of children younger than 6 years of age from divorcing families. A total of 96 sample correlations are used in this analysis; they are not all reported here. Full information on this synthesis is given by Whiteside and Becker (2000).

Arditti and Keith (1993) reported six correlations, among father–child relationship quality (relation), frequency, cooperation, and involvement, for 66 cases. Because warmth and internalizing symptoms were not measured, the data for this study are incomplete. The correlation matrix for this study is given in Table 17.2.

I illustrate computation of the variances and the covariance for the correlations of frequency with relation ($r_{FR} = .561$) and frequency with cooperation ($r_{FC} = .393$). For analysis each correlation is transformed to the Fisher $Z$ metric. Because the sample size is $N = 66$, the variances of the $Z$ values are all equal to $1/(66 - 3) = .0159$. To obtain the covariance between the $z$s for these correlations, I use formula (5) with the variable index labels $s = F$ for frequency, $t = R$ for relation, $u = F$ for frequency, and $v = C$ for cooperation. Frequency appears twice, as index $s$ and index $u$, as both correlations involve scores on frequency of father visitation. Correlations involving the other variable (involvement) are not used in

**TABLE 17.2  Correlation Matrix of Results from Arditti and Keith (1993)**

|  | Frequency | Relation | Cooperation | Involvement |
|---|---|---|---|---|
| Frequency | 1.000 | 0.561 | 0.393 | 0.561 |
| Relation | 0.561 | 1.000 | 0.286 | 0.549 |
| Cooperation | 0.393 | 0.286 | 1.000 | 0.286 |
| Involvement | 0.561 | 0.549 | 0.286 | 1.000 |

this computation. Noting that $r_{FF} = 1$ and $r_{CR} = .286$, the covariance of the two correlations $r_{FR}$ and $r_{FC}$ is

$$
\begin{aligned}
\text{Cov}(r_{RF}, r_{FC}) = &[0.5 r_{FR} r_{FC}(r_{FF}^2 + r_{FC}^2 + r_{FR}^2 + r_{RC}^2) + r_{RC} + r_{FC} r_{FR} \\
&- (r_{FR} r_{FC} + r_{FR} r_{FR} r_{RC} + r_{FR} r_{FC} + r_{FC} r_{CR} r_{CF})]/n \\
= &[0.5(.561).393(1 + .393^2 + .561^2 + .286^2) + .286 + .393*.561 \\
&- (.561*.393 + .561^2*.286 + .561*.393 + .393^2*.286))]/66 \\
= &.00154.
\end{aligned}
$$

The covariance between the two Fisher's $Z$ values for these correlations is

$$
\begin{aligned}
\text{Cov}(Z_{FR}, Z_{FC}) &= \text{Cov}(r_{FR}, r_{FC})/[(1 - r_{FR}^2)(1 - r_{FC}^2)] \\
&= .00116/(1 - .561^2)(1 - .393^2) = .0027.
\end{aligned}
$$

Covariances among the other pairs of correlations were computed in a similar fashion. Thus the data file contains six lines of data for this study, including a $6 \times 6$ covariance matrix, with the variances (diagonal elements) all equal to $1/(n - 3) = .0159$ and covariances computed as above.

## V. APPROACHES TO MULTIVARIATE ANALYSIS

Most meta-analyses begin with the estimation of a set of effects from the series of studies. Whether the effects are viewed as *common* or *average* effects depends on whether the reviewer adopts a fixed-, random-, or mixed-effects model for the data. These models are described below. Choice of a model depends on the assumptions the reviewer is willing to make, and can be made on theoretical grounds before the analysis, or on the basis of empirical data about the plausibility of these models. With an empirical approach, the significance of the predictors and the extent to which predictors account for between-studies differences lead to the choice of a fixed-, mixed-, or random-effects model.

Often the best model is a mixed model—a combination of one or several explanatory variables (the fixed effects) plus an index of remaining "unexplained" variation or uncertainty in the results. A key activity in any meta-analysis is the exploration of variables that may potentially explain differences among the study results, and the type of analysis used depends on the nature of the potential predictors. Methods analogous to analysis of variance can be used to study categorical predictors, and weighted regression methods are used to study continuous predictors or combinations of categorical (dummy) and continuous predictors. The general linear modeling approach discussed here subsumes both types of analysis.

### A. Fixed-Effects Models

The simplest fixed-effects model posits that all observed variation in the study results is due to within-study sampling error. The hypothesis of homogeneity (consistency) of effects concerns differences across studies, thus it does not consider differences between outcomes *within* studies. Adopting this model in the multivariate case is equivalent to assuming that a single set of $p$ population effects underlies every study in the review. This model typically does not hold, because studies vary in a variety of interesting ways. However, it is easy to examine this simple model using a homogeneity test (for more detail see Becker, 1992; Gleser & Olkin, 1994; or Hedges & Olkin, 1985), and frequently this is the first statistical test done in a meta-analysis.

The homogeneity test in meta-analysis is a goodness-of-fit test based on variation in the effects across studies. The homogeneity test statistic (which is a chi-square statistic under the hypothesis of homogeneity) is essentially a weighted variance, which increases when the variation among study results is larger than the dispersion expected due to sampling error alone. In a univariate homogeneity test, each study result is weighted by the inverse of its sampling variance, and for multivariate data each study's set of effects is weighted by its variance–covariance matrix.

When results appear consistent across studies, we may want to test hypotheses about the $p$ effects. We can test the simple omnibus hypothesis that all effects are zero (e.g., coaching has no effect on either math or verbal SAT scores), or we can examine hypotheses concerning individual effects or differences between effects for the $p$ different outcomes. By constructing contrasts of interest among the estimated effects (and using their estimated variances), we can test a wide variety of hypotheses concerning effects in the synthesis, while modeling dependence.

For example, if $\delta_1$ represents the common effect for the SAT math outcome, and $\delta_2$ is the effect for SAT verbal, to test the equality of the math and verbal effects, we would test $H_0$: $\delta_1 = \delta_2$ or equivalently $H_0$: $\delta_1 - \delta_2 = 0$. (This hypothesisis is tested below.) We compute the contrast $C = \hat{\delta}_1 - \hat{\delta}_2$ and its estimated variance $\text{Var}(C) = \text{Var}(\hat{\delta}_1) + \text{Var}(\hat{\delta}_2) - 2\,\text{Cov}(\hat{\delta}_1, \hat{\delta}_2)$. Dependence among the two effect values is accounted for via the covariance term in $\text{Var}(C)$. The test statistic is computed as $z = C/\sqrt{\text{Var}(C)}$, which under the null hypothesis has the standard normal distribution. The test is conducted by comparing the value of $z$ to an appropriate $\alpha$-level critical value from the standard normal table (e.g., for a two-sided test at $\alpha = .05$, use the critical value $z_c = \pm 1.96$). Similar tests can be conducted under random-effects and mixed model assumptions.

## B. Fixed Effects with Predictors

When results do not appear consistent across studies, a more complex fixed-effects model may still apply. If all between-studies differences can be explained by a set of predictor variables, we still have a fixed-effects model.

When population effects may relate to some explanatory variable or variables (say, $X_1$, representing publication date) we can explore that relation using regression methods, incorporating $X_1$ into a model such as $\delta_{ij} = \beta_0 + \beta_1 X_{1i}$, for outcome $j$ in study $i$. The regression model here is distinctive in that it shows no error term—the parameter $\delta_{ij}$ is predicted perfectly, once we know the value of the predictor $X_1$. Any error (say $e_{ij}$) is still conceived to be due to sampling; that is, it reflects only the degree to which each $d_{ij}$ deviates from $\delta_{ij}$. For the sample effect size the model would be $d_{ij} = \beta_0 + \beta_1 X_{1i} + e_{ij}$.

Furthermore, as noted above, outcome-specific $X$s can be used as predictors for different outcome parameters. If $X_{2j}$ represents duration of treatment aimed at outcome $j$ then we can model outcome $j$ in study $i$ as $\delta_{ij} = \beta_0 + \beta_{2j} X_{2ij}$. Separate intercept terms for each outcome can also be used, leading to the models $\delta_{ij} = \beta_{0j} + \beta_{1j} X_{1ij}$, for outcomes $j = 1$ to $p$ in study $i$. If we fit the model with a separate intercept and slope for all $p$ outcomes, we would estimate $R = 2p$ regression coefficients.

Regression analyses in multivariate meta-analysis are typically completed using generalized least squares (GLS) regression estimation (again, see Becker, 1992; Gleser & Olkin, 1994; Hedges & Olkin, 1985, chapter 10; Kalaian & Raudenbush; 1996; Raudenbush, Becker, & Kalaian, 1988; Timm, 1999). Regressions in univariate meta-analysis use weighted least squares (WLS), with weights based on variances. GLS slope estimates are weighted using the inverse variance–covariance matrix of the data points. Raudenbush, et al. (1988) give an extended GLS example for effect-size data.

Once a model has been estimated, regression diagnostics familiar from primary research can be used to examine its validity. Residuals give evidence of poorly fitting effects (large residuals or standardized residuals), and evidence of multicollinearity can be assessed by looking at correlations among the estimated slopes, and at the standard errors of the slopes, comparing the latter across models when possible. Also, as is true for regression analyses in univariate meta-analysis, a chi-square test of specification error for the model is available. That is, we can ask whether each regression model has explained all between-studies variation in the data beyond what would be expected due to chance or sampling error.

Estimated regression coefficients can be used to test hypotheses about the importance of individual predictors or sets of predictors. The importance of specific individual predictors can be tested (via $z$ tests for each slope), and the omnibus hypothesis that all intercepts and slopes equal zero can be tested using a chi-square test. Consider the models $\delta_{ij} = \beta_{0j} + \beta_{1j} X_{1ij}$, with $j = 1$ for SAT-math effects and $j = 2$ for SAT-verbal effects. The omnibus test asks whether $\beta_{01} = \beta_{02} = \beta_{11} = \beta_{12} = 0$, and the test of the effect of duration of coaching on SAT-math would examine $H_0: \beta_{11} = 0$. The importance of added sets of variables can be examined, if additional models are estimated that contain all the predictors already modeled, plus others.

## C. Random-Effects Model

When the overall hypothesis of homogeneity is rejected or if the results of the series of studies are expected to vary because of a multitude of factors, a reviewer may wish to adopt the random-effects model. The random-effects model assumes that studies are sampled from a population of studies, and that the population effects for each outcome may truly differ across studies. In the most extreme case, each study is viewed as having its own unique set of population effects. Each study can show a different *true* effect for each of its $p$ outcomes.

Under the random-effects model, uncertainties due to differences in population effects are incorporated into the variance of each data point and thus into the variances of overall (average) effects. The specific form of the estimator of population variance may differ according to the type of index being summarized. Also in the multivariate case, population covariation may be added to within-study covariances, though when the number of studies is small, covariance estimates may be unstable and are sometimes ignored.

Under random effects, effect estimates across studies represent average, not common, values. Neither the average effects nor the variances of the effects are assumed to be equal across the $p$ outcomes. One limitation of this model is that it assumes that population effects *do not* relate to specific study features such as type of sample, age of students, or treatment implementation. Such relations are neither modeled nor tested, though in reality they may exist.

## D. Mixed Model

Typically the real state of affairs is somewhere between the fixed-effects and the random-effects models—in other words, a mixed model is likely best for most data. Mixed models apply when some explanatory variables relate to the study results, but do not explain all between-studies variation in the data. With a mixed model we explore potential relations between population effects and explanatory variables using regression methods; however, the mixed model with one predictor is written as $\delta_{ij} = \beta_0 + \beta_1 X_{1i} + u_{ij}$, for outcome $j$ in study $i$. The residual component $u_{ij}$ represents between-studies variation in the population effects that is not explained by the predictor(s) in the model.

Regression analyses for the mixed model require methods more complex than the GLS methods described for the fixed-effects case above. Only Kalaian (1994) and Kalaian and Raudenbush (1996) present a mixed model for multivariate data, and for brevity details of multivariate mixed-model estimation are not presented here.

## E. Further Analyses for Correlations

In examinations of effect sizes and proportions, the final goal is typically to estimate and examine variation in effects across studies. With correlations, however, further analyses under any of the three models described above allow us to model complex interrelations among variables in the correlation matrix (Becker, 1992, 1995; Becker & Schram, 1994). The reviewer can specify relations of sets of predictors with intermediate and final outcomes—creating a predictive model that may not have been examined in any single study in the synthesis. The common or average correlation matrix from the studies is used to estimate a set of linear models that specify both direct and indirect relations of predictor variables with outcomes of interest. These methods produce estimated standardized regression equations, and corresponding standard errors for the slopes in those equations, which enable the reviewer to test for partial relations that are not captured by zero-order correlations.

Details of multivariate meta-analysis estimation and tests for correlations are described more fully in papers by Becker (1992, 1995) and Becker and Schram (1994). In short, the procedures are based on generalized least squares (GLS) estimation methods that incorporate information on variation and covariation among the correlations to obtain weighted average correlations and estimates of their uncertainty. In the case of random-effects modeling, between-studies variation must be estimated and incorporated; this can sometimes be accomplished using software for hierarchical data analysis such as HLM (Bryk, Raudenbush, & Congdon, 1996).

## VI. EXAMPLES OF ANALYSIS

Two examples illustrate the kinds of questions that can be addressed in multivariate meta-analyses. One illustration examines data on SAT coaching; this literature has been reviewed by several authors using different datasets and synthesis approaches (e.g., Becker, 1988b; DerSimonian & Laird, 1983) . The first data set includes effects from seven studies, as well as two predictors: study date and the duration of coaching. The second example features data drawn from a synthesis of the relations of various predictors to outcomes for very young children in divorcing families (Whiteside & Becker, 2000).

### A. Scholastic Aptitude Test Coaching

We first estimate the average effects of coaching on math and verbal outcomes under the fixed-effects model and test the homogeneity of results across the seven studies. These analyses follow procedures outlined by Raudenbush et al. (1988). The means of the math and verbal effects weighted by the inverse variance–covariance matrix are 0.26 and 0.08, where the first and larger effect is the math effect. Model 1 in Table 17.3 shows these results.

We ask whether these effects can be viewed as common effects by computing the homogeneity test statistic ($Q_E$). For these 14 outcomes the test value is $Q_E = 46.56$, which is compared to the chi-square distribution with 12 degrees of freedom. The value is highly significant, so all studies do not appear to come from the same population. We conclude that the mean values do not represent common effects from all studies. Some studies may show coaching effects greater than these means, and others may have effects smaller than these means.

Regression models were estimated to explore patterns of variation among the outcomes. In model 2, duration of coaching is modeled as a single potential predictor, and separate intercepts are modeled for math and verbal effects. Model 3 also has two intercepts, and has separate slopes for the relation of duration to the math and to the verbal outcomes, in case the effect of duration is different for the two outcomes. Table 17.3 shows the slopes, their standard errors (in parentheses), and fit tests for Model 1 (with only separate means or intercepts), and for three models including duration.

Adding duration as a predictor common to both effects did not significantly change model fit ($Q_E$) compared to Model 1, and the coefficient for duration in Model 2 has a standard error four times its slope. This alone suggests that duration does not have much value as a single predictor. In addition, adding duration has increased the size of the standard errors of the coefficients for math and verbal intercepts (versus those for Model 1) by over 70%, an indication of potential multicollinearity. Another index of multicollinearity is the matrix of correlations among the coefficients, all of

**TABLE 17.3    Coefficients, Standard Errors (in parentheses), and Fit Tests for Scholastic Aptitude Test Coaching Studies[a]**

| | Intercepts | | Predictors | | | Test statistics | |
|---|---|---|---|---|---|---|---|
| Model | Math | Verbal | Duration | Duration for math | Duration for verbal | $Q_E$ ($df_E$) | $Q_B$ ($df_B$) |
| 1 | 0.26 (.096) | 0.08 (.095) | | | | 46.56 ($df = 12$) | 8.73 ($df = 2$) |
| 2 | 0.30 (.164) | 0.12 (.164) | −0.001 (.004) | | | 46.49 ($df = 11$) | 8.80 ($df = 3$) |
| 3 | 0.04 (.178) | 0.37 (.176) | | 0.007 (.005) | −0.009 (.005) | 29.26 ($df = 10$) | 26.02 ($df = 4$) |
| 4 | −0.15 (.148) | 0.08 (.095) | | 0.013 (.003) | | 33.14 ($df = 11$) | 22.15 ($df = 3$) |

[a] The standard error of each coefficient is shown in parentheses.

which are larger than .80 for this model. The correlation, for instance, between the slopes for the math intercept ($\beta_1$) and duration ($\beta_3$) is $-.00055/\sqrt{(.0270 \times .000017)} = -.81$. These correlations are obtained from the matrix of variances and covariances among the slopes, which is

$$\text{Var}(\hat{\beta}) = \begin{bmatrix} .0270 & .0238 & -.00055 \\ .0238 & .0270 & -.00056 \\ -.0006 & -.0006 & .00002 \end{bmatrix}.$$

The standard errors for the coefficients of Model 2 in Table 17.3 are the square roots of the diagonal elements of this matrix.

To see if the role of duration is better described by using a separate slope for its relation to math and verbal effects, we fit Model 3, where $\delta_{ij} = \beta_{0j} + \beta_{1j}*\text{Duration}_{ij}$ for studies $i = 1$ to 7 and outcomes $j = 1$ (math) and $j = 2$ (verbal). Its predictor matrix includes two columns of zeros and ones for the intercepts and two columns containing values of the duration variable. We estimate four coefficients $\beta_{01}$, $\beta_{02}$, $\beta_{11}$, and $\beta_{12}$; the estimates and their standard errors are given on the third line of Table 17.3. Adding separate slopes for duration has not led to significant test statistics for any individual coefficients, but the overall fit value ($Q_E$) has dropped dramatically. Standard errors for the intercepts are quite large, and evidence of multicollinearity appears in the large correlations between each outcome's intercept and its slope for duration—both are above .80. The model significance test ($Q_B$) is large, suggesting that separate slopes account for some variation, but the model shows considerable multicollinearity.

A last fixed-effects model (Model 4) includes an intercept for each outcome, but includes the duration predictor for only the math outcome. Model 4 shows a relatively small fit-test value ($Q_E$) and a significant slope for math duration (with a smaller standard error as well). The slope suggests that 10 hours of coaching increases the SAT-math scores of coached students by slightly over one-tenth of a standard deviation ($0.013 \times 10 = .13$). However, the math intercept (the predicted effect for 0 hours of coaching) equals $-0.15$, so that at least 12 hours of coaching are needed to provide a gain on SAT-math. The SAT math mean increases by 0.013 for each hour of coaching, thus 12 hours will produce a positive estimated effect, as $12 \times .013 = .156$. (In this data set studies 3 and 4, offering the shortest coaching programs, showed negative coaching effects on SAT-math.) None of the correlations among the model's coefficients are larger than .75, which is high but does not indicate extreme multicollinearity. This is the best of the fixed-effects models, but though it explains a good deal of the variation in coaching effect sizes, even this model does not totally explain the differences in coaching effects, as $Q_E = 31.58$ exceeds 19.68, the $\alpha = .05$ critical value of the chi-square distribution with 11 degrees of freedom.

We next explore a random-effects model for the coaching data. Variance components for the math and verbal effect sizes, estimated using a method-of-moments estimator (Raudenbush, 1994, p. 312), are 0.338 and 0.013, respectively. That is, if the distribution of true SAT-math coaching effects had a variance of 0.338 (and SD = 0.58 = $\sqrt{0.338}$), then 95% of the SAT-math population effect sizes would lie in a range wider than 2 units ($\pm 1.96 \times 0.58 = \pm 1.13$ or 2.26 units wide)—a very broad range of effect-size values. By contrast, 95% of the distribution of SAT-verbal population effects would lie in a narrower interval, less than half of a standard deviation wide ($\pm 1.96 \times 0.11 = \pm .22$ or an interval of width .44).

When these variance components are added to the within-study variances for the coaching effects, random-effects estimates of the math and verbal effect-size means are 0.16 (SE = .25) and 0.11 (SE = .11) standard-deviation units, respectively. These means are fairly similar to the fixed-effects means (from Model 1 in Table 17.3), but both standard errors are larger under the random-effects model. Neither mean differs significantly from zero in the random-effects case, and the math and verbal effects also do not differ from one another. The test statistic for this last comparison of means, computed using components from the random-effects estimate of the variance of the means, is $z = (0.16 - 0.11)/\sqrt{0.061 + 0.011 - 2*(-0.0026)} = 0.18$. Under the random-effects model, the effects of coaching appear negligible.

## B. Postdivorce Adjustment of Children

The first step in this analysis was to compute mean correlations across the samples under the fixed-effects model, and to assess homogeneity of the 21 relations among the seven variables under study (see Table 17.4). Two omnibus tests were performed. The first test addressed the general question of whether any of the sets of correlations show more variation than would be expected due to sampling alone. This test was carried out by computing the chi-square test statistic $Q_E$ from the 96 correlations and their variance–covariance matrix (see Becker & Schram, 1994, p. 369). Here $Q_E$ has $96 - 21 = 75$ degrees of freedom, and the sample statistic $Q_E = 156.13$ is significant at the .05 level.

The test that all 21 population correlations equal zero uses the chi-square statistic $Q_B$, here with 21 degrees of freedom. This test value $Q_B = 190.08$ is significant, indicating that nonzero values are present in the matrix. Eight relations (with correlations shown in bold above the diagonal in Table 17.4) are significant when individual $z$ tests on the means are computed. (The $21 \times 21$ variance–covariance matrix among the means is not reported here, but was used to compute these tests.) However, the fixed-effects model is not totally appropriate for our data because the test of homogeneity suggested that some relations show unexplained between-studies variation. It would be better to adopt the random-effects model to

**TABLE 17.4   Weighted Average Correlations under Fixed-Effects Model (above Diagonal) and Random-Effects Model (below Diagonal)**[a]

| | Involvement | Frequency | Relation | Cooperation | Warmth | Restrictive | Internalizing |
|---|---|---|---|---|---|---|---|
| Involvement | 1.000 | **0.259*** | **0.274** | −.089* | 0.052 | −.000 | −.082 |
| | • | 3 | 3 | 3 | 2 | 2 | 2 |
| Frequency | 0.320* | 1.000 | **0.411*** | **0.235** | −.102 | 0.021 | −.001 |
| | 3 | • | 6 | 5 | 2 | 2 | 7 |
| Relation | **0.338** | **0.345*** | 1.000 | **0.213** | **0.166** | 0.087 | **−.161** |
| | 3 | 6 | • | 7 | 5 | 4 | 10 |
| Cooperation | −.081* | **0.226** | **0.223** | 1.000 | −.049 | 0.064 | −.113 |
| | 3 | 5 | 7 | • | 4 | 4 | 8 |
| Warmth | 0.081 | −.093 | **0.158** | −.067 | 1.000 | 0.141 | **−.198** |
| | 2 | 2 | 5 | 4 | • | 4 | 6 |
| Restrictive | −.004 | 0.054 | 0.089 | 0.062 | 0.134 | 1.000 | −.061* |
| | 2 | 2 | 4 | 4 | 4 | • | 7 |
| Internalizing | −.119 | 0.027 | **−.173** | −.147 | **−.203** | −.032* | 1.000 |
| | 2 | 7 | 10 | 8 | 6 | 7 | • |

[a] The number of correlations averaged is shown below each mean. Relations that showed significant between-studies variation are asterisked. Mean fixed-effects correlations were computed by weighting by the variance–covariance matrix of the sample $z_i$ values. Random-effects means were computed by weighting each $z$-transformed correlation by $1/((n_i - 3)^{-1} + S_Z^2)$, where $S_Z^2$ is the estimated variance component. Boldface means differ from zero with $\alpha = .01$. The correlation between Involvement and Frequency differed from zero under the fixed-effects model, but not under the random-effects model.

account for the unexplained between-studies variation before estimating and testing the mean correlations.

Table 17.5 shows estimated variance components for each relation. A variance component is zero when the population correlations are equal (homogeneous) for a particular relation. These values are presented in the metric of Fisher's transformation, thus comparisons between them are of interest, but their values cannot be related directly to the correlation scale (i.e., 0 to ±1). Four of the 21 values, shown in bold, were significant at the .01 level according to univariate tests (i.e., a homogeneity test for each relation), and three more reached significance at $\alpha = .05$. Some variation may have resulted from differences in reporting (e.g., father involvement was reported by the mother in some studies and the father in others) or differences in conceptualization of variables across studies. Whiteside and Becker (2000) discuss these sources of variation in detail. Here I use these variance components to recompute the means under the random-effects model.

The mean correlations computed under random-effects assumptions are reported below the diagonal in Table 17.4. The four correlations for relations with significant variance components have changed the most from

**TABLE 17.5  Variance Components (in Fisher z Scale) for the Divorce Data[a]**

|  | Involvement | Frequency | Relation | Cooperation | Warmth | Restrictive | Internalizing |
|---|---|---|---|---|---|---|---|
| Involvement | .000 • | | | | | | |
| Frequency | **0.039** 3 | .000 • | | | | | |
| Relation | 0.025* 3 | **0.051** 6 | .000 • | | | | |
| Cooperation | **0.077** 3 | 0.004 5 | 0.025* 7 | .000 • | | | |
| Warmth | −.009 2 | −.011 2 | −.005 5 | 0.002 4 | .000 • | | |
| Restrictive | −.000 2 | −.006 2 | 0.020 4 | 0.019 4 | 0.000 4 | .000 • | |
| Internalizing | −.011 2 | 0.023 7 | −.001 10 | 0.020* 8 | 0.013 6 | **0.060** 7 | .000 • |

[a] The number of correlations used in each computation is shown below each variance component. For further analyses, negative variance-component estimates are set equal to zero, indicating no variation. Four values marked in boldface differ from zero at the $\alpha = .01$ level. Those four values, plus the three asterisked values, differ from zero at $\alpha = .05$.

their fixed-effects values, though only the correlation of father visitation with preseparation father involvement was not significantly different from zero under the random-effects model. Because random-effects models incorporate more uncertainty for each data point, the standard errors for mean effects tend to be larger than those from fixed-effects models. Random-effects means are therefore less likely to differ significantly from zero.

For illustrative purposes I focus now on the correlations involving the outcome internalizing symptoms, shown in the last column or row of Table 17.4. At this point one might want to test whether father–child relationship quality (Relation) or maternal warmth (Warmth) showed a stronger correlation with internalizing symptoms. That is, we could test $H_0$: $\rho$(Relation, Internalizing) = $\rho$(Warmth, Internalizing). The appropriate test uses the means (in the Fisher $Z$ metric), their standard errors, and the covariance between the $Z$ values for the two mean correlations to compute a standard normal deviate ($z$). Under the random-effects model the value $z = (-.173) - (-.203)/\sqrt{(.0023 + .0040 - 2 \times .002)} = 0.63$ is not significant at the .05 level indicating that the two populations correlations do not differ.

In the analysis designed by Whiteside and Becker, the mean correlations in Table 17.4 were used to compute a series of path models linking the six predictors to each other and, either directly or indirectly, to the child's internalizing symptoms. Because population variation was minimal for most relations, and because variance components are quite poorly estimated when the number of studies is small, fixed-effects means were used for these computations.

Figure 17.1 shows one path model for these seven variables. Relationship quality, father visitation, and maternal warmth and restrictiveness were intermediate outcomes, and internalizing symptoms was the final outcome in the model. Slope coefficients ($b$s) and their standard errors are shown near each path, and significant relations (i.e., paths for which $|b/\mathrm{SE}(b)| > 1.96$) are shown as paths with heavier lines. Significant slope values are shown in boldface. Though at this point the statistical analysis described by Becker and Schram (1994) does not include either tests of indirect effects or tests of model fit, such tests are possible. Inferences about indirect effects here are based on finding two adjacent direct effects.

Although the only two predictors with significant direct relations to

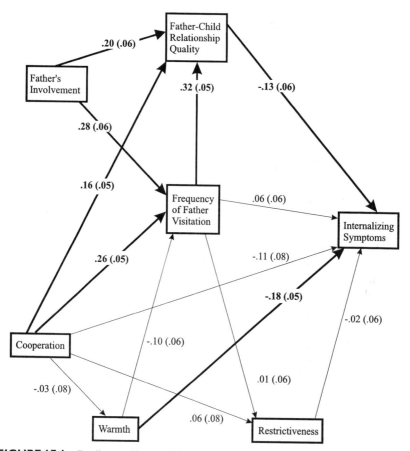

**FIGURE 17.1**    Predictors of internalizing symptoms in young children of divorcing families.

internalizing symptoms were father–child relationship quality and maternal warmth, other factors had influence through connections with father–child relationship quality. (The slopes shown in the figure for these two predictors were estimated after removing the other unimportant predictors from the model). Previous narrative reviews of this literature had concluded that father visitation had little effect on child outcomes, but those reviews focused only on the direct connection between these variables. That relation is not significant either in the correlation matrix or this path model. However, visitation is the strongest predictor of relationship quality in this model, suggesting either that fathers with good relationships visit often, or that fathers who visit more frequently develop and keep stronger positive relationships with their children. (Given the timing of the measures, we cannot infer directionality in these data.) Relationship quality, then, is significantly related to internalizing symptoms. Children who had good relationships with their fathers were low on the internalizing symptoms scales. Preseparation levels of father involvement and parental cooperation work similarly, suggesting a complex constellation of interacting variables related to positive parenting and paternal involvement, ending with positive outcomes for the child.

## VII. SUMMARY

Reviewers have a wide range of options available when faced with multivariate data in meta-analysis. No reviewer should ever ignore dependence among study outcomes. Even the most simplistic ad hoc options are better than pretending that such dependence does not exist.

The specific set of analyses a reviewer selects will depend on a number of factors. First, the reviewer must consider the structure of the data in the review. How many studies present multiple outcomes? How different are the sets of outcomes in those multivariate studies? Are comparisons between the different outcomes interesting and important theoretically, or are some outcomes of minor importance or presented so rarely that the accumulated evidence about them is minimal? If the multivariate data are a small part of the evidence in the review, it may make sense to use a simple approach, such as dropping or combining outcomes or using sensitivity analyses to evaluate the impact of the dependence on the results of the review.

If multivariate data are a major part of the evidence in the review, more sophisticated analyses may be needed. When comparisons among overall effects for the outcomes in the review are not of interest, creating and analyzing independent subsets of data makes sense. If comparisons of effects across outcomes are of primary interest, the reviewer may decide to model dependence. Then he or she must consider how much information is available about the dependence. Do primary research reports give the values of correlations among the multiple measures (in the case of multiple outcome constructs)? Is the outcome a familiar one, so that test technical manuals may provide information on interrelations among the outcomes? Alternately, are there other sources of information on dependence? If so, the approach of fully accounting for dependence enables the reviewer to estimate overall effects and to examine patterns of differences among study results using rigorous statistical methods. These methods should lead to more informative reviews, based in full knowledge of the complexity of the accumulated evidence.

## REFERENCES

Arditti, J. A., & Keith, T. Z. (1993). Visitation frequency, child support payment, and the father-child relationship postdivorce. *Journal of Marriage and the Family, 55,* 699–712.

Becker, B. J. (1988a). Gender and science achievement: A reanalysis of studies from two meta-analyses. *Journal of Research in Science Teaching, 26,* 141–169.

Becker, B. J. (1988b). Synthesizing standardized mean-change measures. *British Journal of Mathematical and Statistical Psychology, 41,* 257–278.

Becker, B. J. (1990). Coaching for the Scholastic Aptitude Test: Further synthesis and appraisal. *Review of Educational Research, 60,* 373–417.

Becker, B. J. (1992). Using results from replicated studies to estimate linear models. *Journal of Educational Statistics, 17,* 341–362.

Becker, B. J. (1995). Corrections to "Using results from replicated studies to estimate linear models." *Journal of Educational Statistics, 20,* 100–102.

Becker, B. J., & Fahrbach, K. (1994, April). *A comparison of approaches to the synthesis of correlation matrices.* Paper presented at the annual meeting of the American Educational Research Association, New Orleans, LA.

Becker, B. J., & Schram, C. M. (1994). Examining explanatory models through research synthesis. In H. M. Cooper & L. V. Hedges (Eds.), *The handbook of research synthesis* (pp. 357–381). New York: Russell Sage Foundation.

Bryk, A. S., Raudenbush, S. W., & Congdon, R. T. (1996). *Hierarchical Linear and Nonlinear Modeling with the HLM/2L and HLM/3L Programs.* Chicago: Scientific Software International.

Chiu, C. W. T. (1997). *Synthesizing multiple outcome measures with missing data: A sensitivity analysis on the effect of metacognitive reading intervention.* Unpublished manuscript, Department of Counseling, Educational Psychology, and Special Education, Michigan State University, East Lansing, MI.

DerSimonian, R., & Laird, N. (1983). Evaluating the effect of coaching on SAT scores: A meta-analysis. *Harvard Educational Review, 53,* 1–15.

Dunlap, W. P., Cortina, J. M., Vaslow, J. B., & Burke, M. J. (1996). Meta-analysis of experiments with matched groups or repeated measures designs. *Psychological Methods, 1,* 170–177.

Federal Trade Commission, Boston Regional Office (1978). *Staff memorandum of the Boston Regional Office of the Federal Trade Commission: The effects of coaching on standardized admission examinations.* Boston: Author. (NTIS No. PB-296 210).

Fisher, R. A. (1921). On the 'probable error' of a coefficient of correlation deduced from a small sample. *Metron, 1,* 1–32.

Glass, G. V. (1976). Primary, secondary, and meta-analysis of research. *Educational Researcher,* *5*, 3–8.

Gleser, L. J., & Olkin, I. (1994). Stochastically dependent effect sizes. In H. M. Cooper & L. V. Hedges (Eds.), *The handbook of research synthesis* (pp. 339–355). New York: Russell Sage.

Greenhouse, J. B., & Iyengar, S. (1994). Sensitivity analysis and diagnostics. In H. M. Cooper & L. V. Hedges (Eds.), *The handbook of research synthesis* (pp. 383–398). New York: Russell Sage.

Greenwald, R., Hedges, L. V., & Laine, R. D. (1996). The effect of school resources on student-achievement. *Review of Educational Research, 66*(3), 361–396.

Hedges, L. V. (1981). Distribution theory for Glass's effect size and related estimators. *Journal of Educational Statistics, 6*, 107–128.

Hedges, L. V., & Olkin, I. (1985). *Statistical methods for meta-analysis.* Orlando, FL: Academic Press.

James, L. R., Demaree, R. G., & Mulaik, S. A. (1986). A note on validity generalization procedures. *Journal of Applied Psychology, 71*, 440–450.

Kalaian, H. (1994). A multivariate mixed linear model for meta-analysis. Unpublished doctoral dissertation, Department of Counseling, Educational Psychology, and Special Education, Michigan State University, East Lansing, MI.

Kalaian, H., & Raudenbush, S. W. (1996). A multivariate mixed linear model for meta-analysis. *Psychological Methods, 1*, 227–235.

Kim, J. P. (1998). *Meta-analysis of equivalence of computerized and pencil-and-paper testing for ability measures.* Unpublished manuscript, Department of Counseling, Educational Psychology, and Special Education, Michigan State University, East Lansing, MI.

Little, R. J. A., & Rubin, D. B. (1987). *Statistical analysis with missing data.* New York: Wiley.

Olkin, I., & Siotani, M. (1976). Asymptotic distribution of functions of a correlation matrix. In S. Ikeda (Ed.), *Essays in probability and statistics* (pp. 235–251). Tokyo: Shinko Tsusho.

Pearson, K. (1904). Report on certain enteric fever inoculations. *British Medical Journal, 2*, 1243–1246.

Piggott, T. D. (1994). Methods for handling missing data in research synthesis. In H. M. Cooper & L. V. Hedges (Eds.), *The handbook of research synthesis* (pp. 163–175). New York: Russell Sage.

Raudenbush, S. W. (1988). Estimating change in dispersion. *Journal of Educational Statistics, 13*, 148–171.

Raudenbush, S. W. (1994). Random effects models. In H. M. Cooper & L. V. Hedges (Eds.), *The handbook of research synthesis* (pp. 301–321). New York: Russell Sage.

Raudenbush, S. W., Becker, B. J., & Kalaian, S. (1988). Modeling multivariate effect sizes. *Psychological Bulletin, 102*, 111–120.

Raudenbush, S. W., & Bryk, A. S. (1987). Examining correlates of diversity. *Journal of Educational Statistics, 12*, 241–269.

Rosenthal, R., & Rubin, D. B. (1986). Meta-analytic procedures for combining studies with multiple effect sizes. *Psychological Bulletin, 99*, 400–406.

Ryan, E. D., Blakeslee, T., & Furst, D. M. (1986). Mental practice and motor skill learning: An indirect test of the neuromuscular feedback hypothesis. *International Journal of Sport Psychology, 17*(1), 60–70.

Steinkamp, M. W., & Maehr, M. L. (1983). Affect, ability and science achievement: A quantitative synthesis of correlational research. *Review of Educational Research, 53*, 369–396.

Thomas, H. (1989). *Distributions of correlation coefficients.* New York: Springer-Verlag.

Timm, N. H. (1999). A note on testing for multivariate effect sizes. *Journal of Educational and Behavioral Statistics, 24*(2), 132–145.

Tippett, L. H. C. (1931). *The methods of statistics* (1st ed.). London: Williams & Norgate.

Whiteside, M. F., & Becker, B. J. (2000). Parental factors and the young child's post-divorce adjustment: A meta-analysis with implications for parenting arrangements. *Journal of Family Psychology, 14*, 1–22.

# 18

# GENERALIZABILITY THEORY

### GEORGE A. MARCOULIDES
*Department of Management Science,*
*California State University at Fullerton, Fullerton, California*

## I. INTRODUCTION

Generalizability theory is a random sampling theory for examining the dependability of measurement procedures that has been heralded by many psychometricians as "the most broadly defined psychometric model currently in existence" (Brennan, 1983, p. xii). Although several researchers can be credited with paving the way for generalizability theory (e.g., Burt, 1936; Hoyt, 1941), it was formally introduced by Cronbach and his associates (Cronbach, Gleser, Nanda, & Rajaratnam, 1972; Cronbach, Rajaratnam, & Gleser, 1963; Gleser, Cronbach, & Rajaratnam, 1965) as an extension to the classical test theory approach.

The fundamental axiom of the classical test theory approach is that an observed score (X) for any individual obtained through some measurement procedure can be decomposed into the true score (T) of the individual and a random error (E) component. The better a measurement procedure is at providing an accurate indication of an individual's true score, the more accurate the T component will be and the smaller the E component. In practical applications of classical test theory, reliability is typically defined as the ratio of the true score variance to the observed score variance (i.e.,

*Handbook of Applied Multivariate Statistics and Mathematical Modeling*
Copyright © 2000 by Academic Press. All rights of reproduction in any form reserved.

the proportion of the observed score variability that is attributable to true score variability across individuals), and the reliability coefficient increases as error decreases.

Unfortunately, the assumptions that follow the fundamental axiom of the classical test theory approach are often unreasonable (Cronbach et al., 1972). In particular, the existence of a single undifferentiated random error in the measurement procedure is quite questionable. In contrast, generalizability theory recognizes that multiple sources of error may occur simultaneously in a measurement procedure (e.g., errors attributable to the test items, testing occasions, and test examiners). Generalizability theory extends classical test theory by providing a flexible and practical framework for estimating the effects of multiple sources of error, thereby providing more detailed information that applied researchers can use in deciding how to improve the usefulness of a measurement procedure for making decisions or drawing conclusions. Thus, the focus of the classical test theory concept of reliability (how accurately observed scores reflect corresponding true scores) is expanded in generalizability theory to address a broader issue: how accurately can researchers generalize about a persons' behavior in a defined universe from observed scores? (Shavelson, Webb, & Rowley, 1989, p. 922).

The use of generalizability theory to study the dependability of measurement procedures has expanded over the past two decades, but most published aptitude tests currently in use continue to provide psychometric information based only on the classical test theory approach. Two major obstacles have hindered the more widespread use of generalizability theory in the social and behavioral sciences: (a) the complexity of the original treatments of the theory provided by Cronbach and his associates, and (b) the lack of tailored computer programs for performing a generalizability analysis (Shavelson & Webb, 1991). Tailored computer programs for calculating a generalizability analysis are becoming more widely available, and some commercial statistical packages now provide procedures for computing the estimates needed in a generalizability analysis. Nevertheless, few nontechnical introductions to the theory have appeared in the literature.

This chapter provides a nontechnical overview of generalizability theory and illustrates its use as a comprehensive method for designing, assessing, and improving the dependability of measurement procedures. Throughout the chapter example cases and situations are reported in sufficient detail to enable the reader to verify calculations and replicate the results. In addition, references are made to computer packages. This chapter will enable readers to begin to use generalizability theory procedures in their own research. The first section of this chapter gives an overview of the fundamentals of generalizability theory. Although generalizability theory is applicable to a number of models, this section concentrates on the simple one-facet model, which is the most common measurement procedure used. Different types of error variance and generalizability coefficient estimates are introduced to illustrate the distinct advantages of generalizability theory over classical test theory. The next two sections give an overview of the basic concepts extended to multifaceted and multivariate measurement designs. In the fourth section, some new extensions to the basic generalizability theory approach are introduced. The final section provides an overview of computer programs that can be used to conduct generalizability analyses.

## II. FUNDAMENTALS OF GENERALIZABILITY THEORY

In their original formulation of generalizability theory, Cronbach et al. (1972) promoted the perspective that observed scores obtained through a measurement procedure are gathered as a basis for making decisions or drawing conclusions. Cronbach et al. (1972) provided the following argument for their perspective:

> The score on which the decision is to be based is only one of many scores that might serve the same purpose. The decision maker is almost never interested in the response given to the particular stimulus objects or questions, to the particular tester, at the particular moment of testing. Some, at least of these conditions of measurement could be altered without making the score any less acceptable to the decision maker. That is to say, there is a universe of observations, any of which would have yielded a usable basis for the decision. The ideal datum on which to base the decision would be something like the person's mean score over all acceptable observations, which we shall call his "universe score." The investigator uses the observed score or some function of it as if it were the universe score. That is, he generalizes from sample to universe. (Cronbach et al., 1972; p. 15)

Based upon the above argument, Cronbach et al. (1972) placed the notion of a *universe* at the heart of generalizability theory. All measurements are considered to be samples from a universe of admissible observations. A universe is defined in terms of those aspects (called *facets*) of the observations that determine the conditions under which an acceptable score can be obtained. For example, the facets that define one universe could be personality tests administered to military recruits during their first week of enlistment. It is possible to conceive of many different universes to which any particular measurement might generalize, so it is essential that investigators define explicitly the facets which can change without making the observation unacceptable or unreliable. For example, if test scores might be expected to fluctuate from one occasion to another, then the "occasions" facet is one defining attribute of the universe, and multiple testing occasions must be included in the measurement procedure. The same is true for the choice of test items and other aspects of the measurement procedure. Ideally, the measurement procedure should yield information about an examinee's universe score over all combinations of facets, but in reality researchers are limited in their choice of particular occasions, items, or other facets. The need to sample facets introduces error into the measurement procedure and limits investigators to estimating rather than actually measuring the universe score.

The basic approach underlying generalizability theory is to decompose an observed score into a variance component for the universe score and variance components for each of the facets associated with the measurement procedure. Although some attribute of persons will be the object of measurement in most situations, generalizability theory also considers the possibility that any other facet in the procedure could be regarded as the object of measurement. In that case the facet for persons is treated as an error component. This feature has been termed the *principle of symmetry* (Cardinet, Tourneur, & Allal, 1976).

## A. A One-Facet Crossed Design

A common measurement procedure is to administer a multiple-choice test consisting of a random sample of $n_i$ items from a universe of items to a random sample of $n_p$ persons from a population of persons (see Table 18.1). This design is called a one-facet design because the items facet is the only facet included in the measurement procedure.

The observed score $(X_{pi})$ for one person $(p)$ on one item $(i)$ can be expressed in terms of the following linear model:

$$X_{pi} = \mu + \mu_p^* + \mu_i^* + \mu_{pi}^* + e \tag{1}$$

where

$\mu$ = overall mean in the population of persons and universe of items,

$\mu_p^*$ = score effect attributable to person $p$,

$\mu_i^*$ = score effect attributable to item $i$,

$\mu_{pi}^*$ = score effect attributable to the interaction of person $p$ with item $i$, and

$e$ = the experimental error.

**TABLE 18.1    Data from a Hypothetical One-Facet Study**

|          | Items |   |   |   |   |
|----------|-------|---|---|---|---|
| Person   | 1     | 2 | 3 | 4 | 5 |
| 1        | 9     | 6 | 6 | 8 | 8 |
| 2        | 8     | 5 | 6 | 6 | 6 |
| 3        | 9     | 8 | 7 | 8 | 8 |
| 4        | 7     | 5 | 6 | 4 | 4 |
| 5        | 7     | 5 | 6 | 3 | 4 |

Because there is only one observation for each person–item combination, the $\mu_{pi}^*$ and $e$ effects are completely confounded. Cronbach et al. (1972), represented this confounding with the notation $pi,e$ (often referred to as the residual effect). Thus, the model can also be written as:

$$X_{pi} = \mu + \mu_p^* + \mu_i^* + \mu_{pi,e}^* \qquad (2)$$

or simply:

$$
\begin{aligned}
X_{pi} &= \mu && (grand\ mean = \mu) \\
&+ \mu_p - \mu && (person\ effect = \mu_p^*) \\
&+ \mu_i - \mu && (item\ effect = \mu_i^*) \\
&+ X_{pi} - \mu_p - \mu_i + \mu && (residual\ effect = \mu_{pi,e}^*)
\end{aligned}
$$

The universe score for person $p$ is denoted $\mu_p$ (i.e., a person's average score over the entire item universe), and is defined as the expected value of a person's observed score across the universe of items or simply $\mu_p \equiv E_i X_{pi}$. Similarly, the population mean for item $i$ is defined as the expected value over persons or $\mu_i \equiv E_p X_{pi}$, and the mean over both the population of persons and the universe of items is $\mu \equiv E_p E_i X_{pi}$ (Brennan, 1983). The basic assumptions underlying the above model are that all effects in the model are sampled independently, and the expected value of each effect over the population of persons and the universe of items is equal to zero. Given these assumptions, the model is considered a random-effects person-crossed-with-items ($p \times i$) one-facet design.

Generalizability theory places considerable importance on the variance components of the effects in the model because their magnitude provides information about the potential sources of error influencing a measurement. The estimated variance components are the basis for determining the relative contribution of each potential source of error and for determining the dependability of a measurement. In actual practice the estimation of the variance components is achieved by calculating observed mean squares from an analysis of variance (ANOVA), equating these values to their expected values, and solving a set of linear equations. Table 18.2 provides the ANOVA results from the hypothetical one-facet study along with the computational formulas for the variance components associated with the score effects in the model.

It is important to note that there is no restriction on the procedures that applied researchers can use to estimate variance components. Elsewhere researchers (Marcoulides, 1987, 1989, 1990, 1996; Shavelson & Webb, 1981) have described numerous other methods of estimation that can be used to provide the same information as ANOVA, including Bayesian, minimum variance, restricted maximum likelihood, and covariance structure methods. These methods often provide more accurate estimates of variance compo-

**TABLE 18.2    ANOVA Estimates of Variance Components for Example One-Facet Crossed Design**[a]

| Source of variation | df | Sum of squares | Mean square | Expected mean square | Variance components |
|---|---|---|---|---|---|
| Persons ($p$) | 4 | 34.96 | 8.74 | $\sigma^2_{pi,e} + n_i\sigma^2_p$ | 1.58 |
| Items ($i$) | 4 | 17.36 | 4.34 | $\sigma^2_{pi,e} + n_p\sigma^2_i$ | 0.70 |
| Residual ($pi,e$) | 16 | 13.44 | 0.84 | $\sigma^2_{pi,e}$ | 0.84 |

[a] The estimated variance components for the above one-facet design are calculated as follows:

$$\sigma^2_p = \frac{MS_p - MS_{pi,e}}{n_i} = \frac{8.74 - 0.84}{5} = 1.58$$

$$\sigma^2_i = \frac{MS_i - MS_{pi,e}}{n_p} = \frac{4.34 - 0.84}{5} = 0.70$$

$$\sigma^2_{pi,e} = MS_{pi,e} = 0.84$$

$$\sigma^2_{\delta} = \frac{\sigma^2_{pi,e}}{n_i} = \frac{0.84}{5} = 0.17$$

$$\sigma^2_{\Delta} = \frac{\sigma^2_i}{n_i} + \frac{\sigma^2_{pi,e}}{n_i} = \frac{0.70}{5} + \frac{0.84}{5} = 0.31$$

nents than ANOVA in cases involving small sample sizes, dichotomous data, unbalanced designs, or data with missing observations (Marcoulides, 1987; Muthén, 1983). ANOVA is much easier to implement, however, and it continues to be the most commonly used method in generalizability theory.

The Venn diagram is a useful visual aid for understanding the decomposition of the observed score variance into the various sources of variability in a measurement design. For example, Figure 18.1 presents a Venn diagram for the above person by item one-facet design. The intersecting of the circles of $p$ and $i$ denotes the crossing of persons and items. Of course, the magnitude of the areas within the circles do not correspond to the actual

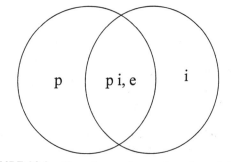

**FIGURE 18.1**    Venn diagram for one-facet ($p \times i$) design.

magnitudes of the sources of variability. Nevertheless, the Venn diagram illustrates clearly the sources of variability observed in a design, and consequently the variance components that must be estimated.

## B. Types of Error Variances

### 1. Relative Decisions

One of the distinct advantages of generalizability theory over classical test theory is that the former allows researchers to distinguish between two types of error variance that correspond to relative decisions (i.e., decisions about individual differences between persons) and absolute decisions (i.e., decisions about the level of performance). Relative error variance (also called $\delta$-type error) is usually of primary concern when researchers are interested in decisions that involve the rank ordering of individuals, so in this instance all the sources of variation that include persons are considered measurement error. For example, in the one-facet design the relative error variance (symbolized by $\sigma_\delta^2$) includes the variance components due to the residual averaged over the number of items used in the measurement. Using the estimates obtained in Table 18.2, this value is 0.17. The square root of this index ($\sigma_\delta = 0.41$) is considered the $\delta$-type (relative) standard error of measurement ($\delta$-type SEM). The $\delta$-type SEM is equivalent (for only the one-facet design) to the SEM in the classical test theory approach. Using $\sigma_\delta$, a confidence interval that contains the universe score (with some degree of certainty) can easily be determined. For example, using the data in Table 18.1, a 68% confidence interval for person #3 with an overall score of 8 would extend from 7.59 to 8.41 (i.e., $8 \pm Z_{\alpha/2}\sigma_\delta$). It is important to note that it is traditional to assume a normal distribution in order to attach a probability statement to the confidence interval. Generalizability theory makes no distributional assumption about the form of the observed scores or the scores effects, but such an assumption is required when using error variances in establishing confidence intervals.

### 2. Absolute Decisions

When dealing with applied problems investigators often must make decisions about issues such as whether an examinee can perform at a prespecified level. In these instances it is the absolute error variance (also called $\Delta$-type error) that is of concern. The absolute error variance reflects both information about the rank ordering of persons and any differences in average scores. For example, in the one-facet design the absolute error (symbolized by $\sigma_\Delta^2$) includes the variance components due to both the item effect and the residual effect averaged over the number of items used in the measurement. Using the estimates obtained in Table 18.2, this value is 0.31. The square root of this index ($\sigma_\Delta = 0.56$), the $\Delta$-type (absolute) standard error of measurement ($\Delta$-type SEM), also can be used to determine a confidence interval that contains the universe score. For example, a 68% confidence interval for person #3 with a score of 8 would extend from 7.44 to 8.56 (i.e., $8 \pm Z_{\alpha/2}\sigma_\Delta$).

## C. Generalizability Coefficients

Although generalizability theory stresses the importance of variance components, it also provides a generalizability coefficient which ranges from 0 to 1.0 that can be used to index the dependability of a measurement procedure (higher values reflect more dependable measurement procedures). Generalizability coefficients are available for both relative error (symbolized by $E\rho_\delta^2$) or absolute error (symbolized by $\rho_\Delta^2$ or $\Phi$—the notation is often used interchangeably). For the one-facet example design these values are:

$$E\rho_\delta^2 = \frac{\sigma_p^2}{\sigma_p^2 + \sigma_\delta^2} = \frac{1.58}{1.58 + 0.17} = 0.90 \tag{3}$$

and

$$\rho_\Delta^2 = \Phi = \frac{\sigma_p^2}{\sigma_p^2 + \sigma_\Delta^2} = \frac{1.58}{1.58 + 0.31} = 0.84. \tag{4}$$

As before, sample estimates of the parameters in the generalizability coefficients are used to estimate the appropriate level of generalizability. It is also important to note that the value of $E\rho_\delta^2$ in the one-facet design is equal to the classical test theory Cronbach's coefficient $\alpha$ for items scored on a metric and equal to KR-20 and Cronbach's coefficient $\alpha$-20 when items are scored dichotomously.

Brennan (1983) indicated that the $\rho_\Delta^2$ (or $\Phi$) generalizability coefficient may be viewed as a general-purpose index of dependability for domain-referenced (criterion-referenced or content-referenced) interpretations of examinee scores. The observed examinee score is interpreted as being representative of the universe of the domain from which it was sampled, and interest is placed on the dependability of an examinee's score that is independent of the performance of others (i.e., independent of the universe scores of other examinees). However, if emphasis is placed on the dependability of an individual's performance in relation to a particular cutoff score (e.g., a domain-referenced test that has a fixed cutoff score and classifies examinees who match or exceed this score as having mastered the content represented by the domain), a different generalizability index must be computed (Brennan & Kane, 1977). The index is denoted by $\Phi(\lambda)$ and represents domain-referenced interpretations involving a fixed cut off score. The value of $\Phi(\lambda)$ is determined by:

$$\Phi(\lambda) = \frac{\sigma_p^2 + (\mu - \lambda)^2}{\sigma_p^2 + (\mu - \lambda)^2 + \sigma_\Delta^2}. \tag{5}$$

For computational ease, an unbiased estimator of $(\mu - \lambda)^2$ is determined by using $(\overline{X} - \lambda)^2 - \sigma_{\overline{X}}^2$. Where $\sigma_{\overline{X}}^2$ represents the mean error variance (Marcoulides, 1993). The mean error variance for the one-facet design represents the error variance involved when using the mean $(\overline{X})$ over the sample of both persons and items as an estimate of the overall mean $(\mu)$ in the population of persons and the universe of items; the smaller the mean error variance the more stable the population estimate (Brennan, 1983). Using the estimates obtained above, the mean error variance is equal to:

$$\sigma_{\overline{X}}^2 = \frac{\sigma_p^2}{n_p} + \frac{\sigma_i^2}{n_i} + \frac{\sigma_{pi,e}^2}{n_p n_i} = \frac{1.58}{5} + \frac{0.70}{5} + \frac{0.84}{25} = 0.49. \tag{6}$$

Assuming that the cutoff score in the hypothetical study were $\lambda = 8$, using Equation 5 the value of $\Phi(\lambda)$ is found to be:

$$\Phi(\lambda) = \frac{1.58 + (6.36 - 8)^2 - 0.49}{1.58 + (6.36 - 8)^2 - 0.49 + 0.31} = 0.92.$$

As indicated earlier, generalizability theory provides a framework for examining the dependability of measurement procedures. Performing a generalizability analysis to pinpoint the sources of measurement error allows the applied researcher to determine how many conditions of each facet are needed (e.g., number of items, number of occasions) to obtain an optimal level of generalizability for making different types of future decisions (e.g., relative or absolute—Marcoulides & Goldstein, 1990). Using the above formulas, several generalizability coefficients for a variety of studies with different numbers of items can be computed in much the same way that the Spearman-Brown Prophecy Formula is used in classical test theory to determine the appropriate length of a test. The difference is that generalizability theory can provide three types of coefficients, whereas classical test theory provides only one (which is identical to $E\rho_\delta^2$). For example, with $n_i = 1$ the generalizability coefficients are $E\rho_\delta^2 = .65$, $\rho_\Delta^2 = .51$, and $\Phi(\lambda = 8) = .67$. In contrast, with $n_i = 10$ the generalizability coefficients are $E\rho_\delta^2 = .95$, $\rho_\Delta^2 = .91$, and $\Phi(\lambda = 8) = .96$. Because the items are a source of measurement error in the example design, increasing the number of items in the measurement procedure increases the generalizability coefficients. Thus, by examining various types of designs, a researcher can readily determine how many conditions of each facet are needed to obtain an acceptable level of generalizability for making future decisions.

## D. Differences between Generalizability Studies and Decision Studies

Generalizability theory refers to the initial study of a measurement procedure as a generalizability (G) study (Shavelson & Webb, 1981). However, it is quite possible that, after conducting a generalizability analysis, a researcher may want to design a measurement procedure that differs from the procedure used in the actual generalizability study. For example, if the results of a G study show that some sources of error are small (e.g., error attributable to items), then an applied researcher may elect a measurement procedure that reduces the number of levels of that facet (e.g., number of items), elect to change the actual design of a study (see discussion below), or even ignore that facet (which can be especially important in multifaceted designs—see next section). Alternatively, if the results of a G study show that some sources of error in the design are very large, the applied researcher may need to increase the levels of that facet in order to obtain an acceptable level of generalizability. Generalizability theory refers to the process in which facets are modified on the basis of information obtained in a G study as decision (D) studies.

Unfortunately, there is much confusion in the literature concerning the differences between a G study and a D study. Cronbach et al. (1972) indicate that the distinction between a G study and a D study is simply a recognition that certain studies are carried out while developing a measurement procedure and then the procedure is put to use. In general, a D study can be conceptualized as the point at which one looks back at the G study and examines the measurement procedure in order to make recommendations for change. In its simplest form, a D study is analogous to an implementation of the Spearman-Brown Prophecy Formula used in the classical test theory approach to determine the appropriate length of a test. In other words, a D study addresses the question, What should be done differently if you are going to rely on this measurement procedure for making future decisions or drawing conclusions? In the case where no changes can or should be made, the G study acts as the D study (i.e., use the same sample of items used in the initial study). A major contribution of generalizability theory, therefore, is that it permits researchers to pinpoint the sources of measurement error and modify the number of observations on relevant facets to obtain a desired level of generalizability.

### E. A One-Facet Nested Design

Another way in which generalizability studies and decision studies may differ is in terms of the actual data collection design. In the above discussion, for example, both the G and the D studies used a crossed $p \times i$ design (i.e., every person was administered the same sample of test items). Nevertheless, it is possible that a researcher might use a D study design in which each person is administered a different random sample of items. Such a situation may arise in computer-assisted or computer-generated test construction (Brennan, 1983). Thus, although the facets that define the universe of admissible observations are crossed, an investigator may choose to use a nested measurement design in conducting a D study. This one-facet nested design is traditionally denoted $i:p$, where the colon implies "nested within." For such a model the observed scores is:

$$X_{pi} = \mu + \mu_p^* + \mu_{i:p,e}^* \tag{7}$$

where

$\mu$ = overall mean in the population of persons and universe of items,

$\mu_p^*$ = score effect attributable to person $p$,

$\mu_{i:p,e}^*$ = residual effect.

A comparison of the above nested design model (Equation 7) with the crossed design illustrated in Equation 2 reveals that the items nested within-persons effect ($\mu_{i:p,e}^*$) represents the confounding of the $\mu_i^*$ and the $\mu_{pi,e}^*$ effects in the crossed design. In the nested design the item effect cannot be estimated independently of the person-by-item interaction because different persons are administered different items (Shavelson & Webb, 1991). Figure 18.2 presents a Venn diagram of a nested $i:p$ design. The variance of the observed scores for all persons and items is the sum of only two variance components (i.e., $\sigma^2 X_{pi} = \sigma_p^2 + \sigma_{i:p,e}^2$). In the nested design $\sigma_\delta^2 = \sigma_\Delta^2$ and only one generalizability coefficient for relative or absolute decisions can be estimated. Table 18.3 presents the variance component estimates for the above nested design using the data from the crossed design (except that each person is now assumed to have been tested on a different set of five items). Using Equation 3 and the estimates presented in Table 18.3, the generalizability coefficient for a five-item test is found to be 0.82.

A major drawback to using a nested design is that it is not possible to obtain a separate estimate of the item effect. Whenever possible, therefore,

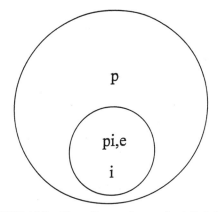

**FIGURE 18.2**   Venn diagram for one-facet ($i:p$) design.

**TABLE 18.3    ANOVA Estimates of Variance Components for Example One-Facet Nested Design**

| Source of variation | df | Sum of squares | Mean square | Variance components |
|---|---|---|---|---|
| Persons ($p$) | 4 | 34.96 | 8.74 | 1.44 |
| Residual ($i{:}p,e$) | 20 | 30.80 | 1.54 | 1.54 |

generalizability studies should use fully crossed designs so that all the sources of variability in the measurement design can be estimated (Cronbach et al., 1972). However, a researcher may elect to change a crossed generalizability study design to a nested design whenever the results from a crossed G study show that the sources of error for a particular facet are negligible. In the case where there is natural nesting in the universe of admissible observations, a researcher has no choice but to incorporate that nesting into both the G study and the D study.

## III. GENERALIZABILITY THEORY EXTENDED TO MULTIFACETED DESIGNS

Generalizability theory provides a framework for examining the dependability of measurements in almost any type of design. For example, Shavelson and Webb (1991) described a G study of the dependability of measures of teacher job performance in which eight teachers ($p$) were observed by three judges ($j$) on their presentation of two subject areas (mathematics and reading; $s$). The design was a completely crossed $p \times j \times s$ design in which each judge rated each teacher on both subjects. The overall quality of each teacher's performance in the two subject areas was rated 1 to 11. For simplicity, the subject area was considered a random facet selected from a universe of possible subjects areas (although the analysis could be conducted with subject areas as a fixed facet—see Shavelson & Webb, 1991, for further discussion). Table 18.4 presents the data from the above study originally compiled by Shavelson and Webb (1991, p. 69).

Several sources of variability can contribute to error in this two-faceted study of the dependability of the performance measures (see Figure 18.3). The teacher variability does not constitute error variation because teachers are the object of measurement. Using the analysis of variance (ANOVA) approach, seven variance components must be estimated. These represent the three main effects in the design (persons, judges, subjects), the three two-way interactions between these main effects, and the three-way interac-

**TABLE 18.4  Two-Facet Example Data Set of Teacher Performance**

| Teacher | Mathematics judges | | | Reading judges | | | Ability estimates |
|---|---|---|---|---|---|---|---|
| | **1** | **2** | **3** | **1** | **2** | **3** | |
| 1 | 4 | 4 | 4 | 5 | 5 | 6 | −.07 |
| 2 | 6 | 7 | 6 | 7 | 9 | 5 | .14 |
| 3 | 8 | 7 | 7 | 4 | 3 | 2 | .25 |
| 4 | 6 | 8 | 7 | 9 | 11 | 7 | .31 |
| 5 | 2 | 1 | 1 | 5 | 5 | 3 | −.88 |
| 6 | 5 | 4 | 4 | 7 | 6 | 5 | −.01 |
| 7 | 4 | 5 | 6 | 6 | 8 | 9 | .08 |
| 8 | 7 | 7 | 6 | 5 | 9 | 9 | .18 |
| Judge severity estimates | −.39 | −.39 | −.39 | .14 | .44 | .58 | |

tion (which is confounded with random error because of the one observation per cell design). The total variance of the observed score is equal to the sum of these variance components:

$$\sigma^2 X_{pjs} = \sigma_p^2 + \sigma_j^2 + \sigma_s^2 + \sigma_{pj}^2 + \sigma_{ps}^2 + \sigma_{js}^2 + \sigma_{pjs,e}^2. \tag{8}$$

Table 18.5 provides the ANOVA source table and the estimated variance components for the above example. Estimation of these variance

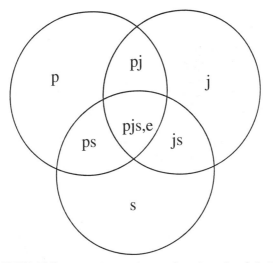

**FIGURE 18.3**  Venn diagram for two-facet ($p \times j \times s$) design.

**TABLE 18.5**   Generalizability Study Results for Teacher Performance Data[a]

| Source of variation | df | Sum of squares | Mean square | Variance component | Variance explained (%) |
|---|---|---|---|---|---|
| Teachers ($p$) | 7 | 111.65 | 15.95 | 1.17 | 22.95 |
| Judges ($j$) | 2 | 4.88 | 2.44 | 0.02 | 0.29 |
| Subject ($s$) | 1 | 12.00 | 12.00 | 0.16 | 3.09 |
| $p \times j$ | 14 | 26.46 | 1.89 | 0.51 | 9.90 |
| $p \times s$ | 7 | 55.30 | 7.90 | 2.34 | 45.82 |
| $j \times s$ | 2 | 2.38 | 1.19 | 0.04 | 0.76 |
| $pjs,e$ | 14 | 12.32 | 0.88 | 0.88 | 17.18 |
| Total | | | | | 100.00 |

[a] The estimated variance components for the above two-facet design are determined as follows:

$$\sigma^2_{pjs,e} = MS_{pjs,e} = .88$$

$$\sigma^2_{js} = \frac{MS_{js} - MS_{pjs,e}}{n_p} = \frac{1.19 - .88}{8} = 0.4$$

$$\sigma^2_{ps} = \frac{MS_{ps} - MS_{pjs,e}}{n_j} = \frac{7.90 - .88}{3} = 2.34$$

$$\sigma^2_{pj} = \frac{MS_{pj} - MS_{pjs,e}}{n_s} = \frac{1.89 - .88}{2} = .51$$

$$\sigma^2_p = \frac{MS_p - MS_{pjs,e} - n_j\sigma^2_{ps} + n_s\sigma^2_{pj}}{n_j n_s} = \frac{15.95 - .88 - (3)(2.34) + (2)(.51)}{(3)(2)} = 1.17$$

$$\sigma^2_j = \frac{MS_j - MS_{pjs,e} - n_p\sigma^2_{js} + n_s\sigma^2_{pj}}{n_p n_s} = \frac{2.44 - .88 - (8)(.04) + (2)(.51)}{(8)(2)} = .02$$

$$\sigma^2_s = \frac{MS_s - MS_{pjs,e} - n_j\sigma^2_{ps} + n_p\sigma^2_{js}}{n_p n_j} = \frac{12 - .88 - (3)(2.34) + (8)(.04)}{(3)(8)} = .16$$

components involves equating the observed mean squares from the ANOVA to their expected values and solving the sets of linear equations (for more details see Shavelson & Webb, 1991, or Marcoulides, 1998). As seen in Table 18.5, the variance component for teachers indicates that there are differences among the teachers' performances. The magnitude of the variance component for the teacher-by-judge interaction (9.9%) suggests that judges differed in their grading of the teacher's performance. The teacher-by-subject interaction (45.8%) indicates that the relative ranking of each teacher differed substantially across the two subjects areas.

The relative error variance associated with this measurement procedure is

$$\sigma^2_\delta = \frac{\sigma^2_{pj}}{n_j} + \frac{\sigma^2_{ps}}{n_s} + \frac{\sigma^2_{pjs,e}}{n_j n_s} = \frac{0.51}{3} + \frac{2.34}{2} + \frac{0.88}{6} = 1.49, \qquad (9)$$

and the generalizability coefficient for a relative decision using Equation 3 is

$$E\rho_\delta^2 = \frac{1.17}{1.17 + 1.49} = 0.44.$$

In contrast, the absolute error variance associated with this measurement procedure is

$$\sigma_\Delta^2 = \frac{\sigma_j^2}{n_j} + \frac{\sigma_s^2}{n_s} + \frac{\sigma_{pj}^2}{n_j} + \frac{\sigma_{ps}^2}{n_s} + \frac{\sigma_{js}^2}{n_j n_s} + \frac{\sigma_{pjs,e}^2}{n_j n_s} \tag{10}$$

or simply

$$\sigma_\Delta^2 = \frac{0.02}{3} + \frac{0.16}{2} + \frac{0.51}{3} + \frac{2.34}{2} + \frac{0.04}{6} + \frac{0.88}{6} = 1.58,$$

and the generalizability coefficient for an absolute decision using Equation 4 is:

$$\rho_\Delta^2 = \Phi = \frac{1.17}{1.17 + 1.58} = 0.42$$

As illustrated in the previous section, generalizability coefficients for a variety of D studies using different combinations of number of judges and subject areas can be estimated. For example, with $n_j = 3$ and $n_s = 4$ the generalizability coefficients are $E\rho_\delta^2 = .59$ and $\rho_\Delta^2 = .57$. Whereas with $n_j = 3$ and $n_s = 8$, the generalizability coefficients increase to $E\rho_\delta^2 = .70$ and $\rho_\Delta^2 = .69$. The subject areas facet is a major source of measurement error, so increasing the number of subject areas in the measurement procedure increases the generalizability coefficients.

Of course, applied researchers must always consider the realistic constraints (especially resource constraints) imposed on the measurement procedures before making final decisions about a D study design (Marcoulides & Goldstein, 1990). For example, satisfying resource constraints while maximizing generalizability is simple in the one-facet design; choose the greatest number of items needed to give maximum generalizability without violating the budget. However, obtaining a solution can be quite complicated when other facets are added to the design. Goldstein and Marcoulides (1991), Marcoulides and Goldstein (1990, 1991, 1992), and Marcoulides (1993, 1995, 1997b) developed optimization procedures that can be used in any measurement design to determine the optimal number of conditions that maximize generalizability under limited budget and other constraints. These optimization procedures generally provide closed-form equations that can be solved by simple substitution. For example, if the total available budget (B) for the two-faceted design is $5000 and if the cost (c) for each judge to observe a teacher in a subject area is \$120, then the optimal number of judges can be determined using the following equation:

$$n_j = \sqrt{\frac{\sigma_{pj}^2}{\sigma_{ps}^2}\left(\frac{B}{c}\right)} = \sqrt{\frac{.51}{2.34}\left(\frac{5000}{120}\right)} = 3 \qquad (11)$$

## IV. GENERALIZABILITY THEORY EXTENDED TO MULTIVARIATE DESIGNS

Behavioral measurements often involve multiple scores in order to describe individuals' aptitude or skills (Webb, Shavelson, & Maddahian, 1983). The most commonly used procedure to examine measurements with multiple scores is to assess the dependability of the scores separately (i.e., in a univariate generalizability analysis), because a multivariate generalizability analysis can be quite complicated, and there are currently no available tailored computer programs for calculating the estimates needed (Marcoulides, 1998). Nevertheless, a multivariate analysis can provide information about facets that contribute to covariance among the multiple scores that cannot be obtained in a univariate analysis. This information is essential for designing optimal decision studies that maximize generalizability.

The previous section examined the dependability of measures of the quality of teacher presentations in two subject areas using a univariate generalizability analysis approach. However, treating the subject area as a separate source of error variance provided no information about the sources of covariation (correlation) that might exist among the two subject areas. That information may be important for correctly determining the dependability of the measurement procedure. In situations such as this, applied researchers should consider conducting a multivariate generalizability analysis and determine whether there are any sources of covariation among subject areas. If no covariation exists, perform a univariate generalizability analysis and base evaluations of the dependability of the measurement procedure on the univariate approach.

## V. GENERALIZABILITY THEORY AS A LATENT TRAIT THEORY MODEL

Latent trait theory (or item response theory—IRT), refers to a family of measurement theories that can be used to describe, explain, and predict the encounter of an individual with a measurement device (for further discussion see Hambleton, Robin, & Xing, Chapter 19, this volume). IRT assumes that an individual's performance on a measurement device (e.g., a test item) can be accounted for by defining characteristics called traits.

Typically, the relation between the unobserved trait and the observed measure is described in terms of some mathematical model (Schmidt-McCollam, 1998). To date, there is no generally accepted model for describing the relation between a given trait and performance on a measure of the trait. Marcoulides and Drezner (1995, 1997a,b) recently introduced an extension to the generalizability theory model that can be used to provide specific information on the reliability of individual ability estimates and diagnostic information at individual and group levels. This extension to generalizability theory can be considered a special type of IRT model capable of estimating latent traits, such as examinee ability estimates, rater severity, and item difficulties (Marcoulides, 1997a).

To illustrate this model, consider the two-facet study of measures of teacher performance presented earlier in which eight teachers were observed by three judges on their presentation of two subject areas (see Table 18.5). The teacher-by-judge interaction (9.9%) suggested that judges graded each teacher's performance somewhat differently, and the teacher-by-subject interaction (45.8%) indicated that the relative ranking of the teachers differed substantially across the two subject areas. The model proposed by Marcoulides and Drezner (1997a) provides three additional important pieces of information that can be used: (a) a diagnostic scatter diagram to detect unusual patterns for each teacher and each judge, (b) an Examinee Index to examine the ability levels of each teacher, and (c) a Judge Index to examine judge severity.

The Marcoulides and Drezner (MD) model is based on the assumption that observed points (i.e., examinees, judges, items, or any other set of observations that define a facet of interest) are located in an $n$-dimensional space and that weights ($w_{ij}$) can be calculated that indicate the relation between any pair of points in this $n$-dimensional space. These weights constitute a measure of the similarity of any pair of examinees in terms of ability level (see Table 18.4), any pair of judges in terms of severity estimates (see Table 18.4), or any pair of items in terms of difficulty (Marcoulides & Drezner, 1993).

Figures 18.4 and 18.5 present plots of the MD weights that show the distribution of teacher performances and the rating patterns of the judges in the subject areas. As can be seen in Figure 18.4, the teachers vary in the ability scores they received and teachers #3, #4, and #5 require further study. The judges rating patterns appear identical in Mathematics (Figure 18.5) but they differ somewhat in Reading. The greatest difference in the composite rankings received by the teachers occurred for teacher #3, who was first in mathematics and eighth in reading (see Figure 18.4). Plots of the MD weights highlight unusual teacher performances and unusual patterns of judges ratings.

The MD model also provides information concerning examinee (teacher) ability or proficiency and judge severity estimates (see Table

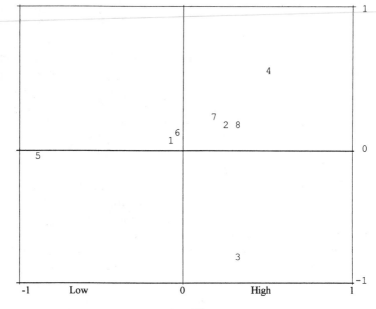

Teacher Ability

**FIGURE 18.4** Marcoulides-Drezner plot for observed teachers in example study.

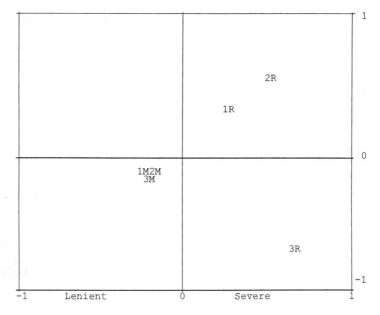

Judge Severity

**FIGURE 18.5** Marcoulides-Drezner plot for judge ratings in example study.

18.4). The MD model can also be extended to include any other latent traits of interest depending on the facets included in the original measurement study. For example, item difficulty estimates could be calculated if each teacher in the above example study had been rated using some set of items. Marcoulides and Drezner (1997a) referred to the ability measure for each teacher ($Se$) as the Examinee Index (the ability estimate is generally represented as $\theta$ in traditional IRT models), and the severity estimate for each judge ($Sj$) as the Judges Index. Because the MD model independently calibrates the examinees and judges so that all observations are positioned on the same scale, the scales range from $+1$ to $-1$. Thus, negative values of the Examinee Index indicate relatively less able examinees and positive values relatively more able examinees. Similarly, negative values of the Judges Index indicate relatively lenient judges and positive values indicate relatively severe judges.

An examination of the teacher ability estimates generated by the MD model in Table 18.4 reveals that teacher #5 ($-0.88$) is the least able and teacher #4 (0.31) the most able examinee (see also Figure 18.5). The judges' severity estimates are identical in mathematics and relatively similar in reading (see Table 18.4 and Figure 18.5). Based on the results of the MD model, it appears that judge #1 is slightly more lenient than judge #2 and judge #3. These results are also corroborated when using the traditional G theory analysis. For example, as presented in Table 18.5, it was determined that judges graded each teacher's performance somewhat inconsistently (the teacher-by-judge interaction was found to be 9.9%).

## VI. COMPUTER PROGRAMS

The computational requirements involved in estimating variance components for use in a generalizability theory analysis can be demanding, especially in multifaceted designs. As a consequence, applied researchers generally must use computer programs to obtain the necessary estimates. Several tailored computer programs (for both Mac and PCs) have been developed for conducting generalizability analyses including GENOVA (Brennan, 1983), GEN, and GT (Cardinet and associates, 1985–97; the programs are available with manuals from http://www.unine.ch/irdp/stat and http://www.cessul.ulaval.ca). These programs are relatively easy to use and they provide excellent procedures for handling most of the computational complexities of generalizability theory. Unfortunately, these programs cannot handle datasets with missing observations.

Appendix A contains examples of GENOVA program setups for the one-facet crossed generalizability study design using raw data, mean squares and variance components as input. One nice feature of the GENOVA program is that any number of D study designs can also be specified for estimation. For example, lines 21–27 specify four crossed D studies with different choices (1, 5, 10, & 20) for the number of observations of the Item facet. In contrast, lines 28–32 specify four nested D studies using the same number of observations of the Item facet.

Other programs that can be used to estimate variance components for use in generalizability analysis include the REML program (Robinson, 1987), the SAS-PROC VARCOMP and SAS-PROC MIXED procedures (SAS Institute, 1994), the SPSS VARIANCE COMPONENTS procedure (SPSS Inc., 1994), and even general-purpose structural equation modeling programs like LISREL or EQS (Marcoulides, 1996). Appendix B contains an example SAS-PROC VARCOMP program setup and sample output for the same one-facet crossed design. As can be seen in Appendix B, the program provides the estimated variance components, but the user must compute the error variances and generalizability coefficients separately.

## VII. CONCLUSION

Generalizability theory is a comprehensive method for designing, assessing, and improving the dependability of measurement procedures. Generalizability theory most certainly deserves the serious attention of all-researchers using measurement procedures. As illustrated in this chapter, the results obtained from a generalizability analysis can provide important information for determining the psychometric properties of almost any type of measurement procedure. Generalizability analysis is also essential for determining what modifications can or should be made to a measurement procedure. Although the example cases illustrated throughout this chapter were purposely kept simple, it should be relatively easy for applied researchers to adapt the discussion to other types of measurement designs that they encounter in practical applications.

## REFERENCES

Brennan, R. L. (1983). *Elements of generalizability theory.* Iowa City, IA: American College Testing.

Brennan, R. L., & Kane, M. T. (1977). An index of dependability for mastery of tests. *Journal of Educational Measurement, 14,* 277–289.

Burt, C. (1936). The analysis of examination marks. In P. Hartog & E. C. Rhodes (Eds.), *The marks of examiners* (pp. 245–314). London: Macmillan.

Cardinet, J., Tourneur, Y., & Allal, L. (1976). The symmetry of generalizability theory: Application to educational measurement. *Journal of Educational Measurement, 13,* 119–135.

Cronbach, L. J., Gleser, G. C., Nanda, H., & Rajaratnam, N. (1972). *The dependability of behavioral measurements: Theory of generalizability scores and profiles.* New York: Wiley.

Cronbach, L. J., Rajaratnam, N., & Gleser, G. C. (1963). Theory of generalizability: A liberization of reliability theory. *British Journal of Statistical Psychology, 16,* 137–163.

Gleser, G. C., Cronbach, L. J., & Rajaratnam, N. (1965). Generalizability of scores influenced by multiple sources of variance. *Psychometrika, 30*(4), 395–418.

Goldstein, Z., & Marcoulides, G. A. (1991). Maximizing the coefficient of generalizability in decision studies. *Educational and Psychological Measurement, 51*(1), 55–65.

Hoyt, C. J. (1941). Test reliability estimated by analysis of variance. *Psychometrika, 6,* 153–160.

Marcoulides, G. A. (1987). *An alternative method for variance component estimation: Applications to generalizability theory.* Unpublished doctoral dissertation, University of California, Los Angeles.

Marcoulides, G. A. (1989). The estimation of variance components in generalizability studies: A resampling approach. *Psychological Reports, 65,* 883–889.

Marcoulides, G. A. (1990). An alternative method for estimating variance components in generalizability theory. *Psychological Reports, 66*(2), 102–109.

Marcoulides, G. A. (1993). Maximizing power in generalizability studies under budget constraints. *Journal of Educational Statistics, 18*(2), 197–206.

Marcoulides, G. A. (1995). Designing measurement studies under budget constraints: Controlling error of measurement and power. *Educational and Psychological Measurement, 55*(3), 423–428.

Marcoulides, G. A. (1996). Estimating variance components in generalizability theory: The covariance structure analysis approach. *Structural Equation Modeling, 3*(3), 290–299.

Marcoulides, G. A. (1997a, March). Generalizability theory: Models and applications. Invited session at the Annual Meeting of the American Educational Research Association, Chicago, Illinois.

Marcoulides, G. A. (1997b). Optimizing measurement designs with budget constraints: The variable cost case. *Educational and Psychological Measurement, 57*(5), 808–812.

Marcoulides, G. A. (1998). Applied generalizability theory models. In G. A. Marcoulides (Ed.), *Modern methods for business research.* Mahwah, NJ: Lawrence Erlbaum Associates, Inc. Publishers.

Marcoulides, G. A., & Drezner, Z. (1993). A procedure for transforming points in multidimensional space to a two-dimensional representation. *Educational and Psychological Measurement, 53*(4), 933–940.

Marcoulides, G. A., & Drezner, Z. (1995, April). *A new method for analyzing performance assessments.* Paper presented at the Eighth International Objective Measurement Workshop, Berkeley, California.

Marcoulides, G. A., & Drezner, Z. (1997a). A method for analyzing performance assessments. In M. Wilson, K. Draney, G. Engelhard, Jr. (Eds.). *Objective measurement: Theory into practice.* Ablex Publishing Corporation.

Marcoulides, G. A., & Drezner, Z. (1997b, March). *A procedure for detecting pattern clustering in measurement designs.* Paper presented at the Ninth International Objective Measurement Workshop, Chicago, Illinois.

Marcoulides, G. A., & Goldstein, Z. (1990). The optimization of generalizability studies with resource constraints. *Educational and Psychological Measurement, 50*(4), 782–789.

Marcoulides, G. A., & Goldstein, Z. (1991). Selecting the number of observations in multivariate measurement designs under budget constraints. *Educational and Psychological Measurement, 51*(4), 573–584.

Marcoulides, G. A., & Goldstein, Z. (1992). The optimization of multivariate generalizability studies under budget constraints. *Educational and Psychological Measurement, 52*(3), 301–308.

Muthén, L. (1983). *The estimation of variance components for dichotomous dependent variables: Applications to test theory.* Unpublished doctoral dissertation, University of California, Los Angeles.

Robinson, D. L. (1987). Estimation and use of variance components. *The Statistician, 36,* 3–14.

SAS Institute, Inc. (1994). *SAS user's guide, version 6.* Cary, NC: Author.

Schmidt-McCollam, K. M. (1998). Latent trait and latent class models. In G. A. Marcoulides (Ed.), *Modern methods for business research* (pp. 23–46). Mahwah, NJ: Lawrence Erlbaum Associates, Inc. Publishers.

Shavelson, R. J., & Webb, N. M. (1981). Generalizability theory: 1973–1980. *British Journal of Mathematical and Statistical Psychology, 34,* 133–166.

Shavelson, R. J., & Webb, N. M. (1991). *Generalizability theory: A primer.* Sage Publications, Newbury Park, CA.

Shavelson, R. J., Webb, N. M., & Rowley, G. L. (1989). Generalizability theory. *American Psychologist, 44*(6), 922–932.

Webb, N. M., Shavelson, R. J., & Maddahian, E. (1983). Multivariate generalizability theory. In L. J. Fyans, Jr. (Ed.), *Generalizability theory: Inferences and practical applications* (pp. 49–66). San Francisco, CA: Jossey-Bass.

**APPENDIX A SAMPLE GENOVA PROGRAMS**

**Program with Raw Data Input**

| | | | | |
|---|---|---|---|---|
| 1 | GSTUDY | P X I DESIGN - RANDOM MODEL | | |
| 2 | OPTIONS | RECORDS 2 | | |
| 3 | EFFECT | *P | 5 | 0 |
| | | +1 | 5 | 0 |
| 4 | FORMAT | (5f2.0) | | |
| 5 | PROCESS | | | |
| | data set placed here | | | |
| 20 | COMMENT | D STUDY CONTROL CARDS | | |
| 21 | COMMENT | FIRST D STUDY | | |
| 22 | DSTUDY | #1 | -PI DESIGN I RANDOM | |
| 23 | DEFFECT | $ P | | |
| 24 | DEFFECT | I 1 5 10 20 | | |
| 26 | DCUT | | | |
| 27 | ENDSTUDY | | | |
| 28 | COMMENT | SECOND D STUDY | | |
| 29 | DSTUDY | #3 | I:P | NESTED DESIGN |
| 30 | DEFFECT | $ P | | |
| 31 | DEFFECT | I:P 1 5 10 20 | | |
| 32 | ENDSTUDY | | | |
| 33 | FINISH | | | |

**Program with Mean squares as Input**

| | | | | |
|---|---|---|---|---|
| 1 | GMEANSQUARES | PI DESIGN | | |
| 2 | MEANSQUARE | P | 8.74 | 5 |
| 3 | MEANSQUARE | I | 4.34 | 5 |
| 4 | MEANSQUARE | PI | 0.84 | |
| 5 | ENDMEAN | | | |
| 6 | COMMENT | D STUDY CONTROL CARDS | | |
| 7 | | | | |
| | same as above D studies | | | |
| 20 | FINISH | | | |

**Program with Variance Components as Input**

| | | | | |
|---|---|---|---|---|
| 1 | GCOMPONENTS | PI DESIGN | | |
| 2 | VCOMPONENT | P | 1.58 | 5 |
| 3 | VCOMPONENT | I | 0.70 | 5 |
| 4 | VCOMPONENT | PI | 0.84 | |
| 5 | ENDCOMP | | | |
| 6 | COMMENT | D STUDY CONTROL CARDS | | |
| 7 | | | | |
| | same as above D studies | | | |
| 8 | FINISH | | | |

## APPENDIX B SAMPLE SAS PROC VARCOMP PROGRAM WITH OUTPUT

**SAS Program with Raw Data Input**
```
DATA EXAMPLE1;
INPUT PERSON ITEM SCORE;
LINES;
1 1 9
1 2 6
1 3 6
1 4 8
2 1 8
:    :    :
:    :    :
:
;
PROC ANOVA;
CLASS PERSON ITEM;
MODEL SCORE=PERSON|ITEM;
PROC VARCOMP;
CLASS PERSON ITEM;
MODEL SCORE=PERSON|ITEM;
```

## Example SAS Output

```
                Variance Components Estimation Procedure
                     Class Level Information
            Class     Levels     Values
            PERSON        5      1 2 3 4 5
            ITEM          5      1 2 3 4 5
         MIVQUE(0) Variance Component Estimation Procedure
                                                Estimate
            Variance Component                    SCORE
            Var(PERSON)                           1.58
            Var(ITEM)                             0.70
            Var(PERSON*ITEM)                      0.84
                     Analysis of Variance Procedure
Dependent Variable: SCORE
Source                   DF   Anova SS   Mean Square   F Value   Pr>F
PERSON                    4     34.96       8.74          .        .
ITEM                      4     17.36       4.34          .        .
PERSON*ITEM              16     13.44       0.84          .        .
```

# ITEM RESPONSE MODELS FOR THE ANALYSIS OF EDUCATIONAL AND PSYCHOLOGICAL TEST DATA

**RONALD K. HAMBLETON, FREDERIC ROBIN, DEHUI XING**

*College of Education, University of Massachusetts at Amherst, Amherst, Massachusetts*

## I. INTRODUCTION

Psychometric theory and psychological and educational assessment practices have changed considerably since the seminal publications of Lord (1952, 1953), Birnbaum (1957), Lord and Novick (1968), Rasch (1960), and Fischer (1974) on item response theory (IRT). Several unidimensional and multidimensional item response models for use with dichotomous and polytomous response data have been advanced, IRT parameter estimation, statistical software, and goodness-of-fit procedures have been developed, and small- and large-scale applications of IRT models to every aspect of testing and assessment have followed (van der Linden & Hambleton, 1997). Applications of IRT models were slow to be implemented because of their complexity and a shortage of suitable software, but now they are widely used by testing agencies and researchers.

These changes in psychometric theory and practices over the last 30 years have occurred because IRT models permit more flexibility in the test development and data analysis process, they have more useful properties than classical test models (e.g., item statistics that are less dependent on

*Handbook of Applied Multivariate Statistics and Mathematical Modeling*
Copyright © 2000 by Academic Press. All rights of reproduction in any form reserved.

examinee samples), and they allow psychometricians to more effectively model the test data they work with.

Today, the development and refinement of the largest and most important assessment instruments in the United States is guided by item response models, including the *National Assessment of Educational Progress* (a national assessment system for grades 4, 8, and 12), the *Third International Mathematics and Science Study* (used to evaluate the quality of science and mathematics achievement in over 40 countries), three of the major standardized achievement tests used in the United States (i.e., the California Achievement Tests, Comprehensive Tests of Basic Skills, and the Metropolitan Achievement Tests), admissions tests used to determine admission to undergraduate universities and graduate schools (e.g., the Scholastic Assessment Test, Graduate Management Admissions Test, Law School

**TABLE 19.1    Classical Test Theory and Three-Parameter Item Response Theory Model Item Statistics**

| Item | Proportion correct | Item discrimination | Three-parameter logistic model item statistics | | |
| --- | --- | --- | --- | --- | --- |
| | | | Difficulty | Discrimination | Guessing |
| 1 | 0.51 | 0.51 | 0.16 | 0.89 | 0.10 |
| 2 | 0.80 | 0.22 | −1.85 | 0.40 | 0.19 |
| 3 | 0.78 | 0.43 | −1.03 | 0.84 | 0.18 |
| 4 | 0.96 | 0.34 | −2.19 | 1.38 | 0.14 |
| 5 | 0.33 | 0.39 | 1.06 | 0.66 | 0.07 |
| 6 | 0.18 | 0.31 | 1.79 | 1.13 | 0.08 |
| 7 | 0.85 | 0.36 | −1.56 | 0.79 | 0.18 |
| 8 | 0.74 | 0.25 | −1.21 | 0.41 | 0.18 |
| 9 | 0.87 | 0.44 | −1.46 | 1.11 | 0.16 |
| 10 | 0.86 | 0.28 | −2.13 | 0.53 | 0.16 |
| 11 | 0.30 | 0.25 | 1.79 | 0.47 | 0.10 |
| 12 | 0.61 | 0.39 | 0.01 | 0.72 | 0.22 |
| 13 | 0.56 | 0.60 | −0.10 | 1.29 | 0.09 |
| 14 | 0.82 | 0.58 | −1.00 | 1.98 | 0.15 |
| 15 | 0.52 | 0.34 | 0.27 | 0.52 | 0.13 |
| 16 | 0.21 | 0.58 | 1.03 | 1.99 | 0.02 |
| 17 | 0.78 | 0.51 | −0.97 | 1.10 | 0.14 |
| 18 | 0.69 | 0.56 | −0.55 | 1.24 | 0.13 |
| 19 | 0.54 | 0.40 | 0.21 | 0.68 | 0.16 |
| 20 | 0.59 | 0.51 | 0.02 | 1.20 | 0.19 |
| 21 | 0.83 | 0.37 | −1.37 | 0.72 | 0.22 |
| 22 | 0.22 | 0.50 | 1.15 | 1.27 | 0.03 |
| 23 | 0.26 | 0.24 | 1.80 | 1.14 | 0.18 |
| 24 | 0.52 | 0.63 | 0.01 | 1.42 | 0.06 |
| 25 | 0.40 | 0.42 | 0.83 | 1.45 | 0.18 |

*(continues)*

**TABLE 19.1** (*continued*)

| Item | Proportion correct | Item discrimination | Three-parameter logistic model item statistics | | |
|------|------|------|------|------|------|
| | | | Difficulty | Discrimination | Guessing |
| 26 | 0.91 | 0.41 | −1.72 | 1.35 | 0.14 |
| 27 | 0.94 | 0.31 | −2.36 | 0.86 | 0.16 |
| 28 | 0.85 | 0.46 | −1.35 | 1.09 | 0.16 |
| 29 | 0.85 | 0.46 | −1.38 | 1.03 | 0.15 |
| 30 | 0.15 | 0.17 | 2.10 | 1.52 | 0.11 |
| 31 | 0.47 | 0.49 | 0.39 | 1.06 | 0.13 |
| 32 | 0.77 | 0.57 | −0.90 | 1.38 | 0.11 |
| 33 | 0.79 | 0.37 | −1.08 | 0.71 | 0.22 |
| 34 | 0.90 | 0.32 | −1.98 | 0.81 | 0.18 |
| 35 | 0.26 | 0.53 | 0.98 | 1.40 | 0.04 |
| 36 | 0.34 | 0.23 | 1.79 | 0.59 | 0.19 |
| 37 | 0.91 | 0.32 | −2.10 | 0.83 | 0.16 |
| 38 | 0.92 | 0.33 | −2.18 | 0.88 | 0.15 |
| 39 | 0.60 | 0.53 | −0.08 | 1.19 | 0.16 |
| 40 | 0.41 | 0.34 | 0.95 | 0.69 | 0.16 |
| 41 | 0.79 | 0.46 | −1.05 | 0.99 | 0.16 |
| 42 | 0.54 | 0.39 | 0.21 | 0.69 | 0.16 |
| 43 | 0.46 | 0.37 | 0.83 | 1.02 | 0.24 |
| 44 | 0.90 | 0.34 | −1.86 | 0.86 | 0.17 |
| 45 | 0.27 | 0.20 | 2.17 | 0.71 | 0.17 |
| 46 | 0.63 | 0.52 | −0.18 | 1.12 | 0.17 |
| 47 | 0.21 | 0.21 | 2.07 | 1.04 | 0.14 |
| 48 | 0.22 | 0.17 | 2.21 | 1.05 | 0.17 |
| 49 | 0.85 | 0.29 | −1.90 | 0.56 | 0.19 |
| 50 | 0.94 | 0.41 | −1.84 | 1.72 | 0.15 |

Admission Test, and the Graduate Record Exam), and the Armed Services Vocational Aptitude Battery (used to assign military recruits to occupational specialties). Numerous other achievement, aptitude, and personality tests could be added to this list of tests in which IRT models are used. Test applications based upon IRT principles and applications affect millions of students in the United States and in other countries each year (Hambleton, 1989).

Clearly, IRT models are central today in test development, test evaluation, and test data analysis. The purposes of this chapter are (a) to introduce a number of widely used IRT models and (b) to describe briefly their application in test development and computer-adaptive testing. These purposes will be accomplished using examples based on a hypothetical 50-item test given to a sample of 750 examinees drawn from a normal ability distribution. The examinee data were simulated to fit the unidimensional three-parameter logistic model. Table 19.1 contains the classical item difficulty (proportion-correct) and item discrimination indices (point biserial correlations) for the 50 items. We conclude this chapter by highlighting new directions for developing and using IRT models and identifying important issues requiring research.

## II. SHORTCOMINGS OF CLASSICAL TEST MODELS

Theories and models in educational and psychological testing specify anticipated relations among important variables such as examinee ability and errors of measurement; they provide the basis for a deeper understanding of the relation among variables; they provide a way for predicting and explaining observable outcomes such as test score distributions; and they are valuable in understanding the role of errors and how errors might be controlled.

Much of classical test theory is concerned with the estimation and control of error in the testing process. The theories and models postulate relations between examinee observable performance on test items, rating scales, or performance tasks and the latent variables describing examinee performance. Error is operationalized as the difference between the latent variable describing ability and examinee performance.

The classical test model assumes a linear relation among test score ($X$), true score ($T$), and error ($E$) ($X = T + E$). Error is assumed to be randomly distributed with a mean of 0 across the population of examinees for whom the test is intended, and to be uncorrelated with true scores and with error scores on parallel test administrations. No distributional assumptions are made about $X$, $T$, or $E$. Based on this reasonable linear model, true score theory has been elaborated into a comprehensive psychometric model that has long been used to guide test development, test evaluation, and test score interpretation (see, for example, Crocker & Algina, 1986; Gulliksen, 1950; Lord & Novick, 1968).

Unfortunately, classical test theory and measurement procedures have several important shortcomings. One shortcoming is that the values of such classical item statistics as item difficulty and item discrimination depend on the sample of examinees measured. The mean and standard deviation of ability scores in an examinee group affect the values of item statistics, and reliability and validity statistics, too. These sample-dependent item and test statistics are useful only when constructing tests for examinee populations that are similar to the sample of examinees from which the item and test statistics were obtained. Unfortunately, field test samples often are not representative of the population of persons for whom the test is being constructed.

A second shortcoming of classical test models is that examinees' scores are largely a function of the difficulty of the items they are administered.

Matching the difficulty of a test to the ability level of the examinee allows the use of shorter tests that produce more precise ability estimates, but, of necessity, test difficulty will vary substantially across examinees. Examinee test scores are not comparable when several forms of a test that vary in difficulty are used (i.e., nonparallel forms).

Another shortcoming of the classical test model is that the best-known and most commonly used form of the classical test model (Gulliksen, 1950) (sometimes referred to as the weak true-score model) requires the assumption of equal errors of measurement for all examinees. This seems implausible (see, for example, Lord, 1984). Violations of the equal error variance assumption are the rule. For example, errors of measurement due to factors such as guessing on a difficult multiple-choice test are greater for low-ability examinees than for average- and high-ability examinees (see Yen, 1983). Although these violations might not be threatening to the overall usefulness of the classical model, models where the assumption is not made in the derivations are preferable. Strong true-score models (Brennan & Lee, 1999; Kolen & Brennan, 1995; Lord & Novick, 1968) represent one excellent solution to obtaining error estimates conditional on test score or ability within the classical test theory framework, but such a solution does not address the other shortcomings of the classical model.

## III. INTRODUCTION TO ITEM RESPONSE THEORY MODELS

Classical test models have served the measurement field well for over 80 years, and countless numbers of tests have been constructed and evaluated with these models. Unfortunately, those models lack the following desired features: (a) item statistics that are *not* group dependent, (b) examinee ability estimates that are not dependent on test difficulty, (c) test models that provide a basis for matching test items to ability levels, and (d) test models that are not based on implausible or difficult-to-meet assumptions.

These desirable properties can be obtained, in principle, within the framework of another test theory known as item response theory or "IRT" (Hambleton & Swaminathan, 1985; Hulin, Drasgow, & Parsons, 1983; Lord, 1980; Wright & Stone, 1979). IRT actually refers to a group of statistical procedures for modeling the relation between examinee ability and item and test performance. Here, the term *ability* is used to describe the construct measured by the test—it might be an aptitude, achievement, or personality variable.

Many IRT models have been formulated and applied to real test data (see Thissen & Steinberg, 1986; van der Linden & Hambleton, 1997). Models have been developed for use with discrete or continuous examinee item responses that are dichotomously or polytomously scored; for ordered or unordered item score categories; for homogeneous (i.e., one ability) or heterogeneous (i.e., multiple) latent abilities; and for a variety of relations between item responses and the underlying ability or abilities.

Typically, assumptions about the dimensional structure of the test data and the mathematical form of the item characteristic function or curve are made in specifying IRT models. Both are strong assumptions in that they may not be applicable to some sets of test data. Fortunately, there are statistical approaches for assessing the viability of the assumptions.

### A. Item Characteristic Functions or Curves

The upper part of Figure 19.1 shows the general form of item characteristic functions (often called item characteristic curves or ICCs) applied to dichotomous data. This model assumes that the ability measured by the test to which the model is applied is unidimensional (i.e., that the test principally

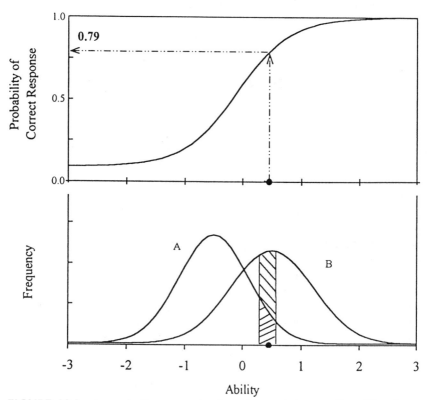

**FIGURE 19.1**   A sample three-parameter item characteristic curve (item 13) and two ability distributions (denoted A and B).

measures a single or unitary trait, and little is lost by characterizing examinee performance by a single ability parameter).

The three-parameter item characteristic functions are of the form:

$$P_i(\theta) = c_i + (1 - c_i) \frac{e^{Da_i(\theta - b_i)}}{1 + e^{Da_i(\theta - b_i)}} \quad i = 1, 2, \ldots, n,$$

which serves as the mathematical model linking the observable data (examinee item performance) to the unobservable variable of interest (i.e., examinee ability). $P_i(\theta)$ gives the probability of a correct response to item $i$ as a function of ability (denoted $\theta$). The symbol $n$ is the number of items in the test. The **c**-parameter (i.e., pseudo-guessing parameter) in the model is the height of the lower asymptote of the ICC. This parameter is introduced into the model to account for the performance of low-ability examinees on multiple-choice test items. It is not needed in a model to fit free-response data. The **b**-parameter (i.e., item difficulty parameter) is the point on the ability scale where an examinee has a $(1 + \mathbf{c})/2$ probability of a correct answer. The **a**-parameter (i.e., item discrimination parameter) is proportional to the slope of the ICC at the point $b$ on the ability scale. In general, the steeper the slope the higher the **a**-parameter. The $D$ in the model is simply a scaling factor.

By varying the item parameters, many "S"-shaped curves or ICCs can be generated to fit actual test data. Simpler logistic IRT models can be obtained by setting $\mathbf{c}_i = 0$ (the two-parameter model) or setting $\mathbf{c}_i = 0$ and $\mathbf{a}_i = 1$ (the one-parameter model). Thus, three different logistic models may be fit to the test data. Figure 19.2 depicts a typical set of ICCs for eight randomly selected items from Table 19.1. The corresponding three-parameter logistic model item statistics appear in Table 19.1. The items in Table 19.1 represent a range of item difficulties (items 4 and 33 are relatively easy, whereas items 22 and 30 are relatively hard), discriminations (items 4 and 16 are highly discriminating, whereas items 5 and 19 are less discriminating), and pseudo-guessing parameters.

### B. Item Response Theory Model Properties

When the assumptions of an item response model can be met by a set of test data, at least to a reasonable degree, the item and ability model parameter estimates obtained have two desirable properties. First, examinee ability estimates are defined in relation to the pool of items from which the test items are drawn but do not depend upon the particular sample of items selected for the test. Therefore, examinees can be compared even though they might not have taken identical or even parallel sets of test items. Second, item descriptors or statistics do not depend upon the particular sample of examinees used to estimate them.

The property of item parameter invariance can be observed in Figure

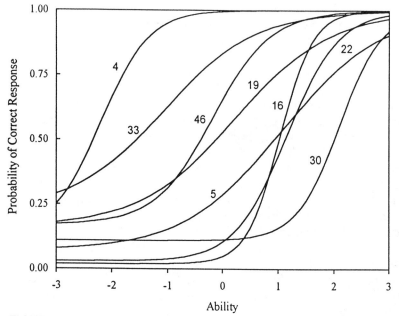

**FIGURE 19.2** Three-parameter item characteristic curves for eight typical items.

19.1. Notice that the ICC applies equally well to examinees in distributions A and B. For examinees at a given ability level, the probability of a successful response to this dichotomously scored item is the same (i.e., 0.79 in this example) regardless of the number of examinees at that ability level in each distribution. For each ability level, there is a single probability of a correct response. Of course, that probability does not depend on the number of examinees in each ability group at that ability level. In that sense, the ICC applies equally well to both groups, and the item parameters are said to be "invariant" across examinee groups. In contrast, classical item parameters such as item difficulty (i.e., proportion correct) are not invariant across examinee groups. In the example in Figure 19.1, the test item would be substantially more difficult in group A than in group B.

IRT models define items and ability scores on the same scale, so items can be selected to provide optimal measurement (minimum errors) at ability levels of interest. They also allow the concept of parallel test forms, which is central to the most popular and commonly used form of the classical test model, to be replaced by a statistical method that permits estimation of different standard errors of measurement for examinees at different ability levels.

IRT models link examinee item responses to ability, and they report item statistics on the same scale as ability, thereby yielding information about where an item provides its best measurement on the ability scale, and about the exact relation between item performance and ability.

## C. Test Characteristic Function

One useful feature of IRT is the test characteristic function (or test characteristic curve: TCC), which is the sum of the item characteristic functions that make up a test. The TCC can be used to predict the test scores of examinees at given ability levels using the following equation:

$$\text{TCC}(\theta) = \sum_{i=1}^{n} P_i(\theta).$$

If a test is made up of test items that are relatively more difficult than those in a typical test, the test characteristic function is shifted to the right on the ability scale and examinees tend to have lower expected scores on the test than if easier test items are included. Thus, it is possible through the test characteristic function to explain how it is that examinees with a fixed ability can perform differently on two tests, apart from the ubiquitous error scores. IRT and classical test theory are related in that an examinee's expected test score at a given ability level, determined by consulting the TCC, is by definition the examinee's true score on that set of test items. The TCC is also valuable in predicting test score distributions for both known and hypothetical ability distributions and for considering the potential effects of test design changes on test score distributions.

## D. Unidimensional Item Response Theory Models for Fitting Polytomous Response Data

Although important technical and application work was going on with the unidimensional normal and logistic IRT models, other developments started in the late 1960s and became a serious psychometric activity of researchers beginning in the early 1980s. Samejima (1969), for example, introduced the very important and useful graded response model to analyze data from Likert attitude scales and polychotomously score performance data as might be obtained from, for example, scoring essays. Her model and extensions of it were the first of many models developed by psychometricians to handle polytomously scored data (see, van der Linden & Hambleton, 1997, for a review of this work).

In Samejima's graded response model, the probability of an examinee obtaining a particular score, $x$, or a higher score up to the highest score $m_i$ on item $i$, is assumed to be given by a two-parameter logistic model:

$$P_{ix}^*(\theta) = \frac{e^{Da_i(\theta - b_{ix})}}{1 + e^{Da_i(\theta - b_{ix})}} \quad i = 1, 2, \ldots, n; x = 0, 1, \ldots, m_i$$

The expression represents the probability of the examinee obtaining a score of $x$ or higher on item $i$. In addition, the probabilities of obtaining scores of 0 or greater, $m_i + 1$ or greater, and $x$ on item $i$ need to be defined:

$$P_{i0}^*(\theta) = 1.0$$
$$P_{i(m_i+1)}^*(\theta) = 0.0$$

Then,

$$P_{ix}(\theta) = P_{ix}^*(\theta) - P_{i(x+1)}^*(\theta)$$

is the probability of the examinee obtaining a score of $x$. It is assumed that each response category or score category has its own "score category function" that provides the probability of examinees at each ability level making that choice or obtaining that particular score. At each ability level then, the sum of probabilities associates with the available responses or possible score points is 1. Very capable (or high ability) examinees would have high probabilities associated with the highest score points and low probabilities associated with the lower score points, and less capable (or lower ability) examinees would have high probabilities associated with the lower score points and low probabilities associated with the higher score points.

Samejima permitted the researcher flexibility in describing these score category functions by their difficulty and a constant level of discrimination across the response categories within a task or problem (i.e., a one-parameter model), or by their difficulty and level of discrimination for each response category within a task or problem (i.e., a two-parameter model). Samejima's model (and variations on it) provides considerable flexibility to the researcher in attempting to fit a model to a dataset.

Figure 19.3 highlights the score category functions for an item with four response categories. Equations for this model, and related IRT models for handling polytomously scored data such as the partial credit model and the generalized partial credit model are found in van der Linden and Hambleton (1997). The discriminating power of the item is 1.00. The item thresholds, denoted $b_{i1}$ (i.e., $-1.25$), $b_{i2}(-0.25)$, and $b_{i3}$ (1.50), are the points on the ability scale where the item score category functions designate a 50% probability of examinees obtaining scores of 1 or higher, 2 or higher, or 3, respectively. For example, at $\theta = -0.25$, an examinee has a 50% probability of obtaining a score of 2 or 3 (and, by extrapolation, a 50% chance of obtaining a score of 0 or 1). For all higher abilities, the probability of obtaining a score of 2 or 3 is greater than 50%, and for all abilities lower, the probability of obtaining a score of 2 or 3 is less than 50%.

The generalized partial credit model (Muraki, 1992) is also popular for

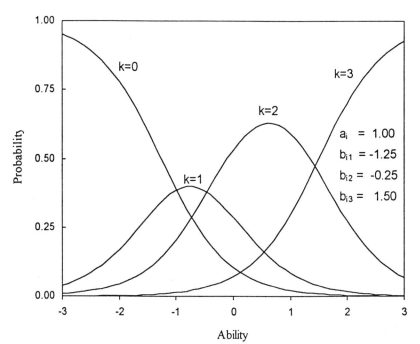

**FIGURE 19.3**  Score category functions for the graded response model fitted to a four-point test item.

analyzing ordered polytomously scored data. The probability of receiving a score of $x$ on item $i$ is obtained directly from the model:

$$P_{ix}(\theta) = \frac{e^{\sum_{k=0}^{x} Da_i(\theta - b_{ik})}}{\sum_{h=0}^{m_i} e^{\sum_{k=0}^{h} Da_i(\theta - b_{ik})}}.$$

The modeling of the response process used by examinees is different from the graded response model, but very similar results to the graded response model are obtained when applied to the same data (Fitzpatrick et al., 1996; Tang & Eignor, 1997). A special case of this model is the popular partial credit model (Masters, 1982; Masters & Wright, 1997), where polytomously scored items are assumed to have equal discriminating powers.

Bock's nominal response model is an even more general unidimensional model than the models that Samejima's work has generated (see, for example, Bock, 1972, 1997). This model can be applied to data where no ordering of the response categories is assumed (that is, nominal response data). For example, instead of scoring multiple-choice responses dichotomously as right (1) or wrong (0) and fitting a logistic IRT model to the 0–1 data,

Bock's nominal response model fits an item response function to each possible answer choice or response category. In this respect, it is like Samejima's graded response model (see Figure 19.3), except that the answer choices or response categories substitute for the ordered score categories 0 to 3. For each answer choice or response category there is a probability function linking the likelihood (or probability) that examinees of different abilities will make that choice.

In one application to multiple-choice test items, the nominal response model allows for the extraction of more information from item responses on a test than simple success or failure on the item. In many tests, some answer choices reflect more partial knowledge than others (e.g., a near correct answer reflects more partial knowledge than an answer choice that represents a serious misunderstanding of the concept being assessed). This model uses information about the degree of correctness of the answer choices in fitting the nominal response model to the item and in the estimation of ability. As a consequence, it yields more precise estimates of person ability and a fuller characterization of the test item and how it relates to the ability measured by the test.

### E. Multidimensional Item Response Theory Models

Multidimensional IRT models were introduced originally by Lord and Novick (1968), Samejima (1974), and, more recently, by Embretson (1984, 1997), Fischer and Seliger (1997), McDonald (1967, 1982, 1989), and Reckase (1997a, 1997b). They represent an excellent approach for handling binary and polytomously scored data that are multidimensional in their character. Multidimensional IRT models offer the prospect of better fitting certain types of test data (e.g., assessments of higher level cognitive skills) because they consider the interrelations among the items and provide multi-dimensional representations of both items and examinee abilities. These models replace the single parameter used to describe ability in the unidimensional IRT models with a vector of abilities. One ability is estimated for each dimension, which is present in the multidimensional representation of the item data. Vectors of item difficulties, discriminations, and pseudo-guessing parameters—one for each dimension—also may be produced (the number of item parameters depends on the model of choice).

It remains to be seen whether parameters for these multidimensional models can be estimated properly, and whether multidimensional representations of items and examinees are useful to practitioners (see, for example, Reckase, 1997a, 1997b). To date, this general direction of model building has been somewhat hampered by software difficulties, but work is progressing and important results can be expected in the future. Examples of software to carry out multidimensional IRT analyses are described by Fraser and McDonald (1988) and Fisher and Seliger (1997).

## IV. ITEM RESPONSE THEORY PARAMETER ESTIMATION AND MODEL FIT

Many approaches for estimating IRT model parameters have been described in the measurement literature (see, for example, Baker, 1992; Bock & Aitkin, 1981; Lord, 1980; Swaminathan, 1997). Variations on maximum likelihood estimation (MLE), including joint MLE (JMLE), conditional MLE (CMLE), and marginal MLE (MMLE), have been the most popular and appear in standard IRT software packages such as BILOG (Mislevy & Bock, 1986) and PARSCALE (Muraki & Bock, 1993).

A common assumption of these estimation procedures is that the principle of local independence applies. By the assumption of local independence, the joint probability of observing the response pattern $(U_1, U_2, \ldots, U_n)$, where $U_i$ is either 1 (a correct response) or 0 (an incorrect response) (in the case of the common IRT models for handling dichotomously scored data) is

$$P(U_1, U_2, \ldots, U_n | \theta) = P(U_1 | \theta) P(U_2 | \theta) \cdots P(U_n | \theta) = \prod_{i=1}^{n} P(U_i | \theta).$$

This principle means that examinee responses to test items are independent of one another and depend only on examinee ability.

When the response pattern is observed, $U_i = u_i$, the expression for the joint probability is called the likelihood, denoted:

$$L(u_1, u_2, \ldots, u_n | \theta) = \prod_{i=1}^{n} P_i^{u_i} Q_i^{1-u_i},$$

where $P_i = P(u_i = 1 | \theta)$ and $Q_i = 1 - P(u_i = 0 | \theta)$. L can be differentiated with respect to the ability parameter to produce an equation. If item parameters are known or assumed to be known, the differential equation can be set equal to zero and solved for the unknown ability parameter. Solving the equation for the ability parameter is carried out using the Newton-Raphson procedure. Item parameters are not known, but available estimates can be used instead to obtain the MLE of examinee ability. The same process is repeated for each examinee, substituting his/her response pattern into the equation, to obtain the corresponding examinee ability estimate. The basic idea here is to find the value of the unknown ability parameter that maximizes the probability of the data for the examinee that was observed.

A similar procedure can be used to obtain item parameter estimates. The likelihood expression (formed for all examinees and items, assuming their independence) is differentiated with respect to the unknown parameters for an item, ability estimates are assumed to be known, and the resulting differential equations are solved using the available examinee responses to the item for model item parameter estimates. The procedure is repeated for each item.

In a typical JMLE process, initial ability estimates are obtained (e.g., converting actual test scores to z-scores and using the z-scores as the initial ability estimates), and then the likelihood equations are solved for item parameter estimates. With item parameter estimates available, new ability estimates are obtained. With new ability estimates available, item parameter estimates are obtained again, and the process is continued until from one stage to the next, the model parameter estimates change by less than some specified amount known as the convergence criterion (e.g., 0.01). The same estimation process can be used with polytomous and multidimensional IRT models.

JMLE has its advantages (e.g., relatively speaking, it is faster than some other estimation procedures with large samples and long tests) (see, for example, Mislevy & Stocking, 1989), but there have remained questions about the quality of the model parameter estimates. For example, consistency of the parameter estimates has never been proven.

MMLE is preferred currently by many psychometricians and is incorporated into current software programs such as BILOG (Mislevy & Bock, 1986) and PARSCALE (Muraki & Bock, 1993). The main virtue of this approach seems to be that the ability parameter is removed from the likelihood equation, which in turn permits more satisfactory item parameter estimates to be obtained. Then, an expected a posteriori (EAP) estimate can be computed by approximating the posterior distribution of $\theta$ based on the available data from the examinee.

Table 19.2 contains basic information about four of the popular software packages today. Mislevy and Stocking (1989) provide a nontechnical review of the properties of one of the standard software packages for fitting logistic test models to dichotomously scored test items. BILOG, to a software package, LOGIST, that was popular in the 1970s and 1980s. LOGIST was developed at the Educational Testing Service (ETS), and for many early researchers of IRT in the 1970s, this was the only software available for fitting multiparameter IRT models. (ETS has recently released a PC version of LOGIST for research use.) LOGIST features JMLE, and although this estimation approach is relatively fast, questions remain about model parameter consistency even for long tests and large examinee samples. BILOG offers JMLE as well as MMLE. With item parameter estimates in hand, BILOG uses Bayesian estimation or MLE to obtain ability parameter estimates. Both software packages produce comparable results when tests are relatively long (45 items or more) and sample sizes are large (bigger than 750). BILOG MMLE procedures appear to estimate parameters somewhat more accurately when tests are short and sample sizes are small.

**TABLE 19.2   Commonly Used Item Response Theory Software[a]**

| Software | Features[b] | Fit indices | Output documents |
|---|---|---|---|
| BILOG-W (Windows 3.x version) and BILOG-MG (DOS Extender) | Dichotomous scoring of response data. IRT modeling using unidimensional 1P, 2P, or 3P logistic models. Single and multigroup marginal maximum-likelihood item calibration procedures for up to 1,000 items (unlimited number of examinees). Scoring using maximum likelihood or Bayesian estimation. Equating, item bias analysis, and drift analyses can be performed with the multigroup version. | Test level likelihood ratio statistic Test and item level chi-square statistics Raw and standardized item level residuals conditional on ability | Phase 1: Classical item statistics Phase 2: IRT item calibration and fit statistics Phase 3: Scoring (examinee parameter estimates) High-quality graphics including: test information and measurement error, item response and item information functions, and item fit |
| MULTILOG Version 6.0 (DOS) | Unidimensional 1P, 2P, or 3P models, graded and nominal IRT models, marginal maximum-likelihood estimation in parameter estimation and maximum likelihood and Bayesian estimation in scoring, multiple category data, multiple group, no strict limitation on the number of items, nor the number of examinees. | Likelihood ratio chi-square goodness-of-fit statistic Observed and expected frequencies Standardized residuals | Internal control codes and the key, format for the data, and first observation Item summary: item parameter estimates and information, and the observed and expected frequencies for each response alternative Population distribution, test information, and measurement error Summary of the goodness-of-fit statistics |
| PARSCALE Version 3.3 (DOS) | Unidimensional, 1P, 2P, 3P logistic models, graded response and partial-credit models, marginal maximum-likelihood estimation, multiple category data, maximum test length and sample size depend on the extended memory. | Likelihood ratio chi-square goodness-of-fit statistic | Phase 0: Log for check up Phase 1: Classical item statistics Phase 2: IRT item and fit statistics Phase 3: Examinee score and ability estimates. |
| ConTEST Version 2.0 (DOS) | Optimal automated test design using pre-existing item banks and linear programming; specifications can be based on IRT or CTT; response data generation for optimal test design research is an option. | | |

[a] These software packages are distributed by Assessment Systems Corporation (ASC), 2233 University Ave., Suite 200, St. Paul, MN 55114-1629 USA. More information can also be found at ASC's website: http://www.assess.com
[b] IRT, item response theory; CTT, classical test theory.

## A. Assessing Item Response Theory Model Fit

The usefulness of IRT modeling of test data is conditional on the extent to which model assumptions are met, and the extent to which an IRT model fits the test data. Local independence of examinee responses to items, unidimensionality of the test, and nonspeededness of the test are among the assumptions of most of the currently used IRT models (Hambleton & Swaminathan, 1985). Hambleton and Swaminathan (1985) classify the types of evidence that are needed to address the adequacy of model fit into three categories: (a) investigations of the violation of model assumptions such as unidimensionality; (b) examination of the presence of the expected advantages of item and ability invariance; and (c) the assessment of the performance of the IRT model in predicting item and test results. Compiling evidence of each type is important because if model fit is poor, desired advantages may not be obtained.

IRT models are based on strong assumptions, some of which are difficult to attain in real testing situations (Traub, 1983). Various procedures have been developed for the investigation of the violation of each assumption. One of the widely used methods of evaluating test dimensionality is to calculate the tetrachoric correlations among the items, submit the correlation matrix to a principal components or common factors analysis, and examine the eigenvalues of the correlation matrix. The eigenvalues obtained from a principal components analysis of the sample test depicted in Table 19.1 were 10.9, 1.9, and 1.7 for components 1 to 3, respectively. Two criteria generally have been used for interpreting the eigenvalues: (a) the first factor should account for at least 20% of the variability, and (b) the first eigenvalue should be several times larger than the second largest eigenvalue (see, for example, Reckase, 1979). The eigenvalues reveal a dominant first factor that accounts for about 22% of the variability. The first eigenvalue is about 5.7 times larger than the second. Factor analysis is not a perfect choice for dimensionality assessment, mainly because linear relations are assumed among the variables and the factors, but this analysis often provides a good approximation for the assessment of the dimensional structure of test data. More up-to-date procedures are identified in van der Linden and Hambleton (1997).

Other assumptions of some IRT models such as the nonspeededness of the test, minimal guessing (with the one- and two-parameter models), and equal discrimination (for the one-parameter model) can be addressed by methods described in Hambleton, Swaminathan, and Rogers (1991). Comparisons of examinee test scores with a time limit and without a time limit, examination of low-ability examinee performance on the most difficult items, and comparisons of classical item point-biserials all can be done to address model assumptions.

Many methods have been recommended for the assessment of the extent to which an IRT model fits a data set (for a review of these methods, see Hambleton, 1989). Examination of the residuals and standardized residuals for a model-data fit study, investigations of model robustness when all assumptions are not fully met, and statistical tests of fit, such as chi-square, are but a few of the many methods used in model-data fit investigations.

Table 19.3 shows the distribution of standardized residuals after fitting the one-, two-, and three-parameter logistic test models (for more on residuals and standardized residuals, refer to Hambleton, et al., 1991). The high percentage of standardized residuals that exceed 2.0 (22.3% with the one-parameter model in the sample data) highlights the poor fit of the one parameter model to the data. Both the two- and three-parameter models fit the data described in Table 19.1 better than the one-parameter model; the three-parameter model appears to fit the data slightly better than the two-parameter model.

Figure 19.4 shows examinee performance on item 13 across 10 ability categories (i.e., the dots represent group performance in ten ability categories) and the one-parameter and three-parameter logistic curves that fit the data best. The residuals and standardized residuals are considerably smaller with the best fitting three-parameter ICC than the best fitting one-parameter ICC. (Recall that the one-parameter model attempts to account for performance entirely in terms of item difficulty, whereas the three-parameter model also uses information about item discrimination and pseudo-guessing in modeling performance.) Residuals should be small and random when the model fits the data, and standardized residuals should be approximately normally distributed (mean = 0, standard deviation = 1) when model fit is good. Figure 19.4 shows that large residuals remain after fitting the one-parameter model to the data, and that there is a pattern to the lack of fit. The best fitting one-parameter model tends to overpredict the performance of low-ability examinees and underpredict the performance of high-ability examinees. The presence of a pattern in the residuals is a sign that variations

**TABLE 19.3   Standardized Residuals Falling in Four Intervals**

| Model | \|0–1\| | \|1–2\| | \|2–3\| | \|>3\| |
|-------|------|------|------|------|
| | Absolute value of standardized residuals (%) | | | |
| 1P | 45.8 | 32.0 | 15.0 | 7.3 |
| 2P | 66.5 | 29.3 | 3.5 | 0.8 |
| 3P | 73.8 | 24.3 | 2.0 | 0.0 |
| Expected | 68.2 | 27.2 | 4.2 | 0.4 |

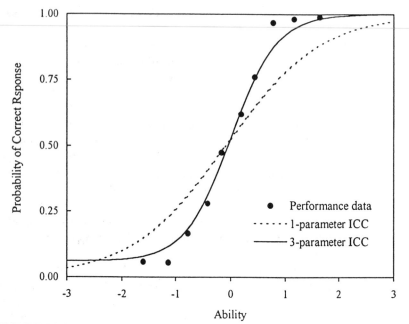

**FIGURE 19.4** Plot of item 13 performance data and the best fitting one-parameter and three-parameter logistic item characteristic curves (ICCs).

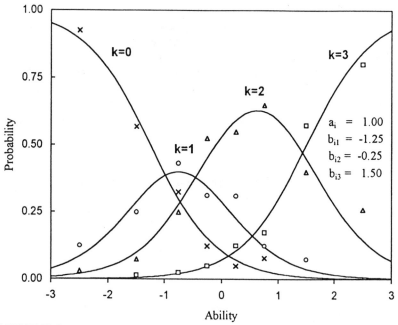

**FIGURE 19.5** Score category functions and residuals for the graded response model fitted to a four-point test item.

in guessing and/or item discrimination have influenced performance, and that a model with more parameters is likely to better fit the data.

Figure 19.5 shows the residuals obtained for the polytomously scored response item shown in Figure 19.3. The residuals appear to be relatively small for each score category. The residuals correspond to the proportion of examinees in each ability category who obtained each of the four possible test scores (0 to 3). The close agreement between examinee performance on the task and the predictions of scores for examinees at each ability level, assuming the validity of the model, is evidence that the best fitting score category functions match the data well.

## V. SPECIAL FEATURES OF ITEM RESPONSE THEORY MODELS

Another feature of IRT models is the presence of item information functions. In the case of the simple logistic models, item information functions show the contribution of particular items to the assessment of ability. The equation below defines the item information function:

$$I_i(\theta) = \frac{[P_i'(\theta)]^2}{P_i(\theta)[1 - P_i(\theta)]}.$$

All variables in the equation were previously defined except one. $P_i'(\theta)$ is an expression for the slope of the ICC (i.e., the discriminating power of the item). Items with greater discriminating power contribute more to measurement precision than items with lower discriminating power. The location of the point where information is a maximum for an item is:

$$\theta_{i_{\max}} = b_i + \frac{1}{Da_i} \ln[0.5(1 + \sqrt{1 + 8c_i})].$$

With the three-parameter model, items provide their maximum information at a point slightly higher than their difficulty (i.e., because of the influence of guessing, the $c$ parameter). The one- and two-parameter logistic models for analyzing dichotomous response data assume that guessing does not influence performance on the item (i.e., $c_i = 0.0$), so the right-hand side of the equation reduces to $b_i$, indicating that items make their greatest contribution to measurement precision near their $b$-value on the ability scale. Similar item information functions can be calculated for other IRT models.

Figure 19.6 shows the item information functions for the 8 items from the 50-item test shown in Figure 19.2. Item 46 is more difficult than item 4, so the item information function for item 46 is centered at a higher ability level than the item information function for item 4. Also, items 16 and 30 are more discriminating than items 5 and 33, so the corresponding item

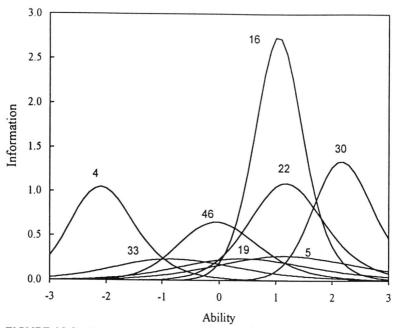

**FIGURE 19.6**   Three-parameter item information functions for eight typical items.

information functions are higher for items 16 and 30. These item information functions are valuable in the test development process, and this point will be addressed later in the chapter.

Another special feature of item response models is the test information function, which is the sum of item information functions in a test:

$$I(\theta) = \sum_{i=1}^{n} I_i(\theta).$$

It provides estimates of the errors associated with (maximum likelihood) ability estimation, specifically,

$$SE(\theta) = \frac{1}{\sqrt{I(\theta)}}$$

The more information provided by a test at a particular ability level, the smaller the errors associated with ability estimation. Figure 19.7 provides the test information function for the 50 items shown in Table 19.1 and the corresponding standard errors (test information is shown by the solid line, and the standard error of estimation by the broken line). Information from item and test information functions allows applied researchers to design a test that will yield desired levels of measurement precision at selected points

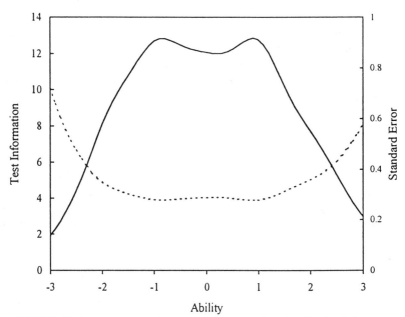

**FIGURE 19.7** Test information and standard error functions for the 50-item test.

along the ability continuum. This point will be discussed in more detail in a later section of the chapter also.

## VI. APPLICATIONS

Item response theory models are receiving increasing use in test design, test-item selection, and computer-administered adaptive testing. Brief descriptions of these applications follow.

### A. Test Development

In test development within a classical framework, items are field tested to obtain item statistics, item banks are established, then items are selected from the banks whenever a test is needed. The content that each item measures, the features of the items such as format, and item statistics are considered in test development.

Test construction methods in an IRT framework are more flexible (Green, Yen, & Burket, 1989). Item and test information functions are used in place of item statistics and test reliability. Item information functions inform test developers about (a) the locations on the ability scale where items provide the most information, and (b) the relative amount of information provided by the test items. A test information function informs test developers about the precision of measurement provided by a test at points along the ability scale. Basically, the more information provided by a test at an ability level, the more precisely ability scores are estimated.

Unlike classical item statistics and reliability, item information functions are independent of other test items so the independent contribution of each item to the measurement precision of the test can be determined. The item parameters, especially the discrimination parameter, determine the information offered by each item at any ability level. The difficulty of an item controls the location of the highest slope of the ICC and hence the location where the item provides the highest information. The explanation of the functioning of polytomously scored items is more complex, but polytomously scored items are often two or three times more informative than dichotomously scored items and often enhance the precision of measurement at the extremes of the ability continuum as well as in the middle of the ability continuum.

The precision of ability estimates is a function of the amount of information provided by a test at an ability level. For example, if a test developer wants to set a passing score at $\theta = -1.0$, items that provide information in the region of $-1.0$ should be chosen. That would increase measurement precision around $\theta = -1.0$ and reduce the number of examinees misclassified (most of whom would have abilities near $\theta = -1.0$). IRT models allow test developers to decide the measurement precision they want at each ability level, and in so doing, to specify the desired test information function. Test items then can be selected to produce the desired test.

An important feature of the item information function is its additive property; the test information function is obtained by summing up the item information functions (Hambleton & Swaminathan, 1985; Lord, 1977). In test development, items are selected that can contribute to the target test information at a prespecified point or range along the ability scale. Those items reduce errors in ability estimation at desired ability levels and contribute to the content validity of the test.

A common starting point in test development is to specify the standard error of estimation that is desired at a particular ability range or level. For example, the desire might be to produce a test resulting in standard errors of 0.33 in the interval $-2.0$ to $+2.0$ and 0.50 outside that interval. Also, the information function of a previous administration of the test, for example, could be used to specify the target test information function. Items that contribute to the test information function at an ability level of interest are selected from a pool. When test items are being selected by the test developer, it is common for test developers to determine their target test information function, compute the information that each item in the pool provides at different points along the ability scale, and choose those items they believe contribute the most information in constructing the desired test. When statistical as well as content and format considerations are taken into account, the process can be time consuming and practically exhausting, though the basic approach is conceptually satisfying.

Figure 19.8 provides a simple example using the items described in Table 19.1. The target test information adopted was that the test should be most effective in the region on the ability scale between −1.0 and +1.0. Then 20 test items from Table 19.1 were selected using random selection and optimum selection strategies. Optimum selection involved choosing items that had b- (i.e., difficulty) values in the general range of −1.00 to +1.00 and generally high levels of a (i.e., discrimination). Clearly, the optimum item selection strategy produced the better fit to the target test information function. In practice, content, format, and other considerations become part of the item selection process.

Automated item selection methods that use IRT models are beginning to receive attention among testing practitioners (van der Linden & Boek-kooi-Timminga, 1989). The development of powerful computers played a role in the inception of the automated test development procedures, and many test publishers are using or considering the use of these approaches in the future (Green, Yen, & Burket, 1989; Stocking, Swanson, & Pearlman, 1990; van der Linden & Glas, 2000). In automated test development, mathematical optimization algorithms are used to select the items that contribute

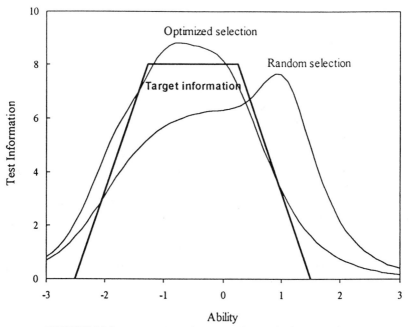

**FIGURE 19.8**   Test design using several item selection procedures.

most to desirable test features, such as measurement precision, content balance, item-format balance, and the length of the test.

Optimal test designs use flexible strategies that optimize some objective function(s) such as (a) minimization of test length, (b) maximization of a test information function, (c) minimization of deviations from the target test information function, (d) minimization of administration time, or (e) combinations of these functions (van der Linden & Boekkooi-Timminga, 1989). The decision variable in the optimization procedure is often the selection of an item, and it takes the value of 1 or 0 for selected and not selected items, respectively. Powerful, flexible, and easy-to-use computer packages for optimal test design are currently available (e.g., ConTEST, see Table 19.2) that allow applied researchers to control many aspects of the resulting test, accommodate the inclusion of many constraints in the test development process, and assemble tests in a short period of time. Basically, the test developer provides the test specifications, and the computer selects the set of items that best meets the specifications. Very often, one of the design specifications is the target test information function. Of course, control of the final test remains in the hands of test developers who can overturn any decisions made by the computer software.

## B. Computer Adaptive Testing

Frederic Lord at ETS became interested in the 1960s in the notion of adapting the selection of test items to the ability level of examinees. His idea was to make testing more efficient by shortening test lengths for examinees by not administering test items that were uninformative—difficult items administered to low-performing examinees and easy items administered to high-performing examinees. The emerging power of computers made computer-based testing possible, and IRT provided solutions to two technical problems that limited the implementation of computer adaptive testing (CAT).

First, an item selection algorithm was needed. Lord (1980) felt that an examinee's ability estimate could be calculated based upon performance on previously administered items in the test. Then, an item providing maximum information at or near the examinee's ability estimate could be chosen from the bank of available test items. Testing would continue until the examinee ability estimate reached some desired level of precision.

Second, when examinees had been administered tests of variable difficulty, statistical adjustments in the ability estimation process would be required to equate the nonequivalent forms. Without equating, examinees could not be compared to each other, a reference group, or any performance standards that may have been set for interpreting examinee performance. The use of an IRT model provides an ideal solution to both problems: items and examinees are reported on the same scale, and items providing maximum information could be chosen to maximize the efficiency of the test. IRT abilities are independent of the particular choice of items, and hence, ability estimates of examinees who may have taken nonequivalent forms of the test can be compared. IRT models allow the computer to be used for item selection, item administration, item scoring, and ability estimation. Without the use of IRT, CAT would be difficult if not impossible to implement. Interested readers are referred to Wainer et al. (1990) and van der Linden and Glas (2000) for more information on CAT technical advances.

Figure 19.9 provides a simple example of a CAT using the test items described in Table 19.1, with examinees having a relatively high, average, and relatively low true ability. (The true ability is known in a simulation study, which is why these studies have been so valuable in advancing IRT methods.) These examinees were administered a CAT-based test on the three-parameter IRT model. What is clearly seen in the example is that these examinees' ability estimates are very close to the final ability estimates by about the 13 or 14th item (high ability), the 18th item (average ability), and the 12th item (low ability). Subsequent items contribute very little extra precision to the ability estimation.

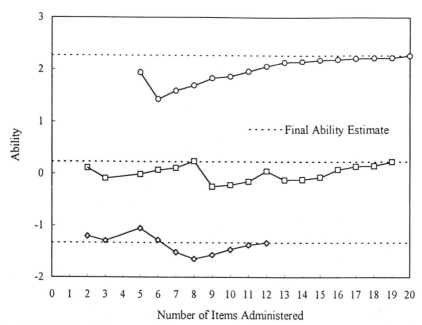

**FIGURE 19.9**    Ability estimation for three examinees across the first 20 administered test items.

## VII. FUTURE DIRECTIONS AND CONCLUSIONS

Three new areas for applications of IRT models and practices are likely to draw special attention in the coming years. First, large-scale state, national, and international assessments are attracting considerable attention (e.g., the Third International Mathematics and Science Study involving over 40 countries). Item response models are being used in many aspects of this work, including the reporting of scores. For example, ICCs, are being used in attempting to attach more meaning to particular points along the ability scale (e.g., a score of 220). At each point of interest (e.g., the average score of persons from a particular country), ICCs are reviewed to find several items on which persons at the ability point of interest have relatively high probabilities of correct responses (e.g., >80%). Then, the skills measured by these items are evaluated to determine the types of skills possessed by persons at or near the ability point. In this way, more meaning can be attached to ability points of interest on the ability continuum over which scores are reported.

Another possibility is to select these points of interest to enhance the meaning of test scores along the ability continuum (e.g., 150, 200, 250, and 300) or to illuminate points of interest defined by the distribution of scores (e.g., 10th, 25th, 50th, 75th, and 90th percentiles). A number of technical controversies may arise from these applications of IRT. For example, it is assumed that ICCs are invariant to the nature and amounts of instruction, but substantially more research is needed on this issue. Also, validation of these enhanced interpretations of ability scores is required to insure that misinterpretations are not made. This requirement is often overlooked with tests designed using classical test theory procedures, but it applies with equal force to all tests, regardless of how they are constructed.

Second, cognitive psychologists such as Embretson (1984, 1997) are using IRT models to link examinee task performance to their ability level through complex models that attempt to estimate parameters for the cognitive components that are needed to complete the tasks. This line of research is also consistent with Goldstein and Wood's (1989) goal to see the construction of more meaningful psychological models to help explain examinee test performance (see, for example, Sheehan & Mislevy, 1990). One potential application might be the systematic construction of test items to insure construct validity. This field of research is known as model-based test development.

Third, educators and psychologists are advocating the use of test scores for more than the rank ordering of examinees or determining whether they have met a particular cutoff. Diagnostic information is becoming increasingly important to users of test scores. Inappropriateness measurement, developed by M. Levine and F. Drasgow (e.g., Drasgow, Levine, & McLaughlin, 1987) uses IRT models to identify aberrant responses of examinees and special groups of examinees. Such information can be helpful in successful diagnostic work. More use of IRT models in providing diagnostic information can be anticipated in the coming years.

The future of IRT as a measurement framework for solving testing and assessment problems is promising. Today, a wide array of IRT models exists for handling nominal, ordinal, interval, and ratio data (see, van der Linden & Hambleton, 1997), and extensive commercial efforts are in place to develop user-friendly software. At the same time, IRT modeling is not a solution to all of the measurement problems in the fields of psychology and education, despite the unfortunately overstated claims of some IRT researchers. IRT is not a "magic wand" that can compensate for poorly defined constructs, badly constructed items, improperly administered tests, and faulty score interpretations. Nevertheless, in the hands of careful test developers and researchers, IRT modeling and applications can be very useful in the construction and evaluation of tests, and in the effective uses and interpretations of test scores.

## REFERENCES

Baker, B. F. (1992). *Item response theory: Parameter estimation techniques.* New York: Marcel Deckker.

Birnbaum, A. (1957). *Efficient design and use of tests of ability for various decision-making problems* (Series Report No. 58-16, Project No. 7755-23). Randolph Air Force Base, TX: USAF School of Aviation Medicine.

Bock, R. D. (1972). Estimating item parameters and latent ability when responses are scored in two or more nominal categories. *Psychometrika, 37,* 29–51.

Bock, R. D. (1997). Nominal response model. In W. J. van der Linden & R. K. Hambleton (Eds.), *Handbook of modern item response theory.* New York: Springer-Verlag.

Bock, R. D., & Aitkin, M. (1981). Marginal maximum likelihood estimation of item parameters: Application of an EM algorithm. *Psychometrika, 46,* 443–459.

Brennan, R. L., & Lee, W. (1999). Conditional scale score standard errors of measurement under binomial and compound binomial assumptions. *Educational and Psychological Measurement, 59,* 5–24.

Crocker, L., & Algina, J. (1986). *Introduction to classical and modern test theory.* New York: Holt, Rinehart, & Winston, Inc.

Drasgow, F., Levine, M. V., & McLaughlin, M. E. (1987). Detecting inappropriate test scores with optimal and practical appropriateness indices. *Applied Psychological Measurement, 11*(1), 59–79.

Embretson, S. E. (1984). A general latent trait model for response processes. *Psychometrika, 49,* 175–186.

Embretson, S. E. (1997). Multicomponent response models. In W. J. van der Linden & R. K. Hambleton (Eds.), *Handbook of modern item response theory.* New York: Springer-Verlag.

Fischer, G. H. (1974). *Ein fuhrung in die theorie psychologischer tests* [Introduction to the theory of psychological tests]. Bern: Huber.

Fischer, G. H., & Seliger, E. (1997). Multidimensional linear logistic models for change. In

W. J. van der Linden & R. K. Hambleton (Eds.), *Handbook of modern item response theory*. New York: Springer-Verlag.

Fitzpatrick, A. R., Link, V. B., Yen, W. M., Burket, G. R., Ito, K., & Sykes, R. C. (1996). Scaling performance assessments: A comparison of one-parameter and two-parameter partial credit models. *Journal of Educational Measurement, 33*(3), 291–314.

Fraser, C., & McDonald, R. P. (1988). NOHARM: Least squares item factor analysis. *Multivariate Behavioral Research, 23,* 267–269.

Goldstein, H., & Wood, R. (1989). Five decades of item response modeling. *British Journal of Mathematical and Statistical Psychology, 42,* 139–167.

Green, R., Yen, W., & Burket, G. (1989). Experiences in the application of item response theory in test construction. *Applied Measurement in Education, 2,* 297–312.

Gulliksen, H. (1950). *Theory of mental tests.* New York: Wiley.

Hambleton, R. K. (1989). Principles and selected applications of item response theory. In R. L. Linn (Ed.), *Educational measurement* (3rd ed., pp. 147–200). New York: Macmillan.

Hambleton, R. K., & Swaminathan, H. (1985). *Item response theory: Principles and applications.* Boston, MA: Kluwer Academic Publishers.

Hambleton, R. K., Swaminathan, H., & Rogers, H. J. (1991). *Fundamentals of item response theory.* Newbury Park, CA: Sage.

Hulin, C. L., Drasgow, F., & Parsons, C. K. (1983). *Item response theory: Applications to psychological measurement.* Homewood, IL: Dow Jones-Irwin.

Kolen, M. J., & Brennan, R. L. (1995). *Test equating: Methods and practices.* New York: Springer-Verlag.

Lord, F. M. (1952). *A theory of test scores* (Psychometric Monograph No. 7). Psychometric Society.

Lord, F. M. (1953). An application of confidence intervals and of maximum likelihood to the estimation of an examinee's ability. *Psychometrika, 18,* 57–75.

Lord, F. M. (1977). Practical applications of item characteristic curve theory. *Journal of Educational Measurement, 14,* 117–138.

Lord, F. M. (1980). *Applications of item response theory to practical testing problems.* Hillsdale, NJ: Lawrence Erlbaum.

Lord, F. M. (1984). Standard errors of measurement at different ability levels. *Journal of Educational Measurement, 21,* 239–243.

Lord, F. M., & Novick, M. R. (1968). *Statistical theories of mental test scores.* Reading, MA: Addison-Wesley.

Masters, G. N. (1982). A Rasch model for partial credit scoring. *Psychometrika, 47,* 149–174.

Masters, G. N., & Wright, B. D. (1997). Partial credit model. In W. J. van der Linden & R. K. Hambleton (Eds.), *Handbook of modern item response theory*. New York: Springer-Verlag.

McDonald, R. P. (1967). *Non-linear factor analysis* (Psychometric Monograph No. 15). Psychometric Society.

McDonald, R. P. (1982). Linear versus non-linear models in item response theory. *Applied Psychological Measurement, 6,* 379–396.

McDonald, R. P. (1989). Future directions for item response theory. *International Journal of Educational Research, 13*(2), 205–220.

Mislevy, R., & Bock, R. D. (1986). *BILOG: Maximum likelihood item analysis and test scoring with logistic models.* Mooresville, IN: Scientific Software.

Mislevy, R., & Stocking, M. L. (1989). A consumer's guide to LOGIST and BILOG. *Applied Psychological Measurement, 13*(1), 57–75.

Muraki, E. (1992). A generalized partial credit model: Application of an EM algorithm. *Applied Psychological Measurement, 16,* 159–176.

Muraki, E., & Bock, R. D. (1993). *PARSCALE: IRT based test scoring and analysis.* Chicago, IL: Scientific Software International.

Rasch, G. (1960). *Probabilistic models for some intelligence and attainment tests.* Copenhagen: Denmarks Paedagogiske Institut.

Reckase, M. (1979). Unifactor latent trait models applied to multifactor tests: results and implications. *Journal of Educational Statistics, 4,* 207–230.

Reckase, M. (1997a). A linear logistic model for dichotomous item response data. In W. J. van der Linden & R. K. Hambleton (Eds.), *Handbook of item response theory.* New York: Springer-Verlag.

Reckase, M. (1997b). The past and future of multidimensional item response theory. *Applied Psychological Measurement, 21,* 25–36.

Samejima, F. (1969). Estimation of latent ability using a response pattern of graded scores. *Psychometric Monograph No. 17.*

Samejima, F. (1974). Normal ogive model on the continuous response level in the multidimensional latent space. *Psychometrika, 39,* 111–121.

Sheehan, K., & Mislevy, R. J. (1990). Integrating cognitive and psychometric models to measure document literacy. *Journal of Educational Measurement, 27*(3), 255–272.

Stocking, M., Swanson, L., & Pearlman, M. (1990, April). *Automated item selection using item response theory.* Paper presented at the meeting of NCME, Boston.

Swaminathan, H. (1997, August). *Bayesian estimation in item response models.* Paper presented at the meeting of the American Statistical Association, Anaheim.

Tang, K. L., & Eignor, D. R. (1997). *Concurrent calibration of dichotomously and polytomously scored TOEFL items using IRT models* (TOEFL Report TR-13). Princeton, NJ: Educational Testing Service.

Thissen, D., & Steinberg, L. (1986). A taxonomy of item response models. *Psychometrika, 51,* 567–577.

Traub, R. (1983). A priori considerations in choosing an item response model. In R. K. Hambleton (Ed.), *Applications of item response theory* (pp. 57–70). Vancouver, BC: Educational Research Institute of British Columbia.

van der Linden, W. J., & Boekkooi-Timminga, E. (1989). A maximin model for test design with practical constraints. *Psychometrika, 54,* 237–247.

van der Linden, & W. J., Glas, C. (Eds.). (2000). *Computerized-adaptive testing: Theory and practice.* Boston: Kluwer Academic Publishers.

van der Linden, W. J., & Hambleton, R. K. (Eds.). (1997). *Handbook of modern item response theory.* New York: Springer-Verlag.

Wainer, H., Dorans, N. J., Flaugher, R., Green, B. F., Mislevy, R. J., Steinberg, L., & Thissen, D. (1990). *Computerized adaptive testing: a primer.* Hillsdale, NJ: Erlbaum.

Wright, B. D., & Stone, M. H. (1979). *Best test design.* Chicago, IL: MESA.

Yen, W. (1983). Use of the three-parameter model in the development of a standardized achievement test. In R. K. Hambleton (Ed.), *Applications of item response theory.* Vancouver, BC: Educational Research Institute of British Columbia.

# 20

# MULTITRAIT–MULTIMETHOD ANALYSIS

**LEVENT DUMENCI**

*Department of Psychiatry, University of Arkansas, Little Rock, Arkansas*

The multitrait–multimethod (MTMM; Campbell & Fiske, 1959) analysis procedure is a powerful measurement design in construct validation research. It offers an uncompromising perspective on the meaning of theoretical variables (e.g., intelligence and depression) that require indirect measurement. Originating in psychometrics, MTMM models now are applied over a broad range of disciplines including business (Conway, 1996; Kumar & Dillon, 1992), medical and physical education (Forsythe, McGaghie, & Friedman, 1986; Marsh, 1996), speech-language (Bachman & Palmer, 1981), political science (Sullivan & Feldman, 1979), nursing (Sidani & Jones, 1995), and medicine (Engstrom, Persson, Larsson, & Sullivan, 1998).

The MTMM is a fully crossed measurement model whereby multiple procedures (methods) are used to measure multiple theoretical constructs (traits) in a large number of persons. In medical schools, for example, the standardized patient examinations are used to measure students' abilities to take patients' medical history (T1), to perform a physical examination (T2), to communicate effectively with the patient (T3), and to provide relevant feedback to the patient (T4). These four types of abilities are referred to as traits in the MTMM analysis. All four traits are measured

*Handbook of Applied Multivariate Statistics and Mathematical Modeling*
**583**

by three types of raters: trained persons playing the role of a patient and faculty in the examination room (M1 and M2, respectively), and trained raters observing students' performances live on a video monitor (M3). In this example, four traits measured by three types of performance ratings form an MTMM matrix, which is nothing more than a correlation matrix between 12 measures (4 traits $\times$ 3 methods). A measure ($T_iM_j$; $i = 1,2,3,4$ and $j = 1,2,3$) refers to an observed variable associated with each trait–method pairing, (e.g., medical history-taking ability rated by the faculty) (i.e., $T_1M_2$). Other examples include the measurements of four competence traits (i.e., social, academic, English, and mathematics) measured by four rating methods (i.e., self, teacher, parent, and peer) (Byrne & Goffin, 1993) and the measurements of three types of attitudes (i.e., affective, behavioral, and cognitive) toward church assessed with four different methods of scale construction procedures (i.e., scalogram, equal-appearing intervals, self-rating, and summated ratings) (Ostrom, 1969).

The objectives of MTMM analysis are (a) to determine whether the measurements of each trait derived by multiple methods are concordant (convergent validity); (b) to show that measurements of different traits obtained using the same method are discordant (discriminant validity); and (c) to estimate the influence of different methods on the measurement of traits (method effect). Traits are universal, and their existence should not depend on the choice of a method to measure them. Therefore, convergent validity requires empirical evidence that different methods can be used to measure a trait. Traits also are unobservable individual difference characteristics. Different labels are used to distinguish one trait from the other to facilitate verbal communications. Consequently, this gives rise to the possibility that different labels may refer to the same trait. The discriminant validity evidence is sought to demonstrate that different labels (e.g., depression and anxiety) indeed refer to two distinct traits. Finally, the measurement of traits unavoidably involves one or more methods. For example, an IQ score of 105 obtained using a performance measure may be an inflated or deflated measure of intelligence because of general or incident-specific biases in the method of measuring intelligence. An IQ score different than 105 might have been obtained if teacher's ratings were used to measure intelligence. Thus, construct validity requires that the scores reflect the magnitude of the trait and be relatively free of method of measurement.

All MTMM analysis procedures provide evidence pertaining to three types of construct validity: convergent validity, discriminant validity, and method effect. However, there are no universally accepted formal tests for construct validity. The analytic definitions of different validity types vary across different MTMM analytic procedures. Therefore, construct validity evidence should be interpreted under the framework of a particular MTMM procedure used in the analysis (Kumar & Dillon, 1992).

The first analytic approach for the MTMM matrix, originally proposed by Campbell and Fiske (1959), relied on the magnitudes, patterns, and averages of zero-order correlations. According to this approach, correlations among the measurements of the same trait by different methods (i.e., monotrait-heteromethod [MTHM] correlations or validity diagonals) should be statistically significant and large in magnitude. The validity diagonals should exceed the correlations among measures of different traits obtained from different methods (i.e., heterotrait–heteromethod) (HTHM) and the correlations between different traits measured by the same method (i.e., heterotrait–monomethod) (HTMM). Returning back to the example of standardized patient examination, a zero-correlation between the history-taking ability measured by the ratings of the faculty and the patient would imply that these two ratings are not measures of the same construct (i.e., history-taking ability). Also, the findings that the correlation between the physical examination and communication measured by the faculty ratings is of the same magnitude as the correlation between the faculty and the observer ratings of physical examination would suggest a substantial method bias in ratings. The method bias also is evident when the patterns of heterotrait correlations within monomethod and within heteromethod blocks are different.

The Campbell and Fiske (1959) procedure assumes that the operation of averaging correlations does not lead to any loss of information in the MTMM matrix (i.e., that all the correlation coefficients in an MTMM matrix were generated by only four correlations). This is a very restrictive assumption and is usually violated in practice. Applied researchers should verify (at least visually) that the range of correlations being averaged is very narrow before reaching any conclusions on construct validity.

The Campbell and Fiske procedure of averaging correlations still is the most commonly used MTMM analytic procedure, but it lacks any formal statistical model, and the subjectivity involved in interpreting simple correlations has led researchers to develop new statistical models for analyzing MTMM matrices. Earlier approaches include exploratory factor analysis (Jackson, 1969), nonparametric ANOVA models (Hubert & Baker, 1978), partial correlation methods (Schriesheim, 1981), and smallest space analysis (Levin, Montag, & Comrey, 1983). With the exception of exploratory factor analysis, a distinguishing characteristic of these descriptive MTMM techniques is that they are relatively free from the problems of convergence and improper solutions. Unfortunately, the measurement structure underlying the MTMM matrix is either implicit or absent in these MTMM techniques. That means that the measurement model implied by these techniques cannot be subjected to the test of model disconfirmability, which is an essential element in scientific endeavor, prior to making any construct validity claims based on the parameter estimates. In sum, these techniques assume that the measurement model implied by the model is correct without offering any way of verifying it. The equal-level approach (Schweizer, 1991) and the constrained component analysis (Kiers, Takane, & ten Berge, 1995), relatively new statistical methods of MTMM analysis, also suffer from this limitation.

The random-effect analysis of variance (ANOVA), confirmatory factor analysis (CFA), covariance component analysis (CCA), and composite direct product analysis (CDP), are statistical models commonly used for analyzing MTMM matrices; these are presented here under the framework of structural equation modeling (SEM). A good grasp of the SEM and CFA fundamentals (see DiLalla, chapter 15, and Hoyle, chapter 16, this volume) are required to implement these procedures with real data.

Artificial MTMM matrices with 12 variables each (four traits measured by three methods) are used to illustrate the models. A random sample size of 1,000 is assumed throughout. Graphical representations of the models and parameter estimates also are provided so that readers can gain first-hand experience analyzing different MTMM models with a SEM software package (e.g., AMOS, EQS, LISREL, Mplus, or Mx) and verify their results. Indeed, readers are encouraged to replicate the examples presented here with a software of their own choice before analyzing a real MTMM matrix. All parameter estimates are significant unless otherwise indicated in the text. Different MTMM matrices are analyzed for different models because it is very unlikely, if not impossible, that all four models fit an MTMM matrix equally well in practice.

A cross-sectional measurement design is assumed throughout this chapter. Some MTMM applications nonetheless adapt a longitudinal measurement design where occasions are treated as methods. An MTMM analysis may not be optimal for longitudinal designs. The issues of measurement of stability and change, measurement invariance, trait–state distinction, and interpretations of convergent validity, discriminant validity, and method effects should be addressed in longitudinal designs (Dumenci & Windle, 1996, 1998). A number of statistical models now are available to address such issues and should be considered before an MTMM analysis for such measurement designs is adopted. Readers should consult chapters on time-series analysis (see Mark, Reichardt, & Sanna, chapter 13, this volume) and modeling change over time (see Willett & Keiley, chapter 23, in this volume).

## I. RANDOM ANALYSIS OF VARIANCE MODEL

A full three-way random ANOVA model specification would allow for estimating the variance attributable to person, trait, and method main effects, three two-way interactions, and one three-way interaction (Guilford, 1954; Stanley, 1961). Four sources of variance are of special interest in analyzing MTMM data: (a) Person, (b) Trait × Person, (c) Method × Person, and (d) Trait × Method × Person (Kavanagh, MacKinney, &

Wolins, 1971). The Person × Trait × Method interaction serves as the error term for testing the significance of the remaining three sources of variance.

The ANOVA approach has early roots in generalizability theory (Cronbach, Gleser, Nanda, & Rajaratnam, 1972; see Marcoulides, chapter 18, this volume) and parallels to the averaging correlations employed by Campbell and Fiske (1959). Following Kenny (1995), SEM representation of the random ANOVA model is depicted in Figure 20.1. Four unique parameters account for the entire MTMM matrix: trait variance (T), trait covariance (two directional arrows in Figure 20.1), method variance (M), and unique variance (e). Note that no subscript is used to distinguish different traits, different trait covariances, different methods, and different errors of measurement in Figure 20.1. For example, the correlation between history-taking ability and communication is the same as the correlations between history taking and the remaining two abilities (i.e., communication and feedback). Also, the error of measurement associated with communication measured by faculty rating is the same as the error of measurement associated with feedback measured by the rater at the monitor. This assumption is known as compound symmetry and applies to all four parameters. The ANOVA model also assumes that the correlation between a measure and a trait factor (all one-directional arrows from traits to measures in Figure 20.1) is equal to unity.

A large Person × Trait effect would signify that the differences among individuals vary in magnitude as a function of the trait being measured (i.e., the rank ordering of participants is not the same across traits). That finding, along with low trait correlations, would provide evidence for discriminant validity. Convergent validity is established when the Person × Method effect is not significant (i.e., people are ranked the same way on a given trait, regardless of the method used to measure it). If the Person × Method effect is significant, it should be smaller than the Person × Trait effect, if convergent validity is to be demonstrated.

The model depicted in Figure 20.1 implies that the traits are correlated (as represented by lines linking the T entries) but that the methods are uncorrelated (i.e., there are no lines linking the M entries). However, two additional SEM possibilities should be considered when the random ANOVA model is interpreted. First, it is possible that the traits are uncorrelated (i.e., the lines linking the T entries would be deleted from Figure 20.1) but that the methods are correlated (i.e., lines linking the M entries would be added). All other features of the model (i.e., the unit factor pattern coefficients, the number of free parameters, and the assumption of compound symmetry) would remain unchanged. Alternatively, three orthogonal (i.e., uncorrelated) common factors (i.e., a general, trait, and method factors) could underlie the MTMM matrix (see Figure 20.2).

Any statistical program with the SEM capabilities can be used to analyze the MTMM matrix. I analyzed the artificial MTMM matrix in

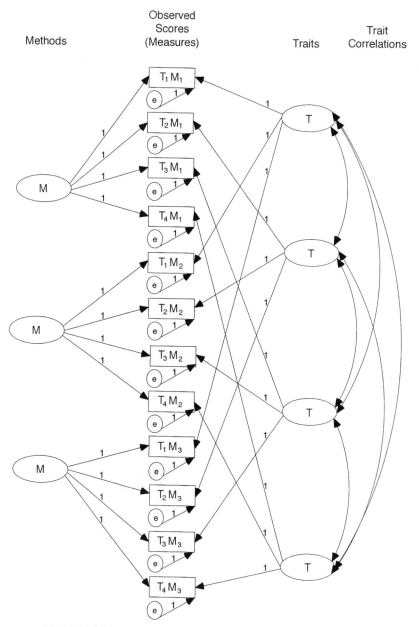

**FIGURE 20.1** Random analysis of variance model: Correlated traits.

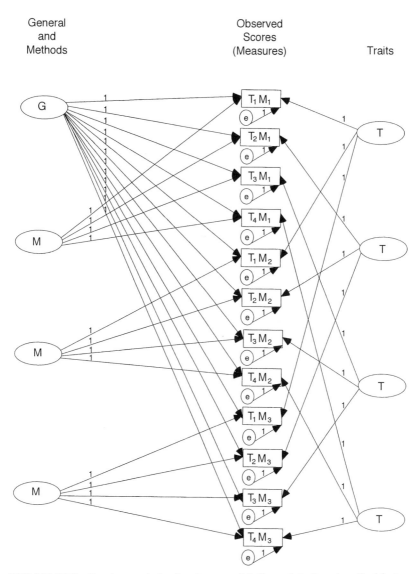

**FIGURE 20.2** Random analysis of variance model: General, trait, and method factors.

Table 20.1 using EQS (Bentler, 1989) to illustrate all three ANOVA models. The unweighted least-square method was used to estimate the model parameters. All three ANOVA models fit the MTMM matrix equally well $(\chi^2_{(74)} = 83.74; p > .10)$. The nonsignificant chi-square statistic provides evidence that each of the three ANOVA models is a plausible representation of the MTMM matrix, hence, the model assumptions (e.g., compound

**TABLE 20.1   Hypothetical MTMM Correlation Matrix for the Random ANOVA Examples**

| Measure | Method 1 | | | | Method 2 | | | | Method 3 | | | |
|---|---|---|---|---|---|---|---|---|---|---|---|---|
| | $T_1$ | $T_2$ | $T_3$ | $T_4$ | $T_1$ | $T_2$ | $T_3$ | $T_4$ | $T_1$ | $T_2$ | $T_3$ | $T_4$ |
| **Method 1** | | | | | | | | | | | | |
| $T_1$ | 1.00 | | | | | | | | | | | |
| $T_2$ | .37 | 1.00 | | | | | | | | | | |
| $T_3$ | .39 | .38 | 1.00 | | | | | | | | | |
| $T_4$ | .40 | .37 | .40 | 1.00 | | | | | | | | |
| **Method 2** | | | | | | | | | | | | |
| $T_1$ | .56 | .05 | .09 | .06 | 1.00 | | | | | | | |
| $T_2$ | .05 | .56 | .10 | .04 | .37 | 1.00 | | | | | | |
| $T_3$ | .07 | .06 | .56 | .07 | .42 | .43 | 1.00 | | | | | |
| $T_4$ | .08 | .07 | .08 | .56 | .37 | .39 | .42 | 1.00 | | | | |
| **Method 3** | | | | | | | | | | | | |
| $T_1$ | .56 | .08 | .06 | .09 | .57 | .06 | .05 | .07 | 1.00 | | | |
| $T_2$ | .05 | .54 | .06 | .04 | .04 | .54 | .06 | .04 | .41 | 1.00 | | |
| $T_3$ | .08 | .07 | .55 | .09 | .09 | .10 | .55 | .08 | .40 | .41 | 1.00 | |
| $T_4$ | .09 | .09 | .08 | .56 | .04 | .06 | .06 | .53 | .40 | .42 | .42 | 1.00 |

Heterotrait–monomethod (method) triangles

Heterotrait–heteromethod (error) triangles

Monotrait–heteromethod (convergent validity) diagonals

Monotrait–monomethod (reliability) diagonals

symmetry) are supported. Parameter estimates and average MTMM correlations appear in the upper half of Table 20.2. It is not coincidental that all three models have exactly the same fit index and that the parameter estimates from three ANOVA models are closely related. All three models are statistically equivalent, hence one cannot choose one or the other, because each model can be reexpressed by the parameters of the remaining two models, as well as by the average correlations. For example, the estimated method variance in the correlated-method ANOVA model (.40) is nothing more than the average correlation (see Table 20.1) between different traits measured by the same method (i.e., mean [cor(HTMM)]). Note that the parameter estimate of .40 from the correlated-method ANOVA model can be reexpressed as the sum of the method variance and the trait correlation (.33 + .07) in the correlated-trait ANOVA model, and as the sum of the method variance and the general trait variance (G) in the general factor ANOVA model. As shown at the lower half of Table 20.2, all three ANOVA models are statistically equivalent, and the same set of parameters can be estimated by simply averaging the correlations (i.e., the Campbell and Fiske approach) in the MTMM matrix.

The random ANOVA model is the most parsimonious approach to modeling the MTMM matrix because it requires the estimations of only four parameters. A meaningful interpretation of ANOVA parameters is difficult to reach, however, for when the assumption of compound symmetry is satisfied, the ANOVA model implies three conceptually different but statistically equivalent models, as demonstrated in Table 20.2. First, the HTHM triangle reflects the correlations between traits, as implied by Figure 20.1 (the correlated-trait ANOVA model); second, it reflects the correlations between methods in the correlated-method ANOVA model; and third, it reflects a common (general) latent variate uncorrelated with trait and method factors, as implied by Figure 20.2 (the general factor ANOVA model). The Campbell and Fiske procedure treats the HTHM correlations as "error" in that they establish a baseline against which to interpret the other correlations. Campbell and Fiske (1959) recommended that maximally distinct methods be chosen in the MTMM analysis so that any interpretation based on the correlated-method model is implausible, but even if that were done, there is no statistical justification for adopting one interpretation or the other because all three models are statistically equivalent.

The only benefit of adopting the random ANOVA model is that it provides a statistical procedure for testing whether the MTMM matrix satisfies the assumption of compound symmetry, which underlies the Campbell and Fiske criteria. When a random ANOVA model does not fit, the interpretations of model parameters would be questionable, as in any other SEM model. For a well-fitting model, on the other hand, the ANOVA method does not provide any substantive information above and beyond

**TABLE 20.2  Parameter Estimates and Model Equivalency among Three Random Random Analysis of Variance Models**

| Parameter | Correlated trait | Correlated method | General factor | Averaging correlations[a] |
|---|---|---|---|---|
| Trait variance (T) | .55 | .49 | .49 | mean[cor(MTHM)] = .55 |
| Method variance (M) | .33 | .40 | .33 | mean[cor(HTMM)] = .40 |
| Trait correlation (T) | .07 | | | mean[cor(HTHM)] = .07 |
| Method correlation (M) | | .07 | | |
| General factor variance (G) | | | .07 | |
| Error variance (e) | .12 | .12 | .12 | |
| Person | cor(T) | cor(M) | var(G) | mean[cov(HTHM)] |
| Trait | var(T) − cor(T) | var(T) | var(T) | mean[cor(MTHM)] − mean[cor(HTHM)] |
| Method | var(M) | var(M) − cor(M) | var(M) | mean[cor(HTMM)] − mean[cor(HTHM)] |
| Error | {mean[var(TiMj)]− var(T) − var(M) −cor(T)} | {mean[var(TiMj)]− var(T) − var(M) −cor(M)} | {mean[var(TiMj)]− var(T) − var(M) −var(G)} | {mean[var(TiMj)] − mean[cor(MTHM)]− mean[cor(HTHM)]} |

[a] MTHM, monotrait–heteromethod; HTMM, heterotrait–monomethod; HTHM, heterotrait–heteromethod.

what already is known from the averaging of correlations, except that the averaging operation is justified. That is, applied researchers do not need to have access to computers to obtain the unweighted least square parameter estimates for ANOVA models. This can be achieved by simply averaging the respective correlations in the MTMM matrix. In sum, interpretive difficulties associated with the problem of model equivalency seem the major obstacle to adopting the random ANOVA model in MTMM analysis. Therefore, despite its widespread use over the last three decades, the model is flawed and applied researchers should avoid its use.

## II. CONFIRMATORY FACTOR ANALYTIC MODEL

For a given set of congeneric measures (Joreskog, 1971), the CFA model can be used to partition the variance of each measure in an MTMM matrix into a common trait and unique components. The unique components can be further decomposed into measure-specific (common method) and random error components. The MTMM can be conceptualized as a replication design in which the measurement of traits are replicated using multiple methods. An important feature of replication designs is that they permit the evaluation of the independent contribution of the measure-specific component to the unique variance. Therefore, the CFA can be used to evaluate trait and method components of measures (Werts & Linn, 1970).

The CFA has been one of the most widely used methods for analyzing MTMM matrices. Its attractiveness is due to the premise that, first, it allows for separating trait, method, and unique components from the measures. Second, it adopts hypothesis-testing strategies to evaluate convergent and discriminant validity and method effects. Third, it estimates within-trait and within-method correlations after the unreliability of measures is taken into account.

The CFA model decomposes the MTMM matrix into three additive components: (a) $t$ common trait factors, each explaining the correlations across different methods; (b) $m$ common method factors, each explaining the correlations across different traits; and (c) $tm$ unique variances (i.e., the variances in measures unexplained by the common factors). From the perspective of construct validity, the most desirable CFA model is the trait-only model, which includes $t$-uncorrelated trait factors and no method factor structure. A variety of CFA models can be specified with different trait, method, and factor correlation structures, ranging from the most parsimonious (trait-only) to the most complex (correlated-trait/correlated-method) model (see Figure 20.3). According to this model, trait factors are uncorrelated with method factors, trait and method factors are uncorrelated with unique components, and unique components are uncorrelated. Note that

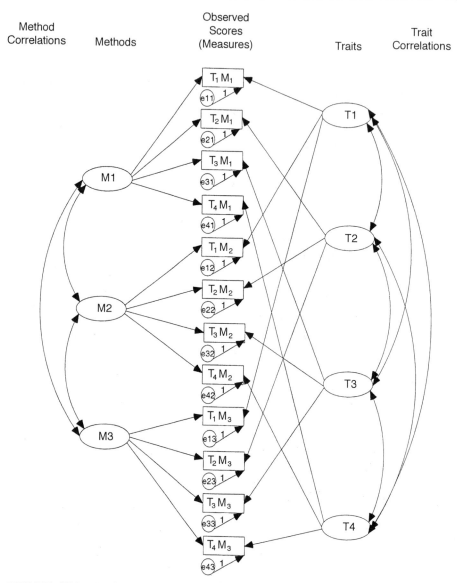

**FIGURE 20.3** Confirmatory factor analysis model: Correlated traits and correlated methods.

this model differs from the random ANOVA model in three respects. First, it allows for correlations within traits and within methods. The ANOVA model can have only one set of such correlations (i.e., either traits or methods are correlated). Second, it estimates correlations between factors and measures (i.e., factor loadings). All factor loadings are set to unity in the ANOVA models. Third, it does not assume compound symmetry, hence, it imposes no equality restrictions on variances and correlations in the model. For this reason, latent variates are identified with trait or method number (e.g., T1 and M3, in Figure 20.3).

The correlated-trait–correlated-method (CTCM) model is illustrated with the hypothetical standardized patient examination using an artificial MTMM matrix presented in Table 20.3. The CTCM model in Figure 20.3 fits the matrix quite well ($\chi^2_{(33)} = 5.63$; $p > .10$), providing evidence that Figure 20.3 is a reasonable measurement structure for the MTMM matrix in Table 20.3. Parameter estimates from the maximum likelihood solution appear in Table 20.4. All parameter estimates are significant, except for the method loading of $T_2M_2$ (.11) (i.e., the faculty rating of the physical examination is relatively free from method bias) and the method factor correlation between M2 and M3 ($-.17$) (i.e., the faculty and the observer ratings).

Large and significant trait loadings in Table 20.4 indicate convergent validity. Providing further evidence for convergent validity is the fact that the overall factor patterns indicate larger trait loadings than the method loadings in Table 20.4. For example, the measure of history-taking ability by the observer ($T_4M_3$) accounts for approximately twice as much variance as the rater type (.86 versus .47). Visual inspection reveals that the trait factor correlations range from .37 to .90 (see Table 20.4) indicating that discriminant validity is mediocre. The discriminant validity diminishes to

**TABLE 20.3    Hypothetical Multitrait–Multimethod Correlation Matrix for the Confirmatory Factor Analysis Example**

| Measure | Method 1 | | | | Method 2 | | | | Method 3 | | | |
|---|---|---|---|---|---|---|---|---|---|---|---|---|
| | $T_1$ | $T_2$ | $T_3$ | $T_4$ | $T_1$ | $T_2$ | $T_3$ | $T_4$ | $T_1$ | $T_2$ | $T_3$ | $T_4$ |
| Method 1 | | | | | | | | | | | | |
| $T_1$ | 1.00 | | | | | | | | | | | |
| $T_2$ | .45 | 1.00 | | | | | | | | | | |
| $T_3$ | .35 | .18 | 1.00 | | | | | | | | | |
| $T_4$ | .62 | .52 | .32 | 1.00 | | | | | | | | |
| Method 2 | | | | | | | | | | | | |
| $T_1$ | .61 | .46 | .42 | .62 | 1.00 | | | | | | | |
| $T_2$ | .46 | .62 | .20 | .53 | .56 | 1.00 | | | | | | |
| $T_3$ | .39 | .19 | .39 | .32 | .51 | .23 | 1.00 | | | | | |
| $T_4$ | .59 | .53 | .34 | .71 | .73 | .66 | .41 | 1.00 | | | | |
| Method 3 | | | | | | | | | | | | |
| $T_1$ | .64 | .49 | .44 | .66 | .60 | .52 | .42 | .62 | 1.00 | | | |
| $T_2$ | .44 | .58 | .19 | .51 | .46 | .70 | .18 | .56 | .59 | 1.00 | | |
| $T_3$ | .42 | .20 | .41 | .35 | .43 | .21 | .42 | .35 | .57 | .24 | 1.00 | |
| $T_4$ | .61 | .54 | .34 | .74 | .58 | .60 | .32 | .74 | .87 | .65 | .45 | 1.00 |

**TABLE 20.4  Parameter Estimates from the Confirmatory Factor Analysis Model**

| Measure | Trait factors | | | | Method factors | | | Unique variance |
| --- | --- | --- | --- | --- | --- | --- | --- | --- |
| | $T_1$ | $T_2$ | $T_3$ | $T_4$ | $M_1$ | $M_2$ | $M_3$ | |
| Method 1 | | | | | | | | |
| $T_1$ | .64 | 0 | 0 | 0 | .41 | 0 | 0 | .41 |
| $T_2$ | 0 | .68 | 0 | 0 | .27 | 0 | 0 | .46 |
| $T_3$ | 0 | 0 | .58 | 0 | .12 | 0 | 0 | .64 |
| $T_4$ | 0 | 0 | 0 | .70 | .51 | 0 | 0 | .25 |
| Method 2 | | | | | | | | |
| $T_1$ | .79 | 0 | 0 | 0 | 0 | .37 | 0 | .24 |
| $T_2$ | 0 | .88 | 0 | 0 | 0 | .11 | 0 | .21 |
| $T_3$ | 0 | 0 | .65 | 0 | 0 | .18 | 0 | .54 |
| $T_4$ | 0 | 0 | 0 | .89 | 0 | .27 | 0 | .14 |
| Method 3 | | | | | | | | |
| $T_1$ | .80 | 0 | 0 | 0 | 0 | 0 | .53 | .08 |
| $T_2$ | 0 | .80 | 0 | 0 | 0 | 0 | .21 | .32 |
| $T_3$ | 0 | 0 | .66 | 0 | 0 | 0 | .23 | .51 |
| $T_4$ | 0 | 0 | 0 | .86 | 0 | 0 | .47 | .04 |
| Factor correlations | | | | | | | | |
| Traits | | | | | | | | |
| $T_1$ | 1 | | | | | | | |
| $T_2$ | .75 | 1 | | | | | | |
| $T_3$ | .85 | .37 | 1 | | | | | |
| $T_4$ | .90 | .80 | .61 | 1 | | | | |
| Methods | | | | | | | | |
| $M_1$ | 0 | 0 | 0 | 0 | 1 | | | |
| $M_2$ | 0 | 0 | 0 | 0 | .63 | 1 | | |
| $M_3$ | 0 | 0 | 0 | 0 | .57 | −.17 | 1 | |

the extent that the trait correlations approach to unity. Evidence for convergent and discriminant validity should be interpreted in the light of measure reliability estimates (i.e., one minus the unique variance in Table 20.4). For instance, the measurement of communication ability rated by trained individuals playing the role of a patient (i.e., $T_3M_1$) has the lowest reliability of .36 $(1 - .64)$, which can also be obtained by the sum of squares of trait and method factor loadings: that is, $(.58)^2 + (.12)^2 = .36$. In other words, the unit variance of $T_3M_1$ is partitioned into trait, method, and unique components (i.e., .34, .02, and .64, respectively). Finally, the magnitude of method factor loadings point to the presence of method effect.

When specified a priori, nested model comparison strategies can also be used to test hypotheses regarding construct validity (Widaman, 1985). Hence, it is possible to express convergent validity numerically by reestimating the model fit without trait factors and comparing the fit statistics with those of the CTCM model. A substantive difference between two statistics indicates convergent validity. From the nested comparison perspective, discriminant validity can be evaluated by comparing the fit of the CTCM model with four trait factors to the fit with one trait factor. The statistical testing of method effect involves a comparison of model fit between the CTCM model with and without method factors. The method effect is absent if both models fit equally well.

The conceptual elegance of the CFA–MTMM model has been challenged in a number of studies. Indeed, separation of trait and method effects is nothing but wishful thinking unless all measures are either tau-equivalent or parallel (Kumar & Dillon, 1992). The CFA model does not make the assumptions of compound symmetry and unit factor loading as discussed for the ANOVA model, so it requires a great number of parameter estimates relative to the number of nonredundant correlation coefficients in the MTMM matrix. This lack of parsimony opens the door to a variety of problems. The empirical and statistical identification is a common problem, which implies the lack of a unique parameter set that can be estimated from the CFA model (see DiLalla, chapter 15, this volume, for model identification). Another common problem is the lack of convergence. That is, the minimizing function (usually maximum likelihood) fails to reach its minimum so no CFA solution can be obtained from the model. When convergence is reached, however, the solution often contains out-of-range parameter estimates (also referred to as Heywood cases), such as negative variance estimates and correlations greater than unity. Finally, difficulties exist in finding a well-fitting CFA model even if a unique solution is obtained. Despite its shortcomings, there have been successful applications of the CFA model (e.g., Bollen, 1989; Dumenci, 1995).

## III. COVARIANCE COMPONENT ANALYSIS

Covariance component analysis is a multivariate extension of the random ANOVA design for completely crossed measurements (Bork & Bargmann, 1966) that offers a viable alternative to conventional models for analyzing MTMM matrices (Wothke, 1984, 1987). The conceptual foundation of CCA follows closely from the generalizability theory of Cronbach et al. (1972). The model assumes that three sets of latent variables account for correlations in the MTMM matrix: (a) a general component factor, which accounts for correlations among all $tm$ measures; (b) $t - 1$ trait contrast factors, which account for differences in traits; and (c) $m - 1$ method contrasts factors, which account for differences in methods. Interpretation of variance component estimates is facilitated by integrating both scaling differences and unreliability of measures into the model.

Despite the fact that all MTMM models address the construct validity issues (i.e., convergent validity, discriminant validity, and relative influence of method variance), and that the meaning of the general factor in CCA framework is somewhat analogous to that in factor analysis, interpretations of trait and method factors differ in these two modeling approaches. The trait contrast factors in CCA capture differences between all possible $(t - 1)$ sets of traits (e.g., set 1: history taking versus set 2: physical exam, communication, and feedback), whereas each trait factor in the CFA captures variability among different ratings of the same trait measures. Similarly, method contrast factors capture differences between two sets of methods in the CCA, method factors capture variability among different traits rated by the same rater type as in the CFA. Additionally, factor pattern coefficients specify the contrasts in the CCA, whereas they are the estimates of the strength of relations between measures and common factors in the CFA. Furthermore, differences in scaling among measures and the characteristics shared by all measures are taken into account in the CCA, but they are not a part of the CFA model. CCA is less susceptible to identification and convergence problems and improper solutions than the CFA models for the MTMM data.

The artificial MTMM matrix in Table 20.5 is used to illustrate a scale-free CCA model with unknown scaling factor and diagonal component factor correlations. Alternative model specifications also are possible. For example, the CCA without the scaling factor provides a parsimonious model when all measures share the common metric. Alternative factor

**TABLE 20.5  Hypothetical Multitrait–Multimethod Correlation Matrix for the Covariance Component Analysis Example**

| Measure | Method 1 | | | | Method 2 | | | | Method 3 | | | |
|---|---|---|---|---|---|---|---|---|---|---|---|---|
| | $T_1$ | $T_2$ | $T_3$ | $T_4$ | $T_1$ | $T_2$ | $T_3$ | $T_4$ | $T_1$ | $T_2$ | $T_3$ | $T_4$ |
| Method 1 | | | | | | | | | | | | |
| $T_1$ | 1.00 | | | | | | | | | | | |
| $T_2$ | .27 | 1.00 | | | | | | | | | | |
| $T_3$ | .19 | .23 | 1.00 | | | | | | | | | |
| $T_4$ | .21 | .24 | .21 | 1.00 | | | | | | | | |
| Method 2 | | | | | | | | | | | | |
| $T_1$ | .34 | −.18 | −.18 | −.19 | 1.00 | | | | | | | |
| $T_2$ | −.18 | .27 | −.15 | −.15 | .14 | 1.00 | | | | | | |
| $T_3$ | −.17 | −.14 | .28 | −.17 | .13 | .16 | 1.00 | | | | | |
| $T_4$ | −.16 | −.15 | −.16 | .31 | .17 | .15 | .12 | 1.00 | | | | |
| Method 3 | | | | | | | | | | | | |
| $T_1$ | .35 | −.19 | −.19 | −.19 | .44 | −.11 | −.10 | −.11 | 1.00 | | | |
| $T_2$ | −.16 | .27 | −.15 | −.15 | −.11 | .35 | −.07 | −.07 | .14 | 1.00 | | |
| $T_3$ | −.18 | −.14 | .34 | −.17 | −.12 | −.07 | .38 | −.09 | .17 | .16 | 1.00 | |
| $T_4$ | −.18 | −.15 | −.18 | .31 | −.17 | −.11 | −.09 | .36 | .21 | .13 | .09 | 1.00 |

correlation structures also may be tested. For the block-diagonal CCA model, within-trait and within-method correlations are estimated. All variance components (i.e., general, trait, and method) are correlated in the full CCA model.

A diagonal CCA model is depicted in Figure 20.4. There are twelve scaling factors corresponding to twelve measures. The scaling coefficients (the coefficients relating the scaling factors to the measures) are free parameters of the model. Disattenuated correlations among the scaling factors are accounted for by three sets of second-order factors: (a) one general factor (F_G); (b) three $(t - 1)$ trait contrast factors (traits 1_123, traits 2_34, and traits 3_4); and (c) two $(m - 1)$ method contrast factors (methods 1_23 and methods 2_3). The contrast matrix contains coefficients that signify the unidirectional relations between the scaling factors and the contrast factors. The contrast matrix entries are predetermined (i.e., fixed parameters), such that the sums of squares for each column are equal to 1, and the sum of product terms is equal to 0 for all possible column pairs. Such matrices are labeled as columnwise orthonormal (the contrast matrix for four traits measured by three variables appears in Table 20.6; see also Wothke, 1996, for three traits measured by three methods).

The model in Figure 20.4 fits the MTMM matrix well ($\chi^2_{(49)} = 57.59$; $p > .10$); hence, the CCA model is a plausible underlying structure of the MTMM matrix in Table 20.5. Maximum likelihood parameter estimates appear in Table 20.6. The scaling variable facilitates interpretation of variance components, especially when a correlation matrix, instead of a covariance matrix, is used in the analysis. A visual inspection of Table 20.6 reveals a relatively low variability in scaling parameter estimates (i.e., ranging from .86 to .1.17), which indicates that measures have a similar metric. The reliability of measures (one minus the unique variance) ranges from .51 to .85. The variance of the general factor is fixed to unity, for identification; the variances of the trait and method contrast factors are free parameters. The size of both trait and method variance components are evaluated relative to unit variance of the general variate (F_G in Figure 20.4). The variance component estimates appear at the lower half of Table 20.6. Variance components for the three trait and two method contrast factors are statistically significant in this hypothetical illustration. For example, the variance component estimate for the trait contrast factor 1 versus 2, 3, and 4 (traits 1_234 in Figure 20.4) is 1.44 and significant; hence, there is empirical evidence that history-taking ability (T1) is discriminated from remaining abilities (i.e., physical exam [T2], communications [T3], and feedback [T4]). The estimate of 1.06 for the method contrast factor 2 versus 3 (methods 2_3 in Figure 20.4) also is significant, meaning that faculty ratings (M2) are discriminated from the observer ratings (M3). Convergent validity is somewhat in question because the average size of trait components is similar to that of method variance components (from Table 20.6, 1.47 and

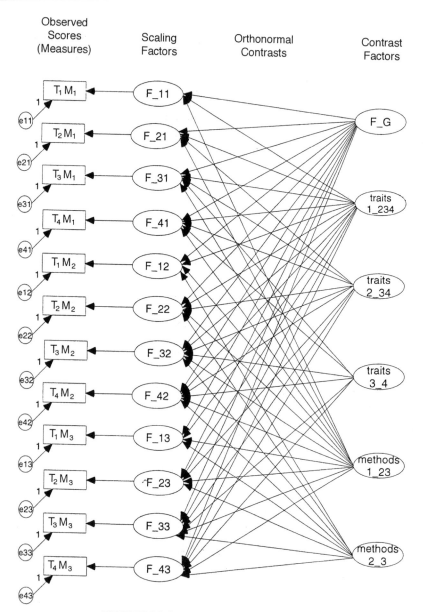

**FIGURE 20.4** Covariance component model.

**TABLE 20.6  Parameter Estimates from the Covariance Component Analysis Model**

| | | | Fixed orthonormal contrast matrix | | | | | |
| | | | Trait contrasts | | | Method contrasts | | |
| Measure | Scaling | F_G | 1_234 | 2_34 | 3_4 | 1_23 | 2_3 | Unique |
|---|---|---|---|---|---|---|---|---|
| Method 1 | | | | | | | | |
| $T_1$ | 1.01 | $+1/2\sqrt{3}$ | $+1/2$ | 0 | 0 | $+1/\sqrt{6}$ | 0 | .27 |
| $T_2$ | 1.01 | $+1/2\sqrt{3}$ | $-1/6$ | $+2/3\sqrt{2}$ | 0 | $+1/\sqrt{6}$ | 0 | .29 |
| $T_3$ | .95 | $+1/2\sqrt{3}$ | $-1/6$ | $-1/3\sqrt{2}$ | $+1/\sqrt{6}$ | $+1/\sqrt{6}$ | 0 | .33 |
| $T_4$ | .95 | $+1/2\sqrt{3}$ | $-1/6$ | $-1/3\sqrt{2}$ | $-1/\sqrt{6}$ | $+1/\sqrt{6}$ | 0 | .33 |
| Method 2 | | | | | | | | |
| $T_1$ | 1.08 | $+1/2\sqrt{3}$ | $+1/2$ | 0 | 0 | $-1/2\sqrt{6}$ | $+1/2\sqrt{2}$ | .26 |
| $T_2$ | .95 | $+1/2\sqrt{3}$ | $-1/6$ | $+2/3\sqrt{2}$ | 0 | $-1/2\sqrt{6}$ | $+1/2\sqrt{2}$ | .43 |
| $T_3$ | .86 | $+1/2\sqrt{3}$ | $-1/6$ | $-1/3\sqrt{2}$ | $+1/\sqrt{6}$ | $-1/2\sqrt{6}$ | $+1/2\sqrt{2}$ | .49 |
| $T_4$ | .98 | $+1/2\sqrt{3}$ | $-1/6$ | $-1/3\sqrt{2}$ | $+1/\sqrt{6}$ | $-1/2\sqrt{6}$ | $+1/2\sqrt{2}$ | .38 |
| Method 3 | | | | | | | | |
| $T_1$ | 1.17 | $+1/2\sqrt{3}$ | $+1/2$ | 0 | 0 | $-1/2\sqrt{6}$ | $-1/2\sqrt{2}$ | .15 |
| $T_2$ | .90 | $+1/2\sqrt{3}$ | $-1/6$ | $+2/3\sqrt{2}$ | 0 | $-1/2\sqrt{6}$ | $-1/2\sqrt{2}$ | .49 |
| $T_3$ | .96 | $+1/2\sqrt{3}$ | $-1/6$ | $-1/3\sqrt{2}$ | $+1/\sqrt{6}$ | $-1/2\sqrt{6}$ | $-1/2\sqrt{2}$ | .37 |
| $T_4$ | 1.00 | $+1/2\sqrt{3}$ | $-1/6$ | $-1/3\sqrt{2}$ | $+1/\sqrt{6}$ | $+1/2\sqrt{6}$ | $-1/2\sqrt{2}$ | .34 |
| Variance components | | | | | | | | |
| | F_G | 1 | | | | | | |
| Traits | 1_234 | 0 | 1.44 | | | | | |
| | 2_34 | 0 | 0 | 1.38 | | | | |
| | 3_4 | 0 | 0 | 0 | 1.59 | | | |
| Methods | 1_23 | 0 | 0 | 0 | 0 | 1.60 | | |
| | 2_3 | 0 | 0 | 0 | 0 | 0 | 1.06 | |

1.33, respectively). A strong convergent validity claim requires much larger trait components than the method components. Any substantive interpretation of the general variate lies outside the CCA and requires further research.

## IV. COMPOSITE DIRECT PRODUCT MODEL

An additive trait–method relation is assumed in the ANOVA, CFA, and CCA modeling approaches to the MTMM matrix, but Campbell and O'Connell (1967, 1982) pointed out that trait and method effects may be combined in a multiplicative fashion. Browne (1984; see also Cudeck, 1988) introduced the composite direct product (CDP) model for MTMM data as a means of formally modeling the multiplicative trait–method effect. The restricted second-order factor model specification of the CDP model adopted from Wothke and Browne (1990) is depicted in Figure 20.5. The free parameters include 12 scaling factors (first-order factor loadings), 12 unique trait components (Dij), and three sets of trait variance–covariance matrices that are set equal across methods (the second-order variance/covariance matrix of F_Ti). The second-order factor loadings are related to the multiplicative method effect, as will be explained below.

The CDP model is illustrated with an artificial MTMM matrix (see Table 20.7) involving four clinical abilities (traits) measured by three types of ratings (methods). The CDP model in Figure 20.5 provides an excellent fit to the MTMM matrix in Table 20.7 ($\chi^2_{(50)} = 43.40$; $p > .10$), suggesting that the CDP is a plausible model for the MTMM matrix. The maximum likelihood parameter estimates appear in Tables 20.8, 20.9, and 20.10. A visual inspection of parameter estimates suggests that all twelve variables are measured on a comparable scale, given the narrow range of scaling parameters from .69 to .72 (see Table 20.8). Correspondingly, the unique trait components are relatively similar in size (i.e., from .98 to 1.10). The size of unique component associated with the faculty ratings (M2), however, is twice as large as the ratings at the monitor (M3) (i.e., 1.24 and .68, respectively).

Convergent validity and discriminant validity are assessed using trait and method multiplicative correlation components. Note that these components are estimated from the CDP model, which are adjusted for measure-specific and random error components (see Figure 20.5). Parameter estimates (see Tables 20.9 and 20.10) and some basic algebra are needed to derive the trait and method multiplicative correlation components, $\Pi_{F\_T}$ and $\Pi_{F\_M}$, respectively. The multiplicative trait correlation components are obtained directly from Table 20.10. The pattern of multiplicative trait components ($\Pi_{F\_T}$) is assumed to be the same across three methods but only those trait components associated with the first method (i.e., patient ratings) are given in boldfaces in Table 20.10 for clarity of presentation. The following steps are needed to obtain the disattenuated multiplicative method components. The elements of $C_{(3 \times 3)}$ matrix are given in boldface in Table 20.9. The pattern of C matrix is set equal across all four traits but only those elements associated with history taking (T1) are given in boldface in Table 20.9. The sum of square of the C matrix (the C matrix multiplied by its transpose, i.e., the C′) yields

$$
\underset{C}{\begin{bmatrix} 1 & 0 & 0 \\ .61 & .28 & 0 \\ .17 & .51 & 1.12 \end{bmatrix}} \times \underset{C'}{\begin{bmatrix} 1 & .61 & .17 \\ 0 & .28 & .51 \\ 0 & 0 & 1.12 \end{bmatrix}} = \underset{CC'}{\begin{bmatrix} 1 & .61 & .17 \\ .61 & .45 & .25 \\ .17 & .25 & 1.54 \end{bmatrix}}
$$

Observed
Scores
(Measures)

First-Order
Factors

Second-Order
Factors

Second-Order
Factor
Covariances

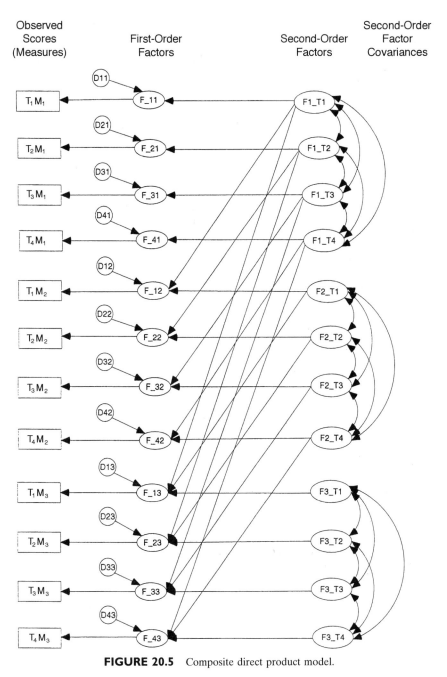

**FIGURE 20.5**  Composite direct product model.

**TABLE 20.7 Hypothetical Multitrait–Multimethod Correlation Matrix for the Composite Direct Product Example**

| Measure | Method 1 | | | | Method 2 | | | | Method 3 | | | |
|---|---|---|---|---|---|---|---|---|---|---|---|---|
| | $T_1$ | $T_2$ | $T_3$ | $T_4$ | $T_1$ | $T_2$ | $T_3$ | $T_4$ | $T_1$ | $T_2$ | $T_3$ | $T_4$ |
| Method 1 | | | | | | | | | | | | |
| $T_1$ | 1.00 | | | | | | | | | | | |
| $T_2$ | .25 | 1.00 | | | | | | | | | | |
| $T_3$ | .15 | .31 | 1.00 | | | | | | | | | |
| $T_4$ | .15 | .08 | .33 | 1.00 | | | | | | | | |
| Method 2 | | | | | | | | | | | | |
| $T_1$ | .30 | .14 | .02 | .14 | 1.00 | | | | | | | |
| $T_2$ | .16 | .29 | .20 | .11 | .11 | 1.00 | | | | | | |
| $T_3$ | .09 | .19 | .30 | .14 | .04 | .16 | 1.00 | | | | | |
| $T_4$ | .11 | .06 | .22 | .29 | .07 | .08 | .12 | 1.00 | | | | |
| Method 3 | | | | | | | | | | | | |
| $T_1$ | .09 | .06 | −.02 | .03 | .11 | .03 | .03 | .03 | 1.00 | | | |
| $T_2$ | .03 | .08 | .03 | .01 | .02 | .11 | .09 | .01 | .39 | 1.00 | | |
| $T_3$ | −.03 | .04 | .09 | .04 | −.04 | .09 | .11 | .11 | .14 | .49 | 1.00 | |
| $T_4$ | .00 | .02 | .00 | .05 | .03 | .03 | .05 | .13 | .26 | .17 | .46 | 1.00 |

**TABLE 20.8 Parameter Estimates from the Composite Direct Product Model**

| Measure | Scaling matrix | Unique methods | Unique traits |
|---|---|---|---|
| Method 1 | | | |
| $T_1$ | .70 | 1 | 1.04 |
| $T_2$ | .69 | 1 | 1.10 |
| $T_3$ | .71 | 1 | .98 |
| $T_4$ | .70 | 1 | 1.06 |
| Method 2 | | | |
| $T_1$ | .70 | 1.24 | 1.04 |
| $T_2$ | .69 | 1.24 | 1.10 |
| $T_3$ | .72 | 1.24 | .98 |
| $T_4$ | .69 | 1.24 | 1.06 |
| Method 3 | | | |
| $T_1$ | .70 | .68 | 1.04 |
| $T_2$ | .70 | .68 | 1.10 |
| $T_3$ | .71 | .68 | .98 |
| $T_4$ | .70 | .68 | 1.06 |

**TABLE 20.9** Parameter Estimates from the Composite Direct Product Model: Second-Order Factor Pattern Matrix

| Variable | F1_T1 | F1_T2 | F1_T3 | F1_T4 | F2_T1 | F2_T2 | F2_T3 | F2_T4 | F3_T1 | F3_T2 | F3_T3 | F3_T4 |
|---|---|---|---|---|---|---|---|---|---|---|---|---|
| F_11 | **1** | 0 | 0 | 0 | **0** | 0 | 0 | 0 | **0** | 0 | 0 | 0 |
| F_21 | 0 | 1 | 0 | 0 | 0 | 0 | 0 | 0 | 0 | 0 | 0 | 0 |
| F_31 | 0 | 0 | 1 | 0 | 0 | 0 | 0 | 0 | 0 | 0 | 0 | 0 |
| F_41 | 0 | 0 | 0 | 1 | 0 | 0 | 0 | 0 | 0 | 0 | 0 | 0 |
| F_12 | **.61** | 0 | 0 | 0 | **.28** | 0 | 0 | 0 | **0** | 0 | 0 | 0 |
| F_22 | 0 | .61 | 0 | 0 | 0 | .28 | 0 | 0 | 0 | 0 | 0 | 0 |
| F_32 | 0 | 0 | .61 | 0 | 0 | 0 | .28 | 0 | 0 | 0 | 0 | 0 |
| F_42 | 0 | 0 | 0 | .61 | 0 | 0 | 0 | .28 | **0** | 0 | 0 | 0 |
| F_13 | **.17** | 0 | 0 | 0 | **.51** | 0 | 0 | 0 | **1.12** | 0 | 0 | 0 |
| F_23 | 0 | .17 | 0 | 0 | 0 | .51 | 0 | 0 | 0 | 1.12 | 0 | 0 |
| F_33 | 0 | 0 | .17 | 0 | 0 | 0 | .51 | 0 | 0 | 0 | 1.12 | 0 |
| F_43 | 0 | 0 | 0 | .17 | 0 | 0 | 0 | .51 | 0 | 0 | 0 | 1.12 |

**TABLE 20.10** Parameter Estimates from the Composite Direct Product Model: Factor Correlations

| Variable | F1_T1 | F1_T2 | F1_T3 | F1_T4 | F2_T1 | F2_T2 | F2_T3 | F2_T4 | F3_T1 | F3_T2 | F3_T3 | F3_T4 |
|---|---|---|---|---|---|---|---|---|---|---|---|---|
| F1_T1 | 1 | | | | | | | | | | | |
| F1_T2 | .52 | 1 | | | | | | | | | | |
| F1_T3 | .21 | .65 | 1 | | | | | | | | | |
| F1_T4 | .35 | .23 | .62 | 1 | | | | | | | | |
| F2_T1 | 0 | 0 | 0 | 0 | 1 | | | | | | | |
| F2_T2 | 0 | 0 | 0 | 0 | .52 | 1 | | | | | | |
| F2_T3 | 0 | 0 | 0 | 0 | .21 | .65 | 1 | | | | | |
| F2_T4 | 0 | 0 | 0 | 0 | .35 | .23 | .62 | 1 | | | | |
| F3_T1 | 0 | 0 | 0 | 0 | 0 | 0 | 0 | 0 | 1 | | | |
| F3_T2 | 0 | 0 | 0 | 0 | 0 | 0 | 0 | 0 | .52 | 1 | | |
| F3_T3 | 0 | 0 | 0 | 0 | 0 | 0 | 0 | 0 | .21 | .65 | 1 | |
| F3_T4 | 0 | 0 | 0 | 0 | 0 | 0 | 0 | 0 | .35 | .23 | .62 | 1 |

The CC' matrix is rescaled into a correlation matrix by dividing each covariance term into the product of two respective standard deviations, that is,

$$.61/[(\sqrt{1})(\sqrt{.45})] = .91,$$
$$.17/[(\sqrt{1})(\sqrt{1.54})] = .14, \text{ and}$$
$$.25[(\sqrt{.45})(\sqrt{1.54})] = .30,$$

which yields the following multiplicative method correlation components:

$$\Pi_{F\_M} = \begin{bmatrix} 1 & .91 & .14 \\ .91 & 1 & .30 \\ .14 & .30 & 1 \end{bmatrix}.$$

The off-diagonal elements of the $\Pi_{F\_T}$ and $\Pi_{F\_M}$ matrices are related directly to the validity diagonals and to the monomethod correlations of the MTMM matrix, respectively. Convergent validity is achieved when the off-diagonal elements of the $\Pi_{F\_M}$ matrix approach unity. The average correlation components of the $\Pi_{F\_M}$ matrix is equal to .45, indicating poor convergent validity in this example. For the discriminant validity, the trait correlation components should be close to zero and should also be less than the method correlation components. From the $\Pi_{F\_T}$ matrix, the average trait correlation components are equal to .43, which not only is much larger than zero but also is slightly smaller than the average method correlation components. Therefore, discriminant validity for this hypothetical MTMM matrix is not supported in this hypothetical example.

## V. CONCLUSION

Definitions of convergent validity, discriminant validity, and the method effect differ to some extent from one analytic technique to another. Extreme caution is warranted when interpreting the results of an MTMM analysis, especially when the analysis does not produce any fit statistic. When a fit statistic is provided, the investigator must first determine if the model fits the data. If not, no claims for convergent or discriminant validity can be made. The finding of a well-fitting MTMM model signifies that model-based interpretations are justified, but not necessarily that support for construct validity was obtained. A well-fitting MTMM model may fail to support construct validity, as illustrated hypothetically in this chapter. When the fit is plausible, then the next step is to interpret the parameter estimates obtained from a particular MTMM model. Conclusions drawn from an MTMM analysis, therefore, should be based on the model of choice.

The method effect plays an important role in the analysis and interpretation of MTMM matrices. Methods long have been considered trait contaminants, but both traits and methods have their own places in the nomological network (Cronbach, 1995; Dumenci & Windle, 1998). The attitude that "traits are good and methods are bad," therefore, should be abandoned.

Independent replication is the essential element in any scientific inquiry. Thus, all information necessary to replicate the MTMM analysis should be reported so that the analysis can be verified independently by others. That includes the correlation–covariance matrix (MTMM) and distributional characteristics of measures. The adoption of the specific MTMM analytic procedure, on which the substantive interpretation is based, should be justified on theoretical, empirical, and statistical grounds. When two or more analytic techniques are used, all analyses should be reported, along with the problems encountered in fitting each MTMM model. Such information not only gives readers the opportunity to draw their own conclusions, but also informs the performance of different analytic models under certain conditions.

## A. Choosing an Analytic Technique

An in-depth understanding of measurement process and an extensive knowledge of substantive theory provide the ultimate guidance in model selection. Familiarity with various MTMM techniques facilitates the choice of analysis. Applied researchers should start thinking about the possible MTMM analyses during the initial stages of research. The statistical model used should provide a sound representation of the measurement process. Research design plays an important role in choosing an MTMM analysis. For example, two traits measured by two methods satisfy the minimum requirements for an MTMM matrix, but a minimum of nine measures (i.e., three traits assessed by three methods) is required to use the CFA, CCA, or CDP models. Statistical tests also should be used to verify model assumptions.

The expected method effect plays an important role in the selection of an analysis. For example, when the methods consist of three self-report batteries supposedly measuring the same set of traits, the choice of analysis should be able to test for the presence of a general method effect, related methods, unrelated methods, and no method effect at all. The presence of method correlations is much less likely when rather different procedures such as self-report, expert rating, and a physiological variable (e.g., reaction time) are used to measure a trait such as intelligence (Wolins, 1982). Such complexities should be taken into account in selecting the model.

Applied researchers typically will face alternative strategies for choosing a statistical model for MTMM data. I recommend choosing the statistical model that best represents the measurement process and the substantive theory. Strong construct validity claims can be made when the model provides a good fit to the observed MTMM matrix. When the model fit is poor, however, conclusions about convergent and discriminant validity cannot be drawn.

Alternatively, two or more analytic procedures can be used to analyze an MTMM matrix. The use of multiple analytic techniques increases the possibility of finding a well-fitting model for the MTMM matrix. The major obstacle of choosing a best fitting model in such a post hoc fashion is that it has a strong likelihood of capitalizing on chance, hence, the findings may not be generalized across different samples. Furthermore, different analytic procedures may yield contradictory construct validity evidence when two or more MTMM models fit the observed MTMM matrix equally well. Therefore, applied researchers who adopt this strategy must be willing to settle for a weak construct validity claim.

A third strategy offers a compromise between the first two strategies. Many analytic techniques (e.g., the CFA, the CCA, and the CDP) offer alternative parameterizations of trait, method, and error effects. For example, alternative specifications of the CDP model allow for testing of a range of error structures. The CDP model without the error parameters in Figure 20.5 corresponds directly to the model proposed by Swain (1975). Nested model comparison strategies for the CDP model also are available to test convergent validity and discriminant validity, as well as the method effect (Bagozzi & Yi, 1990). Thus, applied researchers can specify alternative models a priori, without committing to a single model, thereby avoiding the problem of capitalizing on chance. Applied researchers should realize, however, that this strategy offers a weaker construct validity claim than the first strategy, and the probability of obtaining a well-fitting model for the MTMM matrix is less than with the second. In conclusion, there is no one simple answer to the question of what is the best quantitative technique to analyze an MTMM matrix. Instead, researchers must consider to what extent conclusions drawn from the analysis can be argued successfully against alternative explanations.

## ACKNOWLEDGMENTS

I thank Werner Wothke for his comments on an earlier version of this chapter. I also am thankful to Anne Richards for editorial assistance.

## REFERENCES

Bachman, L. F., & Palmer, A. S. (1981). The construct validation of the FSI Oral Interview. *Language Learning, 31,* 67–86.
Bagozzi, R. P., & Yi, Y. (1990). Assessing method variance in multitrait-multimethod matrices:

The case of self reported affect and perceptions at work. *Journal of Applied Psychology, 75*, 447–460.

Bentler, P. M. (1989). *EQS structural equations program manual.* Los Angeles: BMDP Statistical Software, Inc.

Bollen, K. A. (1989). *Structural equations with latent variables.* New York: Wiley.

Bork, J. K., & Bargmann, R. E. (1966). Analysis of covariance structures. *Psychometrika, 31*, 507–534.

Browne, M. W. (1984). The decomposition of multitrait–multimethod matrices. *British Journal of Mathematical and Statistical Psychology, 37*, 1–24.

Byrne, B. M., & Goffin, R. D. (1993). Modeling MTMM data from additive and multiplicative covariance structures: An audit of construct validity concordance. *Multivariate Behavioral Research, 28*, 67–96.

Campbell, D. T., & Fiske, D. W. (1959). Convergent and discriminant validation by the multitrait-multimethod matrix. *Psychological Bulletin, 56*, 81–105.

Campbell, D. T., & O'Connell, E. J. (1967). Method factors in multitrait-multimethod matrices: Multiplicative rather than additive? *Multivariate Behavioral Research, 2*, 409–426.

Campbell, D. T., & O'Connell, E. J. (1982). Methods as diluting trait relationships rather than adding irrelevant systematic variance. In D. Brinberg & L. H. Kidder (Eds.). *Forms of validity in research. New directions for methodology of social and behavioral science* (Vol. 12, pp. 93–111). San Francisco: Jossey-Bass.

Conway, J. M. (1996). Analysis and design of multitrait-multirater performance appraisal studies. *Journal of Management, 22*, 139–162.

Cronbach, L., (1995). Giving method variance its due. In P. E. Shrout & S. T. Fiske (Eds.), *Personality research, methods, and theory: A festschrift honoring Donald W. Fiske* (pp. 145–157). Hillside, NJ: Lawrence Erlbaum Associates.

Cronbach, L., Gleser, G., Nanda, H., & Rajaratnam, N. (1972). *The dependability of behavioral measurements.* New York: Wiley.

Cudeck, R. (1988). Multiplicative models and MTMM matrices. *Journal of Educational Statistics, 123*, 131–147.

Dumenci, L. (1995). Construct validity of the Self-Directed Search using hierarchically nested structural models. *Journal of Vocational Behavior, 47*, 21–34.

Dumenci, L., & Windle, M. (1996). A latent trait-state model of adolescent using the Center for Epidemiologic Studies-Depression scale. *Multivariate Behavioral Research, 31*, 313–330.

Dumenci, L., & Windle, M. (1998). A multitrait-multioccasion generalization of latent trait-state model: Description and application. *Structural Equation Modeling: A Multidisciplinary Journal, 5*, 391–410.

Engstrom, C. P. Persson, L. O., Larsson, S., & Sullivan, M. (1998). Reliability and validity of a Swedish version of the St Georges Respiratory Questionnaire. *European Respiratory Journal, 11*, 61–66.

Forsythe, G. B., McGaghie, W. C., & Friedman, C. P. (1986). Construct validity of clinical competence measures: A multitrait-multimethod matrix study using confirmatory factor analysis. *American Educational Research Journal, 23*, 315–336.

Guilford, J. P. (1954). *Psychometric methods.* New York: Wiley.

Hubert, L. J., & Baker, F. B. (1978). Analyzing the multitrait-multimethod matrix. *Multivariate Behavioral Research, 13*, 163–179.

Jackson, D. N. (1969). Multimethod factor analysis in the evaluation of convergent and discriminant validity. *Psychological Bulletin, 72*, 30–49.

Joreskog, K. G. (1971). Statistical analysis of sets of congeneric tests. *Psychometrika, 57*, 409–426.

Kavanagh, M. J., MacKinney, A. C., & Wolins, L. (1971). Issues in managerial performance: Multitrait-multimethod analysis of ratings. *Psychological Bulletin, 75*, 34–49.

Kenny, D. A. (1995). The multitrait-multimethod matrix: Design, analysis, and conceptual issues. In P. E. Shrout & S. T. Fiske (Eds.), *Personality research, methods, and theory: A festschrift honoring Donald W. Fiske* (pp. 111–124). Hillside, NJ: Lawrence Erlbaum Associates.

Kiers, H. A. L., Takane, Y., & ten Berge, J. M. F. (1995). The analysis of multitrait-multimethod matrices via constrained component analysis. *Psychometrika, 61,* 601–628.

Kumar, A., & Dillon, W. R. (1992). An integrative look at the use of additive and multiplicative covariance structure models in the analysis of MTMM data. *Journal of Marketing Research, 29,* 51–64.

Levin, J., Montag, I., & Comrey, A. L. (1983). Comparison of multitrait-multimethod, factor, and smallest space analysis on personality scale data. *Psychological Reports, 53,* 591–596.

Marsh, H. W. (1996). Physical Self Description questionnaire: Stability and discriminant validity. *Research Quarterly for Exercise & Sport, 67,* 249–264.

Ostrom, T. M. (1969). The relationship between the affective, behavioral, and cognitive components of attitude. *Journal of Experimental Social Psychology, 5,* 12–30.

Schriesheim, C. A. (1981). The effect of grouping or randomizing items on leniency response bias. *Educational and Psychological Measurement, 41,* 401–411.

Schweizer, K. (1991). An equal-level approach to the investigation of multitrait-multimethod matrices. *Applied Psychological Measurement, 15,* 307–317.

Sidani, S., & Jones, E. (1995). Use of the multitrait multimethod (MTMM) to analyze family relational data. *Western Journal of Nursing Research, 17,* 556–570.

Stanley, J. C. (1961). Analysis of unreplicated three-way classifications, with applications to rater bias and trait independence. *Psychometrika, 26,* 205–219.

Sullivan, J. L., & Feldman, S. (1979). *Multiple indicators: An introduction.* Beverly Hills, CA: Sage.

Swain, A. J. (1975). *Analysis of parameter structures for variance matrices.* Unpublished doctoral dissertation, University of Adelaide.

Werts, C. E., & Linn, R. L. (1970). Path analysis: Psychological examples. *Psychological Bulletin, 74,* 193–212.

Widaman, K. F. (1985). Hierarchically nested covariance structure models for multitrait-multimethod data. *Applied Psychological Measurement, 9,* 1–26.

Wolins, L. (1982). *Research mistakes in the social and behavioral sciences.* Ames, IA: Iowa State University Press.

Wothke, W. (1984). *The estimation of trait and method components in multitrait-multimethod measurement.* Unpublished doctoral dissertation, University of Chicago, Department of Behavioral Science.

Wothke, W. (1987). *Multivariate linear models of the multitrait-multimethod matrix.* Paper presented at the meeting of the American Educational Research Association, Washington, D.C. (Educational Resources Information Center, Document No. TM 870 369).

Wothke, W. (1996). Models for multitrait–multimethod analysis. In G. A. Marcoulides & R. E. Schumacker (Eds.), *Advance structural equation modeling: Issues and techniques* (pp. 7–56). Hillside, NJ: Lawrence Erlbaum Associates.

Wothke, W., & Browne, M. W. (1990). The direct product model for the MTMM matrix parameterized as a second order factor analysis model. *Psychometrika, 55,* 255–262.

# 21

# USING RANDOM COEFFICIENT LINEAR MODELS FOR THE ANALYSIS OF HIERARCHICALLY NESTED DATA

**ITA G. G. KREFT**

*School of Education, California State University, Los Angeles, Los Angeles, California*

## I. MULTILEVEL MODELS FOR MULTILEVEL DATA

This chapter is about the analysis of data collected over individuals nested in contexts or groups. This includes individual data collected over time, as in repeated measurement or longitudinal data, where observations are nested within individuals. I do not cover time series analysis and modeling growth or change, since these are discussed by Mark, Reichardt, and Sanna (chapter 13, this volume) and Willett and Keiley (chapter 23, this volume), but it is quite easy to use the multilevel model discussed in this chapter for the same purpose (see Bryk & Raudenbush, 1987). This multilevel model discussed in this chapter was developed for analyzing hierarchically nested quantitative data. These models became popular a decade ago, in part as a result of the availability of software packages such as HLM (Hierarchical Linear Models; Bryk, Raudenbush, & Congdon, 1996) and MLn (short for Multilevel with *n* numbers of levels, where *n* has a maximum of 9; Rashbash, Prosser, & Goldstein, 1991). For an extensive discussion of this and other software, see De Leeuw and Kreft (1999). The term *linear* indicated that the model is related to the well-known linear models, regression, and analysis of variance (ANOVA), and is labeled *random coefficient* model, continuing the tradition started in De Leeuw and Kreft (1986). Other names for the same model are hierarchical linear models, mixed linear models, and multilevel models.

Before elaborating on the model itself, more needs to be said about hierarchically nested data. A hierarchical data set contains measurements on at least two different levels of a hierarchy, such as measurements of patients and hospitals, where the lower level (patients) is nested within the higher level (hospitals). Other examples are students nested within schools, employees nested within firms, or repeated measurements nested within persons. The lowest level of measurements are said to be at the microlevel or level 1; all higher level measurements are at the macrolevel (i.e., level 2, level 3, and so on). Macrolevels are often referred to as contexts, but in longitudinal and repeated measurement data the context is the individual. Models for analyzing data obtained at micro- and macrolevels are called multilevel models.

Multilevel data can have as few as two levels, as in the case of students (microlevel) nested within classes (macrolevel), and repeated measurements (microlevel) nested within students (macrolevel); or more than two, for example, a three-level model of students nested within classes, nested within schools, or repeated measurements nested within students, nested within school classes. Many more levels can be conceived, such as schools nested within states, and states nested within countries. Hierarchies exist in all aspects of life, and the research question and knowledge of the research field guide the determination of what levels to investigate in an analysis. The best rule is to pick a small number of different levels, as parsimony is always preferable over complexity in a model.

## II. FIELDS OF STUDY WHERE MULTILEVEL DATA ANALYSES CAN BE APPLIED

The examples that follow illustrate what is meant by multiple levels, what multilevel data is, and when applied researchers need to take these levels seriously in analyzing data. The examples also illustrate the broad range of potential applications of multilevel analysis techniques.

### A. Clinical Psychology

Imagine a traditional research situation in which clients are (randomly) assigned to a directive or nondirective group therapy intervention, where each intervention consists of several groups. The response variable is the improvement in the client's level of psychological adjustment, which is operationalized as the difference between the level measured prior to therapy and after completion of the therapy program. It is important to realize that the dynamics of the group are not under the control of the researcher, so that the interactions within each group may develop in different directions, resulting in behavior that is partly due to the type of therapy and partly to specific dynamics unique to the group. If a traditional method such as a fixed effects ANOVA is used to analyze this data, the factor is the type of treatment (i.e., directive or nondirective type of group therapy). Because several groups are nested within treatments, however, combining the groups receiving the same treatment without correcting for the intraclass correlation among observations in the same group confounds the intervention, with variance attributable to within-group dynamics.

Using a random coefficient model in such a situation allows the investigator to take the intraclass correlation into account, and to test whether specific group characteristics account for differences in the outcome variable, above and beyond the effect of the intervention. A two-level hierarchy is present in this example, participants nested within groups. The type of treatment can be considered to be a group characteristic, and a random coefficient model will test for both individual-level and group-level effects. The same effects can be tested using an analysis or covariance (ANCOVA) (see Huberty & Petoskey, chapter 7, this volume) with the restriction that at the group level only the effect of treatment can be assessed.

## B. School Effectiveness Research

School-based drug prevention research is an area in which multilevel analysis is useful for answering questions about the effectiveness of specific prevention programs. Data are obtained from students nested within schools. Some schools receive a specific drug prevention program, whereas other schools that do not receive a program function as the control group. The measure of success is the individual student's behavior in relation to some specific drug (e.g., alcohol). The program is administered at the school class level, but the individual students are the unit of measurement used to evaluate the success of the program.

Students within the same school class share the same environment and the same teachers, so the behavior of the students in the same school class will be more homogeneous than the behavior of students in different schools. In a statistical sense, therefore, the observations are no longer independent, and the assumption of independent observations in traditional statistical models is no longer tenable. If schools are used instead of students as the unit of sampling, we assume that an intraclass correlation exists. The larger the intraclass correlation, the more homogeneous a school (or class) will be, and the more the schools (or classes) will differ from each other. Large variations among the schools in the response (dependent) variable will result in both a large between-school variation and a large intraclass correlation. The intraclass correlation is a measure of homogeneity within groups, and at the same time a measure of heterogeneity among groups in random coefficient models. Ignoring the intraclass correlation and analyzing the data at the student level results in statistical tests of significance that are too liberal, thereby increasing the occurrence of false positives (see Barcikowski, 1981, Kreft & De Leeuw, 1998).

Data from an investigation using this design could be analyzed using a two-level (i.e., student level and school level) or a three-level (i.e., students, classes, and schools) analysis. A theoretical reason for using a three-level analysis is that classes create their own specific class climate, which is a separate level from that of the school level. Of course this is most likely to be true only when students attend the same class all day, as in elementary schools.

### C. Family Research

Family research is another field where multilevel models are useful. The family environment exerts a strong influence on its members, so family members are in many respects more similar to each other than they are to people outside the family. This intrafamily correlation must be considered in analyzing data when families are sampled and members within families are observed. At least two levels are present, the family level (level 2) and the individual level (level 1), and additional levels, such as families nested within neighborhoods, and neighborhoods nested within towns and states, could be included in the analysis.

An example of a multilevel question in family research is what are the effects of family climate on the achievement of children in school? We know that achievement in school is partly determined by IQ. But there are reasons to believe that the relation between IQ and achievement has a within-family as well as a between-family component. The within-family component may be partly due to differences of family members in their IQ (the level 1 variation). The between-family component can be due to two effects: (a) direct effect of the family milieu on achievement, indicated by a between-family (or level 2) variation and (b) an indirect effect. Here an indirect effect is the enhancing (or decreasing) effect of family factors on IQ, which in turn affects achievement. This is a cross-level effect, based on the family milieu and IQ together affecting achievement. In multilevel models a cross-level interaction is created by the multiplication of two variables, in the same way as in traditional linear models. The difference is that in multilevel models the interaction occurs between variables measured at different levels of the hierarchy (i.e., the family climate, measured at level 2, and the student IQ, measured at level 1). Fitting cross-level interactions is one of the attractive features of multilevel analysis. The cross-level hypothesis in this example states that family climate influences the relation between IQ and the achievement of individual children in school.

In this chapter I will present illustrations of cross-level interactions using a real data set. One of the hypotheses is that schools in the private sector have a moderating effect on the relation between homework and math achievement of students. The results of the analysis shows that students in the schools of the public sector need to make more homework to achieve the same math grade as students in schools of the private sector. The cross-level interaction in this case is between the school characteristic "public" and the student characteristic "homework."

### D. Vocational Research

Multilevel modeling can be very useful in analyzing data collected from employees in various occupational settings, such as medical doctors or nurses in hospitals. An intraclass correlation is expected based on the observation that employees within the same hospital act more alike compared to employees in other hospitals. The climate of a hospital can have this homogenizing effect on the behavior of employees towards patients. If the well-being of patients is the response variable, we can hypothesize that well-being can be predicted by the climate and culture of the hospital that determines the quality of care (a hospital effect). A multilevel analysis will include all important patient and hospital characteristics, testing simultaneously the effects of both levels on the well being of patients. Researchers interested in discovering the effect of hospital characteristics (such as type of HMO) on the wellness of patients must consider both the intrahospital correlation and the different levels of the measurement. The variation in health among patients is called the within or level 1 variation, whereas the variation among hospitals is the between-hospital or level 2 variation. The results of a multilevel analysis will show how much each level contributes to the well-being of patients and what characteristics account for that variability.

### E. Summary

Many important research questions in the social sciences are multilevel questions. What part of the criterion is predicted by individual variables, what part is predicted by group variables (school, family, or occupational setting), and what attributes of the group are most strongly associated with the most efficacious outcomes? In multilevel analyses the response variable is measured at the lowest level (e.g., the students, patients, and children in earlier examples). In the following discussions of analyses with random coefficient models, the most frequently used concepts are the intraclass correlation, level 1 variance, level 2 variance, and cross-level interactions.

Only models with two levels will be discussed in this chapter, although more levels are possible. In these models level 2 is the context (e.g., schools, hospitals, families) where level 1 units (students, patients, nurses, or children) are nested within. The criterion variance is partitioned into variances attributed to each level. The ANCOVA concepts of variance within and variance between still apply. The variance within is defined in random coefficient models in the same way as in ANCOVA, while the variance between is a composite of more than one variance, except in the simple random intercept model. In all cases where a random slope is present, several variance components exist at the between level. The number of variances present in a model depends on the number of random coefficients and the number of levels in the data. In models with three or more levels, the variance between is defined as between contexts at level 3, and so on. In models with more than one random coefficient, the variance between has as many variances as random coefficients plus the covariance's, because variances at the same level are correlated.

## III. RANDOM COEFFICIENT MODELS COMPARED WITH FIXED LINEAR MODELS

Fixed coefficient approaches, such as ANCOVA (see Huberty & Petoskey, chapter 7, this volume) and regression (see Venter & Maxwell, chapter 6, this volume) are more restrictive procedures from the same family of linear models as the random coefficient model. ANCOVA is a familiar choice for researchers confronted with grouped data that involve two levels (e.g., the individual employee level and the group industry level). For instance, applied researchers interested in a comparison of salaries of employees working in different industries of the United States can use ANCOVA to compare the average salary in industries after correction for the educational level of workers in those industries. Random coefficient models also are related to multiple regression; using the above example, the random coefficient model provides the possibility to predict income of an employee (given a certain level of education) for each industry separately, because intercepts are allowed to differ. Although the regression part of the model assesses the effects of quantitative factors at the individual level (such as the effect of education on income), the ANCOVA part enables the modeling of qualitative factors at the group level (such as the type of industry in which a person works). The equation for the analysis of covariance is

$$\underline{Y}_{ij} = a_j + bx_{ij} + \underline{e}_{ij}, \tag{1}$$

where $_j$ indicates groups and $_i$ indicates individuals. The $_j$ for the intercept, $a_j$, refers to different estimates for the intercept $a$ for each industry, based on the assumption that some industries have higher starting salaries than others. The underlining indicates that a variable is random (having a mean and a variation around that mean). Using ANCOVA to analyze data where employees are nested in industries requires the assumption that all industries have the same slope (b), which is the same as assuming that the relation between education and income is equal among industries. This last assumption may be too strong, given Kreft and De Leeuw's (1994) findings that gender differences in income are stronger in private industries than in public industries.

The chief advantage of ANCOVA is its predictive power, but it can be used only to answer broad-brush questions, such as whether general differences in income exist among industries. More specific questions about what causes these differences in income cannot be answered with this technique. Furthermore, ANCOVA does not allow for different slopes among groups. The effect of gender on income in different industries can be modeled in multilevel analysis by allowing the slope for gender to be random. A significant random effect of such a slope shows that there is a between variance that could be "explained" by some context characteristic (e.g., whether an industry is private or public).

The traditional regression approach assumes that the intercepts are equal for all groups (i.e., $a_1 = a_2 = \cdots = a_m$), whereas the ANCOVA and random coefficient models assume that they are unequal (i.e., $a_1 \neq a_2 \neq \cdots \neq a_m$). Both the traditional regression and ANCOVA models assume that the slope does not vary significantly among contexts (i.e., $b_1 = b_2 = \cdots = b_m$) whereas the random coefficient model allows applied researchers to assume that the slopes are equal or unequal (i.e., $b_1 \neq b_2 \neq \cdots \neq b_m$). The use of the random coefficient model is indicated when the researcher wants to investigate the possibility that coefficients vary systematically as a function of the context, either by modeling different intercepts and/or different slopes. The assumptions of the random coefficients model are (a) that observations within the same context have correlated error terms, indicating that they are more like each other than individuals in different contexts, and (b) that parameters are random, indicating that they fluctuate over contexts. The assumptions of the random coefficient model may better fit the data when a data set is hierarchically nested, and it represents a sample of an infinite population of contexts, instead of a fixed number of treatments. Random coefficient models are extensions of variance component models (see Searle, Casella, & McCulloch, 1992).

## IV. ILLUSTRATION OF THE RANDOM COEFFICIENT MODEL

In this section a simple random coefficient model for quantitative data is discussed, with two hierarchical levels: the student and the school level. An extension to more than two levels is straightforward, but an extension to random coefficient models with dichotomous or categorical response variables is not to simple and is beyond the scope of this introductory chapter.

It is instructive to compare the concept of random coefficients with the familiar concepts of random explanatory variables (predictors) used in structural equation models (see DiLalla, chapter 15, this volume). The equation for a random coefficient model, with all coefficients random is

$$\underline{Y}_{ij} = \underline{a}_j + \underline{b}_j X_{ij} + \underline{e}_{ij}. \tag{2}$$

Equation 2 resembles a familiar regression model with the following components: a response (criterion or dependent) variable $\underline{Y}_{ij}$, an explanatory (independent) variable $X_{ij}$, and an error term $\underline{e}_{ij}$. The model has two (random) coefficients, the intercept ($\underline{a}_j$) and the slope, ($\underline{b}_j$). The underlining indicates that the coefficients are random, while the $j$ indicates that the coefficients differ among groups or contexts. Random means here that these components are a random sample, assumed to be from a normal distribution, that is, symmetrical around a mean value.

The concept of 'random' coefficients is comparable to the concept of random explanatory variables, as known from linear structural models. The random coefficient model (Equation 2) shows underlining of the coefficients $\underline{a}_j$ and $\underline{b}_j$, indicating that they are random, whereas in (Equation 3) the explanatory variable $\underline{X}_I$ is random and thus underlined.

$$\underline{Y}_{ij} = a_j + b_j \underline{X}_i + \underline{e}_{ij} \tag{3}$$

Because a random part (either of a variable or of a coefficient) represents error variance, the two approaches differ in attributing error variance to a variable ($\underline{X}_i$ in Equation 3) or to the intercept and slope coefficients ($a_j$ and $b_j$ in Equation 2). The interpretation is equally different, where random variables are assumed to differ over measurements (e.g., have a measurement error), whereas random coefficients are assumed to vary over contexts. But both a random coefficient and a random variable are described by two numbers, an average and a variance around that average. In sum, random coefficient models have fixed explanatory variables and random coefficients, whereas structural equation models have random explanatory variables and fixed coefficients.

In random coefficient models not all coefficients are necessarily random. It is up to the researcher to choose which ones to make random and which ones to treat as fixed. That is why this model is sometimes called a mixed model. Nevertheless, a random coefficient model has at least one random coefficient, which by default is the intercept in most software packages.

In Equation 2 $j$ is used to indicate that both the coefficients for intercept and slope differ over contexts or groups. For example, let $\underline{Y}_{ij}$ be the math score of students and $X_{ij}$ the hours of homework per week. The model in Equation 2 allows each school $j$ to have its own intercept and its own slope for homework. A significant random slope for homework may indicate that homework is a strong predictor of math achievement in some schools, but not in others. Using ANCOVA to compare corrected means among schools is similar to testing whether the intercepts differ among schools, but the presence of varying slopes violates the assumption of homogeneous slopes in ANCOVA. The random coefficient model also can be compared to the traditional regression model. Both models estimate fixed effects for intercept and slope, but the random coefficient model also estimates extra parameters for the random components for the same intercept and slope, random at the higher level of the hierarchy. For the model in Equation 2 those parts are the variance of the intercept, the variance of the slope, together with their covariance. Admittedly, Equation 2 does not look like a multilevel model because no second-level (school level) variables are present. The only evidence that we are dealing with multilevel data is in the underlining of intercept (a) and slope (b), which indicates the random variation at the second level. Next I will demonstrate the results of fitting the model in Equation 2 on a real data set.

## A. Illustration and Model Specifications

Data collected by the National Center for Education Statistics of the U.S. Department of Education are used to illustrate the advantages of the random coefficient model. The NELS.88 data are the first in a series of longitudinal measurements on students starting in the eighth grade. Almost 25,000 students from approximately 800 public and 200 private schools in the United States participated in the base year study. The sample is representative for the 3 million eighth graders enrolled in more than 38,000 schools in Spring 1988. From that data I choose math achievement as the response (dependent) variable, and constructed a composite of parents educational level and income, which I call socioeconomic status (SES), as an explanatory (independent) variable. The results of a multilevel analysis with these two variables are reported in Table 21.1.

Table 21.1 contains several parameter estimates that are not familiar from traditional regression analysis, the random components at level 2 and the deviance (defined as $-2$ times the loglikelihood). The value of 11.11 for the variance of the intercept (see Table 21.1) and the relatively small standard error (0.66) of this value indicates that after correction for SES, a significant variation exists among schools in mean levels of achievement. Significance is defined in the usual way, by dividing the parameter estimate (11.11) by its standard error (0.66), which yields a z-score of 16.8. Any z-score larger than 2.00 indicates that an individual estimate is statistically significant. The estimate for the variance of the slope of SES is not significant

**TABLE 21.1    A Random Coefficient Model with Socioeconomic Status and Math Achievement**

|  | Parameter estimates | Standard error |
|---|---|---|
| Fixed effects |  |  |
| Intercept | 50.96 | 0.12 |
| Socioeconomic status | 4.82 | 0.09 |
| Variance Components |  |  |
| Level 1 (individual) |  |  |
| Error | 69.76 | 0.70 |
| Level 2 (school) |  |  |
| Intercept | 11.11 | 0.66 |
| Slope for socioeconomic status | 0.54 | 0.35 |
| Covariance | 1.68 | 0.33 |
| Deviance | 154336 |  |

(i.e., 0.54/0.35 = 1.54). Based on this result, I assume a homogeneous slope among schools for SES in my next model.

Standard scores (z-scores) usually are not provided in the output of software for multilevel analysis. The philosophy behind this nonreporting of the significance level of individual parameter estimates is that the interpretation of analysis results are the responsibility of the user. In the social sciences, which have many correlated variables, the interpretation of complex models, such as the random coefficient model, must be based on total model fit or improvement of fit rather than on individual coefficients. Such a measure of fit is the deviance, defined as $-2*$log-likelihood, derived from the maximum likelihood estimation (MLE) of the parameters in these models. The deviance indicates the difference between the observed and expected values. The difference between two deviances (when comparing two models) behaves like a chi-square distribution and is evaluated in relation to the number of degrees of freedom. I will return to this topic later when this measure of fit is used to compare the fit of two models in a real example.

For a better understanding of the random coefficient model, the intercept and the slope are defined by Equations 4 and 5. The definition for the intercept is

$$\underline{a}_j = \gamma_{00} + \underline{\delta}_{0j}, \tag{4}$$

where $\gamma$ represents the fixed part, and $\underline{\delta}$ the random part of the intercept. The same is true for the slope, as in Equation 5.

$$\underline{b}_j = \gamma_{10} + \underline{\delta}_{1j} \tag{5}$$

The subscripts indicate the function of the gammas and deltas. The first subscripts for the gamma estimates (in all equations) are (0) when related to the intercept, (1) when related to the first slope, (2) when related to the second slope, and so on. The second subscript is used to indicate the relation with second-level (school) variables. If there are no second-level variables in the model, this subscript will be 0, as in both Equations 4 and 5; otherwise the first higher level variable will be indicated by (1), the second by 2, and so on. Note that in Equation 4 the intercept is defined as consisting of a fixed part, $\gamma_{00}$ (with a value of 50.96 in Table 21.1) and a random part, $\underline{\delta}_{0j}$ (with a variance of 11.11 in Table 21.1). Similarly, the slope (Eq. (5)) has a fixed part, $\gamma_{10}$, (with a value of 4.82 in Table 1), and a random part $\underline{\delta}_{0j}$ with a variance of 0.54 in Table 21.1). The grand mean effect in Equation 4 is $\gamma_{00}$, which is the average math achievement over all students, after controlling for SES. The macro-error term, $\underline{\delta}_{0j}$ indicates the deviation (or variation) of the schools' intercepts around this overall or grand mean. The grand slope $\gamma_{10}$ in Equation 5 estimates the overall effect of SES on math achievement, whereas $\underline{\delta}_{1j}$ represents the deviation of the schools' slopes from the overall slope. The random parts $\underline{\delta}_{0j}$ and $\underline{\delta}_{1j}$ are macrolevel errors. Equations 4 and 5 indicate, respectively, that both the intercept $\underline{a}_j$ and the slope $\underline{b}_j$ are modeled as varying over schools. The similarity and the differences between the equations for the random coefficient and traditional regression models can be seen when Equations 4 and 5 are substituted in Equation 2. Replacing $\underline{a}_j$ and $\underline{b}_j$ (in Equation 2) with their new definition yields Equation 6a:

$$\underline{Y}_{ij} = (\gamma_{00} + \underline{\delta}_{0j}) + (\gamma_{10} + \underline{\delta}_{ij})X_{ij} + \underline{e}_{ij} \tag{6a}$$

Rearranging the terms in Equation 6a to put fixed and random parts together, yields Equation 6b:

$$\underline{Y}_{ij} = \gamma_{00} + \gamma_{10}X_{ij} + (\underline{e}_{ij} + \underline{\delta}_{0j} + \underline{\delta}_{ij}X_{ij}) \tag{6b}$$

Note that Equation 6b is actually an elaboration of the familiar regression equation in which the error term (in parenthesis) has been expanded to include more variance components. This is the main difference between a random coefficient model (Equation 6b) and a traditional regression model (Equation 1).

## B. Adding a School Context Variable to the Model

The random parts in a random coefficient model can be used for different purposes. The variance of the intercept can be used to calculate the intraclass correlation, as well as to test if significant variation exists among schools, much like an ANCOVA would do. If the intercept variance is significant, the researcher can proceed by investigating if characteristic of schools can "explain" that variation. In the previous example the intercept was shown to be significantly random, indicating that the schools differ in average math scores, after correcting for SES. The random coefficient model introduced in Table 21.1 and in Equation 6 does not yet contain a school-level variable, but researchers interested in context effects and interactions between the context and characteristics of the individual (cross-level interactions) can expand the model to address such issues. This is illustrated by introducing "percent minorities" as a school-level variable $Z_j$, where the $_j$ indicates that it has different values for different schools. $Z_j$ is introduced in the model as an overall effect that interacts with the intercept $\underline{a}_j$ as shown in Equation 7:

$$\underline{a}_j = \gamma_{00} + \gamma_{01}Z_j + \underline{\delta}_{0j} \tag{7}$$

A comparison with this equation and Equation 4 shows that the variable $Z_j$ is added and has a coefficient gamma 01, where the subscript 0 indicated it is related to the intercept, and where 1 indicates that it is the first school-level variable. In theory the same variable $Z_j$ could be introduced into Equation 5, the equation for the random slope (to create a cross-level interaction between the school-level variable minorities and the student-level variable SES), but the previous analysis revealed that no significant variation exists in the slope; therefore, I have set the variance for SES to zero as shown in Equation 8.

$$b = \gamma_{10} \tag{8}$$

Equation 8 treats the slope ($b$) for SES as fixed (no underlining, no $_j$, no error term). Substituting Equations 7 and 8 in Equation 2 results in Equation 9.

$$Y_{ij} = \gamma_{00} + \gamma_{10}X_{ij} + \gamma_{01}Z_j + (\underline{e}_{ij} + \underline{\delta}_{0j}) \tag{9}$$

Comparing Equations (6b) and (9) reveals that setting the variance for the slope at zero, thus creating a fixed instead of a random slope, simplifies the error term. In Equation 9 the error term contains a within- and a single between-variance component. The results of this model are presented in Table 21.2.

**TABLE 21.2   A Random Coefficient Model with a School Level Variable**

|  | Parameter estimates | Standard error |
|---|---|---|
| Fixed effects |  |  |
| Intercept | 53.16 | 0.19 |
| Socioeconomic status | 4.76 | 0.09 |
| Percent minorities | −0.71 | 0.05 |
| Variance Components |  |  |
| Level 1 (individual) |  |  |
| Error | 69.93 | 0.69 |
| Level 2 (school) |  |  |
| Intercept | 8.78 | 0.55 |
| Deviance | 154184 |  |

Comparing the analyses reported in Tables 21.1 and 21.2 reveals that adding the school variable percent minorities reduced the intercept variance and the deviance. Earlier I discussed that model fit instead of individual parameter estimates are studied in random coefficient modeling to evaluate the improvement of one model over another. The deviance of a model can be compared to the deviance of another model to assess which one fits the data best. The difference in the deviance (i.e., 154336, 154184 = 152) of two models follow the well-known chi-square distribution. Statistical significant differences in fit among models is calculated by comparing the difference in deviance to the degrees of freedom used by each model. A rule of thumb is that an observed difference between models (in their deviances) should be at least twice as large as the difference in degrees of freedom between the same two models, with a slight exception for 1 degree of freedom (consult any chi-square table). The larger the difference, such as 152 observed in my example, and the smaller the degrees of freedom, the more we can be convinced that the difference between models is statistically significant. Another interesting comparison among the models is the school-level variance for the intercept. It shows that this variation is reduced from 11.11 (in Table 21.2) to 8.78 in last analysis. Part of the variation among schools in mean level of math achievement is "explained" by differences in percentage of minorities among schools.

## C. Explained Variance and Intraclass Correlation

It is common practice in traditional regression to use the explained variance or $R^2$ (where $R^2$ = Total Variance − Error Variance/Total Variance) to compare different models for goodness of fit. In the previous paragraph I indicated that in random coefficient models the deviance is used for that purpose. The reason is that in random coefficient models more sources of error are present, and these sources do not sum up to the total error variance in a straightforward way. For instance, we have an error term within and an error term between, which are uncorrelated. Both sources of error are reported in Table 21.2, under the heading level 1 and level 2 variance. If I want to know whether the model in Table 21.2 produces a reduction in variance in level 1 (the $R^2$-within) and/or in level 2 (the $R^2$-between) I need to know how much of the total variance in the response variable math achievement is within-schools variance and how much is between-schools variance. For that purpose, a null-model is fitted to the data. A null model has no explanatory variables, only an intercept and two variances, one at level 1 and one at level 2. The results are in Table 21.3.

The results show a within-school variance of 76.62 and a between-school variance of 26.56, indicating that most of the variation is among students. Still the variance among schools is large enough to justify a further

**TABLE 21.3   A Random Coefficient Model with No Explanatory Variables: The Null Model**

|  | Parameter estimates | Standard error |
|---|---|---|
| Fixed effects |  |  |
|   Intercept | 50.8 | 0.17 |
| Variance components |  |  |
|   Level 1 (individual) |  |  |
|     Error | 76.62 | 0.76 |
|   Level 2 (school) |  |  |
|     Intercept | 26.56 | 1.36 |
| Deviance | 154184 |  |

search of school effects. The variances are useful to calculate a reduction in variance, when variables are added to the model, and to assess the amount of intra-class correlation among students within the same school. The intraclass correlation is defined as the ratio of the between-variance component (26.56) to the total variance (26.56 + 76.62 = 103.18) in the response variable. The resulting value (26.56/103.18 = 0.26) indicates that 26% of the variation in math achievement is attributable to differences among schools.

Comparing the variances in Table 21.2 with these in the null model (Table 21.3) reveals that both variances are reduced by the addition of a student-level variable (SES) and a school-level variable (percent minorities) to the model. The within variance in the null model is reduced from 76.62 to 69.93 in Table 21.2, a reduction of 6.69 or 8.7%, suggesting that the $R^2$-within is about 0.09 and that SES "explains" 8.7% of math achievement. The variance at level 2 in the null model (26.56) is reduced to 8.78 in Table 21.2, a reduction of 17.78 or 67%. The between part of SES (a level 1 variable) and percent minorities (a level 2 variable) "explain" together 67% of the between variation in math achievement. We cannot determine how each variable contributes uniquely to the $R^2$-between because they are correlated ($r = 0.23$). This is small but statistically significant and indicates that the lower the percent of minorities, the higher the SES.

The familiar $R^2$ can still be used for the concept of $R^2$-within, but not for the $R^2$-between, except in the random intercept model. As mentioned earlier, in models with random slopes, the variance between consists of more than one variance component. This results in an $R^2$-between that has a different value for different values of $X$ (see again the error term in Equation 6b). Therefore, the only way to assess the improvement of fit from one model to another in the presence of random slopes is to compare the deviances of the models.

## V. COMPLEX RANDOM COEFFICIENT MODELS

The next example is a model with two student-level and one school-level variable. All coefficients are set to be random, except for the coefficient of the school-level variable. Because school is the highest level here, school-level variables cannot be defined as varying over a higher level and thus they are estimated as fixed. The new student level variable is number of hours of math homework performed per week, whereas the new school-level variable indicates whether a school is in the public (coded 1) or private sector (coded 0). The results of this model are shown in Table 21.4.

The most remarkable feature of this model is the complex random part at level 2, consisting of six variances components, which are the variances of the random intercept, the random slopes for homework and SES, and their covariances. The covariances reflect the fact that disturbances at the second level are correlated. The results show that the slope for SES does not vary significantly among schools (i.e., $0.33/0.47 = 0.70$), but the slope for homework does vary significantly (i.e., $0.35/0.09 = 3.89$). All fixed effects are significant, indicating that SES and homework are significant determinants of math achievement, and the public sector (coded 1) has a negative influence on math achievements as compared to the private sector (coded 0). Students in public schools have a predicted math score that is on average 2.06 points lower than students in private schools.

### A. Models with Cross-Level Interactions

The next model attempts to "explain" the variation among schools of the slope for homework by adding an interaction term to the model, the

**TABLE 21.4  A Random Coefficient Model with Random Slopes**

|  | Parameter estimates | Standard error |
|---|---|---|
| Fixed effects |  |  |
| Intercept | 50.16 | 0.28 |
| Socioeconomic status | 4.35 | 0.09 |
| Homework | 1.24 | 0.05 |
| Public | −2.06 | 0.29 |
| Variance components |  |  |
| Level 1 (individual) |  |  |
| Error | 66.19 | 0.68 |
| Level 2 (school) |  |  |
| Intercept | 10.54 | 0.91 |
| Slope for socioeconomic status | 0.47 | 0.33 |
| Slope for homework | 0.35 | 0.09 |
| Covariance socioeconomic status–intercept | 1.42 | 0.39 |
| Covariance homework–intercept | −0.56 | 0.23 |
| Covariance homework–socioeconomic status | −0.04 | 0.12 |
| Deviance | 153333 |  |

interaction of the school variable *public* with the student-level variable homework. I hypothesize that I will find a significant negative effect for the cross-level interaction with the public sector. I base that on the literature that reports that the private-sector offers, on average, a richer school environment. As a result, doing homework is less necessary than in the public sector. In this same model I set the nonsignificant random slope for SES to zero to reduce the level 2 random part from six to three parameters. The separate equations for the new model are in Equations 10–14, where $Z_j$ stands for the school-level variable public.

$$\underline{a}_j = \gamma_{00} = \gamma_{01}Z_j + \underline{\delta}_{0j} \tag{10}$$

and

$$\underline{b}_{1j} \text{ (homework)} = \gamma_{10} + \gamma_{11}Z_j + \underline{\delta}_{1j} \tag{11}$$

The slope for SES is fixed as in Equation 12

$$\underline{b}_{2j} \text{ (SES)} = \gamma_{20}. \tag{12}$$

Substituting the tree equations in Equation 2 yields Equation 13

$$\underline{Y}_{ij} = (\gamma_{00} + \gamma_{01}Z_j + \underline{\delta}_{0j}) + (\gamma_{10} + \gamma_{11}Z_j + \underline{\delta}_{1j})X_{1ij} + \gamma_{20}X_{2ij} + \underline{e}_{ij} \tag{13}$$

Multiplication and rearrangement of the terms results in Equation 14.

$$\underline{Y}_{ij} = \gamma_{00} + \gamma_{10}X_{ij} + \gamma_{01}Z_j + \gamma_{11}Z_jX_{1ij} + \gamma_{20}X_{2ij} + (\underline{e}_{ij} + \underline{\delta}_{0j} + \underline{\delta}_{1j}X_{1ij}), \tag{14}$$

where the cross-level interaction appears in the term $Z_j X_{1ij}$. The results of this model are presented in Table 21.5.

**TABLE 21.5   A Random Coefficient Model with a Cross-Level Interaction**

|  | Parameter estimates | Standard error |
|---|---|---|
| Fixed effects | | |
| Intercept | 50.9 | 0.32 |
| Socioeconomic status | 4.36 | 0.09 |
| Homework | 0.90 | 0.09 |
| Public | −2.89 | 0.36 |
| Public * homework | −0.44 | 0.11 |
| Variance components | | |
| Level 1 (individual) | | |
| Error | 71.72 | 0.72 |
| Level 2 (school) | | |
| Intercept | 9.94 | 0.86 |
| Slope for homework | 0.30 | 0.08 |
| Covariance homework–intercept | −0.39 | 0.22 |
| Deviance | 153336 | |

It is instructive to compare the differences in deviances of the last two models. The more complex model in Table 21.5 has six second-level estimates for the variance components and a deviance of 153333. The model in Table 21.6 has a variance part at level 2 that is much smaller, and a deviance of 153336. Both models fit the data about equally well, but the latter model contains a cross-level interaction and has a more parsimonious random part. The interaction is statistically significant (i.e., $-0.44/0.11 = -4.0$) and has, as expected, a negative sign indicating that the effect of homework is indeed lower for the public sector. This lower value for the coefficient of homework means that students in the public sector need to make more homework to obtain an equal score in math compared to students in the private sector, supporting the earlier stated hypothesis. It is interesting to note changes in the values for the estimates homework and public as a result of the addition of their interaction. (e.g., the coefficient for public changes from $-2.06$ to $-2.89$). Both coefficients have larger standard errors and smaller coefficients, a common phenomenon in regression models, when interaction terms are added. This occurs because interactions are correlated with their main effects, and together they account for a fixed amount of variation in the response variable. When part of the variance is removed from the main effects and allocated to an interaction, the estimated influence of the main effect is reduced. And vice versa, when all the variance is attributed to the main effects, their influence is magnified. In my example the correlation with the interaction and public is $r = 0.52$, and $r = 0.68$ with homework. Multicollinearity also introduces instability, and hence larger standard errors. This example provides a clear illustration of the fallacy of attempting an atheoretical analysis of multivariate data in the hope that the "numbers will reveal the truth." Hetherington (chapter 2, this volume) and Hallahan and Rosenthal (chapter 5, this volume) discuss these issues in detail.

The fitting of one or more cross-level interactions introduces instability

**TABLE 21.6   The Effects of Adding a Cross-Level Interaction**

| | Model 1 Full model | | Model 2 No interaction | | Model 3 Main effect deleted | |
|---|---|---|---|---|---|---|
| | Parameter estimates | Standard error | Parameter estimates | Standard error | Parameter estimates | Standard error |
| Intercept | 51.43 | 0.70 | 51.52 | 0.61 | 47.90 | 0.19 |
| Homework | 1.52 | 0.18 | 1.48 | 0.05 | 2.00 | 0.16 |
| Ratio | $-0.20$ | 0.04 | $-0.21$ | 0.03 | | |
| Homework * ratio interaction | $-0.002$ | 0.01 | | | $-0.03$ | 0.008 |
| Deviance | 155679 | | 155679 | | 155706 | |

in the solutions, due to multicollinearity. Bryk and Raudenbush (1992, p. 25 e.v.) propose to center student-level variables around their context mean to avoid some of the correlations among explanatory variables of different levels. This strategy does remove the between part of such an explanatory variable, and thus the correlation between the centered (student-level) variable, the second (school) level variable, and their interaction. Centering of student-level variables (e.g., Homework and SES) will remove the correlation of both these variables with the school-level variable (Public) as well as the correlation between public and their cross-level interactions. This procedure, however, may increase the correlation of the centered student variables with their interactions (in this data to respectively $r = 0.88$ and $r = 0.92$). Iversen (1991, pp. 35–72) and Aiken and West (1991) present good discussions of centering in traditional regression models. Because centering in random coefficient models is a much debated and important topic, I will return to it later. First, however, I will discuss in more detail how to choose between a main effect model and a model with cross-level interaction.

The previous example showed that a small change, such as adding a cross-level interaction, can result in different solutions and sometimes lead to different conclusions. I will illustrate with an example the importance of theory in providing direction to choosing between different models. Again I use the NELS:88 data, with math as the response variable, predicted by homework at the student level and class size (student–teacher ratio, labeled ratio in the analyses) at the school level. These variables are used in three different models (see Table 21.6); the first and the last model contain a cross-level interaction between ratio and homework, whereas the model in the middle has only main effects. The theory behind the models is that having smaller classes (as indicated by a smaller ratio) has a positive effect on math achievement. The sign of the coefficient for ratio is expected to be negative, signifying that the larger the ratio, the lower math achievement. The same is expected from the sign for the coefficient of the interaction. I theorize that the lower the ratio the less strong the effect of homework on math achievement. The first model in Table 21.6 tests both theories, the effect of teacher–student ratio plus the effect of an interaction of Ratio with homework. The second model tests only the effect of Ratio; the interaction is deleted from the model. The last model tests only the interaction effect of Ratio and homework; the main effect of ratio is deleted from the model.

All three analyses show a significant positive effect of 'homework' on math achievement, and the expected significant negative effect of Ratio (see Table 21.6). In the full model, however, the cross-level interaction is not significant ($z = 0.002/0.01 = 0.20$), nor does it improve the model fit, as indicated by a comparison of the deviances of models 1 and 2. The coefficients for homework and ratio are very close in magnitude for the two models in Table 21.6, but the standard error of the homework effect drops from 0.18 in the full model to 0.05 in the main effects model, as a result of the removal of the highly correlated interaction. This illustrates again that correlations among explanatory variables introduce instability in the solution. Although both models have the same deviance, model 2 has one more degree of freedom (one less parameter is estimated), making model 2 a superior model in a statistical sense.

The main-effects-only model is least satisfactory given my hypothesis that ratio interacts with homework. Based on that theory, as well as to avoid correlation among variables, a new model is fitted that does not contain a main effect of ratio, only its cross-level effect (see model 3 of Table 21.6). My theory about the existence of an interaction of ratio and homework is supported by the data. In the absence of such a theory, the choice among models will most likely be the model with the best fit and/ or the model that is most parsimonious, which is model 2 in my example. Of course each individual researcher has to defend any choice among models, either by a theory that supports this choice or by model fit.

## B. Effect of Centering Explanatory Variables

Not only does model choice have to be made in a concious way, how to treat the data is another issue. For instance, using centered scores (around the group mean) instead of raw scores for data analysis has consequences for the outcome. Hence this choice must be based on theory or knowledge of the data. I will illustrate this with a well-known example from Raudenbush and Bryk (1986). They noted that the relation between SES and math achievement is stronger in public schools than in private schools and concluded that the private sector is more successful in teaching math irrespective of the student's background. Following their lead, I use a model that predicts math achievement scores from homework and SES at the student level and public versus private sector at the school level. Raudenbush and Bryk (1986) obtained their results by using centering within context (here within schools) of both student-level explanatory variables. Table 21.7 presents results obtained using their model together with two closely related models. The first model uses explanatory variables in raw score form. In the second and third model the student-level variables SES and homework are expressed as deviations from the school mean. In these (centered) models the overall mean of the school the student attends is subtracted from their raw score. For example, the homework score of students doing 5 and 6 hours of homework, respectively, in a school with an overall homework score of 5.4 would be expressed as $-0.4$ (i.e., 5–5.4) and 0.6 (i.e., 6–5.4). In another school with an average of 3.6 hours of homework, the same raw scores would yield centered scores of 1.4 (5–3.6) and 2.4 (6–3.6). The idea is to center part of the student's total score,

**TABLE 21.7   The Effects of Centering**

| | Model 1 Raw score | | Model 2 Centered | | Model 3 Centered with means | |
|---|---|---|---|---|---|---|
| | Parameter estimates | Standard error | Parameter estimates | Standard error | Parameter estimates | Standard error |
| Intercept | 50.16 | 0.28 | 55.06 | 0.35 | 47.53 | 0.47 |
| Socioeconomic status | 4.35 | 0.09 | 3.84 | 0.10 | 3.85 | 0.10 |
| Homework | 1.24 | 0.05 | 1.18 | 0.05 | 1.20 | 0.05 |
| Public | −2.06 | 0.29 | −5.42 | 0.39 | 0.62 | 0.28 |
| Mean socioeconomic status | | | | | 8.14 | 0.25 |
| Mean homework | | | | | 1.65 | 0.20 |
| Variance socioeconomic status | 0.47 | 0.33 | 1.65 | 0.44 | 1.71 | 0.44 |
| Deviance | 153333 | | 153968 | | 153004 | |

contributed to him/herself, and part of it to the school (the average). But centering changes the relation among the scores of students in different schools, and it also removes information about the school mean from the analysis, at least when the subtracted mean is not introduced again in the analysis model, as in Model 2, Table 21.7. If centering is employed, investigators must decide whether to withhold information about level differences among the schools from the analysis (model 2 of Table 21.7) or to make that information available to the analysis by adding school means to the model, (as in model 3 of Table 21.7).

Table 21.7 shows that the coefficients for SES and homework are not influenced by centering, because they change only slightly from model to model. The important effect of centering is mainly present at the second level (the level where the means of SES and homework are either added or deleted from the model). Here it affects the coefficients for the school-level variable public, as well as the second-level variance of the SES slope. Although the public sector effects are statistically significant in all three models, the magnitude and the sign of these coefficients are dramatically different. The second analysis, in which the scores are centered but no means are added to the model (the between-school part of SES and homework, present in model 1 are deleted from model 2), shows an increase of the negative effect of the public sector from −2.06 to −5.42. Reinstating the deleted information about mean differences in SES and homework in the last model of Table 21.7 results in the transformation of the large and negative effect for public to a small positive effect. In general, centering first-level variables removes their between-school effect, thereby making some variance available that could be allocated to a second-level variable in the model. More so if the mean is correlated to that variable, like mean SES and private sector.

The deviances in Table 21.7 reveal that the model fit is also affected by centering. It is not surprising that the model with information removed, the centered model 2, has the worst fit, and the centered model with the information about the level differences reinstated (model 3) has the best. Despite these differences in fit, the best model needs to be picked on the basis of knowledge of the data or theory. In a real situation, researchers pick a model before hand and do not compare different ones, as I have done in this example. A model is chosen based on a certain knowledge, like in my example. I have knowledge that SES influences the achievement of students and that the average SES of the student population in the public school sector is lower than that of the student population in the private sector. That fact indicates that I have to correct, at the second level, for the mean difference in SES among sectors, and thus I would center SES and add the mean SES at the school level. Homework, however, would be left in raw scores (no centering, no mean) since I have no theory about the mean homework as a school-level variable. My preferred model would be a combination of model 1 and 3, a model not presented here. For all purposes, Model 2 would be the worst choice because it deletes important information about between-school variation from the data. It would only fit the purpose of researchers interested in individual within-school relationships among variables only, a most unlikely situation for users of a multilevel technique.

The decision to center or not is an important and informed decision. Cronbach (1976) introduced this method to school effectiveness research to partition the variation of an explanatory variable into a within-school (i.e., due to student characteristics) and between-school (due to school characteristics) component. As a consequence of Cornbach's influence, the software HLM (Bryk et al., 1996) offers this choice, but applied researchers must understand that using context-centered or raw score data is fitting statistically different models, as is illustrated above and shown in Kreft, De Leeuw, and Aiken (1995).

## C. Informed Use of Multilevel Models

### 1. Small Models Are Necessary

It is well established that models having many correlated variables produce unstable solutions, where a small change, such as the addition of a random slope or an interaction variable, can produce greatly differing results. The most parsimonious (and thus more stable) solutions are obtained by fitting random coefficient models with a limited number of variables. A random coefficient model can easily become complex even with a few explanatory variables, as previously shown in Table 21.4 with two student variables (homework and SES) and one school-level variable (public), but eleven parameter estimates. Because random coefficient models become very quickly useless when data sets are large, a relevant theory is needed to guide the choices of (a) what variables to include in the model, (b) what coefficients to make random, and (c) which interactions to create.

## 2. Not All Data That Have Hierarchies Need Multilevel Models

Cronbach (1976) anticipated the need for the random coefficient model when he wrote that the principal flaw of school effectiveness research lies in the failure to understand and deal with the complications that arise from the multilevel character of the data. In the same vein, Raudenbush and Bryk (1986) explain that the social sciences have been plagued by both methodological and conceptual problems, and the random coefficient model creates an opportunity to extend conceptualization (see introduction). It is probably not surprising that when random coefficient modeling software first became available, a little over 10 years ago, enthusiastic advocates began to tout these models as the solution for many data analysis problems in the social sciences. The models do have an intuitive appeal for social science data that very often consist of individuals nested in groups. Indeed it can be reasoned that all data consists of individuals nested within one context or another, and that these contexts are again nested within other, larger contexts. For example, when employees are observed over a period of time, these repeated measurements are nested within employees, these employees are nested within industries, and these industries can be seen as nested within countries. All these levels may contribute to the explanation of a response variable (e.g., why some people produce more money or are more satisfied than others). The realism of the random coefficient model makes it attractive for many data analysis problems.

As we have seen, however, the same models reduce parsimony by fitting nonlinear relations and the introduction of multicollinearity, with more chances that results are merely capitalizations on chance. Cronbach (1976) cautioned about the risk of regression techniques when he wrote.

> There is a lively danger that regression techniques will dramatize relationships that arose by chance; and making hypotheses complex adds to the risk. Nonlinearities may reasonably be explored, but unless there is a rationale for predicting nonlinearity, little credence can be given a non linear relationships the first time it turns up (p. 3).

This danger is even greater when using random coefficient models.

## 3. Enhancing Social Theory

A technique is genuinely useful only if the researcher uses it intelligently with data of good quality, where "intelligently" is defined as knowledge of the technique combined with knowledge of the data. The use of complex models is attractive because they require fewer assumptions regarding the way the data are structured. In random coefficient models assumptions of independently sampled observations and homogeneity of slopes are dropped. However, this appeal can stimulate intellectual passivity, not the least promoted by slogans such as: the 'best' technique or a 'superior' technique, sometimes heard in discussions. As noted by De Leeuw and Van Rijckevorsel (1990), this appeals to many social scientists who are very unsure about the value of their prior knowledge. They prefer to delegate decisions to the computer and they expect techniques to generate knowledge. It is the applied researcher (not the technique or the statistician) who must decide which model to use and how to use it. Multilevel models are important additions to the field, but they cannot help researchers in making decisions when they are faced with conflicting solutions and results, as illustrated in this chapter. Random coefficient models invite the fitting of more and more complex models, but by doing so they place a heavier burden on the applied research to be thoroughly conversant with the relevant theory and research.

## D. A Comparison of the Random Coefficient Models versus Traditional Regression

Random coefficient modeling is assumed to have advantages over traditional regression when data are hierarchically nested, because traditional regression underestimates the standard errors of coefficients. When units (people, animals, plants, or repeated observations on people) are sampled from within contexts, it is expected that those that are closer in time or space will be correlated on some unmeasured context variables, summarized in the error term. This violates the assumption of uncorrelated errors. However, I will present examples where the choice of random over fixed regression models is not so clear cut.

The magnitude of the intraclass correlation gives an indication of the consequences of ignoring the fact that these observations are nested, but its effect (e.g., on the alpha level of the test of significance) is different in different situations. A moderate intraclass correlation has severe consequence for the standard errors in large groups, but the same magnitude will have less consequence in small groups. When dealing with small groups and very large intraclass correlations (e.g., twins and family studies), ignoring the nested design will cause substantial downward bias of the standard errors. However, the intraclass correlation has no effect on the parameter estimates for the fixed coefficients. These are estimated without bias in any situation. Researchers interested in fixed effect estimates only, and less so in hypothesis testing (e.g., in the standard errors of these coefficients) can use tradiational regression equally well.

In some situation traditional regression produces estimates and tests that are as efficient and powerful as those obtained for random coefficients models (see, e.g., De Leeuw & Kreft, 1995). This occurs when first-level observations are nested in very large groups (e.g., hundreds of observations in each group) and intraclass correlations are small or zero. For example, the intraclass correlation is negligible when observations are drawn from large areas, like observations sampled from within states. A low intraclass correlation typically is found in hierarchically nested data where members have infrequent contact or are part of a large and loosely defined group. When analyzing such data using a random coefficient technique for the sole purpose of correcting for an intraclass correlation is not necessary, and more simple techniques will suffice. However, applied researchers will benefit from the application of a random coefficients model when the hierarchical nature of the data adds important information. In such cases the multilevel technique will enhance the power and precision of the analysis.

The random coefficient model is clearly very useful when sparse data are available, and estimations are focused on predictions for each context separately, such as in large data sets with a small number of observations within some (not all) of the groups. One such issue is the question of how minority students are doing in specific graduate business schools. The result of the analysis would be used to establish guidelines for admitting minority students to specific schools, however, the percentage of minority students in some schools is small, making it difficult to make predictions for such schools.

Braun, Jones, Rubin, and Thayer (1983) and Rubin (1983) introduced the concept of *borrowing of strength* for this specific situation. They used a random coefficient model (based on empirical Bayesian estimation procedures) to analyze data in which information based on all graduate schools combined is used to make estimations for individual schools. This allows the investigator to obtain a prediction for each school, including those schools having few minority students. The estimation for the latter schools is strengthened by the information available about minority students and their achievement in other schools. The "prior" information (a Bayesian concept) borrowed from schools with a high number of minority students strengthens the prediction for schools having low numbers of minority students. In the random coefficient method the separate estimates for large schools (e.g., schools having a large percentage of minorities) contribute the most to the overall solution, whereas the small schools (schools having small percentages of minorities) contribute less. That is based on the knowledge that large schools have more stable estimates, because large numbers produce smaller standard errors. It also means that the prediction for small schools is "shrunken" towards the mean, whereas the solution for large schools is not. The concept of "borrowing of strength" is most important when the data contain unequal numbers of observations per group and are too sparse to make separate predictions, whereas the research question requires the investigator to make those separate predictions. The principle of "borrowing" of strength is useful also in family studies where sparse data (e.g., only one child per family) frequently is a problem.

## VI. SUMMARY

Random coefficient models are promising tools for data analysis in fields such as education, health, demography, and animal studies, where data are grouped in meaningful settings and the context is believed to influence the individuals. The method is very useful for prediction when groups are small and group membership is an important predictor (e.g., family and twin studies and repeated measurement studies). Random coefficients models are less parsimonious than traditional regression models, and therefore they yield solutions that are more data specific and, in the case of small data sets, less stable. Furthermore, the properties of the estimation method (empirical Bayesian/MLK) are less well known than those of the estimation method of linear regression (Least Squares). More research is needed on the power of random coefficient models (see Morris, 1995, and Rogosa & Saner, 1995) and on its relation to sample size (Snijders & Bosker, 1994).

The great potential of random coefficient models can work to the users advantage, but also be misapplied. Therefore, I concur with Morris (1995, p. 198) that "fitting hierarchical models is not a substitute for thinking hard about the data and its structure, and for intimate involvement in understanding and in the careful analysis of all components of the data."

## VII. SOFTWARE

Many software tools can be used to analyze hierarchically nested data. A brief overview of these packages follows. See De Leeuw and Kreft (1999) for a more extensive comparison of some of these and other software packages.

**BMDP5-4,** a part of the well-known BMDP package, was written by Jennrich and Schluchter (1986) for the analysis of repeated measurement data. More information can be found at their website: http://www.spss.com/software/science/Bmdp/

**HLM** is based on theory developed by Bryk and Raudenbush (1992), with manuals written by Bryk, Raudenbush, Seltzer, and Congdon (1988; Bryk et al., 1996). The software has a user-friendly question-and-answer format. Their analysis model is developed based on the slopes-as-outcomes model, which is very familiar to educational researchers. Further information can be obtained from http://www.gamma.rug.nl/iechome.html

**MIXOR and MIXREG** are written by Don Hedeker (see Hedeker & Gibbons, 1993a, 1993b). MIXOR does multilevel analysis with an ordinal outcome variable; MIXREG does multilevel analysis with autocorrelated errors. Copies of the programs for PC and Macintosh (in binary format) with manuals can be obtained at the following address: http://www.uic.edu/hedeker/mixdos.html

**Mln** is developed by the Multilevel Project at the Institute of Education, University of London, based on the theory developed by Goldstein (1987, 1995). Mln can analyze up to 15 levels and various crossed and nested structures. Mln is intergrated with the general-purpose statistics package NANOSTAT. Users guides have been written by Rashbash and Woodhouse (1995), and Rashbash, Prosser, and Goldstein (1990, 1991). Their webpage is http://www.ioe.ac.uk/multilevel/. A discussion group regarding multilevel models is made available by this group. The list aims to promote discussion, mutual support, and the spread of information such as conferences, wokshops, and new software. To subscribe send an e-mail with the message: join multilevel (your name) followed by the word *stop* to mailbase@mailbase.ac.uk. After subscription one can post queries and messages to the list. A multilevel Modeling Newsletter is published by the same people at the Mathematical Sciences Institute of Education. University of London, 20 Bedford Way, London WC1HOAL, England. Their e-mail address is temsmya@ioe.ac.uk. Their informative website is at http://www.ioe.ac.uk/multilevel/

**PROC MIXED** is a mixed model component of the SAS statistics system and comparable to BMDP5-V. More information about the options and possibilities of PROC MIXED is available at http://www.sas.com/

**VARCL,** written by Longford (1990), is a program for variance component analysis of hierarchical data. VARCL has extensions to Poisson and binomial response models. More about this package is at http://www.gamma.rug.nl/iechome.html

## REFERENCES

Aiken, L. S., & West, S. G. (1991). Multiple regression: Testing and interpreting interactions. Newbury Park, CA: Sage Publications.

Barcikowski, R. S. (1981). Statistical power with group mean as the unit of analysis. *Journal of Educational Statistics, 5,* 3, 267–285.

Braun, H., Jones, D., Rubin, D., & Thayer, D. (1983). Empirical Bayes estimation of coefficients in the general linear model from data of deficient rank. *Psychometrika, 48,* 2, 171–181.

Bryk, A. S., & Raudenbush, S. W. (1992). *Hierarchical linear models: Applications and data analysis methods.* Social Science Series. Newbury Park, CA: Sage.

Bryk, A. S., & Raudenbush, S. W. (1987). Applying the hierarchical linear model to measurements of change problems. *Psychological Bulletin, 101,* 147–158.

Bryk, A. S., Raudenbush, S. W., & Congdon, R. T. (1996). *HLM, hierarchical linear and nonlinear modeling with HLM/2L and HLM/3L programs.* Chicago, IL: Scientific Software International.

Bryk, A. S. Raudenbush, S. W., Seltzer, M., & Congdon, R. T. (1988). *An introduction to HLM: Computer program and user's guide.* Chicago: University of Chicago Press.

Cronbach, L. J., & Webb, N. (1975). Between class and within class effects in a reported aptitude × treatment interaction: A reanalysis of a study by G.L. Anderson. *Journal of Educational Psychology, 67,* 717–724.

Cronbach L. J. (1976). *Research in classrooms and schools: Formulation of questions, designs and analysis.* Occasional paper, Stanford Evaluation Consortium.

De Leeuw, J., & Van Rijckevorsel, J. (1990). Beyond homogeneity analysis. In. J. Van Rijckevorsel and J. De Leeuw (Eds.), *Progress in components and correspondence analyses,* Chichester: Wiley.

De Leeuw, J., & Kreft, I. G. G. (1999). Software for multilevel analysis. In A. Leyland (Ed.), *Multivariate analysis.* in press.

De Leeuw, J., & Kreft, I. G. G. (1986). Random coefficient models for multilevel analysis. *Journal of Educational Statistics, 11,* 1, 57–86.

Goldstein, H. (1987). *Multilevel models in educational and social research.* London: Griffin.

Goldstein, H. (1995). *Multilevel statistical models.* London: Edward Arnold.

Hedeker, D., & Gibbons, R. (1993a). MIXOR. *A computer program for mixed-effects ordinal probit and logistic regression analysis.* Unpublished manuscript, University of Illinois at Chicago.

Hedeker, D., & Gibbons, R. (1993b). MIXREG. *A computer program for mixed-effects regression analysis with auto correlated errors.* Unpublished manuscript, University of Illinois at Chicago.

Iversen, J. R. (1991). *Contextual analysis.* Thousand Oaks, CA: Sage.

Jenrich, R., & Schluchter, M. (1986). Unbalanced repeated measures models with structured covariance matrices. *Biometrics, 42,* 805–820.

Kreft, I. G. G., & de Leeuw, J. (1994). The gender gap in earnings. A two way nested multiple regression analysis with random effects. *Sociological Methods and Research, 22*(3), 319–342.

Kreft, I. G. G., de Leeuw, J., & Aiken, L. S. (1995). The effect of different forms of centering in hierarchical linear models. *Multivariate Behavioral Research, 30*(1), 1–21.

Longford, N. T. (1990). VARCL. *Software for variance component analysis of data with nested random effects* (maximum likelihood). Princeton, NJ: Educational Testing Service.

Morris, C. N. (1995). Hierarchical models for educational data: An overview. *Journal of Educational and Behavioral Statistics, 20*(2), 190–201.

Rashbash, J., Prosser, R., & Goldstein, H. (1990). ML3. *Software for three-level analysis. User's guide.* Unpublished manuscript, Institute of Education, University of London.

Rashbash, J., Prosser, R., & Goldstein, H. (1991). ML3. *Software for three-level analysis. User's guide for V-2.* Unpublished manuscript, Institute of Education, University of London.

Rashbash, J., & Woodhouse, G. (1995). *Mln command reference.* Unpublished manuscript, Institute of Education, University of London.

Raudenbush, S. W., & Bryk, A. S. (1986). A hierarchical model for studying school effects. *Sociology of Education, 59,* 1–17.

Raudenbush, S. W. (1995). Re-examining, reaffirming, and improving applications of hierarchical models. *Journal of Educational and Behavioral Statistics, 20* 2, 210–221.

Rogosa D., & Saner, H. (1995). Longitudinal data analysis: Examples with random coefficient models. *Journal of Educational and Behavioral Statistics, 20* (2), 149–171.

Rubin, D. B. (1983). Some applications of bayesian statistics of educational data. *The Statistician, 32,* 155–167.

Searle, S. R., Casella, G., & McCulloch, C. E. (1992). *Variance components.* New York: Wiley.

Snijders, T. A. B., & Bosker, R. J. (1994). Modeled variance in two-level models. *Sociological Methods and Research, 22,* 342–363.

# 22

# ANALYSIS OF CIRCUMPLEX MODELS

**TERENCE J. G. TRACEY**

*School of Education, University of Illinois at Urbana-Champaign, Champaign, Illinois*

Guttman (1954) first proposed the circumplex as the circular pattern of correlations in a matrix (formed by two simplexes). Adopting a circumplex representation enables a parsimonious description of the relations among a large set of variables. The many relations among the variables can be understood as inversely proportional to their distance in a circular arrangement. Variables more proximate on the circle are highly related, whereas those more distal are less related. By knowing where variables are placed on the circle, one can generate the degree of relation among all the variables. Also, the circle is defined by two dimensions, and the variables can be thought of as being composed of varying degrees of these two dimensions. Thus the circumplex is an appealingly simple model of many complex relations among variables.

Circumplex models have been applied widely in many psychological domains, including the areas of interpersonal behavior (e.g., Leary, 1957; Strong et al., 1988), personality traits (e.g., Gurtman, 1992a, 1992b, 1997; Wiggins, 1979; Wiggins & Trobst, 1997), emotions (e.g., Larsen & Diener, 1992; Lorr, 1997; Plutchik, 1996), vocational interests (e.g., Holland, 1973,

*Handbook of Applied Multivariate Statistics and Mathematical Modeling*
**641**

1992; Tracey & Rounds, 1996), color perception (e.g., Shepard, 1962), and psychopathology (e.g., Widiger & Hagemoser, 1997). Although very popular as a theoretical representation, there has been less focus on the empirical evaluation of the validity of the circumplex: Is the circumplex an accurate representation of the data? Much of the research on the application of circumplex models to different data sets has used methods such as factor analysis that are not necessarily appropriate for the examination of a circular structure (Fabrigar, Visser, & Browne, 1997) or order relations using inappropriate distributions (Hubert & Arabie, 1987). Several new analytic tools have been proposed for the more appropriate evaluation of the circumplex; I will demonstrate several of these tools in this chapter.

Central to evaluation of circumplex models is the explication and understanding of how the circumplex is being operationalized. Many of the analytic tools have their own implied definitions of a circumplex. There are several means of operationalizing a circumplex that are not necessarily equal or similar. This lack of correspondence among circumplex definitions is often ignored or unrecognized in reviewing research results, leading to inaccurate conclusions.

Guttman (1954) used the term *circumplex* in a general sense to refer to circular models of data, and I will use the term in the same general manner to include all of the more specific models. Within the general circumplex models, Guttman distinguished between two more specific models: the first he termed a *circulant* defined by equal spacing of variables around a circle, and the second he termed the *quasi-circumplex,* defined only by a circular arrangement of variables with no equal spacing. The general pattern of relations for both types of circumplexes is similar. The highest relations appear near the main diagonal of a matrix; the magnitude of the relations decreases as one moves away from the matrix main diagonal, and then the magnitude begins to increase toward the elements farthest from the matrix main diagonal. The circulant and quasi-circumplex models serve as the most basic distinction in operationalizing circumplex matrices. Each model requires the general high-low-high pattern in the correlation matrix as one moves away from the main diagonal, but the circulant also requires that there be equal spacing of variables and that the correlations be equal for similar distances on the circle (i.e., equal radii). The two circumplex models Guttman proposed are not isomorphic, so researchers must be clear about which model they plan to test.

The purpose of this chapter is to review the major methods of examining circumplex models, to highlight the different operationalization models implied in each, and to detail the relative advantages of each. In this chapter, I focus first on the exploratory approaches and then cover the confirmatory approaches to testing circumplex models. To exemplify the methods discussed, I will use the correlation matrix of scale scores obtained from the Inventory of Interpersonal Problems (IIP, Horowitz, Rosenberg, Baer, Ureno, & Villasenor, 1988) that was examined elsewhere (Tracey, Rounds, & Gurtman, 1996). The IIP-C (Alden, Wiggins, & Pincus, 1990) is a rescored version of the IIP that is composed of eight scales, which are hypothesized to exist in a circumplex, and their placement is depicted in the upper left of Figure 22.1. The correlation matrix of the scale scores is presented in Table 22.1.

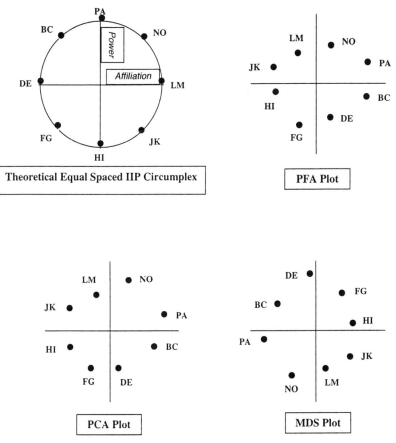

**FIGURE 22.1** Spatial representation of the Inventory of Interpersonal Problems (IIP) circumplex model along with the plot of the IIP data from Table 22.1 using principal factor analysis (PFA), principal components analysis (PCA), and multidimensional scaling (MDS).

**TABLE 22.1** Correlations among the Inventory of Interpersonal Problems Circumplex Scales[a]

| Scale | BC | DE | FG | HI | JK | LM | NO | PA | M | SD |
|-------|------|------|------|------|------|------|------|------|------|-----|
| | | | Correlations (N = 1093)[b] | | | | | | | |
| BC | 1.00 | | | | | | | | .96 | .63 |
| DE | .63 | 1.00 | | | | | | | .89 | .68 |
| FG | .51 | .68 | 1.00 | | | | | | 1.37 | .90 |
| HI | .33 | .49 | .71 | 1.00 | | | | | 1.43 | .82 |
| JK | .23 | .36 | .50 | .72 | 1.00 | | | | 1.41 | .78 |
| LM | .23 | .34 | .40 | .50 | .72 | 1.00 | | | 1.55 | .77 |
| NO | .42 | .24 | .21 | .27 | .42 | .55 | 1.00 | | 1.16 | .68 |
| PA | .68 | .45 | .33 | .18 | .22 | .36 | .59 | 1.00 | .88 | .64 |

[a] Data from Tracey, Rounds, and Gurtman (1996).
[b] BC, vindictive; DE, cold; FG, socially avoidant; HI, nonassertive; JK, exploitable; LM, overly nurturant; NO, intrusive; PA, domineering.

## I. EXPLORATORY APPROACHES TO THE EVALUATION OF CIRCUMPLEXES

### A. Visual Inspection

### 1. Inspection of the Correlation Matrix

One of the most common early methods of evaluating the presence of a circumplex is the procedure of "eyeballing" a correlation matrix for relative fit to the circular model (e.g., Holland, 1973). Do the correlations in the matrix demonstrate the pattern of dropping and then rising in magnitude as one moves away from the main diagonal? This method allows one to get an intuitive feel for the pattern of relations in the matrix, but it is inadequate as a means of formal evaluation and subject to wide differences in conclusions across researchers. To better enable a visual evaluation of whether the data resemble a circumplex model, many researchers have used various spatial analyses of the correlation matrix.

### 2. Inspection of Plotted Factor Scores

Three such spatial approaches to the analysis of correlation matrices are principal factor analysis (PFA), principal components analysis (PCA), and multidimensional scaling (MDS). (See Cudeck, chapter 10, and Davison & Sireci, chapter 11, this volume, for discussions of these techniques.) Each of these analyses represents the data spatially in a manner that permits the detection of a circular pattern if one exists. The relative differences, or lack thereof, among each of these approaches have been debated at length elsewhere (Davison, 1985; Davison & Skay, 1991; Gorsuch, 1990; Loehlin, 1990; Velcier & Jackson, 1990), but the three approaches lead to fairly similar conclusions when used to evaluate the presence of a circular structure. Given this similarity, I will discuss the three approaches collectively; I will refer to the factors (PFA), components (PAC), and dimensions (MDS) collectively as factors for ease of communication. The weights for each of the factors yielded by the analysis can be plotted and the presence of a circular structure evaluated visually. Given that the circular model typically exists in two dimensions, researchers plot the weights of each variable on the first two factors and evaluate the extent to which the data are arranged circularly and look for the presence of any order reversals of the variables. These analyses represent a more sophisticated version of "eyeballing" in which a simpler representation of the data is generated that is less susceptible to variation in interpretation across researchers than is the visual inspection of the correlation matrix.

A thorny issue when using PCA and PFA is the presence of a general factor. In many contexts, PCA and PFA yield results in which the first factor is characterized by loadings that are uniformly positive and generally fairly high across the variables (i.e., a general factor). A general factor can occur even when the circular structure is present in the data (Rounds & Tracey, 1993; Wiggins, Steiger, & Gaelick, 1981). Whenever PCA or PFA yield a general factor, the circular structure typically exists in the space defined by the second and third factors.

The meaning of the general factor—substantive, bias, or nuisance—is still under debate. Several authors have argued that the general factor has substantive meaning and thus needs to be included and examined in any analysis (Rounds & Tracey, 1993; Tracey et al., 1996). Others have argued that the general factor can be considered response bias and thus should be removed (Jackson, 1977; Jackson & Helmes, 1979). Still others view the general factor as a nuisance because it is without substantive or biasing effects, but its contribution must be recognized and accounted for (Prediger, 1982, 1998). Regardless of the researcher's conclusion about the nature of the general factor, the general factor requires explicit attention when performing an exploratory evaluation of a circumplex. If the researcher rotates the factors following initial extraction, the pattern produced will not show a circumplex structure even if one exists. Rotation spreads the general factor across the other factors (Becker, 1996; Hofstee, Ten Berge, & Hendriks, 1998), and the structure that results will not resemble a circle.

Given the possible biasing effect of the general factor yielded in PFA and PCA, I recommend that researchers choose to (a) ignore the first unrotated general factor and inspect the unrotated second and third factors for the presence of the circular structure (e.g., Rounds & Tracey, 1993), or (b) ipsatize the data prior to analysis, thereby removing any variance attributable to the general factor (e.g., Bartholomew & Horowitz, 1991). These methods of dealing with the general factor yield similar results and both remove the effects of the general factor in masking the presence of the circular structure (Rounds & Tracey, 1993; Wiggins et al., 1981).

The issue of the general factor is irrelevant in MDS because MDS centers the data matrix. As Davison (1985) demonstrated, PCA and MDS yield similar solutions with the exception of the absence of the general factor in MDS. A potential problem with the use of MDS in detecting circular patterns is that it is prone to producing curved patterns (Hubert, Arabie, & Muelman, 1998) and thus is biased toward yielding circular structures. Regardless, I have found that plotting the weights obtained from PCA, PFA, and MDS generally yields similar results with respect to the circular structure. To demonstrate this, the correlation matrix in Table 22.1 was subjected to PCA, PFA, and MDS. The variances accounted for by the first four factors were 48%, 14%, 11%, and 4% for the PAF and 51%, 17%, 15%, 5% for the PAC. The first factors of the PCA and PFA were general factors, and the second and third factors defined the area covered by the IIP circumplex. The fourth factors were ignored, as they were much lower in variance accounted for and were not interpretable. The MDS yielded two dimensions that accounted for 97% of the variance: dimension one accounted for 59% of the variance, and dimension two accounted for 38%. The plots of the variables are depicted in Figure 22.1 for each of the analyses (note that the unrotated solution was used for PCA and PFA and that the second and third factors were graphed given the presence of the first general factor). The circular arrangement of the IIP scales is evident in each of the analyses.

Figure 22.1 also demonstrates that the dimensions underlying any circular structure are arbitrary with respect to orientation because the three analyses yielded different factor orientations. Any set of axes can underlie a circular structure. The key defining characteristic of a circular structure is the circle (formed by the location of the variables) itself and not the underlying dimensions. Hogan (1983) noted this distinction in his contrasting of the factor model with the circumplex model. The factor model focuses on defining the most plausible factors (typically orthogonal and approximating a simple structure), whereas the circumplex model focuses on defining the circular structure and ignoring the factors. Although most all circumplex models have theoretically defining dimensions or factors underlying the circle (e.g., power and affiliation in the interpersonal realm, or people/things and data/ideas in the vocational interest realm), these dimensions are arbitrary. Any orientation of dimensions would equally fit the circumplex. The dimensions can be placed using an arbitrary convention (e.g., Gifford & O'Conner, 1987; Wiggins & Trapnell, 1997) or using external criteria to determine the most appropriate placement (e.g., Prediger, 1982; Tracey, 1994). However, Larsen and Diener (1992) demonstrate that different orientations of dimensions yield different results with respect to external criteria. Circular structures are valuable in their capturing the blendings of factors, whereas factor models minimize focus on any such blendings (indeed they strive for simple structure). Although there are overlapping aspects to factor and circumplex approaches, the selection of a circumplex or a factor approach rests upon the purposes of the researcher. Regardless, researchers need to be aware that the presence of any circular structure carries no information on the underlying dimensions.

The problem with such visual inspections of PCA, PFA, or MDS results is that they provide no statistical evaluation of the degree of fit of the circumplex model to the data. Not everyone will come to equivalent conclusions from visual analysis, especially in cases where the fit is questionable. Furthermore, visual inspection does not take into account error variance around the data points. Researchers tend to interpret data points as occupying a certain point in space rather than as lying somewhere within a broad area or region. This makes it difficult to distinguish cases in which a circular structure exists from those in which it does not exist and to determine whether equal spacing is present. Several statistics can be used to further examine the spatial representation of the data provided by the PCA, PFA, or MDS results.

## B. Statistical Tests of the Circular Distribution

Evaluating variables associated with circles requires special statistics, because many of the distributional assumptions underlying the more common statistics are invalid. The main question in contexts where researchers are interested in the circularity of the data is the extent to which the data are uniformly distributed (i.e., equal spacing) around a circle. A wide variety of statistics has been used to examine this issue (see Fisher, 1993; Mardia, 1972; and Upton & Fingleton, 1989, for a complete description of statistics for use with circular data). For example, some researchers have used the Rayleigh test (Fisher, 1993, p. 69; Upton & Fingleton, 1989, p. 233) which evaluates whether the data are distributed in a single direction to determine whether the data can be viewed as circular. Unfortunately, failure to reject the null using the Rayleigh test indicates only that there is no one direction in the data, not that the data are distributed continuously around a circle. Typically, researchers interested in circular arrangements have already ruled out the possibility that the data are distributed in a single direction and are concerned with the extent to which the data are distributed continuously around the circle. Researchers must take care to select an appropriate statistic. The Selkirk-Neave Gap test (Upton & Fingleton, 1989, pp. 248–250) and the Kupier-Stephens K test (Stevens, 1970; and described by Fisher, 1993, pp. 66–68, and Upton & Fingleton, pp. 256–258) are appropriate when the investigator is interested in testing whether the data are distributed continuously around a circle; both are fairly easily calculated. Both tests can be conducted on any data set containing the angular dispersion of the data (i.e., the angles of each data point). In the example illustrated in Figure 22.1, the angles can be obtained by calculating the arc tangent of the ratio of the weights on the two dimensions. Such tests provide information relative to the hypothesis of continuous placement around the circle (i.e., equal spacing of the variables).

The angular gaps in the plot derived from the MDS analysis were 144°, 101°, 67°, 16°, 330°, 300°, 246°, and 187° for vindictive (BC), cold (DE), socially avoidant (FG), nonassertive (HI), exploitable (JK), overly nuturant (LM), intrusive (NO), and domineering (PA), respectively, and application of the Selkirk-Neave Gap test to these data yielded a critical gap value of 22.0. This value was significantly ($p = .01$) less than the two-tailed critical value of 57.0 for eight variables (Upton & Fingleton, 1989, p. 249), thereby supporting the conclusion of uniform distribution of the variables around the circle (i.e., equal spacing). The Selkirk-Neave gap test critical values approximate a $z$ distribution when the number of data points is greater than 30, but the critical values cited in Upton and Fingleton (p. 249) should be used with fewer variables.

Exploratory models such as these all rely on estimates from the data and build error variance specific to the sample into the model. It is difficult to distinguish substantive contributions to the obtained spatial representation from the error contributions when using these approaches. Plotting the spatial representations of the data may help illuminate patterns in many cases, but these spatial representations sometimes fail to reveal circular patterns that really exist. Furthermore, circumplexes can exist even though there are more than two factors underlying the data (Browne, 1992; Farbigar et al., 1997; Guttman, 1954), so relying on a two-dimensional spatial representation to reveal a circular structure may result in missing valid circumplex structures. More powerful tests of the circular structure are confirmatory in nature.

## II. CONFIRMATORY APPROACHES TO THE EVALUATION OF CIRCUMPLEXES

There are a variety of confirmatory techniques for examining the extent to which the circumplex fits a specific data set. All confirmatory approaches require that the researcher specify the pattern that should be present in the correlation matrix, and then the extent to which this pattern accounts for the data is assessed. Hence, the strength of the confirmatory approach is that the *a priori* theoretical specification of the structure to be examined reduces the likelihood that sample-based error variance will be interpreted as substantive in nature. In addition, confirmatory approaches often provide an inferential evaluation of the fit of the specified model to the data, and most confirmatory approaches can be used to evaluate the variance across samples in model-to-data fit.

Confirmatory methods vary in the assumptions required and the specific circumplex model examined. Because these methods are confirmatory, the selection of the specific circumplex model to be examined is central to the selection of the analysis. Guttman's (1954) two circumplex models (i.e., the circulant and the quasi-circumplex) appear to define the extremes of the range of circumplex models examined in the literature. Researchers who use a procedure designed to evaluate the circulant model when they are concerned about the presence of a quasi-circumplex may draw inaccurate conclusions.

## A. Constrained Multidimensional Scaling

As noted earlier, MDS is a common tool used for the exploratory evaluation of the circumplex structure, but it also can be used in a confirmatory analysis in which the data are fit to a specific structure. Gurtman (1992a) suggested using the cosine difference as an index of model–data fit to assess the extent to which results from an unconstrained MDS two-dimensional solution adhere to a theoretical target, but this provides a measure of equal spacing while ignoring the equal radii assumption also contained in a circulant model. The usage of constrained MDS assesses both the equal spacing and equal radii aspects of the circulant model. The theoretical points of a circumplex structure (specifically, the circulant model) in two-dimensional space can be specified, and the extent to which the data fit that structure can be assessed using stress and/or variance accounted for as indices of model–data fit. (See Rounds and Tracey, 1993, for an example of constrained MDS and multisample application). Further, MDS (both constrained and unconstrained) can be applied in multisample contexts to examine the similarity of model–data fit across samples.

For the data in Table 22.1, the unconstrained MDS solution depicted in Figure 22.1 accounted for 97% of the variance. In the constrained analysis, the data in Table 22.1 were fit to a circulant model with dimensions weights of: $-.71$, $-1.00$, $-.71$, $0.00$, $.71$, $1.00$, $.71$, and $0.00$ on dimension 1 and $.71$, $0.00$, $-.71$, $-1.00$, $-.71$, $0.00$, $.71$, and $1.0$ on dimension 2 for the variables BC through PA, respectively. The constrained solution accounted for 94% of the variance. Of course, any constrained model will not fit the data as well as the unconstrained model because of the incorporation of sample error into the unconstrained solution. The relative similarity of the unconstrained and constrained solutions provides information about the adequacy of the constrained solution. Given the good fit of the constrained circulant model and the relative similarity of the variance accounted for by the unconstrained and constrained solutions, it may be concluded that the IIP data provided in Table 22.1 can be adequately described using a circulant model.

## B. Structural Equation Modeling

Structural equation modeling (SEM; see DiLalla, chapter 15, this volume) can be applied to any correlation matrix to examine a variety of different circumplexes. The most important aspect in the application of the SEM method is the specification of the circumplex model to be tested. Three major representations of the circumplex model have been specified: the circulant, the geometric circulant, and the quasi-circumplex. These models vary in their restrictiveness, but each can be examined using SEM.

### 1. Circulant Model

Evaluation of the circulant model using SEM was first proposed by Jöreskog (1974) and first applied by Wiggins et al. (1981). Although SEM approaches to the examination of circumplexes are not common in the literature, the circulant model is the most typical circumplex model examined in a SEM context (e.g., Fouad, Cudeck, & Hansen, 1984; Rounds & Tracey, 1993; Rounds, Tracey, & Hubert, 1992; Tracey & Rounds, 1993; Wiggins et al., 1981). The number of parameters needed to account for the equal spacing requirement of this model is n/2 when $n$ is even and (n-1)/2 when an odd number of variables ($n$) is represented on the circle. Tests of interpersonal octant models (i.e., models such as the IIP-C that have eight variables) require four parameters to represent a circulant model. Tests of Holland's (1973, 1992) vocational personality type model, which has six variables, require three parameters to represent a circulant model.

For the eight IIP-C variables, the four parameters below the main diagonal in Table 22.2 represent the pattern hypothesized to exist in a circulant model of any eight variables. One parameter ($p_1$) represents the eight correlations among the adjacent variables, those one step apart on the circle (PA-BC, BC-DE, DE-FG, FG-HI, HI-JK, JK-LM, LM-NO, and NO-PA). A second parameter ($p_2$) represents the eight correlations among all variables two steps apart on the circle (PA-DE, BC-FG, DE-HI, FG-JK, HI-LM, JK-NO, LM-PA, and NO-BC). A third parameter ($p_3$) represents the eight correlations between variables three steps apart on the circle (PA-FG, BC-HI, DE-JK, FG-LM, HI-NO, JK-PA, LM-BC, and NO-DE). The fourth parameter ($p_4$) represents the four correlations between those variables four steps apart, or opposite, on the circle (PA-HI, BC-JK, DE-LM, and FG-NO). The circulant model further constrains the parameters such that the correlations between adjacent variables are greater than the correlations between variables two steps apart on the circle, which are in turn greater than the correlations between variables three steps apart on the circle, which are in turn greater than the correlations between

**TABLE 22.2** Circulant (below Main Diagonal) and Geometric Circulant Correlation Pattern Model for an Eight-Variable Circumplex[a]

| Octant scale[b] | PA | BC | DE | FG | HI | JK | LM | NO |
|---|---|---|---|---|---|---|---|---|
| PA | 1.00 | $4p_5$ | $3p_5$ | $2p_5$ | $p_5$ | $2p_5$ | $3p_5$ | $4p_5$ |
| BC | $p_1$ | 1.00 | $4p_5$ | $3p_5$ | $2p_5$ | $p_5$ | $2p_5$ | $3p_5$ |
| DE | $p_2$ | $p_1$ | 1.00 | $4p_5$ | $3p_5$ | $2p_5$ | $p_5$ | $2p_5$ |
| FG | $p_3$ | $p_2$ | $p_1$ | 1.00 | $4p_5$ | $3p_5$ | $2p_5$ | $p_5$ |
| HI | $p_4$ | $p_3$ | $p_2$ | $p_1$ | 1.00 | $4p_5$ | $3p_5$ | $2p_5$ |
| JK | $p_3$ | $p_4$ | $p_3$ | $p_2$ | $p_1$ | 1.00 | $4p_5$ | $3p_5$ |
| LM | $p_2$ | $p_3$ | $p_4$ | $p_3$ | $p_2$ | $p_1$ | 1.00 | $4p_5$ |
| NO | $p_1$ | $p_2$ | $p_3$ | $p_4$ | $p_3$ | $p_2$ | $p_1$ | 1.00 |

[a] $p_1 > p_2 > p_3 > p_4$; $p_1$ represents correlation between adjacent scales; $p_2$ represents correlation between scales 1 step removed on the circle; $p_3$ represents correlation between scales 2 steps removed on the circle; $p_4$ represents correlation between opposite scales.

[b] PA, domineering; BC, vindictive; DE, cold; FG, socially avoidant; HI, nonassertive; JK, exploitable; LM, overly nurturant; NO, intrusive.

variables opposite each other on the circle (i.e., $p_1 > p_2 > p_3 > p_4$). For the circulant model to fit the data, these four parameters and the order constraint must account for all 28 correlations in an eight-variable correlation matrix. It is thus a very parsimonious representation of the data.

Gaines et al. (1997) have proposed a slightly different means of examining the circulant model using SEM. They focused on specifying and estimating the parameters of the measurement model (i.e., the factors loadings of the two factors underlying the circumplex) rather than the correlation matrix itself. In cases where there are not only two factors comprising the circumplex (see Browne, 1992, and Fabrigar et al., 1997, for discussions of potential lack of agreement between circular and factor structures) or where there is a general factor in addition to the two factors underlying the circumplex, this factor specification approach using confirmatory factor analysis will not yield an accurate estimation of the circumplex. The more general correlation matrix estimation approach is thus more appropriate in most contexts.

## 2. Geometric Circulant Model

Rounds et al. (1992) proposed a more restrictive version of the circumplex model that assumes that the parameters of the circulant model are proportional and linearly related to the distance around the circumference of the circle. This assumption should hold if the geometric properties of the circle are valid. The correlations between adjacent variables should be four times greater than correlations between variables opposite on the circle, correlations between variables two steps apart on the circle should be three times the correlations between variables that are opposite, and correlations between variables three steps apart on the circle should be two times the correlations between the variables opposite (see Table 22.2). The geometric circulant model is extremely parsimonious in that all 28 correlations in the correlation matrix can be explained using one parameter.

The circulant and geometric circulant models can be examined using any SEM program (e.g., LISREL, Jöreskog & Sörbom, 1993). In all SEM applications, the researcher examines the fit of the specified model to the data. Many researchers (e.g., Bentler & Chou, 1993; Browne & Cudeck, 1993; Marsh, Balla, & McDonald, 1988; Tanaka, 1993) have noted the lack of an unambiguous, unequivocal means of determining whether a model fits or does not fit a data set satisfactorily (see DiLalla, chapter 15, and Hoyle, chapter 16, this volume, for further discussion of fit indices). The common chi-square goodness-of-fit index is greatly affected by sample size and model complexity (Marsh et al., 1988) and is thus not always helpful in evaluating fit. Other indicators of fit are the Goodness of Fit Index (GFI), (Jöreskog & Sörbom, 1993), Bentler and Bonett's normed fit index (BBI, Bentler & Bonett, 1980), Tucker-Lewis index (TLI, Tucker & Lewis, 1973), and the comparative fit index (CFI, Bentler, 1990, also called the noncentralized normed fit index, NCNFI, McDonald & Marsh, 1990). Each of these indices range roughly from 0 to 1 and can be interpreted as proportion of variance accounted for by the model in question. A value of .90 is often used as a cutoff between adequate and inadequate fit (e.g., Bentler & Bonett, 1980; Schumacker & Lomax, 1996). Another frequently used index is the root mean square error of approximation (RMSEA, Steiger, 1989; Steiger & Lind, 1980). This index, a measure of badness of fit per degrees of freedom in a model, takes account of both degree of fit and the relative parsimony of the model (simpler models are better than more complex ones). The RMSEA equals zero when there is perfect fit to the data. Browne and Cudeck (1993) have proposed that RMSEA values of less than .05 indicate very close fit and that values between .06 and .08 indicate good fit. RMSEA values over .10 are viewed as indicative of poor fit. MacCallum, Browne, and Sugawara (1996) view RMSEA values between .08 and .10 as indicative of mediocre fit. Given the different information provided by the different fit indices, it is usually best to use several different fit indices in evaluating model–data fit.

The circulant (4 parameter) and geometric circulant (1 parameter) models were examined for degree of fit to the correlation matrix depicted in Table 22.1 using LISREL (Jöreskog & Sörbom, 1993). The results, derived using the maximum likelihood method of estimation, are summarized in the top half of Table 22.3. All the fit indicators support the circulant model as a valid representation of the data. The BBI, TLI, and CFI all exceeded the .90 criterion of adequate fit, and the RMSEA was below the cutoff of .08 indicative of a "good" model. The four parameters of the model were estimated to be .67, .46, .30, and .24. The fit indicators for the

**TABLE 22.3  Summary of the Fit of the Different Circumplex Models Using the SEM Approach**

| | | | Goodness of fit[a] | | | |
|---|---|---|---|---|---|---|
| Sample | df | $\chi^2$ | BBI | TLI | CFI | RMSEA |
| Circulant (four parameters) | 24 | 178.14 | .97 | .97 | .97 | .08 |
| Geometric circulant (one parameter) | 27 | 536.93 | .90 | .90 | .90 | .13 |
| CIRCUM analyses | | | | | | |
|   Unconstrained | 10 | 46.98 | .99 | .98 | .98 | .06 |
|   Equal communality | 17 | 68.67 | .99 | .98 | .98 | .05 |
|   Equal spacing | 17 | 159.76 | .97 | .96 | .96 | .09 |
|   Equal commonality and spacing | 24 | 239.55 | .96 | .96 | .96 | .09 |

[a] BBI, Bentler and Bonett's (1980) normed fit index; TLI, Tucker & Lewis index (1973); CFI, comparative fit index (Bentler, 1990); RMSEA, root mean square error of approximation (Steiger & Lind, 1980).

geometric circulant model were less supportive than those of the circulant model. Although the BBI, TLI, and CFI all equaled the .90 cutoff of adequacy, the RMSEA was well over .10, which is indicative of a poor model–data fit. The single parameter of the geometric circulant model was estimated to be .13, and this was viewed as not yielding an adequate fit to the data. Thus the circulant model, but not the geometric circulant model, was supported as fitting this data set.

### 3. Quasi-Circumplex

The final circumplex model that has been examined in the literature assumes that the variables are arranged in a circular order but that the spaces among the variables are not equal. This model, first defined by Guttman (1954) and subsequently operationalized by Browne (1992), is valuable in contexts where the researcher cannot specify *a priori* the ordering of the variables (i.e., the relative placement on the circle). The circulant and the geometric circulant models require that researchers specify the general position of the variables on the circle prior to evaluation. The quasi-circumplex model provides a means of assessing the extent to which a circle can fit any set of variables. (Refer to Browne, 1992, for a complete description of this model and method.) Besides examining circular arrangement of variables, Browne's method makes it possible to evaluate changes in fit indices that occur with the introduction of the constraints of equal spacing around the circle and equal communality of the variables (equal radii for each variable, resulting in a circle rather than an ellipse or irregular-shaped circle). Imposing both of these constraints results in a model similar to the circulant model. Browne's approach also enables a more specific examination of the circulant model by breaking it down into its two underlying parts: the equal spacing and equal radii assumptions.

Browne's approach is operationalized in his computer program CIRCUM, which was used to examine the correlation matrix presented in Table 22.1. (Copies of the CIRCUM program can be downloaded from Michael Browne's website: http://quantrm2.psy.ohio-state.edu/Browne/). The program uses both maximum likelihood and generalized least squares procedures to estimate parameters, but in this example they both yielded very similar results so only the maximum likelihood results are discussed. As Browne's approach also utilizes SEM procedures, the models are evaluated with the same fit indices described for other SEM approaches. Fabrigar et al. (1997) and Gurtman and Pincus (in press) present clear descriptions and examples of this approach and of the CIRCUM program.

CIRCUM requires that the user specify the number of free parameters used in the Fourier series, and I chose three free parameters as appropriate for octant data. Models using 1, 2, 4, 5, and 6 free parameters also were examined, but three free parameters resulted in the best, yet most parsimonious fit. Four circumplex models were examined: the quasi-circumplex (i.e., unconstrained) model of circular arrangement, the quasi-circumplex model with the equal communality constraint, the quasi-circumplex model with the equal spacing constraint, and the circulant model, which has both the equal spacing and equal communality constraints (see Table 22.3). The traditional SEM fit indices (BBI, TLI, and CFI) for each model were all well above the .90 criterion, indicating good fit to the data. The RMSEA estimates, however, demonstrated that only the quasi-circumplex model (i.e., unconstrained) and the equal communality quasi-circumplex model were good approximations of the data (both were less than .08). The equal spacing and communality models had RMSEA values between .08 and .10, indicating slightly less adequate fit. The fit indices of the equal communality and the quasi-circumplex models were virtually identical, but the equal communality model was selected as the best representation of the data given its parsimony.

The empirical angle estimates for the equal communality model are summarized in row 2 of Table 22.4 along with the 95% confidence intervals (rows 3 and 4) and the angles associated with the equal spacing, theoretical model (row 1). Although the circulant model did not fit the data quite as well as the other models, the deviation from equal spacing was not great, as evidenced by row 5 of Table 4, nor was the fit of the circulant model poor in and of itself (see Table 22.3). Row 5 of Table 22.4 demonstrates that the JK and LM scales were the scales that had the greatest deviation from the theoretical, circulant ideal. The angular deviation from ideal was small for the other scales.

A valuable feature of the SEM approaches to the evaluation of the circumplex structure in a data set is that they require that the specific model

**TABLE 22.4   Polar Angle Estimates of the Equal Communality, Loose Circular Order Model with Upper and Lower Bounds of 95% Confidence Intervals**

| | Scale[a] | | | | | | | |
|---|---|---|---|---|---|---|---|---|
| | **PA** | **BC** | **DE** | **FG** | **HI** | **JK** | **LM** | **NO** |
| Theoretical angle | 00 | 45 | 90 | 135 | 180 | 225 | 270 | 315 |
| Empirical angle | 00 | 44 | 97 | 136 | 168 | 200 | 232 | 300 |
| 95% confidence interval | | | | | | | | |
| Lower bound | 00 | 40 | 92 | 131 | 163 | 195 | 227 | 295 |
| Upper bound | 00 | 48 | 103 | 142 | 174 | 205 | 238 | 306 |
| Angular deviation of empirical estimate from theoretical | 00 | −1 | +7 | +1 | −12 | −25 | −38 | −15 |
| Communality estimate | .91 | .91 | .91 | .91 | .91 | .91 | .91 | .91 |
| 95% confidence intervals | | | | | | | | |
| Upper bound | .90 | .90 | .90 | .90 | .90 | .90 | .90 | .90 |
| Lower bound | .92 | .92 | .92 | .92 | .92 | .92 | .92 | .92 |

[a] PA, domineering; BC, vindicative; DE, cold; FG, socially avoidant; HI, nonassertive; JK, exploited; LM, overly nurturant; NO, intrusive.

to be tested be operationalized. Then SEM procedure provides a statistical and interpretative evaluation of its fit to the data. Browne's (1992) approach is the most flexible of the SEM approaches in that it enables the easy evaluation of a range of different circumplex models by starting with the least restrictive quasi-circumplex model of circular arrangement and progressively adding the constraints of equal communalities and equal spacing to obtain a test of the circulant model.

## C. Randomization Tests of Hypothesized Order Relations

Another confirmatory approach to the evaluation of the extent to which a circumplex fits a correlation matrix (or any similarity or dissimilarity matrix) is the randomization test of hypothesized order relations (Hubert, 1987; Hubert & Arabie, 1987). This is a general procedure that can be used with any model and applied to most any matrix (this general aspect will be reviewed later). This analysis involves the three separate steps of model specification, statistical evaluation of fit, and interpretative assessment of fit.

The first step requires completely specifying a theoretical model (here a circumplex). The circumplex model examined in this procedure is called the circular order model (Rounds et al., 1992) and is highly similar to the circulant model presented in Table 22.2 (indeed, the model in Table 22.2 serves as the basis of the circular order model). In the circular order model, the relations between each pair of variables are examined relative to all other pairs of variables (with the exception of those relations that are assumed to be equal). In a circular order model of eight variables (as hypothesized to fit Table 22.1), all correlations between adjacent variables (i.e., those one step apart on the circle, such as PA and BC) are assumed to be greater than correlations between variables two steps apart on the circle, such as PA and DE, which are in turn assumed to be greater than correlations between variables three steps apart on the circle, such as PA and FG, which are in turn assumed to be greater than correlations between variables four steps apart on the circle (i.e., opposite such as PA and HI). No comparison is made of those relations that are assumed to be equal (i.e., between correlations of adjacent variables such as PA-BC and DE-FG, between correlations of variables two steps apart on the circle such as PA-DE and FG-JK, between correlations three steps apart on the circle such as PA-FG and DE-JK, or between correlations of opposite variables such as PA-HI and DE-LM). Such a specification of order relations among the correlations completely accounts for the circular pattern. In an 8 × 8 matrix as exemplified here, there are 288 order predictions made among the 28 different correlations in the matrix.

The only difference between the circular order model and the circulant model of Table 22.2 is that in the circular order model there is no constraint of equality among similar types of relations *explicitly* specified. Implicitly these equality relations hold in the circular order model, but they are not specifically examined or constrained. Browne's (1992) quasi-circumplex model (circular arrangement of variables) is much less restrictive than the circular order model in that equality of relations is not even implicitly assumed. So the circular order model is highly similar to the circulant model in that only data approximating equal spacing around the circle are going to be well fit, but the circular order model is slightly less restrictive regarding its constraints than the circulant model.

Using the randomization test, the order predictions of the circular order model are examined for their degree of fit to the data matrix. The number of order predictions met in the data is calculated. Then the order of the rows and columns of the data is varied systematically (i.e., random relabeling) until all possible permutations of the rows and columns have been achieved and the number of predictions met in each is calculated. In an 8 × 8 matrix, this random relabeling of rows and columns results in 40,320 (8!) permutations. The number of predictions met in the data set is compared to the number of predictions met in the permutation distribution to provide the exact test statistic. The number of permutations that meet or exceed the number of predictions met in the data matrix is the numerator, and the total number of permutations is the denominator used to calculate the significance probability value. The significance level provides information on the probability that the model–data fit would occur under the chance conjecture of random relabeling of the rows and columns of the data matrix.

Hubert and Arabie (1987) proposed the randomization test in reaction to the binomial test proposed by Wakefield and Doughtie (1973) because of the inappropriate assumption made in the binomial distribution of independence. The independence assumption implies that the distribution generated by complete replacement as used in the binomial distribution is inaccurate. Complete replacement assumes that every entry (here correlations) in the data matrix is independent of the other entries and thus the denominator should be the full permutation of all items in the matrix (e.g., 28! in the 8 × 8 example used here). Given the lack of independence of the entries in a data matrix, Hubert and Arabie proposed using random relabeling of rows and columns as the basis of the distribution rather than complete random selection. Myors (1996) proposed using Spearman's rho as an alternative test of model–data fit but subsequently found (Myors, 1998) that the randomization test of hypothesized order relations provides a more accurate test because of the assumption of independence.

Hubert and Arabie (1987) proposed the use of the correspondence index (CI) as an interpretive aid to represent the degree of model–data fit. The CI index is defined as the number of predictions met in the data matrix minus the number of predictions violated over the total number of predictions made. The CI is a correlation coefficient (Somers's D, Somers, 1962) and varies from +1.0 (indicating perfect model–data agreement) to −1.0 (indicating complete violation of model predictions). CI values of .00 indicate chance, where 50% of the predictions were met. A CI of .50 indicates that 75% of the predictions were met (75% met minus 25% violated). Thus the CI provides an index of the fit of the circular order model to the data that can be used in comparing fit of different models or across different samples. Rounds et al. (1992) provide a thorough discussion of the randomization procedure, and Tracey (1997) also describes the procedure and his computer program RANDALL that can be used to apply this procedure. (Programs to conduct the various randomization tests of hypothesized order can be downloaded from http://courses.ed.asu.edu/Tracey/).

Using the RANDALL program to evaluate the fit of the circular order model to the data in Table 22.1 results in a $p$ of .0004 and a CI of .94 (278 of the 288 predictions were met, 8 predictions were violated, and there were 2 ties), supporting the circular order model as an adequate representation of the data.

The randomization approach is valuable in its generality and simplicity. It involves no assumptions or estimation of parameters, and it can be used with any type of similarity (or dissimilarity) matrix. For example, even data obtained from a single individual can be evaluated using the randomization test. Furthermore, it enables the comparison of several very different models that cannot be examined using most of the other methods described above. For example, the randomization approach enables examination of the relative merits of the circular order model and a cluster model (see Tracey & Rounds, 1993).

## III. VARIATIONS ON EXAMINING CIRCUMPLEXES

### A. Alternative Matrices

Most of the research utilizing circumplexes focuses on the evaluation of the fit of the circumplex to the correlation matrix. Such evaluations are necessary in establishing the structure of the scales of many personality measures. However, there are several other contexts in which circumplex structures are salient but have not been studied, such as the examination of nonsymmetric matrices.

Not all applications of circumplex models consist of symmetric correlation matrices. In many cases, the data consist of frequency matrices. Two common examples are the presence of high-point scale profiles and the sequence of interactional behavior. As circumplex measures gain more widespread application, using high-point codes or profile types to interpret results probably will become more common. Interpretation of high-point codes is common in the vocational interest area where circumplex models are interpreted relative to the highest two or three scales (e.g., Holland, 1992). Given their usage, it is important to evaluate the circumplex structure of these high-point codes. Another example is the circumplex evaluation of the construct of complementarity. Interpersonal models (e.g., Carson, 1969; Kiesler, 1996; Tracey, 1993) hypothesize that specific sequences of behavior are more frequent and follow certain circumplex patterns (i.e., certain behaviors are more likely to follow other behaviors according to rules generated from the circumplex structure in the behaviors). Complementarity requires the presence of two circumplexes, one describing the behavior of each of the participants interacting, and then further that the two circumplexes correspond to each other in a theoretically defined manner.

The high-point data and the behavioral sequence data typically form nonsymmetrical frequency matrices, which cannot be examined using any of the methods detailed except the randomization approach. Tracey (1994) used the randomization approach to evaluate the circumplex structure of a sequence of interactive behavior recorded in nonsymmetric transition frequency matrices. He found support for the circumplex basis of complementarity, but only after account was taken of the varying base rates of behavior. Two other approaches that can be used to analyze nonsymmetric frequency data are the exploratory approach of correspondence analysis and the confirmatory approach of loglinear analysis.

## B. Correspondence Analysis

Correspondence analysis, also called dual scaling (Nishisato, 1980) and canonical analysis of cross-classified data (Holland, Levi, & Watson, 1980), is comparable to a principal components analysis of a two-way frequency table. It is thus similar to the exploratory techniques of MDS, PCA, and PFA discussed above in that it yields a spatial representation of the relation among the variables, and this spatial representation can be evaluated relative to its resemblance to a circle. Unlike the above spatial techniques, correspondence analysis provides two structures, one for the variables on the rows of the frequency matrix and one for the variables on the columns. These two structures are jointly scaled on the same planes to understand their covariance. Weller and Romney (1990) provide a relatively easy to follow description of correspondence analysis. Tracey (1994) applied correspondence analysis to transition matrices of interactional behavior to examine the presence of complementarity.

## C. Loglinear Modeling

Like the SEM approaches above, the researcher using loglinear modeling specifies a model and applies it to the data, and then assesses fit using chi-squared goodness-of-fit and several other goodness-of-fit indices. The circumplex model typically fit to the data is one similar to the circulant model posited in the bottom of Table 22.2, except that the matrix is full. The researcher specifies only the structure for the interaction term in the loglinear model, as this is where the circumplex structure is hypothesized to manifest itself. Since the frequency matrix is not assumed to be symmetric, the deviation from symmetry is assumed to be a function of varying marginals (i.e., different base rates among the behaviors) and the interaction term is constrained to be full and symmetric. The fit of this model is then evaluated on the frequency matrix. Tracey and Rounds (1992) provide an example of how this loglinear circumplex analysis could be applied to a frequency matrix. (Imrey provides a comprehensive discussion of the approaches to analyzing log-linear models in chapter 14, this volume).

## IV. CONCLUSIONS

There are a variety of models and methods that can be used to evaluate the circumplex, most of which are relatively new. Given our increasing ability to specify and evaluate circumplex models, it is not surprising that extrapolations into circle-like structures that exist beyond the two dimensions of a circle are beginning to appear. Three-dimensional spherical structures have been proposed and evaluated by Tracey and Rounds (1996) in the vocational interest area and by Saucier (1992) in the personality area. Of the above confirmatory approaches, only the randomization approach is general enough to be applied easily in these more complex contexts.

Work by Hubert and Arabie (1995; Hubert et al., 1998) on the evaluation of several structures in an additive manner holds promise as another advance in our search for methods to use in studying the underlying structure. Their method enables the evaluation of both cluster and circular structures as coexisting in the same data set. Formerly, it was possible to evaluate one or the other structure, but not the co-occurrence of each. As interest in both cluster and type models increases (e.g., Bailey, 1994; Meehl, 1992) along with that in circumplexes, such methods could serve as the next set of tools to better understand structure. Our knowledge and interest in different structures is advanced not only by new theory of different structures but also by the development of appropriate evaluation methodologies.

The field is moving toward more confirmatory approaches for the evaluation of circumplexes and away from the early exploratory methods. Several different approaches to confirmatory approaches have been covered, each with its own strengths. The selection of confirmatory approach should rest on the circumplex model of interest. An important benefit of adopting confirmatory methods is that they require researchers to specify explicitly the particular circumplex model that is to be examined. Such specification can only help move the field forward.

## ACKNOWLEDGMENTS

Appreciation is expressed to Cynthia Glidden-Tracey for her helpful comments on this chapter.

## REFERENCES

Alden, L. E., Wiggins, J. S., & Pincus, A. L. (1990). Construction of circumplex scales for the Inventory of Interpersonal Problems. *Journal of Personality Assessment, 55*, 521–536.

Bailey, K. D. (1994). *Typologies and taxonomies.* Thousand Oaks, CA: Sage.

Bartholomew, K., & Horowitz, L. M. (1991). Attachment styles among young adults: A test of a four-category model. *Journal of Personality and Social Psychology, 61*, 226–244.

Becker, G. (1996). The meta-analysis of factor analysis: An illustration based on the cumulation of correlation matrices. *Psychological Methods, 1*, 341–353.

Bentler, P. M. (1990). Comparative fit indexes in structural models. *Psychological Bulletin, 107*, 238–246.

Bentler, P. M., & Bonett, D. G. (1980). Significance tests and goodness of fit in the analysis of covariance structures. *Psychological Bulletin, 88*, 588–606.

Bentler, P. M., & Chou, C. (1993). Some new covariance structure model improvement statistics. In K. A. Bollen and J. S. Long (Eds.), *Testing structural equation models* (pp. 235–255). Newbury Park, CA: Sage.

Browne, M. W. (1992). Circumplex models for correlation matrices. *Psychometrika, 57,* 469–497.

Browne, M. W., & Cudeck, R. (1993). Alternative ways of assessing model fit. In K. A. Bollen and J. S. Long (Eds.), *Testing structural equation models* (pp. 136–162). Newbury Park, CA: Sage.

Carson, R. C. (1969). *Interaction concepts of personality.* Chicago: Aldine.

Davison, M. L. (1985). Multidimensional scaling versus components analysis of test intercorrelations. *Psychological Bulletin, 97,* 94–105.

Davison, M. L., & Skay, C. L. (1991). Multidimensional scaling and factor models of test and item responses. *Psychological Bulletin, 110,* 551–556.

Fabrigar, L. R., Visser, P. S., & Browne, M. W. (1997). Conceptual and methodological issues in testing the circumplex structure of data in personality and social psychology. *Personality and Social Psychology Review, 1,* 184–203.

Fisher, N. I. (1993). *Statistical analysis of circular data.* Cambridge, UK: Cambridge University Press.

Fouad, N. A., Cudeck, R., & Hansen, J. C. (1984). Convergent validity of the Spanish and English forms of the Strong-Campbell Interest Inventory for bilingual Hispanic high school students. *Journal of Counseling Psychology, 31,* 339–348.

Gaines, S. O., Jr., Panter, A. T., Lyde, M. D., Steers, W. N., Fusbult, C. E., Cos, C. L., Wexler, M. O. (1997). Evaluating the circumplexity of interpersonal traits and the manifestation of interpersonal traits in interpersonal trust. *Journal of Personality and Social Psychology, 73,* 610–623.

Gifford, R., & O'Connor, B. (1987). The interpersonal circumplex as a behavior map. *Journal of Personality and Social Psychology, 52,* 1019–1026.

Gorsuch, R. L. (1990). Common Factor analysis versus components analysis: Some well and little known facts. *Multivariate Behavioral Research, 25,* 33–40.

Gurtman, M. B. (1992a). Construct validity of interpersonal personality measures: The interpersonal circumplex as a nomological net. *Journal of Personality and Social Psychology, 63,* 105–118.

Gurtman, M. B. (1992b). Trust, distrust, and interpersonal problems: A circumplex analysis. *Journal of Personality and Social Psychology, 62,* 989–1002.

Gurtman, M. B. (1997). Studying personality traits: The circular way. In R. Plutchik & H. R. Conte (Eds.), *Circumplex models of personality and emotions* (pp. 81–102). Washington, DC: American Psychological Association.

Gurtman, M. B., & Pincus, A. L. (in press). Interpersonal Adjective Scales: Confirmation of circumplex structure from multiple perspectives. *Personality and Social Psychology Bulletin.*

Guttman, L. (1954). A new approach to factor analysis: The radex. In P. F. Lazarsfeld (Ed.), *Mathematical thinking in the social sciences* (pp. 258–348). New York: Columbia University Press.

Hofstee, W. K. B., Ten Berge, J. M. F., & Hendriks, A. A. J. (1998). How to score questionnaires. *Personality and Individual Differences, 25,* 897–909.

Hogan, R. (1983). A socioanalytic theory of personality. In M. M. Page (Ed.), *Nebraska symposium on motivation 1982. Personality: Current theory and research* (pp. 55–89). Lincoln, NE: University of Nebraska Press.

Holland, J. L. (1973). *Making vocational choices: A theory of careers.* Englewood Cliffs, NJ: Prentice Hall.

Holland, J. L. (1992). *Making vocational choices: A theory of vocational personality and work environments.* (2nd ed.). Odessa, FL: Psychological Assessment Resources.

Holland, T. R., Levi, M., & Watson, C. G. (1980). Canonical correlation in the analysis of a contingency table. *Psychological Bulletin, 87,* 334–336.

Horowitz, L. M., Rosenberg, S. E., Baer, B. A., Ureno, G., & Villasenor, V. S. (1988). Inventory of Interpersonal Problems: Psychometric properties and clinical applications. *Journal of Consulting and Clinical Psychology, 56,* 885–892.

Hubert, L. J. (1987). *Assignment methods in combinatorial data analysis.* New York: Marcel Dekker.

Hubert, L., & Arabie, P. (1987). Evaluating order hypotheses within proximity matrices. *Psychological Bulletin, 102,* 172–178.

Hubert, L., & Arabie, P. (1995). The approximation of two-mode proximity matrices by sums of order-constrained matrices. *Psychometrika, 60,* 573–605.

Hubert, L., Arabie, P., & Meulman, J. (1998). The representation of symmetric proximity data: Dimensions and classifications. *Computer Journal, 41,* 566–577.

Jackson, D. N. (1977). *Manual for the Jackson Vocational Interest Survey.* Port Huron, MI: Research Psychologists Press.

Jackson, D. N., & Helmes, E. (1979). Personality structure and the circumplex. *Journal of Personality and Social Psychology, 37,* 2278–2285.

Jöreskog, K. G. (1974). Analyzing psychological data by structural analysis of covariance matrices. In D. H. Krantz, R. C. Atkinson, R. D. Luce, & P. Suppes (Eds.), *Contemporary developments in mathematical psychology* (Vol. 2, pp. 1–56). San Francisco: S. H. Freeman.

Jöreskog, K. G., & Sörbom, D. (1993). *LISREL8: Structural equation modeling with the SIMPLIS command language.* Hillsdale, NJ: Erlbaum.

Kiesler, D. J. (1996). *Contemporary interpersonal theory and research: Personality, psychopathology, and psychotherapy.* New York: Wiley.

Larsen, R. J., & Diener, E. (1992). Promises and problems with the circumplex model of emotion. In M. S. Clark (Ed.), *Review of personality and social psychology: Emotion.* (Vol. 13, pp. 25–59). Newbury Park, CA: Sage.

Leary, T. (1957). *Interpersonal diagnosis of personality.* New York: Ronald.

Loehlin, J. C. (1990). Component analysis versus common factor analysis: A case of disputed authorship. *Multivariate Behavioral Research, 25,* 29–32.

Lorr, M. (1997). The circumplex model applied to interpersonal behavior, affect, and psychotic syndromes. In R. Plutchik & H. R. Conte (Eds.), *Circumplex models of personality and emotions* (pp. 47–56). Washington, DC: American Psychological Association.

MacCallum, R. C., Browne, M. W., & Sugawara, H. M. (1996). Power analysis and determination of sample size for covariance structure modeling. *Psychological Methods, 1,* 130–149.

Mardia, K. V. (1972). *Statistics of directional data.* London: Academic Press.

Marsh, H. W., Balla, J. R., & McDonald, R. P. (1988). Goodness of fit indexes in confirmatory factor analysis: The effects of sample size. *Psychological Bulletin, 103,* 391–410.

McDonald, R. P., & Marsh, H. W. (1990). Choosing a multivariate model: Noncentrality and goodness of fit. *Psychological Bulletin, 107,* 247–255.

Meehl, P. E. (1992). Factors and taxa, traits and types, differences of degree and differences of kind. *Journal of Personality, 60,* 117–174.

Myors, B. (1996). A simple, exact test for the Holland hexagon. *Journal of Vocational Behavior, 48,* 339–351.

Myors, B. (1998). A Monte Carlo comparison of three tests of the Holland hexagon. *Journal of Vocational Behavior, 53,* 215–226.

Nishisato, S. (1980). *Analysis of categorical data: Dual scaling and its applications.* Toronto: University of Toronto Press.

Plutchik, R. (1996). The circumplex as a general model of the structure of emotions and personality. In R. Plutchik & H. R. Conte (Eds.), *Circumplex models of personality and emotions* (pp. 17–47). Washington, DC: American Psychological Association.

Prediger, D. J. (1982). Dimensions underlying Holland's hexagon: Missing link between interests and occupations? *Journal of Vocational Behavior, 21,* 259–287.

Prediger, D. J. (1998). Is interest profile level relevant to career counseling? *Journal of Counseling Psychology, 45,* 204–211.

Rounds, J., & Tracey, T. J. (1993). Prediger's dimensional representation of Holland's RIASEC circumplex. *Journal of Applied Psychology, 78*, 875–890.

Rounds, J. B., Tracey, T. J., & Hubert, L. (1992). Methods for evaluating vocational interest structural hypotheses. *Journal of Vocational Behavior, 40*, 239–259.

Saucier, G. (1992). Benchmarks: Integrating affective and interpersonal circles with the big-five personality factors. *Journal of Personality and Social Psychology, 62*, 1025–1035.

Schumacker, R. E., & Lomax, R. G. (1996). *A beginner's guide to structural equation modeling.* Mahwah, NJ: Lawrence Erlbaum.

Shepard, R. N. (1962). Analysis of proximities: Multidimensional scaling with an unknown distance function II. *Psychometrika, 27*, 219–246.

Somers, R. H. (1962). A new asymmetric measure of association for ordinal variables. *American Sociological Review, 27*, 799–811.

Steiger, J. H. (1989). *Causal modeling: A supplementary module for SYSTAT and SYSGRAPH.* Evanston, IL: SYSTAT

Steiger, J. H., & Lind, J. (1980, June). *Statistically based tests for the number of common factors.* Paper presented at the annual meeting of the Psychometric Society, Iowa City, IA.

Stevens, M. A. (1970). Use of the Kolmogorov-Smirnov, Cramer-von Mises and related statistics without extensive tables. *Journal of the Royal Statistical society, Series B, 32*, 115–122.

Strong, S. R., Hills, H., Kilmartin, C., DeVries, H., Lanier, K., Nelson, B., Strickland, D., & Meyer, C. (1988). The dynamic relations among interpersonal behaviors: A test of complementarity and anticomplementarity. *Journal of Social and Personality Psychology, 54*, 798–810.

Tanaka, J. S. (1993). Multifaceted conceptions of fit in structural equation models. In K. A. Bollen and J. S. Long (Eds.), *Testing structural equation models* (pp. 10–39). Newbury Park, CA: Sage.

Tracey, T. J. (1993). An interpersonal stage model of the therapeutic process. *Journal of Counseling Psychology, 40*, 1–14.

Tracey, T. J. (1994). An examination of the complementarity of inter-personal behavior. *Journal of Personality and Social Psychology, 67*, 864–878.

Tracey, T. J. G. (1997). RANDALL: A Microsoft FORTRAN program for a randomization test of hypothesized order relations. *Educational and Psychological Measurement, 57*, 164–168.

Tracey, T. J., & Rounds, J. B. (1992). Evaluating the RIASEC circumplex using high-point codes. *Journal of Vocational Behavior, 41*, 295–311.

Tracey, T. J., & Rounds, J. B. (1993). Evaluating Holland's and Gati's vocational interest models: A structural meta-analysis. *Psychological Bulletin, 113*, 229–246.

Tracey, T. J., & Rounds, J. (1996). The spherical representation of vocational interests. *Journal of Vocational Behavior, 48*, 3–41.

Tracey, T. J. G., & Rounds, J. (1997). Circular structure of vocational interests. In R. Plutchik & H. R. Conte (Eds.), *Circumplex models of personality and emotions* (pp. 183–201). Washington, DC: American Psychological Association.

Tracey, T. J. G., Rounds, J., & Gurtman, M. (1996). Examination of the general factor with the interpersonal circumplex structure: Application to the Inventory of Interpersonal Problems. *Multivariate Behavior Research, 31*, 441–466.

Tucker, L. R, & Lewis, C. (1973). The reliability coefficient for maximum likelihood factor analysis. *Psychometrika, 38*, 1–10.

Upton, G. J. G., & Fingleton, B. (1989). *Spatial data analysis by example. Vol. 2: Categorical and directional data.* New York: Wiley.

Velcier, W. F., & Jackson, D. M. (1990). Component analysis versus common factor analysis: Some issues in selecting an appropriate procedure. *Multivariate Behavioral Research, 25*, 1–28.

Wakefield, J. A., & Doughtie, E. B. (1973). The geometric relationship between Holland's personality typology and the Vocational Preference Inventory. *Journal of Counseling Psychology, 20,* 513–518.

Weller, S. C., & Romney, A. K. (1990). *Metric scaling: Correspondence analysis* (Sage University Paper series on Quantitative Applications in the Social Sciences, series no. 07–075). Newbury Park, CA: Sage.

Widiger, T. A., & Hagemoser, S. (1997). Personality disorders and the interpersonal circle. In R. Plutchik & H. R. Conte (Eds.), *Circumplex models of personality and emotions* (pp. 299–326). Washington, DC: American Psychological Association.

Wiggins, J. S. (1979). A psychological taxonomy of trait-descriptive terms: The interpersonal domain. *Journal of Personality and Social Psychology, 37,* 395–412.

Wiggins, J. S., Steiger, J. H., & Gaelick, L. (1981). Evaluating circumplexity in personality data. *Multivariate Behavioral Research, 16,* 263–289.

Wiggins, J. S., & Trapnell, P. D. (1997). Personality structure: The return of the big five. In R. Hogan, J. Johnson, & S. Briggs (Eds.), *Handbook of personality psychology* (pp. 737–766). San Diego: Academic Press.

Wiggins, J. S., & Trobst, K. K. (1997). When is a circumplex and "Interpersonal Circumplex"? The case of supportive actions. In R. Plutchik & H. R. Conte (Eds.), *Circumplex models of personality and emotions* (pp. 57–80). Washington, DC: American Psychological Association.

# 23

# USING COVARIANCE STRUCTURE ANALYSIS TO MODEL CHANGE OVER TIME

### JOHN B. WILLETT

*Graduate School of Education, Harvard University, Cambridge, Massachusetts*

### MARGARET K. KEILEY

*Department of Child Development and Family Studies,*
*Purdue University, West Lafayette, Indiana*

In recent years, there have been important developments in the investigation of change. In this chapter, we emphasize three. First, there has been a renewed emphasis on the collection of multiwave panel data because experts and researchers alike have recognized that two waves of data are insufficient for measuring change effectively (Rogosa, Brandt, & Zimowski, 1982; Willett, 1988, 1989, 1994). Second, the analysis of multiwave data has been revolutionized by *individual growth modeling* (Bryk & Raudenbush, 1987; Rogosa et al., 1982). Under this approach, individual changes over time are represented in a statistical model as functions of time, and questions about interindividual differences in change are answered by letting the parameters of this individual growth model differ across people in ways that depend on the person's background, environment, and treatment (Rogosa & Willett, 1985; Willett, 1994, 1997). Third, innovative methodologists have shown how the individual growth modeling approach can be mapped onto the general covariance structure model, providing a flexible new tool for investigating change over time called *latent growth modeling* (see Willett & Sayer, 1994, 1995, for citations to the technical literature).

*Handbook of Applied Multivariate Statistics and Mathematical Modeling*

In this chapter, we use longitudinal data on self-ratings of alcohol use in grades seven and eight for 1122 teenagers (Farrell, 1994) to demonstrate latent growth modeling. There are three main sections. First, we introduce individual growth modeling and show how the multilevel statistical models required for the investigation of change can be mapped onto the general covariance structure model. Second, we show how a single time-invariant predictor, adolescent gender, can be introduced into the analyses to ascertain whether individual growth in alcohol use differs by gender. Third, we demonstrate how a time-varying predictor of change—the pressure to drink exerted on adolescents by their peers—can be included in the analyses. Here, we ask how change in the original outcome, adolescent's alcohol use, depends on change in the time-varying predictor. We close with comments on extensions of latent growth modeling to more complex settings.

## I. LATENT GROWTH MODELING: THE BASIC APPROACH

To investigate change, you must collect multiwave data on a representative sample of people, measuring the status of each member repeatedly over time. In this chapter, we use multiwave data provided by Professor Albert D. Farrell of Virginia Commonwealth University on the self-reported alcohol use of 1122 adolescents (Farrell, 1994). For each adolescent, alcohol use was reported at the beginning of seventh grade, at the end of seventh grade, and at the end of eighth grade. We define corresponding values of the variable, GRADE, as 7, 7.75, and 8.75 years, respectively. Adolescents used a six-point scale to rate how frequently they had consumed beer, wine, and liquor during the previous 30 days. We averaged each adolescent's responses to the three items to create a composite self-rating of adolescent alcohol use in Farrell's original metric. Data were also collected on two predictors of change: (a) time-invariant adolescent gender, and (b) time-varying peer pressure. Gender was coded 1 to indicate when the adolescent was female, 0 otherwise. The peer pressure predictor used an anchored six-point scale to record the number of times each adolescent reported being offered an alcoholic drink by a friend in the previous 30 days (Farrell, 1994). In Table 23.1, we present a subsample of 10 adolescents randomly selected from the full data set. Inspection of subsample information—and the full data set—suggests there is heterogeneity in both entry-level (7th grade) alcohol use and change in alcohol use over the seventh and eighth grades. We present the full-sample mean vector and covariance matrix for all variables in the Appendix.

### A. Modeling Individual Change

As a first step, we must choose a level-1 statistical model to represent individual change over time. This model is a within-person regression model

**TABLE 23.1** Measurements of Adolescent Self-Reported Alcohol Use, Gender, and Peer Alcohol Use over Three Occasions of Measurement[a]

| | Outcome Adolescent alcohol use | | | | Predictors of change Peer alcohol use | | |
|---|---|---|---|---|---|---|---|
| Adolescent ID | Start of seventh grade | End of seventh grade | End of eighth grade | Female | Start of seventh grade | End of seventh grade | End of eighth grade |
| 0018 | 1.00 | 1.33 | 2.00 | 0 | 3 | 2 | 2 |
| 021 | 1.00 | 2.00 | 1.67 | 0 | 1 | 1 | 1 |
| 0236 | 3.33 | 4.33 | 4.33 | 0 | 2 | 1 | 4 |
| 0335 | 1.00 | 1.33 | 1.67 | 0 | 1 | 2 | 1 |
| 0353 | 2.00 | 2.00 | 1.67 | 0 | 1 | 1 | 2 |
| 0555 | 2.67 | 2.33 | 1.67 | 1 | 2 | 3 | 1 |
| 0850 | 1.33 | 1.67 | 1.33 | 1 | 3 | 1 | 2 |
| 0883 | 3.00 | 2.67 | 3.33 | 1 | 4 | 5 | 1 |
| 0974 | 1.00 | 1.67 | 2.67 | 1 | 1 | 5 | 6 |
| 1012 | 1.00 | 1.67 | 2.33 | 1 | 1 | 2 | 4 |

[a] Table includes 10 adolescents selected at random from the full data set.

relating the outcome—here, self-reported alcohol use—to time, and to a few individual growth parameters. If we decide, for instance, that individual change in self-reported alcohol use from seventh through eighth grades has a linear trajectory, then the level-1 model will contain a predictor representing time and two individual growth parameters: (a) an *intercept* parameter representing an adolescent's self-reported alcohol use at the beginning of seventh grade (providing that we set seventh grade as the "origin" of our time axis, as we do below), and (b) a *slope* parameter representing the adolescent's rate of change in self-reported alcohol use over the period of observation. On the other hand, if we hypothesize that change in alcohol use over seventh and eighth grades has a curvilinear trajectory, perhaps quadratic, then the level-1 model would contain an additional parameter representing curvature, and so on.

In addition, recall that classical test theory distinguishes observed scores from true scores. This distinction is critical when individual change is investigated, because change in the underlying true scores, rather than change in the observed scores, is the real focus of analytic interest. Measurement error randomly obscures the true growth trajectory from view, but it is the true growth trajectory that is specified as a function of time and that is the focus of research interest.

How do you choose a mathematical function to represent true individual change? If theory guides your model choice, then the individual growth parameters can have powerful substantive interpretations. Often, however, the theoretical mechanisms governing change are poorly understood. Then, a well-fitting polynomial can be used to approximate the trajectory. When only a restricted portion of the life span has been observed and few waves of data collected, the complexity of the growth model must be limited. Because of this, the use of linear or quadratic functions of time is very popular in individual growth modeling.

One strategy for choosing an individual growth model is to inspect each person's growth record, by plotting their observed scores against time (see Willett, 1989). This type of individual-level exploration is important when latent growth modeling is being used because good data-analytic practice demands knowledge of the data at the lowest level of aggregation so that anomalous cases can be identified and assumptions checked. In our example, initial data exploration suggested that the natural logarithm of self-reported adolescent alcohol use was linearly related to grade level. Thus, we hypothesized the following level-1 individual growth model for self-reported alcohol use:

$$Y_{ip} = \pi_{1p} + \pi_{2p}t_i + \varepsilon_{ip}, \tag{1}$$

where $Y_{ip}$ represents the natural logarithm of the self-reported alcohol use of the $p$th adolescent on the $i$th occasion of measurement ($i = 1, 2, 3$), $t_i$ represents an adolescent's grade level minus seven years (yielding values of 0, .75, and 1.75 for $t_1$ through $t_3$, respectively), and $\varepsilon_{ip}$ are the level-1 measurement errors that distinguish the true from the observed self-reported alcohol use on each occasion for each person.

The shape of the hypothesized trajectory depends on the functional form of the growth model and the specific values of the individual growth parameters. The growth model in Equation 1 contains a pair of individual growth parameters representing the intercept, $\pi_{1p}$, and slope, $\pi_{2p}$, of each adolescent's trajectory of log-alcohol consumption. The intercept parameter $\pi_{1p}$ represents the true (self-reported) log-alcohol use of adolescent $p$ when $t_i$ is equal to zero—that is, at the beginning of grade seven, because of the way we have created $t_i$ from adolescent grade level by subtracting 7. Children whose self-reported alcohol use is higher at the beginning of grade seven will possess higher values of this parameter. The slope parameter $\pi_{2p}$ represents the change in true (self-reported) log-alcohol use per grade for the $p$th adolescent. Children whose use increased most rapidly over grades seven and eight will have the largest values of this parameter. These intercepts and slopes are the central focus of our investigation of change.

### B. Mapping the Individual Growth Model onto the LISREL Measurement Model

Our purpose in this chapter is to show how the individual growth modeling framework begun in Equation 1 can be mapped onto the statistical framework provided by covariance structure analysis, as operationalized in LIS-REL. One key facet of this mapping is that the individual growth model can be treated as the LISREL measurement model for endogenous variables, $Y$. Beginning here, and over the next few pages, we illustrate how the mapping evolves.

In our example, three waves of data were collected on each adolescent's alcohol use, so each individual's empirical growth record contains three entries—$Y_{1p}$, $Y_{2p}$, and $Y_{3p}$—on observed (logarithm of self-reported) alcohol use. Each of these observed values can be represented as a function of time and the individual growth parameters, using Equation 1. By repeating the individual growth model in Equation 1 with suitable values of $Y$ and $t$ substituted, we can represent this vector of observed data in matrix form, as follows:

$$\begin{bmatrix} Y_{1p} \\ Y_{2p} \\ Y_{3p} \end{bmatrix} = \begin{bmatrix} 0 \\ 0 \\ 0 \end{bmatrix} + \begin{bmatrix} 1 & t_1 \\ 1 & t_2 \\ 1 & t_3 \end{bmatrix} \begin{bmatrix} \pi_{1p} \\ \pi_{2p} \end{bmatrix} + \begin{bmatrix} \varepsilon_{1p} \\ \varepsilon_{2p} \\ \varepsilon_{3p} \end{bmatrix} \qquad (2)$$

The introduction of the vector of zeros on the right-hand side of Equation 2 does not change the equation but will facilitate our mapping of the individual growth model onto the corresponding LISREL $Y$-measurement model below. Equation 2 now says, straightforwardly, that every adolescent's alcohol use on each of the three occasions of measurement is constructed simply by bringing the values of their individual growth parameters—their own intercept and slope—together with an appropriate value of time and then disturbing the observation of true alcohol use by an error that arises as a consequence of the measurement process itself.

Now think about the column of measurement errors in Equation 2 more carefully. This equation states that the level-1 measurement error $\varepsilon_{1p}$ disturbs the true status of the $p$th adolescent on the first occasion of measurement, $\varepsilon_{2p}$ on the second occasion, and $\varepsilon_{3p}$ on the third. However, we have so far made no claims about the nature of the distribution from which these errors are drawn. Fortunately, latent growth modeling permits great flexibility in the specification of the level-1 error covariance structure and is not restricted to classical assumptions of independence and homoscedasticity. We do not intend to focus, however, on the flexibility of the level-1 error structure specification at this stage, so we assume in all models that we specify subsequently that the level-1 errors are independently and heteroscedastically normally distributed over time within-person. We can describe this assumption statistically, as follows:

$$\begin{bmatrix} \varepsilon_{1p} \\ \varepsilon_{2p} \\ \varepsilon_{3p} \end{bmatrix} \sim N\left( \begin{bmatrix} 0 \\ 0 \\ 0 \end{bmatrix}, \begin{bmatrix} \sigma^2_{\varepsilon_1} & 0 & 0 \\ 0 & \sigma^2_{\varepsilon_2} & 0 \\ 0 & 0 & \sigma^2_{\varepsilon_3} \end{bmatrix} \right) \qquad (3)$$

where the mean vector and covariance matrix on the right-hand side of Equation 3 are assumed to be homogeneous across children. Supplementary data analysis indicated that this assumption was very reasonable.

Once we have specified the individual growth model in this way, individual growth modeling and covariance structure analysis begin to converge. We have suggested that the individual growth model in Equation 2, along with its associated error covariance assumptions in Equation 3, map onto the LISREL measurement model for endogenous variables, **Y**. In standard LISREL matrix notation, this latter model is simply:

$$Y = \tau_y + \Lambda_y \eta + \varepsilon \tag{4}$$

Comparing Equations 2 and 4, we can force the LISREL measurement model for endogenous variables, **Y**, to contain our individual growth model by specifying the LISREL score vectors in a special way:

$$Y = \begin{bmatrix} Y_{1p} \\ Y_{2p} \\ Y_{3p} \end{bmatrix}, \eta = \begin{bmatrix} \pi_{1p} \\ \pi_{2p} \end{bmatrix}, \varepsilon = \begin{bmatrix} \varepsilon_{1p} \\ \varepsilon_{2p} \\ \varepsilon_{3p} \end{bmatrix} \tag{5}$$

and by forcing the LISREL $\tau_y$ and $\Lambda_y$ parameter matrices to contain known values and constants:

$$\tau_y = \begin{bmatrix} 0 \\ 0 \\ 0 \end{bmatrix}, \Lambda_y = \begin{bmatrix} 1 & t_1 \\ 1 & t_2 \\ 1 & t_3 \end{bmatrix} \tag{6}$$

Notice that substitution of the vectors and matrices from Equations 5 and 6 into the general $Y$-measurement model in Equation 4 leads back to the original individual growth modeling specification in Equation 2. Finally, to represent the level-1 error covariance structure in Equation 3, we let the LISREL error vector $\varepsilon$ be distributed with zero mean vector and covariance matrix $\Theta_\varepsilon$, where:

$$\Theta_\varepsilon = \begin{bmatrix} \sigma^2_{\varepsilon_1} & 0 & 0 \\ 0 & \sigma^2_{\varepsilon_2} & 0 \\ 0 & 0 & \sigma^2_{\varepsilon_3} \end{bmatrix} \tag{7}$$

The advantage of recognizing that the individual growth model and the LISREL Y-measurement model are one and the same is that the covariance structure specification is quite general and can be expanded to cover more complex analytic settings. If more waves of data were available, for instance, then the empirical growth record on the left-hand side of Equation 2 would contain additional elements and we would simply tack rows onto the $\Lambda_y$ matrix, one per occasion of measurement, to accommodate the new waves. Or, if we wished to adopt a more complex trajectory for individual growth, then we would simply add the appropriate curvilinear terms to the individual growth model in Equation 1. This change would add one or more individual growth parameters representing the curvilinearity to the $\eta$-vector in Equation 4 and additional columns to the $\Lambda_y$ matrix in Equation 6.

Notice that, unlike more typical covariance structure analyses, the LIS-REL $\Lambda_y$ parameter matrix in Equation 6 contains only known times and constants rather than a collection of unknown factor-loading parameters that you then try to estimate. Although unusual, this "constants and values" specification forces the critical individual-level growth parameters that are the focus of the investigation of individual change, $\pi_{1p}$ and $\pi_{2p}$, into the LISREL endogenous construct vector $\eta$, which we subsequently refer to as the "latent growth vector." This allows us to conduct between-person analyses of the individual growth parameters—the level-2 analyses of multi-level modeling—by modeling person-to-person variation in the latent growth vector in the so-far unused LISREL structural model. We describe this next.

## C. Modeling Interindividual Differences in Change

Even though we have hypothesized linear change in log-alcohol use over time, everyone need not have an identical trajectory. The trajectories can still differ from person to person because of variation in the values of the growth parameters; some adolescents may have different intercepts; some may have different slopes. These interindividual differences in the growth parameters can be modeled by assuming that each adolescent draws his or her latent growth vector from a bivariate normal distribution, as follows:

$$\begin{bmatrix} \pi_{1p} \\ \pi_{2p} \end{bmatrix} \sim N\left( \begin{bmatrix} \mu_{\pi_1} \\ \mu_{\pi_2} \end{bmatrix}, \begin{bmatrix} \sigma^2_{\pi_1} & \sigma_{\pi_1 \pi_2} \\ \sigma_{\pi_2 \pi_1} & \sigma^2_{\pi_2} \end{bmatrix} \right) \tag{8}$$

Conceptually, this is like bobbing for apples. Equation 8 says that, before they begin to change, each adolescent "dips" into a distribution of individual intercepts and slopes, and pulls out a pair of values for themselves. Although these values are unknown to us, they determine the shape of the individual's change trajectory over time. Because each adolescent draws a different value for their intercept and slope from the same underlying distribution, each person can possess a unique growth trajectory. The parameters of the distribution in Equation 8 then become the focus of our subsequent analyses of change.

The hypothesized distribution in Equation 8 is a simple between-person or level-2 model for interindividual differences in true change. In the current example, in which there are two important individual growth parameters for each adolescent (i.e., intercept and slope), the shape of the distribution specified in the between-person model is determined by five important level-2 parameters: two means, two variances, and a covariance. These parameters describe interesting features of average change and variability in change in the population, and are worthy of estimation in an investigation of change:

- The two mean parameters, $\mu_{\pi_1}$ and $\mu_{\pi_2}$, describe the average population intercept and slope, and answer the research question, What is the population trajectory of true change in self-reported adolescent (log) alcohol use through grades seven and eight?
- The two variance parameters, $\sigma^2_{\pi_1}$ and $\sigma^2_{\pi_2}$, summarize population interindividual differences in initial (grade seven) true self-reported (log) alcohol use and true self-reported rate of change in (log) alcohol use, and answer the research question, Is there between-person heterogeneity in the growth trajectory of adolescent alcohol use in the population?
- The covariance parameter, $\sigma_{\pi_1 \pi_2}$, represents the population association between initial status and rate of change and answers the research question, Is there a non-zero correlation between initial status and rate of change in adolescent self-reported (log) alcohol use, over grades seven and eight?

### D. Mapping the Model for Interindividual Differences in Change onto the LISREL Structural Model

How can we estimate these important between-person parameters in an analysis of change? Fortunately, there is another mapping of the change framework onto the covariance structure framework that helps out. The level-2 distribution of the individual growth parameters in Equation 8 can be modeled via the reduced LISREL structural model, which is

$$\eta = \alpha + \mathbf{B}\eta + \zeta \qquad (9)$$

The important distribution of the individual intercept and slope, $\pi_{1p}$ and $\pi_{2p}$, in Equation 8 can be modeled in the LISREL structural model because this model is responsible for the modeling of the distribution of the $\eta$-vector, which we have forced to contain the individual growth parameters. The specification is straightforward. We free up the LISREL $\alpha$-vector to contain the population average values of the individual intercepts and slopes and constrain the LISREL $\mathbf{B}$-matrix to zero, as follows:

$$\alpha = \begin{bmatrix} \mu_{\pi_1} \\ \mu_{\pi_2} \end{bmatrix}, \mathbf{B} = \begin{bmatrix} 0 & 0 \\ 0 & 0 \end{bmatrix} \qquad (10)$$

This forces the LISREL structural model to become:

$$\begin{bmatrix} \pi_{1p} \\ \pi_{2p} \end{bmatrix} = \begin{bmatrix} \mu_{\pi_1} \\ \mu_{\pi_2} \end{bmatrix} + \begin{bmatrix} 0 & 0 \\ 0 & 0 \end{bmatrix} \begin{bmatrix} \pi_{1p} \\ \pi_{2p} \end{bmatrix} + \begin{bmatrix} \zeta_{1p} \\ \zeta_{2p} \end{bmatrix} \tag{11}$$

and, critically, the elements of the LISREL latent residual vector, $\zeta$, in Equation 11 then contain deviations of the individual intercepts and slopes, $\pi_{1p}$ and $\pi_{2p}$, from their respective population means. This means that the covariance matrix $\Psi$ of the latent residual vector, $\zeta$, will contain the level-2 variance and covariance parameters that represent the interindividual differences in change from Equation 8:

$$\Psi = Cov(\zeta) = \begin{bmatrix} \sigma_{\pi_1}^2 & \sigma_{\pi_1 \pi_2} \\ \sigma_{\pi_2 \pi_1} & \sigma_{\pi_2}^2 \end{bmatrix} \tag{12}$$

By estimating the vector of latent construct means in $\alpha$ and the latent residual covariance matrix, $\Psi$, in an empirical application of LISREL, we can address the research questions about interindividual differences in change cited in the previous paragraph.

Although this application appears complex and mathematical, the central idea is conceptually clear. We can force the multilevel models necessary for the measurement of change into the general framework provided by covariance structure analysis. In Equations 1, 2, 3 and 8, the individual growth modeling framework provides level-1 (within-person) and level-2 (between-person) models to represent our hypotheses about the changes underlying the three waves of panel data in our data-example. Equations 4 through 7 and 9 through 12 show that these models can be rewritten using the format and notation of the general LISREL model. Carefully choosing the specification of the LISREL parameter matrices transforms the LISREL Y-measurement model into the individual growth model (which includes all of our assumptions on the distribution of the measurement errors), and the LISREL structural model becomes the level-2 model for interindividual differences in true change. In analyses that follow, we refer to the covariance structure model for interindividual differences in change in Equations 4–7 and 9–12 as Latent Growth Model #1.

Figure 23.1 presents a path diagram of Latent Growth Model #1. The figure illustrates that, by constraining the LISREL $\lambda$-coefficients (the factor loadings linking **Y** to $\eta$) to take on either the value of 1 and or the values of the times of measurement, $t_1$, $t_2$, and $t_3$, we force the latent construct, $\eta$, to become the individual growth parameters at the center of our change analysis. Then, level-2 relations among these individual intercepts and slopes can be modeled as associations among the latent residuals, $\zeta$, as indicated by the double-headed arrow on the left of the path model.

Completion of model-mapping allows us to use covariance structure

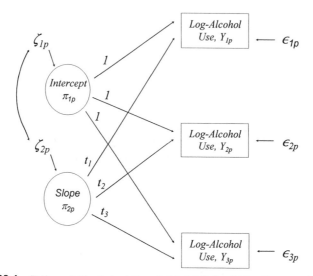

**FIGURE 23.1**     Path model for Latent Growth Model #1, in which self-reported log-alcohol use is a linear function of adolescent grade, and interindividual differences in initial status (intercept) and rate of change (slope) are present.

analysis to obtain maximum likelihood estimates of the parameters of interest in our analysis of change in adolescent alcohol use. All of the critical parameters describing interindividual differences in change now reside in either the $\alpha$-vector, the $\Theta_\varepsilon$-matrix or the $\Psi$-matrix, respectively. We can program LISREL to estimate these matrices and thereby answer our questions about change in self-reported alcohol use for these adolescents. In Table 23.2, we provide a LISREL VIII program suitable for fitting Latent Growth Model #1, in which the $\tau_y$, $\Lambda_y$, $\Theta_\varepsilon$, $\alpha$, $\mathbf{B}$, and $\Psi$ parameter matrices are patterned as described above.

The program for Latent Growth Model #1 has a structure typical of most LISREL VIII programs. Data are input, variables labeled and selected for analysis, and the latent growth model in Equations 5–7, 10–12 is specified. First, examine the level-1 model for individual change. In the model statement (MO) of lines 7 and 8, the LISREL score vectors are dimensioned according to Equation 5. Notice that the $\mathbf{Y}$-vector has three elements (NY = 3), one for each of the repeated measurements on each adolescent, and the $\eta$-vector has two elements (NE = 2), one for each individual growth parameter (intercept and slope, labeled $\pi_{1p}$ and $\pi_{2p}$ in lines 9 and 10) in the linear growth model. The $\tau_y$ vector has its elements set to zero (TY = ZE), as required by Equation 6. The factor-loading matrix $\Lambda_y$ from the Y-measurement model is first declared "full" with elements "fixed" to zero (LY = FU,FI) and is then loaded, in lines 11 through 14, with the values required by Equation 6—that is, the value "1" enters the entire first column

**TABLE 23.2   LISREL VIII Program for Fitting Latent Growth Model #1$^a$**

Raw data input
1. DA NI = 7
2. RA FI = C:\DATA\ALCOHOL.DAT

Variable labeling and selection
3. LA
4. FEM LN_ALC1 LN_ALC2 LN_ALC3 LN_PEER1 LN_PEER2 LN_PEER3
5. SE
6. 2 3 4 /

Model specification
7. MO NY = 3 NE = 2 TY = ZE LY = FU,FI TE = SY,FI C
8.    AL = FR GA = ZE BE = ZE PS = SY,FR
9. LE
10. Pi1 Pi2
11. VA 1 LY(1, 1) LY(2, 1) LY(3, 1)
12. VA 0.00 LY(1, 2)
13. VA 0.75 LY(2, 2)
14. VA 1.75 LY(3, 2)
15. FR TE(1, 1) TE(2, 2) TE(3, 3)

Creation of output
16. OU TV RS SS ND = 5

$^a$ Line numbers and comments are for reference, and should be removed when the program is executed.

(line 11), and the times of the occasions of measurement, measured as deviations from grade seven (i.e., 0.00, 0.75, 1.75) enter the second column (lines 12–14). This ensures that the elements of the $\eta$-vector represent the required individual growth parameters, $\pi_{1p}$ and $\pi_{2p}$. The covariance matrix of the level-1 measurement errors, $\Theta_\varepsilon$, is then declared symmetric and fixed to zero (TE = SY,FI) in line 7, but its diagonal entries are freed for estimation later in line 15, as required by Equation 7.

The shape of the level-2 model for interindividual differences in change is also specified in the program. In line 8, the shapes of the critical level-2 parameter matrices that make up the LISREL structural model are specified. The $\alpha$-vector is freed (AL = FR) to contain the mean values of the individual intercepts and slopes across all adolescents in the population, as is required by Equation 10. The **B**-matrix is forced to zero (BE = ZE) also as a requirement of Equation 10. The covariance matrix of the latent residuals, $\Psi$, is declared symmetric and freed (PS = SY,FR) so that it can contain the population variances and covariances of the individual growth parameters, as required by Equation 12. Notice that we also set the elements of the $\Gamma$-matrix to zero, as it is not used in the current analysis. Finally, line 16 of the program defines the output required.

Execution of the LISREL program in Table 23.2 provides model goodness-of-fit statistics, maximum likelihood estimates of the population means and variances of the true intercept and true rate of change in (self-reported) alcohol use, the covariance of the true intercept and true rate of change, and approximate $p$-values for testing the null hypothesis that the associated parameter is zero in the population (Bollen, 1989, p. 286); see Table 23.3. We have added a brief conceptual description of each parameter. Latent Growth Model #1 fits the data well ($\chi^2 = .049$, df $= 1$, $p = .826$).

The first two rows of Table 23.3 contain estimates of the population means of true intercept (.226, $p < .001$) and true slope (.036, $p < .001$) from the fitted $\alpha$-vector. They describe the average trajectory of true change in self-reported log-alcohol use over the seventh and eighth grades in this population. Adolescents report consuming statistically significant (i.e., non-zero) amounts of alcohol in seventh grade (average log-alcohol use is reported as .226). The positive sign on the statistically significant mean slope estimate suggests that adolescents report consuming more alcohol as they age. Because we have used a natural logarithm to transform the outcome prior to growth modeling, we can interpret the magnitude of the slope as a percentage increase in alcohol use over time. Therefore, given a slope estimate of .036, we conclude that in this population there is,

**TABLE 23.3** **Parameter Estimates for Latent Growth Model #1, in Which the Individual Initial Status (Intercept) and Rate of Change (slope) Parameters of the Alcohol Use Trajectory Differ across Adolescents**

| Parameter | | Estimate |
|---|---|---|
| **Symbol** | **Label** | |
| Population true mean trajectory in alcohol use | | |
| $\mu_{\pi_1}$ | Initial status: average true log-alcohol use in grade seven | .226*** |
| $\mu_{\pi_2}$ | Rate of change: average true change in log-alcohol use per grade | .036*** |
| Population true residual variances and covariance in alcohol use | | |
| $\sigma^2_{\pi_1}$ | True initial status: variance in true log-alcohol use in grade seven | .087*** |
| $\sigma^2_{\pi_2}$ | Rate of true change: variance in true change in log-alcohol use per grade | .020*** |
| $\sigma_{\pi_1 \pi_2}$ | True initial status and rate of true change: covariance of true log-alcohol use in grade seven and true change in log-alcohol use per grade | $-.013$*** |
| Level-1 measurement error variance | | |
| $\sigma^2_{\varepsilon_1}$ | At time-1 (beginning of grade seven) | .049*** |
| $\sigma^2_{\varepsilon_2}$ | At time-2 (end of grade seven) | .076*** |
| $\sigma^2_{\varepsilon_3}$ | At time-3 (end of grade eight) | .077*** |

$** = p < .01$, $*** = p < .001$. $\chi^2(1, 1122) = 0.05$ ($p = .83$).

on average, about a 4% increase in self-reported alcohol use per year of adolescence (the exact yearly increase is actually $e^{.0360} - 1$, or 3.66%).

The third and fourth rows of Table 23.3 give the estimated variances of true intercept and true slope from the fitted $\Psi$ matrix, and summarize population interindividual heterogeneity in true change. Because both variances are statistically significant and therefore nonzero, there are important interindividual differences in seventh grade self-reported log-alcohol use and in rate of change in log-alcohol use in the population. In other words, there is recognizable variation in the trajectories of alcohol use across adolescents in the population, and we can legitimately seek predictors of this variation in subsequent analyses.

The fifth entry in the table is also from the fitted $\Psi$ matrix and provides an estimate of the covariance of true intercept and true rate of change in log-alcohol use across adolescents. The product–moment correlation coefficient between intercept and slope is equal to their covariance ($-.013$) divided by the square root of the product of their variances (i.e., $\sqrt{(.087 \times 0.020)} = \sqrt{(.00174)} = .0417$). Thus, the true rate of change has a correlation of $-.31$ with true intercept—a moderate, but statistically significant, correlation that suggests that adolescents with lower self-reported alcohol use in the seventh grade have more rapid rates of increase as they age.

Finally, the sixth through eighth rows of Table 23.3 provide estimates of the error variances on the three occasions of measurement, from the diagonal of the fitted $\Theta_\varepsilon$ matrix. Although these variances are of similar magnitude across time, there is some evidence of heteroscedasticity, particularly between times 1 and 2.

## II. INTRODUCING A TIME-INVARIANT PREDICTOR OF CHANGE INTO THE ANALYSIS

Once we know that interindividual differences in change exist, we can ask whether this heterogeneity can be predicted by characteristics of the people under study. In our example, for instance, we can investigate whether statistically significant heterogeneity in true change in self-reported (log) alcohol use depends on the time-invariant predictor of change, adolescent gender (FEM). In the following analyses, we investigate whether variation in individual change (i.e., in the intercept and slope of the adolescent alcohol-use trajectory) is related to gender, and address the following research questions: Does seventh grade self-reported (log) alcohol use differ for boys and girls? and Does the rate at which self-reported (log) alcohol use changes over time depend upon gender? Our analysis requires that the gender predictor be incorporated into the previously specified structural model where level-2 interindividual differences in change were modeled.

Within the general LISREL framework, insertion of predictors into the structural model is achieved via the mechanism of the measurement model for exogenous predictors, **X**. You load up the selected predictors of change into the $\xi$-vector via the X-measurement model and then take advantage of the so-far unused $\Gamma$ matrix in the structural model, which contains parameters describing the regression of $\eta$ on $\xi$, to represent relations between the individual growth parameters and predictors of change. For interpretive purposes, we also recommend centering all time-invariant predictors on their averages as part of this process. This can be achieved simultaneously in the X-measurement model, where any predictor can be partitioned into its mean and a deviation from its mean using a mathematical tautology, as follows:

$$FEM_p = \mu_{FEM} + 1(FEM_p - \mu_{FEM}) + 0, \qquad (13)$$

where $\mu_{FEM}$ is the population average of the time-invariant predictor, FEM. Equation 13 says that FEM—and, in fact, any predictor—can be decomposed into two parts: an average, $\mu_{FEM}$, and a deviation from the average, $(FEM_p - \mu_{FEM})$. This representation can then again be mapped directly onto the LISREL measurement model for exogenous variables **X**, which is:

$$X = \tau_x + \Lambda_x\xi + \delta \qquad (14)$$

Comparing Equations 13 and 14, we see that the tautology in Equation 13 containing the predictor of change can be regarded as a LISREL measurement model for exogenous variables **X** with LISREL score and error vectors defined as follows:

$$X = [FEM_p], \; \xi = [FEM_p - \mu_{FEM}], \; \delta = [0], \qquad (15)$$

with constituent $\tau_x$ and $\Lambda_x$ parameter matrices:

$$\tau_x = [\mu_{FEM}], \; \Lambda_x = [1] \qquad (16)$$

and with the covariance matrix $\Phi$ containing only the variance of the predictor of change:

$$\Phi = Cov(\xi) = [\sigma^2_{FEM}] \qquad (17)$$

Notice that, having centered the predictor of change, FEM, on its population average in Equation 13, we estimate that average by freeing the $\tau_x$ vector. Although we do not demonstrate it here, the X-measurement model for exogenous variables in Equations 13–17 can be modified to accommodate multiple time-invariant predictors of change and multiple indicators of each predictor construct, if available. In each case, the parameter matrix $\Lambda_x$ is expanded to include the requisite loadings (under the usual identification requirements, see Bollen, 1989).

Once the X-measurement model has been specified in this way, the

LISREL structural model then lets us model the relation between the individual growth parameters and the predictor of change, by permitting the regression of $\eta$ on $\xi$. The level-1 individual growth model described in Equations 1–7 is unchanged. However, we express the association between the individual growth parameters and the predictor of change by modifying the existing LISREL structural model in Equations 9–12 so that the newly defined vector of exogenous predictors (now containing adolescent gender, centered on its own mean) is introduced into the right-hand side of the model. We do this by utilizing the latent regression-weight matrix $\Gamma$ present in the LISREL structural model for modeling the association between the $\eta$ and $\xi$ vectors. We free those elements of the $\Gamma$ matrix that represent the simultaneous linear regression of true intercept and slope on the predictor of change, as follows:

$$\begin{bmatrix} \pi_{1p} \\ \pi_{2p} \end{bmatrix} = \begin{bmatrix} \mu_{\pi_1} \\ \mu_{\pi_2} \end{bmatrix} + \begin{bmatrix} \gamma_{\pi_1 FEM} \\ \gamma_{\pi_2 FEM} \end{bmatrix} [FEM_p - \mu_{FEM}] + \begin{bmatrix} 0 & 0 \\ 0 & 0 \end{bmatrix} \begin{bmatrix} \pi_{1p} \\ \pi_{2p} \end{bmatrix} + \begin{bmatrix} \zeta_{1p} \\ \zeta_{2p} \end{bmatrix},$$

(18)

which is the general LISREL structural model:

$$\eta = \alpha + \Gamma\xi + B\eta + \zeta,$$ (19)

with constituent parameter matrices:

$$\alpha = \begin{bmatrix} \mu_{\pi_1} \\ \mu_{\pi_2} \end{bmatrix}, \Gamma = \begin{bmatrix} \gamma_{\pi_1 FEM} \\ \gamma_{\pi_2 FEM} \end{bmatrix}, B = \begin{bmatrix} 0 & 0 \\ 0 & 0 \end{bmatrix}.$$ (20)

It is the LISREL $\Gamma$-matrix that then contains the parameters of most interest in an investigation of change. These regression parameters indicate whether the individual intercepts and slopes that distinguish among the change trajectories of different adolescents are related to adolescent gender. In other words, they tell us whether the trajectory of alcohol use in adolescence differs for boys and girls. We refer to the covariance structure model described by Equations 13 through 20 as Latent Growth Model #2.

In Figure 23.2, we extend the Figure 21.1 path model to include FEM as an exogenous time-invariant predictor of change. On the right-hand side, the factor loadings linking $Y$ and $\eta$ remain fixed at their earlier values, forcing the $\eta$-vector to contain the individual growth parameters, $\pi_{1p}$ and $\pi_{2p}$. To the left, the newly selected time-invariant exogenous predictor, FEM, predicts the level-1 intercept and slope constructs, as our research questions require. Finally, the latent residuals, $\zeta$, now capture the variation remaining in the intercepts and slopes, after what can be predicted by FEM has been accounted for.

It is worth thinking about the properties of the latent residuals, $\zeta$, in Latent Growth Model #2. In a regular regression analysis, the residual

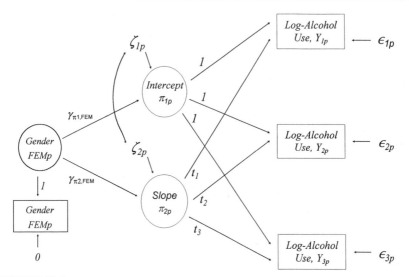

**FIGURE 23.2**   Path model for Latent Growth Model #2, in which self-reported log-alcohol use is a linear function of adolescent grade, and initial status (intercept) and rate of change (slope) depend linearly on gender.

variance equals the variance of the outcome when there are no predictors in the model. Once predictors are added, residual variance declines in proportion to predictor effectiveness. The two elements of the latent residual vector $\zeta$ in Equations 18 and 19 now contain the values of intercept and slope deviated from their conditional means (i.e., from their values predicted by their linear relation with $\mathrm{FEM}_p$). These are the "adjusted" values of true intercept and slope, after partialling out the linear effect of the predictor of change (i.e., those parts of true intercept and slope that are not linearly related to $\mathrm{FEM}_p$). The latent residual vector $\zeta$ is therefore distributed with zero mean vector and a covariance matrix $\Psi$ that contains the partial variances and covariance of true intercept and slope, controlling for the linear effects of the predictor of change. If we predict true intercept and slope by $\mathrm{FEM}_p$ successfully, then these partial variances will be smaller than their unconditional cousins in Equation 12.

Maximum-likelihood estimates of the new parameters in regression-weight matrix $\Gamma$, along with estimates of other unknown parameters in $\Phi$, $\alpha$ and $\Psi$—which characterize our hypotheses about any systematic interindividual differences in change in the population—can again be obtained using LISREL. We provide a LISREL VIII program for fitting Latent Growth Model #2 in Table 23.4, specifying the $\Lambda_y$, $\Theta_\epsilon$, $\tau_x$, $\Lambda_x$, $\Phi$, $\alpha$, $\mathbf{B}$, $\Gamma$ and $\Psi$ matrices as defined above. Lines 1–5 of the program are identical to the program listed in Table 23.2. Specification of the Y-measurement model is also identical.

**TABLE 23.4   LISREL VIII Program for Fitting Latent Growth Model #2**[a]

Variable selection
   6. 2 3 4 1 \
Model specification
   7. MO NY = 3 NE = 2 TY = ZE LY = FU,FI TE = SY,FI C
   8.    NX = 1 NK = 1 LX = FU,FI TX = FR TD = ZE PH = SY,FR C
   9.    AL = FR GA = FU,FR BE = ZE PS = SY,FR
  10. LE
  11. Pi1 Pi2
  12. LK
  13. Female
  14. VA 1 LY(1, 1) LY(2, 1) LY(3, 1)
  15. VA 0.00 LY(1, 2)
  16. VA 0.75 LY(2, 2)
  17. VA 1.75 LY(3, 2)
  18. VA 1    LX(1, 1)
  19. FR TE(1, 1) TE(2, 2) TE(3, 3)

[a] Lines 1–5 and 20 are identical to Table 23.2, line numbers and comments are for reference only, and must be removed when the program is executed.

Differences between the programs begin in line 6, where the predictor of change, FEM, is selected for analysis (and labeled as a construct in lines 12–13). In the model (MO) statement (line 8), the new measurement model for $X$ is defined. First, the $X$ and $\xi$ score vectors are dimensioned to contain a single element (NX = 1, NK = 1), as required by Equation 15. The factor-loading matrix for $X$, $\Lambda_x$, is fixed (LX = FU,FI) and its value set to 1 in line 18 (VA 1 LX(1,1)) as noted in Equation 16. The $X$ mean vector, $\tau_x$, is freed up to contain the population average value of FEM, as required by Equation 16 (TX = FR). The covariance matrix of $X$ is freed (PH = SY,FR) so that the variance of the predictor of change, FEM, can be estimated, as required by Equation 17. Finally, all measurement errors in $X$ are eliminated (as required by the assumption of infallibility embedded in Equation 13) by setting the covariance matrix of the errors in $X$, $\Theta_\delta$, to zero (TD = ZE).

The structural model is mainly defined in line 9. The population means of the individual growth parameters are freed for estimation in the $\alpha$-vector, a vector that contains the means of the endogenous constructs, $\eta$ (AL = FR). The newly important $\Gamma$ matrix is also freed (GA = FU,FR), as required by Equation 20, to describe the regression of the individual intercepts and slopes on the predictor of change. The $\mathbf{B}$ and $\Psi$ matrices have their earlier specifications. The $\mathbf{B}$ matrix is set to zero as it is not being used (BE = ZE), and the $\Psi$ matrix is freed for estimation of the latent residual partial variances and covariance (PS = SY,FR).

**TABLE 23.5   Parameter Estimates for Latent Growth Model #2, a Model in Which the Impact of Time-Invariant Adolescent Gender on the Trajectory of Adolescent's Alcohol Use Is Assessed**

| Parameter | | Estimate |
|---|---|---|
| **Symbol** | **Label** | |
| Population true mean trajectory in alcohol use | | |
| $\mu_{\pi_1}$ | True initial status: average true log-alcohol use in grade seven | .226*** |
| $\mu_{\pi_2}$ | Rate of true change: average true change in log-alcohol use per grade | .036** |
| Population true residual variances and residual covariance in alcohol use, controlling for gender | | |
| $\sigma^2_{\pi_1\|FEM}$ | True initial status: partial variance of true log-alcohol use in grade seven | .086*** |
| $\sigma^2_{\pi_2\|FEM}$ | Rate of true change: partial variance of true change in log-alcohol use per grade | .020*** |
| $\sigma_{\pi_1\pi_2\|FEM}$ | True initial status and rate of true change: partial covariance of true log-alcohol use in grade seven and true change in log-alcohol use per grade | −.012** |
| Population regression of individual change in alcohol use on gender | | |
| $\gamma_{\pi_1 FEM}$ | Regression of true initial status on gender: population difference in true log-alcohol use in grade seven between males and females | −.042~ |
| $\gamma_{\pi_2 FEM}$ | Regression of rate of true change on gender: population difference in true per-grade differences in log-alcohol use between males and females | 0.008 |

~ $= p < .10$, ** $= p < .01$, *** $= p < .001$. $\chi^2(2, 1122) = 1.54$ ($p = .46$)

Table 23.5 contains selected goodness-of-fit statistics, parameter estimates, and $p$-values for Latent Growth Model #2. Estimates of the level-1 error structure have been omitted to save space. The model fits well ($\chi^2 = 1.54$, df $= 2$, $p = .46$). The first two rows present estimates of the population means of true intercept and slope from the $\alpha$-vector, which are similar to those obtained under Latent Growth Model #1. Because we have centered the predictor of change on its own mean in Equations 13, 15, and 16, estimates of the average intercept and slope have their earlier interpretation and describe the log-alcohol use trajectory for the average adolescent (not for a male adolescent, as would be the case if no centering had been employed). This trajectory has statistically significant intercept and slope, indicating that adolescents report nonzero alcohol consumption in grade seven and that this consumption rises about 3.6% per grade subsequently.

Rows 3 through 5 of Table 23.5 contain the estimated partial variances and covariance of true intercept and true slope, controlling for the linear effects of adolescent gender, from the fitted $\Psi$ matrix. Comparing the estimated conditional variances with their unadjusted cousins in Table 23.3, the inclusion of the predictor of change has reduced unexplained variance in true intercept and slope by a very small amount, .8% and 1.5%, respectively, suggesting that gender is a relatively unimportant predictor of interindividual differences in true change in the population. This is further confirmed by statistical tests on the latent regression coefficients, $\gamma_{\pi_1 \text{FEM}}$ and $\gamma_{\pi_2 \text{FEM}}$, listed in rows 6 and 7, where we reject the null hypothesis associated with predicting the intercept at the .10 level but cannot reject the null hypothesis associated with predicting the slope. This suggests that girls report consuming less alcohol initially than boys in grade seven, but that their rate of change in self-reported consumption is indistinguishable from boys.

## III. INCLUDING A TIME-VARYING PREDICTOR OF CHANGE IN THE ANALYSES

Once you recognize that the multilevel models required for investigating change can be mapped onto the covariance structure model, a conceptual portal is opened though which other possibilities come into view. In some research projects, for instance, data are collected on time-varying predictors of change, such as "peer pressure" in our example.

When time-varying predictors of change are available, several different kinds of research questions can be addressed. For instance, the presence of a time-varying predictor allows us to model individual change in both the outcome and the predictor and then to investigate whether changes over time in the two variables are related. In the current example, we can ask, Does adolescents' self-reported use of alcohol increase more rapidly over the seventh and eighth grades if the pressure exerted on them to drink by their peers is also increasing more rapidly? This asks whether the rate of change in alcohol use is predicted by rate of change in peer pressure. Questions like these can also be addressed with covariance structure methods.

To determine whether change in the outcome depends on change in a time-varying predictor, we model individual growth in both the outcome and the predictor simultaneously and investigate whether individual growth parameters representing change in the outcome can be predicted by individual growth parameters representing change in the predictor. In the current example, this requires that we specify individual growth models for both alcohol use and peer pressure. We proceed in the usual way. First, we specify the individual change trajectory of the outcome (self-reported alcohol use) in the $Y$-measurement model, forcing the individual growth parameters that describe the true trajectory into the $\eta$ vector, as before. In previous analyses, we then used the $X$-measurement model to pass the newly centered predictor of change into the $\xi$ vector and ultimately into the LISREL structural model. When incorporating a time-varying predictor, however, we use the $X$-measurement model to represent individual change in the time-varying covariate (peer pressure) and force its individual growth parameters into the $\xi$ vector instead. Then, as before, the relation between $\eta$ and $\xi$ is modeled via the matrix of latent regression parameters, $\Gamma$, which then contains coefficients describing relations between the two kinds of change. (In the example that follows, we have removed the adolescent gender predictor from contention, for simplicity. However, both time-invariant and time-varying predictors of change can be included in the same analysis by combining the two approaches).

In the current example, individual change over time in self-reported (log) alcohol use is modeled in the $Y$-measurement model as before (Equations 1–7), but, now we also represent the natural logarithm of peer pressure, $X_{ip}$, on the $p$th child on the $i$th occasion of measurement by its own individual growth model:

$$X_{ip} = \omega_{1p} + \omega_{2p}t_i + \delta_{ip},\qquad(21)$$

where the child's grade has been recentered on grade seven and $\delta$ represents measurement error. Again, exploratory graphical analyses suggested that Equation 21 was an appropriate model for individual change in peer pressure over adolescence. In Equation 21, slope $\omega_{2p}$ represents change in true self-reported (log) peer pressure per grade for the $p$th adolescent; teenagers who reported that peer pressure increased the most rapidly over seventh and eighth grades will have the largest values of this parameter. Intercept $\omega_{1p}$ represents the true self-reported (log) peer pressure on adolescent $p$ at the beginning of the seventh grade, given our recentering of the time metric; children who report experiencing greater peer pressure at the beginning of seventh grade will possess higher values of this parameter.

As before, we can develop a matrix representation of each adolescent's empirical growth record in both the outcome and the predictor. The former is unchanged from Equation 2, the latter is:

$$\begin{bmatrix} X_{1p} \\ X_{2p} \\ X_{3p} \end{bmatrix} = \begin{bmatrix} 0 \\ 0 \\ 0 \end{bmatrix} + \begin{bmatrix} 1 & t_1 \\ 1 & t_2 \\ 1 & t_3 \end{bmatrix} \begin{bmatrix} \omega_{1p} \\ \omega_{2p} \end{bmatrix} + \begin{bmatrix} \delta_{1p} \\ \delta_{2p} \\ \delta_{3p} \end{bmatrix}\qquad(22)$$

where, as before, we assume that the measurement errors, $\delta$, are heteroscedastic but independent over time:

$$\begin{bmatrix} \delta_{1p} \\ \delta_{2p} \\ \delta_{3p} \end{bmatrix} \sim N\left( \begin{bmatrix} 0 \\ 0 \\ 0 \end{bmatrix}, \begin{bmatrix} \sigma^2_{\delta_1} & 0 & 0 \\ 0 & \sigma^2_{\delta_2} & 0 \\ 0 & 0 & \sigma^2_{\delta_3} \end{bmatrix} \right)\qquad(23)$$

As previously, we note that the empirical growth record in self-reported peer pressure in Equation 22 can be mapped onto the LISREL X-measurement model:

$$X = \tau_x + \Lambda_x \xi + \delta, \tag{24}$$

with LISREL score vectors that contain the empirical growth record, the individual growth parameters, and the errors of measurement, respectively:

$$X = \begin{bmatrix} X_{1p} \\ X_{2p} \\ X_{3p} \end{bmatrix}, \xi = \begin{bmatrix} \omega_{1p} \\ \omega_{2p} \end{bmatrix}, \delta = \begin{bmatrix} \delta_{1p} \\ \delta_{2p} \\ \delta_{3p} \end{bmatrix} \tag{25}$$

the elements of the $\tau_x$ and $\Lambda_x$ parameter matrices contain known values and constants:

$$\tau_x = \begin{bmatrix} 0 \\ 0 \\ 0 \end{bmatrix}, \Lambda_x = \begin{bmatrix} 1 & t_1 \\ 1 & t_2 \\ 1 & t_3 \end{bmatrix} \tag{26}$$

and the error vector $\delta$ is distributed with zero mean vector and covariance matrix $\Theta_\delta$:

$$\Theta_\delta = Cov(\delta) = \begin{bmatrix} \sigma^2_{\delta_1} & 0 & 0 \\ 0 & \sigma^2_{\delta_2} & 0 \\ 0 & 0 & \sigma^2_{\delta_3} \end{bmatrix} \tag{27}$$

When we investigate the association between growth in an outcome and growth in a predictor, both the $Y$- and $X$-measurement models are in use, and we have the additional option to specify not only the covariance matrices of the level-1 errors, $\Theta_\delta$ and $\Theta_\varepsilon$, but also the matrix of their covariances, $\Theta_{\delta\varepsilon}$. In this particular example, this is useful for both substantive and psychometric reasons. In Farrell's survey, both the adolescent's alcohol use and the peer pressure were self-reported on similar instruments and similar scales on each of the three occasions of measurement. We therefore assume that the level-1 measurement errors in self-reported alcohol use and peer pressure covary across adolescents within occasion, as follows:

$$\Theta_{\delta\varepsilon} = Cov(\delta\varepsilon) = \begin{bmatrix} \sigma_{\delta_1\varepsilon_1} & 0 & 0 \\ 0 & \sigma_{\delta_2\varepsilon_2} & 0 \\ 0 & 0 & \sigma_{\delta_3\varepsilon_3} \end{bmatrix}. \tag{28}$$

Having specified the level-1 structure, we can predict the growth parameters describing change in alcohol use by the growth parameters describing change in time-varying peer pressure, by utilizing the LISREL structural model. These predictions are modeled in the usual way in the following level-2 model:

$$\begin{bmatrix} \pi_{1p} \\ \pi_{2p} \end{bmatrix} = \begin{bmatrix} \mu_{\pi_1} \\ \mu_{\pi_2} \end{bmatrix} + \begin{bmatrix} \gamma_{\pi_1\omega_1} & \gamma_{\pi_1\omega_2} \\ \gamma_{\pi_2\omega_1} & \gamma_{\pi_2\omega_2} \end{bmatrix} \begin{bmatrix} \omega_{1p} \\ \omega_{2p} \end{bmatrix} + \begin{bmatrix} 0 & 0 \\ 0 & 0 \end{bmatrix} \begin{bmatrix} \pi_{1p} \\ \pi_{2p} \end{bmatrix} + \begin{bmatrix} \zeta_{1p} \\ \zeta_{2p} \end{bmatrix}, \quad (29)$$

which we again recognize as the LISREL structural model:

$$\eta = \alpha + \Gamma\xi + B\eta + \zeta \quad (30)$$

with parameter matrices:

$$\alpha = \begin{bmatrix} \mu_{\pi_1} \\ \mu_{\pi_2} \end{bmatrix}, \Gamma = \begin{bmatrix} \gamma_{\pi_1\omega_1} & \gamma_{\pi_1\omega_2} \\ \gamma_{\pi_2\omega_1} & \gamma_{\pi_2\omega_2} \end{bmatrix}, B = \begin{bmatrix} 0 & 0 \\ 0 & 0 \end{bmatrix}. \quad (31)$$

It is again the $\Gamma$ matrix that contains the important level-2 regression parameters describing relations between an adolescent's growth in alcohol use and growth in peer pressure (i.e., describing relations among the individual growth parameters describing change in predictor and outcome).

In the previous section, we had to account for the population mean and variance of the predictor of change, FEM. In this expanded time-varying-covariates analysis, we must ensure that there is a place in the model for the population distribution of the intercept and slope of the predictor peer-pressure trajectory. Clearly, trajectories of peer-pressure can differ across adolescents, and so their individual growth parameters will possess population averages, variances, and covariances. These parameters will be forced to be zero, undermining the overall fit of the model, if we do not provide them with a place to reside within the general LISREL model. The natural place to model the population average trajectory of the time-varying covariate—that is, $\mu_{\omega1}$ and $\mu_{\omega2}$—is in the $\kappa$-vector, which records the means of the exogenous construct, $\xi$:

$$\kappa = \begin{bmatrix} \mu_{\omega_1} \\ \mu_{\omega_2} \end{bmatrix} \quad (32)$$

and the associated variability in $\xi$ is modeled in the covariance matrix of exogenous constructs, $\Phi$, which contains the variances and covariance of the individual intercepts and slopes that now describe change in peer pressure over time:

$$\Phi = Cov(\xi) = \begin{bmatrix} \sigma_{\omega_1}^2 & \sigma_{\omega_1\omega_2} \\ \sigma_{\omega_2\omega_1} & \sigma_{\omega_2}^2 \end{bmatrix} \quad (33)$$

In Figure 23.3, we extend the path model of Figure 23.1 to include peer pressure as a time-varying predictor of change. The right-hand side

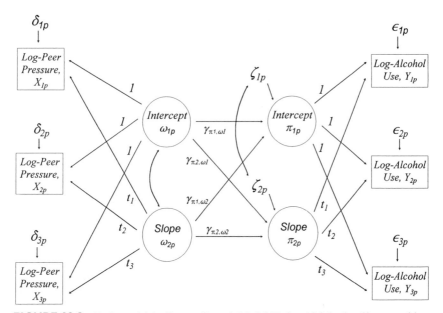

**FIGURE 23.3**    Path model for Latent Growth Model #3, in which both self-reported log-alcohol use and log-peer pressure are linear functions of adolescent grade, and the initial status (intercept) and rate of change (slope) of the alcohol-use trajectories depend linearly upon the initial status (intercept) and rate of change (slope) of the peer pressure trajectory.

of the figure is unchanged from earlier analyses, and continues to represent adolescents' change in log-alcohol use over time. However, a similar path structure has been introduced on the left-hand side of the figure to represent change in peer pressure for those same adolescents. Single-headed arrows linking individual growth parameters for change in peer pressure to individual growth parameters for change in alcohol-use provide the hypothesized links between the two kinds of changes. The latent residuals continue to sop up any between-person variation in the alcohol-use trajectory that remains after what can be predicted by change in peer pressure has been removed. We refer to the model in Figure 23.3 as Latent Growth Model #3.

In Table 23.6, we provide a LISREL VIII program for fitting Latent Growth Model #3 to data. The model is specified as in Equations 1–7 and 22–32. Lines 1–5 are identical to those in the program in Table 23.2. In line 6, three waves of data in both log-alcohol use and log-peer pressure are selected for analysis. The Y-measurement model is specified as before in program lines 7, 10, 11, 14, 16–18, and 19, but modifications have been made to both the X-measurement model and the structural model.

The X-measurement model is specified in program lines 8, 15, 16–18, and 20. In the model (MO) statement (line 8), the **X** and $\xi$ vectors are dimensioned (NX = 3, NK = 2) to reflect the three waves of longitudinal

**TABLE 23.6   LISREL VIII Program for Fitting Latent Growth Model #3**[a]

Variable selection
  6. 2 3 4 5 6 7 /
Model specification
  7. MO NY = 3 NE = 2 TY = ZE LY = FU,FI TE = SY,FI C
  8.    NX = 3 NK = 2 LX = FU,FI TX = ZE TD = SY,FI TH = FU,FI PH = SY,FR C
  9.    AL = FR GA = FU,FR KA = FR BE = ZE PS = SY,FR
 10. LE
 11. Pi1 Pi2
 12. LK
 13. Omega1 Omega2
 14. VA 1 LY(1, 1) LY(2, 1) LY(3, 1)
 15. VA 1 LX(1, 1) LX(2, 1) LX(3, 1)
 16. VA 0.00 LY(1, 2) LX(1, 2)
 17. VA 0.75 LY(2, 2) LX(2, 2)
 18. VA 1.75 LY(3, 2) LX(3, 2)
 19. FR TE(1, 1) TE(2, 2) TE(3, 3)
 20. FR TD(1, 1) TD(2, 2) TD(3, 3)
 21. FR TH(1, 1) TH(2, 2) TH(3, 3)

---

[a] Lines 1–5 and 22 are identical to Table 23.2, line numbers and comments are for reference only, and must be removed when the program is executed.

data that are available on adolescent peer pressure and the pair of individual growth parameters that are required to represent its trajectory of change, respectively. The new intercepts and slopes of the peer-pressure trajectory are labeled in lines 12 and 13. The X-measurement model's factor-loading matrix, $\Lambda_x$, is first specified as "full" and "fixed" in line 8 (LX = FU, FI), but then its elements are given the values required by Equation 26 in lines 15 through 18. The X-mean vector, $\tau_x$, is set to zero in line 8 (TX = ZE), also as required by Equation 26. The mean-vector, $\kappa$, and covariance matrix, $\Phi$, which contain the level-2 parameters that describe the between-person distribution of the peer-pressure trajectories, are freed for estimation in line 9 (KA = FR) and in line 8 (PH = SY, FR), as required by Equations 32 and 33. The covariance matrix of the measurement errors in X, $\Theta_\delta$, is first declared symmetric and fixed in line 8 (TD = SY, FI), and then its diagonal elements are freed for estimation in line 20, as required in Equation 27. Similarly, the matrix containing the covariances of the measurement errors of Y and the measurement errors of X, $\Theta_{\delta\varepsilon}$, is declared full and fixed in line 8 (TH = FU, FI), and then its diagonal elements are freed for estimation in line 21, as required by Equation 28.

Finally, the new structural model is described (line 9). It is similar to the earlier structural model specified for Latent Growth Model #2. The population means of the individual growth parameters describing change in the outcome, adolescent log-alcohol-use, are freed for estimation in the

$\alpha$-vector as before (AL = FR), as required by Equation 31. The $\Gamma$ matrix is freed (GA = FU, FR), as required by Equation 31, to describe the regression of change in alcohol-use on change in peer pressure. Specification of the $\beta$ and $\Psi$ matrices are identical to the earlier specifications. The $\beta$ matrix is set to zero because it is not being used in this analysis (BE = ZE), and the $\Psi$ matrix is freed for estimation of the latent residual partial variances and covariance (PS = SY, FR).

Selected parameter estimates and goodness-of-fit statistics are provided for the Latent Growth Model #3 in Table 23.7, along with approximate $p$-values. We have omitted estimates of the level-1 error structure in order to save space. The model fits reasonably well—although the magnitude of the $\chi^2$ statistic (11.54) is larger than we would like, it is not exorbitant given the 4 degrees of freedom.

A variety of conclusions can be reached by inspecting the table. First, the intercept (.188, $p < .001$) of the average trajectory of log-peer pressure over seventh and eightth grades indicates that adolescents experienced statistically significant pressure from peers to consume alcohol in seventh grade, and the slope (.096, $p < .001$) indicates a statistically significant increase in peer pressure during the period of observation. In fact, because we took the natural logarithm of peer pressure prior to growth modeling, we can interpret the slope estimate (.096) as indicating that adolescents experienced about a 10% increase in peer pressure per year. There is also statistically significant heterogeneity in the peer-pressure change trajectories among adolescents. Rows #8 and #9 of Table 23.7 reveal that the variances of initial log peer pressure (.070, $p < .001$) and rate of change in log peer pressure (.029, $p < .01$) are both statistically significant (i.e., nonzero), but we cannot reject the null hypothesis that they are unrelated (.001, n.s.).

In a similar fashion, rows 1 through 5 of Table 23.7 suggest conclusions about changes in alcohol use. We must interpret these entries cautiously because the estimated values are conditional on the value of the predictor, growth in peer pressure. In other words, the estimated values of $\mu_{\pi_1}$ and $\mu_{\pi_2}$ in Latent Growth Model 3 are those for individuals who have "null" trajectories—trajectories with zero intercept and zero slope on the time-varying covariate. However, the results suggest that change in peer pressure is moderately successful in predicting change in alcohol use. Notice, from rows 11–14 of Table 23.7, that two out of the four regression coefficients linking change in peer pressure and change in alcohol use are statistically significant. These coefficients indicate that the initial level of peer pressure is positively related to the initial level of alcohol use (.799, $p < .001$), and that rate of change in peer pressure is positively related to rate of change in alcohol use (.577, $p < .001$), respectively. We can then conclude that adolescents drink more on entry into 7th grade if they have peers who drink more, and that adolescents report more rapid growth in alcohol usage if they experience more growth in peer pressure to drink. This incontrovert-

**TABLE 23.7  Parameter Estimates for Latent Growth Model #3, a Model Testing the Impact of Changes in Time-Varying Peer Pressure on the Adolescent's Alcohol Use Trajectory**

| Parameter | | Estimate |
|---|---|---|
| **Symbol** | **Label** | **Estimate** |
| Population true mean alcohol use trajectory for those with a "null" peer alcohol use trajectory | | |
| $\mu_{\pi_1}$ | Initial status: average true log-alcohol use in grade seven | .067*** |
| $\mu_{\pi_2}$ | Rate of true change: average true change in log-alcohol use per grade | .008 |
| Population true residual variances and covariance in alcohol use controlling for change in peer alcohol use | | |
| $\sigma^2_{\pi_1\|PEER}$ | True initial status: partial variance of true log-alcohol use in grade seven | .042*** |
| $\sigma^2_{\pi_2\|PEER}$ | Rate of true change: partial variance of true change in log-alcohol use per grade | .009~ |
| $\sigma_{\pi_1\pi_2\|PEER}$ | True initial status and rate of true change: partial covariance of true log-alcohol use in grade seven and true change in log-alcohol use per grade | −.006 |
| Population true mean trajectory in peer alcohol use | | |
| $\mu_{\omega_1}$ | True initial status: average peer true log-alcohol use in grade seven | .188*** |
| $\mu_{\omega_2}$ | Rate of true change: average true change in peer log-alcohol use per grade | .096*** |
| Population true residual variances and covariance in peer alcohol use | | |
| $\sigma^2_{\omega_1}$ | True initial status: variance of true peer log-alcohol use in grade seven | .070*** |
| $\sigma^2_{\omega_2}$ | Rate of true change: variance of true change in peer log-alcohol use per grade | .029** |
| $\sigma_{\omega_1\omega_2}$ | True initial status and rate of true change: covariance of true peer log-alcohol use in grade seven and true change in peer log-alcohol use per grade | 0.001 |
| Population regression of change in alcohol use on change in peer alcohol use | | |
| $\gamma_{\pi_1\omega_1}$ | Regression of true adolescent log-alcohol use in grade seven on true peer log-alcohol use in grade seven | .799*** |
| $\gamma_{\pi_1\omega_2}$ | Regression of true adolescent log-alcohol use in grade seven on the rate of true change in peer log-alcohol use | 0.081 |
| $\gamma_{\pi_2\omega_1}$ | Regression of true rate of change in adolescent log-alcohol use on true peer log-alcohol use in grade seven | −.143~ |
| $\gamma_{\pi_2\omega_2}$ | Regression of true rate of change in adolescent log-alcohol use on the rate of true change in peer log-alcohol use | .577** |

~ = $p < .10$, ** = $p < .01$, *** = $p < .001$. $\chi^2(4, 1122) = 11.54$ ($p = .021$)

ible conclusion most likely could not have been obtained by another method of longitudinal data analysis.

## IV. DISCUSSION

In this chapter, we have shown how individual growth modeling can be accommodated within the general framework of covariance structure analysis. We have explored links between these two formerly distinct conceptual arenas, laying out the mapping of one onto the other, and showing how the new approach provides a convenient way of addressing research questions about individual change. This innovative application of covariance structure analysis offers many flexible data-analytic opportunities.

First, the method can accommodate any number of waves of longitudinal data. Willett (1988, 1989) showed that the collection of more waves of data leads to higher precision for the estimation of individual growth trajectories and greater reliability for the measurement of change. In the covariance structure analyses of change, extra waves of data extend the length of the empirical growth record and expand the sample between-wave covariance matrix (thereby increasing degrees of freedom for model fitting), but do not change the fundamental parameterization of the level-1 and level-2 models.

Second, the occasions of measurement need not be equally spaced. Change data may be collected at irregular intervals either for convenience (e.g., at the beginning and end of the school year) or because the investigator wishes to estimate certain features of the trajectory more precisely by clustering data-collection points around times of greater research interest. Such irregularly spaced data is accommodated by the method, provided everyone in the sample is measured on the same set of irregularly spaced occasions within each domain. When that is not the case, the analyses still can be conducted using multigroup analysis.

Third, individual change can be either a straight line or curvilinear. The approach can accommodate not only polynomial growth of any order but also any type of curvilinear growth model in which status is linear in the individual growth parameters. In addition, because the goodness-of-fits of nested models can be compared directly under the covariance structure approach, one can systematically evaluate the adequacy of contrasting individual growth models in any empirical setting.

Fourth, the covariance structure of the occasion-by-occasion level-1 measurement errors can be modeled explicitly. The population measurement error covariance matrix is not restricted to a particular shape or pattern. The investigator need not accept unchecked the level-1 independence and homoscedasticity assumptions of classical analyses, nor the band-diagonal configuration required by repeated-measures analysis of variance.

Indeed, under the covariance structure approach, a variety of reasonable error structures can be systematically compared and the most appropriate structure adopted.

Fifth, the method of maximum likelihood provides overall goodness-of-fit statistics, parameter estimates, and asymptotic standard errors for each hypothesized model. By using the covariance structure method, the investigator benefits from the utility of a well-documented, popular, and well-understood statistical technique. Appropriate computer software is widely available. In this chapter, we have relied upon the LISREL computer package, but our techniques can easily be implemented using other software such as EQS (Bentler, 1985), LISCOMP (Muthén, 1987), and PROC CALIS (SAS, 1991).

Sixth, by comparing goodness-of-fit across nested models, the investigator can test complex hypotheses about interindividual differences in true change. One benefit of fitting an explicitly parameterized covariance structure to data using a software package like LISREL is that selected model parameters can be individually or jointly constrained during analysis to particular values. This allows the investigator to conduct tests on the variability of the individual growth parameters across people. We can, for instance, fix the value of one parameter to a value common across individuals but permit another parameter to be random.

Finally, the flexibility of the general LISREL model permits extension of the analysis of change in substantively interesting ways. For example, we can *predict* change in one or more domains by simultaneous changes in several other domains. Furthermore, we can introduce additional exogenous variables as predictors of any or all of these changes. The method also enables the modeling of intervening effects, whereby a predictor may not act directly on change, but indirectly via the influence of intervening factors, each of which may be either time-invariant or a measure of change itself.

## REFERENCES

Bentler, P. M. (1985). Theory and implementation of EQS: A structural equations program. Los Angeles, CA: BMDP Statistical Software.

Bollen, K. A. (1989). *Structural Equations with Latent Variables.* New York: John Wiley & Sons.

Bryk, A. S., & Raudenbush, S. W. (1987). Application of hierarchical linear models to assessing change. *Psychological Bulletin, 101,* 147–158.

Farrell, A. D. (1994). Structural equation modeling with longitudinal data: Strategies for examining group differences and reciprocal relationships. *Journal of Consulting and Clinical Psychology, 62*(3), 447–487.

Muthén, B. O. (1987). *LISCOMP: Analysis of linear structural equations with c comprehensive measurement model.* Mooresville, IN: Scientific Software, Inc.

Rogosa, D. R., Brandt, D., & Zimowski, M. (1982). A growth curve approach to the measurement of change. *Psychological Bulletin, 90,* 726–748.

Rogosa, D. R., & Willett, J. B. (1985). Understanding correlates of change by modeling individual differences in growth. *Psychometrika, 50,* 203–228.

SAS Institute. (1991). *User's Guide: Statistics, Version 6.* Cary, NC: Sas Institute Inc.

Willett, J. B. (1988). Questions and answers in the measurement of change. In E. Z. Rothkopf (Ed.), *Review of Research in Education,* Vol. 15. Washington, D.C.: American Educational Research Association, 345–422.

Willett, J. B. (1989). Some results on reliability for the longitudinal measurement of change: Implications for the design of studies of individual growth. *Educational and Psychological Measurement, 49,* 587–602.

Willett, J. B. (1994). Measuring change more effectively by modeling individual change over time. In T. Husen & T. N. Postlethwaite (Eds.), *The international encyclopedia of education* (2nd ed.). Oxford, UK: Pergamon Press.

Willett, J. B., & Sayer, A. G. (1994). Using covariance structure analysis to detect correlates and predictors of change. *Psychological Bulletin,* 116, 363–381.

Willett, J. B., & Sayer, A. G. (1995). Cross-Domain Analyses of Change over Time: Combining Growth Modeling and Covariance Structure Analysis. In G. A. Marcoulides & R. E. Schumacker (Eds.), *Advanced structural equation modeling: Issues and techniques* (pp. 22–51). Hillsdale, NJ: Lawrence Erlbaum Incorporated.

## APPENDIX

Sample Mean Vectors and Covariance Matrices For the Adolescent Alcohol Use Example

| | | | Variance/covariance matrix | | | | | |
| | | | Alcohol use | | | Peer alcohol use | | |
| Variables | Means | Female FEM | Start of seventh grade ALC1 | End of seventh grade ALC2 | End of eight grade ALC3 | Start of seventh grade PEER1 | End of seventh grade PEER2 | End of eight grade PEER3 |
|---|---|---|---|---|---|---|---|---|
| Untransformed data | | | | | | | | |
| FEM | 0.612 | 0.238 | | | | | | |
| ALC1 | 1.363 | −0.019 | 0.516 | | | | | |
| ALC2 | 1.421 | −0.034 | 0.304 | 0.651 | | | | |
| ALC3 | 1.491 | −0.012 | 0.255 | 0.335 | 0.786 | | | |
| PEER1 | 1.347 | −0.018 | 0.304 | 0.200 | 0.165 | 0.860 | | |
| PEER2 | 1.578 | −0.051 | 0.284 | 0.470 | 0.302 | 0.356 | 1.345 | |
| PEER3 | 1.684 | −0.053 | 0.255 | 0.323 | 0.629 | 0.334 | 0.534 | 1.460 |
| Transformed data | | | | | | | | |
| FEM | .612 | .238 | | | | | | |
| ALC2 | .225 | −.008 | .136 | | | | | |
| ALC2 | .254 | −.013 | .078 | .155 | | | | |
| ALC3 | .288 | −.005 | .065 | .082 | .181 | | | |
| PEER1 | .177 | −.009 | .066 | .045 | .040 | .174 | | |
| PEER2 | .290 | −.022 | .064 | .096 | .066 | .072 | .262 | |
| PEER3 | .347 | −.024 | .060 | .074 | .132 | .071 | .112 | .289 |

# AUTHOR INDEX

# SUBJECT INDEX

Agreement, interrater, 28, 111, 116
  chance agreement, computation, 112, 113
  chi-square, 111, 115, 177
  conceptual meaning, 98, 101–102
  criterion-referenced interpretation, 100
  definitions of, 115–116
  distinguished from interrater reliability, 96–101, 117
  errors, cost of, 114–115
  measures, 111
  nominal scales, 112–114
  ordinal and interval scales, 114, 116
  pairwise correlations, 111
  percent agreement, 111–112
  proportion of agreements, 111–112
Aggression, 441, 454, 455, 456, 458–460, 461
AMOS, 586
Anarchism, 55–56,
Analysis of covariance, 18–19
  categorical data, 392, 404, 406

hierarchical nested data, 618–619, 621
Analysis of variance, 18–19, 106–107, 212, 232, 531–540, 586, 587, 591, 593–594, 597, 602
  categorical data, 392, 395–397, 399, 400, 404, 406, 407–410, 424–433
  generalizability theory, 531–540
  relation to canonical analysis, 240–241
  relation to cluster analysis, 316
Archival data, 383, 384, 385
Association, measures for categorical data, 429–431
  crude, 398, 399
  marginal, 398, 399, 400, 405, 410, 420, 422, 424–429, 392, 436
  odds ratio, 397, 399–401, 405, 415, 417, 420, 424, 427–436
  rate ratio, 397–400, 406, 408–410, 415, 420
  relative risk, 399–400, 417
  risk ratio, 397, 399–400
  specific, 398

ISBN 0-12-691360-9

90051